The Dancing Column

Joseph Rykwert

The Dancing Column On Order in Architecture

The MIT Press · Cambridge, Massachusetts · London, England

for Francesco . . .

Publication of this book has been aided by a grant from the University of Pennsylvania Research Foundation, and a grant from the Millard Meiss Publication Fund of the College Art Association.

This book was set in Sabon by Graphic Composition, Inc., and was printed and bound in the United States of America.

Library of Congress Cataloging-in-Publication Data

Rykwert, Joseph
    The dancing column : on order in architecture / Joseph Rykwert.
        p.   cm.
    Includes bibliographical references and index.
    ISBN 0-262-18170-3 (hc : alk. paper)
    1. Architecture—Orders.   2. Columns, Doric.   3. Columns, Ionic.   4. Columns, Corinthian.   5. Body, Human—Influence.   6. Electicism in architecture.
    NA2815.R95   1996
    721′.36—dc20   95-35555
                                                                              CIP

What House more stately hath there bin,
Or can be, than is Man?

. . .

Man is all simmetry
     Full of Proportions, one Limme to another,
And all to all the World besides:
     Each Part may call the furthest, Brother:
For Head with Foot hath private Amitie,
     And both with Moones and Tides.

George Herbert, *Man*

Contents

List of Illustrations   viii

Preface   xvi

I Order in Building   2

Catalogues and recipes ▪ Skeptics, critics, and romantics ▪ The Beaux-Arts and the Ecole Polytechnique ▪ Imitation of the primitive hut ▪ The end of metaphor ▪ Return to sources

II Order in the Body   26

Scriptural columns: John Wood the Elder ▪ Gian Lorenzo Bernini on bodies and columns ▪ Jacques-François Blondel on faces and capitals ▪ Face, character, fate ▪ Psychology in the studio: Charles Le Brun's Conférences ▪ Artificial gesture, natural gesture ▪ Face, character, mood, landscape ▪ Variety of character and fixity of canon ▪ The church as a body ▪ The city as a body ▪ The city as a house

III The Body and the World   68

Love and strife ▪ The body of the first man ▪ Humors, elements, and stars ▪ Astral man, canonic man ▪ The dignity and the misery of the body ▪ Geometry of God and man ▪ Man the measure ▪ The canon realized ▪ Vitruvius' man in motion ▪ Dürer on human variety ▪ Gian Paolo Lomazzo ▪ Michelangelo ▪ Fabric of man, fabric of the world

IV Gender and Column   96

The square and the circular man ▪ *Kanōn* ▪ Polykleitos ▪ A female canon? ▪ Luck and invention

V The Literary Commonplace   116

Girl and boy ▪ The metaphoric animal ▪ The shock of meaning ▪ The primal post ▪ Imitation and manufacture ▪ If nature built a house ▪ Do-alike, look-alike ▪ Tragic building ▪ Persians and atlantes ▪ Caryatids

VI **The Rule and the Song** 142

The sons of Herakles ▪ The first shrines ▪ The heroes' sacrifice ▪ God and king ▪ Altar and temple ▪ Parts of the building ▪ The *demiourgos* at sacrifice ▪ Lions and a column ▪ The holy column ▪ Egyptian Doric ▪ *Djed*: Column and mummy ▪ Egyptians and Greeks ▪ Greek origins

VII **The Hero as a Column** 170

Base ▪ Column: The shaft ▪ Column: The capital ▪ Beams and roof ▪ *Metopai* ▪ Triglyphs ▪ Models ▪ Delphi: The legend ▪ Eretria and Lefkandi ▪ Dreros and Prinias ▪ Thermon ▪ Olympia: The temple of Hera ▪ Olympia: The House of Oenomaos ▪ Delos: The Naxian *Oikos* ▪ Corinth ▪ The awkward Doric

VIII **The Known and the Seen** 210

The temple and its users ▪ Tholoi ▪ Type and project ▪ Six hundred varieties ▪ Tympanum or pediment ▪ The corner triglyph—again ▪ Optical refinements ▪ Theory and practice ▪ Surface

IX **The Mask, the Horns, and the Eyes** 236

Ionic and descriptive ▪ The legend of the first Ionic ▪ Doric territory ▪ Artemis at Ephesus ▪ The worship of Artemis ▪ Artemis and Dionysos ▪ The Ephesian Artemision ▪ Votive columns ▪ The Sikyonian thalamoi ▪ A Persian Ionic? ▪ The Aeolic "order" ▪ Oriental parallels ▪ In Anatolia: Mittani, Hittites, and Urartians ▪ In Anatolia: Phrygia, Lydia, Lycia, Carya ▪ Spirals and trees ▪ Aphrodite at Paphos ▪ Astarte at Kition ▪ The guardians of the doors ▪ Models and shrines ▪ Furniture and fabrics ▪ The horned skull ▪ Humbaba ▪ The column-statue ▪ The reed bundle: Inanna's pillar ▪ Hathor

X **The Corinthian Virgin** 316

The girl ▪ Death and burial ▪ The offering basket ▪ Acanthus—the plant ▪ Acanthus at Delphi ▪ Acanthus and tripod ▪ Delphi: The landscape ▪ The girl again: Persephone ▪ Monuments 1: The temples ▪ Monuments 2: Tholoi ▪ Monuments 3: Monopteroi ▪ The Athenian Olympieion

XI **A Native Column?** 350

The Etruscan arrangement ▪ The disposition ▪ Greek rite, Etruscan rite ▪ Capitolium ▪ Ceres ▪ A Tuscan Ionic? ▪ Columns of honor ▪ An Italic order

XII **Order or Intercourse** 372

The double metaphor ▪ Dorian objectivity ▪ Skin and bones ▪ Sensation and production ▪ Making, imitating, loving

Notes 392

Abbreviations and Ancient Texts 522

Bibliography 532

Index 578

List of Illustrations

I Order in Building

The five orders (Serlio) *5*

Reconstruction of the Temple in Jerusalem (Perrot and Chipiez) *6*

Origin of the Ionic (Chipiez) *7*

Lycian tomb (after Choisy) *7*

The five orders (Normand) *8*

The three faces (Blanc) *9*

The Chinese "style" (Blanc) *9*

Composition of parts (Gwilt, after Durand) *14*

Parc Güell, Ipostilo, of Gaudí: Main entry *16*

Parc Güell, Ipostilo, of Gaudí: Detail of capital *17*

Parc Güell, Ipostilo, of Gaudí: Flank *17*

Woodland Chapel, Stockholm South Cemetery, by Asplund: Exterior (after Ahlberg) *20*

Woodland Chapel, Stockholm South Cemetery, by Asplund: Interior (after Ahlberg) *21*

Woodland Chapel: Plan (Asplund, after Ahlberg) *21*

Woodland Chapel: Sections (Asplund, after Ahlberg) *21*

Chicago Tribune Building: Perspective by Gerhardt *22*

Chicago Tribune Building: Perspective by Loos *22*

Chicago Tribune Building: Sketch plan by Loos *23*

Projected tomb for Max Dvořak by Loos *24*

II Order in the Body

Stanton Drew Academy (Wood) *28*

Stonehenge restored (Wood) *28*

Composite order: Sunrise (H. Vredeman de Vries) *30*

Corinthian order: Morning; Youth (H. Vredeman de Vries) *31*

The Doric order and Hercules (Shute) *32*

The Ionic order and Hera (Shute) *33*

The Corinthian order and Aphrodite (Shute) *34*

The Tuscan entablature of Palladio (Blondel) *35*

The Tuscan entablature of Vignola (Blondel) *35*

Padua, Palazzo della Ragione, Salone: Side wall panels *38–39*

Plato as a dog (della Porta) *41*

Angelo Poliziano as a rhinoceros (della Porta) *41*

Title page from J. Bulwer, *Chirologia* (1644) *42*

The eye, the brain, and the pineal gland (Descartes) *44*

The operation of the pineal gland (Descartes) *44*

Hieroglyphic for the Earl of Shaftesbury's *Characteristics* (1732) *45*

Eagle (Le Brun) *48*

The Eagle Man (Le Brun) *48*

The Eagle Man (Le Brun) *48*

The Owl Man (Le Brun) *49*

The Camel Man (Le Brun) *49*

Illustrations of the passions (Testelin, after Le Brun) *50*

Etonnement (Le Brun) *50*

Attention et Estime (Le Brun) *51*

The angle of savagery (Le Brun) *51*

Corrected perspective construction on a flat plane (Bosse) *52*

Construction of rectangles and parallelepiped (Bosse) *53*

The regulation of interior space (Bosse) *54*

The five orders (Perrault) *55*

Faces on entablatures (Sagredo) *57*

Face on cornice (Francesco di Giorgio)   *57*

The cube mansion (Morris)   *58*

The Corinthian mansion (Morris)   *58*

The Ionic mansion (Morris)   *58*

Man in square and circle (Francesco di Giorgio)   *60*

The canon of proportion with the foot divided into twelve, sixteen, and twenty-four units (Francesco di Giorgio)   *62*

The body in the facade of the church (Francesco di Giorgio)   *63*

Geometrical section of a church (Francesco di Giorgio)   *64*

Section of a church (de l'Orme)   *65*

The city as a body (Francesco di Giorgio)   *67*

III   The Body and the World

Zodiacal man: Miniature   *70*

Phanes in zodiacal egg: Hellenistic stone relief   *71*

Vein man and zodiacal man: Miniature   *72*

Convenientia anni et mundi (Venerable Bede)   *73*

Goddess Nut among zodiacal signs: Sarcophagus lid   *74*

Decan zodiac from the rooftop temple, Denderah   *75*

Microcosmic man: Seasons, temperaments, planets   *78*

Ebstorfer world map of 1234   *79*

Cosmic man: Man and his Creator (Hildegard of Bingen)   *80*

Cosmic man (Pisa, Camposanto)   *81*

Mensura Christi (St. John Lateran, Cloister)   *84*

Man in square (Taccola)   *87*

Man in square (Cesariano)   *88*

Man in circle (Cesariano)   *89*

Eve (Dürer)   *91*

Adam (Dürer)   *91*

Christ holding the Cross (Michelangelo)   *94*

IV   Gender and Column

Ixion on the wheel   *98*

Metrological slab   *100*

Hermes with his lyre: Red figure vase   *102*

Lyre player   *103*

Doryphoros   *105*

Doryphoros   *106*

Doryphoros   *106*

Doryphoros: Head and torso fragments   *107*

Doryphoros: Torso only   *107*

Doryphoros (Apollonios of Athens)   *108*

Polykleitos: Amazon   *109*

Lysippus: Eros stringing his bow   *111*

Euphranor (after). Antinous. Also known as Lansdowne Athlete   *111*

Polykleitos: Amazon ("Capitol" type)   *112*

Apollo Belvedere (Audran)   *114*

Medici Venus (Audran)   *115*

V   The Literary Commonplace

Narcissus mosaic   *117*

Kouros carrying piglet   *120*

Kouros   *120*

Kore poikile   *121*

Kore   *121*

Caryatids (Fréart de Chambray)   *130*

Persians (Fréart de Chambray)   *130*

Atlantes: Temple of Zeus at Agrigento   *132*

Model: Temple of Zeus at Agrigento   *132*

Caryatid porch, Erechtheion: Frontal view   *134*

Caryatid porch, Erechtheion: Side view   *134*

Caryatid   *136*

Caryatid fragment from Treasury of Cnidos   *137*

Caryatid head from Siphnian Treasury   *137*

Hellenistic "Persian" winged supporter   *139*

Tel-Halaf: "Temple-Palace" restoration   *140*

Tel-Halaf: Restored porch   *140*

"Humanized" standing stones from south Brittany   *141*

VI   The Rule and the Song

Priene: Restored columns of temple   *144*

Haghia Triada sarcophagus: Sacrifice scene   *148*

Mycenae: The Lion Gate, exterior view   *152*

Mycenae: The Lion Gate, interior view   *153*

The heraldic ruler figure   *154*

"Ritual or sacred tree" relief: Nimrud   *155*

"Ritual or sacred tree" relief: Nimrud   *155*

Temple of Hatshepsut at Deir-el-Bahari: Exterior   *159*

Temple of Hatshepsut at Deir-el-Bahari: Interior   *159*

Djed character (after David)   *160*

The raising of the *djed* (after David)   *161*

Djed and lily ornaments on a limestone plaque   *161*

Building of the pyramid of Zoser: Section of the stages (after Edwards)   *163*

Stone Heb-Sed pavilions   *164*

Exterior detail of Heb-Sed   *164*

Reconstruction of wooden prototype of Heb-Sed pavilion (after Borchardt)   *165*

Colossus of Rameses III, from Luxor   *166*

Colossi of Rameses II, from Luxor   *167*

Palm leaf columns   *167*

Reconstruction of the Megaron at Tiryns (after Schliemann)   *168*

VII   The Hero as a Column

The constitution of a Greek temple: Heraion at Olympia (after Chipiez)   *172*

The constitution of a Greek temple: Temple of Poseidon at Paestum (after Chipiez)   *173*

Stylobate and column foot: Temple of Aphaia at Aegina   *174*

Column bases: Palace at Pylos, megaron porch   *177*

The narrowing of the Doric flute (after Chipiez)   *178*

Herakles wielding an ax (after Orlandos)   *179*

Carpenter wielding an adze   *179*

Funerary giant pythos   *180*

Reconstruction of painted decoration on echinus and abacus (after Boetticher)   *181*

The "Basilica" at Paestum   *183*

Doric capital: Wooden origin of stone construction (after Choisy)   *184*

Doric capitals: Stone construction of cornice in relation to details (after Choisy)   *185*

Doric temple: Detail of François Vase   *187*

Archaic temple model, from the old Parthenon   *188*

Terracotta cornice detail: A treasury at Selinunte   *188*

Model, from the temple of Hera at Perachora: Present state and reconstruction (after Coldstream)   *190*

Model, from a temple at Argos (after Schweitzer)   *190*

Model, from the Heraion at Samos   *190*

Model, from Chania Tekke   *191*

Model of a round building from Archanes   *191*

Temple of Apollo at Delphi: View from theater   *192*

Eretria Daphnephoron: Plan of site (after Auberson)   *194*

Reconstructed model of hut: Eretria Daphnephoron (after Auberson)   *194*

Eretria Daphnephoron: Excavation   *195*

Lefkandi "Heroon": Excavation plan (after Popham)   *196*

Lefkandi "Heroon": Reconstructed axonometric (after Popham)   *196*

Thermon: Excavations of buildings A and B (after Soteriadis)   *197*

Thermon: Excavation plan (after Sotcriadis)   *197*

Restoration of temple at Dreros (after Beyer)   *198*

Restoration of temple at Prinias (after Beyer)   *199*

Thermon: Temple site (after Soteriadis)   *200*

Thermon: Cornice reconstruction, projection (after Soteriadis)   *201*

Thermon: The metope tablets   *201*

Plan: Heraion at Olympia (after Dörpfeld)   *202*

Elevation: Heraion at Olympia (after Dörpfeld)   *202*

Columns showing cuttings for votive tablets: Heraion at Olympia   *204*

Heraion at Olympia: Central acroterion (after Dörpfeld)   *205*

"Personalized" bronze votive tablet   *206*

Pediment eagle   *208*

VIII   The Known and the Seen

Propylaca of the Acropolis at Athens   *218*

Temple of Poseidon at Paestum   *218*

Temple at Segesta: Corner detail of cornice   *219*

Entasis and inclination of the Parthenon and Propylaea columns (after Penrose)   *221*

The curvature of the Parthenon stylobate   *222*

"Optical" effect and inclination of the columns of a Greek Doric temple (after Choisy)   *223*

The inclination of the columns: The temple at Aegina and the Hephaesteion in Athens (after Chipiez)   *223*

Diagrammatic representation of the Parthenon stylobate curve (after Penrose)   *223*

Example of anathyrosis, Doric: Temple of Zeus at Olympia   *224*

Concave curvature of stylobate: Temple of Apollo at Delphi   *225*

Remains of color stucco: Temple C at Selinunte   *230*

The inlaid eyes of an otherwise vanished statue   *232*

The tonal effect of polychromy: Temple of Artemis at Korkyra (after Dörpfeld)   *233*

IX   The Mask, the Horns, and the Eyes

Bathycles: The throne of Apollo at Amyklae, sections, plan, elevation, and projection (after Martin)   *238*

The temple of Nike Apteros   *240*

Ionic stylobate and wall base: Temple of Nike at Athens   *241*

Ionic base: Erechtheion   *242*

Ionic bases: Didymaion   *243*

Ionic stylobate and intermediate steps: Miletus   *243*

Ionic capital with honeysuckle necking: Delphi, Athenian portico *244*

Ionic volute (according to Porta) *245*

Corner capital: Didymaion at Miletus *246*

Detail of capital: Troad, temple of Apollo Smintheus *247*

Proto-Ionic capital *248*

Capital: Temple of Artemis at Sardis, side view *250*

Capital: Temple of Artemis at Sardis, front view *250*

Temple of Artemis at Sardis: View *251*

Temple of Artemis at Sardis: Plan at three stages of construction (after Fraser) *251*

Temple of Artemis at Magnesia: Elevation (after Humann and Kohte) *252*

Temple of Artemis at Magnesia: Plan (after Humann and Kohte) *252*

Temple of Artemis at Magnesia: Detail (after Humann and Kohte) *253*

Temple of Artemis at Magnesia: Capital *253*

Temple of Artemis at Ephesus: Temple site from the original urban settlement *254*

The great statue of Artemis *256*

The beautiful Artemis *256*

Lenaean vase *258*

Attic skyphos *259*

Development of the Naples Lenaean vase (after Frickenhaus) *260*

Coins of Samos and Magnesia (after Donaldson) *261*

The Aphrodite of Aphrodisias *262*

Artemis and Zeus *262*

The Perge goddess *263*

Votive relief showing Zeus Labraundos flanked by Ada and Idrieus *263*

The Croesan temple of Artemis at Ephesus (after Krischen) *264*

Temple of Artemis at Ephesus: Survey of remains (after Bammer) *266*

The name of King Agesilaos erased from the column base at Ephesus (after Schaber) *266*

Temple of Hera, Samos: The superpositions of different temples on the site (after Gruben) *267*

The Naxian column, Delphi *268*

Aeolic capital from Neandria *270*

Aeolic capital from Lesbos *270*

Votive column of Larisa, restoration (after Boehlau and Schefold) *272*

Aeolic capital: Votive column of Larisa *273*

Tomb of Cyrus (after Ghirshman) *274*

Hattusas: General plan (after Naumann) *276*

Hattusas: Royal palace (after Naumann) *277*

Hattusas: Buildings A and D (after Naumann) *277*

Hittite royal seals with *lugal-gal* character (after Akurgal) *278*

Hittite royal seals with *lugal-gal* character (after Akurgal) *279*

Burned Palace, Beycesultan (after Seton Lloyd) *280*

Megaron 3, Gordion: Interior view *281*

Megaron 3, Gordion: Plan *281*

Türbe near Erkizan on Lake Van *282*

Drawing of mosaic from floor of Megaron 2, Gordion *284*

Mosaic from floor of Megaron 2, Gordion *284*

Xanthos: Theater detail *286*

Xanthos: Harpy tomb *287*

Xanthos: Harpy tomb, details of base *287*

Xanthos: "Sledded" tomb *288*

Xanthos: "Sledded" tomb, constructional details (after Choisy) *288*

Fethiye: Big Ionic tomb *289*

Fethiye: "Sledded" tomb and small Ionic tomb *289*

Doorjamb from Tamassos, Cyprus (after Karageorghis)   *291*

Tamassos, Tomb XI: Plan and section (after Ohnefalsch-Richter)   *291*

Coins showing the temples at Paphos and at Byblos (after Donaldson)   *293*

Basalt relief of two demons with a "sacred tree"   *295*

"Aeolic" capitals: Megiddo   *295*

"Aeolic" capitals: Ramat Rachel   *296*

Relief of Nabu-aplu-iddin   *297*

Cypriot (Aeolic) stele   *298*

Cypriot Hathor stele   *299*

Phoenician Hathor and lotus capitals   *300*

Terracotta Idalion model   *301*

Terracotta miniature shrine in the form of a house   *302*

Cypriot Sphinx stele   *303*

Sumerian foundation nails   *304*

Cypriot miniature shrine (after Tatton-Brown)   *305*

Cypriot bull masks (after Tatton-Brown)   *305*

Columns from Tell-el-Rimah   *306*

Palm column from courtyard   *306*

Inrin temple, Warka   *308*

Osiris columns: Luxor, Temple of Rameses II   *309*

Sumerian "trough"   *310*

Sumerian post-knot sign (after Van Buren)   *310*

Origin of the *djed* column and Sumerian temple post in the reed bundle   *311*

Sistrum handle with Hathor head   *312*

Capitals from Hathor chapel: Tomb of Hatshepsut at Deir-el-Bahari   *313*

Astronomical chapel: Hathor temple, Denderah   *313*

X   The Corinthian Virgin

The setting up of the Corinthian capital (after Serlio)   *317*

The origin of the Corinthian order and the setting up of its capital (Perrault's Vitruvius)   *318*

Invention of the Corinthian order (after Fréart de Chambray)   *319*

White-ground lekythos   *322*

White-ground lekythos   *323*

Development of white-ground lekythos: A warrior deposited in his tomb by sleep and death   *324*

Development of white-ground lekythos: Mourners bringing baskets of pottery to the tomb, which is crowned by an acanthus-like plant   *324*

Acanthus plant   *325*

Palmette decoration of a Campanian funerary vase   *326*

The giant Delphic acanthus and the dancing figures   *328*

Apollo on tripod: Marble relief   *329*

The Delphic Hekation or perirranterion   *330*

The giant Delphic acanthus as a tripod stand: Reconstruction   *331*

Giant marble lebes sustained by acanthus and palmettes   *332*

Palmette and acanthus   *333*

The treasury of the Massalians, Delphi: Capital of column   *334*

Demeter, Triptolemos, and Persephone: Marble relief   *335*

The Eleusinian Telesterion: Plan of the excavations (after Mylonas)   *336*

The Eleusinian Telesterion: View of site   *336*

The temple of Apollo Epikurios, Bassae: Plan (after Dinsmoor)   *339*

Ionic column: Measured drawing by Haller von Hallenstein   *339*

The temple of Apollo Epikurios, Bassae: View   *340*

The Corinthian column: Details (after Cockerell)   *341*

The Corinthian column with its entablature (after Cockerell)   *341*

Tholos in Marmaria, Delphi: Detail   *344*

Tholos in the sanctuary of Asklepios at Epidauros: Plan (after Roux)   *344*

Tholos in Marmaria, Delphi   *345*

Tholos at Epidauros: Interior Corinthian and exterior Doric order   *346*

Tholos at Epidauros: Section (after Roux)   *346*

Tholos in the sanctuary of Asklepios at Epidauros: View of the substructure   *346*

Choragic monument of Lysikrates, Athens: Exterior   *347*

Temple of Olympic Zeus, Athens: Fallen column   *348*

Temple of Olympic Zeus, Athens   *348*

XI   A Native Column?

Tuscanicae dispositiones (after Perrault)   *353*

Cramping of the Tuscan cornice (after Stratico's Vitruvius)   *354*

The Colosseum, Rome (after Desgodets)   *355*

The Amphitheater at Verona (after Maffei)   *355*

Temple of Jupiter O.M., Capitol, Rome (after Gjerstad)   *358*

Temple of Jupiter O.M.: Detail of columns and cornices (after Gjerstad)   *358*

Tomba "dell'Alcova," Cerveteri   *360*

Tomba "dei Capitelli," Cerveteri   *361*

Tomba "dei Rilievi," Cerveteri   *364*

Two coins of the Minucii showing the column of Minucius Augurinus (after Becatti)   *364*

Porta Augusta, Perugia   *366*

Porta Marzia, Perugia   *367*

Santa Maria Novella, Florence: Corner columns   *368*

Medici Villa, Poggio a Caiano   *369*

The orders (Cesariano's Vitruvius)   *371*

Preface

Fascination with the orders of architecture goes back—for me—to adolescence. I had early decided to be a "modern" architect, but I considered that "classical" architecture incorporated some important and timeless rightness, held some lesson that it would only impart to one initiated into its demanding discipline through the long (and, I thought, probably wearisome) apprenticeship of learning the "orders," a lesson that would then serve that architect well when returning to modernity. My sense of inadequacy was only a symptom of the epochal hesitation of modernity before history which had been wistfully and lyrically expressed (two generations before mine) by that great innovator, Guillaume Apollinaire:

> Vous dont la bouche est faite à l'image de celle de Dieu
> Bouche qui est l'ordre même
> Soyez indulgents quand vous nous comparez
> A ceux qui furent la perfection de l'ordre
> Nous qui quêtons partout l'aventure.

That so many of my very unadventurous predecessors and contemporaries had passed through such an apprenticeship without receiving the graces for which I hoped did not deter me. So I followed the routine of drawing the Doric and Ionic column, of throwing shadows on the line drawings, and of rendering them laboriously in Chinese stick-ink, which I would grind myself after the manner of the old Beaux-Arts apprentices. It did not then occur to me (nor did my teachers ever allow me to suspect it) that the routine they taught and I followed had only been formulated late in the nineteenth century, and that earlier architects who learned and used the orders saw and learned them in a quite different way.

Only much later, when I started looking seriously at the architecture of the fifteenth and sixteenth centuries and reading what the architects of the time had said, did I realize that their view of antique columns could only have the most superficial connection with the "orders" I had learned. It led me to think about how the columns and beams of ancient temples had been conceived and constructed by their builders. Although it seemed a central concern of those architects and craftsmen, as is obvious from the most important text surviving from antiquity, Vitruvius' manual, surprisingly little had been written about it, and some of Vitruvius' notions were even dismissed as a late fabulation by his commentators. To me he seemed so much closer to the time of the Greek builders that to dismiss him thus might mean that some valuable aspect of their approach and of its implications might be obscured.

Merely providing a critical (but not a skeptical) comment on Vitruvius might seem too modest an aim for the bulk of the book. And in fact it has grown into something more ambitious, since it seemed to me that any critical (and therefore also historical) account of the orders would have to fulfill at least two conditions—and these conditions inevitably set up an almost irreconcilable conflict.

The first was to provide the context, an anthropological rather than a historical one, within which the columns were formed, showing why they were differentiated and why it was that their configurations had acquired a kind of timeless validity. The second had to be historical and had to provide a genealogy of this idea, show how the columns were constituted and altered in time, how they were worked and perceived by their makers, and what accounts had been offered of their transformations.

The attempt to reconcile these conflicting conditions makes this book truly an essay. In working at it I tried to fulfill the demands both of method and of historical documentation. I hope that readers will experience this conflict as a salutary one, since I did so. They will find the emphasis on method stronger in the first five chapters, on documentation in the next six. And I hope that they will agree that the "orders" of nineteenth-century theory had only the most indirect connection to the *genera* and the *modi* of antiquity.

Although the orders—at any rate as I was taught them—inevitably denied me the grace for which I had hoped, yet as I came to know them historically, they taught me a rather different lesson, since they allowed me to think anew about what people expect from buildings. My ambition then is not merely to provide a hermeneutic vision of Vitruvius' text, but to look through it and use it as a retort in which all we desire and that we rightly expect from our environment might be distilled, if only in a historical form.

A book of this length and scope is inevitably a collective work, even if it carries a single author's name. My thanks are due in the first place to the electors for the Slade Professorship in Fine Arts at Cambridge University who elected me for the year 1980–1981 and allowed me to formulate the ideas on which this book is based. Professor Robin Middleton, one of them, who was also directing General Studies at the Architectural Association at the time, asked me to repeat my lectures there and tried with some success to puncture some of the balloons I floated. The first draft was begun at the Center for Advanced Studies in the Visual Arts in Washington and finished in the Getty Center for the History of Art and the Humanities in Santa Monica: to both these institutions and to their directors, Henry Millon and Kurt Forster, I am deeply grateful; without their help, and (at the Getty) that of Gretchen Trevisan, Kimberley Santini, Daisy Diehl, and Herbert Hymans, as well as Maria de Luca, the book would have taken many more years to complete in far less agreeable circumstances. Between these two terminals, the Graham Foundation for Advanced Studies in the Visual Arts provided the grants that allowed me to visit the sites and museums in Crete, Egypt, and Turkey, while the Getty Institute provided one to visit the Pergamon Museum in (then) East Berlin; the University of Pennsylvania Foundation provided me with the means to make further trips to Turkey and to Greece. Without these visits the book would have been very much poorer. Inevitably, too, I am very indebted to the helpful officers of several libraries: first those of the London Library, without whose forbearance my work would not have been possible, but equally those of Dr. Williams' Library, the Warburg Institute, the Society of Hellenic and Roman Studies at the Institute of Archaeology of London University, and those of the two universities at which I have taught during its making, Cambridge and Pennsylvania. The Kress Foundation awarded me a generous grant that enabled me to collect and commission the illustrations. I am grateful to Lisa Ackerman for her sponsorship.

Many of the ideas in the book have been discussed in seminars at the universities of Essex, Cambridge, Columbia, and Pennsylvania, and I am particularly grateful to those students who questioned my statements and urged me to be more searching. Of course I owe a special debt to those who acted as my research assistants: Victor Deupi, Roy Lewis, Maria Karvouni, Rebecca Williamson, and, at the Getty, Toni Pardi; Taha Al-Douri and Alaa El-Habashi scanned and redrew with great skill and great commitment, while Persephone Braham piloted me through the various straits presented by the word processor. Toby Martinez has provided administrative as well as moral support. Deborah Sandersley took responsibility for many of the photographs.

The seminars on the image of the body summoned by Ivan Illich allowed me to give many of my ideas a new orientation, and the bibliography that Barbara Duden prepared in connection with them has been very valuable. When I began working on these themes, many friends were patient as well as helpful: Michael and Elizabeth Ayrton, Richard Brilliant, Peter Burke, Roberto Calasso, Bruce Chatwin, Edmund Carpenter, James Coulton, Harriet Crawford, Marco Frascari, Moses Finley, Alfred Frazer, Carlo Ginsburg, Alexander Goehr, Remo Guidieri, John Graham, Renate Holod, Theresa Howard-Carter, Charles Kahn, Hara Kiossé, Geoffrey Kirk, Rudolf zur Lippe, Michael Meister, David O'Connor, Roger Norrington, Gregor von Rezzori, Susan Sontag, Leo Steinberg, Cecil Striker, Dalibor Vesely, Michael Vickers, Peter Warren, Irene Winter. Parts of the manuscript were read critically by Andrew Barker, Andrea Carlino, Barbara Duden, William Gass, Ivan Illich, Stella Kramrisch, David Leatherbarrow, Geoffrey Lloyd, Henry Millon, Holly Pittman, Robert Tavernor, Liliane Weissberg, and Richard Wollheim.

The entire first draft was read by my father-in-law, the late Eugene Sandersley, whose patience and knowledge of the ancient languages not only saved me from many slips but also allowed me to develop certain ideas much more fully. Francesco Pellizzi provided editorial as well as moral help throughout this long period. I dedicate the book to him in recognition of many years of close friendship. The final draft was generously but critically read by Walter Burkert, Myles Burnyeat and Ruth Padel, Hans-Karl Luecke, and Anthony Snodgrass, though they must be absolved from any responsibility for my excesses. Some specific suggestions are acknowledged in footnotes.

My association with Roger Conover and Bruce Hunter goes back over many years. I wish to register here my appreciation of a great deal of effort expended on my behalf. I must also thank Alice Falk, whose sharp eyes prevented many an error.

That my wife Anne did not wish to be named as coauthor of this book—which she is in every sense—is a matter of regret to me. Her inexorable questioning made every sentence a collaborative effort, leaving no room for the slovenly or the unformed idea. Moreover, without her intrepid driving on Turkish dirt tracks, her organizational skill, her infinite patience with my chaotic work methods, and her familiarity with the linguistic and bibliographical problems it presented, I could not have brought the work to conclusion.

Philadelphia, November 1994

The Dancing Column

I

Order in Building

▪ Catalogues and recipes ▪ Skeptics, critics, and romantics ▪ The Beaux-Arts and the Ecole Polytechnique ▪ Imitation of the primitive hut ▪ The end of metaphor ▪ Return to sources

For much of the twentieth century the words "academic" and "classical" have been terms of abuse—or at least of disapproval—among artists, particularly when they were linked. They implied stale routine, submission to outworn rules for which universal validity had once been claimed.[1] In architecture these rules were associated with the five orders of columns. During the sixties and seventies there came a renewal of interest, not in the long-despised and long-forgotten rules but in the outward show of classicism, as a protest against the naked economy of commercialized modernism.

In spite of the periodic appeals to return to an "order," a "canon," fustian classicism had long been opposed to an elegant and exhilarating modernism. The promise it held became a wretched actuality as modernism was identified with the exploitation of mechanized building methods. The growing problems and miseries of overdeveloped cities led many people to see even the fustian, as long as it recalled something of the past, as a lost treasure. "Classical" became a blanket term of approval when fragments and caricatures of the orders sprouted in all sorts of "appropriate" and "inappropriate" places.

Everyone knows what an order looks like. From Anchorage to Cape Town, from Vladivostock to Rio de Janeiro there are ministries and cathedrals, institutes, libraries, and even private houses whose entrances are porticoed and walls are scored with columns carrying arches, beams, and pediments that most passersby will recognize as classical. For most purposes, that implies "Roman," "Latin," and/or "Greek"—the last carrying the more refined and respectable associations.

The column and the beam that it carries, when they are combined so as to be recognized (however approximately) as belonging to a definite type, are called an *order*. The type may be established either by the proportions of the thing or by the ornaments which stigmatize it; usually it is determined by a combination of both. There is a large repertory of these recognizable types, both "ordinary" (that is, classical) and exotic, but the classical core is fixed at five orders. Three are Greek: Doric, Ionic, and Corinthian. Two are more or less Roman: the first, of supposedly Etruscan origin, is known as Tuscan; the second, the only one of the five whose label does not refer to a place, is the Composite.

Catalogues and recipes

Four were described and named by the Roman architect Vitruvius in the first century BC; of his *Ten Books on Architecture* much of the second, third, and fourth book are devoted to setting out their detailed prescription and accounting for their origins. Since it was first printed (perhaps in 1486)[2] many have glossed or illustrated that text, and these woodcuts and engravings have had an overwhelming influence on Western architecture.

Several other treatises have included a special section on the orders, which in some cases has virtually become a separate book. In time every systematic writer on architecture was obliged to include such a section in his work, some of which masons and carpenters would use as a pattern for copying, without any reference to the text. Sebastiano Serlio, a Bolognese architect and the first to publish an architectural picture book[3] (as against the more usual thing, an illustrated book on architecture), devoted the fourth of his seven books to the orders. It was he who fixed the list of five orders and also devised the name "Composite" for the additional variant on the Corinthian.

Yet the notion of an order had not been formulated. Leon Battista Alberti, writing about 1450, still uses the word *ordo* to mean the most general, abstract "order": but also to mean "a row of columns" or a contignation of them, "a story"; and much more simply, "a layer" or "course" of stone or brick. The kinds of columns, on the other hand, are the familiar ones: Doric, Ionic, Corinthian, which he details following Vitruvius (but also Pliny). To these he adds the Italic, which is to combine the ornaments (and the virtues) of the other three: the Doric egg-and-dart echinus, the Ionic volutes, the Corinthian leaves and stalks.[4] However, his rationale was not taken up.

Nor was there, when Alberti wrote (and even when his book was printed—in AD 1485/6, after his death), any generally accepted idea of a set or canon of columns. Francesco Maria Grapaldi, author of a compendium of building terms, has no comment on this word in connection with building. Filarete and Francesco di Giorgio have no use for it either. Nor does the first translator of Vitruvius, Cesare Cesariano, a Milanese disciple of Bramante. It appears first in the modern sense in a letter of Raphael. The exact transformation that leads from occasional mentions to the fixing of its sense in Serlio's book has not been traced, though there is nothing improbable about it. Serlio had been the faithful disciple of and had taken over the stock of drawings that had been left by Baldassare Peruzzi, who counts as Sienese (though born in Etruscan Volterra). Peruzzi had been very much a member of Bramante's and of Raphael's circle: the use of the word and the canonic set of five orders must have been fixed sometime between the printing of Alberti's ten books (1486) and of Serlio's fourth (1537), probably by someone or some group in Raphael's circle. It was seen as a formulation from which there could be no appeal after the earliest of Sebastiano Serlio's architectural books appeared: there were five orders, and their arithmetic and geometrical characteristics as well as all their sculptural decoration were fixed.

Serlio's Composite fifth order was also quickly generally accepted, although there was no written antique authority for it. Many "deviant" forms of the Vitruvian orders had in any case been known from antique monuments. There was also an impressive witness to the extra order in the Roman Colosseum, a building whose prestige was as vast as its physical bulk. Each bay of the four-story Colosseum exterior was made up of three arched openings one over the other, framed by half-columns that rose from Doric/Tuscan to Ionic to Corinthian; the top story was a blind wall framed by pilasters with Corinthian capitals, but the shafts were more slender than was usual for that order. In Serlio's Composite, the more elaborate capitals that crowned the columns of the triumphal arches of Titus and of Septimius Severus on the Roman Forum were set on the slender Colosseum shafts.[5] These capitals, which combine the Corinthian leaves with Ionic volutes, gave the order its name.

Serlio's publications show just how things stood in his time: that fourth book—entirely devoted to the orders—was published first, as a kind of *ballon d'essai* for the whole enterprise, and indeed it might as well be considered the first of all order books. It is certainly the first piece of architectural writing in which the five orders were canonized, and the word *order*, a word previously used both quite generally in architecture—much as it is still used in western Indo-European languages (*ordre, ordine, orden, Ordnung*)—and for such generic things as a row or a story, was applied specifically to the proportions and decorations of columns. It was advanced to the specific meaning which has been borrowed in almost every attempt to "systematize" architecture;[6] no other word will quite do for the arrangement of

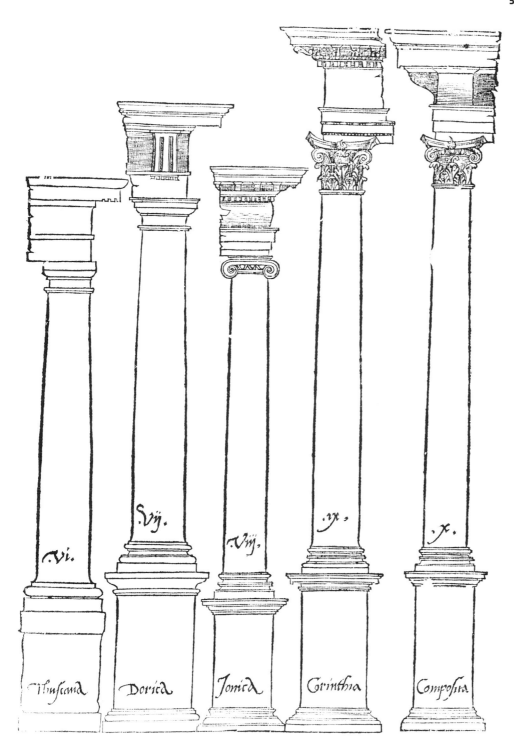

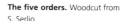

Order in Building

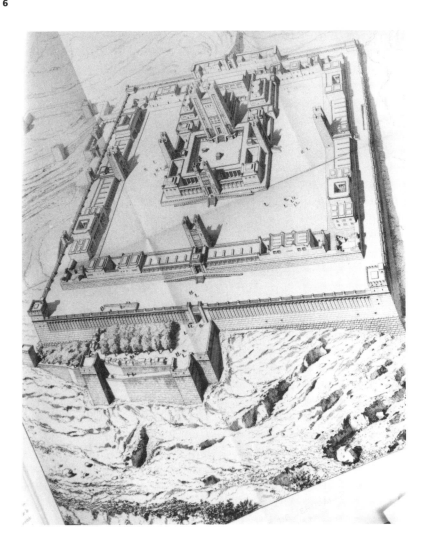

Reconstruction of the Temple in Jerusalem. Photogravure from G. Perrot and C. Chipiez (1882–1889).

beam and column, and no arrangement other than that canonized by Serlio can be accepted in the long run: "l'Architecture n'a que cinq *Ordres* qui lui soient propres."[7]

From 1537 when Serlio's fourth book was first printed until our time there would be a new order book or two published most years. Architects, contractors, masons, carpenters— anyone connected with building—would acquire at least one of them, even if sometimes these would be mere flyleaves for workshop use. Ever since formal architectural training moved into schools during the eighteenth century and replaced the old and haphazard methods of apprenticeship, the schools would lay in stocks of such books for their students and even patronize or publish their own.

Skeptics, critics, and romantics

1 Charles Chipiez   These treatises provided the staple diet of most architectural students, yet surprisingly little attention was paid to the way in which the canon was first constituted. Over a century ago a critical account of the subject was last attempted: Charles Chipiez published the *Critical History of the Greek Orders* in 1876. He was outside the mainstream of academic

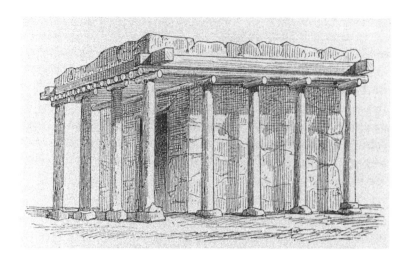

Origin of the Ionic. Wood engraving from C. Chipiez.

French architecture, but something of a disciple of Eugène-Emmanuel Viollet-le-Duc, and also became a teacher at the Ecole Spéciale, the protesting architectural school that for a time stood apart from the French government-licensed system. His maturity coincided with the Franco-Prussian war, and one of his best-known buildings was a memorial of the siege of Paris at the Château de Buzenval. Chipiez's energies were mainly devoted to the work of reconstructing—in large drawings—a series of monuments from the remote past (such as the Jerusalem temple), which he presented in successive Salons and at the World's Fair of 1889. A complement to this work was his collaboration with the historian-archaeologist Georges Perrot on a vast eight-volume *History of Art in Antiquity,* which appeared between 1882 and 1889; Chipiez died in 1901.[8]

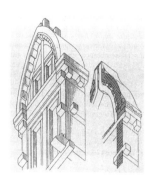
Lycian tomb. Wood engraving after A. Choisy.

   His book on the orders was written at the time when Schliemann had begun digging at Hissarlik (which was to be generally recognized as Homer's Troy) and at Mycenae, though only the preliminary report on Troy had been published[9] and nothing was yet known of the Minoans and very little of the Hittites. Still, Chipiez showed himself aware of the context in which antique (or at least Greek) architecture had grown, particularly of its links to Near Eastern and to Egyptian building. The Perrot-Chipiez joint history, which is familiar to archaeologists and historians, has been persistently ignored by architects, as has Chipiez's book on the orders. It is only at the very end of the century that Auguste Choisy's *History of Architecture* with its marvelously lucid wood engravings interpreted archaeological material as complex as Perrot's and Chipiez's through a constructional rationale. Choisy's notion of "a timber masonry" seemed to reconcile the conflicting accounts of the timber origin of the orders with the demands of stone construction; his book was to become essential reading for Auguste Perret and for Le Corbusier.

   But throughout the nineteenth century, architects were content to use various hand-me-down copybooks, of which the most popular was the one first published by Charles Normand in 1819[10] and constantly reprinted until the 1940s without any substantial modifications. Normand's *Parallèle,* though quite lavish at first, was (like some of his other works) a how-to manual, unconcerned with "why?" or "when?" He took the need for his manual for granted: it was to become the obvious model for all those order books issued in schools of architecture, so that it must have been the best-selling architectural book of the nineteenth

The five orders. Metal engraving from C. P. J. Normand.

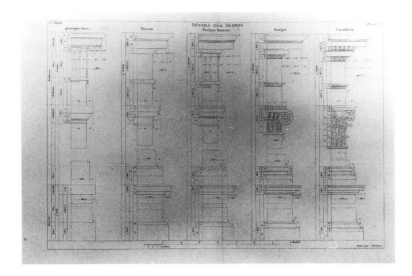

century. Humbler, though almost as popular, was Normand's equally schematic reworking of the famous sixteenth-century book on the orders prepared by another Bolognese architect a generation younger than Serlio, Jacopo Barozzi da Vignola;[11] but this was even more of a copybook, and had no text to speak of. In fact the pedagogic usefulness of the orders was hardly ever questioned during the eighteenth and for much of the nineteenth century. If nothing more, the orders were considered drawing exercises, much in the same way as Latin grammar was thought to be good for the adolescent mind. The orders were taught as a foundation course in which the beginner could learn draftsmanship, presentation, rendering, the casting of shadows, and other such skills while acquiring—by osmosis—the refinements secreted in the models he copied. In the course of their training, architects were constantly presented with the orders as the axiomatic model of historical precedent, and the pressure for an unquestioning acceptance of such teaching (while their real attention was focused on gaining control of the very profitable contractual side of a rapidly expanding and increasingly mechanized building trade) is one explanation for the very limited appeal of Chipiez's work.

2 Charles Blanc   It seems as if any book that discussed the visual arts in general and architecture in particular (and some such books were inevitably read by some architects at least) also had to pay a great deal of attention to the orders. Charles Blanc's very popular *Grammaire des Arts du Dessin,*[12] for instance, discussed the orders as part of a general history of design and had some reflections to offer on the parallel between the horizontality of classical architecture and the lines of a serene human face. To Blanc this parallel seemed to demonstrate that the architecture of beam and column inevitably suggested calm, submission to fate, long duration; his unstated confidence in empathy led him to deduce a universally valid, physiologically based aesthetic theory from this phenomenon. He acknowledged the central idea to the Dutch connoisseur, artist, and cosmographer Pierre Humbert de Superville, who had attempted to develop a system of absolutely fixed signs in works of art, both of colors and linear elements, and who had been only the first of many theorists in Britain, France, and Germany to attempt the construction of a "scientific" or "objective" aesthetic system.[13] For the rest, Blanc was

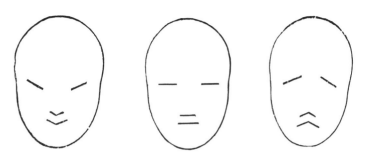

mostly interested in the relation between the timber origins and the stone forms of the orders as they were recounted by Vitruvius.

3 G. W. F. Hegel, Arthur Schopenhauer, and Jacob Burckhardt The same is true of the account of the origin of the orders that Hegel gave in his *Aesthetics:*[14] he was, I am sure, simply reasserting the commonplace when he stated categorically that the three Greek orders—Doric, Ionic, Corinthian—were the most famous, and "that for architectonic beauty and fitness for purpose nothing better had been found, either before or after." The beauty of the orders had a finality in Hegel's aesthetic system, since architecture belonged to the earliest, the symbolic, stage of development in the arts and had long ceased to be of interest to the Spirit whose dialectical realization through physical entities constituted the history of art as far as he was concerned. Strangely enough, his most violent opponent, Arthur Schopenhauer, took an analogous view of architecture, though his aesthetic was genetic rather than historical. The four stages of mineral, plant, animal, and human development constituted the basis for a system in the arts: in Schopenhauer's view (and in his terms) the aesthetic moment occurred when the will was no longer engaged in perception, and the perceiving subject arrived at the still moment of pure representation. Of course, it was not some immanent Spirit whose truth was revealed to the perceiver through the materiality of the work of art. No; what the perceiver obtained through the work of art was insight into Platonic Ideas—archetypes of every category.

As far as the art of building was concerned, these moments occurred at the lowest stage, the mineral. Building (classed with gardening and water management) could only deal with categories such as weight, cohesion, rigidity, fluidity, and light. Schopenhauer's best-known dictum on the matter was that all the arts aspire to the condition of music, since only the higher arts—of which music (in Hegel's system it had been poetry) was the very highest—could allow a view of more exalted archetypes. For him, it meant (in terms of building) that "the column is the very simplest, the purpose-determined form of support. The spiral column is tasteless. The square pier is really not as simple. . . . The forms of the frieze, the beam, the arch and dome are determined by their very structure. . . ." Wood architecture is no architecture; Gothic architecture is based on the fiction of weightlessness and inferior to the antique. The ornament of capitals belongs to sculpture, not architecture. It is the measured and even display of column and beam, of weight and its support, refined through such notions as entasis and the thickening of the corner column in Doric temples, that allows the most immediate aesthetic access to the simple Platonic ideas which architecture can display. The latest measurement taken at Paestum, which found such refinement in Doric temples, confirmed Schopenhauer in his view.[15]

Schopenhauer's senior by a decade (whom he disliked almost as much as he did Hegel), Schelling, took a rather different yet almost equally un-Hegelian view of the orders. In his system, the three "substantial" (*reale*) arts were music, painting, and the plastic arts, of which last architecture was a part. But equally the guiding structure of art was displayed by the "ideal" art, poetry, through its three modes: lyric, epic, and dramatic. Music might analogously be triply articulated according to rhythm, harmony, and melody, and architecture—which he considered a kind of rigidified music—according to its orders: rhythm was represented by the Doric, harmony (rather Vitruvianly) by the Ionic, and melody by the Corinthian. This led him to consider mathematical factors as essential to architecture as they were to music. The echo of his ideas, however little they were to the taste of Schopenhauer (or of Hegel), had a crucial effect on doctrines of geometry in architecture in the nineteenth century.[16]

Another very powerful testimony to the universal authority of the orders (and speculation about them) is the description of the Doric temples the young Jacob Burckhardt had seen in Paestum, which opens his *Cicerone;* he saw the column and its entablature entirely in empathetic, bodily terms.

The Greeks worked through the idealized treatment of form rather than through mass. . . . The first device is the narrowing of the column at the top . . . which assures the eye that the column cannot be overturned; next are the flutes. They signify that the column is condensing and hardening, as it were, gathering its strength. At the same time, they emphasize the upward thrust. . . . The powerful pressure [of the beams] spreads its upper ending out into a swelling—the Echinus. . . . Its profile is the most important measure of strength in every Doric temple; at its base it is edged by three channels, like the folds of a delicate, slack outer skin.

To Burckhardt it seemed evident that the "Doric order is one of the most exalted creations of Man's feeling for form."[17]

That notion of Burckhardt's about the Doric he owed to German romantic historiography, with all its insistent belief in the organic development of nations and of "style," which goes back to the Neapolitan philosopher Giambattista Vico. Burckhardt made this approach his own during his student days in Berlin, where he also absorbed the idea that the Germans, and perhaps the Prussians most particularly, had a kind of inner unity with Greek tribes, especially with the Dorians.

Later, Burckhardt would find confirmation of some of his ideas about Doric materiality in Schopenhauer, whom he admired. But he would see analogous Vican ideas developed sympathetically by French historians, particularly by Michelet and Quinet;[18] Coleridge had given these ideas currency in Britain through his *Philosophic Lectures,* and they had been taken up both by Carlyle and Ruskin. But Burckhardt's exaltation of Doric, written in the 1850s, marks yet another change. Friedrich Schinkel, who had Greekified Berlin and had died some fifteen years before Burckhardt wrote that passage, had used the Ionic order for preference on civil buildings such as the Schauspielhaus and the Altes Museum. The new historiography that had so influenced Burckhardt in Berlin saw the Dorians as the core Greeks and Germany, particularly Prussia, as the Dorian state come again. And that gave the Doric order a new status as the "order of orders."

4 John Ruskin   More surprisingly, John Ruskin also needed to justify much of what he wrote by reference to some such ideas about the orders. There were, as far as he was concerned, only two of them—"and there can never be any more till dooms day"[19]—the Doric and the Corinthian, the convex and the concave. Whatever else may have been more perfect and exalted in Gothic, the elemental alternation of concave and convex "was first invented by the Greeks, and it has never materially been improved. . . ."[20] All the orders beside these two archetypes were debasements, since convexity and concavity were the two dialectical principles of order and no other was necessary.

The Beaux-Arts and the Ecole Polytechnique

During the first half of the nineteenth century, and well beyond it, there was therefore a scission between what was known and discussed among the general public (for whom I have made Hegel, Ruskin, and Blanc token spokesmen) and what was taught in the professional schools. This teaching was dominated by the method of the Parisian Ecole des Beaux-Arts, where the orders were considered the perfect type of historical form. Historical forms did not radiate the exalted clarity of geometry, nor did they display a manifest obedience to natural forces, such as could be read into yet other forms—those derived from the nature of materials and the needs of manufacture. They were only guaranteed by their antiquity, their familiarity.

This threefold division of all architectural forms into geometric, historical, and those derived from the nature of materials was taught from the chair of architecture at the Ecole

Polytechnique by J.-N.-L. Durand for the first thirty years of the nineteenth century. With metaphors he wanted no truck. The orders were ancient and venerable, but any attempt to derive their forms from the human body was only worth holding up to ridicule. With Vitruvius' other "fable," that the details of the orders were derived from ancient timber construction, he did not even bother. Instead, he directed his attention to rebutting a popular account of the origin of the columns and the temple form from a primitive hut.[21] It was from Durand and from his disciples at the new Ecole des Beaux-Arts that postulant engineers and architects—first French, but soon also Italian, German, English, and Balkan (later even American and African, Chinese, and Japanese)—learned the nature of a new schematic and universal classicism. In fact the two Ecoles provided the whole world with a model of architectural education for over a century; their pupils and their pupils' pupils bore this doctrine abroad throughout the world. The method of design and the historical doctrine of the Ecole followed the teaching that Durand had originally proclaimed for the Ecole Polytechnique, and the Beaux-Arts never developed a rival doctrine. Columns were its primary building elements. It made his dogmatic planning method, by which a pupil proceeded from the general to the particular, moving from the main axes to subaxes, and then to planning grids; on the grid, the student would locate the main point-supports—that is, columns—and finally consider the enclosing walls, which left the columns as freestanding as possible. Durand's teaching had to be apodictic, since the insistence on the primacy of column over wall could not be justified either by an appeal to geometry or to statics. It had been formulated—almost exactly as he taught it—some fifty years earlier, and it derived from the conviction that a primitive hut built of tree trunks was the true archetype of all human building. Durand's covert dependence on this teaching gave his irritated dismissal of the argument all its sarcastic virulence.

Imitation of the primitive hut

The teaching about the absolute priority of the primitive hut had been advanced about the middle of the eighteenth century by the Abbé Marc-Antoine Laugier in a brief *Essai* that was to become the architectural ordinance of the Enlightenment. It provided a contractual view of architectural origins, the form of the pedimented temple being the "natural" man's instinctual answer to his physical needs—much as the nuclear family provided the natural sanctioning type for Rousseau's conception of the social contract. Laugier's account of the first hut was conflated with the older legends that Vitruvius and other ancients told about a primitive wooden architecture from which the stone architecture of the Greeks had derived. In the eighteenth century there was hardly any archaeological evidence available against which those legends could be examined, nor were the ancient historians read as assiduously as they would be in the nineteenth. Even the self-appointed guardian of the old teaching against such innovators as Durand, Antoine-Chrysostome Quatremère de Quincy, from his beleaguered position as *secrétaire perpetuel* of the Académie des Beaux-Arts (to which the Ecole was only nominally subject), had to account for the imitation of wooden architecture in terms that—rather reluctantly—took Laugier's primitive hut into consideration. Thus the Doric order did indeed imitate wooden construction, but not some wretched "ouvrage grossier d'un besoin vulgaire"; the exalted stone forms must have derived from an already refined and courtly wooden construction.[22] The schools of architecture had little time during the busy nineteenth century for the subtler teaching of Quatremère.

Though he had no such prejudice against "primitive" origins for the Doric order, the most subtle historian among the architects, Gottfried Semper, rejected both the eighteenth-century myth of the hut as the prototype of the antique temple and the older Vitruvian account of architectural origins: he did so largely because the transition from timber to stone—or rather from primitive, friable materials to more permanent ones such as those used in classical Greek buildings—seemed to him a constant of all monumental architecture. The primitive huts of ancient peoples, of which no trace remains, cannot be reconstructed back from these transformations: even our speculations about them must remain summary. Huts of modern "primitives" offer the only possible substance for our speculations, since to their inhabitants the sacrality of dwelling is incarnate in a twin archetype: the hearth and the double-pitched roof over it.[23] In that sense, of course, the supports, which keep the roof over the hearth and which were the main preoccupation of theorists from Vitruvius to Laugier, have no function at all as part of the archetype, whatever their structural role.

Still, Semper's influence on contemporary architectural thinking (beyond the two schools of architecture at which he taught in Dresden and Zurich) was surprisingly slender. Though architects and historians of art forgot him, anthropologists were to become interested in his ideas a generation or two after his death. The lack of enthusiasm on the part of architects is due, I suppose, to the suspicion with which any teaching that relied on an idea of "imitation" was regarded by many nineteenth-century theorists. Many architects would not sink into the torpid eclecticism implicit in the teaching of the Ecole; and yet all "modern" teaching seemed to reject the old imitative doctrines. The only alternative was an adherence to the belief, also current since the eighteenth century,[24] that a stone monumental architecture had a completely independent development. The arch-rationalist Viollet-le-Duc for one—whatever his differences with Quatremère over Gothic[25]—needed to satisfy himself that the perfection of Greek architecture (the only one that to his mind could rival the ultimate accomplishment of French thirteenth-century Gothic) was achieved by ways quite different from those that ancient writers described. The shape of the column, as he saw it, was imposed by methods of transportation, and its various refinements were introduced as a result of observing the first rude structures under the bright sunlight. The old fables, known largely through Vitruvius even when corroborated by other ancient writers, Viollet treated as later interpolations or as corruptions introduced by Renaissance scribes.[26]

It is therefore very interesting that throughout his critical book on the orders, Chipiez did not find it necessary to refer to Viollet's *Entretiens*,[27] which was by then fifteen years old and one of the most popular architectural books in the French language. Chipiez was perhaps the first modern to take Vitruvius' legend at face value, but then he did not want to frame the orders in a system of his own devising: his concern was to show the slow and majestic transmission of forms, from Mesopotamia and Egypt to Greek lands, and to inquire as to the possible changes of value and meaning through which individual features passed in their wanderings. He ends on a paean to the "bold metaphor"[28] by which the Greeks compared the column to the human body: that bold metaphor, which takes up much of Vitruvius' account of the orders, seems even more vital to me than it did to Chipiez.

The end of metaphor

Metaphors, as I have already pointed out, had been of no concern to Durand, nor to most nineteenth-century architects: it was from Durand's follower, Normand, or one of his coarsen-

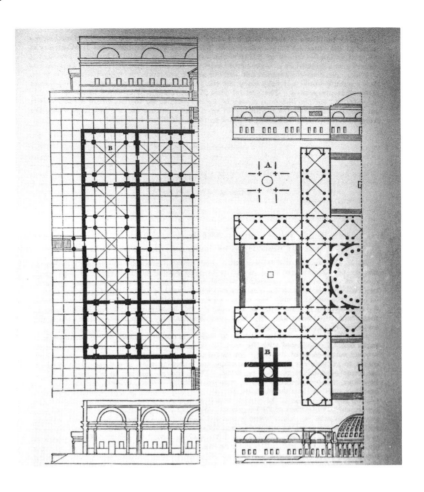

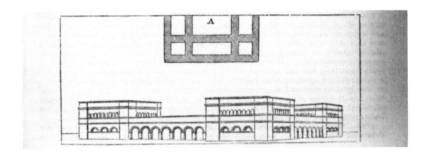

ing imitators that they learned the orders. If called upon to rationalize their use of them, they would probably have provided some form of Durandian argument—as did the last official theoretician of the Ecole des Beaux-Arts, Julien Guadet, at the end of the century.[29]

Strangely, the Anglo-Saxons were rather late in the field.[30] It was not the few Britishers who went as students to the Ecole who proclaimed the Durandian doctrine to the English-speaking nineteenth century, but books. Of these, the most important certainly was the humorless (and grossly overindustrious) architect-polygrapher Joseph Gwilt's *Encyclopaedia of Architecture*.[31] It opened with a long historical introduction, in the course of which he discussed the orders as part of Greek architecture and suggested their Egyptian derivation. But the doctrinal teaching about the orders came in part 3 (Practice of Architecture), which is introduced by an essay on "Beauty in Architecture" in which aesthetics are dismissed as "a silly pedantic term[,] . . . one of the metaphysical and useless additions to nomenclature in the arts, in which the German writers abound. . . ."[32] Gwilt presents himself throughout as the man of good sense, to whom beauty in architecture is, in fact, the true result of a fitness for purpose that is as valid for buildings as it is for machines.

He justifies the special attention he gives to the orders by arguing they are a particular case of beauty resulting from fitness. Gwilt claims to draw his basic ideas from a Scottish philosopher, Archibald Alison, though he probably first learned of them at second hand from J. C. Loudon's *Encyclopaedia,* a popular building handbook for the decade before Gwilt's replaced it.[33] Durand was easy to read in this key. After all, he also maintained that the different heights of columns were simply due to the different requirements of loading. And when it came to the rules of composition, Gwilt deferred to superior authority and quoted Durand extensively—and by name.

The decisive American contribution came very much later; William R. Ware had organized the very first American architectural school at MIT in 1865 and moved to start another one, at Columbia University, in 1881. For the benefit of his Columbia students he composed *The American Vignola,* which remained in print for many years; in 1977 it was republished in a paperback abridgement and dedicated by its editors in the service of a "new Renaissance"—much as the old edition had served the "American Renaissance" that followed the Chicago Columbian exhibition of 1892.[34] Ware's coarse, mechanically printed hand-me-down inevitably claimed Vignola's clear and proud copper engravings as precedent: it is a mark of those first American teachers' servitude to the Beaux-Arts that Vignola is adopted as the model preferred to any other. In France his eminence had become such in the nineteenth century that his name came to stand for any book of rules.

In the preface to *The American Vignola,* Ware tells of an incident from his early career: he had entered the office of Richard Morris Hunt, the very first American pupil at the Ecole, who is now remembered mostly for the Metropolitan Museum building and the socle of the Statue of Liberty (though he was in fact one of the most prolific and successful architects of his day).[35] Young Ware was struggling at his drawing board with details of the Doric order out of one of the many Vignola reprints, when Hunt observed him, took the pencil from his hand, and showed him the "rule" for drawing the capital and cornice by a sleight of hand that makes all moldings and ornaments from the multiples of one-sixth of the column diameter. This shortcut, which Hunt said he had learned as a student at the Beaux-Arts, was of course not devised by Vignola, as Ware (and perhaps also Hunt) seemed to think. It came directly from the very thorough reinterpretation of the orders by Claude Perrault, the author of one of the most ambitious of all the seventeenth-century order books and the best translator-commentator of Vitruvius until that time.[36] Such imprecisions were symptomatic of the philistine and antihistorical character of that early-twentieth-century "Renaissance" that Ware had served.

Yet even at the time of that renaissance, whose heroes on either side of the Atlantic are now seen to have been Stanford White and Edwin Lutyens, there were vast numbers of architects who ruled their designs by the orders, though they would not see them exclusively through eighteenth- or even nineteenth-century pattern books; they had available up-to-date handbooks, most of them shoddy reprints of older ones. The bulk of institutional and colo-

nialist architecture of the late nineteenth and twentieth century was done by them: not only the British in India and South Africa or the French and Italians in North Africa, but even the Japanese in Manchuria produced such building.

Still, there were architects who wanted to appeal to the raw antique, to come to terms with ancient models and rejuvenate them in spite of the bulk of institutional vacuities. Auguste Perret's heroic attempt to pour a classical architecture of reinforced concrete into the primitive timber archetype of shuttering and scaffolding—much as Choisy's Greeks created their stone columns by petrifying timber construction—deserves a more generous treatment than I could afford it here. But other major and innovative architects adapted the existing orders to their own use. Antoni Gaudí was a conspicuous example, perhaps because he made the appeal so sparingly. And of course he was much more "Gothic" than "classical," a self-confessed disciple of Viollet-le-Duc.[37] In one important building, he used the Doric order impressively: the Parc Güell in Barcelona, which was to have been the central open space (called by Gaudí "the Greek Theater") of a new garden-city, or more precisely, of a residential urban park, organized somewhat on the lines of Bedford Park in London but set on the slope of the Muntanya Pelada instead of the flatland of a London suburb. This garden-city did not turn out a success.[38]

Return to sources

1 Antoni Gaudí The main entry into the park was a monumental stairway up to a Doric Ipostilo (hypostyle hall). It was to be the colony's market. Eighty-six Doric columns support a Catalan vault of reinforced brickwork. It is edged by a cornice, which in turn carries the meandering ceramic-covered parapet bench[39] to enclose a wide esplanade (now a place of nursemaids, prams, and ice creams) with a grandiose view over the city and the sea, over the vaults of the hypostyle—just like the stage of many an ancient theater. The esplanade cuts into the hillside, is enclosed by it, so that the natural lay of the land really does resemble that of

Parc Güell, Ipostilo. Antoni Gaudí. Detail of capital: photo by author.

Parc Güell, Ipostilo. Antoni Gaudí. Flank: photo by author.

many Greek theaters. It is not clear whether the idea was Gaudí's or the Hellenophile Güell's, but it is Gaudí's only explicit reference to Greek architecture, and explicit it is not only in the detail but in the context and implications of the project, which is in fact a very personal interpretation of Mediterranean unity. The shafts of the columns are as thin as those in very late Hellenistic examples, and very narrow at the top; the echinus is very widely flared, with a deep necking, like those of some archaic examples, while the abacus of the capitals is octagonal instead of the almost universally used square.

Each column is made up of eleven preformed, hollow concrete drums; the upper eight are fluted, the lower three flat and covered with chips of glazed white tile, as is the vault, so that the interior of the shady hypostyle is lit up by a strange reflected light. The cornice is about a third of the height of the shafts, heavier than most of the ancient examples. While the refined "correct" swelling of the shaft called *entasis* is not used, there are other strange optical devices, thoroughly "unclassical" ones: the outer columns are inclined inward, like flying buttresses, and much of the ornamental detail is improvised.

Gaudí admired Greek architecture for its "simplicity of form and ornamentation. . . . The superposition of the pieces subject only to the action of gravity is achieved with a delicacy of outline and joint, of mass and carving, all of which are minutely studied and corrected." But this admiration does not extend to contemporary "classicists": "the Parthenon is the product of necessity but the Paris opera . . . merely has a stupidly sumptuous facade. . . ."[40]

Of course, Gaudí was not a scientific student of archaeology. The strange arrangement of the Güell Doric was derived from the most obvious sources. The partly fluted shaft has a fairly well known Italian and Hellenistic precedent, the so-called temple of Hercules at Cori, which was the subject of an early plate in Hector d'Espouy's *Fragments d'Architecture Antique,* though Gaudí may well have known it from other publications, such as Piranesi's.[41] His other notable precedents, the temples of Athena ("of Demeter") and of Hera (which had been known as the "Basilica") at Paestum, were also represented in splendid Piranesi engravings. The "Basilica" was particularly important, since it was thought to have been a civil building and was restored as such by Henri Labrouste. That restoration Gaudí certainly knew, since it was published (without acknowledgment) in Luigi Canina's huge *History of Ancient Architecture.*[42] Still, whatever his sources, none of his exemplars looks quite as heavy and fierce as Gaudí's Doric. The barbaric grandeur of the thing was plainly intended to provide as authentic a primitive Greek experience as was possible in moderniste Barcelona.

Gaudí's primitivism has an instructive counterexample: in 1918–1920, some ten years after the Parc Güell, Gunnar Asplund built the little crematorium known as the Woodland Chapel in the Southern Cemetery of Stockholm. It is entered through an extremely widely spaced Doric porch, three columns deep, four columns wide. Because of that wide spacing, the columns (which are in fact canonically proportioned, 1:7) look extremely thin. They support a vast hipped roof of creosoted wooden shingles, more than three times the height of the columns, while the roof covers both porch and chapel within; they carry no cornice, nor do they have bases—the columns rest directly on the pavement. The interior of the rectangular chapel is articulated by a ring of eight columns in a circle supporting a skylit dome, which takes up most of the volume of the roof, turning the whole thing into a toy Pantheon.[43]

2 Gunnar Asplund   Although Asplund's chapel has the high and rather effete finish typical of the best Swedish building of the time, it is obviously intended to look primitive, even primal.

The columns standing in the pine wood are assimilated to the straight, narrow trunks. Their wide spacing, the absence of a cornice, the bulk of the roof all suggest archaic wooden construction; Doric, Tuscan, yes—but also Nordic. It is almost as if in the presence of death Asplund was attempting a regeneration of those forms, which were both national and timeless, classical. However inspired it may now seem, Asplund must have considered this way of designing to have failed him. Apart from one or two tombs, his only use of classical columns after this were the "Pompeian," toylike attenuated herms and Corinthian columns of the Skandia cinema in Stockholm of 1922–1924. Before the decade ended, he had abandoned all classical details, and the 1930 Stockholm exhibition shows him as a fully fledged and highly inventive modernist.[44]

3 Adolf Loos   Another clamorous twentieth-century example never got beyond the drawing stage: Adolf Loos' entry for the Chicago Tribune Building competition of 1922 was in the form of a Doric column in polished black granite. Its shaft was to house offices and stand twenty-one stories on an eleven-story high podium. The building was entered by a portico recessed in the square base, with two Doric columns in antis that carried the appropriate cornice; the column building carried only the stump of a block, invisible from the ground, and no cornice at all.[45] Some critics have bypassed the scheme as a prank, but no one seriously interested in Loos and his work could ever maintain such a view. As a well-known antiornament publicist, Loos had to defend (or so he felt) his frequent use of the orders on public buildings, and his most famous contribution to the Viennese townscape, the Goldmann and Salacz Store (now simply known as the Looshaus), has a rather curious and abrupt ground piano nobile whose bay windows and columns make the sort of Chicagoan reference most of his contemporaries inevitably missed.[46] He made the Chicago Tribune project his most aggressive homage to the Greek column. One other scheme (of 263!) in that competition was to propose a whole building in the form of a column—but it looks vulgar and ham-fisted beside Loos' solemn and grandiose folly.[47]

At the time it might have come as rather a surprise that there were not more columniform schemes entered for that competition. The most influential architectural text about skyscrapers, Louis Sullivan's "The Tall Office Building Artistically Considered," insisted on the tripartition of the building into base, shaft, and capital, even if Sullivan denied invoking any direct analogy to the column.[48] That essay appeared in *Lippincott's* for March 1896, just as Loos, Sullivan's "brother in the spirit"[49] (though very much his junior), was leaving the United States to return to Europe.

Loos almost certainly read that text before he designed the Chicago Tribune building. He had been sent a copy of it in March 1920 in the hope that he would find a European publisher for it; at any rate he considered that he was offering American architects the true solution of the skyscraper problem with which they had still to grapple. When he published the scheme in Vienna after its rejection by the jury, he wrote that when approaching the design he had ruled out the possibility of inventing any new form, since new forms are too quickly consumed and would not answer the promoters' desire to "erect the most beautiful and distinctive office building in the world," as the competition conditions demanded.[50]

He did not really consider alternative solutions to the problem (Loos wrote in a Viennese professional magazine) of offering a type for the American skyscraper. After all, he continued,

Woodland Chapel, Stockholm South Cemetery. Gunnar Asplund. Exterior photo after Ahlberg.

Chapter I

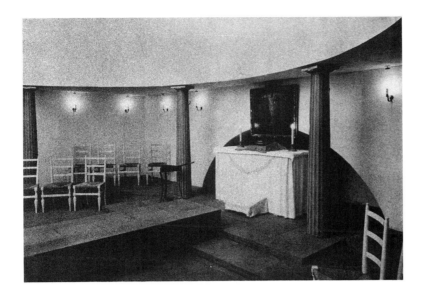

Woodland Chapel. Gunnar Asplund. Interior photo after Ahlberg.

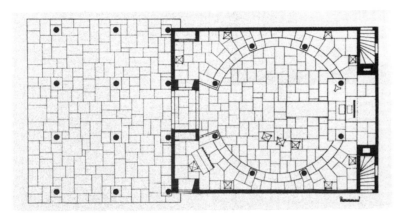

Woodland Chapel. Plan by Asplund, after Ahlberg.

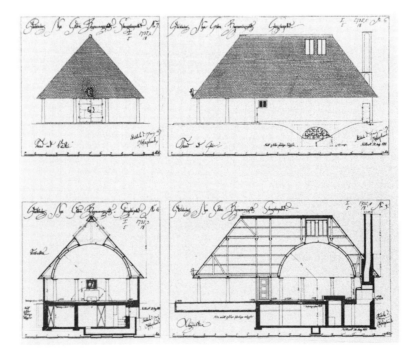

Woodland Chapel. Sections by Asplund, after Ahlberg.

Order in Building

Chicago Tribune Building. Perspective by Paul Gerhardt, from Chicago Tribune (1923).

Chicago Tribune Building. Perspective by Adolf Loos, from Chicago Tribune (1923).

Chapter I

representative examples [of skyscraper building] could, at the beginning of the movement, be distinguished from each other. But now it is already difficult for the layman to tell whether he is looking at a building in San Francisco or Detroit. The writer chose the form of the column for his project. The motif of the free-standing giant column was provided by tradition. Trajan's column already served as the prototype of Napoleon's in the Place Vendôme.[51]

The wordplay on the column of type and the column building for a newspaper, which was offered as an arbitrary justification of the form, seems too weak for this eccentric and powerful project. Loos was very conscious of the anomalous nature of his enterprise. In the apologetic article I have been quoting (headed "Should a Habitable Column be Allowed?"), he refers to other skyscrapers—such as the Metropolitan and the Woolworth buildings, both in New York—which were based on historical types (the Mausoleum of Halicarnassus, a Gothic spire) not originally intended for habitation, and which provoked no objection. Loos was also quite aware how awkward the office accommodation in his cylindrical shaft was bound to be, yet what really mattered to him most was the display of the giant column on its pedestal as an isolated and solemn prophetic object for the Windy City. Indeed Loos deliberately separated the granite-faced base of the building from the other low "functional" brick workshops behind the column to emphasize its splendid isolation, its monumental character. He was convinced that the building of such a column would have a beneficial effect on all the architecture of the future; the envoy of his essay is a vers libre quatrain, which he had printed in English:

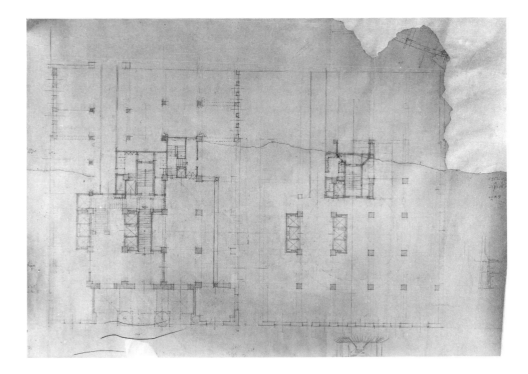

Chicago Tribune Building. Sketch plan, from Loos-Archiv, Albertina. © Verwertungs-gesellschaft bildender Künstler.

The great Greek Doric column must be built.

If not in Chicago, then in some other town.

If not for the "Chicago Tribune" then for someone else.

If not by me, then by some other architect.[52]

What Gaudí had attained by twisting the order to his peculiar missionary and structural purposes, Loos could only assert by isolation and gigantism: the supremacy of value pitted against the city of brute fact. The Doric order appeared to have been *the* ultimate historical form, *the* great human building achievement, unfettered by structural contingency or the base need for shelter. All of them—Gaudí, Sullivan and Loos, and Asplund—saw the Doric order as ultimate, though perhaps only for Loos did that imply the last ever, the last possible.

Gaudí's Doric order became a fragment among other fragments—the broken ceramics, the monstrous animals. Yet he considered all structural forms (and this included "his" Doric columns) as in some way "natural" responses to the ultimate architectural problem of defeating gravity. The Doric order offered that particularly refined response: a civic response, which he saw as ennobling the particular setting.

In any case, for Gaudí the very heterogeneity of the pieces became a kind of guarantee of the homogeneity of the whole. Loos, on the other hand, asserted the otherness of his form against an environment apparently homogeneous. His fragment is a *monstrum*: "monster," but also "exemplar." As for Gaudí, so for Loos, a selection from among existing forms is the unavoidable way, since the history book from which they selected was already complete; all possibilities of ornamental invention were exhausted and his designs would have to be recorded in some quite different way, perhaps not in a history book at all—in a new kind of book, whose compilation had not yet begun.

Yet for Loos certainly, for Gaudí less definitely, the Doric order provided the unique figure that might ennoble the base need for shelter. The grandiose and primal form seems to spring directly from the instinct to build with such vigor that it had to transcend necessity. Each of my twentieth-century masters knew, nevertheless, that the ancient forms were no hand-me-downs, that they would not submit gracefully to the architects' different and contingent demands; all the projects had therefore to violate the limits of the formal convention.

Projected tomb for Max Dvořak. Adolf Loos, from Loos-Archiv, Albertina. © Verwertungsgesellschaft bildender Künstler.

II

Order in the Body

· Scriptural columns: John Wood the Elder · Gian Lorenzo Bernini on bodies and columns · Jacques-François Blondel on faces and capitals · Face, character, fate · Psychology in the studio: Charles Le Brun's Conférences · Artificial gesture, natural gesture · Face, character, mood, landscape · Variety of character and fixity of canon · The church as a body · The city as a body · The city as a house

Asplund and Gaudí and Loos in their very different ways were tributaries to an ancient and grandiose—but apparently buried or broken—tradition: that the Greek orders enshrined and transmitted values of primordial as well as perennial validity. Until the eighteenth century the core notions of that tradition could be taken for granted; from the beginning of the nineteenth, the different historians and architects who wrote about the orders needed to plead and vindicate. That may be why attention clung so insistently to the Greek Doric order, and why my three salient twentieth-century examples are of Greek Doric. It seemed older, nobler—or at least notionally more "primitive" and therefore less "historical"—than the others. The most convincing testimony to this veneration may still be Le Corbusier's account of the Parthenon, both in *Vers une Architecture* (which has had an inestimable influence on several generations of architects), and in greater detail in his travel diary, *Voyage en Orient*.[1]

Constant recourse to the orders had presented quite a different problem for many architects in the sixteenth and seventeenth centuries, since it was their heathen antiquity that required an apologia. They could still be seen adorning the ruins of the temples of pagan gods, as they had been canonically described in the writings of Vitruvius, Augustus' contemporary. An *interpretatio Christiana* in the light of Scripture and the writings of the church fathers was therefore essential, particularly in view of the Puritan rediscovery that the human body was the only true temple the Spirit required. Apologists involved in this endeavor had available to them the texts about the buildings that God had specified to His chosen—Noah, Moses, Solomon, Ezekiel—as well as a few late-antique fragments that could be connected to scriptural people and places and gave physical reality to the texts.[2]

Scriptural columns: John Wood the Elder

Higher biblical criticism was to superannuate that whole enterprise, even if, to many, it could still appear intellectually respectable—though rather old-fashioned—as late as the middle of the eighteenth century.[3] John Wood the Elder, the creator-architect of Georgian Bath, followed the ideas of the two sixteenth-century Jesuits, Juan Bautista Villalpando and Jeronimo Prado, who had maintained that the three Greek orders were descended from an ur-order revealed or dictated by God Himself to King Solomon; they argued that the ancient pagans had in some way stolen it, appropriating the great revelation, which could therefore be read back into Scripture only by way of Vitruvius. These ideas he amalgamated with his own strange musings on the legends of British origins, mostly out of Geoffrey of Monmouth's chronicle, in which the notion of an archaic, just, and almost democratic society was first associated with an ancient pre-Christian British monotheism. Wood imagined that it had been preached by proto-Newtonian Druids in an idyllic age of ancient British social harmony, celebrated in academies in and around the stone circles: Stonehenge, Avebury, and a minor one near Bath at Stanton Drew, which because of its location assumed an enormous importance in his account.[4] The reconciliation of this arcadian antiquity with Holy Writ could only be worked out through an architectural theory in which the orders could be seen as both British and wholly scriptural. In describing the essential notion that underlies all the different orders, Wood offered this definition, much more exalted than the one I formed at the beginning of the previous chapter: "Order is that kind of appearance exhibited to the eye by any artificial object or figure, which by the regularity of its composition is pleasing, and answers to the various purposes for which it is made or intended." Later on in the same passage he continues, "Man is a complete figure

Stanton Drew Academy. Engraving from J. Wood (1765).

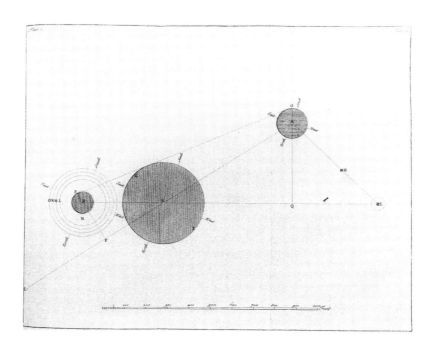

Stonehenge restored. Engraving from J. Wood (1747).

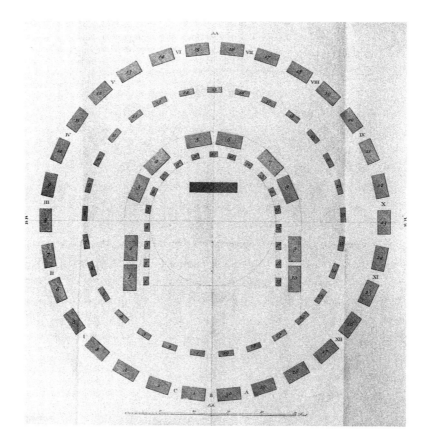

Chapter II

and the perfection of order. . . . And of the infinite number of parts of which he is composed, do but unfold any one of them and what astonishing beauty will arise to the most intelligent eye!"[5]

    To John Wood the human figure seemed the exemplary incarnation of that harmony which also dominated building through the various orders of architecture: not only the human body when alive and moving, but even when it submits to the dissector's knife and shows the internal organization of every member, so that each one separately will be seen to stand to the whole as the body itself does to the ordered world. Hence Wood's insistence on the essential circularity of all bodily organs, a Newtonian conception of corporeal unity which he shared with many of his contemporaries.[6] His use of the architectural orders was therefore as earnest, as fresh, and as individual as all this enthusiasm would lead one to expect. The three-tiered facades that enclose the King's Circus (the first circular space of this kind in Britain) are articulated by coupled half-columns, Corinthian over Ionic over Doric—a regular succession of three orders, each pier carrying a British acorn in place of the biblical pomegranate.[7] Such an analogy between the body and the orders, which is echoed by other parallels between the body and buildings in general, is deeply ingrained in all recorded architectural thinking.

Gian Lorenzo Bernini on bodies and columns

Nearly a century before Wood's book was published, this analogy had surfaced ostentatiously, as it often did in the history of architecture. The incident is familiar to art historians, since it occurred during the only Paris visit of Gian Lorenzo Bernini, who was the most famous architect as well as the most famous sculptor of his day: he was also a playwright and a poet, almost by the way.[8] At the time of his visit he was nearly seventy. He had been summoned to Paris by Louis XIV after he had won the closed competition for a new Louvre palace that was organized among the leading Roman architects of the time when the king and his chief minister decided that no French architect had a grand enough manner for the job. In the event it was Bernini's imperious behavior that made him enemies in Paris; his project (even after being scaled down) turned out to be far too expensive for the king—who in any case had decided to concentrate his efforts on Versailles—and Bernini returned to Rome. However, at the time of his arrival in early June 1665, he was still being feted as the equal of royalty. As the litter in which he had traveled approached Paris, the steward of the king's household, who happened to be a well-known connoisseur and collector (and who also spoke fluent Italian), Paul Fréart, Sieur de Chantelou, was dispatched as the king's special envoy to greet him. Chantelou kept a diary of Bernini's visit, from which my account of the episode is drawn. The diary is in fact a series of long letters that Chantelou addressed to his brother Roland Fréart, Sieur de Chambray, who was himself a writer on art and the author of what was to become, beside Vignola's, the most popular handbook to the orders.[9]

    Chantelou had his carriage drive alongside the litter in which Bernini had made his journey north. Both dismounted, and after the preliminary civilities, both got into the same coach for the entry into Paris. A conversation ensued in which Bernini seems to have done most of the talking, and which Chantelou reports to his brother in some detail, ending:

    as for the matter on which he had come, he told me that the beauty of everything in the world (and therefore of architecture also) consisted of proportion; which might almost be

called the divine part in any thing, since it derived from Adam's body; that it had not only been made by God's Own hand but also in His image and likeness; the variety of the orders arose from the difference between man's body and woman's—because of the differing proportions of each—and added several other things about this which are familiar enough to us.[10]

And indeed some twenty years earlier Chantelou had received a letter from Poussin that included another tantalizing reference to the idea. Chantelou was at that time on a tour of southern France with the court, and Poussin, writing sadly from a wet and sunless Paris, imagines "that the beautiful girls which you saw at Nîmes will have given your spirit no less delight through the eyes . . . than the beautiful columns of the Maison Carrée—since the second are merely agéd copies of the first. I find it a matter of great satisfaction to get such a break during our labors which sweetens the exertion."

The fluted shafts of the Maison Carrée may not seem like even the gentlest erotic image to modern viewers. To Poussin, as to many of his learned contemporaries, the association must have been obvious. Such references may now seem sporadic, yet they do show that

*Post patiens operü, paruoqᶻ aſſueta iuuenta,
Cum subit, ingenias etate parturit artes. .*

.CORÍNTHIA.
.2.

*Prouenit ẽnde animi rerum Prudentia maior,
Et queeüqᶻ ſagax tendando repperit vsus em.*

De ſeize á trente & deux,il erre au Laberynthe
Du Monde perilleux,lors chaſque eſprit ſad on ne,
Pour la vie paſſer á vacation mainte.
Dieu ayde á qui Vertu quiert,par ſa bonté ſainĉte.

Chantelou could merely allude to Bernini's improvised remarks—on the commonplace of the human body as a microcosm, expressing the belief that this notion was fundamental to building—as hardly worth reporting to his brother: they seemed a rather elementary resumé of ideas that the two Fréart brothers had long held in common.[11]

At the time of that incident Chambray's book on the orders was already fifteen years old. Its lasting and steadily growing reputation was due in part to the beauty and clarity of its plates, in part to his digesting the contradictory detailed opinions of his predecessors, but perhaps above all to his new, scholarly precision. With its criticism of Palladio and even Vignola, it became one of the carriers of the new French cultural hegemony over Europe.[12] Of his devotion to the ideal that Bernini had mentioned so casually to his brother there can be no doubt. It even led him to regard the most obviously imitative orders, statue-columns—the "Persians" and particularly the "Caryatids"[13] (of which more later)—with some distaste, since he was a courteous man and therefore found it offensive that women should be made to carry very heavy weights.

The Doric order and Hercules.
Copper engraving from J. Shute
([1563] 1964).

Such literal imitation, which Chambray curtly dismissed, had been enormously popular in the previous century; his contempt was directed mostly at the practice of his contemporaries, who used statues as quasi-heraldic supporters in ways for which there was no antique precedent in fact. Such use had proliferated in the sixteenth century: the earliest book on architecture in English, John Shute's *First and Chief Groundes of Architecture* of 1563, is almost exclusively concerned to illustrate the parallel between each one of the orders and that human body which is its appropriate equivalent.[14] Elaborating on Vitruvian hints, it extends the metaphor to make the orders also signify the five ages of man, each age incarnate in a god or hero—from hoary Atlas as the Tuscan order, to maidenly Pandora as the Composite. Although there are no obvious precedents for Shute's plates, a number of engravers in Flanders about this time (of whom Hans Vredeman de Vries was the most prolific and best known)

The Ionic order and Hera. Copper engraving from J. Shute ([1563] 1964).

were interested in such ideas.[15] Shute however claimed to be following only Serlio and the earliest of the French commentators on Vitruvius, Guillaume Philandrier (or Philander);[16] and indeed his assigning of the figures to the orders is a variation on Serlio's text, though the close parallel of column and figure seems Shute's own contribution (rough as it is) to an understanding of the nature of the orders in a northern land where they seemed quite alien.

About the same time, the idea of the parallel is examined much more searchingly by Vincenzo Scamozzi, Palladio's disciple, who was the most elaborate and painstaking of all the specifically architectural *trattatisti*. His book learning was real enough, but he was also an experienced and successful builder. Scamozzi's insistence on finding a "natural" derivation for every architectural detail is almost obsessive; he was of the opinion, for instance, that the shaft—the least inflected member of the column—should be so specifically characterized that

The Corinthian order and Aphrodite. Copper engraving from J. Shute ([1563] 1964).

even if you saw it deprived of its capital and base, you could instantly tell to which order it belonged, on the analogy of your being able to recognize the features of a friend if you were shown his truncated corpse.[17]

Jacques–François Blondel on faces and capitals

More compelling than the analogy between the torso and the shaft of the column was the much more obvious (and consequently most elaborated) affinity between the head and the capital. *Caput-capitulum-capitellum* is an archaic bit of Latin etymology and is inextricably woven into the vocabulary of the orders;[18] the analogy between the whole column as body, and the cornice as head, is also occasionally if less frequently made. It is therefore odd, but not unexpected, to see it reappear as a kind of critical touchstone in the lectures of Jacques–François Blondel, an exact contemporary of John Wood, who was one of the last representa-

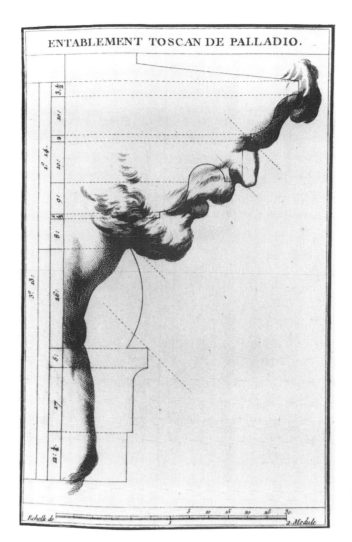

ENTABLEMENT TOSCAN DE PALLADIO.

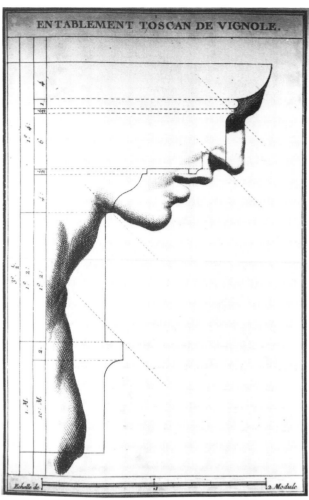

ENTABLEMENT TOSCAN DE VIGNOLE.

tives of the ancient tradition, the ultimate theorist of the ancien régime. Blondel had delivered a course of lectures to the Academy of Architecture every year, and (following precedent) proposed to issue them as a treatise, though he died before the enterprise was quite complete.[19] He illustrates the detail to which I refer in the course of a general discussion of the parallel between body and column, which he very much takes for granted.[20] For Blondel it is not just the best, it is the essential aid to teaching the orders, since the body analogy provides the postulant architect with the one true mnemonic guarantee: only by memorizing the close parallel with the body can he remember accurately the different parts of the various orders. Moreover (and this seems to Blondel to follow closely from the mnemonic guarantee) the face:capital, face:cornice analogy may be used as a critical instrument.

He instances what he means by applying it to a variety of moldings. Comparing the proportions of three select Tuscan cornices by Palladio, Scamozzi, and Vignola, he superimposes the human profile on each of them and finds Palladio's profile to be "like a human face whose parts do not seem . . . to have been made to harmonize. . . . The nose of a twelve year old [is] imposed on the chin of man of eighty, and crowned by the forehead of a man in his middle age." The criticism of Scamozzi is on similar lines; Vignola, on the other hand, is

The Tuscan entablature of Palladio. J.-F. Blondel (1771–1777).

The Tuscan entablature of Vignola. J.-F. Blondel (1771–1777).

Order in the Body

highly commended: "here the three parts seem to present a more acceptable relation between forehead, nose, and chin, which results in the unified profile the previous examples do not show. If these remarks have foundation, they might go some way to justifying our French architects who have preferred the teaching of Vignola to either Palladio or Scamozzi."

Although Blondel warns his readers that the analogy can only serve "as an aid to compare nature and art," he also considers that the character of a building might be influenced and modified by altering the size of the moldings to fit the appropriate human profile "without wounding the rules of good architecture."[21] Blondel does not refer to any of the several precedents that I have so far quoted; but he does claim to have derived this idea from two earlier authors, Diego de Sagredo and Jean Le Blond. Jean Le Blond was a minor seventeenth-century painter-engraver who also wrote a small treatise on the orders, which was dedicated to Charles Le Brun—and owed something to Fréart de Chambray. It hardly alludes to the subject of the analogy between column and body, however.[22] On the other hand, Diego de Sagredo's short book, which was two centuries older, could have been known to Blondel both in French and in the original Spanish.[23] Diego, a sixteenth-century cleric from Toledo who seems to have had a quasi-professional interest in architecture, read a number of earlier works to good purpose: the Philander commentary on Vitruvius, for example, which I have already mentioned. More interesting perhaps was his appeal to the *Dialogue on Sculpture* (*De Sculptura*) by the Neapolitan humanist Pomponius Gauricus, which had a long section on physiognomy that in turn relied on much contemporary speculation as well as on the fragments of ancient works that were known to him.[24]

This part of Blondel's argument approaches the general theme of the parallel between body and column, but also touches on a whole area of speculation that is now treated with scepticism (if not with contempt), though it had been a vital subject of inquiry for centuries: physiognomy. With it comes another matter of the greatest interest to the philosophers and the more or less "scientific" psychologists of the time, character. That last word was to acquire fresh significance in the seventeenth and eighteenth centuries, which led to it becoming, in the nineteenth and even in the twentieth century, the most common and yet the most awkward term of architectural criticism. In fact it even took over some of the meanings of *order,* so that Blondel's use of the comparison between face and cornice is radically different from the predecessors to whom he appealed. I therefore need to consider next the implications and context of Blondel's metaphor.

Face, character, fate

The study of the face, physiognomy, which had in antiquity provided the basis for a scientific psychology and physiology as well as a diagnostic technique for medical practice, came in the Middle Ages to be regarded as a form of divination through its association with astrology. It was consecrated by the authority of Aristotle, to whom a treatise on the subject (in which physical similarity to animals led to revelations of human character) was attributed, and was toyed with by many writers of the time.[25] Later, in the sixteenth century, it was elaborated even further. The most respected and quoted authority on it was the Neapolitan scientific magus, Giambattista della Porta, whose book developing the Aristotelian precept was popular for a hundred years and more.[26] Like many other physiognomists, he had to steer a careful course between physiological determinism and the privilege of human free will. This continued to be a problem of some consequence in the economy of psychological causality: Did the physical character condition, or merely exhibit passion? Did physiognomy express passion or else reveal it? Was the physical exterior merely a codifiable expression of changing interior states, or was it an index of permanent and unalterable psychic formation? A repertory of the different forms of facial configuration in man and animal avoided the tender issue by proliferating the data: by elaborating his taxonomy, della Porta seems to have achieved a scientifically acceptable transmutation of ancient lore while skirting accusations of heresy.

Because it seemed so deterministic, and therefore set close limits on human free will, physiognomy had been under the same condemnation of the church councils and fathers as astrology. In the second half of the first millennium, both disciplines weakened under this ban, though they did not entirely disappear. Things changed after the year 1000. In part this was due to Islamic influence, but there were also strong internal reasons for the change. The new understanding of the social fabric that the Western Empire had promoted, as well as the powerful desire for divine sanction as a manifestation of the corporate health of society, led to a strong current of inquiry into the rather shadowy notion of microcosm taken over from certain writers of antiquity. It was also the time when the notion of judgment assumed a new social prestige, associated as it now was with newly codified and newly written laws. The bound book, *liber,* displaced the scrolled *rotolum* or *volumen* in the imagery as well as the vocabulary of the time.

Liber scriptus proferetur	Lo! the book exactly worded
In quo totum continetur	Wherein all hath been recorded
Unde mundus judicetur	Thence shall judgement be awarded

says a verse of the Dies Irae, a hymn sung or said at all masses for the dead. Thomas of Celano, a Franciscan friar, wrote that hymn or sequence about 1220.[27] The image of that recording book appears interchangeably with the law book or the gospel book in the iconography of the twelfth and early thirteenth centuries, but it is soon displaced in its turn. Christ the judge holds up His wounded hands in the later images to signify His humanity, and the angels approach with emblems of the Passion or with scales on which the fate of souls is decided in pictures of the Day of Wrath. But there was a brief period, in the eleventh and the twelfth centuries—that great age of nascent legalism—when the written judgment had to be recorded in some tangible, showy great book.[28]

Such law-bound thinking seems to imply a reconsideration of the judicial nature of the world fabric as well. Codifying behavior would call for an analogous classifying of human types and the description of the moral nature of each one of them as it related to physical appearance. Around 1140 the most learned astronomer of the time, Raymond of Marseilles, issued the first reasoned Christian apologia for astrology;[29] at the same time Bernard Silvester's epic in prose and verse, his *Cosmography* (which was also known as *De Mundi Universitate, sive Megacosmus et Microcosmus*), which contained much astrological material, was read publicly before Pope Eugenius III.[30] The theme of the microcosm recurred insistently in the thinking of the period, not just among Neoplatonic philosophers but also among the more staid scholastic thinkers. About the year 1160, John of Salisbury, a familiar of the French and English (as well as the Papal) courts, wrote a handbook to statecraft, the *Policraticus,* which he dedicated to Thomas à Becket and which was to prove enormously popular. Although he

was a declared enemy of astrology, the analogy between the created world, society, and body (which he claims to derive from ancient sources) becomes almost the keynote of his text:

> A commonwealth, according to Plutarch, is a certain kind of body which is endowed with life by the benefit of Divine favour. . . . Those things which establish and implant in us the practice of religion and transmit to us the worship of God . . . fill the place of the soul in the body of the commonwealth. . . . Furthermore, since the soul is, as it were, the prince of the body . . . so those whom our author calls the prefects of religion preside over the entire body.
>
> The place of the head in the body of the commonwealth is filled by the prince who is subject only to God and to those who exercise His office and represent Him on earth, even as in the human body the head is quickened by the soul. The place of the heart is filled by the senate, which initiates good works and ill. The duties of eyes, ears and tongue are claimed by the judges and the governors of provinces. Officials and soldiers correspond to the hands. . . . Financial officers and keepers may be compared with the stomach and intestines, which—if they become congested through excessive avidity[—] . . . generate innumerable and incurable diseases, so that . . . the whole body is threatened with destruction. The husbandmen correspond to the feet, which always cleave to the soil, and need the more especially the care and foresight of the head since, while they walk upon the earth doing service with their bodies, they are more likely than others to stumble over stones and therefore deserve aid and protection all the more justly since it is they who raise, sustain and move forward the weight of the entire body. . . . Take away the support of the feet from the strongest body, and it cannot move forward by its own power.[31]

As it was with community, so also with the buildings that housed it, and most obviously with the church, which not only sheltered but also represented the Mystical Body. About a century after John of Salisbury, Guillaume Durand (known as Durandus), Bishop of Mende in southern France, wrote a both learned and popular handbook to liturgical practice in which he guides his reader to an understanding of the plan of a church: "the disposition of the material church answers the human body in the articulation of its parts, since the chancel, where the altar is placed, represents the head; the transepts to the left and right—the arms and hands, while the other part, which lies westward, is the rest of the body."[32] Durandus' handbook echoed the much more extended and searching meditations of earlier theologians, from Augustine to Honorius of Autun and Peter of Celle (who preceded John of Salisbury as Bishop of Chartres), on the same theme, but it runs through much orthodox Western theology of the twelfth and thirteenth centuries.

Michael the Scot, older than Durandus but younger than John of Salisbury, was the court scientist of the Emperor Frederick II and was both physiognomist and astrologer.[33] There was nothing surprising in the conjunction: the astrologer's sky was, after all, crowded with animals—those who guarded the zodiacal houses, and the even more numerous ones into which the constellations were configured. An association of nativity with character through animal resemblance was therefore all too tempting. Nor did the association of character and nativity quite deny free will. But in the following generation, the Paduan physician Pietro d'Abano (reputedly a collaborator of Giotto on the imagery of the Palazzo della Ragione in Padua) made the parallel between physiognomic and astrological determination,[34] so that he shifted the emphasis from the "moral," descriptive tendency physiognomy had had in antiquity to a physical one. It had by then acquired its "decadent" side, associated with the reading of body signs other than the facial (like the lines of the palm) to tell not only character but also the future; physiognomic reading came to be regarded, with palmistry, as a kind of charlatanry and vagrancy in the sixteenth and seventeenth centuries in many countries. However, in seventeenth-century France, where della Porta's rather discursive book was popular, the business of physiognomy was being turned into a modern science at the very time of Descartes and Gassendi.

Whereas della Porta (even if he did not have the reputation of a magus) had proposed physiognomy as a much more powerful and "natural" index to human nature than had been provided by nativities in judicial astrology, its most celebrated "scientific" practitioner and exponent in the seventeenth century was Louis Marin Cureau de la Chambre (an anti-Cartesian, incidentally, in matters of philosophy as well as physics). He was one of the original forty "immortals" of the French Academy, as well as physician to many distinguished and influential men in his time, including both the clamorously disgraced financier Foucquet and his enemy, the Chancellor Séguier.[35]

Marin Cureau may have met the painter Charles Le Brun, who had decorated Foucquet's notorious palace at Vaux-le-Vicomte, either in Séguier's or later in the financier's service. Both men, already distinguished, were protected by Séguier and by Colbert (who, after all, had started his career in the same service) from any troubles at the time of Foucquet's fall in September 1661; their friendship remained constant. Le Brun by then had become chancellor (1648) and was to be rector (1668) of the Academy of Painting (Colbert's instrument in suborning all French art to the royal propaganda machine) as well as (from 1663) director of the Gobelins factory of tapestries—and therefore the most powerful man in the French art world. He designed Cureau de la Chambre's tomb in St. Eustache in Paris, as he had also drawn the frontispieces of Cureau's books.[36]

The relation between the two men was to have momentous consequences when Le Brun set out to restructure the teaching of sculpture and painting in France for the new Royal Academy of Art, which was to have a monopoly of it according to some of the new scientific ideas; this "academic" program was to become a model for the teaching of art all over the world. Previously artists had built up and composed both the voluntary and involuntary characteristics of their figures into poses and gestures that implied a whole repertory of rhetorical precedent. Now Le Brun directed the artist's attention to seizing the involuntary expression of an inner action of the soul.

Le Brun relied on Cureau, certainly, in formulating this program, but he was also familiar with Descartes' physiognomic and "passional" doctrine. The *Passions of the Soul* and the *Treatise on Man* were brief and sharp, while Cureau's works were prolix, diffuse—and unfinished.[37] The difference was not just one of style. The whole of Descartes' space was full: extension was identical (or at least coextensive) with matter. Everything that occurred in extension had to be explained mechanically, as matter acting on matter. It followed that the animal spirits, which regulated both passion and expression by moving through the body, were tiny and swift but definitely corporeal. Their "motor" was the natural heat of the heart. Cureau (who in that, as in other things, was closer to William Harvey's view of blood circulating in the body than Descartes would be) thought that animal spirits were insubstantial, like other kinds of spirit; that they existed in a separate category of being between the body (which, like Descartes, he was content to describe in mechanical terms) and the soul; and that they were the soul's instruments. His views on the whole matter owed a great deal to earlier Neoplatonists.[38] Cureau also thought that spirits were required to impel blood from the heart to the face and the brain, since a liquid such as blood could not carry the pulse of muscular beat throughout the body: that is where he differed from Harvey. He accepted the doctrine of the circulation of the blood wholeheartedly yet he still needed spirits in their chemical (perhaps also alchemical) quality to make Harvey's body-pump work.

Descartes' model, on the other hand, did not require a separate, mediating category of being between body and soul. The soul had a lever or hinge, as it were, on the body; located in the pineal gland (which he also considered to be the "mixer" of our symmetrically received sensations) at the base of the brain, it allowed a passage between perception and the will, as between the body and the mind. In Descartes' view, the soul could command the body to feign expressions and signs of emotion. This made for a crucial difference between him and Cureau, who believed that the involuntary movement of facial muscles would always betray the true state of the soul to an attentive eye. Descartes, though interested in willed and unwilled emotion, had no interest in a "natural" or "semiotic" physiognomy; and since the *res extensa* was completely independent from the *res cogitans,* passions might move the body, but the body could not, in return, condition the thinking faculty, which proceeded free from any physical constraint.[39]

Psychology in the studio: Charles Le Brun's Conférences

However useful Descartes' taxonomy was for him, Charles Le Brun was to rely on Cureau's observations and on his physiognomy;[40] he was, after all, not so much concerned with scientific principle, but was looking for a direct reading of the visible effect of passion on the body. Le Brun's descriptive apparatus, the physiology of passion, was therefore borrowed mostly from Cureau, as were the elaborate analogies between animal and human physiognomy that had not engaged Descartes; and from Cureau also came the description of the workings of passions. From Descartes, who considered that the movement of eye and eyebrow were the most important indices of both permanent character and of passing mood, since both were closest to the pineal gland,[41] Le Brun took over the idea that these visible bodily parts closest to the seat of the soul were also its clearest index, or (as Jennifer Montagu has it) the dial that displayed inner motions; however, he thought that the important lower facial line, that of the mouth, was ruled by the heart.[42] And in that Le Brun followed Cureau.

Plato as a dog. Giambattista della Porta (1586).

Angelo Poliziano as a rhinoceros. Giambattista della Porta (1586).

Engraved title page from J. Bulwer, *Chirologia* (1644).

Much of this could be seen as a reaction against the flood of rhetorical literature in France during the half century before Le Brun.[43] Many of the books dealt not only with speech (declamation) or with facial expression, but also devoted a great deal of attention to *actio*, the language of hand gesture, which was even treated as a separate science—chirology—by some. Not only did *actio* amplify the spoken word for the preacher, the advocate, and the actor: it also provided a vital aid to the deaf and dumb and it could even be scried for clues to the universal language of mankind before the confusion of Babel.[44] Against such a view the skeptics (best represented by Montaigne) maintained that the natural language of the body—of hand and head and eyebrows—in which love and hate, terror and the other passions dictated movement and gesture from instinct (naturally, as it were), had therefore no need of the conventions and artifice of rhetoric: Descartes owed much to this view.[45] To put it in Marin Cureau's terms, however, they were God given, like animal gestures and expressions, while the gestures of rhetoric, like spoken language, were man-made; they were artifacts.

Artificial gesture, natural gesture

The language of passionate gesture, revealed by the scientific study of physiognomy and psychology, had therefore to be entirely different from the highly conventional, artificial one of rhetoric. That is why Le Brun's drawings show patients—men and women who are prey to, are possessed by, their passions—and not the agents of rhetoric, whose gestures are deliberately directed to constructing a discourse. This was Le Brun's first step in the restructuring of the artist's training, and it made his teaching strangely "modern" and original—antirhetorical of course, since in the painter's quasi-rhetorical practice attention concentrated on the importance of willed, even artificial gestures of the body and the hand, *action*, while all that Le Brun catalogued and recorded was largely involuntary, *passion*. By thrusting physiognomy into the center of French academic teaching, he thus effected a brutal and permanent reorientation of the painter's and the sculptor's activity.

Since the representation of the passions had been related to the modes—the musical modes that had so interested Poussin—and beyond them to the great typical system of the orders, such an appeal to the raw immediacy of passion acted as a solvent on that system, and so helped incidentally to loosen the unquestioned authority of the orders on architectural thinking. It prompted an investigation of proper character (the *ethos* of each order), which would determine its appropriate location in a natural setting, and so deprived the columns of their active corporality.

This other matter, the problem of the exact relation between inner character and its outward signs, also preoccupied Le Brun and also required an appeal to antiquity. Another Aristotelian book, the (ethical) *Characters* of his immediate disciple, Theophrastus, became a canonic text in the seventeenth century. The very first near-complete edition was sent to the printers by the Genevan humanist, Isaac Casaubon, in 1592.[46] Theophrastus presented a number of human types by a skillful and fragmentary assembly of external characteristics. The book soon found imitators: the most famous and successful in English was the pious John Earle's *Microcosmographie*, which appeared in 1628. John Aubrey's *Brief Lives* also owe him something. However, the classic "modernization" was to be the *Caractères* of Jean de La Bruyère of 1688, which first appeared as an appendix to his translation of Theophrastus.[47] La Bruyère's short stabs of malice are more brutal than the Greek original: perhaps that is why

The eye, the brain, and the pineal gland (Descartes). Woodcut, after Adam and Tannery (eds., 1897–1909).

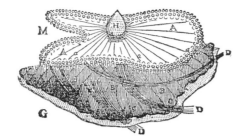

The operation of the pineal gland (Descartes). Woodcut, after Adam and Tannery (eds., 1897–1909).

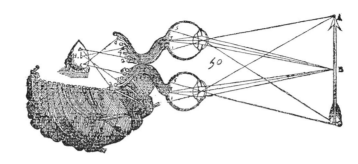

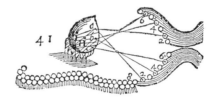

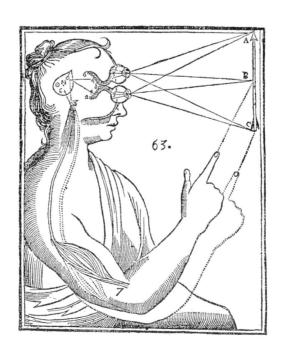

Chapter II

his *ethopoiesis,* his drawing of character, had a very pervasive influence on later European fiction. At the break between the seventeenth and the eighteenth centuries, the word *character* therefore takes on a new importance.

Beside La Bruyère, the writer who most affected the way the word would be used was Lord Shaftesbury, the philosopher-earl; in his main work, the *Characteristics,* he makes no direct mention of Theophrastus, yet clearly his conception of what constitutes character in both people and places was much influenced by the ancient philosopher. Shaftesbury took the word in a radical sense to mean the very marks of writing and painting, as well as what they signify, much as we still use the term in the sense of *letters,* particularly printed ones. Indeed he proposed to reform the use artists made of the universally signifying marks, the *characteristics* of his title, through a search for truth to type in language; the best kind of characteristic would then be a "natural" figure that might convey immediate meaning to the uninstructed. The old apparatus of allegory and learned reference on which artists had long relied seemed to him to be already wearing threadbare, and to be in any case corrupted by "gothick" and "monkish" prejudice.[48] The study of the individually expressive, of a naturally indexed and almost measurable expression (as against a conception of it as either referable to a type or as one of several aspects of an integrated figure), was the only way forward. Shaftesbury was, in any case, another enemy of traditional rhetoric and his work implied the devaluation of all rhetoric and artifice.

Aristotelian rhetoric technique had shown the adept how to bend the assent of an audience by the advantageous presentation of self, or *ēthos,* allied to a thorough advance knowledge of his auditors' reaction, *pathos.* Pathos is indeed concerned with the hearers' or the spectators' emotions, and how these may be influenced. Later textbooks on rhetoric would include a more or less systematic account of human character and passion, of which Aristotle's *Rhetoric* provided a model.[49] But the Theophrastan association of *ēthos* and *charaktēr,* the

Hieroglyphic for the Earl of Shaftesbury's *Characteristics.* Copper engraving by Simon Gribelin for title page of vol. 3 (1732).

Order in the Body

indelible mark or stamp, had already suggested a shift from the way Aristotle, Plato, and their elders had used the word: for them it did not have the sense of cage, of inescapable characteristics, irreversible and determining, but rather all that which in a man (physical makeup, context, custom) guided his actions in much the same way as rational reflection might do. Even that can be read as narrowing the sense the word held for the first philosophers, who had used it to signify the whole experience of man's dealing with the world around him. Aristotle, in particular, discussed the bodily expressions of the soul—or indeed of the soul as the entelechy of a particular body—in his treatise on the parts of animals, though for him as for the older philosophers, ethos, with its implication of choice, could only describe human behavior.[50]

For Aristotle, all human creativity was *mimēsis,* an imitation (or perhaps more correctly, "representation") of human action, of *praxis.* Praxis was never neutral, but had, necessarily, to tend either to good or to bad, and that bundle of possibilities accounted for its ethos; however, in all imitation (and therefore in all forms of art) the imitator's ethos in turn is also a guide to the way he would choose to imitate a person or thing imitated.[51] The study of the rhetor's ethos, of his disposition, atrophies in Hellenistic rhetoric, when ethos comes to be identified increasingly with indelible character. The dialogic aspect of the word is finally suppressed in the Ciceronian translation of *ethikē* as *moralis.* Latin rhetoricians never found the right word to translate the Greek term satisfactorily. Indeed Quintilian was even to suggest in Imperial times that the difference between ethos and pathos is as between comedy and tragedy, between mild and slow (*mites atque compositos*) emotions and *concitatos,* turbulent ones; he added (as he thought) helpfully; *ut amor-pathos, caritas-ethos.*[52] Following Augustine, the theological usage of character as a permanent mark—like a brand burn—conferred by certain sacraments was developed in the Middle Ages. That notion of the indelible mark is conveyed by onomatopaeia, a sound as of a sharp instrument scratching over a hard surface, *charassein:* hence *charagma, charax, charaktēr.* The transmutation shifts the meaning of the word further, so that it turns into an assembly of marks that correspond to a type.[53]

Now to Le Brun the psychology, the ethos, of the artist was of no concern since he was to establish (he thought) an ahistorical and generally applicable doctrine of the art of painting. The primary intentions of his teaching were to discover the symptoms of interior passion and to depict them through facial configurations and the bodily attitudes of each individual personage (or character) within any narrative. Nor was he interested in finding ways of appealing to or affecting the judgment of spectators by passion engaged. The idea, the central character of the painting—its ethos, in fact—would do the persuading. Instead of influencing passion by piecing it together in painting and sculpture through gesture and attitude, he sought to make the visual arts a systematic demonstration or indexing of it; if you will, Le Brun inverted the Aristotelian relationship of ethos and pathos.

And yet Le Brun and his followers were not able to build up a system of notation from direct observation to rival the allusive power of rhetorical conventions. They therefore inevitably turned to the only alternative source, physicopsychological description. The exercises and images that Le Brun developed were first summed up in a set of illustrations to correspond roughly to the passions listed by Descartes. These illustrations by Le Brun were to have a vast influence outside the visual arts.[54] Inevitably the various splendid sets of engravings contributed much to their vogue, but even the texts and drawings that were not widely reproduced inspired a number of students and interpreters.[55] The myology of the face, as it related to the expression of passion, became a favorite subject of anatomists as well as artists. Searching, detailed scientific attention to the face as "revealing," as the key to true character and real (even concealed or masked) passion, diminished the communicative power of the body as a whole. That is why, although he did not wish to devalue antique precedent, the copying of statues and casts, or the example of Raphael, Le Brun's crucial lectures ensured that their place in the artist's formation could never be central again.[56]

Descartes' account of the movements of eye and eyebrow, the line of the nose, and the geometrical relation between the length and position of the mouth and the eye became in Le Brun's teaching a kind of protoscientific physical classification (almost a topology) of all human passion, allied to the older system of parallels between animal and human type—or at least to physiognomy. Le Brun himself went further and attempted to provide a measurable, or at any rate a geometrical, "proof" of the relation of passion to physiognomy. The inclination of the line passing through the center of the eye and the nostrils, as well as the place of one joining the two eyes in relation to the bridge of the nose, would accurately show the degree of ferocity of an animal and could be paralleled in human faces. With that ultra-Cartesian exercise, he opened the way to a new kind of anthropometry, which concerned itself primarily with the face; on it, all later physiognomy as well as the study that came to be known as physical anthropology depended.[57]

It would be quite misleading to foist a consistent philosophical system on Le Brun; certainly, he was not a Cartesian. Insofar as his thinking was at all systematic, it was empirical; his geometrical schemas are always laid over observations. He defeated the one strenuous attempt to make the Academy rational and Cartesian, mounted by the very prolific engraver Abraham Bosse, who in turn was a disciple (and a spokesman for) the geometer-engineer Gérard Désargues. Désargues had been befriended by Descartes and Father Mersenne, was admired by Pascal, and had been a generous teacher to craftsmen as well as other mathematicians, but he published very little.[58] He never occupied any official position, while Bosse— though engravers were not admitted as full Academicians, and he was always described as an engraver—was appointed professor of perspective at the Academy in 1648 and given full membership in 1651; he turned out to be a dogmatic, awkward, and quarrelsome colleague.

Bosse himself maintained that an unfavorable criticism of one of Le Brun's pictures had set the Academy's effective master against him.[59] But their differences were not just personal. The Bosse-Désargues perspective method depended on a very refined version of the *costruzione legittima*. It was a wholly deductive and geometrical procedure that defied experience. In *perspectiva naturalis*, both stereoscopic vision and the curvature of the eye are allowed to correct the constructed perspective from direct observation: that was the method championed by the Academy.[60]

Désargues first, and Bosse after him, had an almost Platonic and certainly Cartesian mistrust of any sensory perception. That is why they (and their colleagues, like the great Father Niceron, the master of anamorphic construction) were so fascinated with ways of deceiving the eye.[61] When it came to making theirs the official doctrine of the Academy, however, and with it a dogmatic assertion of central perspective with buildings laid out parallel to the picture plane, most of the academicians objected. The details of the quarrel have been recounted several times, and they involve a number of the figures whom I have already quoted—the most important being Fréart de Chambray, the publication of whose translation of Leonardo's treatise on painting (with its rather lax proportional doctrine) provided the occasion of at least one battle. Bosse was expelled from the Academy soon after that event, following a nasty row

Eagle. Drawing, Charles Le Brun. © Photo Réunion des Musées Nationaux.

The Eagle Man. Drawing, Charles Le Brun. © Photo Réunion des Musées Nationaux.

The Eagle Man. Drawing, Charles Le Brun. © Photo Réunion des Musées Nationaux.

Chapter II

The Owl Man. Study, Charles Le Brun. © Photo Réunion des Musées Nationaux.

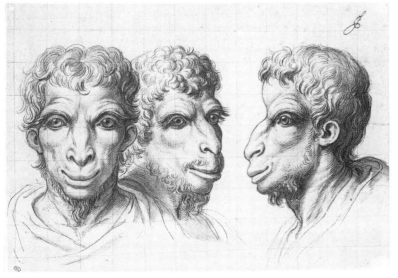

The Camel Man. Drawing, after Charles Le Brun. © Photo Réunion des Musées Nationaux.

Order in the Body

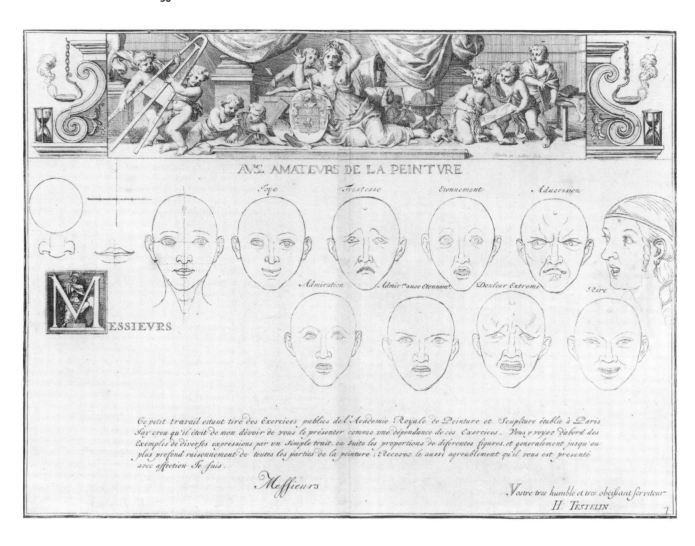

Illustrations of the passions after Le Brun. Dedication plate. Engraving by H. Testelin.

Etonnement, by Charles Le Brun. Paris, Louvre, Cabinet des Dessins. © Bulloz.

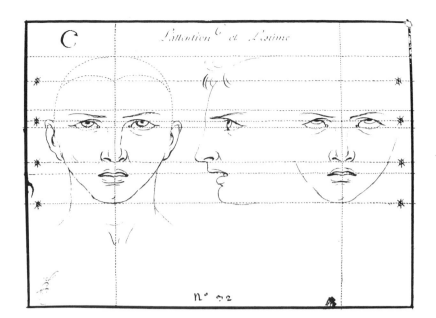

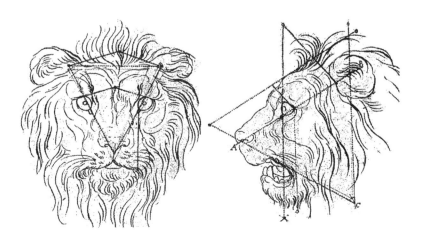

Order in the Body

Corrected perspective construction on a flat plane. Engraving from A. Bosse (1667).

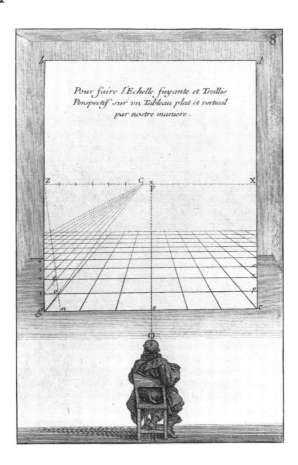

Pour faire l'Echelle fuyante et Treillis Perspectif sur vn Tableau plat et vertical par nostre manière.

about plagiarism in 1661. The independent school he (or his students) attempted to found a year later was suppressed by the police at the instance of Le Brun, and his most loyal students were banished from the Academy.[62]

Bosse seems to have seen himself as a kind of one-man academy in any case, and therefore, inevitably, he also issued a treatise on the orders. Although he explains in a self-deprecating way that his book will not have much text (and refers his readers to Fréart de Chambray's *Parallèle* for greater detail) as it is intended for practitioners, nevertheless he manages one or two disobliging polemical remarks both about Chambray and about François Blondel (the professor of theory at the Academy of Architecture and no relation of Jacques-François).[63] The older Blondel was the main advocate of the "ancient" view of the orders—their timeless rightness and their relation to the proportional teaching about universal harmony—which would in fact be gradually set aside sometime during Blondel and Bosse's lifetime. Of this there is no inkling in Bosse's book on the orders, which merely proposes summary recipes, such as he claims to have drawn mostly from Palladio (who had the better taste) and Vignola (who knew more geometry).[64] The full Cartesian message to the architects had to wait another decade, until Claude Perrault's edition of Vitruvius.

The parallel between Perrault's architectural enterprise and that of his friend, Le Brun, is very striking. Perrault, an architect who was also a famous comparative anatomist, with his brother Charles (a highly placed civil servant) played a part analogous to Le Brun's in French architecture; Claude had demonstrated, on the basis of both his archaeological and

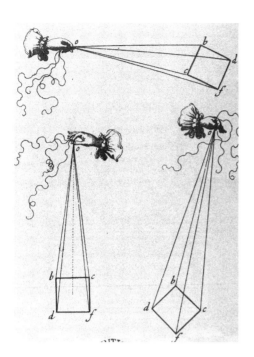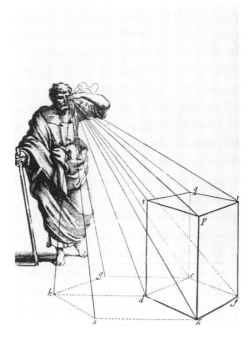

Construction of rectangles and parallelepiped. Engravings by A. Bosse (1665).

anatomical observations, that there was no close correspondence between musical and visual harmony as earlier theorists of both music and of architecture implicitly believed.[65] This led him to propose a division between on the one side commonsense or positive beauty that was "natural" and the arbitrary, culture-dependent beauties of the arts on the other. In spite of his close observation of this distinction, Perrault also upheld the value of ancient precept, since the willful human mind needed that force of example to direct and restrain the fallen, corrupt tastes that depend on the imagination. In Perrault's view (which is in some ways echoed by Le Brun), the value of venerable historical exemplar was in the educating and restraining of that erring human imagination, even if the origin and acceptance of the archetypes were conventions that had been transmitted almost by accident. Perrault was well aware of the parallel between himself and Le Brun, whom he exalted as one of the greatest if not *the* greatest painter of all time.[66] But the revolutionary doctrine proclaimed by the Perrault brothers did not assert itself quite as quickly as Le Brun's had done: it did not have the backing of an academic institution and had to percolate through individual emulators and followers.

The first architectural theorist to make character the essential proposition of his teaching was probably the aged Charles Boffrand in his imitation of Horace's *Art of Poetry.*[67] There was a great demand for verse treatises on the arts in seventeenth- and eighteenth-century France, and Horace provided the most common model for the exercise. But although the form Boffrand chose for his treatise was unremarkable, he did innovate in making it the essential business of architecture to impress the *character* of a building on the spectator, thus making evident its use and its context, by the appropriate adoption of the different orders.

He saw this as justifying the diversity in their proportions and ornaments and saw his precepts about expression and the impression the building is to make as being a simple updating of the Vitruvian rules about decorum. He did not appreciate (or was not interested) in another implication of his teaching: if the choice of an order depends on the location or the use for which the building is intended, then only one order can be appropriate to it. Thus no

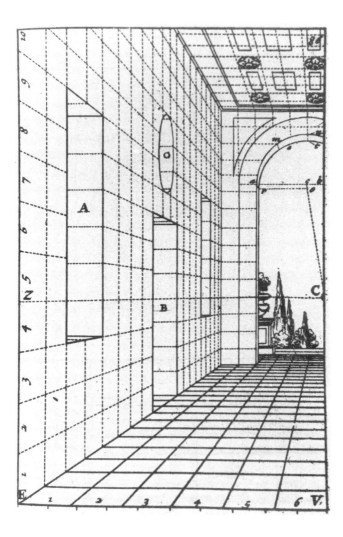

building can ever claim the kind of microcosmic status for which the five-order towers and palaces of the sixteenth century strove so eagerly (and rather naively), which any building that includes more than one kind of column claims in some respect. The way was now open for the nineteenth-century conception of an "order of orders" that I described in the first chapter.

Face, character, mood, landscape

But a more curious and episodic emulation of Horace had appeared in Britain in 1734, eleven years before Boffrand's: the *Lectures on Architecture* by an otherwise little-known English critic, Robert Morris. Writing in alternating passages of prose and (very pedestrian) verse, he recommends recourse to different orders not only according to the use of the building, but also to its situation (town or country, hill or valley, and so on). In the old treatises, Alberti's most explicitly, the situation and name of the site had been regarded as part of the "ornament," the internal and physical presence of the building, while the choice of an order depended on its dedication or on the placing of the column in its fabric. Morris makes the site, the setting, an external determinant of the building's character and therefore of the choice of the proper order. Both Boffrand's and Morris' books established the term *character* as the justification of

Order in the Body

the orders against the post-Perraultian sceptics, and with them it entered architectural parlance, taking on the various further implications that you may find listed for that word in the early nineteenth-century architectural dictionaries of Antoine-Chrysostome Quatremère de Quincy and Francesco Milizia,[68] though it is also discussed in writings bearing on the theory of art more generally.

So I return to Jacques-François Blondel, who first led me into this consideration of character and physiognomy, in the hope of giving a proper context to his apparently strange and isolated exercise. Blondel was familiar with much of the literature I have quoted and with Le Brun's lectures; he knew and admired Boffrand's book.[69] Yet he does not quote Le Brun; by invoking Le Blond and reaching back to Diego de Sagredo, he seems to attempt a reconciliation of his psychologism with an older and independent tradition. For his part, Diego had learned the details of this doctrine in Italy, which he is known to have visited.[70]

Variety of character and fixity of canon

Measure in Diego's title (*Medidas del Romano:* "the Roman's—i.e., Vitruvius'—measurements") is the standard by which the mistakes that his Spanish contemporaries made in the proportions of their columns must be corrected. He begins at the essential instruction with an account of the canon of human proportion, something that was very much at the center of discussion in fifteenth- and sixteenth-century Italian art. His book is in the form of a dialogue between the author (or at any rate a character named Tampeso who seems to represent him) and his questioner and friend, the painter Leon Picardo.[71] He begins by alluding to the notion of microcosm; he underlines the importance of the face as the highest point of creation and provides the outline of a canon, in which he claims to improve on Vitruvius' height for a man, 10 faces, and that of Pomponius Gauricus, who preferred 9 faces to the height.[72] Diego explains his own preference, 9⅓ (which he also states as ²⁸/₃, 28 being a "perfect" number).[73]

In specifying a face: body formula, the various textbooks ranged from the 8⅔ of the early quattrocento painter Cennino Cennini to Vitruvius' 10,[74] while Diego claimed that his particular figure was his own invention. All the books agree, however, in dividing the face into three units (which are sometimes then subdivided into two or three subunits); the length of the nose, which is ⅓ of the face, is often treated as the module by which the rest of the face and body are measured. The Byzantine canon, known from the Athos *Painters' Manual,* also proposes 9⅓;[75] it has been suggested that this particular relation was inherited from a Moslem tradition, which was in turn abstracted from Hellenistic teaching.[76]

There are plenty of obvious references in the *Medidas,* chiefly to Vitruvius and Alberti. As for Blondel, what he borrowed from Diego was the notion of a congruence between face and cornice that Diego seems to have put together out of Fra Luca Pacioli's *De Divina Proportione* and the treatise of the great Sienese painter-architect-engineer Francesco di Giorgio; this last he could have known from one of the several partial copies in circulation then, though he may also have seen and read an original manuscript. There are several that can be called "original" in the sense that Francesco supervised their execution and perhaps helped to illustrate them, and four of them contain two different versions of an architectural and engineering treatise.[77]

From Pacioli, Diego had taken the regulation of the head by a subdivided square; it was also Pacioli who suggested the squaring of the profile, even if Diego's division into nine

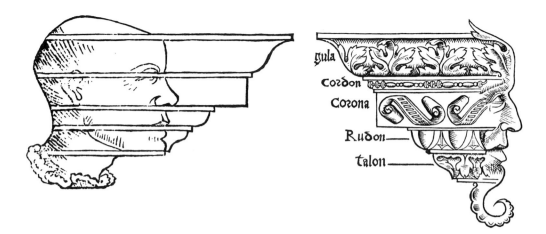

Faces on entablatures. From D. de Sagredo ([1526] 1542, pp. ii, r. and v.).

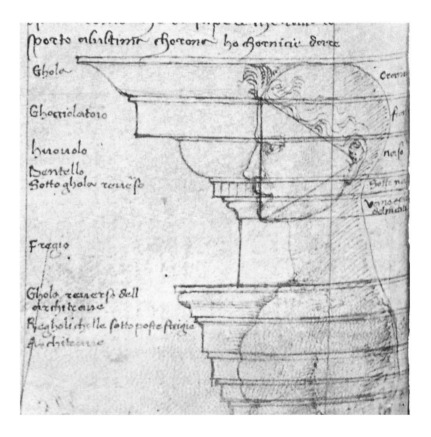

Face on cornice. Francesco di Giorgio, Turin, Cod. Saluzz.

Order in the Body

The cube mansion. Engraving after R. Morris (1759).

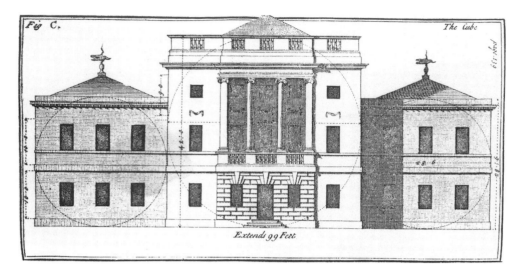

The Corinthian mansion. Engraving after R. Morris (1759).

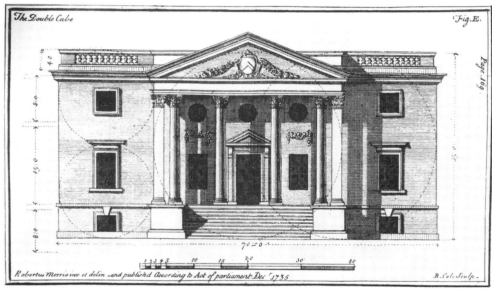

The Ionic mansion. Engraving after R. Morris (1759).

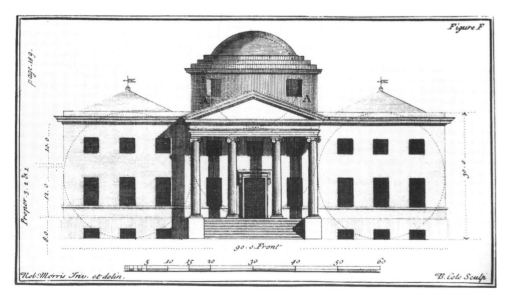

Chapter II

subsquares is much coarser than his. However, the very first to draw the human profile imposed on the section of a cornice to explain the rationale for relating its different members was Francesco di Giorgio; as Pacioli had also done, he regulated it with an equilateral triangle whose corners fall at the edge of the hairline, the chin, and the opening of the ear. Francesco's rationale for it was the inevitable appeal to the ancients:

> several times, considering and investigating whether the proportions of a cornice might not be reduced to those of a human head[,] . . . I saw many which, even if their parts could be given definite shape, yet their proportions could not be ascertained by measurement; but when measuring many other cornices; I have found [over and over again] that they had the same proportions as the head in my drawing shows: epistyle [or architrave] for the shoulders, a frieze for the throat, an astragal for the chin, the dentils for the teeth.

He rather insists on the obvious homonym and goes on to ennumerate the other moldings, ending with "the cyma or throat of the dome, and curve of the head." From all of this, he concludes, it is evident that as the shoulders carry the weight above them, so the epistyle carries that of the cornice, and the rest of the body stands for either a column or the facade of a temple: "On examining the works of the ancients I have found them related in this way. . . ."[78]

This quotation comes from the later and more literary version of his treatise, though in fact, the face/cornice figure in both versions are very similar. The striking difference is the proportion of dentils to the other members. In the first version, they look very much smaller than most on antique examples, since they correspond to the closed mouth of the face; in the later one, the mouth is wide open, and by this rather simpleminded device the dentils become $\frac{1}{5}$ instead of $\frac{1}{8}$ of the cornice.

There is a fairly close correspondence between the arguments in the two versions of Francesco's text, and he seems to have had a hand in illustrating both groups of codices.[79] In both versions Francesco's comparison between cornice profile and human face was part of what now seems an obsessive insistence on the application of the analogy between the whole human body and building. After the cornice, he concerns himself with the capital of the column: that must also be set up in the same way as the cornice, by making it analogous to the three zones of the face, "as the painters divide it." He may have found the rule in Alberti's treatise on painting, in Vitruvius, or taken it for granted as part of workshop practice.[80] If the cornice and the capital are smallest-scale instances of microcosm in building, the plan and facade of a church offer more powerful and binding evidence. The face is only a part of the whole analogy between building and body.

Francesco asserts blandly:

> I will now describe how the facades and doors of temples are to be done. Since the facades of temples are derived from the human body I will set them out according to the methods and measurements that belong to it. You must know that the body is divided into nine parts, otherwise into nine heads, from the edge or inclination of the forehead to the bottom of the chin.[81]

Man in square and circle. Francesco di Giorgio.

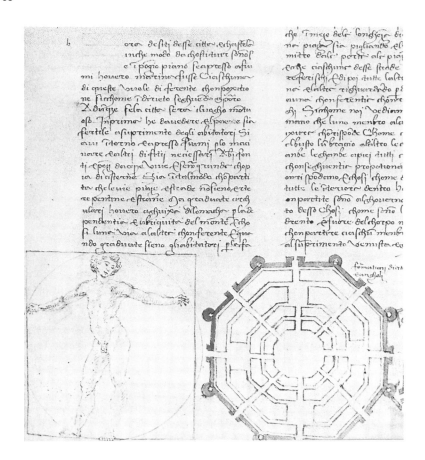

The illustration shows a man with his hand extended, his trunk and head corresponding to the nave of the church, the sloping arms directing the inclination of the aisle roofs. The whole body both of the "Vitruvian" man and of the church is 9⅓ faces high (the extra ⅓ of the formula being the height of the whole head from the hairline down), the nave articulates from the aisles at the elbows, and the main door opens at the knee joints. And of course the head is "in" the pediment. In the later version, Francesco imposed a grid on the figure. The body is now seven heads high, and there is no hairline—it looks bald, as do the bodies inscribed on the plans: they are more schematic, as in the second group of manuscripts the matter is being reduced to a method.[82] When discussing the plans, Francesco is more explicit both about the method and about sources:

> many alert and speculative minds have worked hard to imitate nature in all their activities and from nature have learned their method: as how the divisions and members of the human body, from which the perfect number of which Plato writes, may be found to be derived. Vitruvius [deduces from it] the measurements and proportions of temples and columns, without whose symmetry, so he says, no craftsman can make anything well and reasonably.[83]

The church as a body

Francesco next describes two alternative methods of imposing a plan on a "canonized" human body: one based on seven divisions (that is, on the same grid as the facade described earlier) for a simple building of a nave and two aisles, and an even more elaborate one for a cruciform church with a central dome. They are of course the visual application of the ideas that I have already attributed earlier in this chapter to Honorius of Autun and Peter of Celle, but their concern with overall articulation and with numerology has in Francesco's case been replaced with a reliance on a body-derived geometry.

This implies, incidentally, that rectangular and even square forms had to be reconciled with circular ones. In the later and more theoretical (or at any rate, more explicitly Aristotelian) version of his treatise, Francesco often uses two or three incommensurate prime numbers—5, 7, 11—which have puzzled some commentators. In a diagram early in the book, he explains how to arrive at a geometric relation between 5 and 7, drawing a semicircle into a double square, and a diagonal across one of the squares, to provide the dimension in which he was much interested, the sagitta of that arc which the cross-diagonal cuts off from the semicircle (this turns out to be about $\frac{1}{5}$ of the diagonal, and $\frac{1}{7}$ of the long side of a double square). The diagonal of a square of side 5 is in any case within 3% of 7; and the diagonal of a square of side 7 is 2% short of 10. That long side is the diameter of the semicircle, which is therefore 11 units, since Francesco, like most of his informed contemporaries, took pi at the higher (and more accurate) Archimedean value, $\frac{22}{7}$ or $3\frac{1}{7}$.

Francesco, who follows Vitruvius and Alberti in explaining his design principles on the most exalted of all building types, the temple or church, allowed three forms to it: the circular (or centralized); the long (or basilican); and the mixed (the domed basilica), which he seems to have preferred. The use of the pi-derived numbers allowed him to play on this last type with the grid based on the seven-face human figure that he had adopted earlier, and to take in the medieval method of proportioning *ad quadratum* based on rotating squares. This gallimaufry of arithmetical relationships provides Francesco with the most elaborate of his proportional diagrams, the section through the vault of a basilican (?) church that is the climax, in a way, of his geometrical speculations based on a double grid operated within the same square of seven and of four units. It also allows him to harmonize the human-figure canon with square and circular measurement—to square the circle, as it were, through the human figure, so solving one of the great design problems of his time.[84] And it relates his design procedure to another common conundrum, the perfect number: that is, a number whose factors equal their sum, as $6 = 1 + 2 + 3 = 1 \times 2 \times 3$, or $28 = 1 + 2 + 4 + 7 + 14 = 1 \times 2 \times 14$ or $1 \times 4 \times 7$. Francesco claims to have learned about this kind of number from Vitruvius, though he also refers to Plato.[85] And of course they were among the many numerical conundrums that had fascinated the Pythagoreans. Still, what is remarkable about the formulae that he recommends, for all their numerical and geometrical complexity, is the vivid corporeality with which the setting out of the plan is described. For the smaller plan, the head makes one-seventh of the body: "then set the point of the compass on the umbilicus which is the intersection of the lines [and draw] a circumference from the bottom of the chin to the joint of the knee[,] . . . [after which] you draw a semicircle at the top of the skull ending in AB, and this shall be the place of the image."[86]

The canon of proportion with
the foot divided into twelve,
sixteen, and twenty-four units.
Francesco di Giorgio.

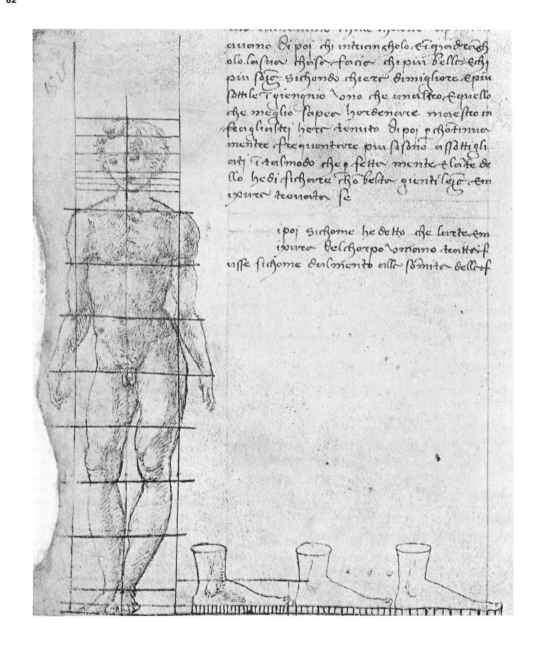

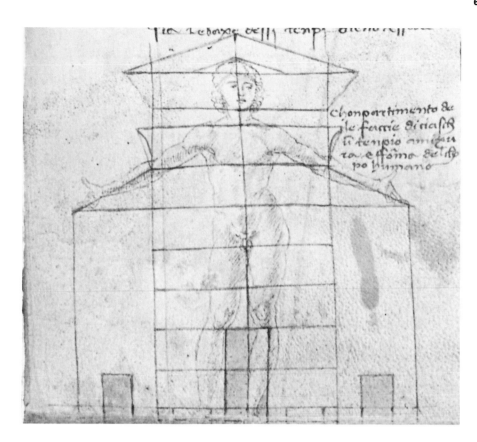

The haptic quality and immediacy—almost as if he were pointing to the relevant member of his reader's body—mark much of the writing about proportions and architecture, subjects that have since become very abstract; but in the particular case of Francesco, the whole of his treatise is built on (if that is the right way of putting it) the image of man incarnate in the very plan of the city. The first page of the earlier version of his treatise has a striking, bold drawing that provides a literal illustration of this idea:

> As Vitruvius has it, all art and its methods are derived from a well-composed and proportioned human body, . . . nature therefore having shown [the ancients] that the face and the head are its noblest part; as the seeing eye can judge the whole of it, so also a castle should be placed in the most eminent part of the city, that it can look over and judge the whole body. It seems to me that the city, fortress, and castle should be formed according to the human body, and that the head should have a proportional correspondence to the appropriate parts; so that the head might be the fortress, the arms its adjoining and enclosing walls, which encircling it on all sides will bind the rest of it into one body, a huge city.[87]

The city as a body

In that first version of his treatise Francesco goes on to specify the details of that fortress; in the second and more developed one, the first forthright illustration has been turned into a much more refined drawing in Dinocrates holding the model of a fortified city (presumably such a one as Vitruvius describes him proposing for Mount Athos, which he had transformed

Geometrical section of a church. Francesco di Giorgio, Magliab. II.I.141, fol. 41 r.

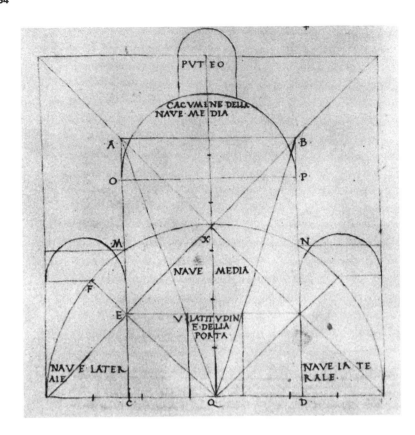

into a colossal statue of seated Alexander, though in the drawing it looks more like one of Francesco's Umbrian fortresses). The written text is even more insistent about the body image as it is applied to the city:

> I therefore say first that the major square [piazza] should be in the middle and center of the city (or as near to it as possible), located as is the umbilicus in the human body, . . . and the reason for this analogy is the following: since human nature first takes all its nourishment and perfection through the umbilicus, so through the [principal] public square all the other places [of the city] are nourished. And there is a justification from nature: that all things which are in common should be equally available to all, as the center [is equidistant] from every part of the circumference. It should moreover be surrounded with stores and honorable business.

Francesco goes on to detail the twenty-two conditions for a correct plan, in which the body returns only in the last rule: "that all the said parts should correspond and be proportionate to the whole city, as each member is to the human body."[88]

The application of the analogy is perhaps clearer in Francesco's drawing on that page than from the text, showing, for instance, that the church is clearly intended to correspond to the heart. By then the idea of the city-building-body microcosm was so widely diffused that Francesco's rather singular exposition of it should not be regarded as eccentric at all. In fact, it has obvious parallels to John of Salisbury's comparison, which I quoted at the beginning of the chapter.

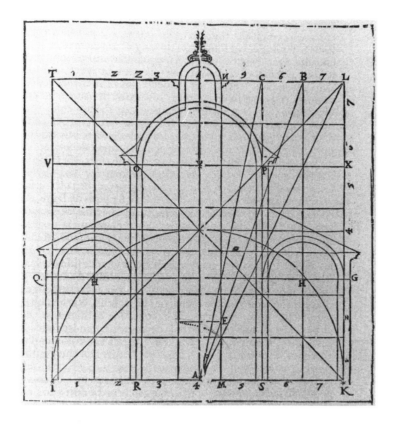

Section of a church. Woodcut from Philibert de l'Orme (1648).

The other major piece of architectural writing of that time, Antonio Filarete's garrulous and diffuse treatise "novel" (written some thirty years before Francesco's books, around 1460),[89] offers a scriptural justification of the basic idea: architecture being an art, it must be a human invention, and indeed we must assume that when Adam was chased out of Paradise by the angel, he put his two hands over his head to make a roof against the inclement weather of the postlapsarian world. Inevitably he extrapolated the rest of building out of his own body: he, Adam, being the paradigm of human proportion. Buildings may in fact be derived from the human body through the columns, which imitate the variety of human physique in their differences. Buildings are like human bodies in other respects: they need to be nourished and maintained throughout their life, they may be wounded, and they have to die in the end. Cities also derive their form from the human body; this is mentioned quite casually and without the circumstantial detail provided by Francesco.[90] Yet Filarete insists on the same idea in several places, and even develops different aspects of it.

The city as a house

But then Filarete and Francesco, as well as other writers on architecture, were only applying an idea so current that it became part of the general and unquestioned deposit of thinking. The idea that the human body is "a little world made cunningly"[91] had repercussions in medicine and alchemy, but also in astronomy and theology—and, inevitably, in law and political theory. For modern times, it had been firmly planted in architectural thinking by Leon Battista

Alberti, although he left the two parts of it separate. "The city is a big house, the house is a small city"[92] is the first part of it. That the building must always be thought of, and indeed analyzed, as body is the other. In this, as in so many other matters, later architectural writers tended to assume a familiarity with Alberti's text on the part of their readers.

As for Alberti, he claimed that he derived his doctrine from antique texts and that he explicitly modeled himself on the one surviving ancient author who wrote exclusively about architecture, Vitruvius. Vitruvius' doctrine on this matter is primarily geometrical, and he begins by assimilating the human figure to the square and the circle. But between Vitruvius and Alberti a different and equally ancient aspect of the analogy, topological rather than geometric, was developed much more luxuriantly, and achieved a complete hold over the popular as well as the scientific imagination: the astrological, to which I alluded earlier, but which I now need to consider more carefully.

The city as a body. Francesco di Giorgio, Turin, Cod. Saluzz.

Order in the Body

III

The Body and the World

• Love and strife • The body of the first man • Humors, elements, and stars • Astral man, canonic man • The dignity and the misery of the body • Geometry of God and man • Man the measure • The canon realized • Vitruvius' man in motion • Dürer on human variety • Gian Paolo Lomazzo • Michelangelo • Fabric of man, fabric of the world

The precious, brilliant, rather effete charm of the zodiacal man in one of the manuscripts that the Limbourg Brothers illuminated for Jean de France, Duke of Berry, about 1400 has often blinded scholars to the commonplace it represents. His (or perhaps her?) double body, the dubious gender, and the startlingly "classical" pose attract so much attention that we are inclined to forget how often a man (less often, a woman) so articulated—or perhaps more accurately, labeled—with all twelve zodiacal signs appears in many books of hours, as well as in many earlier medical handbooks, and persists in herbals or almanacs into our own time.[1] In this almost diagrammatic form he/she stood for a basic notion of all ancient and medieval medicine, that the human body was a topological representation of the universe throughout time as well as in space. "Man himself whom the wise call *microcosmos* (that is, a little world) has a body wholly tempered by qualities that surely imitate the individual humors of which it is composed, as if it were following the season by which it is dominated": so the Venerable Bede, in his calendar musings, which helped to recommend the word to the Western church.[2]

Love and strife

But the notion to which he alludes had a much more generalized currency than the neologism that Bede had coined or borrowed. St. Ambrose of Milan (to take an illustrious example), one of the four fathers of the Latin church, made himself familiar with such doctrines from Hellenistic writers and the earlier Greek fathers, or even such ancient authorities as Aristotle. He explicitly adhered to the teaching of the Stoic philosophers and astronomers that the human body, like the whole world and everything in it, was made up of four elements (*stoicheia*), earth, water, air, and fire. Furthermore, these elements were separated by conflict, *neikos*, and attracted to each other by love, *philotēs*, a word that denoted at once a virtue, a passion, and the force of attraction between inanimate bodies—a universal, almost physical bond. The alternation between affection and repulsion, between *erōs* and *polemos*, sustains the order, the *logos*, of the world fabric and is the cause of the cyclic, recurrent flux of the universe: from the total cohesion of the sphere when love is at the climax, to the total disintegration of the reign of strife.

The teaching about this dynamic constancy and the immortality of the physical world had been first stated clearly about the middle of the fifth century BC by Empedocles of Akragas, and until it was effectively set aside by Robert Boyle, the self-confessedly Sceptical Chymist,[3] it had been a part of mainstream physics, chemistry, and even medicine in the Western and in the Islamic world, in spite of the persistent formation of rival doctrines.[4] Empedocles was considered a physicist, physician, magus, and even a god—and was reported to have worked many cures. The concept that being could only be properly attributed to the four elements and the two forces, love and hate, was part of his teaching, which then passed into the physics of Aristotle and of the Stoa; through the Stoic philosophers it became part of the patristic world picture.[5]

A number of things are implicit in this scheme: that air is a substance; that the world necessarily moves through a series of cycles of varying durations, and our bodies along with it; and further, that the way in which the elements are tempered and distempered in the individual human body can be known—and remedied—by reference to its place and time beneath the constellations. Those constellations that are closest to the ecliptic were given the rule over the movements of the sun and moon and of the principal stars. They were also identified as

Zodiacal man. Très Riches Heures.
Miniature. Chantilly.

Phanes in zodiacal egg. Hellenistic stone relief. Modena, Museo Lapidario, © Galleria Estense.

The Body and the World

Vein man and zodiacal man.
Miniature. Oxford, Bodl. Ashm. 391.

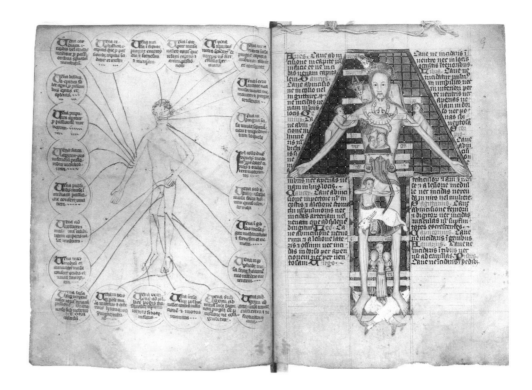

animated identities (hence *zodiac,* the belt of animals) long before any records of these activities were made. It now seems that even in Paleolithic times complex accounts of the regular movements of the planets and stars were maintained. Many different mythologies account for the filling of the sky with gods and heroes, with animals and with mythical creatures, and whatever singular cases may be cited to the contrary, it is most probable that throughout the world people classified celestial phenomena with the same obsessive accuracy as they did earthly ones.[6] The apparently motiveless movements of the heavenly lights demanded narrative explanation in terms analogous to the movements of experience: there was no choice involved in this great work that reason imposed on its possessors. The constellations were animated, always powerful creatures, and sometimes arbitrary ones, whose heavenly movements reflected but also appeared to guide earthly inhabitants. Of all things and creatures in the world, the human body was almost universally recognized as the most imposing case of elemental harmony.

The body of the first man

To the east of the Mediterranean basin—which must remain my main theater—in various Indo-European texts in Persia and in India, the creation of the universe as primal man, or even, the creation of the universe *out of* his body, was also repeatedly sung and meditated. Inevitably, such "scriptures" are mirrored in rituals, such as those ancient ones concerned with the simplest fire altar, the elemental sacrificial ground. The best known of these, the ninetieth

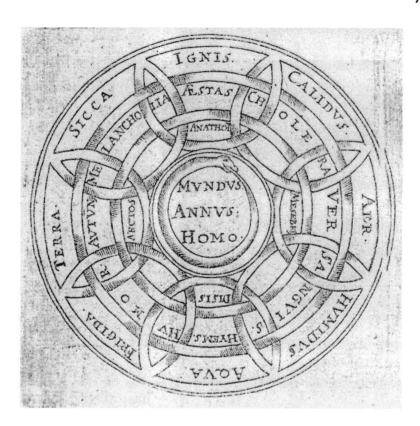

Convenientia anni et mundi. Engraving after a manuscript of the Venerable Bede (1688).

hymn of the Rig-Veda, suggests the creation of the cosmos and of society through a sacrificial dismembering of Purusa, the first man.[7]

The making of a building is, typically, the "raising" or elevating of the "body" of the building on the foundation or the plan of the world diagram, the Vāstupuruṣamaṇḍala. The varieties of mandala, and the rituals associated with it, are among many other things elaborate mnemonics of building forms. The early Indian altars and sanctuaries, before the beginning of Maurya temple building in stone and brick (which corresponds to the dominion of the Hellenistic koiné in the Mediterranean), may well have been raised on the very same plans. The constancy and universal practice of mandala rites in India certainly imply a notional continuity between the larger Gupta stone temples and the temporary earth and wood sanctuaries of pre-Gupta times, which stretch back into prehistory.[8] Similar rituals and notions concerned with analogies between building and body that occur further east and south, in China and Africa, would require too much space to be even outlined here.

Humors, elements, and stars

It is not clear at what point—sometime before the Christian era began—the four-element and the twelve-fold zodiacal systems were associated or overlaid. Since the fifth century BC, physicians agreed that the dominion of any one of the elements over the others in the human body produced varying humors; sometime later each element was seen to correspond to one of the seasons, or one of the world directions, which in turn corresponded to the two equinoxes and

Goddess Nut among zodiacal signs. Sarcophagus lid of Peta-menophis (known as Cailliaud mummy), from Luxor. Paris, Louvre.

Chapter III

Decan zodiac from the rooftop temple, Denderah. Sandstone relief. Paris, Louvre. © Photo Réunion des Museés Nationaux.

the two solstices. Each one of the quarters was further subdivided along the ecliptic ring into three houses, making the twelve houses of the zodiac.[9] The human body was therefore the representation of wholeness in time, as a calendar year, while it also contained the physical structure of the spatial universe, from Aries the ram at the head, in whose house the sun traditionally opens the year at the spring equinox,[10] to Pisces the fish at the feet.

This repertory of sky animals is part of the Mesopotamian heritage. The Kassite kings of Babylonia already thought them powerful enough to guarantee boundaries and had them represented on boundary stones.[11] The lunar calendar and the 360° circle was first formulated in Mesopotamia, though the earliest complete representations of such cycles are Egyptian. In fact, the Egyptians had a much earlier, analogous system of their own: that of the thirty-six decans—a term the Greeks coined, since each animal/star ruled over 10° of the equator, or rather of the sun's path.[12] The body of the Egyptian sky-goddess, Nut, was certainly articulated into decanal calendar divisions by Middle Kingdom times, and perhaps long before. Her worshipper, who was both her double and her brother, in death also identified his own (preferably) mummified body with the decan-divided span of the goddess. As she governed the recurrence of all seasonal events by the passage of the sun through her body, so she also revealed the order of space by spanning the earth: the animals with which parts of her body were identified constituted the whole bodily experience of sentient nature, bridging and mediating between time and space.

The technique of humoral adjustment that depended on these ancient beliefs was for many centuries the most common form of therapy for all diseases. It was operated by such methods as purgings and bloodletting, which inevitably had to be performed at the appropriate seasons. A body-world diagram, such as that of the Limbourg Brothers with which I opened this chapter, was therefore associated with the calendar in many medical documents and could even be read as a simple bloodletting chart. "Simples," too—unprocessed and unmixed plants and minerals used for all forms of chemotherapy—were inevitably selected for their humoral properties. And there were moon-zodiac men as well as sun-zodiac ones. The diagrams had other elaborations, since the sun and planets in the zodiacal houses ruled the outer body, while the moon ruled over internal movements of the humors (and therefore the viscera), much as it controlled the tides on the earth. To some sixteenth-century writers the whole of medicine seemed just a branch of judicial astrology.[13]

Throughout the Middle Ages the fourfold division of the universe and its further articulation into the twelve houses of the zodiac determined the commonplaces of agriculture (of which modern herbals and farmers' almanacs preserve ample evidence) as well as of medical practice; it was also an all-important and all-pervasive mnemonic, an interpretive and computing instrument. But the zodiacal system was not necessarily tied to the notion that the human body was in some way a model of the whole creation, though the association was very old. As I suggested earlier, some ancient philosophers were skeptical about the use of this system for divination, while many church fathers and later theologians were damning: astrology seemed an even more categoric denial of human free will than physiognomy. However, this condemnation was always tempered by the recognition of two crucial gospel references, to the star of Bethlehem and the eclipse at the Crucifixion, which the apologists for astrology always cited.[14]

One of the great Italian thinkers of the fifteenth century, Giovanni Pico della Mirandola, who utterly rejected mantic astrology in the name of free human will, still took the parallel between body and world for granted: "It is a trite saying in the schools that man is a lesser world in whom a body mixed from the elements and a celestial spirit are seen—a celestial spirit and the vegetable souls of plants; the senses of brutes and reason, the angelic intelligence and the likeness of God."[15] Pico knew well that the doctrine was consecrated by a long line of witnesses, which included Plato himself.[16] And that in any case, however skeptical, dismissive, and even damning you may be about it as a tool for forecasting the future, you still had to take the zodiacal-elemental system seriously as a key to certain character traits, an aid to diagnosis and to *pharmakopeia*.[17]

Astral man, canonic man

The many surviving microcosmic medieval pictures, even when they seem to be abstract and geometrical, refer in some way to the zodiacal articulation of the world. In that the medieval Christian and the Islamic world had a great deal in common. Moreover, Islamic astrology flourished in the first centuries of the Arab conquest, since, as I pointed out earlier, it did not fall under the disapproval of theologians as happened in the Christian world, and Moslem scholars eagerly appropriated the Hellenistic works that their Byzantine neighbors sometimes neglected.[18]

At a time when Pico and other thinkers despised or questioned the implications of an astrological-microcosmic system, the revival of Vitruvian studies displaced scientific attention to a different microcosmic doctrine, the canonic understanding of human proportion—a matter also familiar to medieval writers, even if it seems to have been of secondary interest to them. Throughout the Middle Ages, Vitruvius' text had been copied many times; the particular passage on the proportions of the human body was quoted in accurate paraphrase by Vincent of Beauvais (St. Louis' librarian) in his *Speculum Quadruplex,* which was among the most extensive and popular of the many medieval encyclopedias,[19] and it is echoed in many theological and educational works. A notable instance is *Placides et Timéo,* a kind of un-Socratic dialogue, in which the "wise" Timéo explains the world and man's place in it to his disciple. His anthropology is concerned much with sexuality and generation, but ends with an encomium on man, "si digne beste, si haute créature" that he can be called microcosm, because all the four elements are combined in him. He goes on: "know, firstly, that as the world is round, so every man of good measure is round. And in the same way a well-measured man should have the same measure in the spread of his arms as the length of his height to make a proper circle." [20]

Not long after this was written, a Tuscan astronomer, Restoro d'Arezzo, writes about the argument at greater length:

and the body of the world, with its powers, which it has from the highest God, should all be proportioned according to reason, one part with the other, and one member with another[,] . . . and the wise artists to whom nature gave and granted the power to devise and to draw the things of the world, when they draw the figure of a man, divide the measurement in ten equal parts. And from the highest part they make the face and from there nine parts below are counted, and by the face they proportioned the hands, the feet, the chest and the whole body. The well-proportioned form and figure was seen and known by them. And this happened by the nobility and the imagination of the intellectual soul, which was founded in man.[21]

Vitruvius is hardly ever named or even textually quoted in this connection, yet the idea of the human body as microcosm, whether zodiacal or geometrical, recurs constantly, as does the idea that the body is a kind of building or even a kind of temple, to which I alluded in the previous chapter.[22]

Restoro and his contemporaries did not have ready access to the books that even seventeenth-century scholarship demanded, and references to the notion are dispersed and fragmentary. Nor is there any information about how (or even whether) measurements of contingent human bodies were taken and related, and what their microcosmic implications might be.[23] Still, the text of Vitruvius seems to have been familiar enough, treated (if the books it was bound with are any guide) as both a technical manual and a cosmological speculation. Moreover, the canon also had its *interpretatio Christiana,* since the body of Adam before the Fall was the perfect Body (being the image and likeness of his Creator); since we are all descended from him, it was also a guarantee of human unity in spite of all our differences of girth and height and color. Adam's body was also in agreement with that of the Savior, both before and after His Resurrection: and its perfection (on which Scripture seemed in accord with Vitruvius) related it to the canonic proportions of Noah's ark, the tabernacle in the desert, the Jerusalem temple, and other vessels of salvation, as many church fathers—notably Ambrose and Augustine—had asserted and their commentators had reiterated.[24]

Much more explicit than any of these is the fiendishly ingenious picture-poem of Rhabanus Maurus, abbot of Fulda (and later archbishop of Mainz), who wrote at the time of

Microcosmic man: Seasons, temperaments, planets. Ink drawing. Munich.

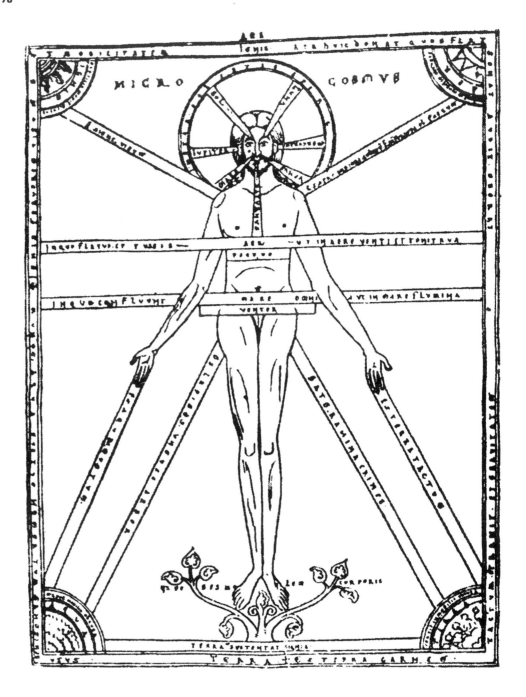

Ebstorfer world map of 1234.
Photogravure (Hannover, 1891).
Photo by Ernst Sommerbrodt.

Charlemagne. Each one of its illuminations is laid over a grid of letters, sometimes square (35 each way), at others rectangular; each square of the grid is a letter and the whole rectangle can be read as one long hexameter. However, the figure within it, and sometimes its separate parts, can also be read as different complete poems, variously scanned. It is dedicated to Louis the Fat, Charlemagne's successor, on the frontispiece miniature. While the opening image of the *Species Salvatoris* provides a meditation on the perfection of Christ's Body, it is the fifth miniature, a square 35 by 35 letters, that makes the message explicit: it is divided into four fields (17 by 17) by a cross (containing the same 35-letter poem left to right as down), and each of them has a little (10 by 10) square within it. The whole diagram declares itself an image of paradise as well as of the Jerusalem temple. It is also an image of the church, since the four inner (10 by 10) squares represent its foundation stones: in each small square the poem has to be read beginning in the top right angle, and these declare that the lower squares represent the prophets and patriarchs, the upper ones the apostles and martyrs, and the whole meditation uses all the scriptural passages on the body as the temple of God that I have quoted.[25]

Such a "moralized" canon is crucial to the account of the mystical visions of divine creation that were granted about the middle of the twelfth century to St. Hildegard of Bingen.[26] The grandest manuscript of her book is brilliantly illuminated. In two of the miniatures, the microcosm man with his hands spread in a quasi-Vitruvian stance is enclosed by concentric world circles; in one of them, the circle of creation enclosing him is in its turn also embraced by a huge world-soul figure whose hands stretch round it. Although Hildegard's man stands upright, his hands are not stretched up like those of Vitruvius but are relaxed lower than the

Cosmic man: Man and his Creator. Miniature, Hildegard of Bingen. Lucca, Bibl. Stat. MS 1942.

Chapter III

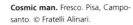

The Body and the World

shoulders, so that while the center of all the circles that surround him is in his crotch (as it is in the case of Vitruvius' square man), yet the place where the top of the head and the arms touch the circle seems composed to make a regular pentagon. This figure is echoed in the Leonardan Codex Huygens, and it was made familiar by the woodcut in Cornelius Agrippa's second book.[27] The inner central circle is one-sixth of the man's height. There are features of this vision that owe something to Bernard Silvester and to the Gnostic literature he had absorbed and transmitted.[28]

A few miles from Lucca, where Hildegard's manuscript was deposited in a monastic library,[29] there happens to be another, vast image of the world soul embracing the universe. It was painted about 1300 on the wall of the Camposanto of Pisa, a century or so after the manuscript was illuminated (probably in the Rhineland), and seems to obey an analogous geometric schema clearly based on a regular hexagon construction.[30] The "square" creator figure—whose equidistant head, hands, and feet represent the four directions—appears on a number of world maps, of which the most famous is the now destroyed Ebstorf *mappa mundi*.

Hildegard and another important contemporary woman writer (though more of an anthologist than a visionary), Herrade von Hohenberg, owe something to the Arab astronomers and astrologers, many of whose works were being translated into Latin about this time. In Herrade's microcosm there is reference to another astral belief, namely that the seven openings of the head refer to the seven planets, the head being a microcosm of the microcosm.[31] As Herrade found the number seven in the body, so other numbers too were verified, counted in it.

This notion of counting in the body is very archaic.[32] Perhaps the oldest explicit text to survive in the Judeo-Christian tradition is a passage at the beginning of the *Sepher Jesirah*, a meditation on the world's wholeness exhibited in the body through the numbers one and ten: "Ten numbers and no more, to count ten fingers, five against five, and the bond set alone: in the word on the tongue and in the cutting of the genital." The author of that book was almost certainly familiar with Pythagorean ideas; he wrote at the beginning of the first millennium AD, but his book became the founding text for much Cabalistic speculation, from which issued the myth of the golem: man remade (or at least simulated) by the operation of letters and number.[33]

The dignity and the misery of the body

As I have already suggested, the status of man as a little universe was considered a self-evident aspect of his dignity in ancient pagan thought. In Christian thinking the idea of man—in the abstract, as it were—being the image of God, an undeniably scriptural notion, was in permanent and irreconcilable conflict with the more familiar complaint of every individual's worthlessness before his Maker and with the conviction that the body was the wretched prison of the immortal soul, which Christian thinking owes to the Jewish tradition. The conflict was mitigated by the perennially powerful idea that the body of Christ as the ultimate and complete Vessel of Salvation was the perfect human body: He was indeed the New Adam. His body had therefore to be depicted and meditated upon as a mirror image of old Adam.[34] In fact, during the later Middle Ages and perhaps earlier in Byzantium, there was a quasi-Vitruvian devotion to the exact measurement of Christ. One muniment of it is still preserved in the cloister of St. John Lateran. Inevitably, it was particularly fascinating to the artists of the fifteenth century.[35]

In spite of the acute interest in Vitruvius during most of that time and the great enthusiasm for the subject of man's dignity, practically no quattrocento representations of the canon of human proportion seem to have survived. Nevertheless, the praise of human dignity was to become a particularly popular subject for humanist writers in the late fifteenth and early sixteenth centuries. Petrarch had written an exemplary exercise on the subject sometime between 1350 and 1360 in his popular book of moral advice, *On Remedies against Either Fortune*,[36] and in a sense it was an answer to a much-read devotional tract, Pope Innocent III's *On the Misery of the Human Condition*,[37] written over a century and a half earlier (between 1190 and 1200).

Pious Petrarch was all too familiar (as he makes it plain elsewhere in the book) with the "misery of the human condition" and with the evanescence of human beauty to attach too much importance to either. Yet he could appeal above Innocent's head, as it were, to a popular work of one of the early church fathers, Lactantius, on the beauty of the human body as a demonstration of the bounty of God.[38] In the course of commentaries on Genesis 1.27 some appreciation of human beauty became an indispensable ingredient in any meditation on the nature of God's image in man, as it had been in many accounts of the argument from design.[39]

The antique, pagan model for such writings was Cicero's series of rhetorical questions in his book *On the Nature of the Gods*,[40] to which (among many other ancient writers) Lactantius also appealed. Of the many orations and essays, the grandest restatement of the doctrine for the fifteenth century was probably made in Nicholas of Cusa's first important book, *On Learned Ignorance*. Nicholas was a figure of power and influence in the church of Nicholas V and Pius II: bibliophile, mathematician, theologian, mystic, politician, preacher and controversialist, diplomat and cardinal. His terminology was more audacious than that of almost any of his contemporaries.

Geometry of God and man

Nicholas' point of departure from current philosophy and theology—which had refined an elaborate understanding of the degrees and chains of being—is that there is no proportional relation possible between finite and infinite, between the world and God.[41] That disjunction or fracture opened the way to a new empiricism in epistemology, to a fideistic tendency in theology, and to a reliance on paradox, riddle, and rite in the approach to the ultimate things. Nicholas hopes that we might yet know something about God (and the world that He created) by way of Man's likeness to Him. The mind's access to number and measure allows the possibility of learning the paradox of our ultimate ignorance, for the relationship between truth and the mind is like that between the circle and its inscribed polygon: however many of the angles may touch the circumference, ultimate identity between circle and polygon is impossible because infinitely remote. The power of human thought can therefore only be metaphoric, and metaphors can only be of human making: they are an artifice built out of our own experience. Only by acknowledging the impassability of the bounds around the human mind can we strive for the unreachable, since the intellect ardently desires knowledge.[42] The world order is an image of its Almighty Creator, who made the world freely. He was not some demiurge under constraint to create, as the ancient philosophers had taught. God was free to make the world in any way He willed and He could have made one quite different from that He did make. Yet

Mensura Christi. Stone construction. St. John Lateran, Cloister. Photos by author.

since the Creator *was* very Being, He was Himself the only possible pattern for any other being.[43] Man is therefore moved by a will which is free—as the will of his Creator is free; and man constantly builds his universe by words, as God had once made the cosmos by the Word. But the converse must therefore be that "Man is God, but not absolutely, since he is man. He is therefore a human God." Man is also the world, and he is therefore a "microcosm or a kind of human world [*humanus quidem mundus*]." It is hardly surprising that Nicholas takes for his own Protagoras' aphorism (which was known through both Plato and Aristotle) that man is the measure of all things.[44]

And indeed, in Nicholas' conception, man is the center and even the justification of the whole world's existence. Man's centrality was demonstrated by the Incarnation. In his *Learned Ignorance* if nowhere else in Nicholas' writings, this centrality is shown to have been the reason (though, of course, not the "cause") of the Incarnation: the final achievement and completion of the microcosm, is—if a scale of value could be permitted in this matter—more important, more essential, than the redemption of original sin.[45]

Man the measure

That aphorism of Protagoras' about man being the measure of all things, often produced as the basic tenet of what is now called humanism, is also quoted with complete approval in a much stuffier, more scholastic source than any writing by Nicholas: the *Summa Theologica* of his near contemporary, St. Antoninus (archbishop of Florence from 1446 to 1459), inevitably in the course of a long passage that St. Antoninus devotes to the dignity of man as microcosm.[46]

For him the continuity of the great chain is not broken: Macrocosmos, the great order, is indeed God Himself; the mediocosmos is the created world, while the *minor mundus,* the microcosmos, is man. All this is consonant with man's excellence, his freedom of will, his sublime dignity of being, and the gravity of his bearing—though we must always remember that he was made of mud, and has only a brief span of life.[47]

Nicholas' restatement of the microcosmic doctrine went a long way beyond the more conventional formulation of the great and saintly archbishop, whose *Summa* is in any case more concerned with social doctrine than with first principles. Nicholas' teaching owed much to German mysticism of the preceding generations and yet attempted to answer the nominalist challenge to scholasticism; he "sailed the narrow channel between the Scylla of nominalism and the Charybdis of scholasticism," writes a recent historian.[48] It is certainly true that after the nominalist revolution the justice of his assertions could only be known by the negative method, or demonstrated by analogies to elementary but highly ingenious arithmetical and geometrical metaphors. Or by yet another recourse to microcosmic thinking: inasmuch as man knew categories by a God-like re-creation of them in each single perceived thing, he assented to the numerical skill of the Creator in his quantitative recognition of number, proportion, and measure.

In this Nicholas was also echoing the ideas of the earlier Neoplatonists, of some of the early church fathers, as well as of Jewish Cabalists. And yet—since he believed that an exact correspondence between the known and the knowable was impossible—there would always be an inevitable and irreconcilable contradiction between the ideal and the empirical measurement, and it would undermine the value of any scholastic cosmological system, though it would remain necessary to seek the utmost precision. Or perhaps put in another way: unless the empirical is in some way referred to the ideal, as the polygon is to the circle, it remains completely incomprehensible.[49] Only in the work of art is this conflict elided, since artists do not imitate nature directly but always follow *symbolica paradigmata* whether making paintings or spoons, and their work must always be directed by the mind. Nothing that is knowable can exist without participating in both art and nature moreover, since the very act of knowing is creative.

Antoninus, the popular and pious archbishop of Florence, died when Leonardo was still a child; Nicholas of Cusa was fifteen years younger than Antoninus, a contemporary of Masaccio and Alberti. What he wrote may well have appealed to some artists, yet Nicholas seems to have had little direct influence on his immediate contemporaries:[50] any contact between his intellectual world and that of Leonardo could not have been direct. Nicholas' contemporaries did indeed think about a canon of the human figure, and about the difference between ideal proportions and empirical measurement, but they seem to have wanted their speculation incarnate in bronze statues and paintings of saints or heroes, and above all, in images of the perfect Incarnation itself. The man in the square stretches his arms out in the gesture of the crucified Christ, and the inevitable similarity did not escape the notice of painters.[51]

The canon realized

The graphic evidence to illustrate this very pervasive idea in medieval architecture seems both scarce and dispersed however. Few architectural drawings (about 2,000 on present count) have survived from medieval Europe, and this has meant that too much has been extrapolated from too little evidence. The best-known collection, the "notebook" maintained by Master Villard d'Honnecourt (and perhaps two successors), was put together about the middle of the fourteenth century. It has been published several times over the last hundred and fifty years.[52] Villard's notebook could hardly have been an isolated collection, and presumably owes its fame to the accident of survival rather than inherent excellence.

Most of the human figures in it seem intended to provide a mnemonic figuration for the geometrical framework that architects and sculptors used, without any over- or undertones, and the text attached to them is not particularly revealing; even if some of the formulae may be derived, directly or indirectly, from Vitruvius—like the tripartition of the face—others owe little or nothing to antique precedents. The few surviving manuals and artists' sketchbooks, as well as records of workshop practice, suggest that various formulae for human geometry may have had wide and independent circulation.[53] They were almost all concerned with the proportioning of figures on the flat surface. Such medieval diagrams, echoed by underdrawings and sinopias of wall paintings, were also devised more as mnemonics or guides than as a systematic presentation of measured, or even numerical, relationships. However they differed from each other, they had certain features in common: so, for instance, the division of the figure into a number of heads or faces (which I described in the previous chapter), as well as a more or less fixed relation between the dimensions of the hand, the foot, and the face, or the division of the face into three zones. They shared some of these features at least with the literary formulae inherited from classical antiquity.

Villard's only real successor appears about a hundred and fifty years later and in a rather different milieu. Mariano Taccola was a contemporary and occasional collaborator of Brunelleschi's. His *Liber de Ingeneis,* composed in the second quarter of the fifteenth century, was also a collection of saws and devices, even if the engineering is very much more elaborate and effective than Villard's and the graphic style much more articulated and pictorial. Like Villard's book, Taccola's is written in a clear scribal hand; unlike Villard, however, his only geometrized body is a full-page drawing (not a gracious one, it is true—almost slipshod, but certainly striking) inscribed with a forbidding claim: "He from whom nothing is hidden created me. And I have all measure with me, both of what is heavenly above as well as earthly and infernal. And who understands himself understands much. He has the book of angels and of nature hidden in his mind. And below, etc."[54]

The Munich Taccola manuscript has been annotated by Francesco di Giorgio, who also copied some of Taccola's machines into his own book. But his "realization" of a passage in Vitruvius, which was all too familiar to him and which I will discuss in detail in the next chapter, was the very graceful, relaxed, and yet impatient—almost balletic—sketchy drawing in the margin of his own treatise, with which, in his turn, Leonardo was familiar. As in almost all the later representations, the figure was shown standing, not lying on the ground (as Vitruvius suggests).[55]

Leonardo's, of course, was to become the most famous of all the pictorial commentaries on human proportion. It was made at the end of the fifteenth century and is now in the Venice Accademia.[56] It is often forgotten that it was drawn before any illustrated edition of Vitruvius had been printed, and that no manuscript illumination to that passage existed (none at any rate has survived). The two drawings are isolated in that they explicitly start from Vitruvius' text on human proportion, but unlike all the later illustrations, attempt to reconcile them by placing the same man with outstretched arms both in the circle and in the square. Moreover Leonardo's circle is not concentric with the square but has the edge of the square as

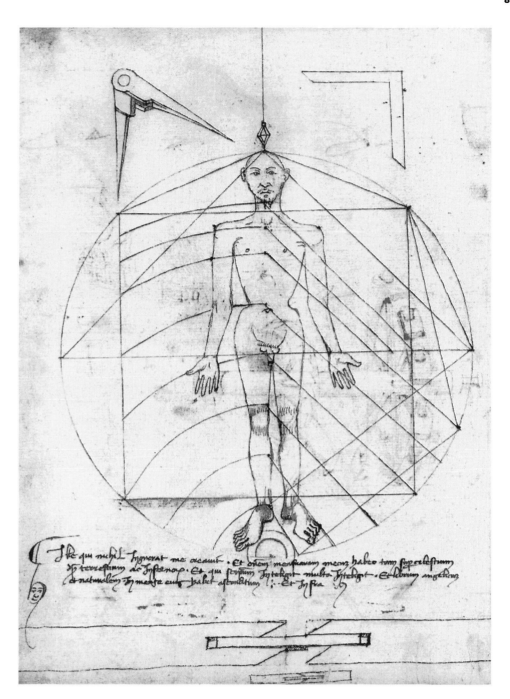

Man in square. Ink drawing by (Pietro Mariano) Taccola. © Munich, Bayer. Staatsbibl.

Man in square. Woodcut from C. Cesariano's Vitruvius (1521).

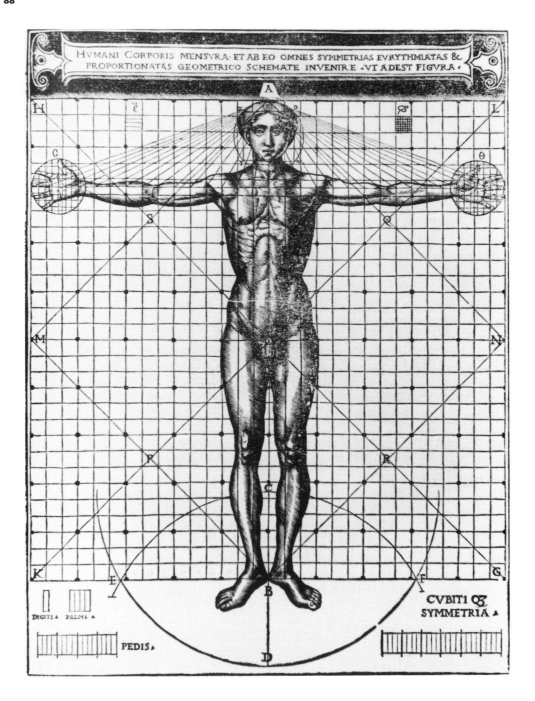

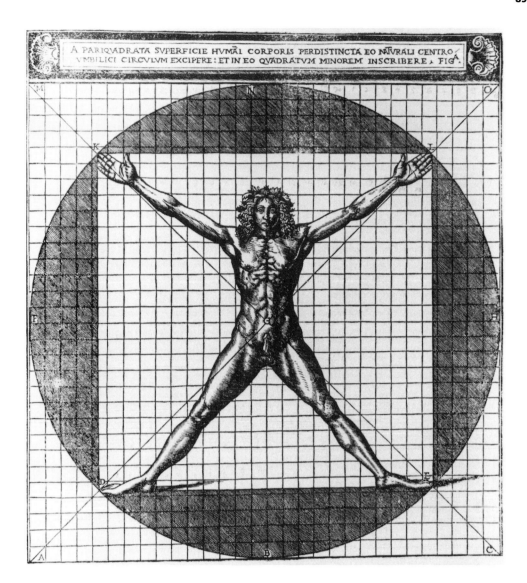

a tangent, so that the center of the figure moves up from the base of the penis (for the figure with legs joined in the square), to the umbilicus of the same figure in the circle, whose feet are now splayed into an equilateral triangle.[57] The scale at the bottom of the sheet is calibrated in cubits, palms, and inches, though the only obvious part of the figure to measure one palm exactly is the generative member. It is essential to recollect that Leonardo's Vitruvian figure stands in close relation to his own more detailed proportional studies, which involved seeking the internal mathematical relationships of the human body from the details of the extremities to the internal sections of the skull.[58]

From the coded, scrappy diagrams in the notebook of Villard d'Honnecourt to Leonardo's isolated, stunning image, only two rather enigmatic speculative drawings investigating the subject have survived, however much Leon Battista Alberti and his immediate followers may have concerned themselves with it. Once Leonardo's drawing emerges fresh and brilliant from the chrysalis of words, many other artists and commentators follow: Piero della Francesca's *De Perspectiva Pingendi* remained in manuscript, but in 1502 the printers Giunta in

Florence published Pomponius Gauricus' book on sculpture, which I mentioned in the previous chapter. Its second section was entirely devoted to physiognomy; in fact it opens with an account of the canon of human proportion.[59] By coupling physiognomy to the canon, Gauricus devised a quasi-empirical method of introducing variation into canonic measurements, which were restated in Fra Luca Pacioli's book *On Divine Proportion* and in the various editions of Vitruvius. The canonic figure was also first illustrated in the Vitruvius edited by the aged Giovanni Giocondo that was also printed by the brothers Giunta in 1511; ten years later appeared the much more impressive Milanese edition by Cesare Cesariano, which indirectly refines Leonardo's first speculation in the matter of human geometry. But while humoral science and physiognomy extended to the whole body offered one way of investigating the validity of the canon, another was provided by setting it in motion.

Vitruvius' man in motion

Leonardo had in fact gone well beyond Vitruvius in his demands on the geometrized body—inevitably, since he not only desired to investigate its inner harmony, but wanted to place the proportioned man in a circumscribing geometry as well as in a rationalized perspective. He therefore needed to take the coordinates of the crucial points on several bodies in order to calculate their exact situation in any perspective. To do this convincingly, he also had to record how these points were displaced in movement. He does indeed take Vitruvius' circumscribing circle and square as his postulate: the center of the circle around the spread-out man is at the navel, while for the cruciform man in the square it is the crotch, so that he rationalizes the two centers to be a ninth of the body-height apart. Leonardo shows his figure-in-the-square man ruled by five equilateral triangles, a pentagon, and an octagon, whereas the figure in the circle (which encloses a man with his arms raised into such positions that he looks very much as if he were posing for Le Corbusier's blue and red *Modulor* series) is also enclosed in a hexagon. His man measures six of his own feet high, and there is also a canon for a woman of eight feet. These studies of human movement are mostly known through the copies that were put together in a notebook now called the Codex Huygens, after its most famous owner, the great Dutch astronomer Christian Huygens.

In that notebook, each of the superimposed "Platonic" figures circumscribes the moving body. Whoever copied these drawings of Leonardo's also thought it necessary to show on another page the simple outline of a man on which were imposed four regularly decreasing circles, each one labeled with one of the elements. Certainly in the medicine of the two following centuries, the analogies between the metric microcosm and the elemental microcosm were to provide a theoretical base for much speculation and experiment, while the astrologers of the time were passionately interested in the geometrical and musical operation of the astrological signs.[60]

Dürer on human variety

Working about the same time in Nürnberg from somewhat different presuppositions, Albrecht Dürer, unlike Leonardo, seems quite incurious about anatomical detail. Yet he also needed a geometry and physiology of movement, even if not one as frighteningly complex as Leonardo's. The changing configuration of different human types, from the tall and thin to the small and

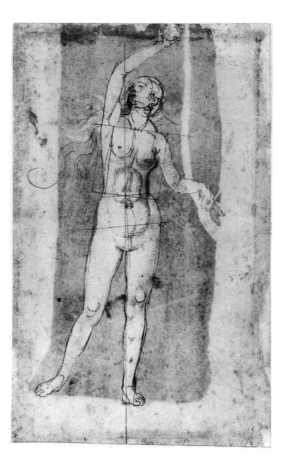

Eve. Construction drawing by Albrecht Dürer. Vienna, Albertina.

Adam. Preparatory drawing by Albrecht Dürer for an etching. Vienna, Albertina.

The Body and the World

fat, from the old man to the baby, was even more important for him; he attempted a kind of algebra of various human proportions, or at least its topology, if the reader will forgive my anachronistic use of this eighteenth-century term. It is nevertheless very striking that his empirically measured figures (which are sometimes even distorted on a curve—there were no distorting mirrors of a large enough size available to Dürer for experiment) are stiff and numb when he records real people as instances of a type, almost as if they were drawings of articulated dolls made only to study their movements, whereas his constructed and ideal figures look bursting with life.

Although Dürer was certainly well aware of and even fascinated by the physiognomic variety of his models, he also implied the possibility of making some systematic relation between body and face, even if it remains fragmentary in his writings. Dürer's conclusion drew one essential lesson from the bulky notes he made on human proportion (and which make up by far the largest part of his copious writings); namely, that the inclusive listing of different types may in itself be the recognition of microcosm: "The Creator fashioned men once and for all as they *must* be, and I hold that the perfection of form and beauty is contained in the sum of all men." [61]

Gian Paolo Lomazzo

Those hints of his were erected into a system of a kind by the blind painter-theorist Gian Paolo Lomazzo, half a century after Dürer's death. Lomazzo systematizes the canon externally, as it were, by moderating its value, since its prescriptions must always be adjusted to the point of view of the spectator. He nevertheless sets out a number of rules relating numerical values to humors and to the work of contemporary artists, who had never been discussed in this kind of text before. A male of ten faces in a square follows Vitruvius' recipe for the "normal" figure, though he also gives a slender formula of ten faces as well: lithe, Mars-like, hot and dry. But another, even more slender male is of ten heads: extravagant, yet essential, because recommended by the great Dürer and included in homage to him. Then Lomazzo describes a young body, of nine heads, as that of Apollo Belvedere and a proportion used by Raphael: this leads him to a parenthesis on the need to vary proportions in the picture, as Parmigianino had always done. Next he considers a male figure of eight heads (the Neptune of Zeuxis) and one of seven heads, the last as of a Hercules. Then he details a woman: one of ten faces, one of ten heads, one of nine faces—the middle proportion appropriate to the Blessed Virgin—and yet another of nine heads, then one of seven heads. Children of six, five, and four heads are considered. The section ends with three chapters on the proportions of horses. This is followed by a conventional account of the orders drawn from Vitruvius, Alberti, and Vignola. What is new in this rather eclectic book of instructions is the introduction of another source, the writings of Cornelius Agrippa, to whom Lomazzo owes much of his temperamental lore and his accounts of gesture and passion, as well as of planetary influence. [62] The criss-crossing counterclaims and counterinstructions had become too prolix to provide an artist with useful precepts. How had the ancients managed this complex business?

Inevitably perhaps, Lomazzo turns to the artist who could speak about this matter with authority, Michelangelo, whom he reports as saying that "modern painters and sculptors should have proportion and measure in the eye . . . and that the marvelous science of the ancients in this matter had been lost by the moderns." [63] It may not have been a helpful precept, but it was an answer.

Michelangelo

The stiff, doll-like appearance of Dürer's measured figures was to offend Michelangelo: "Albrecht discusses only the proportions and varieties of human bodies, for which no fixed rule can be given; and he forms his figures straight upright like poles; as to what was more important, the movements and gestures of a human being, he says not a word."[64] Michelangelo's own book on the subject is one of the great unwritten documents of literature and some traces of it may survive in the work of his friend (and perhaps collaborator) Realdo Colombo, a famous Lombard anatomist,[65] while his obsessive interest in anatomy as well as in the problems of proportioning the human body is attested above all by the nature of his drawings. It is worth noting that one statue of which he is known to have made studies from the skeleton outward is that of the risen Christ at Santa Maria sopra Minerva.[66] Yet Michelangelo also saw this exhausting and anguishing activity as a model for his revolutionary architectural projects, as he once confessed. His unfulfilled obsession was carried into the theory of art in the century and a half that follow by the witness of his pupils and disciples.[67]

It was Leonardo then who had finally transformed the descriptive canon of Vitruvius into an exemplar that became an archetype. If its model seems to have been deduced from empirical observation, the anatomical precision and biting curiosity are Leonardo's own. Even before his time the study of anatomy had become an important accomplishment for any painter or sculptor, and it long continued so. Some of the earliest empirical anatomists in fact were artists like the Pollaiuolo brothers, perhaps just because they were not inhibited by the learned Galenian anatomy current in the medical schools. And the greatest artists of the following generations continued an active and passionate interest in dissection, providing a quite different field of instruction for them that was to last until the revolution in their training operated by Le Brun, which I discussed in the previous chapter.

Michelangelo's friend Realdo Colombo asked his patron for leave from teaching at Pisa to be able to work on anatomy in Rome with "the world's foremost artist."[68] Colombo's teacher Vesalius had complained of the press of artists at his public dissections,[69] but he must have collaborated very closely with his engraver-draftsman, who was probably a pupil of Titian. And in fact Titian himself has been suspected of having contributed to Vesalius' success.[70] The title of Vesalius' most important book, *On the Fabric of the Human Body*, seems to refer to many of the ideas that I have considered; in spite of that title, he was not wholeheartedly committed to an architectural account of the body, and he mocked his disciple Colombo (who was also his successor as anatomic demonstrator at Padua) for paying too much attention to such metaphors.[71]

Fabric of man, fabric of the world

The accident, if accident it is, of its date of first publication, 1543, associates Vesalius' book with another one published that same year by quite a different physician: Nicolaus Copernicus. The coincidence of the date has sometimes seduced scholars into claiming that the works were a two-pronged attack on the old microcosm/macrocosmic system. That it now seems so is most probably hindsight, since Copernicus claimed to be restating Pythagorean notions.[72] As for anatomists, when it came to the study of proportions, even the most passionately empirical among them seemed content to base all speculations on Vitruvius' text, or on other, less

Christ holding the Cross. Marble figure by Michelangelo. Rome, Santa Maria sopra Minerva. © Fratelli Alinari.

elevated writings—which they seem to have used as mere recipes. Yet certainly in Vesalius' figures, and even in those of Claude Perrault more than a century later, the close resemblance between anatomical and built fabric is insistently pointed.[73]

In spite of Michelangelo's carping, the great exception to this general rule was Dürer. He was the first to take another step, to attempt a differentiated survey of various human figures, perhaps with some thought of finding a common algebraical agreement among them. Dürer's book on human proportion, which was not published until several months after his death, became, as its popularity testified, a sourcebook for European artists and indeed an authority for medics as well.[74]

Like Michelangelo and Leonardo, Dürer was familiar with the Vitruvian canon as well as with other formulae of human proportion. But so were many artists before them and throughout the Middle Ages, as I have argued. Although we can only guess at how the canon was transmitted, it is evident that it was held in some esteem by literary people, even if it was only used as a rule of thumb by artists. No known archaeological procedure has revealed quite how or when Vitruvius' or the other various canons were formulated. The one thing that is certain is that by the time Vitruvius wrote it down, the particular formula he recorded had already been circulating for some time.[75]

IV

Gender and Column

▪ The square and the circular man ▪ *Kanōn* ▪ Polykleitos ▪ A female canon? ▪ Luck and invention

In the previous chapter I attempted to reason that the very notion of a canonic order in building, particularly an order based on regulating the proportion between beam and column, drew all juice and nourishment from its being rooted in the analogy between building and body, body and world. Or, if you would prefer, that the validity of the different kinds of column was guaranteed by their being a special way of understanding the building as microcosm.

Some of these roots were cut when the body was read as a system of circulation after William Harvey's great discovery. Circulation of blood, water, and choler was seen increasingly as an analogue of the movement of people and goods in the city. To Harvey the heart was no mere pump, however, but was to the body as the sun to the Copernican world and the sovereign to the royal city; so he wrote in the dedication of his book to Charles I. By the end of the seventeenth century, such analogical guarantees were withdrawn, and in the eighteenth and the nineteenth centuries the orders of architecture came increasingly to be seen as husk, as one kind of historical costume among others, a convenient disguise for an industrial age's startling and threatening architectural forms (such as factories and railway stations), to be preferred in that role over Arab or Indian or Chinese ornamental forms only marginally. In exotic, remote countries they could be offered—and often were—as tokens of European superiority, domination, or both.

However long the inertia of custom sustained the validity of classical orders or colonialism extended their territory, it had become evident by the mid-nineteenth century that if a man-made environment was to be reorganized on one ordering principle, it would have to be grounded in some quite different world picture.

The square and the circular man

And yet in those fifteen hundred years between Vitruvius and Leonardo, when graphic evidence may often seem thin and inconsistent and the appeal to antique ordering principles erratic, the polarity between microcosm and macrocosm had nevertheless remained central to the thinking of the Western world: the doctrine that had been set down by Vitruvius continued, into the nineteenth century, to be regarded as an essential piece of evidence connecting that polarity with building. That is why echoes of the Vitruvian formula were found in literary and liturgical texts throughout the Middle Ages—even if the formula itself is invoked as a general notion rather than examined in any great detail.[2] The assumption behind that formula is the canon of the human body. It is an arithmetical and geometrical model, recorded at the beginning of Vitruvius' third book, where he stated his beliefs about the principles of design (presumably as he inherited them) when applied to temples, which were considered the noblest kind of building:

> It is not possible for any building to be properly designed without symmetry and proportion, that is, if its parts are not precisely related, as are those in the figure of a well-made man. Since nature composed the body so that the face, from the
> **tenth** point of the chin to the top of the forehead (or the hairline) should be a tenth part of it, which is also the length of a spread hand from the wrist to the end of
> **eighth** the middle finger; the head from the chin to its very top an eighth; the distance from the top of the breast bone, where it joins the neck, to the bottom of the
> **sixth** hair is a sixth; [from the middle of the breast] to the crown of the head is a

fourth   fourth. As for the face, a third is from the point of the chin to the underside of
third   the nostrils, and the same to the line of the brows; from that line to the hair of
x six   the forehead takes up another third. The foot goes six times into the height
x four   of the body, the cubit four times, the breast is also a quarter. The other members
have their appropriate measurements: by making use of them famous painters
and sculptors of antiquity gathered glorious and lasting praise.[3]

In the last sentence Vitruvius clearly implies that his canon is only a section of a larger account
of how the parts of the body are internally commensurable with each other. He uses this
formula, an example drawn from nature, to justify the attention that he proposes to devote to
the commensurability of the parts of the temple through the columns and beams; and he insists
that such a canon is the one essential model or pattern respected by all the principal ancient
artists. Some of the body parts, as they are enumerated by Vitruvius, are also the commonplace
measurements used by the Greeks and the Romans. So the whole handstretch is the fathom
or orgyia; half the fathom, or the tip of the nose to the end of the fingers, is the yard of three

feet; the quarter fathom is the braccio or cubit; the foot is a sixth of the fathom and the palm a sixth of the cubit, down to the finger (which in turn is a fourth of the palm) so that the foot is subdivided into sixteen fingers, while the modern foot divides into twelve inches, "inch" being a corruption of the Latin *uncia,* a twelfth.[4] He goes on to imply a more generalized belief:

> Nature so has it that the center of the body is the navel; if you were to lie a man supine on the ground, with his arms and legs spread wide, and set the point of the compass on his navel, the circumference of the circle you could then draw would touch his outstretched fingers and toes. Moreover, as the circular scheme may be found in the body, so also may the square. For if you take the measurements [of a man] from the feet on the ground to the top of his head, and compare this measurement to the full spread of his hands, they will be found to be exactly equal. . . . What is more, from the members of the body [the ancients] derived the standards of measure that seemed necessary for all their work.[5]

No one has ever suggested that Vitruvius found this out by empirical measurement, though for all we know, he may have tried to check the traditional figures. In fact bodily members are the most obvious primary measuring tools, and their internal commensurability must have a part in the most ancient human experience. To cover your face with your hand is about the simplest way to check the formula—and has therefore been said by some to be the most primitive aperçu about corporal symmetry. In many societies people still measure cloth or string with elbow and handspan, and Vitruvius must often have seen it done in markets or on a building site. But there is another aspect to the two texts, at least as they have been strung together, which is not often noted: the dimensions should be drawn from the man-standing-in-the-square, the man-at-rest-in-the-circle, or both, much as Leonardo had attempted to square the circle through the human body. Only a few dimensions could be checked on a body in motion. Certainly, the fifteenth- and sixteenth-century artists who tried to control and elaborate them did learn, as I suggested earlier, that the matter was not straightforward.[6]

Kanōn

There is one object that has survived from antiquity, however mutilated, and seems to have been done to illustrate canonic human dimensions: this is the so-called Metrological Relief in the Ashmolean Museum in Oxford, probably carved about 450 BC or perhaps a little later, somewhere in Ionian lands. It shows the top half of a young man's body, with his arms stretched the full fathom, enclosed in a roughly triangular and pediment-like plaque, while by his head there is an imprint of a foot; the fathom in this case is exactly seven feet. On the other side, which is partly destroyed, there may have been, as was recently suggested, the imprint of a fist or palm. The angle at the top, which seems not to have been exactly symmetrical, is just short of 140°, a little sharper than that of the usual pediment, even if a pediment does seem to be implied by its outline—which has led some authorities to suggest that it was not an independent "work of art" but the ornament of some building. It may well have served to reconcile conflicting measurements or canons, or even to state a canon for which there is no literary documentation. The one thing to which it surely witnesses is the interest in the sort of relation between body and building of which Vitruvius writes.[7]

Metrological slab. Oxford, Ashmolean Museum.

In fact Vitruvius seems to have transmitted a rule that was formulated long before his time. Perhaps his information was drawn from more than one rule or book: so much is suggested by his inconsistencies, the overlaps of the terms he uses, and the allusive way in which he sets out his statement, as if he was appealing confidently to common knowledge. Still, no formula about human proportion earlier than Vitruvius has survived into our time. Such canons certainly existed among Hellenistic artists, who in their turn relied on older traditions. The most famous of all such canons was made sometime about the middle of the fifth century by the sculptor Polykleitos of Argos (or Sikyon) in the form of a statue whose limbs were a solemn proclamation of numbers and their proportions. He almost certainly wrote an explanatory commentary on it.[8] Vitruvius' account of human proportion probably drew on Polykleitos' canon, for the first part of my quotation at any rate. Before I consider the Polykleitan rule further, however, I would like to comment on the general notion of a proportional canon.

The very word *kanōn* was already well-worn by Vitruvius' time, and like so many other words that denote a more a less abstract concept it has a hoary ancestry. Like most such words, it once stood for a tangible thing: it seems to have meant a "straight rod" or "stick," such as a cane, which by metonymy came to mean a measuring rod in classical Greek, or a ruler or even a carpenter's square, a mason's level, and the beam or tongue of a balance; its assumption into musical theory—which inevitably implied number theory—is perhaps more important in this context.[9] *Kanōn* came to mean a monochord, on which the pitch of the plucked string could be altered by moving the bridge. And indeed the word for a clear sound or tone, *phthongos,* may be onomatopeic of the plucked string.[10]

Legend attributed the discovery of the direct relation between audible and "objective" musical harmonies and measurable quantities of weight and length to Pythagoras.[11] It was said to have been the result of a happy accident. The harmonious sounds of hammers striking anvils as he walked past a smithy provoked the revelation, and he is said to have translated the weights of these hammers into elementary harmonies: the *octave,* which corresponded to stopping the string halfway (*2:1*), a *fifth* at three to two (*3:2*), and a *fourth* at four to three (*4:3*); a *major tone* was described as the difference between a fifth and a fourth (³⁄₂ divided by ⁴⁄₃ = ⁹⁄₈). This, of course, meant that the numbers 1, 2, 3, and 4 described or delimited the whole of musical harmony and when added made 10, the number perfect in many other ways;[12] it could—and often was—demonstrated "experimentally" on a one-string instrument, the monochord, though a two-stringed or even three-stringed one was also sometimes used.[13] Yet the importance and excitement of this discovery, perhaps "the first formulation of a natural law in mathematical terms,"[14] is rather difficult to conceive in relation to the performance of music, particularly as the elementary notion of the relationship between distance and tone must have been known empirically to all makers and players of both string and wind instruments.[15] Many-stringed plucking instruments were in any case being played all over the eastern Mediterranean for at least two millennia before the time of Pythagoras, and presumably the method of tuning them was common knowledge long before it was theorized. However, for the complete numerical demonstration of harmony, the first odd and the first even number were multiplied (2×3) and their product 6, another perfect number, produced the demonstrative series 6:8:9:12, which formed another tetrachord: two fourths separated by a tone. The ratio 6:12 makes an octave, 9:8 the major tone, and 6:8 and 9:12 a fourth.[16]

Medieval commentators had already noticed certain similarities between the legend of Pythagoras' discovery of numerical harmony and the paired inventors of music and metalsmithing in the Book of Genesis, the half-brothers Jubal and Tubal-Cain.[17] The total implausibility of the Greek myth about the discovery of harmony (in that there is no relation between the weight of hammers and the sound they produce) may be a pointer to the archaic origin of the account—and perhaps to an association of musicians and metalsmiths on the biblical model.[18]

Like many of his contemporaries, Vitruvius was familiar with such Pythagorean speculations about number: in the canonic passage he glosses over them so nonchalantly that the great cosmic themes are not immediately recognizable, even if he does make a reference to Plato and talks to Pythagoras and the Pythagoreans elsewhere in the book. It is in discussing the perfect number, *teleion,* that Vitruvius makes the most explicit (but there unattributed) reference to Pythagorean teaching:

> The perfect number the ancients established was the number ten, the number [taken from] the fingers of the hands; from the palm they found the foot. While the number called *ten* is found perfected by nature in the fingers of the two hands, Plato was satisfied that the number was perfect since it was made up of separate things, which the Greeks call *monades.* . . . On the other hand the mathematicians maintained that the number called *six* was perfect because it was divisible into fractions that combine into the number six . . . and no less because the foot corresponds to the sixth part of a man's height. . . . [L]ater, when they recognized that both numbers, six and ten, are perfect, they combined both in one and made the most perfect number: sixteen. They found support for this from the foot: since if you take two palms away from the ell, you are left with a four-palm foot; and each palm has four fingers. If it is therefore agreed that numbers are derived from the body . . . [19]

Although the arrangement seems banal—even mechanical—in Vitruvius' text, it was obviously derived from some lost Pythagorean source; but the relation between Vitruvius' "ancients" and his "mathematicians" is obscure. The half-acknowledged debt that he owed to such a source suggests a connection between the forgotten Pythagorean and the post-Polykleitan canon, even if the filiation must remain a presumption.

Although the monochord canon of the original legend may not have been used much for performance, it was certainly used for musical experiments and for tuning.[20] In spite of the elaboration of the numerical, calculable (and therefore rational) teaching of harmony, there was another view, first formulated by Aristotle's cantankerous pupil, Aristoxenus of Tarentum. He held that the numerical harmonies did not always work phenomenally, that they were developed from first principles and therefore could not provide a true account of a listener's experience. Plato's Socrates allows for this problem; but unlike Aristoxenus he considered the calculable harmony superior to the heard one, and therefore ridicules the musicians who "vex and torture" their string on the pegs of harps or lyres to determine by ear the smallest possible audible intervals on which a scale might be based.[21] Whatever the shortcomings of the Pythagoreans, Plato himself would not have accused them of such practical foolishness since he proposed to consult them about the study of harmonics when setting up the right education for the guardians of his republic. That passage in the *Republic* shows the early division among two kinds of Greek musical theorists;[22] and the dispute does touch on a very important strain of my theme, the contradiction between the rational or noetic school of musical theorists and those who wanted to concentrate on what they heard, the phenomenal school. Greek architects constantly had to deal with an analogous division, since the systematic "visual" or "dimensional" formulae of the sculptors and architects had to be modified in a similar way by the demands of optical adjustment and the unavoidable contradictions that were internal to the system itself—which I will have occasion to discuss.[23]

For all his sarcasm and aggression, Aristoxenus did not question the basic calculations of Pythagorean harmony since it seemed irrelevant to music as played and heard. And

indeed by metaphoric extension, *kanones* came to mean any rule or standard of excellence. That sense was extended further so that in the logic of Epicurus and his followers it came to mean the standard of truth, a proof of reality or even plain, reasoned thinking.[24] The notion broadened to literature: so Herodotus was thought to be the canonic writer of the Ionic dialect, as Thucydides was of the Attic.[25] And if visual, plastic artists were also interested in finding a rule or standard of excellence, inevitably the human body was thought to hold the key to it. The word *kanōn* therefore became a technical term for an account of its proportions, and that is the context in which Polykleitos' bronze spear-bearer acquired his name. In late Imperial and into medieval times, the Latin transcription of the word, *canon,* signified anything that must remain unchanged, such as the text of Scripture or the central section of the Latin Mass.

The word was also used for any guide or template on which these relationships were recorded and further was applied to any object that showed or set standards of measurement. Nor is such a conception of the fixed relationships within the human frame limited to Greek and Roman antiquity, although the Greeks may have bound the notion of a canon in terms of linear measurement to an exact analogue in terms of sound. In India, to take one obvious instance, a panoply of canonic measurements exists for the different deities of the Hindu pantheon, though it is a matter still inadequately studied. The whole height of an important figure involved in the building or his palm or even the outer phalange of the thumb seem to have been used often by Indian builders and sculptors to provide the module of a building, even the frames for molding the bricks of a fire altar.[26]

There is a late but very important source for an analogous tradition in Greece. Plutarch (so Aulus Gellius writes at the beginning of his garrulous anthology of various information) praised Pythagoras' subtle reasoning in reconstructing the outstanding height of Hercules

since it was generally known that Hercules paced out the stadium of Olympic Jove in Pisa. While other stadia in Greece which were established later by others were also six hundred feet long, and yet somewhat shorter, he understood straightaway that the size of Hercules, reconstructed from his foot, would exceed that of other men by as much as the Olympic stadium was longer than any of the others.[27]

Whatever Plutarch may have inferred from what he was told or had read in some lost written source, this kind of argument was the stuff of Pythagorean legend, and shows a general familiarity with canonic formulae. In its general outline the legend was ancient: Pindar had already referred to it in his account of the founding of the games by Herakles, who,

. . . Zeus' brawny son,
measured out a holy place
for his mighty father.
He fenced the Altis and marked it off
a pure space[.][28]

As ever, the myth does not record some ordinary transaction but rather a heroic ritual that consecrates and exalts the commonplace. And from such heroic matters, I come down to the particular, as I return to the canon of Polykleitos. He was one of the major artists of the fifth century, the great age of Greek sculpture and architecture, and had probably been a fellow pupil of the even more famous Pheidias in the workshop of Ageladas (also of Argos). The two of them were called wise men, well noted for their *sophia*.[29]

Polykleitos

As I suggested earlier, Polykleitos' canon was incarnate in a statue, the bronze spear-bearer Doryphoros, as well as in an explanatory or descriptive text that he most probably wrote himself. Fragmentary quotations allow only surmise about its nature and contents, and his original sculpture has also perished.[30] Fortunately, it was one of the most frequently copied ancient works of art, both in stone and metal, copied with such concentration and fidelity that even fragments of such copies are not difficult to identify. As I examined them, the Doryphoros became a familiar, almost a friend: the assimilation of some limbless trunks in stone to the type seemed to me to follow Scamozzi's observation, which I quoted earlier.[31]

The copies and the literary references show the admiration for the statue in antiquity, as well as its great authority. Polykleitos was thought to be a kind of new Daedalus, since like that archetypal artist, he made his statues walk. Doryphoros stood in *contrapposto,* the weight of the body taken on one foot, the other moving free. Of course there was some precedent for this freedom of pose, but Polykleitos is credited with the final liberation of male statues, the capacity to make the figures move, which has been associated with the grammatical figure of *chiasmos* (cross emphasis)[32] though the word is not used by ancient critics of the visual arts. Curiously enough, it is when reporting this achievement that Pliny also quotes a much discussed criticism of Polykelitos, which he attributes to an earlier polymath, Varro: that his

Doryphoros. Naples, Museo Archeologico. © Fratelli Alinari.

Doryphoros. Naples, Museo Archeologico. © Fratelli Alinari.

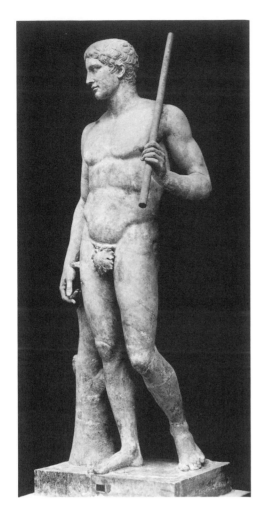

Doryphoros. Vatican. © Fratelli
Alinari.

Doryphoros. Florence, Uffizi.
Photo by author.

Chapter IV

Doryphoros. Head and torso fragments. Rome, Museo Baracco.
© Fratelli Alinari.

Doryphoros. Torso only. Diorite. Florence, Uffizi. Photo by author.

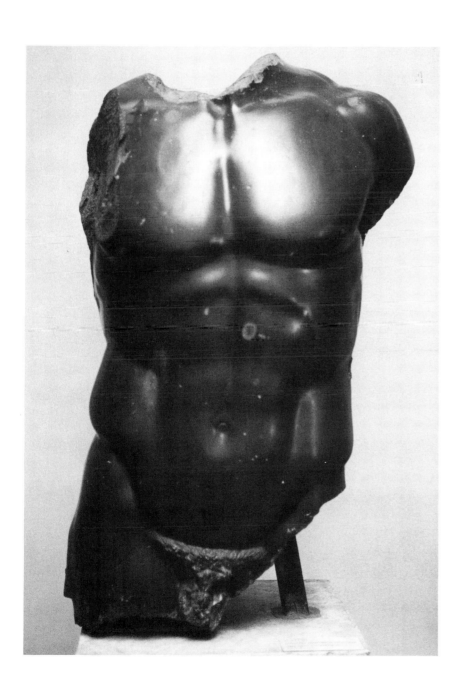

Doryphoros. Bronze herm copy
by Apollonios of Athens (from Herculaneum). Naples, Museo Archeologico. © Max Hirmer.

statues were "square, every single one" (alternatively, "all based on the same model"). Since
he reports this in the very same passage in which he attributes the invention of the relaxed
contrapposto stance to Polykleitos, it must be taken as yet another contradiction in this
puzzle.[33]

That a text of Polykleitos' existed is implied by several purported quotations from it
in later authors. Beside the passage in Vitruvius that I cited earlier (perhaps an indirect quotation, if the longest one), there are other snippets in Plutarch, Philo of Byzantium, Lucian, and
Galen. Galen, in turn, seems to have read them in some Hippocratic writings. Hippocrates of

Chapter IV

Polykleitos: Amazon. Bronze herm copy. Naples, Museo Archeologico. © Fratelli Alinari.

Cos was of the generation after Polykleitos and the best-known ancient physician. He had an almost heroic role, since many writings that bear his name were in fact compilations rather than original compositions. Of his later editors and commentators, Galen, court physician to the Emperors Marcus Aurelius and Verus, was the most famous and prolific. Galen has occasion to quote Polykleitos several times, in one important text coupling him with Chrysippus the philosopher, who had been a systematic compiler of Stoic wisdom in the third century BC. Chrysippus had expanded the familiar doctrine about the cohesion of elements in the body: in discussing mental diseases, he distinguished between the *summetria* that combines the *stoicheia* (its elements) harmoniously so that they constitute the health of the body and the

Gender and Column

analogous but very different *summetria* of *moirai* (its members or limbs), which makes it beautiful. This symmetry, says Galen, "should link finger with finger, all the fingers with the palm, the palm with the wrist [*carpos* and *metacarpion*: perhaps more accurately, 'palm' and 'base of the hand']; these to the ell or forearm, the forearm to the upper arm, and every other part to every other, just as it is written down in the canon of Polykleitos." [34]

Another Polykleitan saying appears as a quotation in Philo of Byzantium's *Belopoeika*, a book in which he also discusses the gradual perfecting of optical refinement by similar small steps. Having reviewed the construction of several siege machines, he concluded: "the right thing [*to gar eu*] is achieved by small steps [*para mikrōn*] through many numbers." [35] It is echoed, though without reference to Polykleitos, in Plutarch's short book on hearing; the beauty (*to kalon* rather than *to eu*) of any human work is achieved gradually, as in Philo's quotation, "through many numbers that come to an accord [*kairos*] in a symmetry and harmony, while ugliness may happen suddenly, from any trivial defect or excess, and can be appreciated [in a lecture] through frowns, and grimaces, roving eyes, lids drooping, yawns, crossing of legs, and indecorous fidgeting." [36] A passage quoted in a different context has an even more enigmatic ring to it: what Polykleitos seems to have said is "that the task is most difficult when clay is at the nail, and the work is drawing to perfection." Whatever he meant by that, it seems to point again to the master's obsession with the perfect detail. [37]

Although Vitruvius refers explicitly to the canon of Polykleitos in his "bibliography," [38] it is not all that certain, as I observed earlier, that he was quoting directly from a text by the sculptor when he specified the proportions of the human body. Later he mentions nine lesser-known artists, each one of whom produced a canon: of these Euphranor of Corinth (also known as the Isthmian) seems to have been the only one whose ideas on the subject were known to other writers of the time. [39] His ideal human figure was probably much more differentiated, accentuated than Polykleitos'. The hugely prolific Lysippus of Sikyon, often considered the last "truly Greek" sculptor (that is how Vitruvius mentions him), who was court artist to Philip II and Alexander the Great, was also famous for his passionate concern with human proportion. He is quoted as having said that he had no other master beside the Doryphoros, [40] though it is clear from the context that he was referring to Polykleitos' statue as a negative example of the kind of "perfection" that he wanted to reject. Lysippus' reputation for the study of the canon is however even more of a quandary than Polykleitos', since there is no trace of his text. And his best-known saying, as puzzling as any of Polykleitos', was that "while they ['the ancients'] represented men as they were, he made them as they appeared." [41]

A female canon?

In spite of the evidence available, the context of all that speculation about an ideal male figure in the fifth and fourth century BC is hazy. Moreover there is no comparable evidence of an interest in the proportions of the female body, perhaps because the naked female figure was a great rarity in Archaic and even Classical Greek art, remaining so until the middle of the fourth century. [42] At one point, however, it became necessary to assert at least the possibility of such a female canon. The occasion was the competition between four or five sculptors for an Amazon statue, to be dedicated in the temple of the Ephesian Artemis. Polyklcitos' was deemed the best (if Pliny's sequence is to be taken seriously), while the others were by Pheidias, Kresilas, Kydon, and Phradmon. [43] Surviving copies of these are not quite as numerous nor as faith-

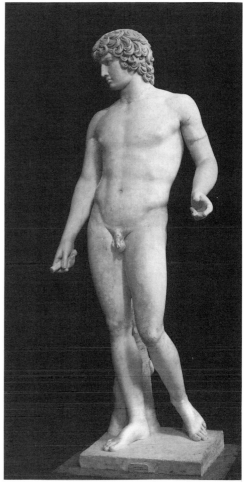

Euphranor (after). Antinous. Also known as Lansdowne Athlete. Copy. Naples, Museo Archeologico. © Fratelli Alinari.

Lysippus: Eros stringing his bow. Vatican. © Fratelli Alinari.

Gender and Column

Polykleitos: Amazon ("Capitol" type). Vatican. © Fratelli Alinari.

ful as those of the Doryphoros, and indeed the art historians' distribution of the different prototypes among the three sculptors is not at all unanimous. The Amazon types are all partly clothed and most have one breast bared, which was the most obvious aspect of their national costume, as it were. That scantily clothed figure may have led to an explicit definition of female bodily proportions on an analogy to the Doryphoros; and indeed in Herculaneum, two lovingly executed bronze herms, one of the Doryphoros and the other of the Amazon, stood in the Villa of the Pisoni. Still, whatever the intentions of the sculptors, there is no evidence for the existence of any written canon here as well.[44]

Luck and invention

The Vitruvian body, both square and circular, was in movement—yet it was timeless; as microcosm it was unique and unalterable. Its assimilation into building therefore required mediation. Moreover the unique microcosm did not seem to apply equally to the Doric and the Ionic columns, and certainly not to the later Corinthian one. Vitruvius thus had a narrative problem, since it was his aim at the beginning of the fourth of his ten books to justify (perhaps more accurately to rationalize or even canonize) the distinction between the different Greek systems in historical terms. He therefore played on the opposition (which he presumably took over from his teachers) between the microcosmic, noetic structure of building, which depends on these principles, and the phenomenal apparatus of adjustable proportions and mimetic ornament, which could be explained mythically, if not historically. He and his readers would have been familiar with the most obvious historical evidence: two of the columns bore the names of the principal dialects of the Greek language, the Doric and the Ionic; the third was called after a city, Corinthian.

The civic column was inevitably shown to be the latest of the three. About the other two he offers evidence in the form of a legendary etiology, set down at such a late date that it is difficult to descry the mythological and perhaps even ritual background from his recounting. However, this is what he writes:

> When Dorus the son of Hellen and the nymph Phtiade ruled over all Achaea and the Peloponnese, he built a temple to Juno in the ancient city of Argos, a sanctuary that fortunately was of this design, and then went on to build more temples of the same kind in the other cities of Achaea, even though the rule of its symmetry had not yet been born.[45]

It is very noticeable that although Vitruvius places the invention of the order at a crucial time in the heroic age, he seems to leave its origin to chance: the temple was "eius generis fortuito formae." Translators rendered the word *fortuito* as "by accident," "par hasard." But since the account of Argos must have come from some Greek source or informant, it is worth noting that *fortuito* is the usual translation of the Greek *kata tuchēn*: and *tuchē* as a cause much interested Aristotle.[46] In fact it may almost be better rendered here by "fortunately," or even— to stretch a point—"as if by design."

Of course the temple in which he has all this happen was a grand and famous one: Hera Argiva had one of the most splendid in all Greece, where Polykleitos (or perhaps Polykleitos the younger, his nephew and also a famous sculptor) made his chryselephantine image of the goddess, to rival the other two great chryselephantine images by Pheidias, of Athena in the Athenian Parthenon and of Zeus in Olympia. And Argos was one of the ancient Dorian settlements, the capital of Agamemnon, who as high king of Argos commanded the Greek fleet and army at Troy.

The absence of a visible prototype, the inevitable disruption of builders' crafts, Vitruvius thought, was to lead to the formulation of a rule:

> When the Athenians later sent out thirteen colonies to Asia all at one time according to the oracle of Delphic Apollo and the common counsel of all Greeks, they appointed leaders for each colony but gave the supreme command to Ion, the son of Xuthos and Kreusa.... [T]hey called that part of the world Ionia after Ion their leader, and having set up shrines to the immortal gods, they began to build temples. And first they decided to build one for Panionian Apollo, such as they had seen in Achaea, which they called Doric since they had first seen that kind of building in Dorian cities. When they wanted to put columns in it, since they did not know the symmetry, and wondering by what method they could make such as would be suitable to carry weight, and in appearance have a beauty to which they were accustomed, they measured a man's footprint and compared it to his height. When they found that in a man the foot was a sixth part of his height, they transferred the same to the column; and however thick they made the bottom of the shaft, they made it, with its

capital, six times as high. That is how the Doric column began to display the beauty and strength of the male body in buildings.[47]

A little further on in the same passage, Vitruvius goes on to set the origin of the Ionic column in a period almost equally remote:

> When later they wanted to build a temple to Diana, and explored a new kind of beauty, they again measured a footprint, but changed to a womanly delicacy, and made the thickness of their first columns an eighth of the height so as to make its beauty slender. A curved base like a shoe was put under the [shaft] and at the head they put volutes like pretty curls hanging down to left and right. The front they decorated with cimasia and festoons and let the fluting fall like the folds of a matronly robe. So they distinguished the two ways of designing columns, based on the human body: one of a naked beauty, virile and without any decoration, and the other female.[48]

The etiological legend, like most such stories, conceals as much as it reveals. Before I consider its implications, I would like to emphasize what the myths, as Vitruvius tells them, show at face value: that the two central orders, the Doric and the Ionic, are identified by name with

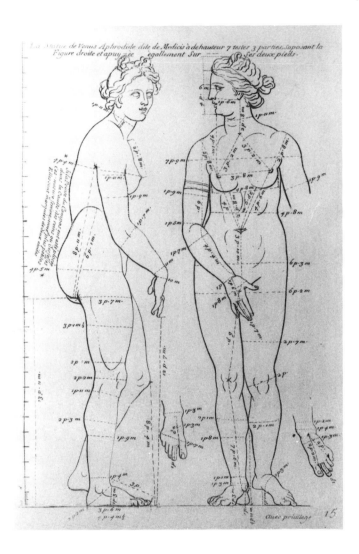

the two main dialects of the Greek language—the Dorian, which was spoken in the Peloponnese and in Crete in historic times, and the Ionian, which was spoken in Attica and by the Western Anatolian Greeks. The literary (as against the spoken) dialects were formulated in the seventh and sixth centuries, as were the rules for columns and beams. They, too, were quite "artificial," and related to literary usage rather than the dialect as it was spoken, even by the poets themselves. Doric was the principal medium of the choric ode—of Stesichorus and Pindar (who was in fact a Boeotian)—and later of pastoral poetry and of the epigram. Ionian was the language of epic—of Homer and Hesiod—as well as of prose, particularly of scientific prose.[49] Dialects were often combined and even mixed in one work. In tragedy, a juxtaposition of Doric for the choral parts and Attic for the main personages was quite usual. Later, in comedy, local dialects were used for caricature.[50]

These two kinds of column are also fully and circumstantially identified with the canonic dimensions of the male and the female human body, and this identification would be constantly reiterated for a worshipper at any Greek shrine by the lines of votive boy-statues, *kouroi* (almost always naked), and girl-statues, *korai* (always draped), which lined the sacred ways in such abundance that no respectable museum of art or classical antiquities would be without an example or two. And it is to kouroi and korai that I turn next.

V

The Literary Commonplace

▪ Girl and boy ▪ The metaphoric animal ▪ The shock of meaning ▪ The primal post ▪ Imitation and manufacture ▪ If nature built a house ▪ Do-alike, look-alike ▪ Tragic building ▪ Persians and atlantes ▪ Caryatids

Narcissus mosaic. From House of Narcissus at Daphne-Harbie. Princeton University, Baltimore Museum of Art.

Girl and boy

Korē and *kouros*, "girl" and "boy," are technical and helpful names for common types of Greek sculpture, which—more than the sculpture of any other culture—concentrated on the isolated human figure, whether clothed and female or naked and male. Since the designers of major buildings were usually famous bronze or stone sculptors, this alone could give the analogy between the represented body and the designed building a certain plausibility.

That analogy, that metaphor, has already been invoked insistently in previous chapters. A repeated recourse to a metaphor in a book devoted to building requires a general apology, since for many readers, metaphor will seem a surface dressing, embroidery on the real business of utilitarian and even of abstract-formal concerns. Such is not my view: I think it an essential part of the business of building, as of all human activity. The particular and neglected body metaphor provides a key to some puzzling features of Greek architecture. I have come to think that it may direct the way all men and women relate themselves to what they build.

The metaphoric animal

But then "natural" language itself, indeed the very possibility of language, has even been considered an extension of some primal metaphor—which is why man has been called the metaphoric animal.[1] Language (in common parlance, at any rate) has always meant articulated sound, eked out by facial grimace and manual gesture to convey intention and matter most exactly and rapidly. Yet grimace and gesture and expressive—even articulated—sound man shares with animals, which make their signs by code in context. However rich and varied their "vocabulary" (as in worker-bee dances outside a hive or the courtship rituals of peacocks), such signs are made in the expectation of the one-to-one reading, while human sound and gesture is always intended for some interpreting other in a constant hope and fear of the polyvalence of the sign.[2] Man shares with other animals not only expressive, articulated sound but play as well; yet the difference between animal and human games is expressed in English by the preposition we use with "play": animals play "with," men play "at."[3] And indeed the play of human beings presupposes a ground and a rule: a reference and an agreement or compact.

The interpretative compact is the primary social bond of humanity, or so it seems to me. It is made in signs and marks and hints for which the body is the only source and its movements the only medium. That is why the body's metaphoric value is my primary human inheritance as it is my reader's. It had to be won through a violent, perhaps even a catastrophic separation of the human body from the undifferentiated flux of nature.

I sense my body as being different from any other bit of creation. The act of my realizing myself both as a separate unique entity and a composite, a manifold, occurs only when I have configured myself in my fellow. It is exclusively human. That is what Nietzsche meant when he wrote that "thou" is older than "I."[4] Yet I also see an other, an uncanny me, looking back from the surface of pools and rivers, from polished stone and bronze: the fate of Narcissus or Narkissos is a warning that the mere act of configuration by reflection may be simple, but it is also hazardous, since the name which that exquisite youth shared with a fragrant, heady (indeed narcotic), springtime moisture-loving flower carries a promise of intoxication but also a threat of stupor and even death.[5] Initiation into the awareness of another and quite different scission, which could have been the first recognition of our likeness in a cadaver—that between our own kind as pure and impure—must have been equally catastrophic and startling. It is a division that defies constant interpretation and speculation. Yet another me also follows my steps, so that I know myself as an upright caster of shadows, and the earth as my sundial.[6]

Recognition, together with the separation that it entailed—the primary estrangement—is the condition of my *knowing* the world outside myself at all, which I can only grasp, only comprehend (if you will forgive the truism) *out* of my body, since that is what I am and that is also all I have for a map or model of my exterior experience. That is the first intellectual "happening," if not quite an intellectual operation. Language, which is the condition and the cage of my thinking, issues out of the body which is the cage and condition of my being.[7] It is the generator of the whole vast project of human culture, its felix culpa.[8] When I *think* of my own body (or the bodies of my fellows), I can only think of it in terms of metaphor; and I must also think of groans and screams, of gesture, of bodily movement, of grimaces even, as the most elementary (or at least the most intimate) artifacts. Perhaps the only possible answer to the primal estrangement is re-presentation,[9] the constantly renewed ordering of the metaphoric experience and its reenactment in the maturation process of the individual. And what representation is more elemental than reflection?

The shock of meaning

Such operations made great intellectual demands on our first parents, which are all too easy to underestimate or even to overlook. Our first ancestors are figures in myth and in revelation, since history knows only the continuity of human polity. Yet the catastrophic passage from the sensed to the thought body and the invention of its language cannot be located in any history or myth or in any other narrative. It cannot even be obviously associated with any other biological conquest of the "human revolution." In the present state of knowledge no agreement seems possible about whether Cro-Magnon people were the first full possessors of language, or if the Neanderthal race had already mastered the techniques of speech.[10]

At any rate, my ancestors and yours achieved and had to consolidate this vast and burdensome intellectual victory of severance by harsh and often very painful means. The process by which their bodies were marked out, signed as being different from—even alien to—their context, had to be both memorable and evident. That is why it almost invariably involved body markings that were mutilations: scarifying, knocking out of teeth, severing of a finger joint, circumcision, or, perhaps most common of all, tattooing.[11] The pain that these involved is understood by both operator and "patient" as a trial by torture: to bring the subject back from an unreflective enjoyment of the world and to realize a new and responsible place in it through an acute concentration on his or her body and its separateness. Paradoxically, it also marks the incorporation of the mutilated individual into the company of (equally mutilated) peers. In many societies, men and women make that reflective act by transforming their bodies painfully into artifacts.[12] Herman Melville described his mysterious Polynesian mariner, Quee-queg, carving a coffin for himself to imitate the tattoos on his body, which, although he cannot himself "read," he knows as a world map, ". . . being the work of a departed prophet or seer of his Island, who by those hieroglyphic marks had written out on his body a complete theory of the heavens and the earth and a mystical treatise on the art of attaining the truth[.]"[13] This is a passage of sharp ethnographic observation, since in many societies tattoo marks act just in the way that Melville cosmically described: as passwords into the underworld or a map for the soul's journey. Tattooing and body painting (and often the mutilations) are also thought to make the subject more acceptable socially, as well as attractive sexually: they are a way of beautifying.[14] However that "beautifying" is understood, it subjects the body to interpretation—it makes the body itself into a story, since my sense of my body is too articulated to allow me to conceive of it as undifferentiated trace.

The technical equipment for the most elaborate body patterning or marking may always have been very elementary: thorns, fish spine, insect needles, and sharp stones had always been readily available, and they went on being used for many millennia before knives were ever flaked in flint. The painful and sometimes dangerous operation of subincision is still performed in Australia with a flint impact blade, of the Paleolithic type.[15]

The primal post

If you accept what you have just read as a premise, I would infer that the isolating of an upright post as an analogue of your body might not seem all that far-fetched, since what is true of speech and gesture and of the decorated body is projected out into the surrounding world and must also be true of all other artifacts. It would seem that one inalienable property of human thinking, as well as the primary mode of animal perception, is orientation in space.[16]

Having abstracted my body from nature and thereby isolated it as an object of attention, I have also established it in a context for its metaphoric interpretation. In consequence, when I carve or even just score an upright post with any mark of intention—as I score (or have scored) my own body to make it sociable—so that any casual passerby may notice that

Kouros carrying piglet. Izmir Museum. Photo by author.

Kouros. From Sounion. Athens, Archaeological Museum. Photo by author.

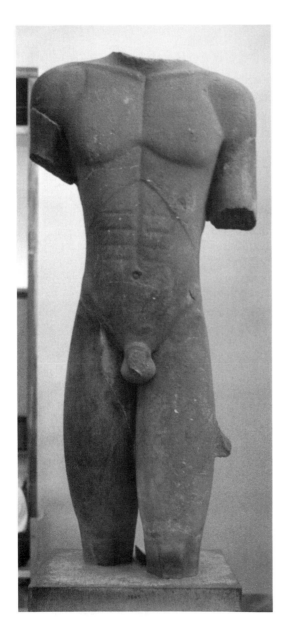

Chapter V

Kore poikile. Athens, Acropolis Museum. Photo by author.

Kore. Athens, Acropolis Museum. Photo by author.

The Literary Commonplace

it has been given its place and shape (or perhaps merely appropriated) by some human agency, I inevitably invite that passerby to see it as a bodily analogue. The planting of a post, let alone a row of them, has therefore always been seen as a metaphoric act and as a type of building in general. It is also an act of taking possession of a ground: every such post implies a circle around itself if only because of the shadow that it casts, whose direction will always be parallel to that cast by my upright body.

The planting of a post is a primal gesture—the ability to orientate ourselves, to know the orthogonality of our body to the ground, is a condition of our being.[17] *Homo* is only *sapiens* because, or even after, he is *erectus*. Before they set up and appropriated the first posts by carving, men may have learned that skill on rooted and growing trees.[18] However many millennia it took, the passage from the coupling of perishable body with growing tree or friable post to a rather different coupling of the body with permanent column seems almost inevitable. Whether such a post does or does not carry a beam is at this point immaterial.

Within this context, the assimilation of the male body to one kind of column, the Doric, and the female to the other, the Ionic, will, I hope, appear as a specific instance of the general metaphor and be as acceptable as it generally was until the eighteenth century. It was one of the essential concepts around which architects would speculate and invent.[19] The extension of the metaphor to three more "orders" and the many further variations on the columns and their garnishes depended (at least in part) on the disagreement about their exact detail: about the proportional minutiae of the rules in ancient literary authority as against those which could be measured in ruins and antiquities, about the best examples to imitate and the exact character of the ornaments, and, above all, about the human figure itself. In spite of the increasing reliance in academic discipline on drawing from the nude, painters and sculptors tended to look at antique statues, not at the live human body, for a model of proportional rectitude.[20] Since the hold of the metaphor loosened at the end of the seventeenth century (as I described in chapter 2), the rationale for using the orders was shaken, and the much-discussed proportional and dimensional structure had to be fixed into an arbitrary system of rules. The generative power of the orders therefore shrank and their hold turned into constriction. The accumulation of conflicting archaeological evidence only intensified the insecurity about them and strengthened the tyranny of the conventional rule.

Although an analogy between body and world had become increasingly improbable, the other one, between body and building, had remained entrenched in everyday speech—and therefore in thinking. But it was not a linguistic leftover marooned as a fly might be in amber. Wittgenstein, so attentive to language's layering, remarked once, "Our speech is an incarnation of ancient myths. And the rites of ancient myth were a language."[21] He would certainly not have considered such linguistic deposits as inert: but then Wittgenstein was also passionately interested in building.

Perhaps it would be more telling to take the evidence of a writer whose influence has grown much since his death, and who detested architecture (particularly contemporary architecture) as he despised his own pruriently indulged body, Georges Bataille. In an antidictionary compiled during the 1930s, he sees architecture as the enemy of the sort of transgressive communication which to him—and to many of his contemporaries—was the state of true human freedom, the only desirable human condition. The willed uncertainty about the future, which Nietzsche had extolled as the ideal human situation, must, he thought, be the enemy of projects, since any project presupposes a hope that sensation will in some way be assimilated to forethought. Of any such project Bataille's mind was impatient: for Bataille the very condition of modernity was established *against* architecture, which is

> the expression of the inner being of societies, much as human physiognomy is the expression of individual being. . . . Great monuments rise like dams, setting the logic of majesty and authority in the face of all disturbing elements. . . . The fall of the Bastille symbolizes this state of affairs: it is difficult to explain this crowd action by anything other than the hostility of a people to the monuments that are its true masters. . . . The mathematical ordering imposed on stone is the pinnacle of the evolution of earthly forms—whose meaning in the biological sphere is proved by the passage from simian to human form; and this last had already implicit within itself all the elements of architecture. In the morphological process men represent nothing but the intermediate step between monkeys and great buildings. . . . Forms have become increasingly static and increasingly domineering. If you are at odds with architecture [therefore] . . . you are at odds with man himself. A whole range of activity—the most brilliant without a doubt, intellectually—tends in that direction, revealing the inadequacy of human domination; thus, although it may seem strange when a creature as elegant as a human being is concerned, a way is shown by the painters toward bestial monstrosity: as if that were the only way to evade the architectural chain gang.[22]

On this view, architecture is only conceivable at all as a mirror of an oppressive social order, and therefore as the "order of orders," redundant in the condition of explosive modernity that Bataille described and accepted eagerly. Yet even he argued the very impossibility of any modern architecture out of the analogy between body and building: a building whose skeletal rigidity is a parody of the soft, pliable, squalid, and lovable body. Architecture must inevitably strive for completion, which constrains humanity, and therefore must deny true liberty, which is the gratifying indulgence of unceasing and insatiable desire—a pious practice that has the Marquis de Sade as its true prophet.

Imitation and manufacture

As in Bataille's contemptuous dismissal, so too in much recent writing about architecture the problem turns on the ancient notion of *mimēsis,* which the Romans unfortunately translated as *imitatio* and we therefore usually limit to *imitation,* meaning verisimilitude. The term has been receiving much attention,[23] though not at all in connection with the idea of imitating the body in building. But then imitation—as an architect might understand the word—can hardly ever be the making of a recognizable likeness, but must involve some process of recomposing a decomposed (or deconstructed, if you will) image by putting it together again as a building. This Bataille had clearly recognized, as had Schopenhauer much earlier.

Insofar as this kind of imitation can be read back into any building—a temple or town hall, or even a home—by the lay public who are the building's everyday users, such imitation will demand repeated explanations or rereadings of the building, and not necessarily ceremonial ones. These readings may be gestural and commonplace, or involve the most homely of rites; they may even require a tacit understanding of those very things that Bernini discussed with Fréart de Chantelou; at the very least they will have to rely on some kind of instinctive, preconscious disposition. However powerful a spectator's unacknowledged empathetic response to building may be (and this is a topic on which there is practically no agreement), an architect who thinks that he can rely on empathy in his designing needs to have some sense of the way in which he is reading and decomposing the corporeality of sensation. The sense that each one of us has of his or her body within the natural order holds an intimation of that whole context. Many of the earlier designers who wrote on (and used) the antique orders relied, more or less consciously, on some amalgam of these factors. The argument from empathy is still favored by some recent writers on proportion, now that the orders no longer attract such speculation;[24] in this tacit form it can almost be taken for a trivialized variant on the old allegory, which taught that love as the child of Need and Resource would guide us to the good and the beautiful, as Diotima had taught Socrates.[25]

Still, this kind of empathy is not at all what Plato meant when he used the word *mimēsis*. For him it still (to put it anachronistically) had the smell of greasepaint lingering about it by its association with *mimos*, "an actor"; a word derived (remotely and by reduplication) from the root **ma*, "to measure," and therefore to imitate by gesture and voice—consequently also to dance. *Mimētikē* was dance and song and could not really be separated from them. Music was therefore almost the Platonic mimetic art par excellence. By the fourth century, a *mimos* was considered of a lower order than the other kind of performer, the *hupokritēs*, "an interpreter," "responder," or "reciter." The term was to acquire a very unpleasant overtone in New Testament Greek; for the Romans, it became *histrio*, a term they derived from the Etruscan.[26]

If, as Socrates (or at least Plato) preached, some absolute and transcendent harmony has any status outside our imaginings following which the creator-craftsman god, *demiourgos*, has made the world by imitation, then it must be part of our humanity to be aware of it, however dimly. For the perceived world will be full of hints and nudges from which even the commonplace artifact derives its dignity (and its maker his or her social standing) through a consensus about the skill that is required to form it, a skill that its inventor would have attained by decomposing the world order and recomposing it through the gestures of the craft by which the artifact was fabricated. For Eros the god or spirit, the uniting daimon prompted the *demiourgoi* to generate all the different works of their minds and hands.[27]

The Greek word for that skill, *poiein*, and its product, *poiēsis*, also means "poetry" in the twentieth-century sense. In antiquity it had a much wider meaning, which included all that we now call the arts and the crafts, all the skills of manufacture. "Creative skills," we are tempted to say, the skills that make the artist so God-like. But remember: Plato's demiurge in the *Timaeus* had worked to make our world, as mortal craftsmen did, by imitation. His creative action was therefore cosmological, not ontological.[28]

Poiēsis was as varied in its forms as love itself, and to make poetry had as many meanings as did "to make love," since poetry is "a general name for the power of calling something into being out of what is not: so that every kind of inventive skill is a poetry, and every artist a poet."[29] Every kind of poetry therefore—shepherding, fishing, farming, pottery, carpentry, tragedy, epic, and inevitably building also—has its *technē*, its way of proceeding, its accumulated experience and know-how, of which Aristotle rather surprisingly says that they are all ways of making imitations of nature: *hē technē tēn phusin mimetai*.[30] In Aristotle's thinking, *mimēsis*, as he makes clear on a number of occasions, is not merely the holding of a mirror up to nature but much more a "doing like," "producing after the fashion of."

If nature built a house

Mimēsis had moved from dance and theater to the fabricating crafts generally about the time of Plato. Aristotle gave a summary indication of what this might consist of in the case of architecture, while demonstrating that as in art, so everything in nature tends towards some end; for

> as in human operations, so in the process of nature, and as in natural process, so in human operation (unless something interferes). Human operations are for an end, hence natural process is also thus. If a house, for example, was brought into being by nature, it would come into being . . . exactly as it now does by art; on the other hand, if natural objects came into being by art rather than by nature, they would still come into being . . . just as they do in the course of nature.[31]

In this matter, Aristotle's formulation is not all that different from what Plato's might have been, but Plato would not have stated it like this, since his intention was inevitably polemic: he advanced his arguments against the "mechanicists" who maintained (according to him) that everything in nature is the product of necessity, whereas Aristotle argued that everything is produced by law and repetition, through regularity.

Chance happenings are not subject to the laws of finality and their end is not set by any intention; they thus become the justification, as it were, of their own ends. But then Aristotle, like Plato before him, was taking up a polemical position against a commonplace. In the *Poetics* he takes up and extends Platonic arguments several times, without invoking his teacher's name,[32] as when discussing the business of *poiēsis* and *mimēsis* that preoccupied Plato at various stages in his development. Throughout the two great dialogues on the nature of the state, the *Republic* and the *Laws*, Plato devotes much attention to the place of the arts in general and imitation in particular within the commonwealth.[33]

Now the word *poiēsis* had changed meaning fairly radically, as I suggested, some generations earlier, when poets began to charge fees at the end of the seventh century. As their activities became less "enthusiastic," less improvised, their procedures turned more solid and craftsmanly and their *technē* became explicit. Poems were considered to bring more lasting fame than did marble, yet at the beginning of the fifth century, Pindar sings of building a stone foundation, a *krēpis*, for his songs—even a golden one—on which he will construct, *teichizomen*, a monument bejeweled with eloquence.[34] He already finds the building/metal-working craft analogy as useful as Plato would in his *Sophist*.

In Plato's *Sophist*, variations of *poiēsis* are sorted out by an unnamed visitor, Xenos, from the Italo-Greek city Elea, Parmenides' city and the home of one of the principal Greek philosophical schools. Although Socrates is present on that occasion, the visiting "Stranger" is the main speaker.[35] In the course of the dialogue, he details the categories of *technē*. There are the acquisitive kind, like hunting or fighting or trade, and the productive, "poetic" kind, like agriculture or shipbuilding. Poetics may even be divine: the bringing forth of stones, plants, beasts, and human beings is the production of divine skill, as is a completely different category of beings: dreams, shadows, reflections. Human skill is analogously divided: on the one hand, *eikastikē* (a conjecture or a representation), which by *mimētikē* imitates some formal prototype in its primary qualities, such as number and measure, is brought to bear in building a house. On the other, *phantastikē* is the skill of secondary depicting, through pictures or through bodily acts, of the original *mimēma,* the primary imitation. The true maker of any one thing is the divine creator of its idea, of which idea the craftsman, the artificer, produces a sensible or even tangible concretization.[36] *Phantastikē* bears the same relation to divine creation as reflections and dreams bear to animals and human beings.

Plato is further (and somewhat differently) concerned with imitation when he considers art and truth in the third and tenth book of the *Republic.* The epic and the dramatic poets, as imagers of the gods and heroes, as imitators who convincingly counterfeit the voice of others and who may therefore make plausible false statements, are scorned by Plato, who was to make the ontological status of the false statement one of the themes of the *Sophist;* and yet the very form of dialogue, of which Plato was the first real master, is mimetic, since it relies on direct speech. Still, the main reason why he proposes to expel the poet from his good city is not for the ontological inferiority of his product, but for the seductive way he may set out bad, wicked examples. On that imputation, all the more winsome modes of music must be rejected from the city, and even the makers of excessively plangent instruments are excluded.[37]

Do-alike, look-alike

There is another kind of imitation removed from truth by a double barrier:[38] that of painters and poets, who imitate not only the objects (such as plants or animals) that are made by the divine craftsman, but also those made by human ones in imitation of the divine prototypes. While the human craftsman turns "*the* bed" into "*this* oak double bed," the painter makes "*this* bed" into "that certain double bed, seen in that certain light and from that certain angle."

The trade of Plato's poets and painters would therefore not seem "real," since it amounts to aping that which others make by spectral images; and, what is worse, they can imitate anything. In that they seem to have something in common with the moneygrubbing sophists whom the Eleatic stranger had found so distasteful. Too much has been made of Plato's disapproval of imitation as *mimēsis* (which he sometimes alternates with a rather more positive notion, *methexis:* "inclusion" or "participation,"[39] since on some occasions he did praise the work of artists highly and compared them favorably with sophists. In fact, there is an ancient tradition that he studied painting; Plato used painters' technical terms and appealed constantly to the words of poets. Xenophon, for his part, described Socrates' visits to artists when he teased them as to whether it was the visible and measurable appearance they imitated or the inner soul.[40] On that showing, the work of the architect, or of any *demiourgos,* would certainly be on a higher plane than that of a painter—as would that of the carpenter. The carpenter, who uses many measuring instruments to ensure accuracy, and who like the shipwright works with certainties, is at the opposite end of the range of *technē* from music, where the best results are achieved by guesswork.[41] The distinction between the carpenter's manual (or servile) work and the architect's "liberal" art belongs to a later time; yet much of the eighteenth-century argument on imitation in architecture, which has had a determinant influence on later architectural theory, still turned on the assumption that the architect who imitates a body in a column is doing something analogous to Plato's "fantasizing." But it is not any specific human body (or even the aggregate of them) that the architect is expected to imitate: it is the human body as a type, or even as an idea. The body is to the fifth-century column as the divinely created idea of it is to the bed.

In any case, Aristotle had a somewhat more benevolent view of the business of imitating than Plato; he appealed to the teachings of an older philosophy—perhaps even an older and less systematic way of thinking—and when in his *Poetics* he considered the way in which the poet imitated, he extended it, as Plato had done, to all the productive trades. Like any other artist, Aristotle's poet must necessarily have objects as models for his imitation. They may be of three kinds: (1) things as they once were, or as they are now; (2) things as they are said or thought to be; and (3) things as they ought to be.[42] This triplication is one of the most frequently quoted passages in the *Poetics;* the things to be imitated are as much processes as tangible objects. The last of the three imitable categories is perhaps most interesting in this context, since it seems to refer not to some "ideas" of the objects (as Plato would have done), but rather to supposing that the process of their growth has an aim, which the imitator can fulfill.

Since the time all this was written (and perhaps because it was written), painting and sculpture, as well as poetry, have come to be known as the mimetic arts—the arts of imitation. In late antiquity, Neoplatonic speculation exalted the soul at the expense of the body, therefore emphasizing the negative aspect of the work of the demiurge who made the material world. His *mimēsis* was also scorned, its original Platonic dignity obscured. The Aristotelian criteria, which were taken up by many medieval thinkers, were often colored by Neoplatonic disapproval. However, imitation took on a new luster in the fifteenth century, when the notion of choosing and following an exemplar became one of the guiding ideas of that mentality which is called Renaissance (one of the most popular devotional works of the time is the Imitation of Christ). At the same time, a new respect for works of art led to a reinterpretation of antique precedent, not only by artists but also by philosophers, of whom Nicholas of Cusa seems to be the most interesting.[43]

Between the fifteenth and the eighteenth centuries, Aristotelian criteria were not, in any case, applied uniformly to all the arts. Critics and theorists would sometimes confuse imitation with verisimilitude. The passionate nineteenth-century discussion about whether music could be considered an imitative art or not inverted the Platonic categories, which had made music the prime exemplar of *mimēsis*. About architecture the dispute had started earlier—in the seventeenth century—but it had also become involved in quite a different category problem: that of a fine as against a servile or mechanical art. In order to look back at Aristotle's scheme, therefore, and glimpse Plato's radical gesture beyond him and the pre-Socratics in the further distance, I have to defy that general acceptance of the hard division between the fine and the useful arts, which Kant had set up in such a way that the classifying of architecture as a fine art was made increasingly difficult.

As all great thinkers do, Kant also rationalized the commonplaces of his time; he distinguished fine from useful art by his general acceptance that what is fine in art is a *free* activity, in the sense that the fine artist acts without any external constraining end in view, and the spectator takes pleasure in that free activity or its precipitate without any thought of appropriation. This apparently echoes but actually inverts the categories of Plato's *Sophist*,[44] since in antiquity the division between what was fine and what was coarse about the arts was seen in quite different terms.[45] The distinction between the fine arts as free activity without any constraining end and the finite applied arts was not thinkable: an art could only be termed free if it was fit for free men to practice, as against one fit only for slaves. The notion of an art not bound by any rules would have defied all ideas of what constituted an art—and therefore (in the first place) a *technē*. It was only at the end of the eighteenth century that it became imperative to teach that "Nature was only beautiful when she looked just like art; and art can only be called fine when we are conscious that she is art and yet she looks quite like nature to us...."[46] This teaching incorporated a view of mimesis which was almost diametrically opposed to that of Greek thinkers, at any rate as they are represented by Plato and Aristotle. In particular, if you turn back to Aristotle's notion of a house generated by nature looking just like one made by art, you will see how the notion has become transformed.

Tragic building

Both Plato and Aristotle had made their discussion on *mimēsis* highly polemical, since it was evidently a matter much discussed in their time. The close association between the body and the column especially was a staple of literary metaphor. To take an obvious example: one of the most frightening, dire shrines in all Greek literature was the Taurian temple of Artemis in the Crimea to which, according to one legend, Iphigenia was confined after the goddess rescued her from the sacrificial altar.[47] She was the daughter of Agamemnon, that high king of Argos who commanded the Greek army and fleet during the Trojan War. As the fleet lay at Aulis waiting to sail for Troy, the contrary winds continued to blow. The consulted diviner demanded the sacrifice of the king's daughter to angry Artemis. The king complied, the winds changed, and the fleet was finally able to sail. That compliance was said to have so enraged his queen Clytamnestra that she murdered the king on his triumphal return ten years later,[48] so forging another link in the fateful chain of guilt and expiation that is the theme of Aeschylus' *Oresteia*.

As the victorious king returns to his palace from Troy after ten years, his faithless queen, who has already planned his murder, heaps metaphors on each other in a sustained welcome of transparent insincerity:

> welcome to your house, my lord,
> as a watchdog is welcome in the sheepfold, the rigging
> to a ship, the strong main post to the roof, the only
> child to a father, the spring to the parched traveler,
> welcome as landfall to the despairing sailor[.][49]

Her analogy between the savior-king and the sustaining pillar was meant to sound as commonplace and conventional as her other comparisons. That night she was to revenge her daughter's sacrifice.

Euripides took up the alternative version of the legend according to which Artemis—selectively merciful—had replaced the girl on the altar with a doe and spirited her away across the Black Sea to the distant Taurians, there to minister at the fearful shrine where all straying Greeks were sacrificed to the goddess. Iphigenia did not kill them with her own hands, as she explains in the prologue of the play, but only prepared them for the sacrifice inside the temple; others did the killing at the altar outside. As the play begins, her brother Orestes arrives in Taurus with his companion Pylades. They look with terror at the bloodstained altar and see the skulls of the Greeks who had previously been sacrificed hung as trophies under the cornice.[50] In the same prologue, Iphigenia gives a moving account of a dream she had, in which

> I had escaped this land
> And lived in Argos. There I slept among my girls:
> Suddenly an earthquake shook the black earth.
> I fled and stood outside: the cornice
> Of the house broke up, and all the roof
> Tumbled from its height down on the ground.
> One single pillar (it seemed to me) was left
> Of all my father's house. From the capital
> fair hair streamed, and taking human form it spoke:
> Then I, faithful to my duty done to strangers
> Sprinkled him with water that prepares for death
> Myself in tears. I construe my vision thus:
> I was dedicating Orestes who now is dead—
> For the male children are the pillars of the house[.][51]

The metaphor equating column and male progeny did not need any gloss for Euripides' audience. It has become, as it must have been long before that time, proverbial in many languages. After all, it is merely a particular application of an all but universal metaphor. Figures bearing a weight, either upright and dignified, like the handles of Egyptian, Etruscan, or indeed Greek mirrors, or oppressed and bent (mostly in Hellenistic and Roman bronzes), are found in furniture and reliefs,[52] yet both literary and archaeological sources provide surprisingly few instances of their most obvious application to building. In spite of the grandeur and the familiarity of the metaphor, as well as the importance of mimesis, Greek architects seemed very reluctant to substitute sculptured figures—whether male or female—for pillars. Vitruvius himself mentions only one such example for each gender of column.

Persians and atlantes

In their agora, so Vitruvius begins his account of the male sculpture-columns, the Spartans erected a portico in which a row of Persian slaves stood in for Doric columns.[53] This is a very different association of body and column from the honorable one that I have culled from the Attic tragedians, and he explains it by reference to the well-known story about how a small number of Spartans defeated the vast Persian army at the battle of Plataea. With the spoils of that battle, they put up a "Persian portico" in their agora, which was to overawe strangers by recalling that act of bravery while it encouraged citizens to defend their liberty. Pausanias, who saw the portico a century and a half after Vitruvius (who also described it, presumably unseen), seemed more interested in the statues as portraits of historical figures than in their structural function: he reports that one statue represented Mardonius, the Persian commander at Plataea; the other he noted in spite of the gender distinction was of the Carian Queen Artemisia of Halicarnassus, who had fought with the Persian navy at the great sea battle of Salamis. His row of portrait-statues is not quite the portico of slaves in barbarous dress bending under the weight of the cornice that Vitruvius described. In any case, Pausanias' statues seemed to be standing *in front of* the columns, not standing in *for* them.[54] No trace of this famous portico has been found so far.

Caryatids. Fréart de Chambray
(1702).

Persians. Fréart de Chambray
(1702).

Chapter V

On the other hand (to remain with the Doric), there is little mention in ancient literature of another monument, the temple of Olympian Zeus in the southern Sicilian town of Akragas or Agrigento.[55] It was the very largest of all Sicilian temples, and indeed of all Doric temples anywhere. For many centuries it lay outside the town as a conspicuous heap of stones. When some of these stones were first fitted together by archaeologists early in the nineteenth century, the telamones or atlantes were reconstituted. They had stood high on a ledge, which was about two-thirds of the column height, and were as tall again as the very thick cornice. There was one figure to every intercolumniation. They alternated in turn—or so it would seem—between bearded and clean-shaven figures. All seem to have been nude and stood with their backs to the wall, straining to carry the cornice on their bent necks and arms. Their exact proportions are very difficult to ascertain (the surface is abraded, and all of their feet seem to be missing), while the reconstructions of the temple have left many things in doubt. It is certain, however, that the attached half-columns stood over twenty meters high on a stylobate whose elaborately molded top edge projected in a semicircle under each one, almost providing them with bases—unique in a Doric order of that date. They seem to be in a proportion of 11:4 with the cornice. Within, the half-columns projected into flat piers through the outside walls, while two-spine internal walls, which also seem to have been reinforced with square piers, ran along the whole length of the interior. The structure was almost certainly strong enough to carry a very heavy roof, which it probably never received, although the pediments had their complement of sculptures: to the east there was a gigantomachy, while the western gable showed the fall of Troy, both subjects highly suitable for a temple that commemorated a victory over the barbarian Carthaginians. Like the telamones, these sculptures were probably done before 450.[56]

All too little is known about the building history of this Agrigentine temple. It was probably founded when the neighboring town of Himera fell, after its protectors, the Carthaginians, had been decisively defeated by an alliance of two Greek tyrants: Gelon of Syracuse and Theron of Akragas. That battle was said to have taken place in 480, on the very same day as the battle of Salamis. The atlantes may therefore plausibly be taken to represent Carthaginian (and Himeran?) slaves. The Carthaginians, always a threat to the Greeks of southern Sicily, were to take their revenge nearly a century later, in 408, under Hannibal.[57] According to Diodorus the temple was being built by Carthaginian slave labor, and the association of the statues of the slaves sculptured on the walls with the slaves on the scaffolding of the building must have been a fine demonstration of triumph. The temple was almost certainly never finished.

Akragas was an opulent and showy Dorian city; all the temples which have survived there are Doric. With the one exception of the Olympeion, they conform to type. The huge Olympeion, however, is not only remarkable for its telamones (which may well turn out to have been unique in their time) but also for its very heavy and clumsy ordering. Such an unusual configuration is often blamed on the Sicilians' tendency to coarsen and confuse, their provincial archaism; but the sheer size of the temple seems to confirm the reputed extravagance of the Akragans, their love of display. The unique composition of columns and atlantes has even been attributed to Carthaginian and Egyptian influence, as has the coarsely molded base that runs continuously under column and wall.[58] At any rate, although it had the usual "nave" and two side aisles, the temple also had an odd number of columns on the front—seven—and therefore could have no central door. In fact, this is one of the very last temple buildings with the archaic odd number of columns on the short sides.

Atlantes. Temple of Zeus at Agrigento. Photo by author.

Model: Temple of Zeus at Agrigento. Photo by author.

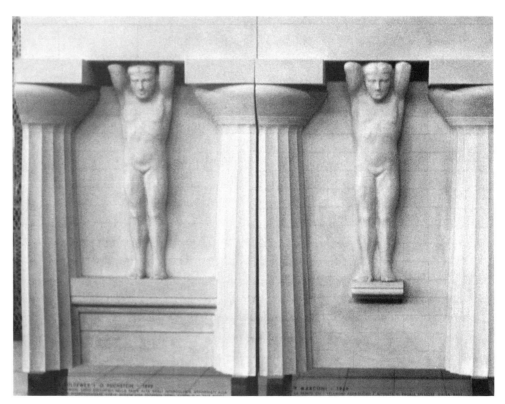

Chapter V

Caryatids

Agrigento provides the archaeological, Sparta the literary example unique in physically juxta-posing Doric columns and the male body. However, with the possible exception of that statue of warlike Artemisia on the Spartan agora, there is no evident counterexample of Doric columns coupled with a female body; in Greek architecture, the female body is on the whole associated with Ionic columns or Ionic entablatures. The most famous and obvious instance are the korai of the porch that is now known as the Caryatid porch of the Athenian Erech-theion, though in antiquity that temple was also known by its main dedication as the shrine of Athena Polias, the ancient olive-wood agalma that was offered the star-spangled peplos at the Panathenaia.[59]

Still, "caryatid" is what Vitruvius called the female equivalent of telamon, and the parallel between them seemed so close that he wanted to establish an analogous etiology of male and female column-statues. He makes no mention of the Doric alternation of column and telamon (as at Agrigento), even though there are parallels to it in a funerary, chthonic context for the Ionic order: in the Nereid monument from Xanthos and in the "mourners'" columnar sarcophagus in the Istanbul Museum.[60] Caryatids, Vitruvius explained, were named after the women of the small town of Carya in the Peloponnese. The town had sided with the Persians in the wars, or at least had remained neutral. After the Persian defeat their enslaved women were shown carrying the heavy weight (of the cornice) to give the same warning the Persian telamones conveyed in Sparta.[61]

A number of difficulties are inherent in Vitruvius' account. The only other mention of caryatids in Greek literature alludes to an attitude quite unlike that of the Athenian korai: in a jokey aside of Athenaeus, an Athenian wit, invited to a tumbledown house, is supposed to have complained that "in this house you have to dine with your left hand supporting the roof, like the caryatids."[62] Athenaeus wrote about AD 200; two centuries earlier, caryatids had been represented steadying a casket on their head with their left hand in the inner propylaea of the Telesterion at Eleusis. Fragments of two of these figures survive, one showing the ornate casket clearly enough; while the inner propylaea were vowed in 54 BC, they were finished over the next generation.[63] Pliny also wrote of the caryatids that a Diogenes of Athens had done for the Pantheon in Rome, and they may also have had one hand helping in support of their burden. They might even have been—like the caryatids of the Forum of Augustus, or those of Hadrian's Villa—direct copies of the Athenian statues.[64] The Romans liked such gracious fe-male supporters as much as they liked brawny and burdened male ones; they were found in decorative reliefs and were also popular in theaters.[65]

Yet there is nothing browbeaten or slavish about these Roman caryatids, certainly not about the Athenian korai; on the contrary, to the naive, uninstructed eye, these last look serene and poised, sandaled and walking—almost dancing—with a light step. They are dressed in a fine woollen, belted Dorian chiton, with a folded himation on the back. In one hand they probably carried shallow offering bowls, while the other restrains the end of their himation. They bend the right foot on the left side of the building, the left foot on the right side: the eight caryatids are divided into two mirror-image groups of three, plus one each. The cornice does not rest directly on their heads, but on a molded, square abacus, in turn set on a circular egg-and-dart echinus or capital, which some have called a *polos,* though others

Caryatid porch, Erechtheion.
Frontal view. Photo by author.

Caryatid porch, Erechtheion.
Side view. Photo by author.

thought it a kind of basket. There are those who rightly point out that it is more like the kind of small round bolster that modern Greeks still use when carrying things on their heads.[66]

That south-facing Athenian caryatid portico was done sometime between 420 and 410, more than half a century after the alleged Caryan defection.[67] Vitruvius, who is the only ancient author to mention it, does not concern himself with another detail about that porch: that it was (or it adjoined) the heroon of Kekrops, the mythical, earth-born, serpentine founder and first king of Athens, much as the tall Ionic northern porch of the same shrine seems connected with the worship of his grandson, Erechtheus. Some forty or fifty years later, King Pericles of Limyra, an ally of the Athenians, built a heroon—which may have served as his father's tomb—in remote Lycia; it also had caryatid porches front and back.[68] These caryatids of Limyra are smaller and coarser than their Athenian prototypes: their heads are veiled and their hair therefore less elaborately dressed.

Both caryatid porches are commemorative, and both carry friezeless archaizing Ionic cornices, though neither conforms recognizably to the Vitruvian legend. It is quite possible that Vitruvius has telescoped two accounts, so that the Carya after which the caryatids are named is not the Peloponnesian town but the Ionian province.[69] Such an anecdote as his might have been commemorated by a single building or its dedication, but it could hardly justify a rare and recurrent building form. A strong tradition must have persisted about the large group of korai that probably stood close by the ancient Erechtheion, since they were either buried very near it by the Athenians when they fled before the Persians, or (as seems more likely) found mutilated when they returned.[70] In any case, the obvious contradiction between the evidence of the works of art and that of the text has been resolved in a number of ways, most often either by refusing the title of caryatids to the Erechtheion women or by supposing that in spite of appearances they are indeed punished slaves.[71]

Yet even if it were quite legitimate to construe these figures as temple "servants" similar to the *arrhēphoroi,* the virgin bearers of wrapped, secret objects to whom they are often compared and who are represented in a number of surviving antique statues, they are not oppressed slaves.[72] These quasi-priestesses were lodged on the Acropolis of Athens (very near the Erechtheion) at state expense. Theirs was an honorable role, not unlike that of the *kanēphorai,* the bearers of holy baskets, who had an important part in the Eleusinian procession and whose burden of enslavement, shame, and punishment must therefore have been sacred and honorable, quite different from that of Vitruvius' caryatids.

The puzzle set by Vitruvius' account is due partly to his choice of a puny Arcadian hamlet to name the slavish quasi-order and so represent all the Peloponnesian towns that were allied to the Persians—or at least neutral in the national war effort, such as Argos. The hamlet Carya was in fact famous throughout Greece for something quite distinct from its disgrace in the Persian war: it was home to the cult of Artemis Karneia or Caryatis, whose main ritual was a stately dance of women devotees round a sacred nut tree. Indeed the term *karuatisein,* meaning "to dance in a stately way," "to do a round dance," was used as a common verb.[73]

In fact, there are "caryatids" that antedate the Persian wars by many years and were therefore already standing at the time of the legendary humiliation of the Caryans. Two of the treasuries at Delphi, that of the Aegean island of Siphnos (probably modern Siphanto, in the Cyclades) and of Cnidos, the Dorian city in Caria (Dorian in dialect, Ionian in style—modern Tekir in Turkey), were small distyle in antis buildings with caryatids for columns. Their cornices are innovative, since they are the first mainland Ionic buildings with a continuous frieze.

Caryatid. Istanbul, Archaeological Museum. Photo by author.

Caryatid fragment. From Treasury of Cnidos. Delphi Museum. Photo by author.

Caryatid head. From Siphnian Treasury? Delphi Museum. Photo by author.

The Literary Commonplace

As for the caryatids, they are archaically smiling figures bearing what appear to be pomegranates, more like independently standing korai than like any telamones. Their dress and their hair braiding are elaborate, and remains of metal ornaments, earrings, inlays, and wreaths adhere to them. The figures in both treasuries carried a cylindrical, sculptured box on their heads, secured to the elaborately dressed hair by a ribbon passing over it (Cnidos) or under it (Siphnos). A wide, flaring echinus-like "bolster" (which in the Siphnyan treasury is highly figured, showing two lions attacking a deer, and in the Cnidian one is divided into petals) carried an unmolded square abacus, so that these Delphic figures have more explicit, more "structured" capitals than those of the Erechtheion caryatids;[74] they bring an explicit Ionian (even Asian) grace into the most Dorian of sanctuaries.

The fact remains that evidence about the *nature* of the interchange between and substitution of figure and column is rather scarce in Greek architecture. If you go further afield, into Asia Minor and Mesopotamia, into Egypt, you will find many instances of the column supporting or displaying a figure, even of supporter-statues. These may also be found in the architecture of nonliterate peoples, and of course fairy tales and legends are forever telling of posts that speak, warn, or even prophesy. I will have occasion to review these accounts later, when I shall discuss non-Hellenic evidence about such matters. Here I would just like to record my view that such an analogy seems already to have been familiar to megalithic builders, whose upright stones clearly allude to it.[75]

The atlantes, the "Persians," and the "caryatids" may therefore all be read as illuminating instances of the one central analogy between column and body, an analogy that the Greeks could take for granted but that may have needed more explicit and insistent representation in cultures in which the making of monumental architecture was still an exciting discovery. In view of what I have argued about the nature of mimesis in building, the reluctance of Greek architects to expose the core of the metaphoric process and their insistence that it was a special case requiring historical justification may seem reasonable after all. Indeed the explicit modeling of such figures may have been mistrusted by the Greeks as too anecdotal, perhaps too particular for the great generalities of building.[76]

However, the natural analogy between column and figure is complicated by further mythological factors: first, the Doric columns clearly, the Ionic more obliquely, tell of early, legendary forms of construction; and secondly, they also tell the history of the two nations whose names they bear. And that I propose to consider next.

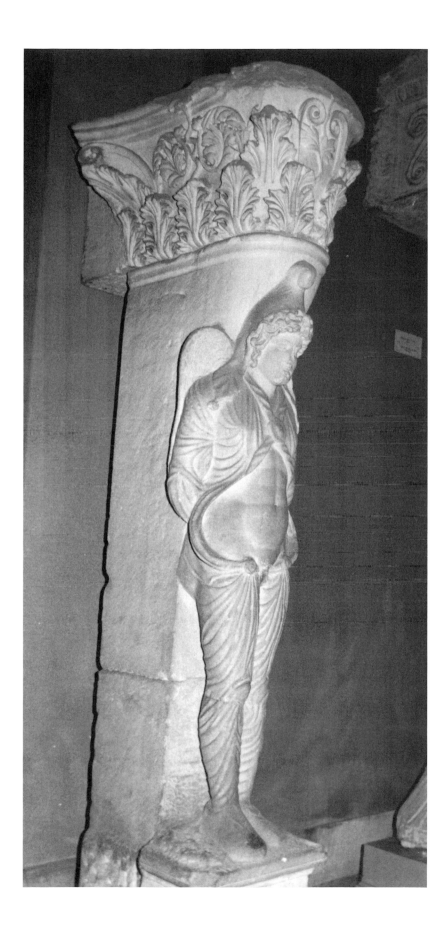

Hellenistic "Persian" winged supporter. Istanbul, Archaeological Museum. Photo by author.

Tel-Halaf: Restored porch. Berlin, Tel-Halaf Museum (now destroyed). After M. von Oppenheim (1931).

Tel-Halaf: "Temple-Palace" restoration. Projection and section, after M. von Oppenheim (1931).

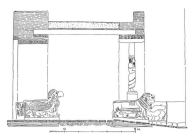

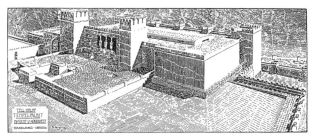

Chapter V

"Humanized" standing stones.
From south Brittany. Museum of
Locmariacquer. Photos by author.

The Literary Commonplace

VI

The Rule and the Song

• The sons of Herakles • The first shrines • The heroes' sacrifice • God and king • Altar and temple • Parts of the building • The *demiourgos* at sacrifice • Lions and a column • The holy column • Egyptian Doric • *Djed:* Column and mummy • Egyptians and Greeks • Greek origins

The sons of Herakles

Some generations after the Greek heroes killed each other off at the Trojan War, the sons of Herakles "came down into Greece"—alternatively, Greece was conquered by the Dorians. That would have been sometime between 1200 and 1000 BC.[1] Thucydides' brief account seems to sum up the situation as we now know it:

> after the Trojan war, the Greecians continued still their shiftings and transplantations, in so much as never resting, they improved not their power. For the late return of the Greeks from Ilium caused not a little innovation, and in most cities there arose seditions. . . . [A]nd the eightieth year after the war, the Doreans together with the Heraklides seized the Peloponnese.[2]

The people who called themselves Dorian spoke a dialect of Greek that spread over the south and eastern Peloponnese, Crete, the Cyclades, and southwestern Turkey.[3] The Heraklidae had to be sturdy and rough like their hero-father, and that is also how the Dorians, who claimed to be their followers and descendants, saw themselves.[4] The Herculean character of their columns must have had some reference to this legendary paternity.

About the time the Olympic games were first held, the arrangement of the column and beam that came to be called Doric must have been constituted in wood and earthenware, then translated into stone not much later. Those first Olympic games—or at least the first reformed games—were held in 776, and the Greeks counted their history from this foundation.[5] The earliest games took only two days, sunset to sunset, as their calendar was counted: a ceremony in honor of Pelops, the hero-founder of Olympia, the first evening, the hecatomb sacrifice to Zeus the next morning, and then the foot race. Other events were gradually introduced, such as horse races at the 25th Olympiad.

The population of Greece probably trebled during the eighth century and the polis took shape. In that period, the stone Doric "order" was fully formed and was used constantly until the time of Alexander the Great. It then fell into disuse until the Romans contrived their modified version of it.[6] In late antiquity Doric columns also came to be used in many other types of building: in public halls, stoai, and occasionally even private houses.[7] Yet it was devised for, and perfected in, temples. In spite of its relatively short life span, it has always fascinated theorists of architecture; and when its true antique shape was rediscovered about the middle of the eighteenth century when Greek monuments became accessible again, it was taken up enthusiastically for current use in all sorts of buildings, particularly those to do with law and the military.

The temple of "classical" Greece was one of those radical innovations mentioned by Thucydides. It was quite different from any earlier Mediterranean building. In Minoan and Mycenaean sites before the first millennium, there are virtually no buildings that could be called temples in the sense in which the Greeks were to understand the term *naos*: a god-house, a chamber sheltering some statue, memento, or fetish of the god as a mark of his or her presence, surrounded on all sides by rows of columns. The god-house was a type of building known throughout the ancient Mediterranean, but the shrine standing in the open, its columns raising the sculptured gable above all the buildings of the city—or over the countryside—was a Greek device.[8] Invariably, such a naos stood within an enclosure, a *temenos*. But

a temenos did not need a temple, a naos. Indeed, it was the other way round: there could really be no temple building without its temenos.

The first shrines

In the Bronze and the Dark Ages, before the first archaic temples were built, Greeks worshipped their gods in caves and hilltops as well as in flat fenced-off ground. Vitruvius may be condensing centuries of prehistory in his account of the origin of columns. When the Greeks first arrived in Asia, he says, "templa deorum immortalium constituentes, coeperunt fana aedificare[.]" Vitruvius distinguishes (and his translators mostly fudge the distinction) two stages: first the setting out of enclosures, *templa* (here the word should be read technically to mean an inaugurated and sacrally enclosed piece of ground, almost literally *temenos*), and then the placing in them of *fana*, "holy things," "holy artifacts" (not necessarily buildings, though the word was often used to translate the Greek *naos*). Those were the first altars and god-houses.[9] There is another Latin word often used to designate a temple: *aedes*, "building," "habitation." Commonly it is used together with a dedication—*aedes Vestae, Jovis,* or *Martis*—and this usage is like that of the Greek *naos;* less obviously, it meant *megaron,* "inner chamber." But *megaron* also had another Latin equivalent, *adytum,* in the sense of shrine, the inner part of the temple. This was a direct transliteration of the Greek *aduton,* the inaccessible, forbidden, sacred part of the temple; the term is used particularly of the inner spaces of oracular temples, as at Delphi. But *aduton* and *megaron* were used interchangeably in one context: to describe the ritual pits, clefts, or caves down which piglets were thrown at the Thesmophoria (a general Greek October festival of Demeter Thesmophora, bringer of laws and rules).[10] No explanation has been offered for the coupling of these two dissimilar words, a coupling which suggests that there were places to which both words could be equally appropriately applied.

Though the aduton might house a statue or a sybil, sacrifice was the main function of both naos and temenos. Sacrifice was what divided humans from the gods, the sacrificers from those to whom sacrifice was made. The great divide between humankind and the gods was fixed when cunning, far-sighted Prometheus, who was himself a Titan, instituted sacrifice to bridge it. That institution is told in detail by Hesiod,[11] and it involved fraud. Prometheus slew a great ox and cut it up to make two portions: one had the meat and innards wrapped in the animal stomach, the white tripe; the second was of bones and gallbladder wrapped in white fat. He offered both bundles to Zeus with expressions of submission. The great god saw through the trick, yet imperiously and willfully chose the fat and bones, to nurture his anger against men in general and Prometheus in particular.

Sacrifice may have divided men from the gods, but it also united them again as partakers in the one sacrificed body. Prometheus' second ruse, which was perhaps the consequence of the first, was to steal fire back from the gods—from Zeus' thunder, from the chariot of the Sun, or from Hephaestus' forge—by hiding it in a fennel stalk when Zeus withdrew fire from defenseless and naked man.[12] For both his crimes, he was punished by being chained to the Crimean rock (from which he was in the end rescued by Herakles), while men were punished by the special creation of Pandora who brought all the curses of mortality down in her box, curses from which they will never have relief.

Since the choice that Zeus made at the Promethean institution was binding once and for ever, the portion of the Olympian gods at all sacrifices was to be the victim's bones wrapped in fat and consumed by fire. The edible part of the animal was eaten, often at a common meal, by the sacrificers, or it was sometimes sold for the benefit of the temple.

The heroes' sacrifice

Homer's heroes were forever sacrificing to the gods or to the spirits of the dead, yet in the epics, at any rate, Greeks hardly ever do so in a regular building. But then, Homer was describing a besieging army, which had to do its sacrificing at open-air altars among the tents. In Troy itself, there was a temple of Pallas Athena, where she was present as the Palladium, an object of uncertain nature (said to have fallen from heaven), while Apollo had a temple outside the walls, south of the city. To that Apollo Smintheus, his priest Chryses dedicated (or perhaps even built with his own hands) a thatched temple, an act of which he reminds the god when he invokes Apollo's help for the return of his daughter Chryseis, whose kidnapping and imprisoning by Agamemnon starts the quarrel that is the thematic pretext of the *Iliad.* On another occasion, Agamemnon consults the oracle at Apollo's "rich" shrine at Pytho or Delphi, a shrine that already then had a stone threshold and that seems to have held a cult statue larger than life.[13] By the time the *Iliad* came to be written down, about the same time as the Doric column and the structure it carried had first been translated into stone, there were many such temples in Greece.

Many of the older shrines, and many simple later ones, were just open-air enclosures. *Temenos,* "a portion," was an old word, found in Linear B documents, though in them it refers to a royal estate, not the enclosure of a sanctuary.[14] The Mycenaean king was not a *basileus,* as the princes and grandees of Hellenic Greece were called, but *anax, wanax,* a term that by the Archaic period had been downgraded to mean "lord," "highness"—a general honorific.[15] Mycenaean Greeks used it for "overlord," "high king." Agamemnon was *anax andrōn* to the Greeks before Troy, "Lord," "leader of men": the two titles could be coupled, showing how they differed. *Anax basileus* probably meant something like "supreme lord and tribal chief," since *basileus* probably connoted inherited honor and even wealth within a tribal structure ("noble prince" might be the closest English equivalent).[16] *Anax* referred to a constitutional (perhaps elective) role. In any case the history and geography of the word are very difficult to chart. A number of *basileis* claimed to be descended from the sons of Herakles—not just Dorian ones, of Argos and Messenia or Sparta, but (by dint of some word juggling) those of Macedon and even barbarian Lydia.[17]

Basileus came to distinguish those with a hereditary (and therefore more legitimate?) title in Archaic Greece from the tyrant whose constitutional role survived into republican times as the title of a state officer. In Athens he was one of the nine archons who were the collective head of state.[18]

God and king

The *wanax* of the Linear B tablets had his temenos, as did some of his immediate inferiors. It was the source of his income: temenos of the divinity incarnate in a king or hero. A temenos might be granted to such a hero or to a great lord as a reward for exemplary courage and skill. They seem also to have been granted as a homage for the divine status of a hero or a king.[19] In any case, a king often received hero worship, even sacrifices, and many would inevitably take place in the king's hall, at his hearth, or in its court, which therefore also acted as a kind of temple.

As for the worship or the heroizing of a great ruler in Crete, the kind who might have occupied the palace at Knossos, there is no definite evidence for it—though the legends about Minos and his brother Rhadamantes, Cretan lawgiver and judge of the dead, or Sarpedon[20] (lover of Zeus and founder of Asian cities) witness to the memory of such a person. The related figure, Minos-tauros, Minotauros, or even Minos the bull, were interpreted at the beginning of the twentieth century in anthropological guise as referring to Minos—or some person close to him (perhaps his son, Asterion) wearing a bull-mask. Although such specula-tions are now less favored, the enormous popularity of the bull's head in Minoan art will not allow that they be dismissed.[21]

The division between god and king had not been as clear in that early Greek world as it was to become once history began. There was a time, if legends were to be believed, when all humans dwelt in community with immortals; mortal women and men certainly coupled with gods and goddesses and had children by them, children of ambiguous or uncertain status, half-mortal, half-divine—heroes and nymphs—and from such unions kings were often said to be descended.

Altar and temple

In the Iron Age, at any rate (which came to be called the Hellenic age in Greece), sacrifice and the meal that followed it were the principal use of the temenos. The familiar classical temenos of the temple seems to have had an almost exclusive religious purpose, and it was separated out, "cut off" from pasture or arable (or even waste) land, to enclose a site that would have been sterile without the constant anointing with libations of water or oil or blood of the original rocky outcrop or of other stones, placed there by worshipers.

The great majority of such enclosures identified by archaeologists have been more or less rectangular.[22] They usually contained a tree, a water source, and an outcrop of rock, or else a boulder: tree and stone.[23] The fire altar was usually the boulder, or a structure connected to it. This outcrop of rock seems to have been important, in certain cults at any rate. It stood for the bones of the earth, much as the bones of the beast stood for the incorruptible and therefore divine part of the sacrifice. It was as if the creature earth offered her bones as a foundation to the builders. The bones of Ge, of Mother Earth—therefore her immutable part—were constantly enlivened by the fat and blood of sacrifices. The piece of ground that was cut out of the city or of arable land to be moistened and greased, the temenos, was the bones of the earth: rocks. And sacrifice was what the temple was for.

Not all sacrifice was bloody. Liquids such as milk, oil, and wine were offered. Also flour and breads of all kinds. Fruit and flowers. Some sacrifices involved alienating an object from secular use by vowing it to the god. Others involved its destruction by drowning or burning—holocaust. But sacrifices of animals were the most spectacular, and the most expensive. Larger beasts were usually killed by an ax blow from behind, their throat then cut with a knife, and the blood caught in a dish for sprinkling on the stone or altar; smaller animals were held over it so that the blood spurted directly on the stone. The blood on the altar had to be renewed constantly as a fertilizing, repairing, resurrecting action. Moreover the butchering—in the sense of cutting up, jointing, and distributing the flesh and the inner organs—was an important part of the rite. Indeed the word for the appropriately adjudicated part of the animal, *moira,* came to mean fate, or the role in life apportioned by fate.[24]

The fat, bones, and skin were often reassembled, the animal reconstituted. Fueled by the fat, the fire then burned it to shapeless ash. At Olympia the ash of thigh bones was mixed with water from the nearby (sacred) river, the Alpheios, and the resulting slurry used as a cement to build up the fire altar of Zeus, which by the second century AD was over twenty feet high.[25] If Pausanias is to be believed, there was another such altar for Hera at Samos, and smaller ones were common in Attica. The famous great altar at Pergamon was also built around such an ash heap.[26]

The ancients found the parallel between cremation and sacrificial rite very striking, though the burning of the sacrificial bones was quite different from the cremation of a familiar. While a cremation fire was spent with wine, and the bones of the corpse were gathered together, "read," and assembled, like those of the sacrificed animal before burning (an analogy made by Orphic writers),[27] they were then wrapped in a cloth and put in an urn for burial. In both sacrifice and cremation, the bones were the incorruptible, the divine or immortal part of the creature, a token of that which is sent up in the smoke of sacrifice; in burial they became the perpetual relic of the dead person, that which made him or her into a kind of deity—to whom, in turn, the immortalizing and fertilizing offerings of moisture, of oil and blood, must be made.[28]

Haghia Triada sarcophagus: Sacrifice scene. Iraklion, Archaeological Museum. Photo © Alison Franz Collection, American School of Classical Studies in Athens.

The act of sacrifice was a kind of multiple trespass, and by analogy, so too was the separating out of a temenos: the first against the integrity of the edible animal, the other against the integrity of the soil. It was a necessary trespass, because without killing man cannot eat his meat, nor can he cultivate land without making some separation between virgin, untamed nature and the order of his ploughing. Sacrifice is to cattle farming what the sacring of a temenos is to agriculture—a reparation—and presupposes a theology, however coarse, however diffuse and unfocused.[29] A theology is implicit in the action of ritual and the recounting of myth, even if it does not become explicit until much later and may never be systematic.

Parts of the building

The earliest buildings of the Greeks, after the return of the sons of Herakles, could not have incorporated a defined type like the temple, nor could their columns have followed the pattern of any recognizable genus. Yet there were posts to support a porch before even the shabbiest of those little shrines appear at the beginning of the iron age. Inside, in a dark chamber with a central hearth, two columns sometimes stood on the long centerline of the shrine on either side of that hearth.[30]

The first sacrifices of the "Hellenes" of which evidence remains were carried out in the straight confines of just such frail and shabby shrines. If these were animal sacrifices, they could only have been of sheep or goats; even the most determined drover would not get a live cow into the earliest Dorian temples—those at Prinias or Dreros on Crete, for instance. The sacrifice was therefore almost certainly begun outside, since only parts of the animals (or less refractory offerings) were burned on the hearth. There is no definite evidence available of external altars at which the sacrifice could be offered, to be butchered there and carried within to be burned. In some buildings, there is a shelf or bench that carries small statuary and other nonperishable offerings such as horns, a practice that harks beyond the Mycenaeans, to Minoan custom.[31] It is also highly probable that these very simple structures—which must have required constant maintenance—and the temenos within which they stood were intentionally related to a proto-urban fabric. The cults that were practiced in them almost certainly required a form of solemn approach, a semipermanent route for the sacrificial victims that must have inflected the pattern of the settlement. But of that we unfortunately know very little.[32]

What separates these "proto-Doric" shrines from both later and earlier buildings, however, are the nature and use of the fireplace hearth. In some cases, it is not merely a piece of raised pavement circled by a decorative border but a kind of stone cist. From analyses of the ashes that they contained, it is clear that organic material (presumably sacrificial) was burned in them, which means that when they were used for their original purpose, these small, smoke-filled chambers had their walls caked with rancid soot. A smoke hole would therefore have been essential, and most restorations provide for one. Presumably the clerestory lanterns of Mycenaean palaces performed that role during the heroic Bronze Age in the much roomier and often two-story megara. On the other hand, all the temple models (with one exception, the flat-roofed house found near Knossos) have steeply pitched roofs, suggesting thatch, the only material available which could allow that pitch: it was how the priest Chryses, in the Homeric account I quoted earlier, covered the temple of Apollo Smintheus near Troy. What remains of the actual buildings is too fragmentary to allow any confident reconstruction of the relation of column to beam, however.

To talk of "order" (as of column-and-beam) or of "type" (as of plan) in such rude architectural circumstances may therefore seem willful and far-fetched. Building techniques varied, even the most rudimentary ones. After the Mycenaean palaces, during the Dark Age, builders did not even seem to be certain how posts should best be planted in the ground: in some quite elaborate buildings, they are set into a hole;[33] in others, they are set on a stone or a terracotta base (as they would have been in a Minoan house, and perhaps in imitation of them), though it is equally possible that these builders still had skills that survived through the disturbances.[34] How were such devices transmitted by preliterate, or nonliterate, people? Inference from scanty archaeological evidence, bolstered by hazardous readings in Homer or Hesiod, cannot quite settle such problems. Clearly even the simplest building techniques, especially when practiced by nonprofessionals (or at any rate nonspecialists) in a farming or even a hunting-gathering community, which depends on a common stock of rule-of-thumb procedures and some form of memorizing and transmitting them, does presuppose a system dependent on stable social relations: these are almost always corroborated from the misty account of legend. With the refinement of construction in larger or at least richer settlements, the existence of a trade organization of specialist builders—of *demiourgoi*—seems confirmed by the repetition of certain plan forms and procedures. The wide diffusion of building types and structural techniques in the first century of the Olympic era could even indicate that these builders were wanderers, like the smiths. Metalsmiths inevitably enjoy a privileged position in all societies that pass from a wood-stone-pottery technology to a metal one.[35]

The *demiourgos* at sacrifice

Long before any of the classical discourse about column types (on which Vitruvius was to draw) could be set down in writing, there would already have been various craft organizations. Such organizations of masons, plasterers, carpenters, and joiners—as well as painters, potters, and sculptors—must already have existed on Crete and Thera, where the level of skill achieved by the end of the third millennium could only have been due to full-time artists-craftsmen working over many generations who had an organized procedure for teaching and transmitting their techniques; this is particularly striking in the matter of motifs in their paintings.[36] There were also goldsmiths and bronze workers. Daedalus became the archetype of these crafts, as he was of the founders and of the masons.[37]

The surviving documents of the Mycenaeans that have been read so far are too dry and clumsy to provide evidence for any written (or perhaps even drawn) constructional practice, yet the rapid elaboration and the similarity of the buildings leads one to suppose, if only by analogy, that the practice must have been codified. The most common form of such codification is a body of work songs accompanied by some sort of heroic etiology. Without such features no craft in the ancient world would have found cohesion possible. Yet long before any craft achieved social standing, never mind proper organization, the builders of Neolithic Greece must already have transmitted certain building procedures orally for generations, in mnemonics dependent on number and rhyme as well as gesture, through which their building types were perpetuated.

As with techniques, so with types: the Iron Age builders inherited certain shapes from their precursors in the Age of Bronze. The central hearth of the earliest two-column Iron Age shrines echoes the central hearth of the four-column megaron, the "great" hall of the Mycenaean *wanax*'s palace. Remains of such fireplaces also survive in the megara at Mycenae, at Pylos, at Tiryns, and perhaps in Athens and Thebes. The very ordering of Mycenaean palaces is eloquent: at that central hearth, at his *eschara,* the Mycenaean king received offerings (which could almost be called sacrifices); by its fire, his jewels gleamed, and he displayed his majesty. It really was the "focus" of his palace.[38]

Around that hearth there usually stood the four columns that held up the roof, with a clerestory or smoke hole over the fireplace. The Mycenaeans seem to have taken over two forms of column from the Minoans: the first was of one piece, usually wooden, wide at the top and tapering down to a narrower base, the inverse of the Hellenic column;[39] the other was the square pillar, almost always made up of stone blocks, with a bulky capital.[40] The square pillar, familiar in Minoan building, seems much rarer in the full Bronze Age—the Late Helladic or Mycenaean period—perhaps because the ordering of the Minoan "palaces" was based on an entirely different way of life. The Minoans on their island lived in open, unfortified buildings, often on flat land, and their palaces were heated by braziers, which all suggest that their alliances were Syrian, Anatolian, and southern. By contrast, the Mycenaeans, although they took over many Minoan building features and procedures, centered their very different fortified, hilltop buildings on a fixed, column-surrounded hearth. Some fragments of round Mycenaean stone columns have survived (most of them charred), as have slightly more distinct remains of a very few stone pillars. Yet there are a number of representations of such columns that allow a tentative restoration and—perhaps more important to me—give some idea of how they were considered by their builders.

Lions and a column

The most explicit of these images urges its own problem. This is the relief over the Lion Gate at Mycenae itself, the largest of all Mycenaean sculptures and also considered the earliest surviving monumental sculpture in Greece.[41] It shows a single column standing on a molded base, usually identified as a double altar. On either side of it, two lions have their forepaws on the altar and hind ones on the ground, making a truncated triangular composition. But while the lions have lost their heads, the column has retained its abacus and cornice. If you observe it closely, you will note that it is in fact a column-and-beam, or even a column-and-roof. The abacus supports a block made up of two horizontal strips: the lower one is a continuous line of discs, almost as if they were the sawn-off ends of tree trunks or of thick branches; the top one is too close to the upper edge of the block to have remained distinct. However, on the gateway of the Treasury of Atreus, where the two columns on the door jambs are a similar shape and support similar ornaments, the upper strip is covered with a spiral wave ornament, usually known as the wave or *chien courant*.[42] The shafts of the treasury columns are zigzagged with chevroned strips inscribed with spirals. The *torus*, the swelling part of the capital, has the same ornaments as the shaft, while the hollow molding underneath it is adorned with a crown of leaves. By contrast the much smaller and dumpier Lion Gate column is quite smooth. Such differences in the proportions and the ornament (fluted or not, flowered or smooth) of columns as between the Lion Gate and the Treasury of Atreus were quite common, while the moldings may well have been of two or three types only.[43] The shaft may also be occasionally fluted.

But at this point, it is not the Mycenaean column in itself (which will have to be considered later) that concerns me, as much as the composition of the sculptured slab and why it has been set over the gate. One sense of it seems obvious at once. It stands guard: a column is the standing object par excellence.[44] At Mycenae it is the central column that seems to be the guardian, though it is also guarded in its turn. The composition of the relief, as has often been observed, is a variant on the familiar device (often called "heraldic"), which has an inert object between two animated ones: usually a tree between two animals. More often than not, what seems a tree is a composite object, contrasting stiff artificiality with the bustling, wriggling verisimilitude of the supporting animals or humans. The triple theme is sometimes varied to show a human figure, usually male, imposing its will on two beasts, bulls or lions. It appears early in glyptic art, in the first Uruk cylinder seals even before the first metal tools and the devising of writing in Sumeria, and about the same time in small sculptures. Although it is also found in Egypt during that period, it is much less common there and is sometimes regarded as a trademark of Mesopotamian art. The device soon spread to stone reliefs, to bronze and copper sculpture and vessels, ivory for furniture, and embroidery. Common on the orthostats of Assyrian palaces, it is taken up by the Hittites, the Phoenician, and other Syrians, who also emulate Mesopotamian cylinder seals, as (for a while) do the Egyptians, the Rhodians, and of course the Minoans.

Glyptic art travels. Sealed jars were the most common object of trade. The seals themselves are small, and merchants carried them, sometimes even as personal jewelry.[45] That particular heraldic motif was therefore as familiar to a Mycenaean sculptor as it had been to the Minoan ones. However, I think it true to say that no artists will ever imitate such a generic motif gratuitously, as an abstract pattern: they will only take it up if they can find a use and a context for it. In the case of the Lion Gate, the use would not be so much for the protection of the gate, since this was signified much more graphically (if not more successfully) around the Eastern Mediterranean (and beyond, as fairy tales suggest) by setting two brutes at the doorjambs—as the Assyrians, the Persians, and the Hittites did—for the protection of the city.[46] They might be lions or sphinxes: in Hittite lands, even the gods themselves; in Babylon and Assyria, gryphons and human-faced monsters. Much earlier, this seems to have been done by setting two divine columns or posts on either side of a gate or door. Since at Mycenae the

lions were set over the gate and guarding the column, it may well be that the column on its double altar was a sign of the divinized city, guarded by its fierce protectors.[47]

Alternatively and, it seems to me, more probably, the whole slab is an apotropaic representation of the guarding of the city, in which the visitor or aggressor is invited to make the identification and to beware. This idea raises a question that has exercised a number of scholars: whether this column may in some way be divine, not (as some have suggested) merely the token for a building, for the palace of the Atreidae on the hill above it and the circuit of the walls, or even, by metonymy, for the Atreid tribe and dynasty. By extension, it suggests a more general question that cannot be answered directly: are not columns in some sense always the holiest part of a building?

What requires more comment is the way in which the material (and therefore the craft) involved changes. Most of the tapered columns in the Minoan and Mycenaean palaces

The Rule and the Song

The heraldic ruler figure. New
York, Metropolitan Museum.
Photo by author.

The heraldic ruler figure. New
York, Metropolitan Museum.
Photo by author.

were wooden. For the first time on the scale of real building—rather than the tiny seals and
only slightly bigger rhytons and wall paintings—the tapering column is represented in stone
at Mycenae: in the Treasuries of Atreus and Clytamnestra, as well as on the Lion Gate. Since
that column is usually represented carrying its beam, another problem arises: the triangular
hole that is corbeled out to carry the slab over the Lion Gate is the same shape as the opening
over the door of the Treasury of Atreus (and that of other tholos-tombs) and may have carried
similar decoration, whether sculptured or painted. The triangular hole is often justified empirically as *lightening* the load over the beam that spans the door opening below. The shape en-

"**Ritual or sacred tree**" relief:
Nimrud. London, British Museum.
Photo by author.

"**Ritual or sacred tree**" relief:
Nimrud. Ivory. New York, Metropolitan Museum. Photo by author.

closed by the corbeled stones is almost exactly the same as that of the smoke holes of the few surviving clay and stone models,[48] which seem to be of shrines rather than houses, and that in turn suggests that the relation of column, beam, and triangular opening was not an invention of the Greek Geometric period, was not a Dorian importation, but may already have been formulated by Mycenaean craftsmen into a *topos,* a transmitted and taught technical commonplace. A pitched roof, presumably a thatched one, also appears on one of the very few glyptic representations of a building.

The formula certainly has to do with "real," "experienced" buildings in wood and terracotta, and perhaps metal as well—not "ideal," painted ones. Unlike many holy pillars of the Greeks and of other civilizations, which are represented unburdened, in the Lion Gate at Mycenae we have for the first time perhaps an explicit representation of a consecrated or quasi-sacred column and beam.

The holy column

That columns were regarded as holy in certain circumstances is well-known. What is perhaps not so evident is that they were considered quite apart from the more enigmatic baetyls, or holy stones.[49] I must therefore discriminate between the burdened column and the baetyl on the one hand, the wooden column and sacred tree on the other. A tree may be designated as holy, chosen as an epiphany quite arbitrarily. However, there is a whole class of artificial objects, such as the maypole, that are treated as venerable in many cultures. I have already pointed out that sort of tree in Assyrian representations. It seems to have been a common cult object, of a similar kind to the Egyptian *djed,* that is, an upright, fabricated of reeds or wood and twisted or woven fiber, perhaps decorated with ribbons and streamers as well as flowers.[50] In some representations, it may also have been covered by a net or cloth. In Assyrian reliefs its artificial nature is underlined by the gryphons and monsters that surround it, who are usually shown carrying a small bucket in one hand and a cone with which they touch the "flowers" of the maypole in the other. This has been read by some Assyriologists as showing the fertilization of the female date palm with male flowers, though in fact the date palm found the climate of Assyria too cold to bear edible fruit there. It must represent an old farming custom of the more southern Sumerians (which they, for their part, never depicted), remembered only in the form of a ritual procedure.[51]

The construction of such ritual or cult objects from different plants, the *asheroth* or *asherim* of the Semites, and their representations in stone and majolica are echoed by the relief surface of the Mycenaean-Minoan capital, which reveals its composite nature. The metal (presumably bronze) leaf crown attached to its hollow molding is often shown.[52] The shaft of the column stood on a flat base, when it did have one, and was crowned by a heavy torus molding with a concave necking that joined it to the shaft. Such a ring necking is visible in the fragments of the lower and bigger Treasury of Atreus columns, and also in several of the small surviving ivory models. What of the torus above it? It will surely not have been turned on a lathe with the capital, and must also have been molded either in terracotta or bronze and fastened or nailed to it.

Whatever its construction, there can be little doubt that the column over the Lion Gate is meant to be seen as holy, even dreadful—and male. Minoan male statuettes and wall paintings tend to show the male figure with feet close together, the hips narrow, and the shoulders very wide. These figures have an almost rhyton-like aspect, which is particularly striking on the so-called cupbearer fresco in the Knossos palace. Was that intended to assimilate them to the outline of the inverted-tree column? Female figures have narrow shoulders and wear conical, voluminous, flared skirts: and that makes them unlike any known column shape.

My suggestion that there may have been an analogy intended between the male figure of Minoan art and the shape of Minoan and Mycenaean columns is the mere outline of a hypothesis; there is yet another matter that I must address here. Exceptionally in Mycenaean art, you will find cylindrical columns with parallel sides: the little model recently found at Arkhanes near Knossos has them, so has the slightly earlier cult scene from the tholos at Kamilari near Phaestos; and among the strange collection of terracottas of the end of the Middle Minoan period which were in the Knossos palace, in the "Loom Weight Basement," there are three such columns, complete with common base and separate capitals. They carry token cylindrical beams like the Lion Gate column, and on two of them (presumably, also on the third) birds are perched.[53]

Scholars have attached a great deal of importance to the presence of birds among the relics of Minoan sacrality; but there is another image that might be recalled here, that of the repoussé gold-sheet shrine model found in one of the shaft graves at Mycenae, with bird "acroteria,"[54] used by Arthur Evans as the basis for his restoration of that part of the Knossos palace which he believed was its sacred area. In fact all too little is known about how the different chambers in the palaces were used—whether some were more or less "sacred" than others, or indeed if the whole palace should not be read as a kind of temple.[55]

A prime piece of evidence in this connection is the sarcophagus from Haghia Triada, which carries the most explicit surviving representation of a Mycenaean ritual.[56] It is not for me to decide whether it represents the cult of the dead or the cult of the gods: what concerns me is how the physical objects in the scenes are displayed. The two long sides of the sarcophagus show processions and sacrifice. On one side, a half-naked woman is pouring what seems to be a libation from a large bucket into an even larger vessel, which is like a Greek *skyphos;* she stands between two columns, each one crowned with crossed double axes on which a bird is perched. The double axes and the birds are colored yellow, suggesting (in conjunction with some material objects) that the whole "capital" of this column is of bronze, that it may be a gilt bronze crowning element. The lady celebrant is followed by two other women in long, fringed dresses; the nearest carries two more buckets suspended from a stick laid across her shoulders as a yoke, while the further one plays a lyre. Moving the other way are three half-naked young men; the first two are carrying leaping calves in such a lively stance that they must picture artifacts, while the third carries a crescent, usually interpreted as the model of a boat. The figures are moving towards a stepped construction behind which stands a stylized tree and a male figure in a stylized dress, considered by some a statue of the dead man (or even his mummy), by others a god: he has even been identified as Zeus Velchanos, an adolescent and bearded Zeus, who later had a temple at Haghia Triada.

The other long side of the sarcophagus shows three female figures in long dresses (the one whose head survives wears a headdress), standing in an offering posture. Before them is the offering table, on which lies a bound bull or ox, whose blood seems to be flowing into just such a bucket as is being used on the first side of the sarcophagus. Behind the offering table, a young man is playing an aulos; a half-naked woman, like the earlier one pouring the libation, seems to be offering a basket and a jug on an altar that stands in front of a building or fence crowned with "horns of consecration," within which is a tree, perhaps an olive. Between the altar and the building stands yet another pole carrying a double ax on which a black bird (a raven?) perches.

A number of commentators have noted certain features of these scenes that relate them to later Greek custom. The pouring of libations through bottomless vases into tombs was common in Geometric and Archaic Greece. The aulos player and the basket in which the sacrificial knife was usually carried were standard features of Greek blood sacrifice. The statue wrapped in a cloak, the feature that has led some to identify it as an Egyptian mummy, is in fact more like some acrolithic statue. More curious from my point of view are the poles that carry the double axes and birds, the usual marks of Minoan-Mycenaean sacrality. They must therefore be read as yet another form, perhaps a semipermanent one, of the sacred tree.

The columns painted on the Haghia Triada sarcophagus were certainly reversed with respect to Mycenaean ones, narrowing rather than widening to the top. If they were to be read as tree-trunk pillars, therefore, they had to be set upright, the way the tree might grow, and not reversed. Were they ceremonial posts erected for the occasion of some ceremony (perhaps a funeral) and then taken down? Or were they already harbingers of the new kind of columns that were to appear in the temples of the Greeks, especially the Dorians?

Egyptian Doric

In spite of the whole weight of legend I have quoted to support the identification of the Doric genus with the heroic male figure, many scholars have been reluctant to discuss, never mind accept, either a ritual or a corporeal origin for the formation of as obvious, pervasive, and matter-of-fact a form as the Doric column. Architectural historians have preferred to look for some formal, positive, even material precedent: for the Doric there was one to hand in Egypt, which offered such an obvious example that the whole body-column business could be relegated to the study of mythology.

At Beni-Hasan, on the east bank of the Nile, were the tombs of the great men of the city of Menat-Khufu (just north of Tell-el-Amarna), which Champollion found opening westward onto the river, cut into the living rock.[57] Many of them had one or two polygonal pillars supporting a kind of abacus. In some of those tombs, notably those of the Nomarch Amenemhat (no. 2) and of Khnum-Hotep (no. 3), the "mayor" of the town, polygonal columns also "supported" the roof inside the tomb. Champollion was very excited by his discovery and suggested that the Greek Doric order must have been derived from the Egyptian precedent. Because he considered—wrongly, as it happened—that the Beni-Hasan tombs dated from the ninth century BC and that some of the wall paintings represented Greek prisoners, a direct derivation seemed all too plausible, particularly as he was in fact following the eighteenth-century architectural historians who had speculated on the subject. Since his time, these and similar (though much later) Egyptian columns came to be known as proto-Doric.[58]

The most spectacular use of them came some centuries later and hundreds of miles further south, at Deir-el-Bahari outside Thebes, in the vast cenotaph of Queen Hatshepsut. There they appear in much greater numbers and with a great deal more refinement. They appear, too, in the buildings of Tuthmosis III at Karnak. In these later constructed (as against the earlier rock-cut) examples, they are mostly built up of drums. Many of them are cut into eight faces, more into sixteen, which suggests that the form was arrived at by chamfering each corner of a square-plan pier: once for eight sides, and twice for sixteen. Often they stand on a low and flat circular disk for a base, while the top of the column is usually square on plan and flush with the cornice, which in turn is also flat and unrelieved in its lowest member. When you look at these columns from the front, therefore, you can read the abacus block as belonging to the columns, as a dip in the cornice, or as an excrescence on its soffit.

The obviously lithic character of these surviving examples prompted one or two commentators to suggest that these earliest square-plan piers (such as may be found in the granite Valley Temple of Chephren at Karnak, to take a famous Old Kingdom example) evolved into the faceted columns at Beni-Hasan to allow for more light in the tomb chambers.[59] Moreover, the square-plan prototype continued to be used well into the Hellenistic period, while round, slender, fluted columns were used both much earlier and much later than the chamfered pier—the earliest use being the tall pilasters of the pavilions in the enclosure of the first large pyramid, that of Zoser at Sakkara. In any case, such evolutionary theories tend to gloss over the most telling of the difficulties: on the cornices at Beni-Hasan small ribs project along the top edge of the main beam, as if they imitated the rafters of a wooden construction, while the ceilings of the tombs of Khnum-Hotep and Amenemhat are painted with a checkered pattern, clearly intended as representations of rush matting; some of the stone beams are even painted to simulate wood. The imitation of wooden construction in stone thus seems as much a feature

Temple of Hatshepsut at Deir-el-Bahari. Exterior photo by author.

Temple of Hatshepsut at Deir-el-Bahari. Interior photo by author.

The Rule and the Song

Djed **character.** From the temple
of Seti I at Abydos. After R. David
(1981).

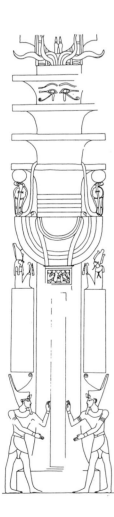

Djed **character.** From the temple of Seti I at Abydos. After R. David (1981).

of Egyptian as of Greek architecture; the Beni-Hasan rock tombs are worth exemplifying here simply because they were so often quoted as an instance of Egyptian pure-stone architecture that might have served as a prototype for the Greeks.

It is in any case a very long time and a long way from Deir-el-Bahari, let alone Beni-Hasan, to the most primitive forms of Greek Doric. As evidence now stands, it would seem that such columns were used only for a limited period between the Eleventh and the Eighteenth Dynasties and that the last Egyptian examples are therefore a millennium earlier than the first Greek ones.[60] They are, moreover, rather unlike the columns used by contemporary Minoans and Mycenaeans.

Djed: Column and mummy

However, I mention the rock tombs here for another reason, since the square pier is very often joined to a male human body, specifically that of the god Osiris, in his common form of the embalmed pharaoh. Wrapped tightly in his mummy bands and wearing his double crown, his hands crossed on his chest and grasping the flail and crook of Egyptian kingship, this figure makes for a different conjunction from the vital six-footer of the Vitruvian text. It—or perhaps he—leans on the pier; it is bound to it, but not identified with it. This conjunction appeared at Deir-el-Bahari and at Karnak, two shrines I have already mentioned, as well as in the various temples round Thebes, like the Ramesseum and the shrine at Medinet-Habu; most overwhelmingly, perhaps, it is seen in the rock-cut shrine at Abu-Simbel. The power of the figure, colossally reiterated, is hypnotic and shows, if showing is needed, that the association of column and human body is no invention of the Greeks.

That association of body and post in Egyptian architecture is again obscured by its very commonplace, as it is in the Greek case. To my mind the association of the mummy-body of Osiris and the *djed* (or *tet*, now more commonly transliterated as *dd*) column is particularly telling. Some have even suggested that the *djed* was already an independent holy object, an archaic ritual implement, before it was identified with Osiris.

Physically, this object was very friable, even perishable: a bundle of such reeds as grow abundantly along the Nile, bound with fillets at top and bottom. The inner tufts would be pulled out and flattened round into a crown or halo three or four times, to make a kind of composite capital.[61] As a hieroglyph it signified the god Osiris, much as the stand crowned with a knot, which was a different kind of ritual implement, signified Isis. There are myths in which the *djed* was the central feature of a temple in some previous cycle of creation, sometimes as an incarnation of the god himself, sometimes as a perch on which the god rested. In the Theban cosmogony it also signified one of the origins of the world: the protruding tuft of reed on which the Horus-falcon settled in the world morning. The *djed* was ritually raised in many Osiris temples: that is, it was made up lying down, and then pulled up by ropes.[62] In some temples one of these was made and set up and raised frequently, in others it was reserved for special occasions. In Busiris (Abusir) it was used to support the Osiris-mummy; Busiris, "the home of Osiris," was the center of the ninth nome of lower Egypt, called *Dd.w.* The hieroglyph was part of many words to do with stability and justice.

The formalized, almost hieroglyphic, *djed* appears early as a decorative device. In the wattle screens, simulated in faience, which decorated the temple in the Zoser enclosure of Sakkara, it is already cramped into arch supports in a way that shows it to have been a familiar

The raising of the *djed*. Fom the temple of Seti I at Abydos. After R. David (1981).

***Djed* and lily ornaments on a limestone plaque.** New York, Metropolitan Museum. Photo by author.

The Rule and the Song

and easily recognizable shape. Its various transformations, from small faience amulet to larger-than-life column-as-god, have never been adequately chronicled.

But then there is no canonic repertory of Egyptian column forms. Most histories of architecture provide one or two familiar instances: the straight bell-capitaled, the swelling reed-bundle shape (perhaps related to the *djed*), the palm leaf and Hathor-capital column, and several other minor variants. Different forms may be used within a single building, and all of them may be combined with square-plan pylons. The most slender—to Western eyes perhaps the most "refined" of them all—are the fluted pilasters that support the gently arched roofs on the Heb-Sed court in the King Zoser pyramid complex. Again the form and the detailing are so accomplished that it is difficult to imagine them as unique or initiating examples of a completely new architectural manner. Zoser was the founder of the Third Dynasty, and his reign is usually dated about 2650 BC; his pyramid complex is the first of such giant Egyptian buildings, one moreover for which the name of the architect has been recorded: Zoser's vizier-healer, Imhotep. For some time now a search has been going on for Imhotep's tomb under the Serapeum at Sakkara, where in Hellenistic times Asklepios, whom the Greeks were to identify with Imhotep, was particularly worshipped.[63] While Asklepios is identified with Imhotep the healer, his skill as an artist and as a deviser of crafts suggests a quite different Hellenization, which is presumably why Diodorus Siculus says that Daedalus was worshipped "on the island of Memphis." According to Diodorus, Daedalus learned one of his most famous skills, that of making statues, from the Egyptians and had also worked in Egypt as an architect. Indeed some Greeks believed that he had imitated Egyptian sculpture.[64]

There is no agreement about what Daedalus' art was supposed to have been like. According to some ancient authorities, his statues were so lifelike that they rolled their eyes and moved their limbs; yet the figures that were shown to Pausanias as Daedalus' work were in his view rather graceless, rather like sculptures that he describes elsewhere as Egyptian.

Whatever the (no doubt complex) origin of the Greek kouros figure, which I have already described in the previous chapter, early kouroi have explicit Egyptian affinities. Some of them could even be mistaken for Egyptian sculptures: the legs set apart in a walking stance, left leg forward; the hands clenched into a fist (in Egyptian sculpture usually grasping a short cylinder); the arms rigidly stretched along the flanks; the long curly hair combed into a chignon, which falls like the headdress of a pharaonic portrait. But the apron that usually concealed the Egyptian midriff is stripped off: curiously enough the Egyptians, who were not shy about female nudity, practically never represented a male figure completely naked.[65]

Egyptians and Greeks

Apart from the legends about Daedalus, there is another tradition which confirms Greek dependence on Egyptian precedent: that of the Samian brothers who made a statue in two separate parts in different quarries, using an Egyptian method of coordinates (rather than of proportions) so as to join the parts exactly when they were brought together.[66] The Greeks' account of their debt to the Egyptians (even of Egyptian superiority in some things) was a familiar story in Plato's time.[67]

The Egyptian influence on Greek art, which I have noted in Greek sculpture primarily in representations of the body, was in any case generally acknowledged by the Greeks. Nor can the influence of Egyptian architecture be reduced to the dumb copying of isolated formal devices, as nineteenth-century theorists and historians suggested. What the Greeks certainly owed to the Egyptians—or perhaps merely had in common with them and with many other ancient peoples—is the custom of representing older and more perishable buildings through stone forms.[68]

The consistent Egyptian usage of reproducing timber and even reed and mud construction in glazed terracotta and in stone was narrative in intention. While no accounts exist of how the procedure was adapted, the buildings of Zoser's Stepped Pyramid provide an obsessional representation in stone of an ephemeral architecture. Well into the Hellenistic period, forms of this friable building of straw, sun-burned brick, and carpenter's work for the living coexisted and challenged comparison with the heavier stone architecture of tombs and temples.

This procedure is really obtrusive in one part of the outbuildings that make up the enclosure of the Stepped Pyramid. The part known as the Heb-Sed court was a stone monumentalization of a very special temporary construction, a ritual "field" usually put up for the thirtieth anniversary of the pharaoh's "coronation" or enthronement, the *Sed* festival (the festival of the throne or of settlement, the jubilee), when he was expected to run or dance around it as if it were his kingdom by metonymy—in miniature—since each nome or district had a pavilion of reed, leaves, and mud edging the field. At Sakkara that field solidifies into carved stone: each pavilion has a door, but it allows you entry only into a shallow niche carved into the solid rock. The arched roofs seem supported by the slender and fluted columns that I mentioned earlier; the wedging and webbing of the temporary construction are shown, even demonstrated, in sculptured limestone. And yet the whole thing is carved out of a mass built up of ashlar blocks: none of the buildings have any interiors. Imhotep must have painstakingly preserved all the appearance of the Heb-Sed festival that Zoser had once celebrated for perpetuity.[69] The Heb-Sed pavilion pilasters represent only one of several very different types of column, all with vegetable references, which may be found elsewhere in the enclosure. It is not known how that strange monument was used ritually, if at all, in the funerary cult of the pharaoh; since it was part of the very first big pyramid complex (and the first building in ashlar stone, according to the chronicler Manetho), it might be expected to have exceptional features. Later pyramid complexes, all of which retained and displayed the different features of the pharaonic ritual, did not have the elaborate articulation of the Zoser pyramid.

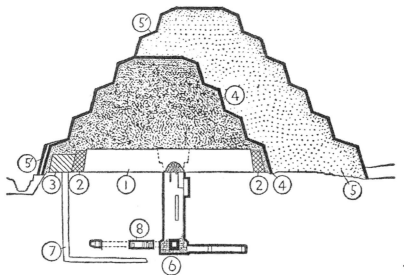

Building of the pyramid of Zoser. Section of the stages, after I. E. S. Edwards (1967).

Stone Heb-Sed pavilions. From pyramid complex of Zoser at Sakkara. Photo by author.

Exterior detail of Heb-Sed. Photo by author.

Chapter VI

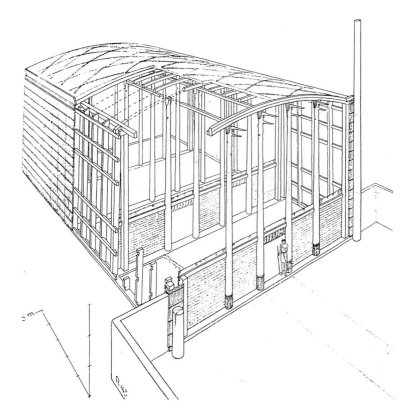

From Zoser until the last pharaohs (who were also Roman emperors), there is a continuity in Egyptian building that lasts for three thousand years. Yet the Egyptians have left us only fragmentary accounts of their building lore; ritual texts and sacred books contain the most extensive passages, usually allusive, and always difficult to interpret. For civil architecture, even that of royal palaces, the writings are even less helpful, and the archaeological evidence about the brightly painted wood, reed, and mud buildings pitifully scarce. We have to read the architects' and the builders' intentions off paintings and models since there is so little left of the buildings themselves.

Greek origins

Something was certainly known in Greece about the vast Egyptian buildings. Still, it is difficult to trace direct formal architectural influences from Egypt to mainland Greece, though it is easier to find contacts and relationships on the coast of Asia Minor and on the Aegean Islands. But neither association can quite provide evidence for an alternative account of the origin or the derivation of the Doric column and its ornaments to bypass the only extended theoretical commentary available for Greek architecture, which claims to set out the methods and explain the intentions of their builders: that of Vitruvius. I propose to discuss this in the next three chapters.

In spite of all the analogies between them, there were also enormous differences between the Greek-speaking Neolithic and Bronze civilizations on the one hand, and those of the Iron Age on the other. Three centuries passed during which a whole way of life seems to

Colossus of Rameses III. From Luxor. Photo by author.

have been forgotten, while a new was yet to be quite formulated, one that would involve both building methods and the ways in which buildings were inhabited. When the earlier period was recalled, it was as an age of gods and heroes to whom the relics of their buildings bore witness. I have already alluded to such a possibility at Eleusis. A crucial piece of evidence about the physical connection between the two epochs, as well as the disjunction between them, is offered by the remains of the old shrine on the site of the main megaron in the palace complex (or "acropolis"). The Mycenaean megaron there seems to have been made over to a proto-Doric temple shape sometime in the eighth century, perhaps earlier.

The temple at Tiryns (if it was that—it has so far turned out to be unique) shows a phenomenon familiar from other sites with almost anatomical precision. The site and ruins of the Mycenaean palace is absorbed by a Greek cult. The megaron may have been ruined for some time when the Archaic temple was built, but new arrivals, new settlers, used its east wall and two out of the four column bases to build a chamber about two-thirds the area of the megaron but much narrower. They also used the eastern column bases and its porch to make an interior central colonnade. Perhaps even earlier, the circular altar in the courtyard was enlarged into a rough square and coated with the same stucco with which the floor of the courtyard had been laid. It was poorly built when compared with the circular, radially cut stone pit of the Mycenaeans, but this makeshift building may well have been that first temple of Hera that the Argives were to destroy in 420, from which they took an ancient image of the goddess to their much grander Heraion.[70]

An overlay of the novel, processional temple type over the abandoned megaron with a circular hearth at the center of the square made by four columns witnesses to the violent

Colossi of Rameses II. From Luxor, temple of Amun. Photo by author.

Palm leaf columns. London, British Museum. Photo by author.

The Rule and the Song

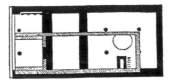

The reconstruction of the Megaron at Tiryns as an Archaic temple. After H. Schliemann (1886).

change in custom, in religious practice, in social structure—in the whole mentality. Even more than in the earliest Geometric shrines, as the methods of building become more elaborate a clear type of temple develops. Several of the more imposing ones (at Samos, at Eretria) have the one-hundred-foot-long plan, *hekatompedon,* with a central row of columns. Such a structure presupposes that whatever functions went on within the building required a movement around that central colonnade, a processional function that may have meant that whatever gifts were brought had to be deposited in the apse, or at a point furthest from the door. As in the smaller shrines of the Dark Age, these could not have been the oxen of the *hekatombeus*—the words are obviously related—but were either the "incorruptible" parts of the animals or some quite different objects of offering. At any rate, these buildings show that the use to which such temples were put was institutionalized, and that the superimposition of the new Greek type on the old Mycenaean base allows for a change in ways of worship and in ways of building—and of thinking.

VII

The Hero as a Column

▪ Base ▪ Column: The shaft ▪ Column: The capital ▪ Beams and roof ▪ *Metopai* ▪ Triglyphs ▪ Models ▪ Delphi: The legend ▪ Eretria and Lefkandi ▪ Dreros and Prinias ▪ Thermon ▪ Olympia: The temple of Hera ▪ Olympia: The House of Oenomaos ▪ Delos: The Naxian *Oikos* ▪ Corinth ▪ The awkward Doric

What Vitruvius called *templum* and *fanum* or *aedes*, the Greeks had called *temenos* and *naos* or *oikos*. Naos, the temple building proper, was the dominant Greek building type, not only because it dominated the fabric of the city, as its tallest—or its highest—building, but also because its most prominent feature, the column and beam (and the exact relation between them) in their different genders, was developed in and for the temple building as the core of Greek architectural thinking and practice. Vitruvius claimed that he set out the methods and explained the intentions of these temple builders faithfully. In spite of his systematic pretension, however, and the corroborating evidence of other Greek and Latin writers, his account remains perplexing, if only because he wrote long after much of his material was first formulated, and in a different language. Thus he sometimes provided contradictory rationales for the same feature without any real apology for the irreconcilable nature of the evidence.

In that most famous of his formulae, the temple was measured or spaced in terms of the male human body to establish the relation between the human figure and the geometry of the world fabric. More specifically, the male was also the model or the *idea* of the Doric column. And in the making of the particular Doric relation, as will also be the case in the female Ionic, the ancient makeup of the column before it was canonized in stone is explained by reference to older materials and building practice—most obviously carpentry, but also pottery and, to a lesser extent, weaving.

To put it another way: the body analogy guaranteed the notional integrity of the order, its corporate nature, its numerical and dimensional identity, the cosmogonic and legendary-historical aspect of the architecture; but anecdotal evidence was more appropriate to the fastidious detail, which was also a reminder of the destructive working of time. To petrify that which time would have destroyed was to defy the elements. Whatever this might have meant to Egyptians or Indians, for the Greeks such defiance was heroic. Vitruvius' dual account of the sources of architectural forms are complementary: as in epic, cosmogony and theogony flower into heroic legend.

To comment usefully on the terminology of the different features of the temple, if only for the purpose of describing it, I will have to treat each member historically, since Vitruvius was quite dogmatic about the fidelity of the stone arrangements to the earlier building techniques that they imitated and represented. Having, as he thought, exhibited the body analogy quite adequately, he goes on to comment on the other aspect of the genera that seem to require justification, beginning with the Doric cornice, to tell "how they were generated, and by what principles they were first devised"; he concludes the detailed reasoning to which I will often return by announcing his general rule that "what could not have really been made [in practice], they [i.e., the ancients] thought could not really be justified as an image." [1]

In describing the temple—and indeed in designing it—the Greeks articulated it frontally, and proceeded from the base up; Vitruvius followed them, and I will be faithful to him. Seen frontally then, the main zones of the temple building were obvious: the base; the range of columns and walls; and the beams with their roof structure, which included the pediments.

Base

The base itself was divided into two parts: the rubble foundations (the *krēpis*, usually invisible), and the stepped platform wrapped around it on which the columns stood, which raised the temple above the level of the approach.

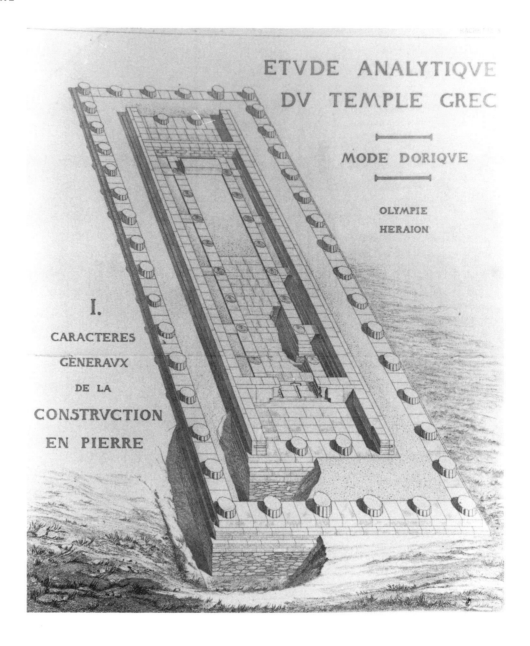

ETVDE ANALYTIQVE DV TEMPLE GREC

MODE DORIQVE

OLYMPIE
HERAION

I.
CARACTERES
GENERAVX
DE LA
CONSTRVCTION
EN PIERRE

The elevated temple platform emphasized and restated the bony, stony outcrops on which so many temples were set, the rock from which they sprang. It also had something to do with another idea pervasive in the ancient world: raising the house of god to a level higher than the daily commerce of men, putting the god on a higher plane, as it were. It was therefore closer to the artificial mountains of Mesopotamia or the "high places" of the Syrians than to the range of steps that leads up to some "classical" bank or town hall. In the Etruscan (and therefore also in the Republican Roman) temples with which Vitruvius was familiar, this effect was all the more obvious since such temples—which went on being built in timber and brick long after the Greeks had opted for stone—stood on a molded stone base, usually a higher

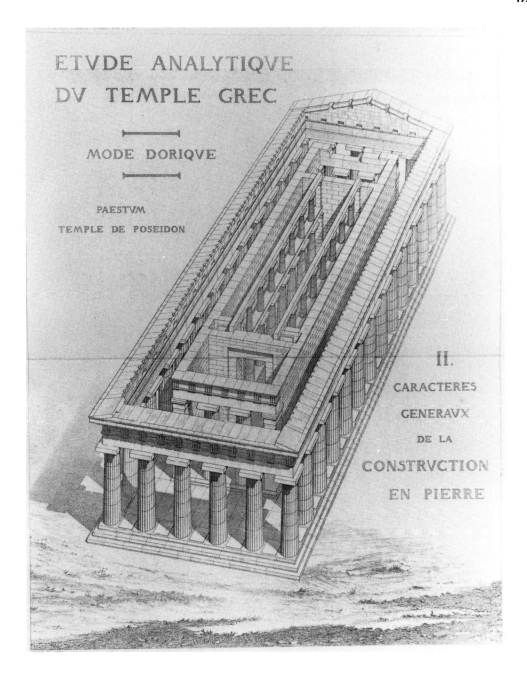

one than the analogous Greek temples. It was plain on three sides, opening on one side only in a flight of quite usable steps.

At the risk of wearying the reader, I must insist that the very steps of the platform in a Greek temple were primarily considered a representation by their builders, which is why they were almost always made both too high and too deep for human ascent. The visitor who wanted to shelter under the colonnade (or to enter the cella) had to walk up by intermediate blocks that were set into the steps at selected points, or by a ramp.[2] Naked and unshod, the Doric column stood on the top step of the *krēpis,* on the stylobate. When the Romans domesticated the column (following Hellenistic usage during the late Republican period), they often

Stylobate and column foot:
Temple of Aphaia at Aegina.
Photo by author.

set the shaft on a base, a more elaborately molded one than those used by earlier Roman builders for native Etruscan or Tuscan columns.[3] However in almost every Greek example, the columns came sheerly down to the temple platform. It was the top step of that platform which was the stylobate proper: from the Greek words *stulos,* "a post," and *bainein,* "to stand upright." A Latin word *stylobatum* seems to have been coined by Vitruvius, which like most of his other technical terms is a Graeco-Latin neologism, transliterated out of one of the many Hellenistic treatises on which he drew.[4]

Taken strictly, *stylobate* applies only to the upper surface of the top step, but by extension the name was given to the whole platform.[5] Moreover the word was—in Hellenistic Greek, at any rate—used almost interchangeably with *krēpis* or *krēpidōma*, which means a base in building terminology, though in common speech it means a shoe or a sandal, especially the popular strung sandal.[6] *Krēpidoma, krēpis,* or even *crepis* can therefore mean the whole flight of steps, as Hesychius points out in his dictionary.[7] It did not stand directly on the foundation, the *themeilia,* and there is usually another layer, the *euthuntēria,* of stone finer than the foundation, and jointed like the *krēpis* proper, which establishes a datum for the building.[8]

Column: The shaft

Between the surface of the stylobate and the bottom of the column shaft, no molding or other visible bedding intervened. The columns that stood on it were not standardized, though Vitruvius' rule can be interpreted through those that remain in the surviving major Greek temples. His tentative approach suggests that he was providing a model rather than a rigid prescription. That is why "type" seems to translate the text much more fittingly than "order," and why there are so many variants in the proportions as well as in the ornament of the Doric column and its cornice during the four centuries of its greatest vigor.

The bare and rough character of the shaft may of itself have carried an allusion to roughly worked tree trunks, standing on the stone base and held in place only by the great weight that they carried. After all, the use of such tree trunks fitted in with the Dorian self-image as it was reflected by their customs. In Sparta, which might be considered the Dorian heartland, the use of the adze was actually forbidden: wood could only be worked with a saw or an ax. The Spartan view of overrefined carpentry is illustrated by an anecdote that Plutarch tells about their famous king, Agesilaos II, who, as he was dining in an opulent house in Anatolia (where he was on campaign), looked up at the ceiling and in mock surprise asked whether trees really grew square in that part of the world. On being told that they did indeed grow round, he asked again: "And if they grew square, would you want to make them round?"[9]

In pre-Hellenic Greek buildings, the shafts of columns do not seem to have been fluted: instance those in the so-called Treasury of Atreus in Mycenae, or in the relief over the Lion Gate there. At the treasury, the smooth surface of the shafts carried a chevron-and-spiral pattern all over, while that of the Lion Gate column seems to have been plain. However, close to the Lion Gate, there is yet another and less impressive (but almost equally well-preserved) beehive tomb, known as the Treasury of Clytamnestra, where fragments of the lower part of a half-column and of one capital survive on one side of its door.[10] These gypsum half-columns are fluted, with thirteen flutes to the half-column, and were fixed to the rougher main framework of pudding stone with metal cramps. There are a few other survivals of this kind in Mycenaean building, including a rather dubious instance of convex instead of concave fluting.[11]

Such Mycenaean buildings are of stone. There is a half-timbered example, however, not usually cited in this connection: the so-called Palace of Nestor, at Epano Englianos on Navarino Bay (which may have been the ancient Pylos). The palace was built toward the beginning of the fifteenth century and was burned late in the thirteenth, perhaps the last Mycenaean palace to be so destroyed by invaders or rebels (or both). Many of the wooden columns, which were irretrievably consumed in the fire, left a negative impression on the column bases: in the main megaron, in the forecourt, and in the outbuildings, there are socket holes baked into bases of clay and potsherds as the building burned. That very slender evidence suggests that they tapered downward (like most Minoan and Mycenaean columns) and that they were worked with some kind of curved blade that furrowed the wood, though the furrows may have been concealed by paint or decorated with a smooth paste finish. The columns were almost certainly made of entire tree trunks; indeed, the Minoan-Mycenaean column is a tree trunk stood on its head.[12]

The Doric shaft and the Hellenic column are generally tapered upward in the "natural" direction, while the Minoan-Mycenaean column follows the custom of many rural communities who set stakes head down. It was an obvious device, some have thought, to prevent rooting. Of course, the same problem of rooting trunks must also have existed for the builders of the first Iron Age as well as the Dorian ones, but then the Iron Age builders used posts that were simply hammered into the ground, even if more substantial columns would be set in postholes and rammed in. All of them must have presented the same rooting problem that the Minoans and Mycenaeans avoided, one which would become much less pressing once the custom of setting the wooden column on a stone (rather than beaten earth or clay) stylobate or base became the rule; in doing so, these later builders may have been following some imported and alien technique, though they may have avoided some of the consequences of such a "natural" rooting tendency by using hardwood such as cypress, which does not root once cut—and whose resistance to rot made it a highly valued timber, as Pliny and Vitruvius insist.

There is yet another obvious distinction between the Mycenaean column and the Hellenic, Iron Age one: the shaft of the latter, even when it was made in marble, stuccoed, and painted, seems to have been designed in all its detail to make insistent reference to the way the wooden construction had once been put together. No such constructions have survived intact in Greece, but in the main hall of the treasury of the palace in Persepolis, which was part of Darius' original foundation that employed many Greek masons about the year 500, there was virtually a forest—ninety-nine of them. The base, a single torus, carried a tree-trunk shaft wound with reed bands to make a key for the stucco casing, which was highly patterned and brightly painted; the stone columns (of which there were several groups) of the main buildings seem all to have been elaborately fluted.[13]

Epic and tragedy carried that kind of detail equally insistently. In the twenty-third book of the *Odyssey*, there is a surprising reference to tree roots in a building when, after her unwelcome suitors have been slaughtered, Penelope challenges Odysseus in a final test of true identity. After she tells a servant to move the bed out of their marriage chamber, Odysseus reacts indignantly to her apparent lack of trust, and in Chapman's version says:

> "Who is it can move
> My bed out of its place?
>   . . .'t was done
> By me, and none but me, and thus was wrought:
> There was an olive tree that had his growth
> Amidst a hedge, and was of shadow proud. . . .
> To this I had a comprehension
> To build my bridal bower; which all of stone
> Thick as the tree of leaves I raised, and cast
> A roof about it nothing meanly graced. . . .
> Then from the olive every broad-leaved bough
> I lopp'd away; then fell'd the tree; and then
> Went over it with my axe and plane,
> Both govern'd by my line."[14]

It is clear that Odysseus knew the bed to be immobile because some part of its frame was still rooted. And, although Chapman added a plane to Homer's *chalkōi*, he is surely right (against some other translators) in making Homer's bronze an ax, not an adze.[15]

The Spartan distrust of high finish, which I mentioned earlier, was canonized by Lycurgos' laws, which did not allow the use of anything finer than the ax for structural elements, though saws could be used to cut boards.[16] When Odysseus was making his bed (and no one has yet convincingly shown which part of it was the tree stump, nor the shape of the room that was built round that olive tree),[17] all the rough work of course had to be done with

an ax; presumably the silver and ivory inlays that he vaunts were later worked with a knife or chisel. His ax must have been the standard double ax, *pelekus*,[18] the kind of ax with which heroes fought in pictures and priests used in sacrifices. No self-respecting hero can be imagined using an adze. Unfortunately, the only instance of the double ax used as a working tool in Greek painting represents Herakles throwing down a Doric column,[19] though in a Euripidean fragment, an unnamed figure describes the holy temples whose cedar beams are closely fitted with the help of a steel ax and glue made from a bull's hide.[20]

The usual curve of an ax blade is an arc on a chord, so that the outline of the flute seems like its negative. In the general fit of a wooden structure (though the actual wood of columns was more usually of oak or pine—and Odysseus' bed was fig) such as that described by Euripides, the space between the hollow of the flute and the curve of the ax would have been very fine. The top and bottom of the trunk shaft may have been cut with a saw, of course, but as he worked the column-trunk, an ax wielder may well have ringed it and cut a notch all the way around, or perhaps marked it with a number of notches to give it a finish. In the stone column, these notches are represented by the *hupotrachēlion*, literally, "under the throat."[21] It is almost as if the body metaphor overlaid the description of a feature that refers to carpenter's work. Those notches would show a carpenter where to complete his flutes. The top of the trunk, above the finished flutes, would have been left rough, to be concealed inside the capital.

Where the projecting spur of a wall corresponds to a column, the wall stops on an element whose three faces reproduce the contignation of that column and support the same cornice. That wall end is called an *anta* or *parasta*: an echo piece, a face piece. It can be read as a stone replica of the wooden boards that protected the end of a brick or rubble wall—an orthostat, as it were, extended up to cover the whole height of the wall at an exposed point. The whole terminology of *templum in antis* or *naos en parastasin* speaks of spur walls that can be imagined as being of sun-dried brick, protected on the exposed ends with wooden boards, which are matched to the columns and which together support a pediment.[22]

Column: The capital

The very word *capital* obviously reintroduces the terminology of the body image: *capitelum* as a diminutive of *caput,* "the head." [23] It is a common enough word, used even as an endearment, and although Vitruvius does not give the Greek equivalent here, it is mostly a variant on *kranion,* "top of the head or skull": *epikranis,* "the brain,"—but also *epikranion,* "the coping or headdress," as well as *kiokranion, kionokranon.* Sometimes the other word for head, *kephalē,* is also used,[24] though the analogy is so obvious that most commentators do not bother to gloss it.

Inevitably, the part of the shaft between the hypotrachelion and the capital was called the *trachēlos,* "necking." [25] In fact, as far as the Hellenistic theorists were concerned, the Doric capital proper started at the hypotrachelion.[26] The necking quite often involved the completion of the flutes in a curve before it met the flared part of the capital, the *echinus.* Where the column shaft entered the echinus, its curve wrinkled up into a number of rings (which are sometimes replaced by an ornamental molding), conventionally labeled by the Latin term *anuli,* "annulets"; most commonly there are four lipped wrinkles that cover or emphasize the joint between the two members.[27]

The fourfold ring convention becomes general sometime before 500 BC. Earlier than that there are several examples in which the molding is a hollow one, internally carved in a

The narrowing of the Doric flute. After C. Chipiez.

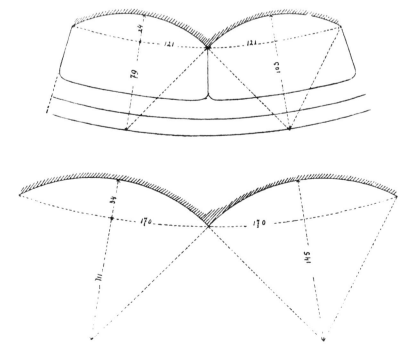

Herakles wielding an ax. After
A. Orlandos (1966–1969).

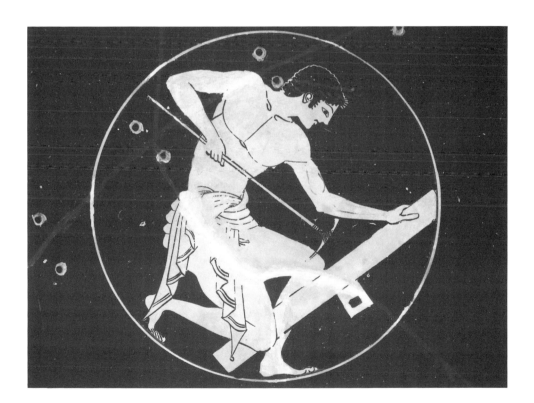

Carpenter wielding an adze.
London, British Museum.

The Hero as a Column

Funerary giant pythos. From Athens. Metropolitan Museum, New York. Photo by author.

relief with floral designs. These are found in the "new megaron" at Tiryns, in the Great Temple of Artemis at Korkyra (Corfu), in the Old Temple at Syracuse—and at Paestum, both in the temple of Demeter and in the "Basilica," where the palmette relief in the necking is particularly crisp.[28]

If the details of the shaft metamorphose and transform timber worked with an ax, then the echinus details, with their hollow moldings and annulets, seem to recall pottery or molded clay, perhaps as cleaned and cut with a string. Various forms of the palmette ornament, such as those at Paestum, are commonly found in fictile bits and pieces from various Geometric and Archaic buildings.[29] The narrow, deep undercut molding may be separated by a fillet from the wide swell of the echinus, or it may pass sharply from one profile to another. It is the kind of relationship between curves and moldings you might commonly find in any number of amphorae or kraters in most large museums with a collection of Greek vases.

As for *echinus*, it means almost any kind of curved and spiny thing in nature: the husk of a chestnut for instance, or a hedgehog. Most frequently, it refers to the various species of sea urchin, which lose their spikes as they dry to a thin beige or pink shell out of water and were often used as cups: "Hippocrates" recommends a dry urchin-shell for mixing medicines, because of its purity.[30] The word came to mean a pottery cup, and later almost any vessel or plate; in its Latinized form, it is almost any bit of pottery. I therefore suppose that the name of the molding refers to a ceramic, cup-shaped piece, which covered the joint between the post and the bracket that it carried. At any rate ancient writers, Vitruvius included, have little to say about it. It is one member in the Doric arrangement that develops consistently: in almost all the early examples, the flat, shallow echinus corresponded to a sharp curvature of the stone column shaft and a tighter narrowing of the shaft at the *trachēlion*. The later narrowing and steepening of the echinus is almost a smooth progression in Archaic to Hellenistic building, until at the end it becomes a near-vertical element.

An examination of the pottery of the Geometric period might almost sanction the idea that the molding was not a purpose-made piece, but that the name recalled instead the use of an existing cup through which the trunk or column shaft was thrust. The analogy is with the great Geometric krateres of the late ninth and eighth century BC, which were used as monuments (sometimes man-sized, that is, about 1.75 m high) and had the bottom broken out, so that libations poured into them would soak directly into the ground. And certainly, many flat drinking cups without any feet were made, such as might have been used for a Doric echinus.[31]

Strings or thongs, tied around the shaft just under it, held the echinus in position. These were the same kind of reed thongs that acted as a key for the plaster on the columns of the treasury at Persepolis and were perhaps commemorated or imitated by the anuli. In all later discussions, the ceramic character of the molding, with annulets acting as an edging or a raking paste marking of the joint, is forgotten.

However, the way in which the echinus makes the transition between the narrow end of the shaft and the plate on which the beam is carried remains constant. This tablet, the

Reconstruction of painted decoration on echinus and abacus.
After C. Boetticher (18/4).

abacus, helps to steady the beam on the post and provides a transition element that ensures that it lies flat. In almost all existing examples, the plan of the upper end of the echinus is a circle inscribed in the square of this abacus. The abacus is a carpentry member again. In Latin, *abacus* can mean any kind of board: gaming board, counting board, bench, counter, sideboard.[32] The plain square board on top of the column seems to be the only member of the Doric to have survived, unchanged both in shape and use, since Minoan times. On wall paintings and gemstones, on reliefs as well as in the remains of the stone columns, the Minoan-Mycenaean "order" is crowned by a swelling molding and topped by a flat board.[33]

The articulation between abacus and echinus must have been echoed—if the surviving architectural fragments are taken as representative of general Greek practice—by the colors that covered most Greek stone moldings, to which I will return later. But while many were carved and molded as well, the echinus and abacus were only painted. From the fragments of color that have remained, it would seem that drooping leaves or tongue ornaments, sometimes conventionalized almost to look like an egg-and-dart patterning, were proper to the echinus, while the abacus carried a rectilinial fret (also called a "broken meander"). This relation of the patterns—spiky leaves on the lower part of the vessel, and the broken fret on one of the upper registers—is also about the commonest manner in which Geometric and Archaic pots were painted.[34]

Beams and roof

All that belongs to and happens above the Doric columns, together with the whole cornice or entablature they carry, is a confused and confusing representation of a wooden roof structure. The lowest member seems simple enough: it is most commonly a flat and unadorned surface, the beam lying directly on the columns, as it were. It is called *epi-stilium* in Latin, by analogy to the stylobate—from the Greek words *epi,* "on top of," "over"; and *stulos,* "a column or post." Yet the word does not seem to have been used at all in Greek texts.[35] Sometime in the fifteenth century, the Italians devised the neologism *architrave* to denote the main beam, and that is the term most commonly used nowadays.[36]

Between the epistyle and the next main member above it is a thin projection called *tainia,*[37] or "fillet," which looks as if it bound the whole structure together. In the prescription for the orders, the epistyle and the cross-beams that rested on it, as well as the filled-up spaces between those cross-beams, called *metopes,* counted as one condignation. The name of the molding therefore bears no particular relation to any form of construction, either of wood or earthenware, but since the woollen fillet called tainia was commonly used by both men and women to bind hair, it returns the terminology to the image of the body, with reference now not only to the column but to the whole arrangement.

The modern name of the high and elaborate member made up of the metopes, which I mentioned earlier, and the sawn-off beam ends that they separate is *frieze.* The word is a late coining, a corruption of the Latin *phrygiones, phrygium opus:* "an embroidered dress," or just "an embroidered hem," imitating the kind of work the women of Phrygia in Central Anatolia were considered particularly good at. It moved from Latin to Italian early on: Dante already used *fregio* to mean "hem" or "ornament," and even *fregiare* "to decorate," and its opposite *sfregiare,* "to dishonor." [38] In architecture, it appears in the sixteenth century. The antique technical term was *zōphoros:* that which carries animals, figures. And *zophoron* is what Vitruvius called it, transliterating the Greek directly into Latin.[39]

The "Basilica" at Paestum.
Photo by author.

The Doric frieze (to use the modern word) consisted of alternating members. One kind, the metope, carried figures in relief on a square ground, and was separated from the other, narrower forms. This second kind, which projected beyond the metopes out of the plane of the frieze, was called a *triglyph*. The word means "three marks" (glyphs). The triglyph slips a short tongue, called a *regula*, " rule," down through the tainia onto the epistyle and frieze.[40] To the regulae were attached some small cylindrical blocks, which were sharply outlined against the plain surface of the epistyle; there were almost always two of these for every shoulder between the glyphs. The little blocks were called in Latin *guttae*, "drops"; their Greek technical name does not seem to have survived.[41]

Vitruvius comments on the Doric cornice made up of triglyphs and metopes and the details of all their ornaments to insist on the general rule about the faithful imitation of archaic construction in stone, enlarging on the passage I quoted earlier:

Each member must be kept to its proper place, category, and arrangement. In view of this, and imitating carpenters' work, when artists build temples of stone or marble, they reproduce such [carpenters'] details by carving them in stone, and believe that the wooden arrangements should be closely followed. . . . For instance, ancient craftsmen who were building in a certain place had set beams in such a way that they projected beyond the wall to the outside, and filled the spaces between these beams with masonry; . . . then they sawed the beams off flush with the wall. As this did not look at all pleasing, they set tablets cut into such a shape as we now make triglyphs and painted them with a cerulean encaustic,

Doric capital. Wooden origin of stone construction. After A. Choisy.

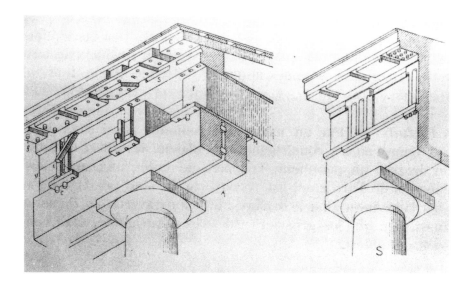

so that the rough end-grain should not offend the view. . . . This is how the ordering of the roofs into triglyphs with metopes between them originated in Doric architecture.[42]

One difficulty is evident from the first reading of the text: Vitruvius seems to be describing a building whose parallel outer walls are bridged by beams. There is no word here of any outside columns, never mind colonnade, nor of any pitch to the roof. The walls are leveled off and the beams set on them—on some kind of wall plate, perhaps. Then between the voids left for the beams, the masonry is built up to the underside of the soffit, or perhaps even higher, to form a parapet around a flat roof.

    Although Vitruvius is not concerned in this passage with first principles—he takes the arrangement of column and beam for granted—yet he wants to demonstrate the etiology of singular details to an inquirer, and he hints at the disagreement among his authorities:

Still other workmen later projected rafters at right angles to the [line of] the triglyphs, and edged them with a cyma molding [or gutter]. As the triglyphs were derived from an arrangement of the beams, so the mutules were from those rafters projecting on the underside of the top member of the cornice. That is why the mutules are, as a rule, carved at an incline to the horizontal in stone buildings [much] as the rafters would have been set, which need to slope in order to throw off the rainwater.[43]

As the triglyphs represent beams lying horizontally, so the mutules represent rafters lying at a slope; of course rafters lying over horizontal beams would be at right angles, not parallel to them—and could not, *pace* Vitruvius, appear in the same cornice.

    One way of dealing with such contradictions is to assume that Vitruvius refers to two quite different types of building. This may be due to the fact that he is drawing on two separate Greek sources that he has not troubled or perhaps has been unable to harmonize. If you look closely at the detail, you will see that the mutules do not in fact connect structurally

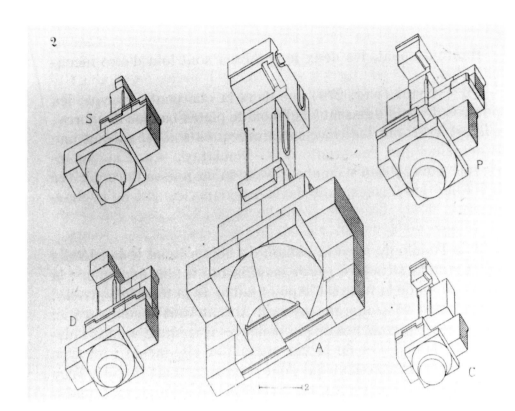

Doric capitals. Stone construction of cornice in relation to details. After A. Choisy.

with any other member of the cornice. In our post-Ruskinian (and post-Viollet-le-Duc!) days this does, in any case, seem rather eccentric. And as Vitruvius (in the tone of one who has explained all) continues, it becomes increasingly clear that his account is not altogether satisfactory, not even to himself. Indeed he goes on to dispute an opinion that he is the only one to mention in writing, though it must have been well-known and well defended by some earlier grammarian or architect: "It is impossible for the triglyphs to have represented windows as some have maintained. . . ."

Metopai

This curious aside, which seems to contradict Vitruvius' earlier explanation of triglyphs as coverings for beam ends, is an attempt to provide a genealogy for the word *metope*. Now *opaios* can stand for any kind of hole or opening, and *metopaios*, or *methopaios*, for anything that is between such holes.[44] The word definitely carries the implication of being between two voids, two holes-in-the-wall when it is used architecturally. In this particular case, the holes could only be for beam sockets, to be covered by the triglyphs. Vitruvius regards this explanation as completely unacceptable, even offensive:

> triglyphs are placed at the angles [of a building] and over the column centers, where it is impossible to put window-openings on any account whatsoever. . . . For the same reason, dentils in Ionic buildings would have to be interpreted as being set in window spaces, since the intervals between dentils, like those between triglyphs, were called metopes. For the

Greeks call *opai* the holes for beams and rafters that our people call *columbaria,* "pigeonholes."[45]

The problem seems to be caused by conflating two terminologies that describe quite different constructions: on the one hand, Vitruvius is commenting on a walls-only building, in which the beam sockets are indeed like openings in a masonry wall, like pigeonholes. This description would also be accurate if the masonry were continued above a wall plate in the same material, particularly if it were done before the beams were set in the sockets or a flat roof put on. On the other hand, if he were commenting on a timber-framed structure, then the long beams carried on columns would carry cross-beams, and inevitably therefore the spaces between the cross-beams would appear as openings.

That is just how they do appear in early representations of Doric temples, such as the one on the François Vase from Chiusi, where the beam ends, the triglyphs, are clearly shown solid like the rest of the structure, while the *metopai* (the spaces between beams) appear as voids. The François Vase was probably made about 560–550 BC, about the time of the first surviving "classic" stone Doric temples, like the Artemision of Korkyra.[46] Like some other representations in vase painting, the François Vase shows a wooden building, such as might have survived into their time. But then the actions represented on vases, and the buildings that provide their background, are not meant to be contemporary but instead represent the doings and dwellings of heroes, even of the gods in some remote or timeless epoch, though for many of the props and backgrounds the painters relied on their own experience.

Let Euripides again serve as witness to illustrate Vitruvian obscurities: in the passage of *Iphigenia in Tauris* quoted in the previous chapter, Orestes and Pylades are planning not only their escape, but also the theft of the Artemis image from the Taurian temple (it was what they had come for in the first place). As they look up the temple walls, Pylades remarks:

Look, in the cornice the gaps between the triglyphs [the *opai*]
are wide enough for us to climb out.[47]

Although Euripides' tragedies also told of mythical events, their date in the days of the heroes can be determined. In the case of the Iphigenia tragedies, it was the generation after the Trojan War; the buildings Euripides described are therefore already ancient, half a millennium away.

To sum up: Vitruvius was talking both about a kind of building in which the beam ends corresponded to voids and about an older walled architecture in which columns played no role at all. In such an astylar architecture the voids would be visible as *opai* before the beams were laid, and the bits of wall between them could quite correctly be called *metopai*. Would Vitruvius have considered astylar buildings as temples, in any case? Is he even concerned to elucidate the passage in such terms? I think not. But then again he seems also to be accounting for a timber or half-timber architecture in which the beam ends are seen as solid and are separated by voids as part of a framed construction that could be set either on posts or on walls.

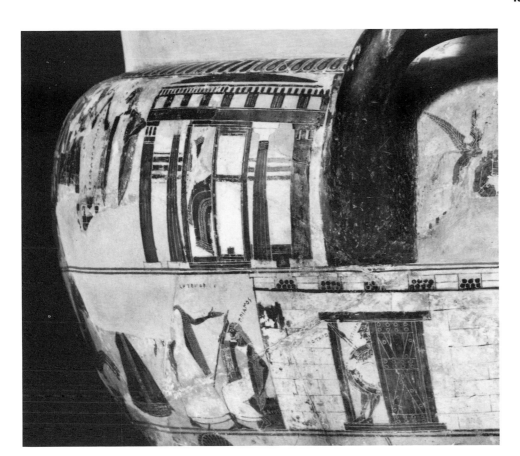

Doric temple. Detail of François Vase. Florence, Museo Archeologico. © Fratelli Alinari.

Triglyphs

These beam ends, the triglyphs, show the problem of representation in architecture up very sharply. The word *triglyph* refers to the grooves or incisions, the glyphs. There are two full cuts in the surface of the panel or tablet that is chamfered at the edge; and the two chamfers, considered as half-glyphs, add up to a third one—or so sixteenth-century commentators thought. In fact, the number of glyphs can vary, but three is preferred for the rule, since three is the number of "wholeness." The raised bits between them are *femur* in Latin, *meros* in Greek—"a bone," "a thigh," or "a block."[48] The one near-unbroken rule about the relation of column and frieze is that one triglyph should fall over the middle of the column. In the stone version the droplets or guttae that project down below the triglyphs appear like wooden pegs wedging the tablet in place through the tainia, which makes it a representation of a wall plate. That at any rate is consistent both with a possible wooden construction and with Vitruvius' text. As for the presence of a tablet at this point in the timber building, there is really no coherent explanation; the one provided by Vitruvius is just not articulated enough. There is no doubt from the text, however, that the sawn-off beams cosmetically represented by the tablet were composite beams—that is, *trabes,* beams made up of several planks—and it is a firm rule that the planks in a composite beam had to be separated by interspaces. A four-planked composite beam would be required to produce three such *laxationes,* "interspaces," with corresponding glyphs. However, if the beam ends were all chamfered, only three planks would have been needed for a prototype of the common arrangement.[49]

Archaic temple model. From the old Parthenon. Athens, Acropolis Museum. Photo by author.

Terracotta cornice detail: A treasury at Selinunte. Agrigento Museum. Photo by author.

When these unsightly beam ends were first masked by the tablets (which also held the decorated metope panels in place) can only be guessed. The stone representation of wooden construction was a second-level representation: the wooden triglyph tablet, slotted into the wall plate and counterwedged by wooden pegs, already stood for the chamfered ends of a composite beam. This explanation has not satisfied many authorities; some have suggested that they were another kind of column, one that derived more or less directly from the Minoan-Mycenaean Kyanos or triglyph frieze, or that they (or the metopes) were the windows of an attic story.[50]

The top of the triglyph was quite often thinly projected out, as a band that might even sometimes continue over the metopes. A thin flat molding was laid immediately above the frieze on which triglyph and metope alternated. As the triglyphs fixed the metopes into place, so they were in their turn fixed between this flat molding and the tainia below;[51] occasionally (though only rarely) the molding between the frieze and the cornice is made more elaborate. The cornice above projects forward very sharply and is inclined down, to make an acute angle with the next flat molding above it and provide a drip. The underside of the soffit in this projection belies the simple drip function. It is usually carved with panels in relief, one per triglyph and one per metope, called *mutuli* in Latin, usually "modillions" in English (from an Italian corruption); these are in turn studded with small cylinders, usually three rows of as many pegs as there are guttae under the triglyph, which are called by the same name.[52] If they do represent rafters, as Vitruvius maintained, the contradiction mentioned earlier is compounded, since the pegs presumably served to fasten thin purlins (or even boards) to the sloping rafters (they would not have been needed over the flat beams).

Higher yet, the flat band of the cornice, sometimes separated from the moldings above and below by thin fillets, is called *geison* (though not by Vitruvius), a "band" or "coping"; or in Latin *corona,* a "wreath" or "crown."[53] Like tainia, geison and corona are terms of the (scantily) clothed body. Both words for the flat molding imply the binding of the head: the wreath and the fillet. The word *corona* turns into "cornice" through yet another Italian corruption, but its implication is clear: it is the crown of the arrangement.

In most cases there is yet another topmost molding, the *cyma*: it is sometimes counted as one element with the geison.[54] *Cyma, cimatium,* and *sima* all derive from the Greek *kuma,* anything billowing or swollen, commonly a "wave." This particular swelling, which presumably returned to the terminology of terracotta construction, acted as a gutter and therefore accompanied a sloping roof. Often it was pierced with discharging lion's-head gargoyles, with palmettes or honeysuckle trailing between them.[55] Not properly part of the column-and-beam arrangement, it completed the building. The roofline was decorated with acroteria at the corners and at the ridge. The cornice should have no geison or cyma on the narrow side of the building if there was to be a pediment over it.[56]

Such were the column and beam of the pedimented Doric temple. The relation between it and the earliest hipped temples at Isthmia and Corinth has not yet been satisfactorily explored. Moreover, even if this whole arrangement was given general currency in the course of the seventh century (as is now generally thought most probable), the exact origin and implication of each molding seem already to have been a matter of speculation at that time, though their names recalled and celebrated time-honored building practice, and some of the techniques to which they referred were still in use.

Model from a temple at Argos. Athens, Archaeological Museum.

Model from the Heraion at Samos. Samos, Vathy Museum.

The separate members of the column and beam have so far been discussed independently. They have been identified and labeled—but their original relations and interconnections can only be appreciated within the context of the scarce evidence that survives in models, in some cultural usages, and in fragments on the various archaeological sites.

Models

Of the relevant surviving models, most are of more or less rectangular buildings with sharply pitched roofs, the pitch much steeper than that of any classical pediment. The two best-known examples come from Argos and from Perachora. They have open porches and open gables and are also quite elaborately painted. Both have small window holes high up in their walls.

The Perachora model has a hairpin plan, an overhanging roof (presumably of thatch), and its walls seem of rendered mud brick; the porch is supported on twinned round posts. The Argive model is square-ended and its porch is framed with squared timbers. The roof may have been of clay on reeds; at any rate, it was painted, like the walls. The porch has a separate flat roof that carries on into the interior as a kind of wall plate on which the roof seems to rest; the walls are ribbed, indicating a half-timbered construction.

Other limestone models—presumably votive—of both flat-roof and thatched, pitched-roof houses, though very damaged, were found at Samos, in the precinct of the Hera temple. The restoration of the oval-ended houses whose foundations were found in Old Smyrna, with steep thatched roofs hipped over the apses and a gable vent between the curve of the hip and ridge, has followed these models.[57]

A flat-roofed house model of terracotta was found at Chania Tekke, near Heraklion in Crete. The walls are ruled into squares, perhaps indicating regular masonry. There are small square side windows, while the heavy doorframe has a fanlight (or smoke hole?) over and molded into it. The flat roof overhangs the walls, and there is a prominent chimney. This may be the type of flat-roofed building Vitruvius had in mind when he wrote of the ancient Doric cornice. The leaf of the door is ornamented with the concentric circles familiar from late Geometric pottery.[58]

While in Italy many elaborate circular house models (and some rectangular ones) were made as cinerary urns, there are virtually none of that kind in Greece; on the other hand, a number have been found on Crete, one of which—probably late ninth century—is of painted terracotta and better preserved than the others. Its conical roof rises to a chimney or smoke hole; two figures on the roof (accompanied by a dog?) seem to be looking down it. In the side of the miniature building is a niche that rises the full height of the wall, and contains an *orans* female figure, protected by a removable door, which was fastened by a crossbar. The type, it has been suggested, is an oriental importation and most could have been used for domestic cults: that they were used as shrines from which the barred door was removed to worship the goddess would, at any rate, explain why each one of them, whatever their size, had that feature;[59] it has also been suggested that it may represent the discovery of a buried tholos tomb.

The Cretan models seem to have had no direct successors, nor are they obviously related to any surviving buildings. Indeed the development of the building (as of the column type) is not played out in the circular but in rectangular buildings of the Dark and Geometric Ages, many of which are known only from legend.[60] Of these the legends which surround

Model from Chania Tekke.
Iraklion, Archaeological Museum.

Model of a round building from Archanes. Iraklion, Archaeological Museum.

The Hero as a Column

the building of the Delphic shrine are, in their enigmatic way, perhaps the most intriguing and evocative.

Delphi: The legend

When he visited Apollo's shrine, Pausanias the traveler was told a number of legends. According to one of his informants there, the shrine was first "in the form of a shanty" and was built from laurel or bay brought from Tempe, the Thessalian valley where the main festival of Apollo Daphnephoros, the bearer of laurel, was celebrated. It was presumably constructed of thin sticks and withies; sweet bay will not grow into a tree and is not much use for building.[61] Sweet bay or laurel, *daphnē,* figures much in the worship of Apollo. The best-known legend about him tells of Daphne, a girl he amorously pursues, turning into a sweet bay tree to escape his embrace.[63] And at Delphi, if Pausanias is to be believed, a regular burning of a laurel hut took place. A young man would be confined in it, but saved to lead a Daphnephoric procession with his contemporaries to Tempe (the valley of the river Peneios, Daphne's father).

That ritual commemorated the first shrine. Two alternative legends give details of the second shrine. Some held that it was made by bees of wax and feathers and sent away to the Hyperboreans by Apollo; according to this version, Olen (a Hyperborean singer) was the founder of the oracle (the Hyperboreans have always been attributed various supernatural powers, as in the legends about Pythagoras).[63] According to the other version told to Pausanias, the shrine was built "by a man called Pteras," who is not otherwise known to mythology but whose name means "wing-y." *Pteron* means a "flap" or "fin" or "wing"—but also a "rudder" or "oar" or "sail." It also meant "shelter" or an "awning"; most importantly here, its derivative *pteroma* was the technical term for the colonnade round the temple—and that is how Vitruvius uses it. *Pteris,* however, also means "fern," the feathery plant; and in yet another account of the same Delphic legend, Pausanias dismisses the suggestion by one of his informants that ferns were the original material of the shrine. If that second shrine was anything like the small building that the archaeologists restored at Eretria, fern stalks (or perhaps fern grass?) could have been twisted into a lashing rope in its construction, or platted into mats for the walls. The fern-feather-wing-colonnade conjunction is another instance of the Greek partiality for wordplay.[64]

The third shrine was to have been of bronze, and Pausanias reports that Hephaestus was reputedly its builder or smith; he does not accept the story fully, but sees no reason to disbelieve the existence of the bronze building and indeed cites parallels.[65] It may well be that some parts at least of these buildings—doors, roof elements—were in fact of bronze, whether hammered plates or lost-wax pieces. Most of the ambitious early Greek bronzes were made by the lost-wax (cire-perdue) technique. The association of bees and wax with a bronze building is therefore plausible even if the wax model was inevitably destroyed in the process of casting.

Yet as a building material bronze was not common; presumably here bronze sheet was used over a wooden core, as in so many surviving door leaves (the ones at the entrance of the Roman Pantheon still retain their original wooden core as well as their bronze casing, as do the somewhat later ones at St. Sophia in Constantinople).

That kind of construction, much used in Mesopotamia for things like city gates, was an obvious intermediate stage between a terracotta-and-wood finish and the hard permanence of stone; and, of course, bronze could be used to ornament a wooden structure, much as terracotta was. It would presumably have had a wood core overlaid with beaten bronze sheet—though the ornament, such as the acroteria, might have been cast separately. Pausanias, however, dismisses as implausible Pindar's praise of the "six charmers all of gold" who sang over its gable; presumably they were merely gilt (over bronze?) acroteria figures.[66] Still, since bronze building elements were usually melted down and reused, and thus were expunged from any future archaeological record, reconstruction is notoriously difficult.[67]

Even if he would not accept that there were gold acroteria on the temple, Pausanias had no difficulty with the bronze part of the story: and indeed he cites quite a number of bronze buildings, both legendary and actual. The bronze Delphic temple had by then inevitably vanished, and if Pausanias did not allow that it might have been deliberately melted down, he was equally uncertain whether it had been swallowed by the earth or melted in a fire. At any rate, it had been replaced by a stone temple designed by two named and celebrated hero-architects, Trophonios and Agamedes. That first stone temple was also burned down in the first year of the 58th Olympiad (544 BC). The great and famous second stone temple, which Pausanias in fact visited and the archaeologists have since reconstructed, was the Doric building designed (so he was told) by one Spintharos of Corinth, of whom nothing else is known.[68]

Eretria Daphnephoron: Plan of site. After P. Auberson (1968).

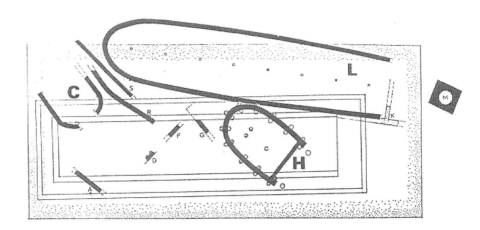

Reconstructed model of hut: Eretria Daphnephoron. After P. Auberson (1968).

Eretria Daphnephoron: Excavation. Photo by author.

Eretria and Lefkandi

Apollo Daphnephoros was also the god worshipped in the ancient (perhaps eighth-century) hairpin building, which was replaced in the sixth century by a squared temple, in the city of Eretria on Eubea, not far from Thebes. That hairpin building, a hecatompedon, had an antaed porch and a central row of columns, like most Geometric and Archaic period temples whose plans have survived.

Hard by it, at a lower level and set at an awkward angle, there was a much smaller hut, proportioned almost exactly like the Perachora model: a short, squarish hairpin, almost a horseshoe. The wall of the temple building proper has a sinking to take up the edge of this tiny hut, perhaps to show its priority on the site. Only at Eretria the structure of the smaller building—which seems to have been light and of timber lashed together with rope—was probably set on either side of a dwarf edging wall, perhaps of rush mat, or even of pisé.[69] This unique arrangement suggests that it may indeed have been designed like the models, with three posts around a hearth (perhaps a sacrificial hearth on the Mycenaean precedent), while the smoke issued through a pedimental hole, or perhaps a triangular hole between the top of the porch and pitch of the main roof. The space that was occupied by the revelation of the god in the classical temple was simply left open.

Another discovery made a few miles away in 1980 has considerably modified scholars' assumptions about the Dark Age. A substantial building, now sometimes called a *heroon*, was found near the Eubean village of Lefkandi. That place does not seem to have earned a

Lefkandi "Heroon." Excavation plan, after M. Popham (1993).

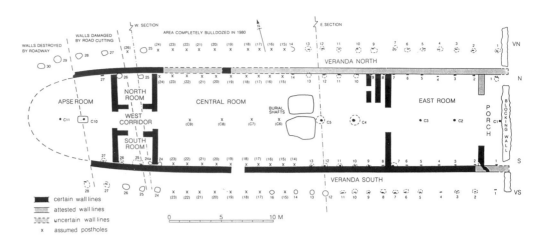

Lefkandi "Heroon." Reconstructed axonometric, after M. Popham (1993).

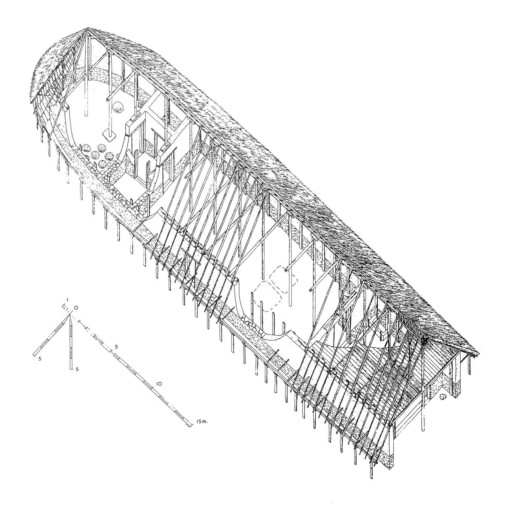

Thermon: Excavations of buildings A and B. After G. Soteriadis (1901).

Thermon: Excavation plan. After G. Soteriadis (1901).

The Hero as a Column

**Restoration of temple at
Dreros.** After I. Beyer (1976).

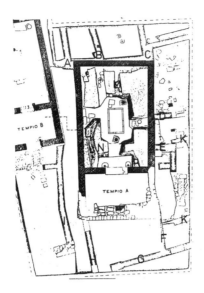

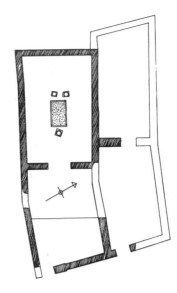

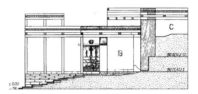

**Restoration of temple at
Dreros.** After I. Beyer (1976).

mention in any ancient text, nor yet been given a name by modern scholars. This heroon is an oriented hall, some fifty meters long, subdivided into more compartments than was usual at the time it was built, which was probably about or just after 1000 BC. Its thatched roof rested on a center line of round posts, on walls of mud brick on stone footing, and there was a double line of much smaller, rectangular posts—one range inside the walls and around them, the other at a distance of about a meter—forming what the excavators call a "veranda" around the building. All the posts are in special pits, and secured with stones—unusually carefully for the period. The eastern porch opened into a squarish front room (perhaps open to the porch), then followed the main chamber; the apse, of which little remains, was closed off by two smaller rooms, presumably used for storage.

The main chamber contained the oddest feature, two undisturbed burial shafts on either side of the center colonnade. In one were the skeletons of four horses. In the other were buried the cremated remains of a man, wrapped in a purple cloth and contained in a (probably Cypriot) bronze urn, which must have been more than a century old when the burial took place. Adjoining in the shaft were the inhumated remains of a considerably younger woman, whose chest was covered with a gold pectoral, the largest piece of gold dating back to the Dark Age of Greece; there was other jewelry, a sword in a scabbard, and other objects. This "Homeric" burial seems to have been that of some important figure or potentate: of the many tombs found in that part of Eubea (mostly of the Geometric period), it is by far the most impressive.[70]

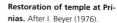
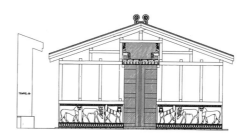

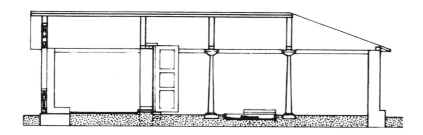

The Lefkandi building seems to have stood above ground for about forty years, after which it was deliberately razed. The roof fell inside the walls but the burials were not touched, which suggests that it was not destroyed by hostile, looting enemies. The ruin was covered by a tumulus. It is none too clear whether this was a cult building in which burials were made, or whether a cult celebrated the burials. However, what is material to my purposes is the two-stage construction: first the walls of rubble and sun-dried brick, then the stockade, which seems to have been almost an elementary peripteron. This has suggested to some that the pitched roof of some later Dark Age or Geometric building was added to a flat-roofed, perhaps Mycenaean-type structure. At any rate, one plausible sequence of events has the walls built before the cremation of the "warrior" took place, and the roof built after the burials. A mile or so away there is a Bronze Age settlement known as Xeropolis, not sufficiently excavated at the time of writing, which may hold some key to the significance of the Lefkandi building. It was certainly occupied both during the Iron and the Bronze Age.[71]

Eretria and nearby Chalkis played their part in Greek legend: it is from there that the Greeks sailed to conquer Troy, and even if not much is known about the early history of the island, it is nevertheless hardly surprising that unchronicled remains there have proved important.

Dreros and Prinias

Crete, on the other hand, which had been the site of the greatest European buildings in the late Stone and Bronze Ages, contributed little to the mainstream during the great half-millennium of Greek history. Nothing of much consequence was built there during the classical period, yet in the Dark Age and the Geometric-Archaic period that followed it, Crete had a period of brilliant legal and constitutional innovation. Two small temples of that early time,

**Thermon, Temple of Apollo:
Temple site.** Photo, after G. Soteriadis (1901).

one in northern central Crete, the other just north of the Idean cave, have suggested to some archaeologists that they are in fact dealing with a meeting of conflicting tradition on the by then impoverished island.

The main shrine at Dreros (near the bay of St. Nicolas) overlooks a substantial agora of some thirty by forty meters, from which it is reached by a stairway. The evidence shows that it was dedicated to Apollo. It was so very well built that its dry masonry walls still rise over two meters. Inside there is a rectangular stone eschara to one side and two column bases at either end that probably supported a flat roof. A small limestone gorgoneion is its only piece of architectural sculpture; but three bronze statues found within have been interpreted as Leto and her two children, Apollo and Artemis.[72] The god is bare headed, naked, and striding; the two women static, clothed, and hatted. The statues were made from hammered and riveted plates, and they were set on a bench inside the shrine room. Presumably they were the cult statues. They may well be later than the building proper, perhaps placed there after the building had already been in use for some time. The original excavator, Spiridon Mari-natos, inspired by the Perachora model, has suggested that the porch had a flat roof, while the temple chamber had a steeply pitched one, which allowed a large triangular smoke hole between porch and shrine. Recent restorations show it looking more like the Chania-Tekke model, with a flat roof over the whole building, and perhaps a Mycenaean-type clerestory vent or chimney over the eschara.[73]

The temple at Prinias (located about the middle of the island, halfway between the two great palaces of Phaestos and Knossos) was probably built about fifty years after Dreros, and the ruins offer a jumble of stone and terracotta architectural sculptures. It has been restored by its excavators as a flat-roofed stone chamber with heavy square stone pillars forming a porch; the animal reliefs were arranged as a frieze. More recently the restoration has been revised and the building shown as a half-timbered construction, gabled or pedimented at the front, hipped at the back, the double-pitched roof a kind of replicated queen-post frame based on Cypriot parallels. Instead of being set up on the frieze, the relief plaques have been restored as orthostats edging the mud-brick walls (which link the building to the orientalizing tenden-

Thermon: Cornice reconstruction, projection. After G. Soteriadis (1901).

The four surviving metope tablets from Thermon, restored from fragments: gorgoneion; hunter carrying quarry; chess(?) game between two men; Perseus with the head of **Medusa.** Athens, Archaeological Museum.

The Hero as a Column

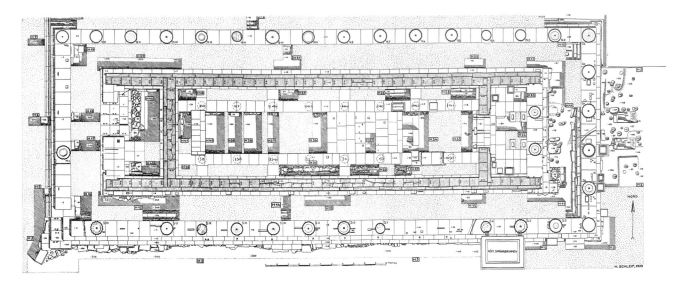

Plan: Heraion at Olympia. Drawing by H. Schleif, after W. Dörpfeld (1935).

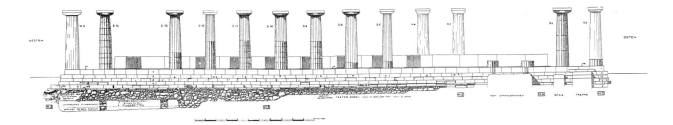

Elevation: Heraion at Olympia. Drawing by H. Schleif, after W. Dörpfeld (1935).

cies in contemporary pottery). Within there is a stone *eschara*, as at Dreros, and the columns may have been two on either side or else set around it in a triangle, like the posts in the Eretrian shrine.[74]

Thermon

However, the remains of one entire Doric peripteral temple that had not been reconstructed in stone do survive, at Thermon on Lake Trikhonis, to the northwest of Delphi. Not only are the traces of a wooden colonnade a prime piece of supporting evidence for the Vitruvian account, but its relatively light construction has not disturbed the older buildings at deeper levels. Thermon seems to offer a particular documentation of the development of the Doric arrangement.[75] The site was probably settled in Mycenaean times and already had some kind of sanctuary then, since a small group of much older buildings surround the foundations of a seventh-century temple. The longest (known as building A) is a columnless hairpin hall, with its usual porch at the entrance, or so it would seem. Other buildings of uncertain nature were ranged around it at the same level, presumably constructed at the same time. Later, but still in the Bronze Age, another and more ambitious building, B, was set up almost exactly parallel to building A. It was wider, also had a columnless porch, but was squared, with a separate chamber (a "treasury"?) opposite the porch. The corners and ends of the walls may well have been protected with wooden antae.

Perhaps a century or more after it was built, this flat-ended (and presumably flat-roofed) later building was surrounded by a line of posts that reproduced or echoed the hairpin plan of the original building A (which may by then have disappeared). These posts have been variously interpreted as a stockade marking out a small temenos, or as the supports of a new and steeply pitched (and therefore presumably thatched?) roof. It has even been improbably suggested that they were oblique struts that buttressed the thrust of the roof, as in some northern European thatched buildings. Perhaps the most likely interpretation is that an apsidal building, which has been called B1, using substantial wooden posts, replaced the rectangular building B and in its turn provided the base for C. And indeed the roof of building B, like that of most apsidal buildings, would have been pitched. As for its use, large deposits of carbonized organic material have been found at the southern ends of both buildings, in a stone-edged well nearby, and in pithoi inside and outside the building, suggesting years of burned sacrifices.

In any case, this new colonnaded megaron was razed in the seventh century to make place for that much larger and newer Doric temple. This was a wooden structure on a stone stylobate with elaborate terracotta revetments: it was never transformed into a stone temple. And yet it was a rich and important shrine, which acted as the center of the Aetolian league until it was destroyed in a savage attack by Philip V, the last independent king of Macedon.[76] What survives from the destruction wrought by Philip's troops is the stylobate itself and on it the footings of some of the columns, as well as fragments of four terracotta metopes, terracotta shards from the cornice. As in many Archaic shrines, the ridge of the roof was supported by a central row of columns, which were presumably very like those of the peripteron, the exterior colonnade; the two series of columns seem to have been of the same diameter. That, at any rate, was the arrangement followed in the much later archaizing temples (such as the "Basilica" at Paestum) where the row of central columns carried a dwarf wall to support the ridge. Because nothing remains of either row of columns, all this must be left to surmise.[77] The succession of buildings on the site and the similarities of location, dimension, and shape suggest continuity between Mycenaean and Hellenic occupation.

A similar building, of which fewer traces remain, was built about the same time a little further south, at Kalydon. The few surviving terracotta fragments are so like the ones from Thermon that it has even been suggested that they came from the same Corinthian workshop, even if the detail of the fixing of the metopes to the main structure is different in each: by projecting lugs molded as part of the panels at Thermon, by countersinkings for swallow tail pegs at Kalydon.[78]

Although Thermon was rich and politically important as the capital of the Aetolian league, it was not central to the life of the Greek nation. The survival of its wooden temple into Hellenistic times can be regarded as a provincial archaism. Much more important, though much more enigmatic, was the temple of Hera at Olympia, the sanctuary of the great Panhellenic games and one of the universally revered shrines of the Greek nation.

Olympia: The temple of Hera

The Olympic Heraion was the most famous temple of mixed construction—using stone, terracotta, and timber—still in existence at the time of Vitruvius. Two centuries after him, when it

was described by Pausanias the traveler, it was still one of the great sanctuaries of the Greek-speaking world, more venerable than the much bigger temple of Zeus nearby that housed one of the seven wonders of the world, Pheidias' statue of the god.[79]

Pausanias visited the temple in AD 176. And he found, in the back porch (or *opistho-domos*), one oak column still standing among the stone ones. This report had puzzled earlier commentators, but was confirmed and explained by the excavations of the nineteenth century. The first surprise came when the columns of the temple were reassembled, since the capitals and shafts, diameters and profiles were all entirely different, even in their construction; most columns were made up of drums, but three were monolithic. Altogether eighteen have survived; there is no trace of that single wooden one. Of course, it could have been replaced in stone after Pausanias' time, since each wooden column of the original seventh-century temple was substituted for a stone one as an occasional votive gift of some potentate or city, and the temple was in use until the fourth century AD.

This process of replacing seems to have started soon after the temple was built. In some of the column shafts there are sinkings that cut across several flutes, each a recess for

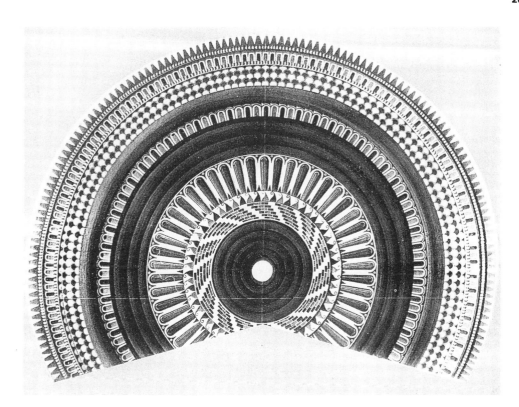

housing a dedicatory tablet.[80] The earliest stone column is about contemporary with the foundations. Of the cornice nothing remains, and it has been assumed that it was of wood, though what is left of the acroteria and other revetments shows that it was decked out in terracotta like those of other early shrines; the most splendid relic of it is the great dish-acroterion, which was mounted on the ridge, facing the altar. At any rate, the temple became a repository of the Greek nation's (and its allies') devotion to the Olympian site, as were the rich treasuries that overlooked it. The altar, around which a rite that was a crucial part of the Olympic games was acted out, stood in front of the temple. It is still the place where the torch is lit for the modern games.

The Heraion whose remains we now see was probably built about 600 BC; like many other famous temples, it stood on the site of two successive older shrines and incorporated parts of all of them. About these older shrines we know little, though Pausanias' informant told him that the first one had been built by Oxylos, the mythical Bronze Age king over Elis, who was said to have turned up in the Peloponnese with the sons of Herakles: this would put its foundation to somewhere between 1100 and 1090 BC.[81] On the other hand Herakles is credited with the instituting of the games themselves,[82] which would put their beginning earlier still, even on Pausanias' chronology. As I noted in chapter 6, the games claimed a variety of mythical ancestors: but the historic ones, which became the datum of Greek chronology, were first recorded in 776 BC. At the time the oldest shrine of Hera and the great open-air altar of Zeus must have seemed venerable enough to act as the consecrating building for the greatest Panhellenic event.

The seventh-century temple is in fact the oldest surviving one to have a double internal colonnade instead of the archaic single one. It is the oldest, too, of whose cult figure (a limestone statue of Hera, about twice life-size) any relics remain.[83] Its material fabric was still more like that of the temple at Thermon than of any great classical shrine. The base and walls were of stone with some brick, but they were also half-timbered, while the peripteral columnar structure was wooden, and the decorations were all of terracotta.[81]

Olympia: The House of Oenomaos

Pausanias wrote about yet another, even more venerable wooden column at Olympia, whose exact location and shape have not been recovered. From his description we may infer, however, that it stood about halfway between the "great" altar (in front of the temple of Hera), and the "new" temple of Zeus. The Eleans, the local people, called it the pillar of Oenomaos. This Oenomaos was a hero-king of Pisa in Elis (very near Olympia) and was the son of Ares, the god of war: according to one version of his legend, he was also the son-in-law of the giant, Atlas. He determined to give his daughter Hippodamia as a prize to any suitor who could beat him in a chariot race: but if they lost, they died. Oenomaos won every time and killed them all. Their skulls were exhibited on the roof of his palace, much as the Taurians exhibited those of the sacrificed Greeks. Pelops, the hero-son of Tantalos, whom the gods revived after he had been slaughtered by his father and served up to them in a stew, won the race by a low trick. Oenomaos' horses dragged him to his death, Pelops married Hippodamia, and he gave his name to the Peloponnese; however, this being a Greek myth, no one lived happily ever after.[85]

The House of Oenomaos, which must presumably be thought of as a megaron complex of the Mycenaean type, was destroyed by Zeus, and the one column that Pausanias still

"Personalized" bronze votive tablet. Athens, National Museum. Photo by author.

saw at Olympia was all that remained of it, bound in iron hoops and covered by a four-poster shelter to keep it from falling apart. Pausanias adds:

> There is a small bronze plaque on it which carries this elegiac inscription:
> I am all that is left of a famous house, stranger;
> I have been a pillar in the house of Oenomaos,
> And now by Kronian Zeus I lie in bands.
> Now I am venerable: dreadful fire did not devour me. . . . [86]

This tablet may have been seen as analogous to the *vota* on the columns of the Heraion. At any rate, Pausanias gave his conventional idea of the age of this column, though the inscription implies that it was about the oldest thing on the site. It would in any case have predated Hera's first shrine (and even her cult), and may in fact really have been a fragment of a Mycenaean building. Whenever it was dedicated, it must already have been identified as a relic.

Such antiquities of timber building were not common, but they did exist elsewhere. Near Mantinea, there was a wooden shrine of Poseidon the Horse around which the Emperor Hadrian had a stone outer casing built. Pausanias records that he did not see inside: it was only accessible to the initiates. He had been told that the original building had been put up by Trophonios and Agamedes, the legendary builders of the first stone temple at Delphi. There were a number of such "primitive" buildings of mixed construction still in use at the end of paganism. However, some were deliberately put up in an archaizing way, while others were maintained unaltered and rebuilt regularly. Ritual thus kept the memory of archaic construction alive and therefore in the popular imagination.[87]

Delos: The Naxian *Oikos*

The progression from small, half-timbered temples with a central row of columns to the standard stony "Hellenic" plan of nave and two aisles is still too orderly. Inevitably there is an example which reverses the progression: the *Oikos* of the Naxians at Delos shows the opposite transformation. Of the original building on the site (which some consider the primary temple of Apollo on the island), the external granite walls enclosed a double wooden colonnade built about 650 BC or soon after, about the time of the wooden Doric temple at Thermon and perhaps a century later than the transformation of the Tiryns megaron. The building, which seems never to have had a pteroma, was given a tetrastyle porch. What appears to have been a marble libation basin and two rows of eight postholes in the rock are the relics of the internal plan. The colossal kouros, some twenty-five feet high—one of the very first (as well as the largest) Greek monumental stone sculptures—was set up next to, indeed almost touching, the north side of the building soon after its construction. About a century later, perhaps about the time of the "purifying" of Delos by Peisistratos in 540, these internal posts were pulled down and a new stone pavement placed over the beaten earth floor of the original building, whose outer walls were probably left intact. Additionally, the double colonnade was replaced with a central row of Naxian marble Ionic columns to support a new pitched roof.[88] It is not at all clear why, at the same time as the Athenian ascendancy over Delos replaced that of Naxos, the Naxians chose to make their "advanced" seventh-century building into an "archaic" one. It may well be that at Naxos the modification of internal plan marks a change of use, from temple into treasury. The main Naxian temple of Apollo was built about that time a few feet away.

Corinth

Of course the inventor—if that is quite the word—of the central-nave, peripteral Doric temple, whose earliest surviving example seems still the Olympian Heraion, has never been named. Pindar, who might be considered the hero-founder of professional flattery, praised Corinth as the home of peace and justice and added that the city was also the home of the Greek *technical* achievements: Corinthians created the Dionysian dithyramb and the horse-bit, and "on the gods' houses [they] planted the double royal bird." [89] The "double royal bird" has been taken by many commentators to refer to the twin pediments of the Doric temple. The royal bird is almost always the eagle, *aetos;* and it is also the word for pediment, because the pediment has the shape of a bird with outstretched wings. [90] The pediment was further taken to stand for the whole temple by metonymy, so that Pindar's elliptic verse has been read as an assertion that a (if not *the*) Doric "order" was devised in Corinth in the middle of the sixth century, either as the culmination of an experimental period or as a new creation. [91] It is an attractive suggestion, since seven columns of the temple of Apollo still stand picturesquely in the city, and they display familiar archaic characteristics: they have no entasis, but the end intercolumniations are narrowed, the end columns slightly thinner, the stylobate curved for optical correction. It is a long temple, since it has the usual cella for the cult image as well as a rear chamber (which has been interpreted as another shrine or as a treasury). The temple was probably built just after 550, some decades after the Olympian Heraion.

There was until the sixteenth century another, somewhat later, Doric temple on the site, which was destroyed by the Venetians. In fact even the "regular" temple of Apollo was not the first shrine: there had been an early-seventh-century temple to house an image, which as fragments of the roof indicate was a hipped building, like the nearby Isthmian temple of Poseidon with brightly painted terracotta decorations. It may have been destroyed in the passage of Corinth from tyranny to oligarchy at the end of the seventh century, after which the surviving Apollo temple was begun.

Pliny would have us believe that the whole apparatus of terracotta revetments, the edge cyma with its gargoyle masks and relief sculpture in clay (as well as the *fastigium*), were all invented in Corinth by Butades (or Debutades) of Sikyon. He tells the charming legend of Butades' daughter drawing the outline of the shadow thrown by the head of her departing lover, and his inventing portraiture by making a clay relief on it. [92] It may equally well be that the perfecting of the roof tile was the work of Corinthian potters. Its general employment from the early sixth century onward led to a lowering of the roof line and therefore—whatever Pliny meant—to the formation of the figured triangular pediment as an essential element of the temple.

The awkward Doric

The link of the mature Doric constructions to some of the examples I mentioned earlier, like the little hut-shrine at Eretria, or the Perachora, Samos, and Acropolis models, remains puzzling. Most of them had porches with corner posts or columns, over which beams had to cross; anyway they implied buildings whose roofs were partly double-pitched and partly flat, perhaps the result of the transmission of building techniques that may well have been carried through from an earlier age.

Whatever the nature of the accounts—legendary, pseudo-historic, topographical, archaeological—the transition from impermanent wood and clay to the seemingly indestructible stone, sometimes mediated by terracotta, bronze, or both, is common to all of them. And yet there is something scandalously violent in a metaphor which demanded that a structural form devised for whittling and fitting together with pegged joints and wedges, like any rough carpentry, be transformed into sculpture in the hardest of crystalline building stones, and the beautiful and miraculously worked stone covered with a stained stucco slurry or painted so as to imitate the wooden original more closely.

That violence is not altogether camouflaged by Vitruvius' bland manner. Vitruvius himself was uneasy with the Doric arrangement, although he gives such an extended and detailed account of it. In the end he advises (or at least cautions) against using it at all. And indeed, the Romans hardly ever used it, while Hellenistic architects, who made it very slender and reduced the cornice to a fraction of its original depth with respect to the columns, were not altogether happy with it either. "Several architects of antiquity," says Vitruvius, did not think it fitting to build temples in the Doric order, because of its "faulty and discordant [*mendosae et disconvenientes*] proportions." [93]

The reason why these fifth- and fourth century architects rejected the Doric order will only become apparent when the way it grew is contrasted with the fully conventionalized canon, in which the glaring historical contradictions that mark Vitruvius' account are exposed. What follows from them, however, are all those awkwardnesses that Vitruvius mentions: the impossibility of taking the metope-and-triglyph system around a corner, of spacing column and frieze with some regularity. All those discords between architectural canons and the demands of optics as well as of construction (rather like the discords of music) could be sweetened only by the distance of a legendary past.

VIII

The Known and the Seen

• The temple and its users • Tholoi • Type and project • Six hundred varieties • Tympanum or pediment • The corner triglyph—again • Optical refinements • Theory and practice • Surface

The genuine discrepancies that so embarrassed Vitruvius in his account of the Doric order may well have been an irritant as essential to its vitality as the comma, the lemma, was to the challenge presented by the Greek musical scale; and these discrepancies were, in any case, the inevitable consequence of the long and haphazard development that I have described. They resulted in a double bind, which harnessed architecture to a dialectic between the timeless canon of human proportion on the one hand and the historical narration of the details on the other.

This dialectic seems to have been completely internalized by the end of the Archaic period. The genera of columns provided the dominant note of the man-made environment, and the terminology in which they were articulated or discussed had meshed with daily speech. Were you to ask a Greek man of the sixth or fifth century what a temple should be like, he could probably give you a description of it that might seem almost like an outline specification. He would start by having it set in its temenos. He would almost certainly say that it should be a rectangle: irregular or circular plans were not the rule for temples, though they were also important.[1] He would then take you through the entry in the enclosure wall, a porch, itself usually columnar, the propylaea. Through it you would see a stone altar: a construction large enough to accommodate at least one whole carcass of a cow or bull, though some were large enough to accommodate several. The altar usually stood in view of the central door of the temple building, though not on its centerline, or even square with it. What my hypothetical Greek guide would certainly describe most carefully, what would be most striking to him about the temple building (as indeed it still is to us), is the forest of columns that supported its cornices and roof. Although a visitor to any Greek town might have caught sight of it from various points, or even seen it rising above the landscape from a distance, the first full sight of the temple on entering the temenos would be an oblique view (the propylaea were almost never axial) of the narrow side facing the altar. My fictitious ancient Greek would almost certainly expect the pediment that crowned the front to contain a sculptured epiphany of the god to whom the shrine was dedicated, though, as I pointed out earlier, the pediment over the back of the building might be plain. The triangle, of course, implied a double-pitched roof, whose ridge ran between the apices of the two pediments.

There would be an even number of columns on the two shorter fronts, and this again could be taken for granted by my ancient interlocutor—even if there were a few notable archaizing exceptions, of which more later. The even number of the columns allowed the middle intercolumniation to correspond to a central door, and the cult statue would then be set against the back wall of the inner temple to face the entering worshipper.

The cella would be lit inside by that huge door, which would sometimes be wider than the middle intercolumniation; it might have illuminated the cella sufficiently for someone standing outside it to see within. A few temples were open to the sky, and in some the light from the door might have been supplemented by a number of smaller roof lights.[2]

It had taken about a century for the model or schema of the temple to be set. The Doric form most probably evolved in the Peloponnese, the Ionic (of which much more later) on the eastern Aegean seaboard. The two patterns may have been devised without reference to specific building materials, although the spread of the fine-masonry stone temples turned out to be very quick, once the technique was adopted. The extent and the speed of the diffusion argues for an early verbal and numerical formulation of the rule and of the type, since it is unlikely to have happened through any graphic medium. But within the type there were also many contradictions and incoherencies, and this in turn suggests that it must have been devised in more than one place and by several groups of people—even if all were working to that end at the same time.[3]

A number of the contradictions were due to the change from wood and ceramic to stone; yet others reflected another change, at the end of the seventh century, from an odd to an even number of columns. The new, even-numbered-front scheme also implied an internal rearrangement of the whole temple, and raised many problems for its builders; and it remains a worry (of a different order) for their historians. Indeed, during the sixth, and perhaps into the fifth century, important (if provincial or secondary) buildings—such as the "Basilica" at Paestum or the Hermes/Aphrodite and Apollo/Artemis temples at Samos, as well as that Sicilian marvel, the Agrigentine Zeus—continued to be planned according to the old arrangement such as that of the wooden temple at Thermon. There, the pediments on the narrow side of the temple are supported by an odd number of columns, of which the middle one corresponded to a central colonnade inside the cella that presumably bore a dwarf wall.[4]

Even more important, the central colonnade plan presupposed a form of cult in which the main action would involve a procession that entered the temple by one door, proceeded along and around the center line of columns, and went out by the other door. It implies, too, that whatever the cult object or place, it would be relatively small, and would be placed asymmetrically in the plan—as in the Cretan shrines I described in the previous chapter. And although there are archaeological data that could allow such an interpretation, very little is known about its implication for ritual practice. At any rate, the fixity of the type did much to mask the contradictions I mentioned, but without resolving them.

By 600 the fully developed wooden temple structure is adapted to a new kind of worship, which requires an internal space centered on a dominant divine image. Hera at Olympia may have been offered the first such temple, with a central aisle and a central door opening to show a cult statue of colossal dimensions. This was the temple type that was to dominate the stone architecture of the Greeks for four centuries and more. We must therefore assume that in the important centers of the proto-Geometric and Geometric periods, Argos, Corinth, Athens, and Thebes, as well as on the islands (Samos most obviously), there were earlier instances of "experimental" Doric buildings of which no legible trace remains. It is hardly conceivable that the builders of a temple in provincial Thermon, who seem so accomplished, so assured in their manipulating of the Doric type, should also be the first to formulate it. Thermon seems much more like the realization of a well-tried scheme rather than an initiating and originating building.

The temple and its users

The early center-colonnade temples coincided with the formation of the essential Greek institution, the polis: perhaps paradoxically, they became the quintessential buildings of the Greek city. First came the hairpin apsidal shrines with a central row of posts (such as the hecatompedon of Apollo Daphnephoros in Eretria). The center-colonnade temples become squared later, as at Thermon B and at the Archaic temple of Artemis Orthia, which was the national shrine of the Spartans.[5] Both of these were near contemporary with most of the models I discussed in the previous chapter. The Spartan shrine was built of pisé reinforced with wooden posts, and faced in the lower parts of the wall with stone orthostats—a form of construction common all over Mesopotamia and Anatolia, which it may well share with some Cretan buildings such as the temple at Prinias.

Many of these new temples were "paced" or measured out at a hundred feet in length. The number one hundred, which implied a great but calculable size, had particular sacrality: great temples were *hekatompeda,* great sacrifices *hekatombē* (hundred-beasters), as I have already pointed out. *Hekatompedon,* "a hundred-footer," is an adjective used by Homer about ships, and it was presumably used of temples that had that measurement; the term survived into classical times as the name of the naos, the inner shrine, of the great Athenian temple of Athena Parthenos, popularly known as the Parthenon. It commemorates the old wood-and-brick hundred-foot in antis Doric temple that had stood on the adjoining site until the tyrant Peisistratos replaced it with a peripteral Doric building about the middle of the sixth century.[6]

Many of the earliest peripteral temples were hecatompeda, but also the second-generation ones: most probably the Geometric Heraion at Olympia and the cella of the Corinthian Poseidon temple at Isthmia, as well as the oldest temple at Ephesus, which lay somewhere between that later temple and its altar, at right angles to the main direction of the Artemision. The roofs of these temples, with the crowning elements of their structures, were inevitably the parts that perished most completely: if number games were played within them, the clues may be lost forever. That this was part of the Greek builders' preoccupations is clear: for instance, the exact hundred-foot pacing, which had been played over their length in apsidal buildings, was later transferred to the width, and primarily the front. Moreover, it recurs in the measurements of the cella walls of many subsequent Doric temples: Athena Alea at Tegea, Apollo Epikurios at Bassae. It was also applied to Ionic ones, such as the Athena temple at Priene, and Hermogenes' much-lauded shrine of Artemis at Magnesia.[7] Their metric dimensions will, of course, change, since they will have been measured in a local foot, which varied from town to town.[8] The first squared hecatompeda, that of Hera on Samos or of Poseidon at Isthmia, belong to the eighth and seventh centuries.

These temples had porches and presumably two doors, or just open entrances, one on either side of the central column line. It is not known whether that central colonnade was usually higher than the peripteron, or if it carried a dwarf wall over which the roof ridge was to be set. At any rate, no evidence survives of a cult station in such an apse.

Still, between the building of these first hecatompeda or the Spartan Artemis Orthia temple (say about 850–800) and the earliest fully Doric or Ionic buildings (which went up about 650), the whole apparatus of detail must have been codified—if not in words, then in the practice of the building crafts, which must already have been organized by this time in some rudimentary way. There is no record of names or places more specific than that which the archaeologists and Vitruvius have provided and I have here outlined, following them. The change I have described from the central colonnade to the central chamber with a cult statue presupposes a transformation of ritual and religious practice, or, at any rate, of religiosity.[9]

The juxtaposition of the two small Cretan temples and the variations of the model types, as well as the succession of events at Lefkandi, suggest that Vitruvius' account of the two constructional forms does indeed follow some remote, and by his time forgotten, historical sequence. Something of the kind seems also to have governed the construction and the cult of the main shrines at Eretria. But there is another, neglected precedent: in the north of Greece—in Epirus and Macedonia—a structural technique developed that depended on sound timber and the ready supply of mud and clay for bricks. The Neolithic and Bronze Age or Helladic builders separated the structural elements from the screening ones. The results were buildings not unlike the little shrine of Apollo Daphnephoros in Eretria, except that the posts, and probably the beams as well, seem to have been made of larger timbers. In such houses, again virtually nothing is known about the roof. It does seem as if it was flat in the earlier structures, but by the end of the Bronze Age, apsidal buildings, presumably covered with double-pitched roofs, also appear.[10]

Tholoi

For all the dominance of the rectangular temples, the Greeks had always built circular structures above as well as below ground. By far the earliest as well as the largest of these must have been the vast brick rotunda that once stood on the hill on which the much later Mycenaean citadel of Tiryns was built. It dates back to the early Bronze Age, before 2000, and is therefore about contemporary with the first Minoan palaces. The diameter would have been some twenty-eight meters—perhaps one hundred feet? Nothing is known about its roof, whether it was flat or domed, and its use—as "ceremonial" complex or a granary—remains equally uncertain, though it is the largest single construction in mainland Greece to have survived from that time.[11]

Roofed circular pits, presumably for storage, were built at Knossos and Phaestos,[12] and there were many beehive tombs at Mycenae, Pylos, and Orchomenos.[13] Other tombs, beehives under a kurgan-type tumulus, have been found at Vrana near Marathon and at Thorikos in southeast Attica. At Mycenae the royal grave circle was surrounded by a covered passageway, though the whole group may well have been buried by such a tumulus, for which the orthostats of the passageway may have acted as a retaining wall. It was merely the grandest of grave circles, even if there was yet another in Mycenae itself; and additional ones have been found in Pylos and as far north as modern Albania. There are also a number of rather mysterious circular building models, perhaps used as household shrines, which have been found on Crete and which I mentioned earlier.[14]

There is no direct connection between such buildings and models on the one hand, and the tholoi from the Archaic period onward on the other[15] (of which very few have survived). Perhaps unique is a relic from the Geometric period, four columns supporting a roof within a circular wall, in the small settlement of Lathuresa in Attica. The Archaic sanctuary of the Cabiroi near Thebes had a circular altar; and there may, of course, have been others, but these are all the traces that have remained. Tholoi become buildings of importance again toward the end of the fifth century: it is easiest to discuss them in the context of Corinthian building, so I will therefore return to them in chapter 10.

Type and project

The public space of the Greek city, with its orthogonalities, welcomed the self-contained circular building. As for the columns within and without it, the fifth-century tholoi followed the precedent of rectangular buildings. Variations of column type presumably also paralleled those of such buildings. The type of wood, brick, or stone temple building to which they apply appeared about the same time all over Geometric Greece: at Thermon in the northwest, at

Eretria on Eubea, in Corinth and on the Isthmus, and perhaps most impressively, on the easternmost Aegean island, Samos. During the time when the polis came to be dominated by a new kind of ruler thrown up by a rising middle class, the tyrants, [16] the type is technically known as peripteral. Since the columns surround the cella like wings, *ptera* came to dominate Greek building.[17]

Yet, precisely because of the familiarity and persistence of the type, the architect of such a building would be largely concerned with refining and elaborating the apparatus of column and beam to match it to the variant of temple that he was asked to design, any demands of site organization and practice apart. He would of course also have to concern himself with the precise relation of the inner shrine to the peripteral columns, although in outline the "normal" sixth- and fifth-century type was so simple as to seem diagrammatic. In fact, the first design decisions—about the number and "gender" of columns, the intercolumniations, the depth of the column rows—did not need to be made by an architect, nor did they require any drawings or even models; they could be verbal and may not even have been recorded.

If Vitruvius may be taken as a reliable guide, once the number of columns (between two and ten, but never more than ten) on the front was established, and the nature of these columns and their intervals were fixed, the next crucial decision could be taken: the setting out of the total size of the building. To follow Vitruvius once more, it would seem that it was determined at the outer length of the front colonnade, presumably measured at the top of the stylobate, since according to him its "whole length" is to be divided to obtain the module— and two of those modules make the bottom diameter of the column. What he does not tell his readers is how this original length was established. Almost certainly in some cases it was measured in whole-number units of the local foot or yard—hence the hecatompedon—but many other factors must have played a role, from the limits imposed by the funds available, to the promptings of oracular responses or some ritual involving the founder.[18]

At any rate, from the module—which is to architecture what *prōtos chronos,* "primal time," the indivisible short measure of rhythm, is to music and poetic recitation—the other measurements followed by subdivision. The module will be the first feature of the building for which something drawn would actually be essential, even if until then everything could have been decided verbally. It should therefore not really be surprising that, with the one exception of the Didymaion graffiti, no Greek working drawings have yet been recovered; but a number of detailed working specifications, *sungraphē,* lettered on stone and tablets and later monumentally affixed to the buildings, have been recorded and discussed.[19]

The prescription which Vitruvius offers follows their pattern; every part is dimensioned in relation to ones that have already been put in position,[20] though the other elements of the column and beam are not necessarily stated in terms of the module but are proportioned progressively upward. My account attempts to follow the craftsmen as they erect the building, deriving the dimensions from the larger parts and adding their fractions rather than multiplying the smaller units.

It is notable (particularly as modern proportion-freaks look at that first) that the height of the building is not proportioned directly to the width of the stylobate in any of the formulae, but it follows from the way in which the detailing has been worked out. As for the length of the temple, Vitruvius gives only one (and that a summary) formula: the length is to be twice the width and it is to be arrived at by doubling the intercolumniations. If the number of columns is doubled instead, says Vitruvius, it will cause the error of producing one intercolumniation too many. The cella, on the other hand, is to be 3:4 in proportion. Very few temples follow this formula exactly.[21] This does suggest that there may well have been other canonic recipes which Vitruvius did not know or which he chose to ignore—or else that the canonic formula was not a rule, but the account of a type to be emulated and varied.

The width of the temple at the stylobate may not have been the first but it was the essential dimension to be determined, and every major part of the building could be set out as its fraction. It locked dimension to the type, since the number by which the length was divided to find the column diameter and the intercolumniation would vary according to the species of column and the type of temple. For the Doric order, Pliny records that in remotest antiquity columns were one-third of the temple width high, which seems to have been true—in some cases, at any rate.[22] For his part, Vitruvius provides two rather different rules. In the third book, he cites reverentially the formula that his intellectual hero Hermogenes devised for Ionic temples, calculated in diameters. The stylobate is to be divided into 11½ parts for a tetrastyle, a front of four columns; 18 for a hexastyle, six columns; and 24½ for an octostyle, eight columns. The "part" thus worked out is the module. In his fourth book he sets out a somewhat different (and perhaps an older) rule and gives two module numbers: 27 for tetrastyle and 42 for hexastyle. The first is an odd number, suggesting a wider central intercolumniation, while the second would suggest uniform distancing. At any rate, the basic type of column arrangement in both cases would be diastyle, three diameters: a type Vitruvius likes, though he finds it rather risky (structurally, as I will have occasion to mention later). Still, I need hardly insist on the numerological importance of 27, the cube of three, in the Pythagorean scheme of things, while 42, six sevens, was also favored. Commentators have had many problems with this particular specification and some have even (to my mind, improbably) suggested that these numbers are "abstract" or "ideal," arrived at by numerological speculation and relating neither to workshop practice nor to the measurements that might have been taken from existing buildings.[23]

Intercolumniation introduces another variant into the type: there are five varieties calculated in terms of the diameter of the bottom of the shaft. These range from the narrowest, pyknostyle, 1½ diameters, to the widest, araeostyle, 3 diameters and over—so straggly that it could not be spanned by any stone beams, which would have cracked over such a distance. I will discuss the latter in connection with the Tuscan column. However, another factor was involved in these variations, since the module height of the (presumably Ionic) shaft was also related to the intercolumniation. It varied from the 10-module shaft in pyknostyle to the 8½ of the araeostyle: commentators have noted that the governing number obtained by adding the height modules to the intercolumniation was always 11½.

The exception and ideal among these relations is inevitably called eustyle—*eu* being the prefix of approval. It was of 2¼ diameters, while the column remained at 9½, so that the sum of the modules was 11¾: in any case, it was widened to 3 diameters in the middle. If Vitruvius is to be credited, it was another device of Hermogenes'.[24] A quite different set of possible variants is in the relation between column and wall, from the two-column porch between anta walls (distyle in antis) to the ten-column wide, two-column deep range of columns surrounding the cella on all four sides (dipteral decastyle). All this provides for some thirty-five variants combining column and column, columns and wall, which seem to apply indifferently to all three Greek genera.

Six hundred varieties

Some commentators have noted, however, that the twenty-seven modules of Vitruvius' four-column temple coincide roughly with the divisions of the great and famous sixth- and fifth-century hexastyle temples: Aphaia at Aegina, the Theseion, the Propylaea of the Acropolis at Athens, the temples of Demeter and Poseidon at Paestum, or of Apollo Epikurios at Bassae. Moreover, since Vitruvius arranges two triglyphs between the columns (where the classic temples have only one), his tetrastyle and the classic hexastyle have the same number of triglyphs (eleven) and of metopes (ten) to the temple width. The choice of the module seems therefore to have been closely related to a numerical fix of the type.[25]

I have already noted Vitruvius' advice that the awkward Doric is best avoided in temples. That is why he relegated its modular recipe to his fourth book. Proceeding rather differently from him, I have considered the temple type as the bridging—or dividing—element between the Doric (which I discussed in the last chapter) and the Ionic (which I will consider next). For Vitruvius, the asymmetrical dialogue between the two older kinds of column—the tough, stocky Doric and the graceful but matronly Ionic—is adjusted by the slender, maidenly Corinthian. Yet Ionic is obviously the *eu*, the best: just because it is the median column between the extreme ones. Whatever later writers may have suggested, Vitruvius' method (or scheme, or whatever you call it) only allows of three Greek columns, to which the Etruscan or Italian arrangement is a kind of appendage or perhaps a reflection. The formula or schema for the Ionic and Corinthian buildings follows directly on Vitruvius' discussion of the proportion of the human body in the third book. It is also much more detailed.

However, when he comes to specify the details of an Ionic temple, Vitruvius introduces a different method of calculating ratios from the Doric, dependent not on abstract proportions but on the dimensions of columns. He prescribes different proportions for the reduction of the column diameter between top and bottom, as well as for the changing epistyle-to-column ratio. His Ionic arrangement is therefore not really modular; measurable size takes precedence over proportional relation.[26] But differences of size apart, and in spite of the fixity of the type—and assuming, too, that most variations could be carried out using all the four kinds of columns he proposes—the variants listed by Vitruvius could produce something like 600 possible temple fronts.

Tympanum or pediment

Another relic remained from the older, more makeshift buildings: the triangular void between the inclines of the double-pitched roof and the flat covering of the porch came to be curiously called "pediment" in modern parlance (at least in English),[27] and *aetos* or *tympanon* (*-um*) in Greek and Latin. Pindar's claim that it first appeared in Dorian Corinth in the reign of his patron, the tyrant Kypsellos, has already been considered in my discussion of the alleged invention of the stone Doric columns in Corinth.

The pediment was the highest member not of the column genus, but of the temple as a type. Although it is very much a part of that symmetrical temple type, it was regarded as more important and perhaps even more sacred on the face that turned toward the main altar in the temenos.[28] In hairpin plans, which often underlie the long and narrow hecatompeda, there would only have been one tympanon in any case, since the apsidal roof had to be hipped.

Propylaea of the Acropolis at Athens. Photo by author.

Temple of Poseidon at Paestum. Photo by author.

In Archaic temples, it is always decorated, while the tympanon facing away from the altar may be left blank, as seems to have been the case in the temple of Athena at Corfu: Korkyra.

You will appreciate the technical implications of this form if you also consider that Greek temple builders did not use triangulation in roof construction. They seem to have laid their sloping roofs directly on the columns and walls,[29] and the persistence of that wasteful, cumbersome usage gave additional plausibility to the traditional account of the origins of the Doric order. It does not allow for another feature, however: the porch that may have butted up against the end wall of the structure.

The corner triglyph—again

That complex scheme or type detailed by Vitruvius (and by myself, following him) dominated every form of building wherever Greek was spoken, and beyond. While the ordering of beams and columns was primarily worked out in temples, civil buildings of all kinds—palaces, council halls, and gymnasia, and yes, circular buildings as well—were articulated and ornamented with columns as well as beams that had probably all been devised for temples. But the cornice, the final, edging element of the arrangement, was often used to top a wall within a temenos. For instance, a cornice would run over all the columnar buildings—temples, propylaea, and treasuries; but cornices would also edge the altars and even crown the enclosure walls. The scheme was dominant in spite of (or perhaps because of) its interior inconsistencies.

The Known and the Seen

I have had occasion to consider one of these inconsistencies in some detail: the triglyph-and-metope frieze which seems to imply that the main beams in the building, those covered by the triglyph tablets, lie in one direction only, across the narrow length of a temple. The difficulty of turning the corner with triglyphs was therefore inescapable when the representation came to be worked out in terms of stone construction; if the main beams are laid parallel to the line of columns and directly on top of them, then the epistyles (which were effectively wall plates) had to be thicker than the wide-spanning cross-beam that they supported. The hierarchy of beams required that the epistyle, being the lowest, should also be the heaviest beam, while the higher ones would become progressively lighter.

The real difficulty is a consequence of that first rigor: the corner triglyph, which represented the thinner beam, could therefore not both come to the edge of the construction and lie over the center of the corner column below it. As Vitruvius pointed out, there are three ways of solving (or of eliding) this problem, all unsatisfactory because they break the regularity of the alternation as well as the rule of composition:

1. you may lengthen the corner metope;
2. you may leave a piece of metope on the edge of the frieze to turn the corner, thus weakening the structural representation; or
3. you may draw in the corner column so that the end intercolumniation is shorter than the other ones.

Although there are examples of all three usages, and sometimes the methods were mixed, the last one was adopted most frequently by the architects of the great Doric temples. The corner triglyph problem has often been regarded as the paradigm of the difficulty of turning the corner in any constructional system. However, in Greek architecture the narrowing of the corner intercolumniation and the consequent displacement of the corner triglyph was linked to some other and more elaborate devices by which the main dimensions of the building were corrected or tempered through minor adjustments.

Optical refinements

When Vitruvius recommends optic "refinements" for Ionic columns, similar to those which the corner triglyph-metope problems suggested for the Doric, he does so in much more detail. The shafts of columns, he directs, must thicken as the distance between them grows:

> In an araeostyle, if the columns are a ninth or tenth part of their height, they will look slight and mean . . . because the air consumes and diminishes the thickness of the shaft. . . . If, on the contrary, it is an eighth part of the height in a pyknostyle, it will look swollen and graceless for the closeness and crowding of the interspace. . . . Corner columns must [therefore] be made one-fiftieth thicker [than the other columns] in diameter because they are surrounded by air, and will seem thinner [than they really are] to spectators.

And he concludes: "Where the eye fails you therefore, reason must remedy the defect." [30]

I write more than sixty years since this particular feature of Greek architecture was last studied in book form. [31] There were, of course, mentions of the practice, however cryptic,

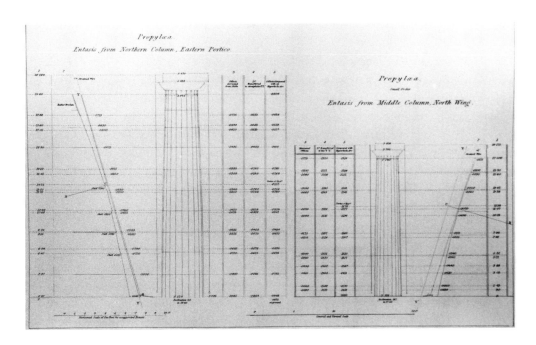

Entasis and inclination of the Parthenon and Propylaea columns. After F. C. Penrose (1888).

in Greek literature. Plato himself refers most explicitly to such optical correction as practiced by sculptors in the *Sophist*, when the Eleatic Stranger and Theaetetus discuss the "reality" of images, so as to decide what kind of maker, or rather "poet," the Sophist might be.

> Stranger: The perfect example of *eikastikē* [the art of making a likeness] . . . consists in creating a copy which conforms to the proportions of the original in all three dimensions and moreover gives the proper colour to every part.
>
> Theaetetus: Is not that what all imitators try to do?
>
> Stranger: Not those sculptors and painters whose works are of giant size. If they were to reproduce the true proportions of a well-made figure, the upper parts would look too small and the lower too large (as you know), because we see the one at a distance, the others from close to. . . . So artists, leaving the truth to take care of itself, do in fact put into the images they make not the real proportions, but those which will appear beautiful. . . . The first kind of image, being like the original, may fairly be called a likeness [*eikōn*]. . . . And what are we to call the kind which only appears to be the likeness of a well-made figure, because it is not seen from a satisfactory point of view, but to a spectator with eyes that could fully take in so large an object would not even be like the original it professes to resemble? Since it seems to be a likeness, but is not really one, may we not call it a semblance, *phantasia*?
>
> Theaetetus: By all means!
>
> Stranger: So the best name for an art which creates not a likeness but a semblance will be dissembling, *phantastikē*. . . . And there are two forms of image making, the making of likenesses and the making of dissemblance.[32]

Plato made many other slighting references to the unreliability of sensory perception and to artists who adjusted or falsified the true measurements and colors of things that they repre-

The curvature of the Parthenon stylobate. Photo by author.

sented to make them appear "correct," much as he derided the acoustic experiments of some of his contemporaries.[33] On the other hand, he extolled building and carpentry above other *technai*, other arts (e.g., medicine, music, rhetoric), because it involved exact calculation and relied on "scientific" instruments. This kind of craft, in which the play of number was a *technē*, was closest to another kind of activity, *tiktein*, "generating"—or even *teknopoiein*, specifically "bringing forth children" (something toward which Eros urges every living being, human and animal, so that they may through reproduction approach immortality). But they never wish to bring forth in ugliness, since love urges us to "reproduce and give birth in beauty," *tokos en kalōi*. Diotima taught Socrates about a *technē* in which the imitation—whether it is the work of god, man, or animal—is not inferior ontologically to the thing imitated.[34]

For all that, the relation between sensation and measurement was very interesting to many early "psychologists."[35] As for artists, it was well-known that Pheidias had a reputation for a particular skill at this kind of adjustment. The story of his competition against Alkamenes (perhaps a pupil of his) was told by John Tzetzes, a twelfth-century Byzantine man of

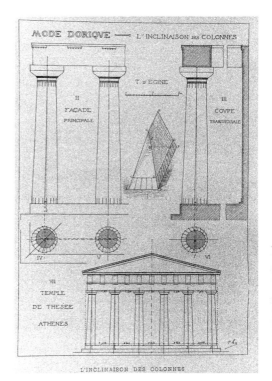

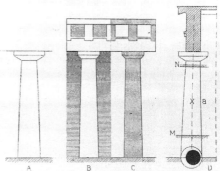

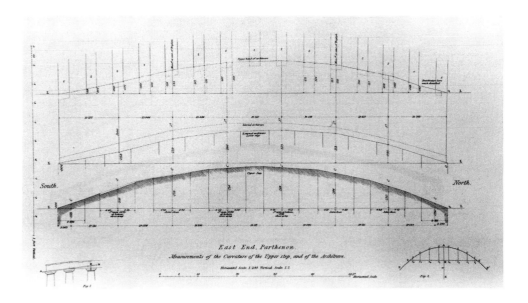

"Optical" effect and inclination of the columns of a Greek Doric temple. After A. Choisy.

The inclination of the columns: The temple at Aegina and the Hephaesteion in Athens. After C. Chipiez.

Diagrammatic representation of the Parthenon stylobate curve. After F. C. Penrose (1888).

Example of anathyrosis: Doric.
Temple of Zeus at Olympia. Photo by author.

letters, on two separate occasions, though he does not mention the source from which the story came. Alkamenes, it seems, made

> very graceful human figures, but was unskilled in optics or geometry. He was a great frequenter of public assemblies where he picked up hangers-on, lovers, and companions. Pheidias, on the other hand, shunned public occasions and reserved love and devotion for his art. . . .
>
> On one occasion the Athenians decided to set up two statues on high pedestals to Athena, and those two sculptors started work submitting [them] to the popular will. Alkamenes made the statue of the young goddess both delicate and womanly at the same time; Pheidias made his according to the rules of optics and geometry, knowing that the higher things are, the smaller they seem: so he gave the statue protruding lips and prominent nostrils. . . . At first his competitor's work seemed so much better than Pheidias' that people were preparing to cast stones at it. As the statues were raised up on their columns, however, the work of Pheidias appeared much more accomplished, while Alkamenes' statue and he himself were derided.[36]

Concave curvature of stylobate. Temple of Apollo at Delphi. Photo by author.

Tzetzes' account of the optical corrections in giant or highly placed statues is, apart from the Vitruvian discussion, the most explicit parallel in ancient criticism of art to the adjustment of dimensions and the correction of horizontal lines in building. There were, it is true, other applications of optical rules in the arts, such as the perspectives of stage scenery or the adjustment of letters on high inscriptions. But the correction of building parts and of horizontal lines were practiced generally and yet were so subtle that in spite of Vitruvian promptings, not even entasis was suspected by careful eighteenth-century surveyors: it was only gradually realized through the 1820s and 1830s. The curving of the horizontal lines of the whole building was the most elaborate and spectacular of all these adjustments, and also the last to be discovered.[37]

As is by now generally recognized, the stylobates and the epistyles or architraves of many Doric and some Ionic buildings (stoa as well as temples) were curved, usually in parallel, often on elevation and sometimes also on plan.[38] This curve, like that of the entasis, was not always convex, as had seemed to the first writers on the subject. Moreover, in many temples the corner columns were not only drawn in, to narrow the intercolumniation, but also slightly inclined inward. In some of the buildings of the highest finish, the whole colonnade would even have a very slight inward inclination, while the centers at the top and bottom might both lie on such a curve.[39]

The painstaking adjustment of detail to cater for optical effect was a great technical achievement, since it required both accurate calculation and precise measurement in execution. It is in glaring contrast to the evident lack of interest in what in the twentieth century seems the principal aim of technology: structural economy. In the columns made up of drums, anathyrosis—the method of jointing stones by cutting them very precisely at the edges while most of the faces were left recessed and rough—meant that only about 10% of the stone surface had any structural function.[40]

Theory and practice

These adjustments appear, moreover, to be in theoretical conflict with the elaborate canonic regulation of the column and beam, though the Greeks certainly did not seem to consider these two approaches as being in opposition, in spite of Plato. It is almost as if these refinements were no part of the *poiēsis* of the building, no part of its theoretical typology, of the *rhuthmos*,[41] but belonged to another order of concerns: as if they corresponded to the praxis of the architect, much as the speed of his fingers on the lute strings or the timbre of his voice were part of a musician's praxis and had no direct connection to the intricate but majestic structure of musical theory. It was Aristotle who first clearly distinguished "making," *poiēsis*, from "dealing with," *praxis*, although such a distinction was already clear to Plato. As I have already pointed out, Plato opposed skill in music, which relied largely on guesswork, to skill in carpentry, which relied almost entirely on measurement and precision.[42]

Musical performance did not call for any theoretical training, after all, but relied on learning by gesture and by ear. Praxis would have been taught by one generation of musicians to the next, through the usual channels of transmitting know-how: rule of thumb, gestures, mnemonics.[43] Analogously, as Philo of Byzantium suggested, the process of making building parts look different from what they actually are is not achieved by precise calculation but by experiment: "by a process of trial and error, adding to the measurements and then subtracting, by inclining planes and all other methods until the form is agreeable to the view and harmonious to sight [*euruthma phainomena*]."[44]

The feature of ancient columns that might be considered their most obvious optical correction is entasis. The word itself is of doubtful origin, as I pointed out earlier; Vitruvius speaks about it obliquely, promising a diagram showing how to work it out at the end of the book, which unfortunately does not appear in any surviving manuscript. Wags have even suggested that he found it too difficult to do and so just left it out. Various exaggerated forms of swelling were practiced by sixteenth- and seventeenth-century architects, but the true curvature of Greek columns was first noted and measured by C. R. Cockerell at Aegina. Still, the method by which the very slight curve was set up was unknown until Lothar Haselberger found the diagram of the construction scribed on the inner wall of the shrine of Apollo at Didyma near Miletus. The method relies on drawing a quadrant on a gridded surface and taking dimensions from it in such a way that they are maintained in one direction and multiplied by a constant in the other, to provide a curve that to modern eyes looks like a gentle ellipse. It has been suggested that the very plausible method used at Miletus has provided a clue for the construction of Greek optical curves in general.[45]

In Vitruvius' Graeco-Latin vocabulary, symmetry or commensuration seems very closely related to *posotēs*, quantity and number, while eurythmy determines the *poiotēs*, the quality of the building.[46] Philo seems to use similar terminology when discussing the skill of the builder of war engines; in the same way as buildings, he maintained, their design could not be wholly reduced to calculation but had to be constantly improved by experiment.[47]

Presumably that is why the device of optical correction, unlike the invention of canonic proportion, is not attributed to any divine revelation, nor even (in the written records, at any rate) dignified with a hero-inventor. It may well have been learned by the Greeks from the Egyptians, yet in the only extended treatment of the matter, that which is offered by Vitruvius, there is not even any attempt to provide a quasi-historical legend about its discovery.

Indeed Vitruvius' advice on this matter is dispersed, and given more in hints and nudges than in any continuous account. It is not always consistent—perhaps because he relied on several handbooks and did not attempt to reconcile their recommendations. He sums up his own general attitude most clearly at the beginning of his sixth book; you may detect an echo, however feeble, of what Plato had been asserting in the remark that

one is the beauty which is close to hand, another that which is high above us; nor is the same valid for enclosure as for the open air, so that you need great ingenuity to take the right decision. The eyes, moreover, do not give a correct idea of things, but will deceive the mind in its judgments . . . . Since things which might seem false are true, while the eye, on the contrary, will accept things which are quite different from reality, I put it beyond doubt that something must be added or taken away according to the requirements and nature of their situation . . . and for this theoretic knowledge is not enough, but acute ingenuity is also required. . . . First therefore the measure of the symmetries must be established, from which surely the modifications may be deduced; then the unit of the outer length is fixed on the site of the future building; once that size is fixed, there will follow the working out of proportions in such a way that observers will not have any doubt about its eurythmy.[48]

The laws that governed optical phenomena as well as the nature of vision had been amply discussed long before Vitruvius' time. Euclid had formulated the laws of optical deformation in his usual limpid form. Aristotle and his pupils speculated on the theories of the pre-Socratic philosophers, some of whom maintained that vision came out of the eye, while others held that it entered the eye from the outside world.[49] But a view quite contrary to all of theirs was advanced by Epicurus and his followers, to whom the *sensa* were the only possible reality and for whom noetic "corrections" were irrelevant to the builder and even the spectator—to whom the squareness and therefore the solidity of the building mattered above all.

In beautifully (but unsuccessfully) urging Epicurean doctrine on all educated Romans in the generation before Vitruvius, Lucretius explained how the account the senses gave of phenomena must always be preferred to any "measured" one: "Proinde animi vitium hoc oculis adfingere noli"[50] (Never allow the eyes to comply with such weakness of the mind). Later in the same book he compares reliance on the senses to orthogonality in building. As the one guarantee of truth, and therefore also of stability, total reliance on the evidence of the senses made speculation about any such matters as the *scamilli impares,* which rely on the disparity between the measured and the sensory, quite irrelevant:

> As in a building, if the first rule is misshapen, and crooked the set-square, if it deviates from the straight line, or if there is a flaw in any part of the level, then all the construction must be at fault, everything awry, crooked or sloping, leaning forward, leaning backward, so that some things seem about to fall, others do collapse, ruined by the first mistaken measurements. So the cause of anything must appear perverse and false, which has arisen from falsity to the senses.[51]

However, the teaching of the rhetoricians and philologists who formulated the Latin vocabulary for later philosophers on a Greek model—Cicero and Varro were the most important—owed more to the New Academy and to Stoic thinkers. Vitruvius' advocacy of Greek-inspired optical refinement, however little it may have been practiced by his contemporaries (and Vitruvius quotes no examples), was consonant with what Octavian, as a patron of architecture, would have wanted to learn about, even if he did not have builders with the skills necessary to apply it.

Vitruvius came to know all this either directly from scientific literature, or perhaps only from secondary sources, from the lost architectural writers. He certainly took it for granted and saw no contradiction with the canonic rules; the evidence is the passage I quoted. Like Philo of Byzantium, he describes the achievement of eurythmy as a secondary procedure, which followed the fixing of size, type, and detail and the establishing of proportions. It consists, in his and Philo's words, of adding on to and of taking away from some modular dimension, assuming therefore that there was something to take from and add to, a datum. The datum was of course achieved in *posotēs*—numerical and proportional harmony—which eurythmy will allow to register adequately in the fallible sense of the beholder. Vitruvius defines *eurythmy* among the essential critical terms at the very beginning of his treatise:

> it is the graceful appearance and the commensurate look of the members of the whole composition when they correspond to each other. This results from the agreement of the parts of the work among themselves: as height with width, width with length—in short, when everything is a harmony of commensuration.[52]

The harmony of which Vitruvius wrote was not the primary numerical, musical harmony of number and quantity, but another one, more difficult to define—and hence his commentators' problems—of quality. This was achieved by the process of sensitive adjustment according to absolute size, place, and position and the coloring and the elaboration of ornamental detail, for all of which no positive rules have survived, if they had ever been formulated. And yet procedures of the kind to which Vitruvius refers may already have been known in the seventh century, though it is impossible to say at what point such practical skill is isolated within a "professional"—as distinct from a "craft"—discussion of theory. Not only are there no written documents about such matters, but even the structure of the buildings that archaeologists have recovered provide no reliable clue for it, since accurate measurements of earlier buildings, where terracotta and wood as well as stone were employed, are almost impossible to ascertain. Yet even in some coarsely built and unsanctified constructions, stylobates had a convex curvature that can be readily observed.

In building entirely of stone, such detailing involved the accurate shaping, perhaps even the separate setting out, of every single block of stone that appeared to the visitor and of many that did not. Starting with the crepidoma: while all its vertical joints were kept perpendicular, horizontal lines were so inclined as to make the curve convex in elevation, and sometimes (perhaps) to make a convex or concave curve on plan. The slope of the flooring between the temple wall and the columns was sometimes also inclined. The curvature affected both the stylobate and the soffit of the epistyle over it, and in many cases also the pediment and its inner wall. In elevation the chord of the curve was almost always parallel to the horizon; inevitably, the curve was not continuous but was made up of straight sloping lines. This was true of the long and the short sides of the plan. On the other hand the short sides were often also curved on plan, and on Egyptian precedent (if Pennethorne is to be believed) this curve was concave. It would therefore be fair to say that as a general rule the curves tended to be convex in elevation and sometimes concave in plan, though the criteria for doing either have not been systematically investigated. Columns were curved to the entasis as I described earlier; the abacus blocks were cut on a slight bias on the front, orthogonally on the sides. The surface of the column and sometimes also the wall inside the pediment were slightly bent forward. All this required precision cutting and abrasion of the stone: the inclinations were sometimes as little as one degree from orthogonal. There is still no general survey of these phenomena, and it has only recently been realized that it affects not only temples but also other public buildings.[53]

Allowing for the very different relation of cost and material in modern building, it may be worth quoting a recent parallel. When the London Cenotaph was designed by Edwin Lutyens in 1919, as a memorial to the dead of the Great War, the builder's estimate was for only one-tenth of the sum voted for the purpose by Parliament. In order to spend the whole sum without altering the outline design with which Lutyens was completely satisfied, he had the stones cut individually to take up the plan curve of the street and the camber of the road. In calculating these curves as well as the general proportions of the Cenotaph, he seems to have relied entirely on the rather questionable publications of Jay Hambidge, the prophet of "dynamic symmetry," while being unaware of the research that had been published on the matter in the preceding decades. And yet, when he designed the vertical lines of the Cenotaph to meet a mile above the center of the building, he was convinced that he was following Greek example.[54]

Surface

Such high and precise finish could be expected in buildings that were in fair-face masonry. But Greek buildings were finished in stucco, all of which was colored. These colors were usually bright and pure—almost heraldic—and always laid flat, never graded. The Greeks liked strong contrasts, as on red and black pottery, and they also liked strongly colored objects "found" in nature: witness the blue marble of the altar outside the temple at Delphi and on the frieze below the Nike temple in Athens. The harsh and eye-catching effect of such contrasts seems to a modern observer to distract attention from the optical and geometrical corrections on which great ingenuity and care had been lavished, and that suggests that eurythmy was not merely intended to produce a smooth impression in the observer's view, an impression that some modern historians have considered the principal aim of the optical refiners.

As for the colors themselves, we know most about the dyes used from the obvious sources: Vitruvius and Pliny. They both wrote at a time when painted stucco was used more on the inside than the outside of buildings. Yet Vitruvius is clear about Greek usage: the triglyphs of wooden temples had been waterproofed by being coated with blue wax, *cera caerulea*; hence blue is the color of triglyphs on Doric temples.[55] However, Vitruvius, like Pliny, is more interested in the use of color for pictures than for the sharply contrasting flat planes of Greek temples.

Like buildings, sculptures were painted. Sculptures were even made of more than one material—wood and terracotta, stone and metal; marble statues might have bronze wigs and belts, wear bronze armor and gilt jewelry. All-bronze statues were varnished with a thin bitumen to keep the surface from oxidizing, and this may have given them the prized "liverish" color described by Pliny the Elder, whose encyclopedic book provides so much of our information.[56] The naturally reddish-gold color of bronze with high copper content was also appreciated. In buildings it was much used for grilles and doors, and legends—which I quoted in another connection—told of whole bronze temples with gold acroteria.[57] Even the most expensive and showy finish, gilding, was not despised. The smooth architrave of the Athenian Parthenon, which some people regard as the archetype of white Greek architecture, was decorated with a series of circular gilt bronze shields. Bronze was occasionally deliberately patinated. There is a record of bronze acanthus leaves for a Corinthian capital in Rome that were marinated in urine to take on a bright green patina before being fastened to the column.[58]

But whether of stone or bronze, Greek statuary almost always had some applied gilding and coloring, and sometimes inlays as well. Eyes of shell and glass can be found on the Delphic Charioteer or the Marathon Kouros and even more conspicuously on the Riace bronzes. Red lips and blood on wounds might be of glass paste; bits of it remain on several statues (for example, the Sitting Boxer in the Museo delle Terme in Rome). Silver eyebrows, hair, and fingernails were common enough and have left traces on many bronzes. Such inlays do leave telltale marks, but painted colors usually fade or flake off if they are not renewed and maintained. The blank-eyed stare of so many Greek statues is due simply to the color being washed out or having fallen away.

The two most famous statues of the Greek world, the Zeus of Olympia and the Athena Parthenos, which were both several times life-size, were made up of many different materials. Olive-wood cores were covered with gold plaques for clothing and with baked ivory (to simulate the color and texture of human skin) for the naked bits; shell eyes, of course; bronze armor; real jewels; and so on. Such chryselephantine statues glittered in the dark sanctuaries (no windows, only some top openings, and a few oil lamps did not exactly provide blinding light), an incense altar before them sending up the occasional puff of smoke. And the stone walls, the columns as well as the timber roofs and ceilings, were finished with stucco prepared for painting with bright, harsh colors, which must have been refracted by the gold plates in spite of the incense smoke.

The Minoan and Mycenaeans and their northern contemporaries, much more refined builders than the "first Greeks," had also colored their sculptures and buildings. When the "classical" Greeks began to use stone instead of brick or wattle and daub, it was quite often protected by decorated and molded plaques of kiln-baked earthenware that would be fastened to the soft tufa with bronze nails. Such revetments were at least as elaborately enameled as the Geometric and orientalizing pottery vases of the period.

When, at the end of the orientalizing period, sometime before 600 BC, the Greeks passed from softer, coarser tufas and limestones to hard, crystalline marble, colored stucco replaced ceramic facings; even on the softer stones, stucco was used as a ground for color, and the colors became more brilliant. The stains or dyes were mineral ochers—oxides and carbonates of iron and copper—burned earths, soot, vegetable (like saffron) or animal stuff (the purple dye from shellfish, which was very volatile and expensive), and ground stones like malachite, hematite, cinnabar, or lapis lazuli. Since ground stones are stable, they provide the surest indication of certain hues used by the ancients.[59]

Two basic techniques were used. The first, the encaustic (preferred by many sculptors, according to Pliny), is effected by tinting hot wax, which may be applied directly to wood or stone when it is liquid; when it sets, it can be scraped away with a hot tool, leaving the color as a stain on the marble or wood. This has the advantage of not affecting the sculptor's surface, but it is wearisome to apply over a large area. The alternative approach was painting on stucco. On dry stucco, the Greeks would mix pigments in egg white, gum, or fish glue, but

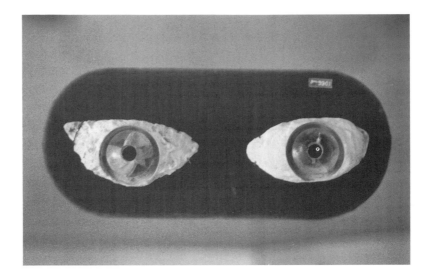

sometimes they waterstained the plaster while it was fresh and still damp, in the technique later called *fresco;*[60] for very large areas, the stain would even be mixed directly into the stucco, before it was applied. More volatile colors were occasionally varnished with oil-bound wax: this process, known as *ganōsis,* was even carried out ritually on some statues by the Romans.[61]

The Greeks and Romans were, of course, not the first to color their buildings. In Mesopotamia, the Sumerians, the Babylonians, and the Assyrians had covered sun-dried brick buildings with glazed and highly colored earthenware. The Egyptians covered theirs with a white stucco, which they also painted with bright, heraldic colors, but the Greek combination of glazed terracotta and stucco suggests that they brought together in their temples the colored terracotta techniques that had been used to decorate Mesopotamian brick buildings and the painted stucco over stone manner commonly used by the Egyptians.

However, unlike the Egyptians the Greeks hardly ever allowed white stucco to show through; on the contrary, it was always colored and used for "breaking" the staring white gleam of the marble. This could be done by mixing the stucco with milk in which saffron had been steeped. Pliny tells of a temple where you could still smell the saffron on your fingers if you wet them with spittle and rubbed them on the stone.[62] The steps, capitals, and the roof structure, as well as the walls of the temple chamber both inside and out, were all painted, as were the reliefs on the cornice and the figures in the pediments—and, of course, all the details of the roofs: acroteria, anthemia, gargoyles.

If the colors used in buildings did not refer directly to natural phenomena, they seem to show the way things were made in older times, recalling wood construction, wood carving, and textile patterns—as well as the terracottas of the Geometric and orientalizing epochs. Plant forms, the palm or honeysuckle, were always as conventional in color as they were in outline, and some were borrowed directly from Near Eastern or Egyptian art.

That bright copper-oxide green I mentioned earlier would only have been a rough approximation of the deep leaf green of the true acanthus: Greek coloring was not naturalistic. In a famous passage in the fourth book of the *Republic,* Plato exemplifies the balance of happiness of different classes within the state in the story of the sculptor approached by a naive visitor, who

The tonal effect of polychromy.
Temple of Artemis at Korkyra.
After W. Dörpfeld (1935).

censured us for not applying the most beautiful pigments to the most beautiful parts of the image, since the eyes, which are the most beautiful part, had not been painted with purple, but with black. We would think it a reasonable enough justification if we replied: don't expect us, quaint friend, to paint the eyes so fine that they will not be like eyes at all . . . but observe whether by assigning what is proper to each we render the whole beautiful.[63]

In Doric buildings, the columns and beams, if in yellowish stone—tufa or poros—might be left untouched; if they were of marble, they would be stuccoed or just tinted yellow. In the frieze, each metope enclosed a rather agitated and brightly colored sculpture, which was not necessarily painted over its entire surface: the "naked" parts might be untreated, but clothes, armor, hair, and so forth were colored in, and the colors were much more varied than the mere reds and blues of the ornaments. Metopes were separated by the "cerulean"-colored triglyphs, while the horizontal moldings on the cornice and those enclosing the tympanum were sharply patterned in red and blue. The background of the tympanum, against which some of the grandest and most famous Greek sculpture (the Elgin Marbles, the Olympia Apollo, the Aegina Athena) has been outlined, was bright blue, since it indicated the heavenly vault in which the divine figures that were revealing themselves in the tympanum had to be seen, much as the dead were seen against the color of the underworld. The pediment figures were again brightly colored, and sometimes ornamented with bronze pieces. Above the main roofline, the acroteria and other ornaments were often colored red and blue but were occasionally of bronze and gilt. A certain amount of gilding on the skyline was, in any case, not unusual in more opulent buildings.

On Ionic and Corinthian columns, the bases were often picked out in bright reds and blues, sometimes with additions of green, black, and some gold. The tops of the columns were bound with a brightly colored necking under the capital, and the spirals of the Ionic volute were heightened with red and blue to make it appear springy, almost metallic; and much the same was done to some of the roof ornaments. The outbuildings, the votive statues, and the propylaea of the temple were all similarly colored.

Monochrome architecture and the monochrome sculpture that came with it belong to the late Roman Empire. That may be the change that Virgil was prophetically regretting when he wrote of the advent of a *discolor aetas,* a monochrome age of war and greed that was to follow the golden age of Saturn.

There is another surface matter which most observers usually note parenthetically (if at all) about Greek buildings: they were inscribed. Not as modern buildings might be, with plaques or labels—or even in large monumental characters on a frieze or in a panel—but right across, as it were. Steps, walls, and even columns might be covered with characters, sometimes contained within a specially smoothed piece of wall, or perhaps attached on a bronze plaque to the wood or stone of a column. They could be grand and enigmatic, like the *E* in the temple of Apollo at Delphi, which prompted Plutarch to a whole book as a commentary; or gnomic, like the *gnōthi sauton,* "know thyself," which Socrates (and/or Plato) keeps on quoting.[64] But most inscriptions tell of dedications, great events, donations, make explanatory comments on the sculptures, and even provide specifications of the building itself. The stylobate and walls of the Athenian treasury at Delphi could be read as a book; the steps of the temple of Apollo at Syracuse carry an Archaic inscription in monumental characters of the kind you might expect on a cornice. The letters of such inscriptions, unless heightened with color, would be illegible or virtually so. Red would have been obvious for the purpose: however, red was often used to mark out the lower parts of a building, the steps especially, since it was the color associated with the earth itself and the underworld (the backgrounds of funerary monuments on which a dead person was represented were almost always red for this reason).

All the inscriptions, the bronze ornaments, the gilding, and the harsh bright colors did not mask the Doric discrepancies or distract architects and observers from the subtleties of optical refinements. By the fourth century the only way architects could cope with the discrepancies was to reduce them literally: columns became very slender, capitals shrank in proportion, and the cornice became so thin that there were sometimes three triglyphs between the columns instead of one. As for the columns themselves, they were curved upward in the tense movement of the flutes and pinched in, to appear even more slender than they actually were.[65]

In his enthusiasm for the power of such effects, Vitruvius even associates them with the very origin of temples surrounded by columns, when he says that the surrounding colonnade "was invented to make the buildings dominant because of the plasticity of contrast, and besides, it gave leisured shelter round the shrine to a crowd of people who may have been surprised and hindered by a shower of rain."[66] The reader may well find this explanation partial, even a little lame, but it is worth quoting as a witness to the *asperitas* that Hellenistic architects and critics attributed to the plastic quality of their building; they considered it a primary attribute of temples, an explanation of their isolation within the city fabric and their exaltation over the houses and the civic buildings of any town.

The Mask, the Horns, and the Eyes

• Ionic and descriptive • The legend of the first Ionic • Doric territory • Artemis at Ephesus • The worship of Artemis • Artemis and Dionysos • The Ephesian Artemision • Votive columns • The Sikyonian thalamoi • A Persian Ionic? • The Aeolic "order" • Oriental parallels • In Anatolia: Mittani, Hittites, and Urartians • In Anatolia: Phrygia, Lydia, Lycia, Carya • Spirals and trees • Aphrodite at Paphos • Astarte at Kition • The guardians of the doors • Models and shrines • Furniture and fabrics • The horned skull • Humbaba • The column-statue • The reed bundle: Inanna's pillar • Hathor

"Temples should not be made to every god according to the same pattern," Vitruvius insists, "since every god brings about a different holiness in his worship."[1] These differences he had already outlined at the beginning of his first book, when he explained the meaning of the term *decor.* As a noun the word has become nearly obsolete in English, though two rather different adjectives derived from it, *decorated* and *decorous,* give an accurate enough notion of the Latin sense. Vitruvius considered that the requirements of decor were met when the appropriate features were set together according to tradition. This follows from *statio,* "which the Greek call *thematismos,*" and which he rather vaguely identifies with custom and nature.[2] But he does go on to exemplify: temples to certain gods,

> such as Jupiter Fulgur [Lightning], the Sky, the Sun, and the Moon should not be roofed; those to Minerva, Mars, and Hercules, because these gods are wielders of much power, should be Doric, without anything fancy; those to Venus, Flora, Proserpine, Fountains, and the Nymphs will be seen to attract the right worship if buildings for them are in the Corinthian manner, since those columns are slenderest and most flowery. . . . The middling situation of Juno, Diana, Liber Pater [the Roman opposite number of Greek Dionysos], and of gods who are like them will have been respected by building Ionic temples to them, so as to keep the just measure between the severe manner of the Doric and the tenderness of the Corinthian.[3]

It is clear from this passage that for Vitruvius the nature of the deity and the cult he or she was offered regulated the formal arrangement of temples in general. The most obvious determinant was the choice of the "appropriate" kind of column. He makes this recommendation insistently for two of them (perhaps a little wearisomely) in terms of gender. Doric is masculine, naked, tough, heroic. Ionic is equally explicitly feminine:

> first they took an eighth of the column for its thickness, so that it should look more slender. They put the swell of the base underneath it, as if it were a sandal; at the capital they attached braided coils, hanging down like hair to the left and to the right, and arranged rounded moldings and garlands as tresses dressed over the forehead, while over the whole body they let the grooves fall as if they were the folds of a matronly robe.[4]

Vitruvius' description makes the Ionic column much more of an explicit icon of the human figure than his Doric. He almost describes it as a figural sculpture. Insistent though he is on gender, neither the Doric nor the Ionic are given any genitally sexual characteristics. The Greeks were not in the least shy about naming or showing them, so that if they are not mentioned or shown, they cannot have been important in this connection. Gender—the assumed cultural characteristic—is being displayed here, not any explicit generative function. As if to emphasize it further, Liber Pater[5] is mentioned as a suitable dedicatee of an Ionic temple, together with Juno/Hera or Diana/Artemis, while Minerva/Athena is to have a Doric one. This distinction between sex and gender is important: the femininity of certain male deities, and the corresponding masculinity of the female ones, will be discussed later in this chapter.

Familiar though he was with the texts from which this conception of the appropriate form of temple for certain deities was derived—and which have since been lost—Vitruvius was yet so remote from Greek religious practice that he does not remark on some of the most

Bathycles: The throne of Apollo at Amyklae. Sections, plan, elevation, and projection, after R. E. Martin (1987).

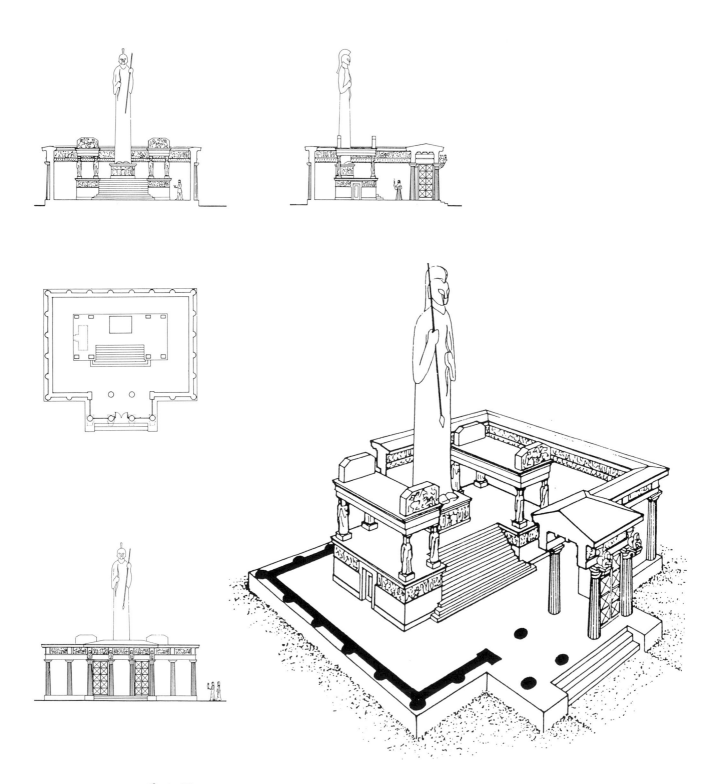

Chapter IX

obvious breaches of decor: in the Olympian sanctuary, both Hera and Zeus had Doric temples, while in the equally famous one on the Athenian Acropolis, Minerva/Athena was given both an Ionic and a Doric temple within the one temenos, dedicated to her different aspects (Ionic as Polias, protectress of the city and giver of the olive; Doric as Parthenos, virgin warrior). Even within one temple the rules could be neglected: the Doric Parthenon had Ionic features. The mixing or combination of genera was much more common than the plain Vitruvian doctrine might suggest. At Amyklae, outside Sparta, in the Doric heartland, Bathycles of Magnesia with a group of Ionian craftsmen designed columns for the throne of Apollo that were Doric on one side and Ionic on the other, making it one of the most quirky shrines to have been described by Pausanias.[6]

Vitruvius' distance—historical as well as geographic—from his Greek mentors, and from the monuments they built, may be why he is unaware of (or merely uninterested in) the great rarity of the third kind of column in earlier Greek building practice, even in the classical period. Setting it aside therefore (for the next chapter), I will not follow Vitruvius in placing the Corinthian after the Doric but work chronologically, discussing next his "middle-ing," the Ionic.

Ionic and descriptive

As it was described by Vitruvius then, and as it was usually built from about the beginning of the sixth century at least, the Ionic column and its beam were articulated into the same principal parts as the Doric, so that some technical terms were common to both. To start again at the foundation: the Ionic also stood over a crepidoma and a stylobate, but the Ionic stylobate differed from the smooth Doric in that it was, as often as not, marked by a deep, square-cut joint between the steps.[7] The first really important difference between them was that the shaft of the Ionic column was "sandaled" in its base, while the Doric shaft stood barefoot on the stylobate.

The base　The rule for that base, as it was set down by Vitruvius, demanded a square plinthos. *Plinthos* means a "brick" or "tile"; since he worked in multiples or fractions of a module for the Ionic, as he did for the Doric (the module being the whole of the diameter of the underside of the shaft, not half, as it was in the Doric),[8] Vitruvius specified that the plinthos was to be one-sixth of the module thick, and the rest of the base was to be one-third of the diameter; the whole base was therefore half the module high. The circular molded base, the base proper, rested on the square plinth. Again, as in the Doric, much of the terminology of the different parts of base, column, and beam has to do with ceramic work or woodworking. But in addition, Ionic has a number of textile, or at any rate "knotted," elements.

Vitruvius specified two variants of the base for this kind of column; he called the first "Attic." It was articulated into three moldings: a convex torus (called *spira* in some texts, a "swelling"; alternatively a "twisted rope," or a "thong"); a concave skotia or trochilos; and another torus. In Latin, *torus* may be any knot or bulge, while in Greek the word—like many "tor-words"—referred to carpentry, to the lathe, to drilling or piercing.[9] These three moldings were separated by flat fillets, though Vitruvius has base and shaft separated by an astragal (a knuckle, a joint), a thin concave strip usually filled with a bead-and-roll ornament, to which, as to a string of knuckles or beads, the term could well refer; in the same way *spira* might be

a twisted or plaited rope, caked with clay or covered with stucco. At any rate, the sequence of torus-skotia-torus had both "working" implications and provided a description of the sequence of the base: light-dark-light.

Although the lathe was a relatively recent invention, tor-words are as old as the Greek language itself. By the same token *torus* meant "thrilling," "piercing" (as of the voice), and "shining" (as of light). This last meaning may be the most relevant, since *skotia* is a darkling, concealment, or shadowing: indeed, "shadow" comes from the same root. *Trochilos* means a rut: either a wheel rut or the groove of a pulley.

Vitruvius' other Ionic base was similar; but its moldings (skotia-skotia-torus) were divided into thirds and sevenths, while the Attic one was divided into thirds and fourths. Moreover, the Attic base seems to have been one-sixth wider than the diameter of the column, while the Ionic one was three-eighths wider; either widening also has its technical Greek name, *ekphora*, literally a "jutting out."

Chapter IX

Ionic stylobate and wall base.
Temple of Nike at Athens. Photo
by author.

The two numerical recipes provided by Vitruvius are meant to be typical rather than normative: they are by way of example. Many more complex arrangements than these two bases survive in existing monuments: the plinth might be octagonal, as at the temple of Apollo at Milesian Didyma; single moldings might be repeated, or the whole sequence duplicated; one or both toruses might be fluted or fretted. Almost invariable was the semicircular sweep at the bottom of the column shaft, where it came down to the astragal that separated it from the top molding of the base. In most surviving examples the shaft was fluted, though the ending of the flutes in the sweep differed in execution.

The shaft   Now the flutes, as Vitruvius insists, are to recall or even represent the folds of a woman's dress, the Ionic chiton. They are therefore not drawn to an arris, like the Doric flutes, which represent heroic ax working, but are gouged out from the wooden trunk of the shaft—presumably with a semicircular chisel—and are therefore to be semicircular on plan, separated by fillets.[10] There are to be twenty-four of them (as against twenty of the Doric ones). The enthesis on the Ionic column is calculated in a way similar to that of the Doric; it was to be demonstrated by Vitruvius in a diagram at the end of the book—of which there is no trace in the surviving manuscripts.[11]

Quite often, at the top of the shaft just under the capital there is a flat necking (for which Vitruvius provides no technical name), decorated with honeysuckle pattern, palmettes, or both.

The capital   Another diagram that has been omitted from all surviving manuscripts, which was essential to supplement the verbal exposition, was the one for the drawing of the volute—the most distinctive and the most puzzling feature of the column. The scrolls that Vitruvius described as representing a woman's curly, braided hair might be imagined carved or scribed into such a bracket as is inserted between the shaft and beam, the bracket inevitably lying parallel to the line of that beam, an arrangement often illustrated in "evolutionary" histories of architecture.

In actual examples the scroll or spiral was worked out of the eye (the oculus) or central circle (which is often convex—a boss—but it can also be a piercing, a round hole). To construct it, a square was inscribed into the circle of the oculus, and the quadrants of which the spiral was made up were drawn from centers set on the diagonals of that square. The spiral was virtually continuous, since each quadrant was diminished by half the diameter of the oculus, while the center from which it was drawn moved along the diagonals.[12] The geometry of the whole capital was based on a small subunit: one-eighteenth of the module. The oculus was one such subunit in diameter. The centers of the two oculi were 16-units apart, and level with the top of the shaft. They were 4½ units from the bottom line of the abacus. The whole volute capital was 16 units deep and 24 units wide.

Vitruvius goes on to explain the other details by imagining plumb lines let down over the face of the block from which the capital is to be carved, though he continues to use the

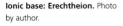

Ionic base: Erechtheion. Photo by author.

Ionic bases: Didymaion. Photo by author.

Ionic stylobate and intermediate steps: Miletus. Photo by author.

The Mask, the Horns, and the Eyes

same subunit to tell his reader—as if he were specifying instructions to a workman—how to calculate the different features. The widest point of the capital is 24 units. The deepest dimension is 8 units (excluding the abacus), giving the simplest possible outline proportions: (1) height, (2) depth = distance between the centers of the oculi centres, and (3) total width.[13]

In the canonic capitals, the line of the spiral was made up of one or more fillets, while the surface between its coils was concave. The swelling spiral drawn from an oculus was not specific to the Ionic capital: it occurred on altar corners and on brackets, sometimes in multiples; or it was merely painted, as on the tainia of the Parthenon cornice. In the painting of the capital (like the Doric, the Ionic was almost always brightly painted), the "eye" quality of the oculus was emphasized by special color, and sometimes by a glass or metal boss fastened to it.[14]

The spiral does not lie directly over the column shaft however, but on a cushion or couch—which was in fact called *pulvinar* in Latin (its Greek name, *proskephalaion*, is known

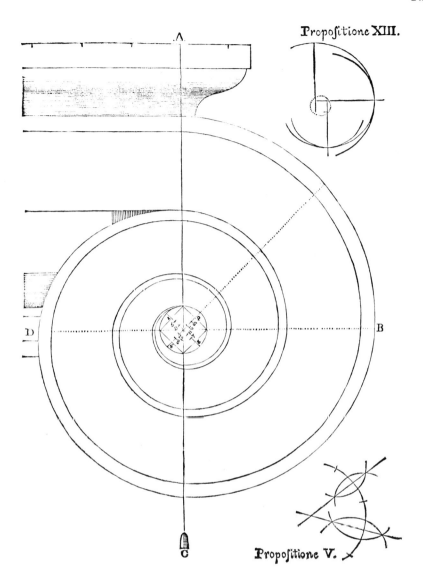

Propoſitione XIII.

Propoſitione V.

Ionic volute. According to Giuseppe Porta in G. Selva (1814).

only from inscriptions[15]—which Vitruvius wanted 2½ units deep. Since it was roughly a quadrant in section and belonged to the shaft rather than to the bracket, it projected farther from the surface of the capital than the abacus, and its swelling was emphasized by the egg-and-dart ornament that usually scored its full height. Another detail that Vitruvius' specification omitted is the drawing of the spiral down to the pulvinar over the center of the column. This created a hiatus at the meeting between three curves, which was usually covered by a small palmette curling from the fold, the tuck-in of the spiral towards the pulvinar. All of this was colored in complex patterns. The spiral, for instance, quite often had three colors: one for the raised flat molding, one for the central spine, and a third for the countersunk background.

Nor were the capitals always equally finished on both sides of the block. Sometimes, it seems, the spiral was carved on one side, and only painted on the other.[16] However, there is a further detail that this matter introduces, and that Vitruvius' specification also omitted: namely, the treatment of the curved side surface, which joins the two faces of the spiral. Seen

The Mask, the Horns, and the Eyes

sideways it looks as if the spirals were edges of a rolled sheet. This "sheet" was almost always concave, and almost always decorated with a central fillet in early Athenian examples. The Croesan capitals at Ephesus have three of them, but by the fifth century an even number was common: four at the temple on the Ilissos, eight at the Erechtheion. Later the whole surface might be covered with a reticulated pattern, as of bay leaves, which really did seem to represent the spiral as a kind of rolled, flattened garland, and there were even more fanciful treatments.

As in the Doric capital, the abacus was a square tablet lying directly under the beam. But the Ionic abacus was much thinner, only 1½ subunits thick; nor was it a simple block, but articulated, usually with a thin fillet over a cyma reversa molding, often cut with an egg-and-dart pattern. However, the obvious formal distinction (which the width of the volutes projecting beyond the square of the abacus implies) was its one-directional bias, quite unlike those of the Doric and Corinthian capitals.

The capital was, therefore, of its very nature one-directional. Moreover, as far as can be ascertained, the Ionic capitals on the sides of the temple did not run parallel to those on its front and back; all of them faced outward, creating a facade. This meant that a formal problem was presented by the capital at the corner of the temple—not by the cornice, as in the Doric temple.[17] In the mature capitals, this was resolved by cramping the two corner volutes into one, which projected at 45° to the faces of the building, while at the back the two volutes were mitered into each other at right angles. The earliest clear example of this was the temple on the Ilissos near Athens, which was built by Ictinus and Callicrates, the architects of the Parthenon, in about 450. Some twenty years later this arrangement was repeated in that of Nike Apteros on the Athenian Acropolis—though between these two dates lies the rather surprising attempt, in the temple of Apollo Epikurios at Bassae, to design Ionic capitals that had four (virtual!) faces. The rather low-slung volute and its very moderate projection provided a type of capital that many Hellenistic architects were to emulate.[18]

The cornice As the Doric, so the Ionic beam is articulated into three members. A three-stage trabs or epistylium rests on the abacus. It is to be the width of the top of the column, or 16

Corner capital: Didymaion at Miletus. Paris, Louvre. Photo by author.

Detail of capital. Troad, temple of Apollo Smintheus. Photo by author.

subunits of the capital.[19] Above it runs the wide band of the zophoron (figure carrier), now usually called the frieze (a term I discussed in connection with the Doric beam), and over that the top member, the cornice proper.

The three stages or fascie of the epistylium are usually at different height (as 3:4:5 in Vitruvius) and sometimes separated by a decorated beading; the epistylium widens upward by as much as the column narrows. At the top, it is to be edged by a cymatium, one-seventh of its whole height. For the zophoron, Vitruvius rules that its height is to depend on whether it is to carry figures (as its name suggests) or to be blank. It is to be in the ratio of 3:4 to the epistyle if it is bare, and 4:5 if it is figured, but in either case it is to have a cymatium the same proportion to it as the epistyle: 1:7. The zophoron seems to have been a later importation into the Ionic arrangement. All sorts of important Ionic buildings—the Ephesian Artemision notably, but also the Mausoleum at Halicarnassus, or the Athena temple at Priene, and even the Caryatid porch in the Athenian Erechtheion—have no frieze at all between cornice and architrave.

Some scholars have suggested that such zophora echo orthostats of earlier Anatolian buildings, and that they were analogously displayed at the foot of the walls of Ionic ones; these orthostats were echoed by relief panels in some Asiatic rooftop structures, and in Ionic ones

Proto-Ionic capital: Corner. Athens, Agora Museum. Photo by author.

following them. They argue that the third element was introduced into the Ionic entablature to give it a tripartite division similar to the Doric one by pulling the rooftop panels down between the cornice and the architrave, probably toward the end of the fifth century.[20]

Above the frieze, dentils are to project, representing the wooden purlins of a roof. Presumably this prototype roof was a flat one, since the dentils are at right angles to the vertical surface of the wall, unlike the slanting mutules under the Doric cornice. Dentils are to be as high as the middle stage of the epistyle and project to the same dimension. The space between the dentils is called the metope, like the Doric interspace of the triglyphs, but it is tiny in comparison; if the breadth of the dentil face is taken as one unit, it will be two high, while dentil is to metope as 2:3 and the proportion is again 1:2:3. The dentils are also edged by a cymatium, which rises to one-sixth of their height. One more fascia, equal in height to the middle molding of the epistyle, crowns the cornice with one more cymatium. As in the Doric, the top member of the cyma or gutter is a separate element, not properly part of the cornice.

The reader will have noticed that it is virtually impossible to give a specification for the capital that is both exact and readable. Some of its curious and entrancing details are susceptible of historical interpretation, however, and I will attempt this in what follows.

The legend of the first Ionic

For all the obvious differences between the two columns and the many disparities between Doric and Ionic temples (such as the prevailing westward instead of eastward orientation of the naos), the typology of bomos, eschara, and temenos were much the same in the two kinds of sanctuary, so that the writers about building did not need to concern themselves with such matters.[21]

According to Vitruvius' account, the very first shrine the Ionian Greeks ever established was a temple to Apollo, and it was of the Doric kind. The new settlers on the Asian shore wanted it to be like the venerable Argive Heraion, but since they had no precise record of the earlier temple's proportions, they chose a fine young man's footprint to serve as its module, and adapted its proportion to his height. That was in the time of the heroes; by Vitruvius' time, a stone Doric temple had long replaced the old wooden temple at Argos, which had been destroyed by fire in 423 BC—and which in turn stood on one of the most ancient foundations in the Peloponnese.

Vitruvius' account is marred by misunderstanding. The Panionian shrine was not dedicated to Apollo, but to Poseidon, more particularly to Poseidon Heliconius, tamer of horses and savior of ships.[22] The Panionion stood on the northern slopes of Mount Mykkale, on a piece of coast that was contested between Priene and the Samians. But Vitruvius' misapprehension is natural enough, since in the account of Greek colonization that he set down to explain the origins of (and the differences between) various kinds of column, Apollo receives the colonists' primary dedication, while the second (Vitruvius does not say where) was to his sister Artemis. The column-and-beam arrangement was called Ionic, Vitruvius adds, since it had first been devised in Ionia, as the Doric took its name from the Dorian city of Argos, where it had first been recorded. Indeed, the archetypical Greek temples with extended mythical and epic building histories are that of Ephesian Artemis for the Ionic genus, Delphic Apollo for the Doric.[23] Doric and Ionic could therefore be thought of as brother and sister, like Apollo and Artemis—divine *enfants terribles,* hunters, sharpshooters, but also the sun and the moon. They were the twin children of that Lato, Latona (whose main shrine was near Xanthos),[24] who was also known on the Aegean coast as the Great Mother of the Gods, and whose own realm would overlap that of Artemis rather than that of Apollo.

Doric territory

The territory of the Doric column was not only that of the Dorian dialect, it was also the Olympic realm where Zeus was the supreme ruler: he knew everything, and he saw through all subterfuge. Like his father, Apollo was clear-sighted, the clairvoyant. First in the whispering oak leaves at Dodona, then in the Sybil's incoherent cries at Delphi, both were the lords of prophecy, both declared the will and wisdom of the gods. Themis, Zeus' second wife and the mother of peace and order, who guides men to know and to do the gods' will, was present at Apollo's birth. Just because they were so alike, Apollo was the most likely god to dethrone Zeus. To avert it, Zeus took the stone that his mother had once given Chronos as a substitute for the son he wanted to devour—the stone Chronos could not digest—and set it as "a sign and a marvel" in Apollo's own sanctuary at Delphi.[25]

The passage of building from wood to terracotta to bronze to stone was recounted in legend at Delphi and commemorated by rite.[26] Enough early Doric columns in wood and terracotta are known from travelers' tales, as well as from excavations, to allow a more or less credible fleshing out of such a development. There is no shortage of early and "primitive" stone columns from which to trace the development from wood and terracotta into stone, as there is also no real difficulty about isolating the male and heroic character of the Doric order. The more "effeminate" character of the Ionic is equally explicit. The much more detailed account that Vitruvius gives of the figural nature of Ionic ornament (as well as its close precedents) is perhaps best examined in the temple that had the same archetypal character for the whole of Ionic building that Delphi had for the Doric: the Ephesian Artemision.

Capital: Temple of Artemis at Sardis. Side view. Photo by author.

Capital: Temple of Artemis at Sardis. Front view. Photo by author.

Temple of Artemis at Sardis.
View. Photo by author.

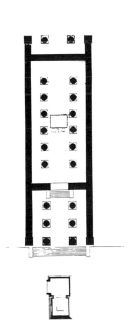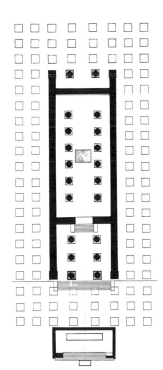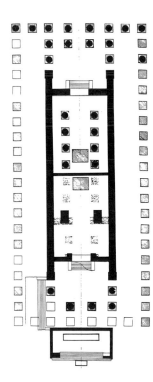

Temple of Artemis at Sardis.
Plan at three stages of construction. After K. J. Fraser in G. Hanfmann and J. Waldbaum (1975).

The Mask, the Horns, and the Eyes

Temple of Artemis at Magnesia: Elevation. After C. Humann and J. Kohte (1904).

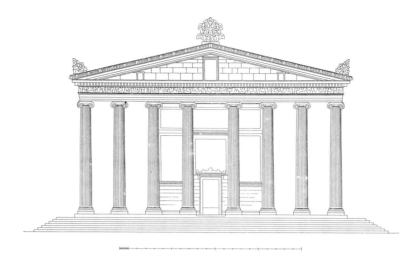

Temple of Artemis at Magnesia: Plan. After C. Humann and J. Kohte (1904).

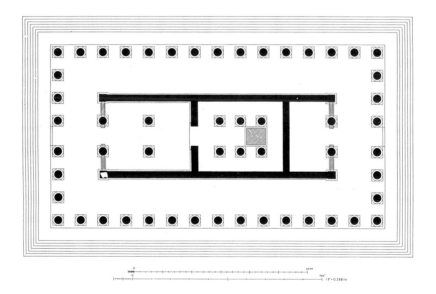

Temple of Artemis at Magnesia: Detail. After C. Humann and J. Kohte (1904).

Temple of Artemis at Magnesia: Capital. Photo by author.

The Mask, the Horns, and the Eyes

Temple of Artemis at Ephesus.
Temple site from the original urban
settlement. Photos by author.

Chapter IX

Artemis at Ephesus

The Artemision at Ephesus was the largest Ionic temple apart from the Samian Heraion (which was never completed and was not nearly as tall) and the Didymaion near Miletus (which was not roofed). It was also one of the Seven Wonders of the World. Ephesus was a much older shrine than the building history of the monuments on the site can show, as is often the case. Epic and myth supply accounts for the archaeologists to dismiss, dispute, or check against surviving ruins and fragments.

Foundation legends for the city as well as the shrine are extremely confusing. The eponymous founder of the city was either an autochthonous (and male) Ephesos or Koressos, son of Kaystris (this father and son also provide the names of a mountain, now Bülbül-Dagh, and a stream, now Kuçuk Menderes); another eponymous figure is mentioned sometimes, the "Lydian Amazon" Ephesos. However, the divinized hero-founder who received a cult in the town was not its eponym but the Ionian Greek *ktistēs* and colonist Androklos: son either to Kodros, the last (or last-but-one) king of Athens, who died at the hands of the sons of Herakles, or to his brother Neleos. The sons of Kodros, the Kodridai, were revered as founders in several Ionian cities, of which the most important in this context was Claros, a little to the north of Ephesus.[27] An oracle shrine to Apollo stood there, famous enough to rival the other great Apollonian oracular shrine in Ionia, at Didyma outside Miletus. Claros was a smaller brother to Ephesus; perhaps just because of its closeness and its connection, the Apollonian shrine had to be—unusually for Ionia—a Doric one, to distinguish the brother's temple from that of his great sister and to make the family connection with Ephesus.[28]

According to yet another legend, long before Androklos arrived to drive away the native Carians and Lelegians, the cult of Artemis had been brought to Ephesus by the Amazons,[29] who were said to have founded several other Aeolian towns—notably Smyrna. The memory of their queen Samorna or Smyrna was also venerated at Ephesus.[30] Legend had settled these fierce, Ares-worshipping women in the north of Anatolia, perhaps even in the Crimea and the Caucasus, though there seem also to have been other groups of them, as in Libya. They figured much in Greek myth and epic; they tangled with Dionysos as well as with Herakles and Theseus, besieged Athens (where they had a cult), and fought against the Greeks in the Trojan War.[31] At the end of the second millennium, they were said to have ruled much of the Ionian coast "by the favor of Herakles." For all that, little is known about their historical antecedents, nor have any remains associated with them been uncovered by archaeologists. However the conflict of the various Ephesian legends is resolved, the Androklos version indicates that an oracle ordered him to settle where he found a sacred oak (perhaps that very one where yet another Amazon, the priestess-queen Hippo [probably identical with Hippolyta], had set up the first xoanon, a stick or board that stood in for god or goddess),[32] as well as a spring of sweet water. In all of these legends, the feminine nature of the cult, and the importance of women in the worship of Artemis, is clearly emphasized. The tombs of these main mythical figures—Hippo, Samorna, and Ephesos, as well as Androklos—were located within easy reach of Ephesus.[33]

As for the great temple, it stood in its vast enclosure on that same seashore, on the marshy, reedy plain on which Androklos had landed and which was a proper ground for a shrine to Artemis. It faced west (as was customary in Asia Minor), over the harbor and the sea, and arriving pilgrims could moor their boats at the edge of the temenos. The town first overlooked it from the hill, from the acropolis now called Ayusoluk.[34] Later, it expanded

The great statue of Artemis.
Ephesus Museum. Photo by
author.

The beautiful Artemis. Ephesus
Museum. Photo by author.

Chapter IX

downhill gradually around the temenos. The silt from the Little Meander has been pushing the coastline westward (it has moved some ten miles in 2,500 years), but already at the time of the building of the Hellenistic temple, in 286 BC, the Diadochos Lysimachos moved the whole town to the new harbor emplacement at the foothills of Bülbül-Dagh. A double processional way, some fifteen stadia (over 2.5 km) long, went from the Magnesia Gate of the new city toward the great Hellenistic temple (the one Vitruvius knew and St. Paul saw), which was burned by the Goths in AD 263.[35]

In the second millennium, however, in the age of heroes—and perhaps even earlier—there had, from time immemorial, been an object at the center of the cult: a xoanon, a statue, known from replicas in stone, bronze, and terracotta, which was indeed very much Homer's *Potnia therōn Artemis agroteri,* "mistress of wild animals, Artemis the Hunter"—particular patron of suckling animals, *Polymaston, multimammia.*[36] That large mummylike figure must have stood in the *sēkos,* the inner shrine of the temple, her feet protruding from under a kind of apron which bound her legs and torso: xoana were practically never naked.

In the larger versions of the statue, the straight edges of the stiff apron are clearly visible on the flanks, and the folded drapery of the neck and back of the Ionian chiton are also clear. The lower edges of the chiton are also pulled out from under the apron, over the feet. The apron itself is divided into heraldic-style rectangles decorated with deer, lions, horses, and gryphons in the central ones; bees, genii, and rosettes on the edges. Above the blanket, the statue's chest is festooned with (most often) four rows of pendulous protuberances, usually called her breasts. On her stiffly extended arms, she carries lions, while her hand probably grasped staves. On the chest, above the breasts, are necklaces of acorns and of zodiacal and other figures, as well as garlands. Behind her head is a halo disk (sometimes a hood), sprouting winged lions. She is coiffed with a high *kalathos* hat, a polos or diadem ornamented with sphinxes and gryphons—and in the largest figure of the type, the one found in the prytaneion of Ephesus, the hat is crowned with a lacy colonnade. These elaborate attributes were of course added to: they were as clothing thrown over what may have been an amorphous (or only a roughly shaped) piece of wood or stone, that original fetish Queen Hippo had first worshipped in the Ephesyne oak. The technical term for the stiff apron, *ependytes* (literally, a "pullover"), is applied to the cult statues of other deities, and also to the ceremonial dress of their officiating priests.[37]

The worship of Artemis

Artemis was, in any case, a powerful and popular goddess in Asia Minor. In Ephesus itself, she was also worshipped as Protothronie (she of the first throne), and her seat was probably carved into the rock on Bülbül-Dagh.[38] At Perge, the idol of Artemis Anassa (the Lady) was clearly a baetyl, dressed in even stiffer and more "detachable" clothes than her Ephesyne namesake, as was the Artemis at Sardis; while Artemis Leukophryene (of the white brow) of Magnesia on the Meander looked very like the Samos-Ephesus type of the goddess, to judge by the coins of the city. At Aphrodisias of Carya, a little inland from Ephesus, the large Ionic temple also sheltered a similar statue of a somewhat different (though equally matronal for her city), Aphrodite; her statue also wore a heavily embroidered apron, if the surviving copies are to be believed.[39]

The images on coins and seals show that the cult statues of the other great Aegean goddesses—of Hera in the temple at Samos and the Athena that was the Palladium of Ilion—

were vested like the Artemis and Aphrodite statues. The odd one out in this group of cults is the image of Zeus Stratos of Labranda (much worshipped by Mausolos of Halicarnassus), which was even incongruously garnished with mammae, and which was associated with the better-known Jupiter Heliopolitanus.[40]

Heliopolitan Zeus is usually assumed to be a Hellenized avatar of the Syrian sky-god Hadad, a close relation of the Semitic Ba'al; his ependytes was covered with heads modeled in relief, representing the seven planets personified. Indeed the embroidered ependytes, stiff with metal appliqué, is a thoroughly Asiatic device, known from Anatolian, Syrian, and Mesopotamian art, as was the polos, the fortified headdress of Cybele—who, like her namesake, the Syro-Hittite Kubaba, often acts as the protectress of cities as well as the promoter of crops and of all fertility.[41]

Cybele, Nin-Astarte, Aphrodite, Hera, Athena, and of course, Latona—it is all too facile to identify each separate and individuated cult with a general one of the Great Mother in Asia Minor, which is sometimes hypothesized as a late development of the immemorial and archaic cult of the first farmers. Yet the differences between these deities may be as interesting as their similarities. So for instance, some have attempted to account for polymammism by a

Lenaean vase. Naples, Museo Archeologico. © Fratelli Alinari.

commonplace—a necklace of ostrich eggs or a garland of some kind of fruit—yet in fact the obvious features of Ephesyne Artemis are not everyday at all, but must refer to specific ritual instruments, to some deliberately significant feature of the xoanon's costume. What these ependytes-wearing deities share, in any case, is not their gender, but the focusing of their cult on an image, fallen from heaven—*diopetus:* blocks of wood or (as in the case of the Great Mother of Pessinus)[42] stone, sometimes roughly shaped, sometimes formless but doubly draped like Zeus Stratos of Labranda, in an undergarment like the chiton, and with the stiff, heavily embroidered ependytes "pulled over." It was the most obvious and effective way of "personifying" them.[43]

The faces of the roughly shaped xoana must inevitably have also been ritual or quasiritual additions: masks. Those worn by wooden xoana may have been merely painted or they may have been much more elaborate, even of silver or of gold. The rigid frontality of the statues of Ephesian Artemis probably relies on the identification of the face as a mask.[44] Practically all the surviving copies of the statue are late Hellenistic or Roman ones and many are more or less heavily restored, though great quantities of them must have been made over the centuries

Attic skyphos. London, British Museum.

The Mask, the Horns, and the Eyes

in various places and in different materials, since the Ephesian cult, like the Delphic and the Delian, had long been missionary.[45]

It was tempting once to see the Ephesian cult as peripheral to "true" Greek religion, and this statue as an oriental idol enshrined in a Greek temple. Yet that very idol, and the similar one of Hera at Samos, were culturally central to Greek identity. Homer, though claimed more insistently by Smyrna, was intensely studied at Ephesus, which was also the home of Greek elegiac poetry. Hesiod came from Cyme, on the coast facing Lesbos; Sappho and Alkaios came from the island proper. Heraclitus was a hereditary magistrate in Ephesus. Xenophanes of Colophon founded the school of Elea. Pythagoras was reputedly a Samian, and so on. The importance of the shrine, beyond its vast size and universally recognized splendor, also depended on the place of the cult in the structure of Greek religious life.[46]

Artemis and Dionysos

The Artemis xoanon, like some of the other statues I mentioned, was almost certainly masked, yet the special Greek deity of the mask, the masked god par excellence, was not Artemis but Dionysos. Legend had him come as an immigrant, either from Phrygia or Thrace. Most popular as the god of wine and of drunkenness, he was worshipped in a rustic Athenian festival, the Lenaea, as a mask hung on a tree above an Ionian chiton.[47] The mask was bearded, but the chiton, usually embroidered, was also ornamented (where the chest might have been) with two cups turned to the viewer to look uncommonly like breasts. Amphorae and kantharoi of wine were offered to him, as well as bowls of porridge and barley scones; and the strange composite idol was garnished with sprigs of vine and ivy.[48]

Whether he came from some geographically localized region or from the legendary land of Nyssa, Dios-nyssos was already present to the Mycenaean Greeks,[49] though his secondary name, Bacchos (which is also the name that his initiates take—bacchoi), was indeed Lyd-

Coins of Samos and Magnesia.
After T. L. Donaldson (1859).

ian, or perhaps Phrygian.[50] Dionysos Dendros, the leafy one, the masked form I described, was known throughout Greece and was shown on a series of Attic pots.[51] In some places, as on Naxos, there were two masks, one of the furious god made of vine wood, another of the gentle god made of the fig tree,[52] while at Corinth there were xoana of the god made of the fir tree from which Euripides' wretched Pentheos had watched the Bacchic orgies.[53] In that Bacchic cult, as in those of Artemis and Aphrodite that I described, the "clothedness" of the god was an essential part of his worship since not only was the image masked and draped, but the actual putting on and off of the clothes may well have been part of the rite. In that he really does seem to represent the diametric opposite of Delphine Apollo's dazzling nudity.

Dionysos was always the wayward god. His worshippers were possessed; "outside" or beyond themselves, some of them shunned urban and even rural society. There were several cases of proceedings against his mystai, which in the Roman Republic provided one of the rare instances of religious persecution (in this case, for public immorality) before that of the Christians.[54] Artemis was also a goddess of the "exterior," of wild things—the powers of nature untamed—of everything beyond the scope of city and village, though in a very different way; yet she and Dionysos were both Olympian gods, and both ruled over parts of human life that the clarity of state religions did not seem to contain. But each of the different parts was essential to the religious life in ancient Greece.

The Greeks seem deliberately to have apportioned sections of what we might consider obviously matriarchal territory to masculine deities, of whom Dionysos is the most remarkable instance. Unlike several Syrian and Phoenician gods, he does not make an appearance as the consort of the Great Mother in one of her many manifestations, since he is the central deity of his own extremely popular cult and mystery.[55] He is moreover often an ostentatiously male, phallic god. Reputedly he was the first idol of Athens, the erect phallus known as Orthos (the standing one) set up in the temple of the Horai, the goddesses of ripening. The procession with a phallus was a normal feature of the Dionysiaca.[56] Although he is a phallic god (especially dear to women) in one aspect, in another, as in the many mask images that were erected in his shrines, he appears as "unmanned," hermaphroditic or bisexual.

As Apollo and Dionysos are summer and winter, they are also paean or dithyramb. Their alternation governs the Delphic calendar. Apollo has another, equally important

The Aphrodite of Aphrodisias.
Aphrodisias Museum. Photo by
author.

Artemis and Zeus. Relief plaque.
Ephesus Museum. Photo by
author.

Chapter IX

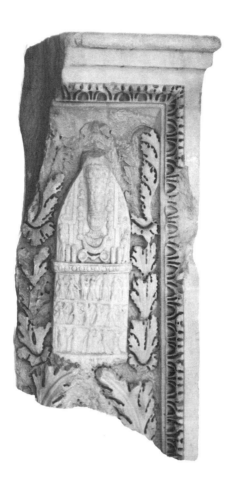

The Perge goddess. Antalya Museum. Photo by author.

Votive relief from Tegea showing Zeus Labraundos flanked by Ada and Idrieus. London, British Museum.

The Mask, the Horns, and the Eyes

coupling, since, as I said earlier, Apollo and Artemis are brother and sister, children of Latona, sometimes called the Great Mother of the Gods.

Apollo may also have been a "young" (in the sense of being a "newish") god of the Greeks, like Dionysos,[57] but at Delphi he almost certainly supplanted an older deity, as his sister had also done at Ephesus. Whatever his marginal (and chthonic, as mouse or snake) manifestations in the Olympian religion, Apollo was the lord of harmony, of justice, of clear and far sight—therefore also of mantic prophecy. The Doric order was thus appropriate to him, as Vitruvius had said and the temple at Claros has already shown. That Ionic should be appropriate to both his sister and to his counterpart Dionysos—even though Ionic columns are so clearly individuated by Vitruvius as female—seems to me to require further comment, which is best done by returning to the buildings and columns of the great Ephesian temple.[58]

The Ephesian Artemision

That last temple was said to be the eighth on the site and built exactly over the ruins of the seventh, which had been largely funded by Croesus of Lydia and burned by one Herostratus, an arsonist (who has no other claim on our attention), in the year 356, reputedly on the same night on which Alexander the Great was born. And the seventh had at its center a series of even older shrines—going back perhaps to the Bronze Age. When Alexander took Ephesus without battle or siege in 334, he offered to pay for the rebuilding of the famous and venerable shrine on condition that he became its dedicating patron. His offer was rejected, and the last temple was built by public subscription. Pliny says that its columns were offered by several kings.[59]

That famous penultimate, seventh temple had been "designed" and built by Theodoros of Samos, who had also been the architect of the Heraion of his native island, together with a kinsman (perhaps his father), Rhoecos. The Ephesians, who had intended to outdo the Samians, employed Theodoros together with another architect, Chersiphron of Cretan Knossos; and Chersiphron's son, Metagenes, is said to have completed the Artemision. Work on the new temple may have begun as early as 580; in 562 Croesus of Lydia captured Ephesus, exiled its tyrant Pindar, and paid for many of the columns of the temple, before he himself was deposed by Cyrus of Persia.[60]

According to Vitruvius, Rhoecos and Theodoros wrote a book on the Samian Heraion, while Chersiphron and Metagenes wrote one about the Artemision.[61] Theirs would have been the first literary, even canonic, statements about Ionic building. Since these huge temples are also the two earliest such dated buildings to survive in the Greek world, the history of Ionic building properly begins with them.

However many temples had stood on the site before Rhoecos and Theodoros' great enterprise, the shape of the seventh shrine is—in the present state of the excavations—very difficult to determine. Many remains as well as inscriptions from it have been found and they, on the whole, confirm the ancient historians' account: the last two temples had the same number of giant columns, probably 127 of them, as Pliny says.[62] It is almost certain, moreover, that both the seventh and the eighth temple had an octastyle front, three rows deep, with a wide central intercolumniation on the altar side. At the back of the building, a row of columns filled the wider central intercolumniation, which gave the rear a nine-column facade. This much seems certain, even if the ordering of more columns within the temple has not been fixed. Furthermore, since a channel ran along the floor from the block on which the statue must have stood toward the main entrance, the assumption has also been made that the shrine was at least in part hypaetheral.

If the back had the archaic nine columns, the front eight, while the sides two rows of twenty columns, it was a dipteral octastyle in Vitruvius' terms. The same layout was followed at the Heraion at Samos.[63] At Ephesus, some of the columns were inscribed with Croesus' dedication on the astragals at the bottom of the shaft; of these, a few were bilingual, in Lydian and Greek. One column, presumably also surviving from the seventh temple, carried a much later dedication, by Agesilaos—perhaps the Spartan king (whom I mentioned earlier) and who visited Ephesus in the years 396/5.[64]

Before Croesus' temple was begun, however, a seventh-century building of ashlar masonry had stood between the stylobate of the later temple and the site of the future great

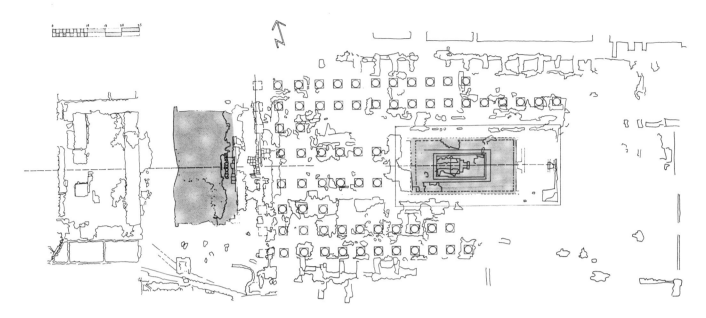

Temple of Artemis at Ephesus.
Survey of remains, after A. Bammer (1984). The two hecatompeda are shaded.

The name of King Agesilaos erased from the column base at Ephesus. London, British Museum. After W. Schaber (1982).

altar. Like the altar, that shrine, of which nothing but the foundation layer remains, was oriented not west-east, but north-south. Its character is uncertain, though it seems, like many early temples, to have been a hundred Ephesian feet long: another hecatompedon. It may well have been the shrine that the Cimmerians destroyed and that Croesus' temple replaced, though no convincing explanation is available for the change of direction.

The strange usage of an odd number of columns on the back of the temple, while the number is even on the front, bears witness to the persistence of archaic features of the cult of the Ephesian xoanon, as does the south-facing altar. Of course there is no telling what the style of the columns might have been before Croesus' reform, since the meager architectural fragments that survive from the pre-Croesan buildings do not include any moldings. It seems that some of the columns of the seventh temple, presumably those on the front and perhaps also on the sides, had elaborately figured drums, as well as necking ornament at the top of the shaft, just under the capital. As I said earlier, the entablature of this temple, like those of most Asiatic Ionic shrines, had no frieze, but a richly carved crowning cyma, with particularly splendid lion-head gargoyles. It may well also have had reliefs, orthostat-like, on the lower parts of the walls: at any rate, some surviving fragments seem to belong to corners of such

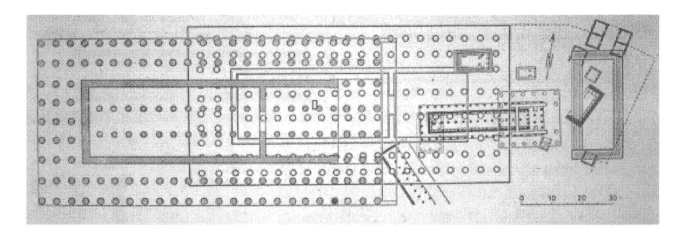

Temple of Hera, Samos: The superpositions of different temples on the site. After Ernst Buschor.

lower friezes.[65] The capitals of the Croesan Artemision, like many early Ionic columns, have no "eye": their spiral wound itself into a point. In these columns (or perhaps only in those of the main front), the spiral of the capital was entirely covered by an eight-petaled rosette.[66] No convincing attempt has yet been made to restore the capitals of Rhoecos' Samian temple.

The evidence about the first maturing of the Ionic order is therefore both ample and perplexing. But about the same date as the generally accepted one of the Ephesian temple, Ionic buildings dedicated to Artemis appear in their canonic and obviously familiar form in Marseilles, in Epizephyrean Locri in Italy, at Naucratis in the Nile Delta. At this time, too, a full Ionic order is used on a tomb in remote mountain Persia. Many of these examples are not structural; their columns do not actually carry a beam but are carved into rock as part of a facade, or cast in bronze, or yet stand isolated, carrying a statue. Votive columns are almost always Ionic: such Doric examples as there are, as for instance on the Athenian Acropolis or at Delphi (of which fragments only survive), seem to have been orientalizing in one way or another.[67]

Votive columns

If at the end of the seventh century Ionic columns also appear in conspicuous isolation, they almost all have volutes without an oculus, and they all seem to have been used as bases for a sphinx: at Aegina probably first, then at Delphi and Delos, where the islanders of rich and fertile, Dionysos-haunted Naxos offered Apollo sphinxes on Ionic columns at both shrines.[68]

The association of the sphinx with the Ionic column indicates other concerns. The sphinx arrived in Greece—from Egypt via Syria and Anatolia, probably—and changed sex on the way, perhaps in Syria; it/he/she became female: a death dealer and a man-eater (as in the story of Oedipus) and therefore also an apotropaic being, who is represented on shields as well as on many Lydian, Cypriot, and Greek tombs. On Greek tombs it is sometimes set on a base of quasi-Ionic (though more elaborate) scrolls, like the roof ornaments of a temple or treasury, rather than the capital of a column.[69] At Delphi the Naxian column carries a sphinx over the very spot where Apollo was said to have slain Python. This again associates the voluted column with another nexus of Greek religion: changing ideas about death and the soul, which of course had a powerful impact on architecture, as I will have occasion to point out

later, particularly in the next chapter. The Naxians offered votive columns, but did not have any treasuries at Delphi or Olympia. These treasuries at Panhellenic shrines were smallish buildings, sometime quite plain on the outside, sheltering stores of metal shaped into tripods and statues that were also regarded as bullion and security: of bronze primarily, but also of silver and gold, they played an important part in Greek economic life.

The Sikyonian thalamoi

It is not entirely surprising therefore that when Pausanias visited Olympia, he was shown two bronze thalamoi—the word is usually translated as "chambers"—in the treasury of the Sikyonians, one Doric, the other Ionic; they have long since disappeared. He was told that they were dedicated by the Sikyonian tyrant, Myron, to commemorate his victory in the chariot race of the 33rd Olympiad (648 BC). Pausanias assumed that the Sikyonian, since it stood at the west end of the platform of treasuries, was also the oldest of them; but excavations have shown that what remains, and what he would have seen, was built partly of local stone and partly of a reddish sandstone imported from Sikyon, and was in fact two centuries or so later than the date he supposed. However, even in mid-fifth century, an Ionic building in the western

Peloponnese would have been a rarity; if Pausanias had been right about the date, it would have been unacceptable on the current account of these developments.

Sikyon was a harbor town on the Gulf of Corinth, where later Dorian arrivals seem to have come into conflict with an earlier Ionian population: the two different thalamoi may have referred to that. At any rate, Pausanias was very impressed, and even notes that the smaller of the two weighed 500 talents (about 18,000 kg!). Since the surviving and excavated treasury is a single-chamber Doric building, the surmise of earlier commentators on this text—that the bronze was used to panel two inner chambers—is untenable, and the alternative hypothesis, that they were "detached" models (the term "portable" hardly seems to apply) has been generally adopted. This introduces the paradox of a first Ionic order having been found on Dorian (or at any rate Aeolian-Arcadian) land. However, these models cannot be reconstructed—not even notionally.[70]

A Persian Ionic?

A remote and quite different "earliest" Ionic building, one with columns (or at any rate half-columns) that support brackets ending in scrolled-spiral capitals, is not strictly speaking a building, but a tomb cut into the rock at Da-u-Dukhtar, near Kurangun (about 50 km north of Kazerun), in southwest Persia. Although it was robbed and emptied centuries ago (as were the adjoining tombs), their discoverer assumed that Tespes (Chishpish) and Cyrus I (Kurash) were buried there; father and son are mentioned as rulers of the southern Persian kingdom in a document of Assurbanipal dated 639. Thus the approximate date for the monument (before 640) would place it several decades before the Croesan temple at Ephesus, though a few years later than the presumed date of the Olympian thalamos. This ascription has been so much questioned recently however that it is extremely difficult to locate the building in relation to the other evidence.[71]

Such are some of the earliest recorded instances of more or less properly recognizable Ionic columns, with or without their beams. Others are mentioned in texts; some are located on awkward sites, not in the context of normal, urban temple building, and these may well have been earlier than the great Ephesian and Samian ones. However, while throughout Greek history the Artemision is the mythical archetype of Ionic building, a rather different kind of capital is associated with the origin of the Ionic volutes.

The Aeolic "order"

The earliest columns built by the Greeks with volute capitals do not provide a direct precedent for the canonic Ionic capital, nor do they fit Vitruvius' legendary account of its origin in any way, which is why some scholars have raised the arrangement to the dignity of a separate "order." A number have been found on the islands (Lesbos, Samos) and in Asia Minor (Smyrna, Neandria, Larissa); and there are some similar later ones in mainland Greece, particularly in Attica.[72] The one characteristic common to these capitals and to the later Ionic ones is the double spiral. In the Ionic capital however it seems drawn out of, or imposed on, the face of a bracket and it works horizontally. On the Aeolic capitals, the spirals rise vertically out of a ring molding, as if they were something tied to the top of the shaft. The capital therefore seems too springy, too flowery, to provide any notional bearing for a beam. The

Aeolic capital from Neandria. Istanbul, Archaeological Museum. Photo by author.

Aeolic capital from Lesbos. Istanbul, Archaeological Museum. Photo by author.

leaves have sometimes been identified as those of the water lily, but they look more generic. Moreover, their form, fixed over several centuries, seems too set and regulated to have been derived from the impression of a three-dimensional flower drawn in profile.

To return to the description of the Aeolic columns: the oculus is often hollowed out, and the triangular space between the divergent spirals filled in with a palmette. Such capitals were certainly colored, though practically no traces of paint have appeared on them. On the other hand, several of them are still pierced for the fixing of metal ornament, perhaps for complete sheathing. The abacus and the beam over it would have rested either on the palmette alone, or sometimes on both the palmette and the tops of the volutes together, making the capital work as a bracket.[73] Aeolic capitals are, unlike the Ionic columns to which they are related, a provincial phenomenon, and a relatively short-lived one at that. Later knowledge of this form by the Greeks is attested by many representations of "Aeolic" columns on painted pottery where they represent buildings of the time of the heroes, often supporting a "Doric"- type triglyph cornice.[74]

Moreover, long after this form of double spiral was replaced by the canonic Ionic capital in stone building, it remained a standard form in furniture decoration—at any rate as it is recorded in vase painting. Aeolic capitals first appear in the context of building activity that had much more in common with building types further east at the break of the millennium or even before: in Phrygia and Lydia, Urartu, Syria, and Mesopotamia.[75] Because the spirals in these capitals rise vertically out of a binding ring molding, the "natural" forms that they look most like are horns, rather than hair curls. But even that resemblance is very indirect. That ring molding from which they rise is usually ornamented with conventionalized leaves (sometimes two such rings), and a botanic connection has therefore seemed more obvious than a zoological one. Vitruvius is in any case explicit that garlands were invoked by the molding attributed to the capital.

Garlands are a constant feature of Greek worship, particularly of sacrifice, so that it almost seems as if the ring of leaves under the Aeolic horns was there to make such an allusion.[76] The heads of animals about to be sacrificed were often garlanded, as were the horned skulls that appear commonly in the metopes of the Doric cornice, though they are not a common ornament of the Ionic beam or column.[77] But I will have occasion to speak of the bucranium later, without reference to garlands.[78] If horns they are, however, those on the Aeolic capital would seem to resemble a goat's, a mouflon's (or even a ram's) rather than those of a bull. In the art of the Near East, two goats are often represented—heraldically, if not ritually— climbing and eating from either side of a tree: usually the composed or artificial tree of Mesopotamian reliefs, which I mentioned earlier.[79] Such themes are common enough eastward, in Hittite lands or Luristan. Indeed the splendid spirals that decorate the early Bronze Age pottery from the Elamite highlands are often the vastly enlarged horns of stylized mouflons, ibexes, and goats. Elamite deities, such as those on the stele of Untashgal or the rock relief at Kurangun, wear conspicuously horned headgear. The horned hat or helmet is a mark of divinity or of sovereignty all over the ancient world, of course, but horns seem to have had a particular importance for the people of Eastern Mesopotamia and the highlands that are now between Iraq and Iran. Even in the late twentieth century, the Lurs (as well as other modern peoples of northern Iraq), have maintained the practice of incorporating a horned and partly modeled skull of a mouflon, a ram, or an ibex into the roofs of their houses, particularly at the corners.[80]

Votive column of Larisa. Restoration, after J. Boehlau and K. Schefold (1940–1942).

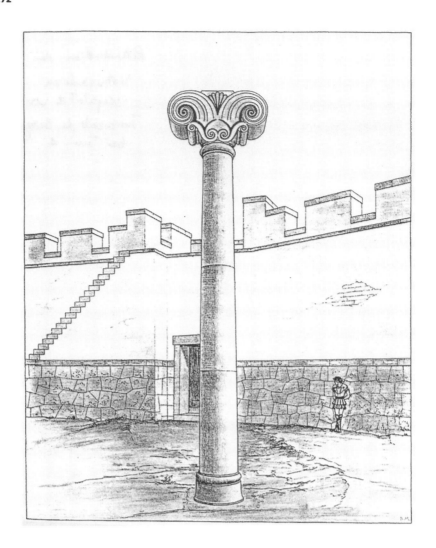

Oriental parallels

Goat and bull protomes are a very common feature of the capitals in Persian palatial architecture. The tall, rather slender columns of the palaces at Susa and Persepolis stood on bases that could be described as elaborated versions of Vitruvius' Ionic moldings. Much of the detailed carving was probably executed by Greek craftsmen.[81] Yet the striking thing to anyone looking for parallels between Ionic and Persian columns is that these protomes on the capitals are not carved into the bracket, like the spiral volutes of evolutionary historians, but are actually at right angles to the beams they carry. This is perhaps most obvious in one of the earliest surviving examples of this kind of column, which again is not built up but carved into the rock of Naqsh-i Rustem in the tomb of Darius the Great (522–486). The bracket may be clearly seen to pass between the protomes, as it probably always did in the palace at Persepolis. About the end of the sixth century, therefore, the Persians, with or without some Greek help, were using such forms rather differently from the Greeks, suggesting not filiation but perhaps some common origin. Hellenistic critics, who recognized the similarities between Greek and Persian buildings, described them as rather more opulent than Greek ones.[82]

Aeolic capital: Votive column of Larisa. Istanbul, Archaeological Museum. Photo by author.

Persian stone architecture had been a rapid (even a sudden) growth, about the middle of the sixth century, in the reign of Kurash II, Cyrus the Great (553–528). His presumed cenotaph or tomb near the city that he founded, Pasargadae, might indeed be considered the monument that opens the sequence of vast Persian construction. The tomb has no identifying marks or inscription and is not known ever to have housed a burial, though tradition had Cyrus killed in battle against Caspian tribesmen but buried in his new city.[83] The grandly sober, ziggurat-like tomb is a strange, specifically Persian amalgam of Ionic and Mesopotamian features.

Long before Cyrus, a highly developed stone-and-timber architecture had been practiced by the Hittite kings westward, in Central Anatolia. By the time the Persians arrived in what we now consider their homeland, Iran, the neighboring Hittite Empire was only a confused memory, a legend; and its monuments were already in ruins. However, the Persians knew the monumental architecture of Mesopotamia and of Syria, which Cyrus had conquered, and of course his son Cambyses was to subdue Egypt.

The stone architecture of the Persian palaces is to some extent an emulation of the monarchies to the west and the north that they had defeated—the Urartians and the Neo-Hittites—rather than of the Elamites, whose land they settled. Some of its forms, particularly the great hypostyles of the palaces, recall those of the Hittite lands; what may have been even more influential on the Persians was that they recruited their craftsmen in Anatolia rather

Tomb of Cyrus. After R. Ghirshman (1964).

than among Mesopotamian brick builders.[84] They also employed other Anatolian peoples as builders: Urartians, Phrygians, and Lydians, and even the remoter Ionians of the Aegean coast.

In Anatolia: Mittani, Hittites, and Urartians

The most impressive stone architecture in Anatolia before the works of the Greeks and their allies were the buildings of the Hittites and the Urartians. The people of the Hittite empire spoke a number of languages,[85] some of them Indo-European, some Caucasian, which they wrote in cuneiform (borrowed from Mesopotamia), in their own hieroglyphs (of uncertain origin), and later also in a linear syllabic script. Their court and religious ceremonies seem to have been almost programmatically polyglot. They had vigorous and complicated trade relations with both Mesopotamian powers (which they recorded in their own languages, in Sumerian, and in Akkadian, the lingua franca of eastern Mediterranean diplomacy as well as with Egypt. Eastward expansion brought them into conflict with the Caucasian-speaking Mittani, an elite or a ruling caste who had set up a strong state between Anatolia and Assyria. The Mittani obstructed the Assyrian trade routes and caused the economic crisis that contributed to the decline of the first Assyrian empire. When the Mittani buffer state was finally ground down by both its powerful neighbors, the Hittite-Assyrian confrontation became direct. But the Mittani had never managed to dominate the Mediterranean seaboard, and there Egypt and the Hittites exercised dominion throughout the second millennium.

A certain amount of dynastic intermarriage between Hittite Great Kings and pharaonic families or Syrian princes marked their alliances. Syria, Canaan, and the Phoenician lands were influenced by both Hittites and Egyptians and were the obvious terrain of their rivalries. A number of wars were fought between the two superpowers: and the famous battle of Khadesh of 1288,[86] when the Hittites were commanded by the Great King Muwatallish and the Egyptians by Rameses II, though undecided, has been fascinating to historians, since it is circumstantially recorded as a victory by chroniclers of both sides. Some years later, Rameses II married a daughter of another Great King, Khattushilish III.

The Hittite metropolis, Hattusas, near the modern Turkish village of Böghazköy, had within its compass at least one big, many-vaulted-chamber Mesopotamian-type temple and a number of smaller ones. Its "acropolis" was fortified separately and is now called Büyükkale; the walls enclosed a hilltop palace with a large diplomatic library and a hypostyle hall that was an artificial forest of thin columns quite unlike the riverine hypostyles of the Egyptians—and unrivaled until the late classical Greek ones (the Mystery Hall at Eleusis being the most impressive) or the palaces of the Persian Great Kings.[87] Hattusas was enclosed by an enormously long, buttressed, and battlemented wall; the gates were guarded by armed gods, lions, and monsters.

For the three hundred years the Hittite empire lasted, the Great King's title, *Lugal.-gal*, was written on the seals and commemorative reliefs as a composite character: one was a roughly conical bundle tied with thongs in two or three places, *lugal*, and the other, written over it as a double spiral, *gal*.[88] Two of these characters would be put on either side of the king's name and would hold a winged sun-disk over it like a beam, so that the whole thing looked surprisingly like an Ionic aedicule.

About 1175, the Great King Shuppiluliumash II fell, and with him Hattusas and the central Hittite power—to unknown invaders, not to their traditional enemies, the Assyrians. Their empire never recovered.[89] Unfortunately these victors over Shuppiluliumash and the later builders of nearby villages have left very little stone in place for the historians of material culture to consider, which makes the establishing of any parallel between the characters on the seals and the building forms an unresolved and teasing problem.

Much of the second millennium Anatolian building was—to adapt a term applied to western European construction—half-timbered, as was much Cretan and some Mycenaean architecture. One of the largest of that time in Asia Minor is the "burned" palace of Beycesultan, on the upper reaches of the Meander, upstream from Miletus and Priene in the lake country. That palace may well have been destroyed when the Arzawa were defeated by the Great King Labarnash, founder of the Hittite Empire, and the palace may well have been an Arzawa metropolis.[90] Their language was related to the Hittite one, as were their building methods: stone foundations, half-timbered walls using sun-dried brick and sometimes also stone. However, the closest parallel, not only in methods of construction but also in arrangement (compact, many-chambered plans, with internal courtyards and light wells), seems to be the contemporary buildings in Crete.[91] The Hittites had extensive trade agreements with Cretan rulers and seem to have practiced a form of preferential import taxation in their favor.[92] All of the Aegean states must have learned something of Hittite diplomatic skills, of the value of negotiation and preferential trade as well as of interstate dynastic marriages. The Arzawa maintained trade links with Egypt, and although the details remain confused, they seem also to have had relations with Mycenae. How exactly the Trojan War fits into this pattern remains exasperatingly obscure. At any rate, the Hittites knew of two other powerful states on the western coast of Anatolia, Millawanda and Apsas; and although the resemblance of these names to Miletus and Ephesus is thought to be accidental, most scholars find it too close to disregard.

The power vacuum left in Anatolia after the fall of the Hittite empire was not filled by the Neo-Hittite cities of northern Syria, but by the Urartian alliance. The people who lived around Lake Van and east of it spoke a Caucasian language rather like Hurrian, which they

Hattusas: General plan. After R. Naumann (1955).

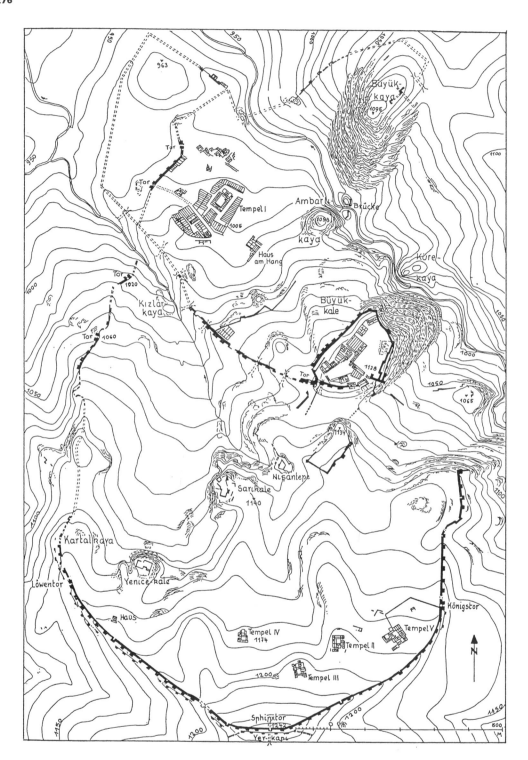

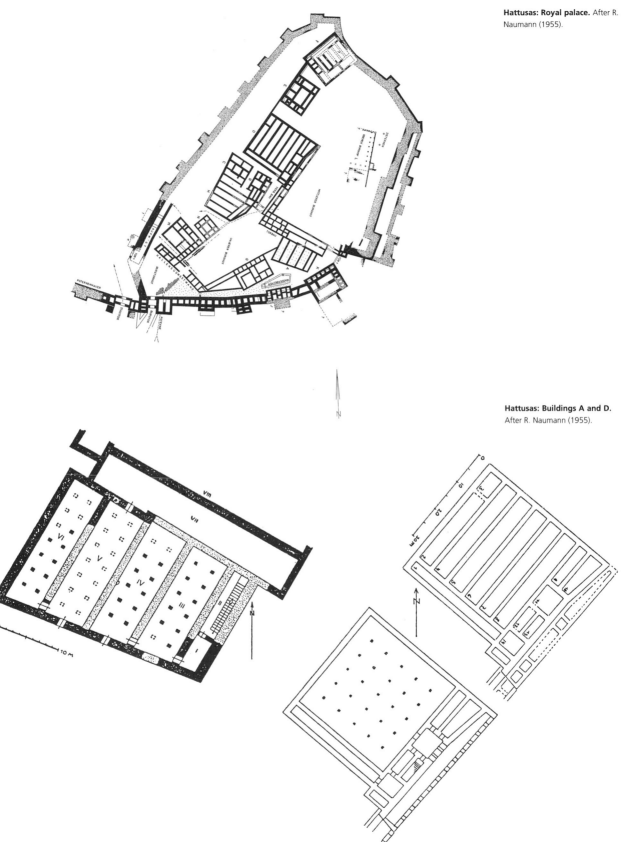

Hattusas: Royal palace. After R. Naumann (1955).

Hattusas: Buildings A and D. After R. Naumann (1955).

The Mask, the Horns, and the Eyes

wrote in Assyrian-type cuneiform. About the beginning of the first millennium, they transformed themselves into a powerful monarchy eastward of the old Hittite centers. They were also fine builders. Unfortunately their architecture has not been surveyed adequately, partly because the sites are divided between the Soviet Union, Turkey, Iraq, and Iran, and many of them are in rather inaccessible terrain. However, it is clear that their settlements were defended by powerful stone fortifications not unlike the Hittite ones, whose appearance is known from Assyrian reliefs, as well as from Urartian bronze models, which were perhaps votive offerings. They also built substantial columnar—perhaps even hypostyle—halls; and they may have planned the first fully orthogonal settlements, though they are most famous as very skilled bronze casters.[93] Indeed they manufactured bronze for export, and even seemed to have exported it to Archaic and (inevitably) orientalizing Greece. The fragmented Hittite empire had been taken over by those princes who were their immediate neighbors to the south and east.

Urartian power was weakened by the constant pressure of the nomadic Cimmerians from the Caucasus and of the Elamites from the east. It finally fell to Sargon II of Assyria, attacking from the south. King Rusas I committed suicide on being told of the fall of the Urartian holy city, Musasir, in 714, as Midas II of Phrygia, son of Gordias, was to do a few years later, in 696 or 695, after another Cimmerian attack.[94] The Urartian state continued for another century or so, during which time some of their most splendid fortified palaces—with

Hittite royal seals with *lugal-gal* character. Yazilikaya, Chamber B; Tudhaliyas IV with his protective god. After E. Akurgal (1962).

their columnar halls, predecessors of the Persian hypostyles—were built. Yet by the end of the seventh century, Urartu had ceased to exist independently; it came to be known as Armenia from its new lords, and under that name it became a satrapy of the Persian empire.

During the centuries of Urartian power, Hittite skills and organization and even a form of the Luvian language had been maintained in a number of the provincial centers to the south of Hittite lands that had not always been loyal to the imperial rule: Zinçirli, Karatepe, Tilmen-Huyuk, Tarsus, Kharkemish, and several others. These principalities—sometimes in alliance with the Urartians—were to give the Assyrians a great deal of trouble.[95]

Some of the smaller states had builders as fine as the imperial Hittites or the Urartians who were their contemporaries. At Tel-Halaf (once the Aramaic-speaking Guzana), halfway between the Tigris and Euphrates and on the Syrian side of the present Turkish border, its ninth-century prince, Kapara ben Chadianu (ca. 850–830), built or extended a palace (the "Western Palace")—or perhaps a temple—whose main entry, guarded by the usual monster-sphinxes, was supported by three column-deities: from left to right, a god on a lion, a god on a bull, and a goddess on a lion. The iconic device is obviously Hittite, but its architectural use seems an idiosyncratic development of the columnar porch, which had its origin in Syria in the second millennium. At Tel-Halaf, statues in the round had each an animal stand for the base; the body of the god is, as it were, the shaft, while the high polos-hat stands in for an

Burned Palace, Beycesultan.
After H. Seton Lloyd (1967).

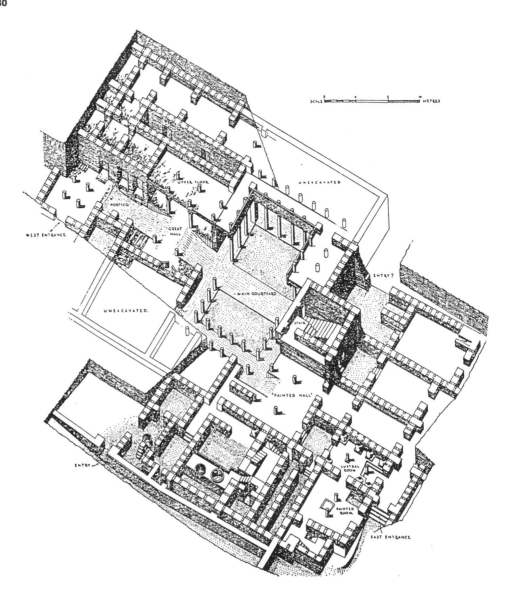

outsize abacus—all of which makes these anonymous deities into the indirect and improbable ancestors of the Ionian Caryatids.[96]

This palace is a prime instance of a *bit-hilani*, an Assyrian expression that has become a technical term. It denotes a palatial building (sometimes but less often a temple), whose porch of one to four columns is enclosed by pavilions on either side and leads to a long public room, usually across the axis. It is not clear where this form originated. It is first identified in northern Syria at the palace of King Niqmepa of Alalakh (a protégé of the Pharaoh Tuthmosis III, 1504–1445) upstream from Antioch on the Orontes, which was probably built soon after 1450.[97] The bit-hilani became very popular with Neo-Hittite builders: at Zinçirli, Tel Tayanat, and Sakjagözü, there are further examples counting from one to three columns. Unfortunately only at Tel-Halaf have the entire sculptured columns actually survived.

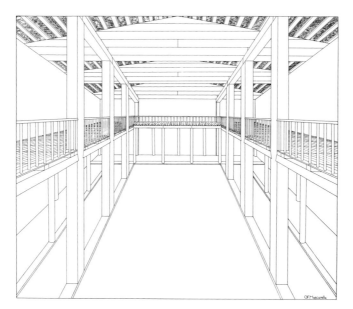

Megaron 3, Gordion. Interior view. Restoration drawn by G. F. Muscarella. © University of Pennsylvania Museum, Philadelphia.

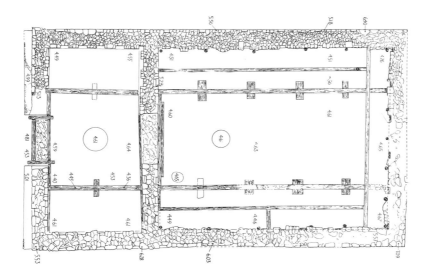

Megaron 3, Gordion. Plan drawn by J. Last. © University of Pennsylvania Museum, Philadelphia.

The Mask, the Horns, and the Eyes

Türbe, near Erkizan on Lake Van,
Turkey. Photo by author.

In Anatolia: Phrygia, Lydia, Lycia, Carya

Of the westward political configuration that the defunct Hittite empire considered as rivals, particularly of the kingdom of the Arzawa, little is known, nor is it clear who exactly were the Arzawa's westward neighbors whom the Hittites called "Luqqa" and "Karkissa." As it happened, the establishment of these Aegean centers and the time of their prosperity and expansion coincided with the beginning of Greek colonization on the western coast of Anatolia—where the people whom they called Lycians, Lydians, and Carians lived, whom some have identified as the same Luqqa and Karkissa.

While the Urartians filled the eastern power vacuum, a very different group of marauders, originating probably in the Balkans, whom the Assyrians called "Mushki" and "Tabal," seem to have welded themselves into the state which we know as Phrygia, which in a very different way became the power westward of the old Hittite empire.

The Phrygian monuments were quite different from those of the supposed Arzawa or of the Hittites. The royal residence of their capital, Gordion, seems to have been much simpler than the complex palace at Beycesultan—alignments of megaron-like halls, so common in Anatolia. At least one of these probably had a timber frame standing independently within its brick walls to support a second floor gallery and a roof that may in turn have been supported by a primitive form of truss. Unfortunately all too little survives of the pre-Cimmerian timber architecture.

The largest (MM) of many tumuli at Gordion may indeed have been the tomb of Midas II, the near-mythical great prince of the Phrygians. It seems to have been built according to an imported model, generally known by its Russian or at any rate Altaic term, *kourgan*: the burial was in a single-chamber log hut, which was then covered by an artificial mound. The grave goods were splendid, though in the case of the alleged tomb of Midas there were many textiles, elaborately carved and inlaid wood (witness to their extraordinary woodworking skills), and ivory but very little silver or bronze, supposedly because it had been looted by the Cimmerians before the king's burial.

On the other hand, the shrines that the Phrygians carved into the rock faces in their lands are their most splendid architectural relics and are mostly rock-face representations of buildings. Usually they are facades with gently pitched roofs; many have a central acroterion in the form of two volutes, which may stand for conventionalized horns. Often there is a central post in the end gable, sometimes guarded by two animals. There are other types of building shown, but all their walls are fretted in a way that is evocative of weaving, even more than their very repetitive marquetry—as if they represented tents made up of elaborately patterned carpets hung on a light wooden frame, as if they were still meant to recall the tents of remote nomad ancestors, as (much later) the türbes of Islamic saints in Seljuk lands were intended to reproduce in solid stone the perishable forms and surfaces of embroidered or woven tents.[98] Many rock-cut Phrygian facades have an empty niche and the suggestion of a public space before it, as if to act as background to some cultic act in the course of a procession in which an image was carried and deposited in the niche. A few of the reliefs are connected with tombs: as at Arslan Tas in the Köhnüs Valley, where two lions rear on either side of a phallic-looking pillar, their forepaws on a box that is both column base and the entrance of the tomb, in a dilated version of the two lions on either side of the column over the Lion Gate at Mycenae.

Mosaic from floor of Megaron 2, Gordion. Drawing by J. Last and Christopher Polycarpoi. © University of Pennsylvania Museum, Philadelphia.

Mosaic from floor of Megaron 2, Gordion. © University of Pennsylvania Museum, Philadelphia.

Chapter IX

I have already noted that *opus phrygium* is the root of the word *frieze;* another Lydian or Phrygian word, *tapete,* "carpet," has infiltrated many European languages through Latin: *tapestry, tapis, tapizzieren.* Textiles were eminently portable and were undoubtedly the carriers of many themes and of styles: the embroideries and carpets in the permafrost kourgans and graves at Pasaryk owe something to Asia Minor, to Persia, perhaps even to Mesopotamia.[99] And indeed the most neglected single Phrygian contribution to the fine arts may have been the invention of the pebble mosaic. Its highly patterned but random effect may well have been devised as a replacement of textiles thrown over a rush or beaten-earth floor.[100]

Apart from their vast stone carpet tents, the Phrygians represented another kind of construction on their tombs: carefully fitted, heavy wooden stave framework. Anatolia was once very rich in building timber—oak, pine, larch—almost like Lebanon was in cedar. The shrines and tombs of the Phrygians are less explicitly "wooden" than the later ones of the Lydians and Lycians, whose funerary architecture covers extended areas of rock face all over the southwest of Anatolia. All these stone tombs are in imitated wooden construction. Whether freestanding or rock cut, most of them represent two analogous (and perhaps archaic) types of carpentry construction: they are either boxes covered by a pointed vault, sometimes with a central post, or else they are facades paneled with elaborately recessed, carefully fitted squared wooden frames.[101] Occasionally these boxes are relieved with sculptured panels, which exceptionally, as in the Harpy tomb at Xanthos, become the dominant element.

South of the Phrygians, the Lycians, who lived in very hilly country, were also a people who recorded their doings on rocks and the walls of buildings. The great inscribed stone pillar at Xanthos, probably the tomb of their dynast, is bilingual, Greek and Lycian, which last is an Indo-European language related to Arzawan and Luvian—and to Hittite; the tomb (though constructed a century later about the end of the sixth century) was probably crowned with a Neo-Hittite throne flanked by two lions. A later inscription at the neighboring Letoon adds Aramaic as a third language on a commemorative stele.[102]

The Lycian builders must have raised carpentry to a level as high as the joinery of the Phrygians. Their country is still relatively little known to archaeologists, and evidence about their achievement has to rely on what can be found above ground. No timber buildings of any antiquity have survived, and the modern parallels, which are invoked to show the similarity of some recent wooden building forms to those which must have existed to provide precedent for the stone tombs, are too miserably put together to be at all convincing. In particular the setting of the whole frame or case on sleepers curved at both ends as if they were a kind of sled seems to have been a local invention. The timber structure is moreover not only carved in representation but in some cases the stone pieces are actually cut separately to represent each wooden member. These representations are much more logical than they are in Greek architecture.

As for the rock-cut tombs, their facades represent a kind of sectional elevation of a heavy timber structure that seems also to have carried a roof of clay and straw laid over a layer of round log purlins. But whereas in the entablature of classical buildings the details are carried around the corner as if there were no distinction between the front and side construction of a wood-and-clay roof, on the Lycian tombs the directional nature of the construction is an important aspect of the ornament, while the superstructures, triangular or ogival, seem to indicate that such independent roofs were a prominent part of their timber building.[103]

Xanthos: Theater detail. Photo by author.

Xanthos: Harpy tomb. Photo by author.

Xanthos: Harpy tomb. Details of base. Photo by author.

The Mask, the Horns, and the Eyes

288

Xanthos: "Sledded" tomb.
Photo by author.

Xanthos: "Sledded" tomb. Constructional details. After A. Choisy.

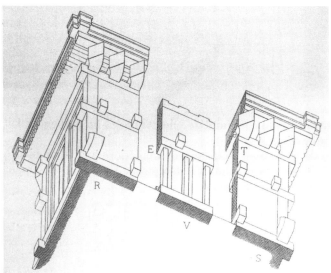

Chapter IX

Fethiye: Big Ionic tomb. Photo by author.

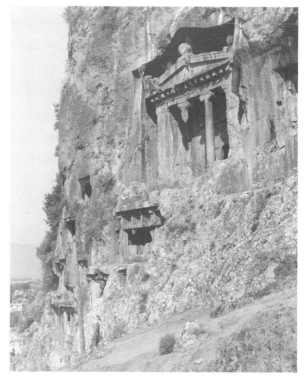

Fethiye: "Sledded" tomb and small Ionic tomb. Photos by author.

The Mask, the Horns, and the Eyes

Spirals and trees

All over the Hittite empire hieroglyph inscriptions appeared side-by-side with cuneiform script,[104] which gives place later to cuneiform alphabet (first recorded at Ugarit, Ras Shamra) and finally to Western Semitic and Phoenician script. Already then, linguistic and epigraphic variations are clearly reflected in the art of the region, including its architecture. The technique of protecting the lower parts of mud and brick walls against splash and flood with hard stone uprights, now called by the Greek-derived term *orthostats* (which I mentioned in connection with early Ionic building), seems to have appeared in Hittite lands and in Syria about the same time, a century or two after 2000, and became an essential part of north Syrian and Assyrian architecture.[105] Orthostats were usually carved with reliefs as well as inscribed, reliefs that are often of a narrative kind, though some were also emblematic and celebrative. It is these last that display a motif of great interest in this context: the "madeup" or artificial tree.[106]

I have already noted this theme and suggested its connection with Egypt, since in Babylonian and even in Sumerian art, both the *djed* column and the "Isis knot" seem to have some of the characteristics of an artificial plant, perhaps of a ritually fabricated one. At Kharkemish the sacred or perhaps more accurately the "ritual tree" appears on a number of reliefs, sometimes flanked heraldically, as the older Babylonian ones were, by two figures, usually warlike and male. The tree sometimes grows out of two spirals at its foot. All of these images display between one and three double spirals of the kind I have described in connection with the capital of the Aeolic column. They do rise one above the other, but not vertically from the stem of a trunk. They cross over to form a triangle, and the two spirals grow out of the triangle almost as if they were horns rising out of the forehead of an animal skull—though no such obvious reading seems intended there.[107]

Aphrodite at Paphos

At the entrance of two princely tombs at Tamassos in Cyprus (modern Politiko), such capitals butt into the doorjambs of the main chamber, which was carved into rock in imitation of a timber house. The butting is so abrupt that some archaeologists have thought that these stelae were reused older monuments. No inscription celebrates the princes, their families, or their fellows (who were buried there sometime in the eighth or seventh century), although forms of writing had by then long been in use in Cyprus.[108]

Cyprus may contain some of the most undervalued of Mycenaean survivals. The dialect was part of the Arcadian Cypriot family and by the Archaic period (when the Tamassos princes flourished), it was written both in the new alphabetic script and in the older syllabary. Unlike Linear B, to which it was probably related, the syllabary survived the Dark Age and was used in a slightly modified form to record the new Greek names and words (in some cases in biscript records), alongside the newer alphabet.[109] Mycenaean Greeks had colonized the western coast of Cyprus during the fourteenth century, perhaps peacefully. Who the "original" Cypriots had been, or where they had come from (if come they did), remains a matter of surmise.[110] In the second millennium the island certainly fell under alternating Egyptian and Hittite rule—or, at any rate, influence.

Greek culture did not suffer eclipse in Cyprus during the wanderings of the Sea Peoples—that predatory barbarian alliance well known from all around the eastern Mediter-

Doorjamb from Tamassos, Cyprus. After V. Karageorghis (1973).

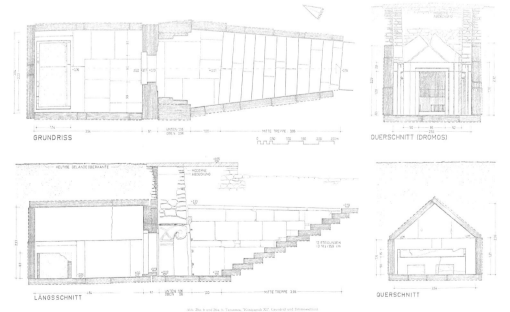

Tamassos, Tomb XI: Plan and section. After M. Ohnefalsch-Richter (1903).

The Mask, the Horns, and the Eyes

ranean—as did that of the Greek homelands. Sometime after the "settlement" of those Sea Peoples, Phoenicians established themselves on the eastern Cypriot shores; a little later the Assyrians asserted a tribute-exacting lordship over the island, which the Egyptians took over again after a century or so.[111] The island is therefore one of the few continuously settled areas where the Greeks lived in close contact with Phoenicians, as well as with Egypt and Mesopotamia. It is almost the only place where the tangled strands of their relationship during the Dark and the Archaic Ages may be traced and even separated.

Greek myth situated an essential theogony in Cyprus. Kronos castrated his father Uranos, and flung the severed genitals into the sea. They fell into the foam near Paphos. From that foam, pinked and fertilized as it was by the blood and semen that spilled from the divine genitals, Aphrodite rose; near the spot the most famous shrine of the goddess was built, around which a city grew.[112] Legend has the sanctuary and the cult founded either by Agapenor, King of Tegea (who led the Aetolian troops in the Trojan War), or by another, probably earlier *ktistēs*-colonist Kinyras, who gave his name to the dynasty of the Paphian priest-kings.[113] Archaeological remains appropriate to such a foundation as Paphos, where occupation seems to have been almost continuous since Chalcolithic times, were only identified with certainty some thirty years ago at Paleopaphos.[114]

While at Paphos a cult was offered to the goddess who was already called Aphrodite in the Archaic period, at Kition—only some fifty miles east, on the southern coast of the island—another (or perhaps analogous) goddess was worshipped by the Phoenicians as Astarte/Ishtar from the middle of the ninth century, on a site that had been sanctified for some centuries before their arrival.

Astarte at Kition

The masonry techniques of the Bronze Age temples at Paphos and Kition were similar: ashlar with drawn joints. The technique, which had also been developed by the Hittites and the Urartians, was used by the Phoenicians after them. No detailed reconstruction of the Bronze Age temple at Paleopaphos has been offered so far, though it is quite clear from images of it on Paphian coins, as well as from the remains, that it was not in any way like a Greek shrine, but more like a Phoenician or even a Syrian one.[115] Kition has been much more explored. The earliest temples there, probably built about the year 1300 but abandoned before the millennium turned, seem to have been of a common Syrian-Mesopotamian type: oblong, single-chamber buildings with double walls.[116] No evidence is available about what deities were worshipped there in the Bronze Age or on the nature of the cult, though the enclosure seems to have included a sacred garden. A temple explicitly dedicated to Astarte seems first to have been consecrated about the time of Itba'al (or Etbaal; ca. 890–860), who had been a priest of the goddess at her temple in Tyre; after his accession, he united that city with Sidon. His much more famous daughter, Jezebel, married King Ahab of Israel with consequences about which the second book of Kings (and the Prophet Elijah) had a great deal to say.[117]

Itba'al certainly promoted the cult of Astarte in the Phoenician colonies. His foundation at Kition is interesting here just because it represented a type of building well-known throughout Syria and Mesopotamia: a temple with a columnar approach. It was considerably larger than earlier buildings on the site. There were two rows of seven columns on either side of the "nave" at Kition, ending in a square, shallow apse. This apse was separated from the

columnar forecourt by a two-leaf door. A type of temple consisting of an approach court (or chamber) and an apsidal shrine is associated in the Near East with city gates and with the ground-level outbuildings of a ziggurat.[118] As often happens in Mesopotamia and in Syria, the pillars of the forechamber or forecourt are squared timber posts socketed into flat, square stone bases, while the stone pedestals built of ashlar blocks guard the shrine door. It is not certain whether they supported statues, or whether they were the tall (and nonstructural) door-guarding pillars that appeared on the coins of Kition and of other Cypriot (as well as Phoenician) shrines.

This practice was widely diffused and sometimes translated into rather different circumstances. The Urartians, for instance, placed leaf spears (which they seem to have bunched or branched into artificial plants, perhaps equivalent in some way to the Mesopotamian sacred trees) in stone bases on either side of the entrances to their shrines—presumably as guardians of the door. That is how they appear in the Assyrian relief celebrating the capture of the Urartian temple at Musasir, as well as on either side of the majestic relief of the weather god, Teisheba, who stands between them on a bull and holds a leaf blade in one hand and a bowl in the other in a gesture reminiscent of the Assyrian fertilizing genies.

The guardians of the doors

At Cadiz (or Gades) beyond the Pillars of Herakles (as the Greeks called the Straits of Gibraltar) in the temple of Melquart (the Phoenician Herakles), there were, according to Strabo, two bronze columns eight cubits high.[119] Two such columns, one of gold, the other of emerald, were also seen by Herodotus in the temple of Melquart in Tyre.[120] Perhaps equally famous (though in another way) were two columns before the inner shrine in the Jerusalem temple, the two that were specifically named Jachin and Boaz.[121]

The elaborate ornaments of Jachin and Boaz, as they are described in the first book of Kings, do not look like the columns and stelae from Cyprus and the Phoenician mainland, which were carved a few generations later. However, since nothing physical remains from the Solomonic temple, something about the structure and the ornaments can only be deduced from surviving buildings of the period—at Megiddo, Hazor, Ramat Rachel, and even in Jerusalem—or perhaps also inferred from representations of various Assyrian and Phoenician monuments. Was the "lily work," *ma'aseh shoshana,* of Scripture the double spirals of the familiar columns? Were the chains (or garlands) and pomegranates like the ornaments of Assyrian canopies? Like others who have tried to fit the archaeological fragments into the text, I can only speculate that they might have been so.

At any rate, perhaps the highest achievement of Phoenician craftsmanship was the Jerusalem temple, and it was earlier than the Cypriot examples. Moreover, it had an as yet all too little known context: at Megiddo, to take one instance, doorjambs, rather like the ones at Cyprian Tamassos, guarded the doorways of building 338 (the "Governor's Palace"), of which extensive fragments remain, and which is probably older than the Cypriot stelae by a century or more. In both of them, again, the "horns" of the fleur-de-lis volutes project from either side of a triangle. At Ramat Rachel similar doorjambs, a little later than the ones at Megiddo (but still earlier than those at Tamassos), have a kind of atrophied palmette in place of the triangle.[122] In several cases the triangle encloses an emblematic ornament: a crescent wrapped around a disk, the moon sign associated in the eastern Mediterranean with Astarte and by the

Basalt relief of two demons with a "sacred tree." Ankara Museum. Photo by author.

"Aeolic" capitals: Megiddo. Jerusalem, Israel Museum. Photo by author.

The Mask, the Horns, and the Eyes

"Aeolic" capitals: Ramat Rachel.
Jerusalem, Rockefeller Museum.
Photos by author.

western Phoenicians with Ta'anit—goddesses so similar that the two have often been thought identical.[123] As at Tamassos, these capitals were not entirely standardized: single doorways may be flanked by two of rather different pattern. Perhaps more curious is another property of the Jewish capitals: each one is carved of a single piece of stone, and some are very highly finished. Some even have surviving painted decoration, which makes them notable members in an architecture of much rubble construction, reinforced with timber and finished in plaster, even if its quoins and cornerstones were also often laid in ashlar.

In Phoenician lands and in Cyprus, double spiral columns were also reproduced on stelae, outside any structural context; sometimes freestanding, sometimes as a pair flanking an aedicule. In this relief, not only do the usual "proto-Ionic" volutes spring downward from the triangle, but upward tendrils spring up from it, framing a field on which a tree is flanked by two tiny sphinxes. As in most of my other examples, these fragile tendrils, which look more acceptable on jewelry and furniture, are crowned by a molded abacus, as if the stela was indeed intended to carry a beam. One of these stelae, from the tophet at Monte Sirai in Sardinia, has two such columns—not facing into the opening as in Tamassos or the Israelite examples, but shown frontally. These pilasters hold up an Egyptian-style cornice on which a winged sun-disk (a sign of divinity?) is displayed, crowned by a row of cobra heads. Within this decidedly Egyptian aedicule, a stylized woman, presumably a goddess, holds a disk (moon?) in her hands: another figure familiar from Phoenician sculpture.[124]

A much more "classical" naiskos in the Louvre has proper Ionic-capitaled columns flanking a scene of a priest addressing the goddess, though on one or two other such stelae the columns are versions of the Doric.[125] Such stelae featuring frontal aedicules are found scattered throughout the various Phoenician territories: North Africa, Sardinia, Sicily, Spain. The Archaeological Museum in Palermo has several dozen of them, as has the museum at Cagliari.[126] On the funerary rocks outside Byblos (in the valley of the Adonis river, now Nar Ibrahim), two such more or less "Ionic" and pedimented aedicules containing indistinct deities are carved in flat relief.[127]

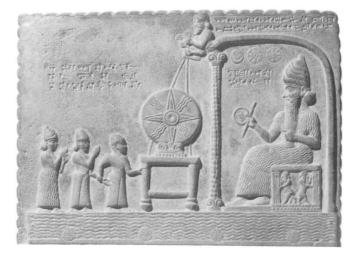

Relief of Nabu-aplu-iddin, the "Sun-God Tablet." London, British Museum.

The Mask, the Horns, and the Eyes

Cypriot (Aeolic) stele. New York, Metropolitan Museum. Photo by author.

Cypriot Hathor stele. New York, Metropolitan Museum. Photo by author.

The Mask, the Horns, and the Eyes

Models and shrines

Aedicules are not straightforward models, though they can be read as contractions, reduced in features as well as in scale from an aedes. In the Phoenician homeland, at Amrit (Marathus) north of Byblos, a similar shrine of uncertain Iron Age date has a cella more or less intact, though the other ornaments and columns have perished, except for the cornice.

At least in one case, in the archaic temple at Tharros on the eastern coast of Sardinia, enough of a Phoenician building has remained standing to allow of a restoration in the form of a magnified aedicule with a sanctuary niche flanked by Phoenician-Ionic pilasters. The cella of such a temple might contain the statue of a deity or baetyls (upright stones—any number up to five have been found) or yet other symbols: the so-called Phoenician caduceus, for instance, or a schematic incense burner, or a dolly figure made up of a large triangle surmounted by a small circle with a horizontal line bent upwards between them, to make *orans* arms, also known as the "Sign of Ta'anit"; yet another form is a curious bottle shape, possibly representing a headless body, or a mummy, or even some combination of some or all of them.[128] Quite often the columns flanking the niche are detached. In the temples at Amrit, as at Tharros, the columns were probably attached to the walls of the antae. Twin columns guard-

ing the doors of a shrine, whether attached or freestanding, seem to have been an essential element of Phoenician building.

Idalion in central Cyprus also had a famous shrine of Melquart/Herakles, where one of the rare examples of a complete architectural model in Cyprus was found, though it seems to refer to the cult of Astarte rather than to Melquart's. It shows a square building, with a woman-faced and -breasted bird looking out of the door, and women looking out of each window. The walls are pitted with what have been presumed to be "pigeonholes" for Astarte/Aphrodite's doves. The front door is sheltered by a narrow plank by way of a porch, and flanked by two columns that barely touch the little roof. They are a three-dimensional version of the stelae type, a bud surrounded by four volute-leaves, and they do not support anything.[129] They are often described as being Egyptian in style, but similar columns in fact recur throughout the Near East. However, these tiny columns offer the only three-dimensional prototype for the flat bracket capitals in full-size buildings.

Furniture and fabrics

An early and much-quoted example of such three-dimensional columns is the visible one (of two—the relief shows only a side elevation), which carries a flexible or floppy canopy over the statue of Shamash in a building dedication of Nabu-aplu-iddin, king of Babylon about 880–855.[130] The slender palm-trunk shaft of the column ends in a palm-bud capital, which is nonstructural, as in the Idalion model, but it has an almost identical motif at the base, while a widened version of it supports the sun disk before the statue.

About the same time that very motif appears on an orthostat relief of Nabu-aplu-iddin's enemy and rival, Assur-nasir-pal (883–857), king of Assyria, where it supports a tent canopy. In this later relief (there are a number of more or less schematic parallel examples) the motif makes no pretence at all of carrying weight: it is merely a finial, and the beam is attached to the column below it. This has led some historians to generalize that the volute capital, even the Ionic capital specifically, originates as a finial, whose setting conceals the joint between beam and column.[131] But Assur-nasir-pal's "tent" shown in that relief is perhaps more of an

honorific canopy than a dwelling. And a number of such canopies recall the "standards"—bronze or stone-tipped staffs—that Assyrian charioteers are sometimes shown carrying in the earliest orthostat reliefs of the ninth and eighth centuries. At the same time double volute capitals appear in representations of permanent buildings, where they are definitely shown as the supports of a roof.[132] Indeed these capitals seem closer to the made-up sacred trees that are such a feature of Assyrian palace reliefs than to the etiolated tipped staffs on which the canopies were supported.

The double spiral motif, still explicitly dependent on the sacred tree, becomes very pervasive early in the first millennium. The most impressive image of it may be the polychrome glazed brick wall from the "Throne Room" of King Nebuchadnezzar II of Babylon (604–562). One wall is covered with a panel of these tall "trees"—four of them, each one crowned with three sets of volutes one above the other, of the kind that spring from a triangle. They are linked by tenuous, ribbonlike bands knotted with lotus flowers, while the whole "capital" is crowned with a palmette. These trees are clearly not weight bearing.

The panels are edged by patterned borders that are not continuous, but are butted as if they were embroidered strips sewn together.[133] In fact the whole of this decoration evokes a textile hanging. There are precedents for it in Mesopotamia: a number of stone thresholds are clearly carved in imitation of carpets.[134] Imitation of textile in stone throughout the ancient Near East recalls another matter (on which I remarked earlier): the use of the double spiral of such "Aeolic" design in the ornament, and probably in the actual wooden and bronze fabric

Cypriot Sphinx stele. New York, Metropolitan Museum. Photo by author.

of ancient furniture. It survived as a decorative theme in wood and ivory long after it had disappeared from stone building.[135] Even the columns on Nabu-aplu-iddin's shrine of Shamash may, in a way, be more like pieces of furniture than of building: the thin palm-trunk shaft must be of metal or hardwood, not of stone or brick; in any case, the palm trunk is obviously artificial, as are the much older, fragmentary palmiform half-columns of the temple at Tell-el-Rimah.[136]

The analogy between plant and column is explicitly invoked in many representations that I have mentioned, even if the plant aspect is almost always mediated by human action, particularly by ritual, as in the making of the sacred tree. There is no reference to all this in ancient architectural writings, or at any rate not in Vitruvius' text about the Ionic column.

The horned skull

Double spiral columns recur on earlier hut-shaped model buildings, which were relatively common burial urns or household shrines, in many places in the Near East. Several of them inevitably display a variant of it: one of the most interesting has two pairs of spiral on each column, one sprouting up, the other one down.[137] The theme of the double curve is most explicit on two terracotta shrines from Jordan, linking it to many double volute capitals and ornaments in Phoenician lands, as well as to the Cypriot stelae.

A quite different analogy between head and capital was very much closer to Vitruvius' teaching. Palmette finials alternate in Mesopotamian imagery with animal, particularly horned animal, finials. As I have already observed, horns, bulls' horns especially, have sacrality in the Eastern Mediterranean region, as they have for any people who are tied to cattle for their livelihood. The "horns of consecration" of Minoan and Mycenaean art may not have any direct animal reference, but authentic animal horns of consecration (or something like

them) seem to have figured on top of some of the pillars in front of Cypriot and Syrian shrines. In some cases, these finials may have been actual bull or oxen skulls, though, of course, no such monument would have survived intact. On the other hand, there are two small but impressively credible models of such a structure, in which bull skulls are the finials of the three upright posts that frame the wall of a shrine.[138] If the human figure that seems to be offering a vessel before them is taken as an indication of the scale, these buildings would have been some nine to ten meters high. In Cyprus, too, there is some evidence that suitably cleaned bull and ox skulls were used as masks in temples—at Kition, for instance—and in circumstances of which we still know very little. One explanation of the Cretan Minotaur was that he could only be understood as a celebrant masked in just such a way, though this is only one aspect of a ritual which was very common in the ancient world.[139] Archaeology (as well as literature) can supply many instances of their use—ritual and quasi-ritual.[140]

But for masks or otherwise, horns were quite often used as a sign of divinity (as I have pointed out in discussing Elamite uses of them). Indeed heroic or divine personages wore horned helmets, while detached horns quite often adorned the hats or helmets of various deities: Assyrian, Hittite, and Cypriot.[141] Wholly artificial bull heads, hollowed to make vessels—rhytons intended perhaps for libations, with painted or gilded horns—are familiar representations in Aegean art. The gods were well known to be interested in bulls' horns, and to have a special preference for them when they were gilded:

> Each year
> Gifts with my own hands I shall bring
> To thy temple and stand a lustrous bullock at thy altar,
> with gilded brow

Cypriot miniature shrine. Nicosia Museum. After V. Tatton-Brown (1979).

Cypriot bull masks. Nicosia Museum. After V. Tatton-Brown (1979).

Columns from Tell-el-Rimah.
Paris, Louvre. Photo by author.

Palm column from courtyard. Istanbul, Archaeological Museum. Photo by author.

Chapter IX

Virgil has young Ascanius vow to Jupiter.[142]

Human builders were also known to have used animal skulls, suitably modified, as a building element, or at any rate an ornamental one, since the earliest days.[143] There is an attested continuity of ideas all over the late Neolithic ancient Near East between horned skulls and the tops of posts and columns. Even the Dorian Greeks worshiped Apollo Karneios as a ram-headed pillar, a kind of animal herm.[144] I have already commented on masks that divinize posts in connection with Dionysos, and he, of course, could also be horned. However, bucrania have appeared in many building contexts. A late Hellenistic commentator on Homer noted that the bull is a moon animal, presumably because of the crescent shape of his horns. This association has suggested to recent scholars that Neolithic peoples, who practiced committal of the dead by excarnation, were familiar with the shape of the womb as it was joined to the fallopian tubes and saw it as an image of the bucranium, setting the metaphor even more firmly in the Mother's realm than that suggested by the crescent image of the horns.[145]

Humbaba

On the other hand, the mask appears as a building element early in Mesopotamia. The custom of making foundation deposits, which almost always included a half-human peg or nail, is current in Sumeria and echoed in all Mesopotamia from prehistoric times until the Persian conquest. The figure, first made of baked earthenware and later cast in copper, was a sharp cone at one end, and a human figure (whether male or female) inscribed by the dedicator at the other; though it could also be an animal, such as a bull. The object has been interpreted as representing a godly version of the glazed cone that was a common finishing element in early Mesopotamian building. Alternately, and perhaps more justifiably, it represents the divinized doorpost as a guardian of the building; it served to nail the temple foundations ritually to Apsu, the water on which the world was based.[146]

Another common guardian, which grew to elephantine proportions in Assyrian building, is the winged, animal-bodied, human-faced demon. Often he was represented on a much smaller scale as a talisman or guardian. A different guardian was the deity Humbaba or Huwawa, warden of the forest cedars of Lebanon and patron of divination from intestines, who was also the infamous enemy of the first epic hero, Gilgamesh. His head, which Gilgamesh cut off, had some of the properties of the gorgoneion, and like it is commonly found as an amulet.[147] At an Iraqi site, Tell-el-Rimah (perhaps the ancient Karana), at least one mask of Humbaba, which has an insistently swirly and linear aspect, was mounted on the pillars before the gates of the Ziggurat temple.[148]

Though as column capitals they are unique—for the time being, at any rate—the Tell-el-Rimah masks are crucial: they are among the very few capitals of any design to survive from Mesopotamian buildings, and indeed they may have survived by a fluke. Yet they presumably represent a common building practice.

The column-statue

It is not clear where and when the column was "invented," or at any rate substituted for the wooden prop. In Egyptian and Sumerian Early Dynastic building, the half-column is already an important articulating device. Anthropomorphic posts are common enough in northeastern

Europe, in Southeast Asia, or in Africa, wherever wood is the major building material. The constitution of column, whether of stone blocks (as in the pavilions that surround the pyramid of Zoser at Sakkara) or of brick (as in the cone-studded walls of the White temple at Warka), required a further step in abstraction.[149]

How the idea of a capital as head relates to that of column as a body, whether conceptually or historically, is impossible to determine. In Mesopotamia, the alternating statues of gods and goddesses, each one pouring out water from some kind of globular vessel, are fitted into the recesses of the walls of the temple of Inrin at Warka, which the Kassite king, Karaindasch, built about 1440.[150] Deities pouring streams from vessels is a well-known type, but what makes the Inrin temple at Warka unique is that the statues alternate with pillars and actually form elements of a building.

In Egypt the earliest surviving "columnar" statues are in rock-cut tombs from the Fifth Dynasty.[151] No evidence of such statues survives from the more perishable Mesopotamian constructions, if indeed they ever existed there. And yet both civilizations have a veneration for and rituals connected with a divinized bundle made from reed, straw, or both. In Egypt it seems to have been male—the *djed* bundle, of which I have said something above; though there was also another and less common device, the so-called Isis knot, a floppy construction of leaves, which sometimes alternates with the *djed*, particularly in Early Dynastic works.[152] In Mesopotamia similar bundles appear on reliefs of ritual scenes flanking both shrine entrance and the figure of the deity, known to commentators as Inanna bundles; they are made of reed and finished at the top with a knot or a roll. "Building element" may not be quite the right term for them. When they guarded the door, they marked the sanctity of an

enclosure. When they stood on either side of the goddess, they may well have been signs of her divinity and her authority, and that is how they appear on an alabaster vase from Warka on which the goddess receives offerings, probably for her marriage feast. A contemporary and stylistically related gypsum trough shows what seems a *srafa* building in the center, from which a lamb emerges on either side toward a ram and a ewe. Bundles project from the building, overhanging the lambs, and other bundles frame the scene.

The device reappears constantly in connection with deities and buildings in the earliest Sumerian art, particularly on cylinder seals. Sometimes coupled bundles are shown beside the door or on either side of it, but occasionally they seem to be sticking out in the middle of the roof. Moreover there are a number of variants on the basic bundle; it could have a single streamer and one knot, three or two pairs of knots and no streamer, and so on: clearly, it is not regarded as a building element but something more like a sign, or even a character, a letter form, though its value, whether as a unit of meaning or of sound, is uncertain.[153] Common though it is in the earliest Sumerian reliefs and inscriptions, it seems to disappear during the Early Dynastic period, soon after 3000.

The reed bundle: Inanna's pillar

The various explanations offered of the bundle and its sacrality are almost inevitably unsatisfactory. The great German archaeologist, Walter Andrae (who wanted them to be the generators and the direct ancestors of the Ionic columns), thought that they were the two monumentalized doorposts of a reed hut. The knot on the reed post represented the eyes

Sumerian "trough." London, British Museum.

Sumerian post-knot sign. After E. D. Van Buren (1945).

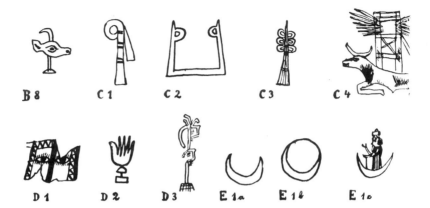

B 8 C 1 C 2 C 3 C 4

D 1 D 2 D 3 E 1a E 1b E 1c

through which a rod was passed, on which the mat (which served the shrine as a door) was rolled up. This seems to me a rather difficult argument to sustain, in spite of the vast learning Andrae brings to it. The device appears in the earliest Sumerian art in connection with both buildings and deities, yet seems to vanish from imagery gradually, though it does so more than two thousand years before even the "Solomonic" Aeolian capitals were first devised.

Nevertheless, the constitution of the column from the reed bundle, as Andrae has charted it, has a number of credible antecedents. Andrae gives great importance to the constructional methods of the Marsh Arabs, whom he saw using the huge reeds of the Tigris-Euphrates estuary (they grow up to six meters, or over eighteen feet high), much as the Sumerians had done, in an unbroken transmission of building technique over five thousand years. The most impressive are the tunnel-like *mudhifs*, "mens' houses." They were—before the Gulf War—built by a technique called *srefe*: bundles of reeds were set in the mud in two parallel rows and inclined toward each other to form a tunnel. In the end arches, the flowering tufts are crossed over to provide a decorative finial. The skeleton is covered by woven reed mats and sometimes plastered with mud. In the larger huts two upright bundles act as reinforcement and columns, and the end "walls" are fretted with more elaborate matting; the floors are carpets and mats over beaten mud.[154] Where Andrae seems to me to fudge the evidence is in wanting the curved, mat-holding reed bundle to act structurally from its very beginning, as it might (though in fact it does not) in the srefe huts of the Marsh Arabs.[155]

The origin of the *djed* column and of the Sumerian temple post in the reed bundle. After W. Andrae (1930).

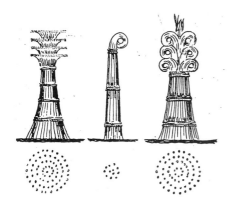

On the other hand, in the earliest dynastic Egyptian buildings of brick and wood, even of stone, one detail is insistently repeated: a representation of binding twine or thong. It is carved with obsessional repetitiveness and accuracy in reliefs and sculptures, and its importance for Egyptians is borne out by the *sem* hieroglyph, which signifies "join": a schematic version of the knotting of the bindweed and papyrus plants, the vegetable symbols of Upper and Lower Egypt.[156] Most monumentally, it is carved in the details of the Heb-Sed court of the first large Sakkara pyramid, that of Zoser; it appears repeatedly in the palm capital, which from the time of the Fifth Dynasty recurs throughout Egyptian architecture. These capitals are obviously "made-up"; they are not palm trunks as found in nature, but trunks stripped and squared at the top end to make an abacus—usually inscribed—for the beam to rest on, with palm leaves fastened to the crown to mask the joint (though they make no pretence to any structural role). The thong that ties the leaves to the column shaft is always carved, showing the end tucked in for fastening. In spite of its name, therefore, it does not represent a real tree, but an artificial one, perhaps a stripped palm trunk, its squared top protruding from the palm leaves tied around it so as to "reconstitute" it, through some building ritual. In the same way, the most common lotus capitals usually have a square abacus protruding from the flower, while the even more common bud capitals, with their double swelling and the buds bunched for a capital, also have a square abacus, which is supported by the post within the bundle and not on the frail flower buds. Egyptian columns therefore seem to me to be consistent representations of a post to which floral or other ornaments are fastened; and the nature of their ornaments plays on this process of binding.

Sistrum handle with Hathor head. New York, Metropolitan Museum. Photo by author.

Capitals from Hathor chapel.
Tomb of Hatshepsut at Deir-el-Bahari. Photo by author.

Astronomical chapel. Hathor temple, Denderah. Photo by author.

Hathor

Another important piece of evidence is the form of Egyptian standards. They were long staves with carved finials; in historic times the standards of all the forty or so nomes into which Egypt was divided accompanied the pharaoh on ceremonial occasions.[157] They are thus shown on the Narmer palette. Narmer, or Horus Nar, is identified with Menes, the founder of the dynastic rule of the pharaohs. Of these standards, that of Diospolis Parva (about halfway between Thebes and Abydos) was a cow skull mounted on a pole.[158] Diospolis was the nome of the goddess Hathor, an archaic cow deity, to whom several pharaohs (notably Mykerinus, the builder of the third Gizeh pyramid, ca. 2500) had a special devotion.[159] Her particular interest to me is that she is the only Egyptian deity whose head also appears as a column capital—analogously to that of Humbaba, and on present evidence somewhat earlier.

She is also one of the very earliest deities to have an identifiable surviving image: the human-faced cow head on the same victory palette of King Narmer.[160] Sometime later, she begins to appear either as a woman with cow's horns between which the sun is mounted, or a woman with cow's head, or a cow with a human face, or else a human being wearing a cow mask.[161] As a cow she represented the sky (which was the underside of her belly), and from her udder the Milky Way spurted, and her legs were the four corners on which the world stood. This majestic image of the sky cow may sometimes obscure her venereal nature, explicit in the mythical account of her seduction of her father, Re, her excessive partiality for beer, and her kindness to humanity.[162] The center of her cult was a little upstream from Diospolis, at Denderah; although the temple now on the site is Ptolomaic, there was certainly one on the site since the time of Chephren, Mykerinus' father, and perhaps earlier.[163] The twenty-four vast columns in the porch of the present Ptolomaic shrine are Hathor-headed, with one face for each direction, while six of them face the visitor on the main front. Like all those images, they show the youthful goddess, with her cow's ears peering through thick hair. That thick hair is probably not the goddess's own hair, but a wig. In Mesopotamia, where beard and head hair seem to have been regarded as ritual accoutrements, they were often curled, or even lacquered and covered with gold leaf or powder; the Egyptians, on the other hand, wore both wigs and false beards, sometimes as a part of ceremonial dress. Indeed a tightly platted beard seems to have been one of the pharaoh's insignia. Moreover, statues as well as people wore wigs of stone or metal, and also of wood and real hair.[164]

Hathor's wig sometimes falls on either side of the face straight; at other times it rolls into a heavy coil on her shoulders and the wide necklace-pectoral that covers them. This is often the case in Hathor-headed capitals, whether sculptured or painted. Two such full-size coiled-wig capitals from Cyprus are in the Louvre; two miniature ones (though with the goddess provided with very prominent horns instead of the cow ears) were found at Paphos and Amathos. All are quite late "classical"; there is also an earlier instance, on a stele of the late seventh or early sixth century.[165] The image is reproduced in various media, at different scales: a Hathor-head with its voluted wig appears as a repoussé image on a piece of bronze plate from a grave neighboring the ones with the volute columns.[166] This last is configured exactly like the head on the stele and in turn is to the stele much as the double curve "palm" is to the capital. The arguably "provincial" or "colonial" Cypriot images replicate the outlines of much earlier Hathor-heads, such as are sculptured in the temple of Rameses' II queen, Nefertary, at Abu Simbel, and painted in the nearly contemporary tomb of Vizier Senmut at Thebes. Their most conspicuous appearance is in the Hathor chapel of Queen Hatshepsut's temple at Deir-el-Bahari.[167]

However it is clear from that Cypriot stele that her head is almost interchangeable with and can be read as analogous to the volutes of the "lily work" of the Aeolic capitals. Is this perhaps *also* what Vitruvius was recalling when he identified Ionic volutes as the curls of a woman's hair, that the head and heavy wig of Hathor are a kind of realization of the Vitruvian metaphor? To put it another way: it may be that the great achievement of Ionian craftsmen (and of their patrons) is to have created or—to term it more accurately, perhaps—to have selected from existing prototypes (probably sometime between 700 and 600, at the very time when the Doric arrangement had matured) a form that would both be abstract, generalized enough not to refer directly to any of the specific themes to which I alluded, and yet maintain a sufficient contact with them: with composed "sacred trees," with masked, bewigged, and gowned idols, and with skull (or shell, or horn) carrying-posts. All this may seem remote from the refinement and geometric ingenuity of the Ionic capital, yet when Vitruvius wrote about its creation six or seven hundred years later, these themes were still present to him—as they must have been to many of his contemporaries, who knew the column as female—because it spoke to that realm of Greek experience that mused over beasts and their fertility, over seed growing in the earth, over death and rebirth.

X

The Corinthian Virgin

▪ The girl ▪ Death and burial ▪ The offering basket ▪ Acanthus—the plant ▪ Acanthus at Delphi ▪ Acanthus and tripod ▪ Delphi: The landscape ▪ The girl again: Persephone ▪ Monuments 1: The temples ▪ Monuments 2: Tholoi ▪ Monuments 3: Monopteroi ▪ The Athenian Olympieion

For his third kind of column, the Corinthian, Vitruvius does not set out detailed instructions for its design; as far as he was concerned, it was almost a kind of Ionic since it had much the same base and shaft.[1] Only the capital is different—and much taller, a whole diameter high, whereas the Ionic capital is only one-third of the diameter of the shaft.[2] Those two-thirds added to the capital do make the Corinthian column seem disproportionately slimmer than the other two.

For the rest, although the underside of the capital has to have the same dimension as the top of the shaft from which it grows, all its other measurements are determined, as in the other columns, by the module, which is derived from the diameter of the shaft at the base. In the module-high capital, the top one-seventh is given to the square abacus, while the remaining six-sevenths subdivide conveniently into three equal zones. The lower four have two superimposed rings of acanthus leaves and the top two are taken up by the stalks and flowers of the plant which curl on the diagonals of the abacus.[3] The abacus, like those of the other columns, is a square tablet in outline, though in the Corinthian capital the sides curve inward and the corners are usually chamfered. Its diagonal on plan is twice the height of the capital and the elevation of the capital on the diagonal is therefore one to two, but orthogonally (as it is usually shown) it is contained in a 1 to $\sqrt{2}$ rectangle.

Its composition and geometry make the Corinthian capital symmetrical about two axes, and that is much more useful than the Ionic one, because it presents no corner problem: if you set a Corinthian capital under an Ionic entablature, those perplexing riddles proposed by the corner of the entablature in the Doric, or the corner capital in the Ionic, simply will not arise.

Vitruvius goes on to say that this novel column can carry either a Doric or an Ionic entablature; the monuments tell a different story, however, since practically every surviving Greek Corinthian building has an Ionic one.[4] Vitruvius motivates this rather curious precept about alternative entablatures by saying that the third kind of column was "procreated" by the two older ones and could therefore "relate" to either;[5] she was their child. But a gender duplication had to be explained. Vitruvius distinguishes the two female columns by saying that as Ionic was womanly, so the more slender Corinthian was girlish, virginal. By way of comment, Vitruvius resorts to anecdote:

A Corinthian girl of good family, just old enough to think of marriage, fell ill and died. After the funeral her nurse gathered the pots and cups, of which the girl had been very fond when she lived, into a basket that she carried to the monument and laid on top of it. She covered the basket with a tile so that the things she put inside might survive that much longer than if it had been left open. As it happened, she had put that basket directly over an acanthus root which—being pressed by the weight—sprouted rather stunted shoots and leaves when the next spring came round. The shoots clung to the sides of the basket as they grew, and being forced outward by the weight of the tile were bent into curls and volutes at the corners. Callimachus (who for the elegance and refinement of his carving was called *catatechnos* by the Athenians) passed by the monument just then and noticed the basket and the tender leaves. Pleased with the whole thing and the novel shape, he made some columns for the Corinthians based on this model and fixed the rule of their proportion.[6]

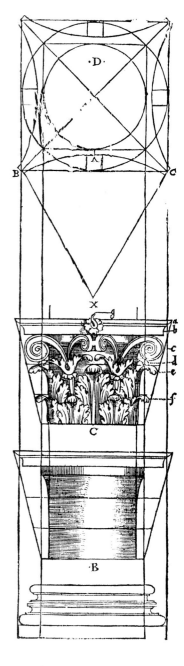

The origin of the Corinthian order and the setting up of its capital. From C. Perrault's Vitruvius (1684).

The Corinthian Virgin

Though a rather thin explanation in any case, this touching anecdote differs radically from Vitruvius' accounts of how the Doric and Ionic were devised, since it is explicitly set in recent history, not in the days of the heroes. It even involves a well-known artist whose signed works could be seen in collections and in public places in Vitruvius' time. Although some commentators find the anecdote trivial,[7] to me it seems to show all the characteristics of a myth—a myth reflecting ritual usage and harnessed to changing religious demands.

The girl

There are five elements in the story: the girl, death and burial, the offering basket, acanthus, and spring and reflowering. The girl (emphasized in Vitruvius' account) was virginal, even if a Greek reader might have found that almost paradoxical—since she was Corinthian.

Korinthos aphneios, "Corinth the wealthy," the Greeks commonly called the city: rich as well as orientalizing, and therefore also corrupt. Virginity had great curiosity value in Corinth, as it has in many licentious or permissive societies. At one of its several sanctuaries of Aphrodite—she was the patronal deity of Corinth as Athena was of Athens, and her armed statue stood (with those of Ares and Eros) on Acrocorinth—her priesthood included temple prostitutes on the Syro-Phoenician model.[8] To the poets Aphrodite was *Kuprogeneia,* "Cyprus-born": in Corinth it was generally believed that her cult had indeed come from Cyprus, though some thought that it had been imported from another island settled by the Phoenicians, Cytherea. It was certainly seen as "oriental."

Death and burial

Whether virgin or merely virginal, the girl died from causes unnamed and was buried as was customary, presumably by a roadside, outside the city gates. She was not cremated but inhumated, which was the most usual kind of burial in Greece, though cremation was intermittently practiced.[10] Being laid in earth, however, seems essential to the anecdote.

A tombstone was put over her grave, according to custom: *monumentum,* Vitruvius calls it. What Greek work was he translating here? *Mnima,* for which "monumentum" was the usual translation, served for anything that acted as a reminder. A sumptuary decree of Demetrios of Phalerum (317–307), which he enacted in emulation of Solon's decree on the same subject (promulgated nearly three centuries earlier), limits tomb markers to three kinds: a column (no higher than three cubits), a stele, or an amphora.[11]

All these forms of monument had a long history, going back to the Dark Age. Many amphorae, krateres, and loutrophores with funerary paintings, some six feet high or more, had already been tomb markers before the Archaic period in the very way Demetrios regulated; and many of them were pierced so that libations poured into them would fall directly on the grave.[12] By the sixth century, marble amphorae and lekythoi were taking the place of the ceramic ones on the tombs of the rich. The stele and the vase seem older forms of grave marker than the column; even older was the heap of earth or stones, *tumba* (of which Homer often speaks), so that the grave was often conceived as being of two elements: tumba, the heap; and stele, an upright slab. That was already how the graves in the royal circle at Mycenae were marked. However, to provide the appropriate model for the finicky, literal-minded Callimachus, the monument that Vitruvius mentioned must have been such a column as Demetrios' decree was to specify, for Athens at any rate.

The offering basket

Offerings at the tomb—from animal sacrifices to the sort of small gift in a basket that the Corinthian maid's nurse had brought—were very common. The word Vitruvius uses here for the basket round which the acanthus grew, *kalathos,* he transliterated from Greek; Latin grammarians usually translate it by "quasillus," a small circular hamper, wider at the top.[13]

Acanthus—the plant

The Greeks associated many plants with funeral rites. In Athens marjoram or oregano was usually strewn on the bier and four broken branches of vine were put under the corpse during the laying out, the *prothesis.* Cypress boughs were nailed to the doors of a house of mourning. A laurel branch was dipped in a bowl of water at the entrance for lustral sprinkling. The head of the corpse was garlanded or crowned with *selinon,* a name variously translated as "parsley" or "celery."

The bier was also hung with fillets—woollen bands that were later taken to the monument, wound round it, and usually tucked in or tied into a knot. Plants and wreaths were sometimes put on top of the tomb, but also stuck into the knotted fillets; selinon was commonly used for this, but acanthus was also placed at the foot and the head of monuments after the funeral.[14] Asphodel (which grew plentifully in the underworld, according to the poets) was often planted on or around graves.[15] What mourners had in mind when making such tributes they did not make explicit: those today who place flowers on tombs would probably say that they are keeping the memory of the dead person fresh and green.[16]

That a basket should have been placed *in summo,* at the top of a tombstone, seemed awkward enough to commentators on this passage; but a root of acanthus on top of the tomb, under the offering basket, was a real puzzle. In fact the most popular illustration of it, in Fréart de Chambray's *Parallèle,* shows Callimachus making a drawing of the tile-covered basket that had been placed on a small earth-mound (at the foot of a much larger pyramid), with a bushy acanthus growing round it; and that engraving was often emulated.[17] However, as the details of Greek custom and their painted pottery became familiar in the course of the nineteenth century, what had once been perceived as a conundrum to be explained away turned out to be a commonplace illustrated by the paintings that decorate many of the lekythoi. These vases are tall and thin, with an extra narrow neck, and were used for ointments since archaic times; the decorated ones I have in mind were developed in Attica near the beginning of the fifth century, and were discontinued after 390 or thereabouts.[18]

Lekythoi had sometimes been placed at the head of the corpse during the laying out, or under the bier, but the white-ground ones were developed specifically as a grave offering; they were used to pour a trickle of scented oil over the corpse, and several of them might be put around the base of the monument. Most were glazed black at the foot and the lip, while the belly and neck of the vase were covered with a white slip, on which figures were brush-drawn in black glaze, sometimes heightened with white glaze or colors, blue, green, russet—usually sketchy, like 1920s Picasso drawings. The scene represented had (almost always) to do with burial: the dead person is carried to the monument or mourners sit or stand by it. On many lekythoi the monument, if it was columnlike, carried a basket, sometimes surrounded by a cabbagy plant, which could well be read as a rough drawing of an acanthus; it was certainly too lush for parsley or celery.[19]

White-ground lekythos. New
York, Metropolitan Museum.
Photo by author.

White-ground lekythos. New York, Metropolitan Museum. Photo by author.

Development of white-ground lekythos: A warrior deposited in his tomb by sleep and death. After E. Pfühl (1923).

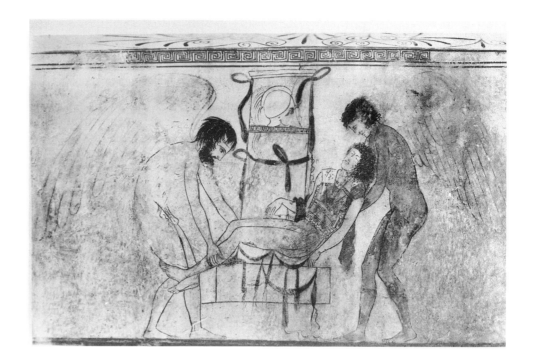

Development of white-ground lekythos: Mourners bringing baskets of pottery to the tomb, which is crowned by an acanthus-like plant.

Acanthus plant. Photo by author.

Another ambiguity is inherent in the customs and the legend (of which many commentators complain) because the word *akanthos* means something like "flowering spike" in Greek, and several different unrelated plants bear that name botanically in Mediterranean countries: such as *Acanthus spinosus,* a large thistle; or *Acanthus mollis,* which is also known as *Brancus ursinus.* Vitruvius' plant must be the second of them, the soft one.[20] In any case, both plants have floppy, serrated leaves and straight, long, simple and smooth stalks. Both are tenacious weeds that will grow almost anywhere, though only *mollis* was cultivated as a garden ornament.[21]

In Greek tomb sculpture, stone acanthus appears long before the Corinthian column, primarily in the acroteria that crown tombstones, or sometimes in combination with an elaborated form of the palmette. In Attica especially the palmette, a common finial of Archaic grave stelae, is replaced in later times by an acanthus-and-palm-frond relief.

Now the palmette is a very stylized bunch of leaves (making a kind of pom-pom), tied by a band at their base. The origin of this device is uncertain: some have suggested that it was Cretan, Minoan.[22] It certainly appears on the made-up trees of Assyrian and Neo-Babylonian reliefs, and there takes a shape much more like the representation of date palms in Assyrian sculpture than like the palmetto on the volute capitals that I described in the previous chapter.[23] The Hittites and Urartians also knew it.[24] Sometimes it is used in conjunction with the lotus flower, which is a conventionalized form and may be of Egyptian origin. At any rate the Assyrians liked fringes of lotus flowers alternating with buds on their dress, their furniture, and their relief sculptures. The Greeks of the Archaic period, on the other hand, seem to have transformed this kind of fringe into a common palmette-and-lotus pattern. Such alternating ornamental devices were run along the cyma moldings crowning the cornices of classical period temples.[25]

Archaic stelae in Attica had palmette acroteria to crown them, and occasionally the palmettes were framed by double spirals that supported the tomb's guardian sphinx,[26] but sometimes they were independent crowning ornaments, usually in flat relief, or even painted in encaustic on a flat, framed ground. In the later fifth century they become aggressively specific and even three-dimensional—or at least swell into high relief. In all of them two spiraling and stereated stalks of some undefined and frilly-leaved plant seem to issue unrealistically from a low and wide acanthus, while palm fronds rise between and above the stalks to weave themselves into the palmette shape. The stalks are always fluted, suggesting the reeded appearance of celery rather than the smooth acanthus. It is worth emphasizing the unlikely conjunction of leaves, because the ornament does not represent any single plant, but seems to consist of three different ones: leafy acanthus at the base, a spiraling ribbed stalk (perhaps of celery or some such plant), and palm fronds rising between them.[27]

The palmette acroterion rising out of an acanthus plant was not merely a tomb ornament. It was a common enough finial even in Archaic buildings: the Doric temple at Assos and the temple of Aphaia at Aegina had central acroteria that were very like—if more elaborate than—the finials of stelae. Such acroteria grew in size and complexity during the Hellenistic period. At Tegea, in the temple of Alea (of which more later), the ridge tiles were probably a series of acanthus-and-palmette upstands, and the central acroterion over the pediment was a fantastic confection in which three palmettes sprouted from an elaborate acanthus-and-spiraling-branch artificial plant. At Nemea nothing remains of the acroteria, but the Archaic temple (which the building covered) also had modest palmette ridge tiles.[28] Another such acroterion crowned the summit of the tholos at Epidauros, while yet another is Hermogenes' pseudo-dipteral Ionic Artemision in Magnesia on the Meander. The monument of Lysikrates in Athens had a truncated version of such a finial to support the tripod.

Acanthus at Delphi

The most impressive of these plants was perhaps the limestone one, over twelve meters high, that stood just above the main entrance of the temple of Apollo at Delphi, by the temenos of Neoptolemos, its base probably built into the structure of the small shrine. Neoptolemos/ Pyrrhos, son of Achilles, is a secondary figure in the accounts of the Trojan War, even if it was he who actually killed King Priam. His importance at Delphi was quite out of proportion to his shadowy role in epic. According to the tragedians, he was in his turn murdered by the jealous Orestes at the hestia, the fire altar of the Delphic temple, which was, in a sense, the fire altar of all Greece. He is also identified with a pre-Apollonian deity of the place.[29] His body, originally buried under the temple threshold, was moved to a shrine of its own by King Menelaus when he visited the oracle.[30]

The name *Neoptolemos* can be translated as "young warrior" or "new recruit," while *Purrhos* connects him both with fire and with Deukalion (the Greek Noah), who had Pyrrha for consort and whose "ark" landed on Parnassus. The altar at which he was killed, within the sanctuary, is close both to the omphalos Zeus fixed as the world center and the tomb of Dionysos, by the Pythia's tripod. But his shrine and the acanthus column—if the reconstructions are correct—faced all visitors to the Delphic sanctuary before they turned the corner of the blue Chian altar to move into the temple proper.

Fragments of the great acanthus tripod-stand were found by the archaeologists strewn down the temple stairway: only the "capital" and the foot of the sculpture have been reassembled in the Delphi Museum.[31] Over an inscribed square plinth, a tall, composite (and celery-like, rather than acanthus-like) stalk rises out of three vast floppy acanthus leaves. Four rings of smaller leaves act as binders for the reeds that make up the stalk; from the fifth, the

The giant Delphic acanthus and the dancing figures. Photos by author.

Apollo on tripod. Marble relief. Athens, Archaeological Museum. Photo by author.

top one, three more large acanthus leaves spread out. Standing back-to-back on each one, life-size girls are dancing. Each is crowned with a diadem or garland, or perhaps what looks like a diadem is a palm leaf wound around the basket filled with such good things as the girls who walked in the sacrificial procession in honor of Neoptolemos carried on their heads.[32]

Notwithstanding, many interpreters of this monument have taken them to be dancer Thyades, the three Dionysian maenads named after Thyia, the daughter of a legendary autochthonous priest who, according to some ancient authors, introduced Dionysian worship there. They danced ecstatically on Parnassus, sometimes quite losing their way.[33] The Dionysos they worshipped, and whom they woke with their dancing, was called Liknites, after the winnowing basket in which they would "discover" him, within a cave up on the mountain: the *liknon,* a flat-bottomed basket, seemed to be used both for winnowing and as a cradle for the newborn and it was quite different in shape from the *kalathos* of the Corinthian capital.

Acanthus and tripod

From the top ring of the floppy leaves the stalk sprouts to support the bowl of a bronze tripod, whose legs once rested on the tips of the acanthus leaves, between the girls, while its bowl rose above their heads. The tripod, which disappeared long ago, would therefore have been over two meters high, so that the whole group must have stood over fifteen meters. A much bigger tripod—the biggest bronze one known and some six meters high, cast from the Persian armor captured at Platea in 479—stood about twenty meters away.[34] The battle between Herakles and Apollo for the Delphic tripod was one of the foundation legends of the shrine.[35] Tripods of various sizes proliferated in the sanctuary. By far the most important of them was the golden

tripod inside the temple on which the Sybil sat for the god to possess her when she prophesied. For this the bowl was covered by a concave lid, the *holmos* or *epithēma,* which (like the tripod kettle) has a recurrent mythological association with Dionysos.[36] The omphalos, which stood before the tripod, was also one of his reputed tombs.[37]

Display and ritual tripods were *apuroi,* "not meant for the fire" or cooking, much as modern athletic trophy cups are not really meant for drinking. But of course the tripod is originally a cauldron or kettle, a *lebēs,* on a minimal, three-prong stand.[38] No monument quite like the great Delphic acanthus has been found or described anywhere else. But the association of lebes, underworld powers, and plants—particularly the acanthus—seems very ancient. At Delphi itself, in the museum, there is an Archaic stone lebes-stand which almost looks like a primitive model or prototype for the Delphic acanthus column. Three Hekatean figures, back-to-back, stand around a plant stem, and the bronze lebes is balanced on the (rather generalized) flower springing from that stem. Nothing is known about this curious object, which was found in the temenos, except that it was probably made late in the seventh (or early in the sixth) century.[39]

In Greek ritual usage, the tripod proper had assumed a curious role: in its votive form it was bullion, and such bronze, electrum, gilt, and even golden tripods crowded temple treasuries (and could be used as collaterals for bank loan purposes) or stood on stone stands at temple approaches, as many did at Delphi. Tripods were also tokens of power in diplomatic exchange. They could be trophies as well as ritual implements (such as the one used by the Sybil). Yet another association also seems material here: in some myths the bodies of the aged were boiled to be rejuvenated; the dead, to be revived. Medea, who had a particular devotion to Hekate (perhaps as her daughter), boiled the body of an old deer experimentally with her magic herbs and brought it back to life rejuvenated to convince the daughters of King Pelias of Iolkos to do the same with their father—though that second operation failed because she withheld the magic herb from them.[40]

Lebes associated with acanthus is another recurrent theme. A lebes on a stand is one of the most impressive Athenian funerary monuments from the Hellenistic period: the whole thing is carved out of marble, and the lebes, with gargoyle-head handles like an Archaic bronze vessel, is set on palmettes over huge acanthus leaves.[41] The golden acanthus plants that grew between the Ionic canopy columns of Alexander the Great's funeral chariot described by Diodorus Siculus must have been even more opulent and impressive.[42] In fact, lebes and acanthus, lebes and tripod, and lebes and Hekate make obvious funerary associations; in the fifth and fourth centuries they seem to allude not only to death but also to resurrection, or at any rate to a life after death.

The transformation of the typical dragon-handled bronze lebes into a limestone one points up another problem: the first acanthus-and-lebes monuments could well have been entirely of bronze. It might be relevant that Corinth's bronze was almost as reputed as its perfumes.[43] "Corinthian" may have referred as much to the bronze of the ornaments as to the place of its origin. The ostentatious transformation from bronze to stone, from the malleable and riveted metal to hard monolithic marble, may also have represented the immortality offered by the monument.

Delphi: The landscape

To return to Delphi: visitors who had made their way up from the lowest Athena shrine at Marmaria, though the great temenos of Apollo and past his temple, past the theater and up to the Cnidan Lesche with its Neoptolemos paintings and its vision of the underworld, could have looked back across the remains of the Mycenaean settlement, where a secondary omphalos, the "Stone of Kronos" stood, to the tholos of Neoptolemos and the great acanthus column. Below these, they could have seen the treasuries of the Thessalian city of Akanthon and beyond it, that of Corinth, to make a verbal as well a visual play of the associations with which I have been concerned.[44]

Below the main temenos of Apollo, in the lower shrine now called Marmaria, was the Massalian treasury, where the flowering or leafy column first appears in Greece, though it does not bear acanthus leaves. The small treasury was an ex-voto of the people of the harbor city now called Marseilles. Marmaria was the first enclosure the pilgrim entered on his way up to the main sanctuary, and was dedicated to Athena Pronaia.[45] The Massalian treasury stood between the small Doric temple of the goddess and the tholos. Like most other treasuries at Delphi or Olympia or Delos, its one hall was prefaced by two columns in antis. These columns carried a simplified and boldly ornamented Ionic entablature and had twenty flutes separated by arisses, not by fillets as is usual in Ionic shafts. The capitals were made up of

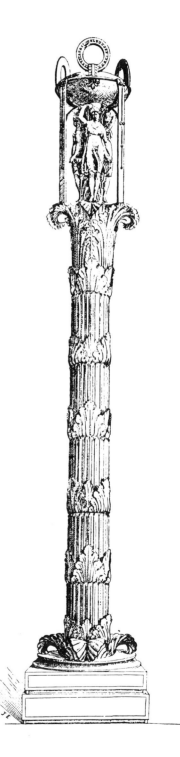

Reconstruction of the giant Delphic acanthus as a tripod stand. After J. Pouilloux and G. Roux (1963).

The Corinthian Virgin

**Giant marble lebes sustained
by acanthus and palmettes.** Athens, Archaeological Museum.
Photo by author.

Chapter X

bunched and highly conventionalized leaves, shaped rather like the petals of the Mesopotamian palmette, each leaf in the capital corresponding to a flute on the shaft, from which they were separated by a bead-and-roll molding. The occasion of the Massalian votum and the generally admitted oriental origin of the columns remain a puzzle.[46] The hard stone, the proportions, and the ornament invite comparison with the treasuries of the Cnidans and the Siphnians, which I discussed in a previous chapter.[47] They stood a little higher in the main temenos and their entablatures were carried by caryatids—also in antis, like the Massalian columns—whose repertory of moldings had obvious community with the Massalian treasury. This association of size as well as ornament would have made another set of connections, which may have been as obvious to a visitor then as they are oblique and fragmentary to us.

Ever since their discovery, the Massalian columns have been called proto-Corinthian, even if they are about a century earlier than Callimachus or the temples at Bassae and Nemea: no monument has yet been found to provide the missing link, to show the displacement of the synthetic-looking palmette leaves by the acanthus, as had happened on the Attic tombs. Even the Corinthian half-columns inside the neighboring tholos are more like the single column at

faraway Bassae than the adjoining Massaliot ones. Their form and presence raise a nagging attendant problem about Corinthian origins. The palmette ornament was of course familiar to the Greeks; moreover, throughout dynastic history until Roman times, the Egyptians used a stone column to represent a palm trunk, with palm-leaf fronds tied to it with a knotted thong to make a capital in funerary buildings. The Massaliot columns, designed and built for a flourishing trading port, have been linked to both Egyptian and the less obvious Phoenician precedents, but it took a century before the Greek masons made the connection between the Delphic "prototype" and the new column type. Rapidly then, at the beginning of the fourth century, it became an accepted, almost a canonic, element in their architecture.

In the context of the Delphic shrine, the association of funerals and offering basket, rejuvenation, acanthus and tripod, and the immortality of the soul does seem to turn around, or even arrange itself about, the Corinthian column. This is, of course, not local to Delphi and involves all the elements of Vitruvius' tale—except the girl.

And the girl? The most obvious Greek word for girl, *korē*, was also another name of a goddess, Persephone "of the trim ankles."[48] Violent Hades dragged her off to the underworld unwilling, to reign over the dead. Her mother Demeter was distraught and searched for her day and night—even by torchlight, especially by torchlight. Resting finally at Eleusis, Demeter drank her holy draught (malted barley water flavored with mint),[49] *kukeōn*, which was to become her sacrament. But she was still enraged at her loss, and she would have deprived the earth of seed grain to reduce humanity to starvation had Zeus not intervened to obtain the return of Persephone to her mother—though cunning Hades had seen to it that before she went back to the world of the living and of the gods she ate some pomegranate seeds. These grains had such powerful charm that she had to go down to the underworld for the three winter months each year, to be its queen.

The Eleusinian Telesterion. Plan of the excavations. After G. E. Mylonas (1961).

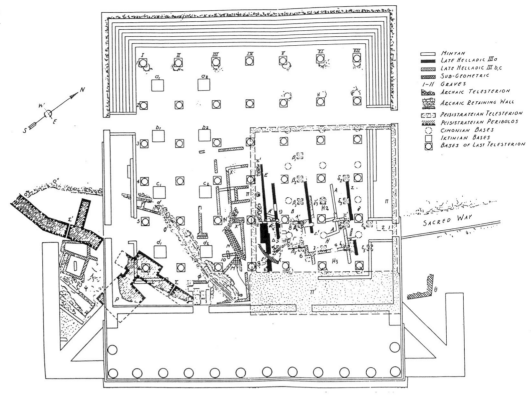

MINYAN
LATE HELLADIC IIIa
LATE HELLADIC III b,c
SUB-GEOMETRIC
I–II GRAVES
ARCHAIC TELESTERION
ARCHAIC RETAINING WALL
PEISISTRATEIAN TELESTERION
PEISISTRATEIAN PERIBOLOS
CIMONIAN BASES
IKTINIAN BASES
BASES OF LAST TELESTERION

The Eleusinian Telesterion. View of site. Photo by author.

Chapter X

The girl again: Persephone

Mētēr kai Korē, "Mother and girl," were the cult names of Demeter and Persephone. The Homeric *Hymn to Demeter* begins with an account of the famous rape, and goes on to tell of the consoling of the mother and of the institution of the cult at Eleusis where Demeter had first taught her mysteries to mankind, and where they were celebrated by torchlight.

Persephone's rape by Hades was not the only one she had to suffer in myth. The Orphics believed that she was also abused by her father Zeus in serpent (or dragon) form; the offspring of the union was Zagreos, the child who was destined to inherit world rule.[50] Jealous Hera incited the Titans against him; they seduced the infant god with mirrors and toys, tore him apart, cooked him on a tripod and on spits, and devoured him. Only his heart survived, and that Athena saved; she gave it to Zeus who swallowed it before uniting himself with Semele to beget Dionysos. Dionysos is therefore Zagreos reborn. But the Titans were burned up by lightning and from their ashes humanity was made; accordingly Zagreos/Dionysos is the divine part in men, while from the Titans they have inherited savagery and guilt.[51]

Zagreos was also a hunter, as his name (*za-agreios*) suggested: the divine, the great hunter. Aeschylus identified Zagreos with Hades, or at least had him be the son of Hades rather than of Zeus. Their roles were so parallel, however, their parts so similar, that one suggestion has been that the passage from Zagreos to Dionysos marks the passage from hunting to viticulture as the generator of ecstatic experience.[52] Zagreos never quite fused with Dionysos in myth. When the two gods were seen as acting in the same way, it was because Zagreos was also chthonic Dionysos—though chthonic Dionysos (if Heraclitus is to be believed) was only Hades in another guise.[53] Like Demeter and Kore, Dionysos also had cults that tied him explicitly to agriculture. In the Eleusinian mystery, the first initiate and priest, Triptolemos (the "Thrice plougher"), was both the inventor of the plough and the missionary propagator of agriculture. As for Hades/Pluto, he was not only the god of the underworld and king over the dead, but also the daunting guardian of the seed grain.[54]

The recital and the acting out in ritual of such myths of origin may not only have attended and even helped to impart skills in ploughing, planting, and harvesting to the young, but, enhanced as they were by the terror of darkness and the glimmering torchlight, they also invested that work with spiritual meaning.[55] And they were also seen to hold out another, a very powerful and much more general promise. It is not possible to trace the metamorphosis of an agricultural legend into a token of immortality. Whatever the original stimulus for the formation of mystery cults and of their organization in Greek society, by the time Socrates spoke about initiates, they seem to have been promised purity, joy, and immortal divine vision—not an abundant harvest.[56]

Eleusis was merely the best known of the shrines at which the promise was made; for the Athenians the Eleusinian mysteries were mysteries par excellence, and there may have been a cult there since Mycenaean times.[57] Yet north of Athens, at Flia (modern Chalandri), much closer to the city than Eleusis, there was another mystery shrine, and another near Messene at Andania; near Megalopolis, the ancient Arcadian megaron of Demeter-Kore of Lycosura claimed the remotest antiquity. Equally archaic were those of Zeus on Mount Ida in Crete, and the halls of the Cabiri on Samothrace. There were other mysteries not dependent on a specific shrine, but imparted by wandering initiators—as the Mithraic mysteries were in the Roman world. Such unlocalized cults might claim foundation by a god, Dionysos, or by a hero, Orpheus, or yet by a historical figure, Pythagoras.[58]

All of these rites seem to have in common the initiatory pain and suffering in imitation of those of a god; then, preoccupation with fertility—both individual and sexual as well as social, that of the soil; and lastly a promise of eternal life, or at least freedom from the fear of death. In the initiations of the itinerant "masters," this last element became increasingly important. Such mysteries never replaced—or even quite displaced—the civic religion into which all Greeks were born; they existed side-by-side with it. But initiation was entered on by an individual, by his or her own choice; mysteries were therefore increasingly seen as the means of personal salvation.

In the sixth and fifth century, initiation to one of the older mystery cults might have satisfied an individual's desire for salvation. In the later fifth century, the desire, the thirst, had become so general, so diffused and powerful, that the old cults of Demeter and Kore or of Dionysos, as they were celebrated in the public sanctuaries, were found wanting—and the mysteries became much more diversified. Building also required a more specific allusion than could be provided by the time-honored, maternal Ionic style or manner.

The goddess who was a girl—Kore—was also the column herself. She is the daughter of the Doric and the Ionic ways of building, and she was born when the Greeks required a new sign. For four or five hundred years, the two older ways of interpreting the world in building had stood for the two sides of Greek religion: the exterior, the masculine side of nomos and the sky; the interior female side of earth and fertility, of death and of rebirth. That bond strained when the Greeks "invented" the immortal and individual soul: a momentous invention that Plato (who was probably a generation younger than Callimachus the sculptor) celebrated in the *Phaedo* and the *Phaedrus*.[59] Inevitably, Greek architects could no longer rely on their conventions. Kore had to be given a commanding position in their works.

Monuments 1: The temples

Of the last ten hexastyle Doric temples in Greece, Italy, and Asia Minor, three included Corinthian columns. In the temple of Apollo Epikurios at Bassae, south of Olympia and in the mountains overlooking the Gulf of Kiparissa, the first Corinthian column appears in the context of a temple building. It is a single column in the cella, and it is framed by two Ionian colonnades; the peripteron is Doric. Its debut is therefore chaperoned by the older columns: the temple at Bassae has everything.

It was probably built soon after the Athenian Parthenon, and Ictinus (who had collaborated with Pheidias on the Parthenon in Athens and perhaps also in Olympia) is credited with its design by Pausanias, who thought it the most beautiful, most "harmonious" temple in the Peloponnese except for that of Athena Alea at Tegea in Messenia.[60] In spite of Pausanias' recommendation, Ictinus' Doric columns had no entasis; the shafts were tapered but straight. Yet the design of the whole reveals many subtleties on close examination: the north front columns are much thicker than those on the sides. Their intercolumniations are also somewhat narrower. Less obvious—it was noted only a few years ago—was the shaping of the plan into a kind of irregular trapeze, perhaps in order to reduce the apparent length of the sides for the worshipper approaching through the propylea.

The temple at Bassae faces north, unlike most Doric temples: towards Delphi, some have suggested. It is also rather longer in plan than usual, six by fifteen columns.[61] The materials are local limestone for the bulk of the work, and marble (perhaps from Paros?) for some

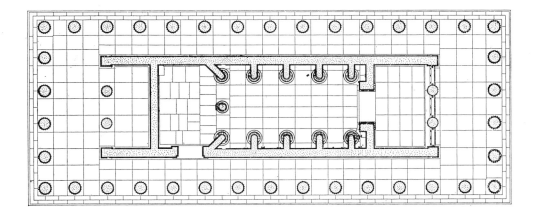

The temple of Apollo Epikurios, Bassae. Plan. After W. B. Dinsmoor (1950).

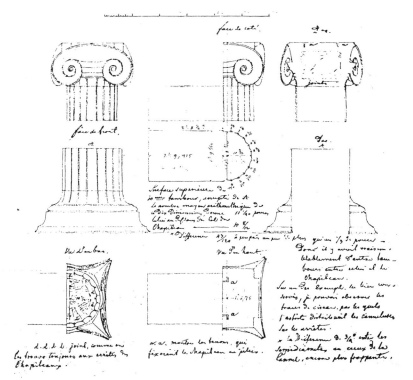

Ionic column. Measured drawing by Haller von Hallenstein.

The Corinthian Virgin

of the ornaments. The cella (which may have been hypaetheral) had five Ionic three-quarter columns projecting into it on short spur walls.[62] Between the last pair, closing the cella, and separating it from the adyton stood, archaically, a single, central column—that earliest Corinthian column to have survived in a temple.[63] The adyton had another curious feature: an eastward door, which some thought was intended to allow the statue of Apollo to look to the rising sun.[64]

Since its discovery, and its almost immediate destruction, the Bassae Corinthian capital has presented problems.[65] Scholars, puzzled both by its unheralded appearance and by its prominence—standing archaically central to the Apollonian shrine—have preferred a formal explanation of this innovation. Yet the suggestion that Ictinus had devised it to solve the perennial directional problem of the Ionic capital seems unsatisfactory not only for the contextual reasons I have suggested, but also on the evidence of the monument itself: the Ionic columns that surround it are (potentially) four-face in their modified Bassae form. Although the central position of the column within the cella would have made a sideways view of the usual Ionic capital wholly unsuitable, yet the way the interior frieze runs over the columns would have allowed even a normal Ionic column to fit frontally under the entablature.[66] Setting a Corinthian column in a row of Ionic ones also posed an additional problem, that of relative height. At Bassae this is resolved by giving the Ionic shafts a deep base mold and an extra large flare, while the Corinthian is set on a much thinner base, so that its slenderness is emphasized by the contrast. The reason for its sudden appearance therefore cannot be explained as the solution of the Ionic corner problem.

Some years later another famous master, Skopas of Paros, built the temple of Athena Alea at Tegea, where her Archaic shrine had burned down ca. 390. The ancient, probably pre-Hellenic goddess Alea seems to have had an ancient cult thereabouts and been assimilated to Athena rather late in her history.[67] Skopas' temple had an external Doric peripteron, shorter than the one at Bassae: six by fourteen. Its Doric columns were very slender, which was almost

The Corinthian column. Details.
After C. R. Cockerell (1860).

The Corinthian column with its entablature. After C. R. Cockerell (1860).

ostentatious in a very large temple.[68] The cella had a door in the side, as had the one at Bassae (but facing northward, since the temple had an east-west axis), which was lined with Corinthian half-columns, seven to each side, carrying a shorn cornice.[69] Its front pediment enclosed a sculpture group showing the killing of the Calydonian boar, while the rear one showed Achilles wounding Telephos.[70]

The last of the three, the temple of Zeus at Nemea, was built half a century later, perhaps as late as 330. Nemea lies about twenty kilometers southwest of Corinth. Although most readers will associate it with the first of Herakles' labors, that of killing and skinning the invincible (presumably the last European!) lion,[71] in antiquity it had another and equally venerable tradition, concerning the royal child, Opheltes, who would only attain happy maturity (according to the Delphic oracle) if his body never touched the ground until puberty. His nurse, Hypsipile, being met by the Seven who were on their way to Thebes and asked for a drink, put the child down on a selinon plant, from which he fell to the ground and was killed by a serpent guarding the spring. In memory of Opheltes—who may have been celebrated in an Archaic heroon nearby—biannual Panhellenic games were founded (though an alternative account has them founded by Herakles after his first labor); the surviving stadium installations are more or less contemporary with the temple.[72] Unfortunately, there was not much of a town at Nemea. It seems to have been primarily a ceremonial center, and although fifth-century Christians were to build a basilica there, the shrine was already ruined when Pausanias saw it, the temple abandoned and the cult image moved to some other place.

On the exterior the temple of Zeus was a peripteral hexastyle with the usual in antis pronaos. But it had no opisthodomos and was therefore even shallower than the temple at Tegea—six by twelve columns. Like the temples at Bassae and at Tegea, it stood on the site of an Archaic shrine. The Doric shafts are unusually slim, but they do have a gentle entasis. The stylobate and crepidoma are "refined" both in vertical curvature and in the downward slope of the top step from the wall to the edge. On the interior, six by four detached Corinthian columns, about two-thirds of the Doric ones high, enclose a central space; the four end ones, against which a cult statue may have stood, separate the cella (as the three do at Bassae) from a narrow adyton, where a sinking and a crypt served for some as yet undiscovered and perhaps oracular ceremony. The bases of these columns look like rather etiolated versions of the Ionic bases at Bassae, while capitals are versions (again refined) of the ones at Tegea. Each Corinthian column supported a smaller (in a ratio of 2:5) Ionic one over its rather shorn entablature.

Monuments 2: Tholoi

Early Corinthian columns appear in circular buildings, which the Greeks called tholoi, as well as the temples. *Tholos* is a Homeric word meaning a circular or a vaulted chamber such as the treasury tombs at Mycenae discussed in chapter 8 or at Orchomenos,[73] though by Plato's time it was applied—particularly in Athens—to the circular civic buildings where the prytaneia received ambassadors, and even took their common meals.[74] Late in the fifth century, circular buildings appear with Corinthian columns on the inside, but Doric (as at Delphi and Epidauros) or Ionic (at Olympia) ones on the outside. While some of the temples of that period include Corinthian columns, all tholoi of consequence have them.[75]

About 470–460 the most famous tholos, the Prytaneion, was built on the Athenian agora. Its outer wall was indeed circular, about 17.5 meters in diameter. The six interior columns (or posts) are set out in a quasi-regular hexagon, and there were no exterior ones.[76] Sometime in the fifth century, a circular building with an exterior fencing or colonnade about the same diameter as the outer wall of the Athenian Prytaneion was built in the southern part of the city of Eretria, on Eubea.[77] It had a central circular sinking, perhaps an altar—but nothing is known about its superstructure or its use.

The first true peripteral tholos to have survived, the small (6.5 meters in diameter) Archaic one at Delphi, was presumably a treasury or an anathema at the entrance of the main enclosure, since the nearby treasury of Sikyon was partly built of its fragments. It is almost the only Doric building whose triglyph frieze was organized without regard to the spacing of the columns that bore it, though all its metopes are square; this cornice was extraordinarily oppressive, since it was half the height of the columns.[78] More than a century later, sometime between 415 and 380, a much more refined tholos (which was twice as big as the Archaic one, being 13 meters in diameter), was built close to the Massalian treasury that I described above, in the Marmaris enclosure at Delphi.[79] Twenty quite robust (6½ diameters high) Doric columns surround the cylindrical shrine; within are ten (?) Corinthian columns, the height of the Doric shafts and completely cylindrical, only just detached from the wall. They stand on a low black marble base and the floor is laid in the same marble; the structure is of white marble but only where fine finish is required, and in limestone for the rest. The Corinthian capitals have their volutes centered on a palmette and a single row of acanthus leaves, with a sharp-cornered abacus, an arrangement that recalls the capital of the single column at Bassae.[80] Unfortunately, nothing has survived of the roof or ceilings within. Again, the nature of the building is unknown; it has been suggested that it, too, is an anathema of the Athenians to their patron goddess, whose temple is nearby.

The second of these peripteral buildings is in the enclosure of Asklepios at Epidauros. It is the biggest of them, just over twenty meters in diameter.[81] Pausanias, who admired it and thought it the work of Polykleitos (like the theater),[82] called it tholos, while the inscriptions recording the building work call it *thumelē*, the "altar," the "sacrificial place." The general arrangement is like that at Delphi, except that its Corinthian columns are fully detached. It is also built of a mix of marble and limestone, though it has many more traces of color. Like the building at Eretria, it had a central circular opening in the floor, which however looked down into a cylindrical chamber, around which ran three ring walls separated by interconnected passages. No explanation of all this is offered by ancient sources, though it has been suggested that it was used for shock therapy by enclosing the patients in the passages with the (harmless) yellow snakes that were sacred to Asklepios. The exterior Doric and interior Corinthian columns are the same height; the Corinthian is near-Vitruvian, with double acanthus leaves. The superstructure—like the rest of this tholos—is much better preserved, including the remains of the extravagant palmette finial of the roof.[83]

The last of these circular peripteral buildings is the shrine in Olympia that Philip II of Macedon vowed in 338, after the battle of Chaeronea, where he defeated the allied armies of Thebes and Athens and became the effective master of all Greece—indeed he had been proclaimed Captain-General, *Stratēgos autokratōr,* in view of a new Persian war, which was delayed by his assassination. This heroon overlooked that of Olympia's founder Pelops: it was the record and witness of the Macedonian dynasty's claim to their Greekness and also laid a claim to their heroic descent from Herakles, who first establish the Olympic games.[84]

Tholos in Marmaria, Delphi. Detail. Photo by author.

Tholos in the sanctuary of Asklepios at Epidauros. Plan, after G. Roux (1961).

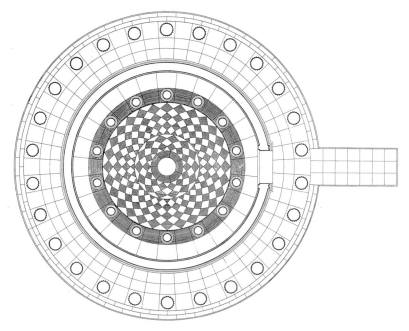

0 1 2 3 4 5 6 7 8 9 10 MS

Chapter X

Tholos in Marmaria, Delphi.
Photo by author.

Monuments 3: Monopteroi

The Philippeion is not the last tholos. There were several later ones, of which the Arsinoeon at Samothrace is perhaps the most famous, but they were no longer relevant to the development of the Corinthian column.

On the other hand, there is another class of circular buildings that does play a role in it. They were stone bases (even monuments) on which bronze tripods, such as competition prizes, were often set up in public places. The most impressive surviving one is the choragic monument of Lysikrates in Athens, which commemorated his choir's winning of a Dionysian singing competition in the year 335. It is the only survival of several such monuments lining Tripod Street, which curved around the eastern end of the Acropolis, from the square of the Prytaneion to the eastern entrance of the theater of Dionysos.

Lysikrates' monument was a cylindrical monopteros,[85] and it was elaborately decorated with acanthus: six half-columns, some of the earliest fully developed Corinthian ones—carved some eighty years after their Vitruvian legendary origins—surround the stone cylinder.[86] It is their very first appearance on the exterior of a building. The reliefs of the frieze show the story of Dionysos and the pirates, as told in the Homeric *Hymn* and often repeated in Greek and Latin literature. The wall of the cylinder, articulated just below the column capital, carries a row of tripods in relief and thus restates the associations—acanthus, tripod, and Dionysos—that I have already described.

Tholos at Epidauros. Interior Corinthian and exterior Doric order. Photo by author.

Tholos at Epidauros. Section, after G. Roux (1961).

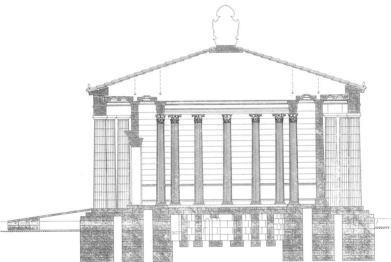

Tholos in the sanctuary of Asklepios at Epidauros. View of the substructure. Photo by author.

Chapter X

The Athenian Olympieion

The last of the great Greek temples, one of the tallest and biggest, was wholly Corinthian, the only true peripteral Corinthian temple in mainland Greece. It was built over the base of the vast and unique dipteral octastyle that had been begun by Peisistratos toward the end of the sixth century but was more or less abandoned just above the stylobate level when his son, Hippias, was driven out of Athens in the year 510. Much bigger than any other Athenian temple, it was regarded with disfavor: Aristotle bracketed it with the pyramids and the Heraion at Samos (begun by another tyrant, Polykrates) as the kind of public work designed to prevent rebellion by keeping citizens poor and busy.[87] Themistocles' removal of some column drums from the site to build up the neighboring city gate in 479 incurred no blame.[88]

Aristotle's association of the Olympieion with the Heraion at Samos draws attention to the fact that both temples were huge dipteral octastyles. Although the dimensions of the recovered Peisistratid drums, as well as their intercolumniations, suggest a Doric temple, their refined and very varied column spacing seem to belong to an Ionic one.[89]

348

Temple of Olympic Zeus, Athens. Fallen column. Photo by author.

Temple of Olympic Zeus, Athens. Photo by author.

Chapter X

Construction on the site began again in the year 179, at the behest of Antiochus IV Epiphanes of Syria. The new design was arrived at by competition, which was won by a Roman citizen (probably of Greek origin), Cossutius.[90] His vast Corinthian columns are regarded as "classic," as normative. When Sulla took Athens in the year 86, he transported some columns to the Roman Capitol,[91] to stand in the temple of Jupiter Optimus Maximus; from there, in turn, they seem to have had an irreversible effect on Roman architecture and sculpture. The Olympieion was finally finished and dedicated by the Emperor Hadrian in AD 129 (or 131), 300 years after its foundation. In fact, Antiochus' forebear, Seleucus Nicator, had ordered a quasi-Corinthian temple of Zeus Olbios at Diocaesarea in Cilicia, and Antiochus' undertaking in Athens was part of his campaign to make Olympic Zeus the supreme god of his empire in imitation of Seleucus. Antiochus is now chiefly remembered for his profanation of the Holy of Holies in Jerusalem, and for setting up an altar—of Olympic Zeus, of course—there: one of the measures that provoked the Maccabee rebellion. The Athenians had no parallel cause: yet on Antiochus' death, construction stopped and, after two desultory starts, had to wait for Hadrian's decisive action.

Now if what Vitruvius said about the relation between the different kinds of column and the dedication of the temple is right, then surely even a late Hellenistic architect such as Cossutius would not have thought it decorous to build a Corinthian temple to Olympic Zeus, whose Doric shrine in Olympia was one of the wonders of the world. But the Athenian temple was not merely dedicated to the deity of its cult image. It was also a shrine of the old chthonian gods, of Gea and Kronos. Most memorably, it contained the chasm in which the water of the flood Zeus had sent (and Deukalion survived) was swallowed by the earth. Deukalion was said to have founded the very first sanctuary on the site, which has been identified with the limestone walling under Peisistratos' stylobate.[92] Even under the patronage of Olympic Zeus, the Corinthian columns must commemorate or celebrate the underworld and its powers: they mark the place given over to thoughts of destruction and regeneration.

The Corinthian column then was to become the column of late Republican Roman and of Imperial architecture. In view of the development of religion in the Roman Empire, this can hardly be surprising. Its popularity may be due to the Roman fondness for elaborate and flowery ornament (as most authorities have suggested), but it may also have something to do with the way this last column isolated and represented a new and essential aspect of religious life: the emphasis on the immortality of the soul.

XI

A Native Column?

▪ The Etruscan arrangement ▪ The disposition ▪ Greek rite, Etruscan rite ▪ Capitolium ▪ Ceres ▪ A Tuscan Ionic? ▪ Columns of honor ▪ An Italic order

Although it was to become the most popular with the Romans, the Corinthian column—like the Doric and the Ionic—was a native of the Aegean, an import or an immigrant in Italy. Vitruvius devoted so much of his descriptive power to those foreign columns that when he came to the native one, which he called Tuscan, he dealt with it very summarily. The Ionic arrangement (which he favored) had occupied most of his third book, while the Doric and Corinthian took the first six sections of the fourth. He finally turned to the Tuscan temple in the seventh section, before dealing with circular temples and with altars.[1] However much he attended to native building traditions and techniques elsewhere in his work, he treated the Tuscan as a poor relation here, since even in his time it was more often carried out in wood than in stone and was usually spaced so widely that the beam it bore had always to be in wood. He seems rather patronizing about the wooden, overspaced buildings: he calls them "straggly," "top-heavy," "lowly," "dumpy."[2] Their detail is always extraneous, since such a wooden construction required a great deal of protection against the weather, and it was almost always loaded with a display of terracotta facings as well as ceramic sculpture at the corners and over the ridge.[3]

The Etruscan arrangement

The rules Vitruvius gives for these columns and their trabeations are much more cursory than for the three Greek ones. Moreover, he reports no legend, not even a historical anecdote, to explain the origin either of the Etruscan temple or of its column. But then the Etruscans, who had become literate between the eighth and seventh centuries—a little later than the Greeks—have left us a language which is still barely comprehensible, and very little of what we now call literature.[4] And indeed the Etruscan paintings and reliefs that illustrate recognizable epic or tragic scenes usually refer to Greek plays and paeans, particularly to stories about the Trojan War or the Seven against Thebes.[5] The accessible Etruscan writings that have survived are elaborate ritual texts, quoted in Latin translation by several Roman historians and antiquarians.

In Vitruvius' treatise the first reference to the Etruscan temple comes near the beginning, when he sets out the Greek customary teaching about the decorum of choosing an appropriate site for every deity, and continues with a gloss drawn from the rules of the Etruscan "discipline": temples of Venus, Vulcan, and Mars should be sited beyond the city boundary. The reason for this, he explains, is that

> adolescents and mothers of families should not become accustomed to the spectacle of sexual licence; buildings would be free of the danger of fire if prayers and sacrifices are offered to Vulcan outside the walls; finally, if devotion is offered there to the divinity of Mars, armed conflict among the citizens will be avoided, while the god will shelter the same walls from enemies and the dangers of war.[6]

How much of that passage was Vitruvius' rationalizing gloss, and how much was the Etruscan rule, is not clear; nor is anything else known about other details of temple siting that might have been regulated in this way. Still, even if he calls these rules "Etruscan," "Tuscan" is the word he later uses for the temple and the column. Tuscan or Etruscan: the words might seem interchangeable, yet Vitruvius uses them a little differently. *Tusci*, from which *tuscus, tuscani-*

cus, was mostly adjectival; it may have been a corruption of an old Italian form (as the Oscan *turskum,* in turn perhaps a corruption of the Greek *tursēnoi*) and not of any known Etruscan term. *Etruscan* was for more exalted usage: Cicero used it when referring to the *disciplina* or to gold wreaths. For their part, the Etruscans called themselves *Rasenna.*[7]

At any rate, when he came to specify his Tuscan "disposition," Vitruvius began by setting out the plan, as he had done for Greek temples. Its outline, the locus (not the area of the Greek temple), was almost always formed by a tall base, molded at the top and bottom (though he does not say so), and never by the stepped stylobate of the Greek ones.[8] The base was to be five units broad by six long—nearly square. He halved the units of width next, to set up a ten—"module" elevation. Four of these (half-)units were given to the central and most important chamber—Vitruvius' Etruscan temple, unlike his Greek, has three shrine chambers—and three to each of the side ones.[9] The whole length was also to be halved, the front part for the portico, the back for the shrines. The corner columns were to be aligned with the antae of the outer walls, the intermediate ones with the walls that partitioned the shrines, and there was to be a second row of columns between those on the front of the building and the walls of the shrines: this rule produced a porch at least two columns deep to preface the windowless cellae. It is therefore quite a different *kind* of building from the peripteral temple of the Greeks. Still, to assimilate it to the Graeco-Vitruvian typology I discussed earlier, it could be called a dipteral prostyle tetrastyle, although it has also been called *peripteros sine postico.*[10] Indeed, Vitruvius himself points out that it was easy enough to replace the wooden "Tuscan" columns with stone Ionic or Corinthian ones and thus make the temple a taller and therefore also a tighter building, more like a Greek one.[11]

As for the columns, Vitruvius directs that their height was to be one-third of the whole width of the base[12] and that—exactly as in his "modified" Doric—the shaft was to be seven diameters. Pliny the Elder also has a little to say on this matter (though only in an aside) when considering the composition of stucco. He, too, knew four kinds of cylindrical columns: a Doric of six diameters, a Tuscan of seven, an Ionic of nine, as well as a Corinthian, whose shaft is of the same proportion as the Ionic; even if (like Vitruvius'), it has a capital two modules high. Pliny also knows a square-based "Attic" column, about whose proportions he has nothing to say, however.[13]

The disposition

To the shafts of their columns the Tuscan builders, unlike the Greek devisers of the Doric, gave not only a capital but also a base when it became terracotta and stone; the process of changing wood into stone was very different (and happened much later) in Italy than in Greece.

To start at that base: it and the capital are to be given a height of half a diameter each. Of the base, half the thickness was the plinth and the other half a thick torus with its fillet. The Tuscan plinth, unlike the square Ionic and Corinthian ones, was to be circular, while the torus of the base is a simplified clay or ceramic shoe for the shaft—which was not to be fluted.[14] The shaft was to be smooth, and this gave it an even greater appearance of solidity than the vertically grooved Greek columns were meant to convey. Moreover the diminution (the top of the column being ¾ of the diameter at the foot) was much more pronounced than the Vitruvian Greek entasis, though, of course, in Archaic Doric shafts it was even grosser.

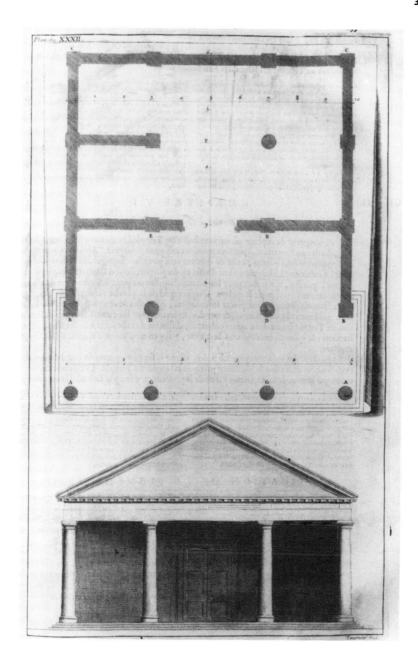

The capital is divided into three equal parts: a hypotrachelion separated from the shaft by a bead; a quadrant echinus supporting a square abacus, whose sides equal the diameter of the column at the base; and the abacus proper.[15] Again, some of the archaeological evidence does not quite illustrate the text. An isolated painted terracotta casing for a column of the archaic temple of Mater Matuta on the Forum Boarium in Rome was found in the St. Omobono excavations there. The wooden structural post inside it was plain, but the casing is molded to represent a fluted shaft with a quite elaborate capital and base. Of the shaft little is left; but the base is a highly decorated torus, while in the capital, the echinus—which is a thin quadrant, carrying an equally thin square abacus—is supported by a ring of schematic leaves.

Cramping of the Tuscan cornice. After S. Stratico's *Vitruvius* (1830).

Each leaf corresponds to a flute and has a small tongue sticking down into it through the striped roll-molding that seems to fasten them. This detail produces a section familiar from other Etruscan stone examples, from Vulci and Falerii, and is associated with Mycenaean as well as several proto-Doric ones.[16]

Vitruvius goes on to specify the details of the beams that such columns were to carry. They were to be wooden and they were to be composite; as thick (he presumably means no thicker than) as the shaft at the hypotrachelion, they were made up of two boards secured to each other with swallowtail cramps in such a way as to leave two fingers'-breadth in between for ventilation, and so prevent dry rot.[17] They are to be "as high as the building requires." The text does not specify any frieze at all, though the most famous "monumental" example (the "ground floor" of the Roman Colosseum) already had one. The frieze was inserted into the formula in the sixteenth century, when the disposition was made into an order.[18]

Vitruvius is explicit in his prescription for the ornate roof that crowns the *dispositio*: above the beams and above the walls, the joists of the roof are to project one-quarter of the height of the column as mutules.[19] The ceramic facings, *antepagmenta*, are to be fixed to their ends.[20] Even higher was to be the tympanum of the pediment, which may be in masonry or carpentry. At the very top was to go the (projecting) ridge beam, the *columen* or *culmen*,[21] and the central acroterion associated with it.

The columen was a much more substantial and prominent building member than any Greek ridge beam. It would almost certainly have been composite, like the beams of the porch, and therefore presented the same "unsightly" divisions and end grain that the Greek builders were said to have masked with a triglyph. In many temples this masking may well have been the principal pedimental sculpture. Something of the kind was already suggested by some of the fragmentary models, particularly the terracotta temple from Nemi.[22] It shows a deeply inset pediment (into which a dwarf roof with elaborate antefixae is set) as a kind of atrophied porch within the pedimental space. There is an elaborate figure-sculpture on its antepagmentum as well as on the ends of the side beams. Less familiar, because recently excavated, is the brilliantly intricate (and relatively well-preserved) pedimental sculpture in the tripartite temple A at Pyrgi—the port of the city of Caere—showing a scene from the Theban cycle. It was just that kind of acroterial columen-cover. In the temple enclosure of Pyrgi also, the first bilingual Etruscan inscriptions were found on gold tablets recording the dedication of a temple there to Uni-Hera-Astarte in both Phoenician and Etruscan. Since the language of the dedication as well as the name of the goddess point to Syro-Phoenician rather than Carthaginian connections, this very surprising find gave a new coloring to Etruscan orientalism while it confirmed the specifically Syrian links of Caere.[23] The patrons and worshippers in the port seem to have wanted both an Italiot-type and a Greek-type wood-and-ceramic Etruscan temple in the temple compound. Adjoining the three-cella temple A was the single-cella temple B. It was peripteral and had a full Greek-style ceramic pediment, even if its intercolumniation was araeostyle.[24]

Certainly in Vitruvius' Etruscan temple, the tympanum (as the word he used itself implies) was closed and ornamented. On the inner side the pediments may also have been figured—that is, if the shallow triangles between the sloping ceilings and the straight painted walls of the tombs at Tarquinia and Chiusi (presumably imitating house interiors) may be taken as a precedent. In several of them, the heavy column carved into the rock is "supported" by the painted representation of a stumpy column, though its relation to real building is very difficult to establish.

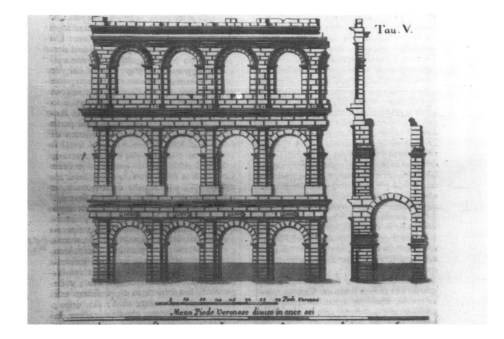

As for the sides of the building, mutules were to project one-quarter of the column height, while the overhang with its gutter is to be one-third of the finished roof.[25] Enough statues and acroteria survive in fragments to allow a general idea of what such wooden and terracotta Etruscan temples looked like, from the outside. On top of the roof, over the ridge, there would be set quite elaborate figural sculptures. That much is clear from the Veian shrine outside that city, at Portonaccio, where is found a small tripartite temple about 18.5 meters square, just over 10 meters high from the top of the podium to the ridge beam. Nothing much is left of the pediment. But on the ridge of the temple stood at least four figures, somewhat over life-size, which are regarded as outstanding works of Etruscan ceramic sculpture and have even been compared to the Ionian korai of the Athenian Acropolis.[26] They are a specifically

A Native Column?

Etruscan feature and have been regarded by some scholars as figurative and highly refined descendants of the huge horned thatch-anchors that appear often on the house urns of the ninth and eighth centuries.[27]

That the early buildings of the Latins and Etruscans were thatched is well-attested, not only by the slopes and surface of the hut urns but also by the stone imitations of thatch-roofed huts in the earlier tombs; the "della Capanna" and the "della Nave" tombs at Cerveteri show it clearly. The columen also referred to a familiar structural feature of early Etruscan and Latin building. Huts, whether oval (as at San Giovenale) or rectangular (as on the Palatine in Rome), are known to have a central row of posts, which supported a ridge beam; on the front of the hut, this ended (as is shown by the hut urns) in a triangular smoke-and-light hole and a porch, while the other end was presumably apsidal, as in the hut urns and many tombs.[28] Vitruvius has no word about a second pediment at the back of the building. If my reading of the evidence is right, then the roof would have been apsidal in the earliest ones, but hipped later: appropriate in the directional Etrusco-Italic temple in a way that would not have satisfied in a peripteral Greek one.[29]

How the Etruscans worshipped in their sanctuaries is known only through analogies with Roman religious practice. Yet this is not always clear, since Romans and Etruscans had ambiguous attitudes toward each other. The Etruscans were a wealthy and powerful confederation when the Rome of the kings and the early Republic was a small (if rising) city-state on which a group of villages depended. For some time the Etruscans seem to have ruled Rome, with intermittent opposition to their rule; at other times the Romans were at war with them—witness Lars Porsena of Clusium.[30] At yet other times, some Etruscan cities were more or less willingly allied with the Romans. By the time they entered history, however, the Etruscans were already weakened by Roman and Greek pressure from the south and by the Gaulish invasions from the north. Their ill-timed (but repeated) attempts to form an alliance with the Phoenicians of Carthage inevitably hastened their political decline. Believing themselves in any case to be doomed to extinction after ten *saecula* ("generations," rather than 100-year centuries) of prosperity, they were more or less voluntarily absorbed into Roman society.[31] Their language faded into oblivion, while many of their social and religious practices were taken over by the Romans, so that what Vitruvius considered the Tuscan-type temple (and many now consider Italic more generally) became a part of the Roman heritage.

Roman ambivalence in this matter has something to do with their sense of political and cultural inferiority to the Etruscans, which—when they transferred it later to the Greeks—Horace's tag expressed perfectly:

> Graecia capta, ferum victorem cepit et artis
> intulit agresti Latio.[32]
> Once conquered, Greece captured her savage victor
> and brought the arts to loutish Latium.

Horace had no such pithy formulation for the debt to the Etruscans, who, according to Roman tradition, at some point ruled not just Rome, but all Italy; in fact, they never dominated more than a third of the peninsula. They had claimed descent from an elite of Asiatic emigrants.[33] According to legend the Romans, too, were descended from a Trojan—and therefore an Asian—hero: Aeneas, and his companions. Indeed, in Roman "society" it was counted a definite social advantage to be able to claim Etruscan ancestry.[34]

Greek rite, Etruscan rite

To the Romans the Etruscans had seemed an unusually pious people:[35] their antiquities were accordingly considered venerable, so that both "Etruscan" and "Tuscan" were extended to mean native and ancient generally, rather than specifically Etruscan. That more general sense may well be what Vitruvius meant by his *Tuscanicae dispositiones:* ancient, venerable, native ways of arranging things.[36]

In religious practice, at any rate, the Romans differentiated between "Etruscan" and "Greek" rites. The Etruscan discipline was a whole revealed religion, set out in books allegedly dictated by two deities: an earth-born child-god Tages and the nymph Vegoia. Its rules are now only known from fragments preserved by various Latin writers, and they regulated the organization of the state, the founding of towns, and the division of land; they also governed the whole business of divination from sacrificial innards, from lightning, and from birds.[37] The Greek rites dealt with other kinds of sacrifice and with consulting oracles (as against interpreting omens): the elected priesthood dedicated to them was called—successively, as their numbers grew—*Duom-* or *Decem-* or *Quindecem viri sacris faciundis.* They consulted Sibylline books, written on rolls (linen, or perhaps leather ones) in their custody, for which purpose they even had to have paid translators from Greek.[38]

The Romans were deliberately eclectic in their religious practices. They were even proud of appropriating foreign deities and ceremonies, to the extent of inviting the gods who protected enemy towns and nations, during a siege or before a battle, to Rome with promises of temples and worship.[39] These deities would arrive in the form of their statues, which meant that the Romans made piety out of stealing the cult statues of their enemies. Yet like the other Latin peoples and their relatives—Oscans, Umbrians, even the Etruscanized Faliscans—the Romans did have their own aboriginal gods and rites, which they owed neither to the Greeks nor to the Etruscans or their other neighbors, and they therefore did not group them under any special label.[40]

Capitolium

Of all the Roman temples, the most famous as well as the largest was built in the early years of a flourishing and quickly growing Rome by Lucius Tarquinius Priscus, the fifth Roman king.[41] When the first recorded Roman shrine was moved from the Quirinal to the Capitol Hill and the new temple vowed to Jupiter Optimus Maximus, the king inevitably summoned Etruscan craftsmen to execute this huge work, which (omens declared) was to become *caput mundi.* He did not live to finish it. The building was continued by his successor, Servius Tullius, and finally completed by his son, Tarquinius Superbus, who—in his turn—imported Etruscan carpenters for the final stage, presumably to build the roof.[42] Tarquinius Priscus also commissioned Vulca of Veii (the only Etruscan artist whose name has been preserved by historians) to make the main cult statue of Jupiter Optimus Maximus, with his regularly vermillioned face and gold-embroidered *toga picta*—which the Roman triumphing general would imitate. The quadriga that formed the main acroterion over its pediment was the subject of another legend confirming the oracular destiny of Rome to become ruler of the world. It also was commissioned from Veian ceramic sculptors, this time by Tarquinius Superbus.[43] The temple was dedicated in 509, just after he had been expelled from Rome, as one of the very first public acts of the Republic.[44]

Temple of Jupiter O.M., Capitol, Rome. After E. Gjerstad (1953–1973).

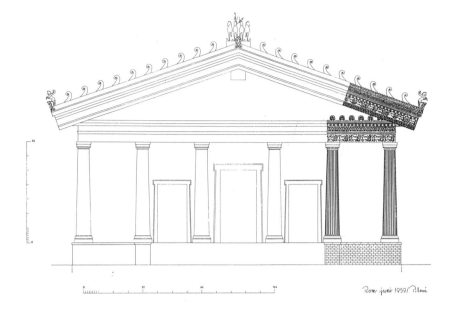

Temple of Jupiter O.M.: Detail of columns and cornices. After E. Gjerstad (1953–1973).

The very size of the Capitoline temple made it an exceptional building at the time of its construction. Later Roman writers comment on its size, disproportionate to the modest "primitive" city, since it is wider than that of Zeus in Olympia and not much narrower than Artemis at Ephesus.[45] It is difficult to see what Italic rivals it may have had for several centuries; very few three-cella temples earlier than the third century have survived, even in the most fragmentary state. Those at Orvieto, Pyrgi, Marzabotto, and even Veii are all smaller.

The Capitol was much like the sort of building Vitruvius described, with a deep portico and set on a high podium. Its base was not exactly proportioned according to the formula he gave,[46] and the building also differed in another, more obvious way: it was a hexastyle, not a tetrastyle. Those two extra columns on the front corresponded to a range of columns lining either side of the cellae, whereas in Vitruvius' formula the walls of the cellae were not porticoed. These porticoes end on antae, which complete the back wall of the building projected to either side. The general configuration gives the impression that the nucleus of a Vitruvian-type tetrastyle has been surrounded on three sides by an extra rank of columns— typologically if not historically.[47]

On the platform was a shrine to a deity who had been worshipped there before Jupiter and the Triad: Terminus. In token of his prior claim to the site, his altar was inside the temple, but it was out of alignment with the walls of the newer and grander building. Since his cult could only take place in the open air, a hole was made in the roof of the porch so that his altar was not sheltered.[48] This opening in the roof and the unaligned altar to another deity was certainly echoed in other Etruscan and Roman buildings.

One such altar stood on the platform of the temple known as the Ara della Regina in Tarquinia. It was at an angle just over 65° to the main building, but was itself fairly accurately oriented north-south; it consisted of a small platform and a molded block, which was presumably the altar proper. Some pavement stones orthogonal with it lie under the base of the main building, suggesting that the quirky orientation was done for the same reasons as those given for the siting of Terminus' shrine on the Capitol in Rome. The temple proper was a large and majestic building, with voluted columns, and was built in the mid-fourth century. The group of winged terracotta horses from its columen-cover is almost as famous an Etruscan sculpture as the Apollo of Veii, and was reproduced on Italian airmail stamps during the 1930s.[49]

Another instance of a "disoriented" shrine is in the Capitoline temple of Cosa, a Roman colony on the Etruscan borders in the territory of Vulci, which had been founded in 273. There, an altar at 45° to the orthogonals of the temple building also stood outside the portico, but on the platform. It was aligned to an earlier shrine on the site (probably the very first), which had been buried within the podium of this new temple, under the geometrical center of the middle cella; the altar stood out in the open. Yet the portico also had a large opening in the roof, an impluvium like that of a Roman or Etruscan house, with a substantial stone cistern underneath (covered by pitched stone blocks), to collect the rainwater funneled into it from the roof. The Etruscans, as well as the Romans, seemed to cultivate such irregularities as tokens or even assertions of the primitive and rustic sanctity of their urban temples.[50]

The anniversary of the consecration of the temple of Jupiter Optimus Maximus on the Ides of September was the beginning of the Roman state year, and on that date the praetor would hammer a bronze nail into the outside wall of the temple, on the side of the Minerva shrine. This calendar nailing seems also to have been an Etruscan rite.[51] The soft-stone, timber-

framed building on its high podium, with its elaborate, very polychrome terracotta decorations and its nail-studded wall, definitely had an Etruscan look. It had to be thoroughly restored a number of times, since its high, exposed position meant that it was liable to be struck by lightning; moreover it was irreparably burned by an unknown arsonist on 6 July 83 BC in a fire in which the original cult statue also perished.

When Sulla decided to reconstruct it in stone, he brought capitals from the temple of Olympic Zeus in Athens to provide models for his masons, as I reported in the last chapter. But very little is known of Sulla's building. It was finished by Julius Caesar in 46 and restored by Augustus (that was the temple Vitruvius knew, of course). It still had a wooden superstructure—presumably with ceramic ornaments—which burned in the civil war that led to Vespasian's accession.[52] Domitian finally rebuilt it as a Corinthian temple of Parian marble, and that last one was plundered and wrecked by the Vandals of Genseric in AD 455.[53] Sulla's importing of Corinthian capitals from Athens and Domitian's use of them do rather suggest that even in the first Capitoline temple, which might almost be considered the archetypal "Tuscan disposition," the column capitals were perhaps not quite the rudimentary ones that Vitruvius described.[54]

Ceres

About twenty years after the Capitolium had been consecrated, during the famine of 496 and after consulting the (Greek!) Sybilline Books, the thrice-consul Spurius Cassius (Vecellinus) vowed a temple to another triad, Ceres, Liber, and Libera, which was dedicated in 493.[55] It

burned down in the year 31, when Augustus decided to replace it with a stone building, in turn dedicated by Tiberius in AD 17.

Though she was a Latin deity, Ceres had a good deal in common with Greek Demeter. This temple of hers near the Circus Maximus stood—perhaps in obedience to the rule reported by Vitruvius—outside the Republican pomerium, which ran just beyond the Circus Maximus, on the lower slopes of the Aventine. Spurius Cassius was himself plebeian, the last plebeian consul for a century and a half. While the Capitol was the shrine of monarchy and later of the patricians, the cult of Ceres was associated with the Republic and was most popular with plebs. Vitruvius counts that shrine of Ceres among the araeostyles of the Etrusco-Italic type, along with the temple of Hercules that stood nearby.[56]

In spite of its different constituency, the temple of Ceres seems to have been planned on the Capitoline model; certainly an Etruscan-type building, it was considered great and famous and was much admired, notably by Cicero, even if denigrated by Vitruvius.[57] It was decorated with terracottas by artists whose names have survived because they signed their work with a Greek epigram (which claimed that Damophilus had done the right side of the building, Gorgasos the left).[58] Was this employment of Greek artists a tribute to the Hellenizing character of the worship of the goddess? The authorities are silent on the subject, though Dionysius of Halicarnassus, who gives the most circumstantial account of its history, Greekifies its dedication to Demeter, Dionysos, and Kore. The terracotta sculptures were carefully preserved, the decorative panels from the sides of the cellae being dismounted and framed, perhaps to be displayed in the reconstructed temple.[59] All this did not result in a Greek peripteral building, which would probably have been too alien to the Romans of the early Republic.

Vitruvius mentions a number of other Italic or Tuscan-type temples in Rome, which he had certainly seen. The three-cella temple, the only one that he specified, is however not the only Etruscan or Italic one, even if the triple type had an obvious dignity, exalted Etruscan house building, and sheltered a number of important cults. Less common and unlisted in antiquity was a type with a single oblong cella, whose back wall projected on either side, while three sides were enclosed by colonnades: for example, the temple of Diana at Ariccia. Moreover, the most common Roman temple type (of which a splendid instance was that of Mars Ultor on the Forum of Augustus) descends directly from it. A third variant, which can also be recognized as a separate type, has a long cella, with two side chambers projecting forward into wings, and an in antis porch between them. The most important (and biggest) such temple was the Ara della Regina at Veii, though much earlier examples seem to follow the same plan. Restorations have not really provided this last type with a full architectural configuration.[60]

A Tuscan Ionic?

The only surviving monumental stone instance in Rome of the column-and-beam Vitruvius called Tuscan—although marginally more slender, yet with a base only slightly more elaborate, and with the smooth cornice he demands—is in the Colosseum, which was begun, long after he died, by Vespasian.[61] Even that arrangement is often called Doric rather than Tuscan. Both Serlio and Scamozzi saw the Colosseum orders as a regular succession of Doric, Ionic, and Corinthian and relied on Vitruvius and the antiquarian descriptions of vanished buildings for their Tuscan. For all that, Serlio first (and other writers following him) modified Vitruvius' account to obtain a simple mnemonic: the proportion of diameter to shaft in the five orders was a simple sequence from Tuscan 1:6 to Composite 1:10. Palladio did find monuments to legitimate his Tuscan, but he had to go further afield, to the amphitheaters of Verona and Pola, for his precedent.

Yet his pupil Vincenzo Scamozzi assumed that there had been temples and public buildings with such columns, even wooden theaters in earlier times. Unfortunately what remains of the Italian temples compares neither in antiquity nor in quantity with those of the Greeks; they were so perishable that their most conspicuous relics are the surviving stone podia.[62] Something more may be learned about them from the few scaled-down votive models of Etruscan temples and the earlier hut urns, which provide the safest if not always unambiguous evidence about full-scale buildings.[63] Tombs offer a different kind of evidence, since many of them were cut out of the rock. Inevitably, the enduring stone architecture of the dead is a primary witness (as it was in Egypt) about other ways of construction.[64] Apart from those tombs and some powerful polygonal stone fortifications with arched gates, Italic peoples, much like the Etruscans, do not seem to have produced a specific monumental and enduring architecture in which rites might be celebrated and enshrined. Of the tombs many have atriumlike central chambers, with a ceiling carved into a representation of a timber or—as in some of the cases I mentioned earlier—even a thatched roof;[65] many also have three door openings to alcoves or internal chambers out of the central one, an arrangement that echoes the ancient Etruscan house as well as the temple that Vitruvius described. Some tombs and a very few hut urns have displuviate roofs arranged to include a (simulated) impluvium, the most famous being the della Marcareccia tomb in Tarquinia.[66] Occasionally details are based, however remotely, on Greek and even oriental precedent.

Such details are common enough in tombs carved into soft tufa, though the richest group is the large and crowded necropolis—the Banditaccia of Cerveteri (Caere)[67]—which was in use over several centuries. It is hardly surprising to find "Tuscan," even in some cases quasi-Doric, faceted columns in many of them.[68] But what has not been stressed or studied enough is the presence of voluted and flowered capitals there. In one of the earlier tombs, known as Tomba dei Capitelli (dated to the beginning of the sixth century), two octagonal columns have cubic, low-relief capitals of the Archaic Greek Aeolic type: doubled pairs of volutes (with a palmette between them) support a thin abacus and a beam over.[69]

When cremation took over from burial as the main funerary custom of the Etruscans toward the end of the fifth century, tombs declined in importance as a witness to the architecture of the living.[70] Nevertheless, the largest tomb in the Banditaccia necropolis, dei Rilievi, was built and decorated late in the fourth century. It owes its name to a spectacular series of reliefs, which were certainly stuccoed and painted to represent armor and various household objects, hung on its walls and pillars. Its large central chamber is even more atriumlike than that of most other tombs. The two square central piers and the broad pilasters that articulate the walls have rather narrow, strictly Aeolic capitals attached to them, not unlike the much older ones found on Cyprus.[71]

Very similar, if simpler—more "Tuscan"—capitals are represented on a fourth-century, rectangular house-sarcophagus from Chiusi on whose narrow front an arched entrance gate is flanked by two Aeolic pilasters, which support globular vases. It seems almost like the façade of a Phoenician-style temple with its guardian pillars. Did that urn represent a house or palace, or even a temple for the soul? Since it is all but unique an answer is almost impossible.[72] All these "pseudo-Aeolic" capitals are, of course, later by some centuries than the Aeolic capitals of the Greeks or the Cypriots. Yet their presence at Caere, along with the presence of the Phoenician-Etruscan inscriptions in its port, Pyrgi, which I mentioned earlier, gives substance to many speculations about Etruscan orientalism.

As in the buildings, so in isolated honorific pillars, there might have been Aeolic or other figured capitals representing a more general Italic *consuetudo* than Vitruvius' dogmatic Tuscan *dispositio*. No columns with Aeolic capitals seem to have survived except in the tombs, even if their representations on figured bronze mirrors and on pottery are very common, echoed in Etruscan (as in Greek) furniture. This might imply that there had perhaps been buildings with something like Tuscan-Ionic columns in addition to the well-recognized Tuscan-Doric type found in Vitruvius and the later handbooks.

Columns of honor

The most famous and ostentatious of these Tuscan columns of antiquity are—inevitably—Roman and Imperial: the two vast ones of Trajan and its imitation, that of Marcus Aurelius in Rome. Each one carried the statue of the emperor (replaced by St. Peter and St. Paul, respectively); each is topped by a simply molded abacus over a quadrant echinus; each is decorated with an egg-and-dart relief; and each has a shaft wound around with a spiral band declaring the emperor's achievements, while it stands on a garlanded torus-base.[73] The twin columns were one of the landmarks that identified the city, as did the Colosseum and the Pantheon.

Inevitably, there was much precedent for them. "Columnarum ratio erat attoli super ceteros mortales," wrote Pliny the Elder to explain the use of commemorative columns, while

Two coins of the Minucii showing the column of Minucius
Augurinus. After Becatti.

(if the poet Ennius is to be believed) the Romans would put them up quite routinely.[74] Reputedly the earliest had been set up by King Tarquinius Priscus in honor of the Etruscan augur,
Attius Navius, and it overlooked the Comitium near the Curia Hostilia on the Roman Forum.
The statue on it was said to be of bronze and smaller than life, but nothing is known about
its base, the column proper, except that it vanished in a fire of AD 52.[75] About the more interesting column of L. Minucius Augurinus, outside the Porta Trigemina—at the foot of the
Aventine, and near the temple of Ceres—we are much better informed. Minucius was a popular benefactor during the famine of the year 439, and the statue was vowed to him by a grateful
plebs. Coins minted by his descendants show a column with an Aeolic capital and a shaft
made up of horizontal bands;[76] from the corners of the capital bells were suspended on chains.
Such bells were much loved by Etruscan monumental builders. The fantastic, huge tomb of
Lars Porsena, which Varro (as quoted by Pliny) seems actually to have seen, was hung with
bronze bells. That enigmatic description has always fascinated artists and commentators[77] (in
Etruria even potters liked chains and bells, and the use of such ornaments on a commemorative
column is not at all improbable). Still, the coins offer rather indistinct evidence, and the texts
are not very explicit either. The only thing certain is that the column of Minucius had some
form of Aeolic capital, perhaps with a human head at the springing of the spirals (the detail
of the coin is not clear enough).[78]

Quite different evidence is provided by the arched gates of Etruscan cities; at Falerii,
Saturnia, and Volterra, the gates are spanned by full voussoir arches, their keystones and im-
posts carved in relief. In the "Arch of Augustus" at Perugia (also a voussoir arch),[79] squat Ionic
pilasters stand on a cornice that superficially (but perhaps intentionally) looks like a Doric

one: recessed squares containing circular shields, divided by triglyphlike and even squatter colonettes, garnished with stunted Ionic bases and capitals. Over another gate at Perugia, the Porta Marzia, two Ionic pilasters frame the actual arch: between them stretches a gallery that passes over it, but under the cornice. This gallery is made up of four fluted and (again) stumpy colonettes, while a fretted balustrade reaches halfway up the columns. Statues, perhaps of divinities—Tinia flanked by the Dioscuri is one suggestion—stand in three of the five "windows," with horses leaning out of the end ones, making an almost ironic comment on Vitruvius' notion that the metopes of a cornice may be read as windows.[80] Stone architecture in these gates show columns and beams almost as quotation from a whole different preoccupation with building.

Outside Etruscan lands, there is some puzzling (and as yet little known) archaeological evidence that seems to support my conjecture: a kind of mixed Aeolic-Corinthian type of capital appears first around Taranto in Apulia. The volutes spring from a single circle of acanthus leaves and are separated by a human (usually female) mask; notable examples are those on the ruined Greek cross church of San Laucio outside Canosa.[81] Even earlier—mid-fourth century—are those in the hypogeum in the garden of Palazzo Palmieri in Lecce: capitals with acanthus leaves, eagles, and female heads. They were used, about the same time, in rock-cut tombs of the third and second centuries: at Volsinii, in the Campanari tomb; at Sovanna, in the Ildebranda cave. There are some isolated capitals in museums;[82] in complete, surviving buildings they are less familiar. However, one temple with just such capitals overlooked the Forum (and overlapped the circular comitium) of Paestum, the old Greek colony Poseidonia. Although it had first been founded from Sybaris, it was captured by Oscan-speaking Lucanians early in the fourth century; incorporated into the Latin confederacy in 273, it became a Roman colony some time later. The circular comitium was probably begun early in the third century, while the relatively small temple building (the podium is about 32 by 14 meters, including stairs and altar) was probably begun about a century later.[83]

The cornice of this temple was a mixed Doric-Ionic arrangement: a metope-and-triglyph frieze with dentils over, of the kind Vitruvius had damned. Although the three-diameter setting out of the columns (the shaft is about 1:8) made it an araeostyle, the beams and all the superstructure were stone, not wood. The proportions of the plan were more Hellenistic than Italo-Etruscan, yet one detail is telling: the back wall of the cella projected on either side, to finish in antae. The peristyle runs around the front and the sides, and the colonnades finish at those antae, according to the Capitoline precedent.

Lucanians had become notoriously "Tuscanized." This temple of theirs was a very different, a much rougher building than the earlier Greek monuments of Paestum/Poseidonia. It is not clear whether it is a belated, provincial, and awkward Hellenistic-Italic confection, a mere echo of Tarentine usages, or if it might be taken, like that rather earlier and much more refined Tomba dei Rilievi at Caere, as a witness to the much wider use of Aeolic capitals in Etruscan building.

In the absence of any definite evidence about their appearance, it is at any rate possible to assume that the grandest and perhaps the largest of them, the Capitoline columns (in a temple so elaborately ornamented), were not of a Vitruvian "Tuscan order," as they are usually shown in the restoration, but would have been something more Aeolic. Since the time of Sulla, after all, they were to be replaced with Corinthian marble ones. And I see no positive reason not to think of them as leaved and voluted.[84]

An Italic order

Vitruvius' Tuscan disposition must therefore not be read as a narrow interpretation of the archaeologically ascertained evidence of Etruscan building. His Tuscan column—and consequently that of the later Renaissance theorists—cannot be derived with any assurance from a general survey of their monuments; his brief account was Vitruvius' bid to set native architecture on the same footing as that of the Dorians and Ionians. In that he was echoed by the elder Pliny.

Yet later Roman architects tended to prefer their Italianized version of the Doric column to Vitruvius' authentically native one. The lower story of the Roman Colosseum may come nearest to it in Rome proper, though even that is not clear; but it does appear (with the circular bases that Vitruvius demanded) in the amphitheaters at Verona and at Pola. Its common use in amphitheaters may be a deliberate attempt to emphasize the "Italic" origin of the amphitheater-type against the temple and the theater, which had come from the Greeks.[85]

All these examples are, however, sufficiently isolated to justify Vignola's observation that he had to make up his Tuscan order entirely from Vitruvius' description, since he could not measure any example of it in Rome.[86] The much more precise, even pedantic (as well as scholarly) Roland Fréart de Chambray (1650) refused the title of order to the Tuscan, thinking it was only appropriate to entablatureless columns, and quoted Trajan's as its prime instance.

Already by the time of Serlio and Palladio, any account of the Tuscan order was filtered through a rather bewildering historicomythical account of the Etruscans. In the Middle Ages there had been a good deal of uncertainty about them. The end of paganism had meant

that the little that survived of the Etruscan language and their discipline would vanish. By Diocletian's time Etruria had become Tuscia; and the more solemn name was only a remote and rather dim memory. For a thousand years the trope of Etruscan power, cultural superiority, and a dispersive devotion to freedom from central authority was only occasionally recalled— since Virgil and Livy, at any rate, were read. But information about them was rather generalized, as it was for Dante, for instance, who recalls their place in the genealogy of Aeneas and his progeny in *De Monarchia* and confines a number of Homeric and Virgilian heroes together to the first circle of hell as unbelievers.[87] Boccaccio included Lucretia among his *Famous Women,* and she proved a useful exemplar of heroic fidelity to many moralists.[88] Otherwise the Etruscans are only mentioned sporadically in medieval letters.

However, the myth of Etruscan brilliance and greatness was revived enthusiastically during the period of great Florentine prosperity in the thirteenth century. Florentine historians and eulogists—Giovanni Villani, Ricordano Malespini, Brunetto Latini, Coluccio Salutati, Gherardi da Prato, and later Machiavelli—all allude to the great Etruscan past, to their superior culture and liberty-loving politics against the imperial pretenses of both Rome and Milan.[89] The words *Toscanicus, Etruscus,* return to the Italian language late in the fifteenth century, and *Etruria* was to be revived both as a geographical term and as a title for the Medici dukes.[90]

As for architecture, Leon Battista Alberti, who had great faith in the Etruscans as painters, was quite convinced that the art of architecture had been brought to an even greater perfection in Italy than in Greece, and Italic columns (which had features of the other three) therefore had a special place in his dispensation.[91] His own practice shows him playing with such a notion: the capitals in his first important building, the Tempio Malatestiano in Rimini,

Santa Maria Novella, Florence.
Corner columns. Photo by author.

correspond to his description of that Italic column. He used a version of the Vitruvian Tuscan—which, as far as he was concerned, was an earlier and more perfect version of the Doric—in the lower story of the rusticated Palazzo Rucellai; and he seems to have developed a square-shafted Attic-Doric-Italic capital (for the flanking piers of Santa Maria Novella), which was much emulated. The "Italic" column was also developed, quite independently of any Vitruvian rules, by Francesco di Giorgio. It shows up, fully elaborated, in the table of the orders with which Cesare Cesariano illustrates the third book of Vitruvius.[92]

Other architects have used these rules in building; Bernardo Rossellino (who may in this, as he was in other things, have been directly influenced by Alberti) used a rather slender (1:7) Tuscan on the lower story of the papal palace, and a curious Tusco-Italic on the main piers of the cathedral at Corsignano/Pienza, for Pope Pius II. Giuliano da Sangallo, in the villa he rebuilt in the 1480s for Lorenzo the Magnificent at Poggio a Caiano, applied an "unclassically" wide, pedimented stone "temple front" to the facade to make a colonnaded porch. The columns can be read as a version of Alberti's Tuscan-Attic; and they are even smaller than the one-third of the width Pliny and Vitruvius recommend for their height. The glazed ceramic frieze of their cornice seems intended as an Etruscan-style *antepagmentum*; Andrea Sansovino, who modeled it, had also made a large ceramic figure of Lars Porsena of Clusium for Montepulciano, which claimed him as its mythic founder—a figure Vasari certainly claims to have seen.[93]

The buildings of the Etruscans, just because their style was not linked to any definite pagan legend nor detailed and elaborate enough in the texts, acquired the status of a proto-architecture. Reliable evidence about Etruscan building was very thin before the beginning of

systematic archaeology, and mythographers had to embroider on it. Their complex relation to the Romans was used to establish the historical priority—and the moral as well as political superiority—of the Etruscans. The legend that Dardanus, the founder of Troy, had come from Etruria (more specifically from Fiesole, though his father Corythus was also taken to be the eponymous founder of Cortona) was refurbished out of Servius' commentary on Virgil. Atlas, for his part, was regarded as the ancestor of Corythus and the founder of Fiesole (Giovanni Villani already believed that). He, it was said, had been turned to stone by Perseus, using Medusa's head, into the African mountain that held up the fabric of the sky.[94]

All this was mixed in the late fifteenth century into a very intoxicating bit of etymology by the theologian-mythographer and faker Annio da Viterbo: the mysterious Etruscan language could be read (he maintained) as Aramaic. He used this notion, together with some very peculiar wordplay, to fuse Scripture with medieval chronicles and legends to suggest that the Italic god-king Janus (after whom the Roman hill Janiculum was named) was the same person as the Patriarch Noah, who was "the second Adam" and therefore the true ancestor of humanity. He had come to Italy after the flood and died in his final home there, a cave on the Janiculum. That hill was therefore not only associated with a pagan past, but had a scriptural connection to provide Christian respectability for the legends of Etruscan and Roman origins.[95]

His "chronicle," for which he claimed remote antiquity but seems to have written himself, proved to be very attractive not only to the Medici and some other Italian princes, but also to French royalty (if only through the Medici connection). The Albertian account of the superiority and archaic grandeur of the Italic-Tuscan columns was very useful to the Medici court, which is why it figures so prominently in the headquarters that Giorgio Vasari designed in 1560 for the family who were soon to become Grand Dukes of Etruria: the Uffizi Palace in Florence. Tuscan, too, was the ornament of the monolith granite shaft (from the Baths of Caracalla), which Duke Cosimo I set up in the Piazza Santissima Trinità, and to which he added a bronze base and capital. The Medici association with the order was further extravagantly displayed when Catherine de Médicis, as the widow of Henry II, Queen Mother of France, built her Paris palace in which she set up a Tuscan column twenty meters high—with an internal staircase, like the two Roman columns—for use as an astrological observatory.[96]

These concerns appealed beyond partisan Medici interests: the twin columns of Hercules, with their motto *Plus ultra,* which were the personal device of Charles V, are therefore inevitably Tuscan columns—associated with Mauretanian Atlas, as well as with Hercules, the perfect device for the world ruler on whose empire the sun never sets.[97] Many other liberties were taken with the Tuscan name. It is perhaps not surprising that the Swiss architect, Hans Bluom or Blom, could maintain quite seriously that the toughest of the orders, which he derived from the Doric, was the Tuscan; "Tuscan," he added, is a corruption of *Teutsch,* and this also recommended it highly to the Germans.[98]

For all Vitruvius' by then familiar condescension, therefore, Tuscan columns and trabeations appear in many late-fourteenth- and fifteenth-century buildings. Bramante and Peruzzi use them constantly. They appear often in the baser buildings, which their rustic association and their obvious rudeness encouraged: in stables and outbuildings, in city gates and fortifications. Their triumph is in the forecourt of St. Peter's, which Gian Lorenzo Bernini made into a vast, dubiously Tuscan, amphitheatrical colonnade.[99] Fifteenth- and early-sixteenth-century columns could be quite fanciful. After 1537 and the Serlian clampdown, the Tuscan style became as recognizable as it was standardized.[100] The genera of the columns, which from antiquity until then had been varied and permutated, had become the Orders of Architecture.

Since the other "orders" were definitely associated with Greece, and even particularly Greek locations that at the time were too remote, too far off the travelers' routes, and even inaccessible, Vitruvius' rough Tuscan became attractive as the archetype of the column, precisely because of its lack of definition. And indeed it was the most elementary of the orders: certainly in Serlio's list, though for him (as for Alberti), the Doric and the Tuscan capital were virtually identical. He gives a number of possible applications for it, all of them involving rusticated walls, keeping the 1:3 ratio of column height to building width: they are utility buildings, such as fortresses and city gates. But this rustic view of the Tuscan column, though common, was not general.

In the nineteenth century, historians and architects raised the Dorians to the status of quintessential (Indo-European and proto-Prussian) Greeks, and the newly discovered Greek Doric order to the status of the "order of orders." Four hundred years earlier, much the same was done by the mythographers and panegyrists of the Florentine Renaissance who exalted the Tuscan column to a similar status. Yet in the end, there can be no order of orders. The ancients, wiser in their ways, could not have conceived such an exclusive notion. Columns and their cornices were used to articulate and even differentiate, not to unify, a building. It is precisely in the wide range of changes I have discussed—both of proportion and of ornament—that the notion of column as gender and as temperament continued to fascinate the Greeks and the Romans as it would the artists of the fifteen and sixteenth centuries.

A Native Column?

XII

Order or Intercourse

· The double metaphor · Dorian objectivity · Skin and bones · Sensation and production · Making, imitating, loving

A simple post, a strut, a column—and the joist or beam it carries—may not seem the best framework in which to manifest the complexities of temperament and gender or unravel the intricacies of Greek identity. Yet the metaphor that brackets notions about these matters— matters as much of mind or soul as of body—with a plain and inert thing of wood or stone has persisted over millennia. It elaborates the primal identification of the standing human body with an upright post.

The double metaphor

The column-and-beam element is, in itself, a constituent of the man-made, of the artificial world. It is also part of an all-encompassing metaphor that makes human shelter an embodying, an in-corporation. The tenacity of that image, which cannot quite be reduced to a plain, "conceptual" statement of any particular clarity or interest, should only be superficially surprising. As I registered its many transformations, it has come to seem irreducible, atomic. However, wherever, whenever that image was first devised, the organization of any building since has evoked, even imposed, some comparison to the human body with its articulations, varieties, mutations—all its inevitable historicity. The metaphor itself, consciously or unconsciously willed by the builders, assumed an absolute character for me in spite of all the slipperiness and lack of definition, its almost inevitable fluidity. There seemed to be no appeal from it. Nor could it be narrowed to any definite shape, since it is much more a clutch of connections, of relations. I can only deduce or intuit this archetypical metaphor through all the bewildering varieties of palpable, visible historical instances. Observing and registering these varieties can provide me with a history more deeply grounded and radical than any account of a rationally formulated concept ever could. Human understanding is, after all, always enclosed—even constrained—by the boundaries that its own time must set for it, while the various transformations of a metaphor can sometimes be a trace of the concealed currents that run through the intellectual landscape of a given period; they can throw into outline those dominant attitudes that may shape, confine and even distort the most rationally formulated notions and concepts. Their very jumble and ferment is the ground in which meaning is founded: they nourish all systematic thinking. The variety and richness of these transformations may also show the daring with which the human mind is able to transcend reason through the very manipulation of image.[1]

Metaphor is generally understood as a figure of speech limited to two terms only: *this* is *like that*.[2] But the metaphor with which I have been concerned is more extended—a double one—in that it involves three terms: a body is like a building and the building in turn is like the world. That metaphor returns in a more global similitude: the whole world is itself understood as a kind of body. It is this very condition that inspired Walter Benjamin's meditations on the origins of astrology, the paradox of transcendental—or at least non-sensual— mimesis: a part of what he called the "mimetic heritage" of mankind.[3]

The condition for any person "finding himself" in the man-made world must therefore be that buildings should be like bodies in the first place, and in the second, like whole worlds—far-fetched though this may sound. More explicitly, there is a way in which all the different parts of my metaphoric equation make up a "function": they bring a number of variables together with a fixed element. On one side is a perpetually changing awareness of the body, the slower modification of building methods, and the differing visions of world order (or disorder); on the other is the sure expectation of the metaphoric commerce between them against which all the changes may be rung.

The condition, perhaps the only necessary condition under which architecture may be produced at all, has always depended on that double metaphor, since architecture is the essential *parlar figurato* (the speaking in figures) of building. The human capacity for figurative (as against expressive) speech has given us all poetry—it is the true being of poetry—and the metaphoric process is its primary mode of operation. I have discussed the decay and marginalizing of that figured speech (or at least its decomposition and its replacement at the center of human communication by the plain statement) in the course of the seventeenth century, when rhetoric came to be seen as the "father of lies."[4] Nature then ceased to offer lessons, the stars paled in their effect on our character and our fate; stones and plants lost any remarkable affinities to human affects; animals no longer seemed to have any virtues to teach us nor any vices to warn us against. In the visual arts the consequences of this change were drawn gradually, as happened, for instance, when the composition of gesture was replaced by the expression of passion in French academic teaching.

As for architecture, notions of harmony and composition were displaced by those of taste and character. By the nineteenth century, the absence or marginality of any figuration was taken for granted as part of an intellectual atmosphere shared and to some extent even generated by architects and other artists as well as their critics. Lewis Mumford, to take an illustrious and recent example, drew a distinction between the "prose" and the "poetry" of architecture in an essay published over forty years ago. Echoing but severely modifying Ruskin, he suggested that we should expect architectural poetry only from the few great masters of any generation, who force the frontiers of the discipline by transforming the received types of building. Others, the lesser lights, follow on a few years or decades later by developing the poetic statements of the masters in the more humdrum but less demanding work that satisfies the everyday needs of patrons and public, thus making up the "plain speech," the journalism of architecture.

Was Mumford right? Should the world, in any case, expect only plain speaking of its builders now? Does not the promise of prosiness provide the guarantee under which all twentieth-century building of any size is financed, produced, received? Even if some figuration may be allowed as a top dressing in luxurious and important corporate headquarters, must not the bulk of building (as the bureaucrats and the developers at any rate have long insisted) be reliably prosy? I have a great reverence for Mumford, but this distinction of his may have turned out to be as unhelpful as Nikolaus Pevsner's much more often quoted claim that architecture is to building as Lincoln Cathedral is to a bicycle shed; for, he continues, "the term architecture applies only to buildings designed with a view to aesthetic appeal."[5]

Mumford's and Pevsner's remarks (made half a century ago), blunt, reductive, and stingily binary though they are, have been very influential. Although it is surely not what they intended, they have been read to justify the shunting of anything said about a matter so patently aesthetic (and—by implication—superfluous and "ornamental") as architecture to a sideline, so that most of those who called themselves architects could feel dispensed from thinking that their work would ever be subjected to close economic or political attention, or that it mattered at all if it were. The straightforward, the economically and technologically determined "art" of building, which aimed to satisfy basic human needs without pretension or frills, could therefore be considered as quite independent from—and in some ways even superior to—the "aesthetic" one of architecture. This was a crucial shift in the thinking of the time, which had a powerful effect during the rush of reconstruction after World War II; it seemed justified by the urgent need for mass housing, for school and hospital building in ruined Europe. But when those problems were no longer so immediate, commercial developments in North America were able to benefit from the same withdrawal of "aesthetic" constraint. Inevitably, any issues of history and theory were then removed from the center of architects' attention.

Dorian objectivity

The arguments for breaking all aesthetic constraints were older and more deeply rooted. Directly after World War I there was a small but very influential avant-garde group in Germany with a distaste for everything that smacked of artifice and form. The emotional distortions of Expressionism were to be replaced by a return to the object itself. In the austere and troubled Weimar Republic, art lived a paradox, since it was to be sober and unaffected, a literal statement, a document.[6] From painting, the slogan *Sachlichkeit* was quickly adapted to architecture. Heavy, ornate "Wilhelmine" art had been buried with imperial ambitions. Art Nouveau, too (echoed in the work of some contemporaries, notably Erich Mendelsohn), with its obsessive interest in ornamental invention, had been discarded.

Neue Sachlichkeit had started as the title of an exhibition of painting and sculpture but was immediately adopted by architects. Mies van der Rohe, by far the most important figure in the group, proclaimed in 1923 that "the art of building is the will of the present formed in space"; he explicitly rejected any attempts at innovation not directly dependent on new materials and tersely declared his contempt "for all aesthetic speculation, every doctrine, every formalism." The very term "architecture" was unacceptable to him and he continued to speak of his work as *Baukunst,* the "art of building." "Our work is to shape forms out of the essence of our task through the means which our time offers," he claimed—though his "forms" might more accurately be called "shapes," since in Mies' "objective" terminology they do not have a responsible and intentioned "author," but are dictated by the impersonal "present." The forms are, in any case, direct and untampered products of an industrial manufacturing process, quite innocent of any metaphorical or even referential implication; nor are they shaped by any search for proportional or syntactic refinement. Transmuting such "objective" elements and composing them into buildings worthy of "aesthetic" attention—in Pevsner's terms—would therefore have to rely on a subsequent "receiver," a user or critic; an observer-consumer might even construe them as the unconscious bearers of Mumfordian "poetry," although hopes or expectations of such a sanction should not govern any of the designer's decisions. His part is to "stick to his last," to satisfy all the contingent demands of the job. Aesthetics and poetry are the recipient's business. Art had therefore to eschew any semblance of the stylish and the ornamental, not only of the aesthetic. Paradoxically, the object presented, an unvarnished statement of things as they are, was to be self-sufficiently the work of art. Mies van der Rohe made this formulation the center of his teaching when he moved to Chicago in the late thirties, and he always remained faithful to it.

The doctrine of *Sachlichkeit*—in architecture, at any rate—echoed the conviction, furthered by a number of important thinkers throughout the nineteenth century, that art was dead. Not quite dead, as Hegel (who first formulated the idea in his lectures on *Aesthetics*) kindly allowed, but of no real further importance, since the work of the Spirit had, by his time, been freed from the indirect and figurative ways by which artists could invest the truth with sensory manifestation; it was directed toward the plainer declarations of the philosopher and the statesman. Much as the Enlightenment philosophers had proclaimed the demise of myth and superstition (by which many of them meant organized religion, as well as the narrow sectarian interests that depended on the churches), so Hegel and many of his followers proclaimed the death—or at least the irrelevance, the unimportance—of art. In France a rather similar effect was achieved by a different argument, built on different premises. It was the argument of the Ecole Polytechnique teachers, chiefly Jean-Nicolas-Louis Durand and his follower, Charles Normand (which I outlined in chapter 1): all historical ornament (of which the Orders of Architecture were the most important by far) was a convention that did not need to be related to any account of its development or to any conception of metaphor in architecture. The convention merely furnished those middle forms that served to bridge the demands of material and manufacture on the one hand, the grandiose purities of geometry on the other. A century later *Neue Sachlichkeit* would extend the argument so as to demonstrate the futility of any such middle forms.

In any case, on Hegel's showing, architecture was even deader than the other arts; since it was the first, the most material and the most earthbound of them, it would (of course) also be the first to atrophy. Poetry, because it is the least dependent on sensuous experience, remained as the most enduring of all the arts, the only one for which the Spirit would have further use. However much or little they took from Hegel, many later thinkers were fascinated by this notion of his, even if his own concern with beauty (as the revelation of the Idea) left a whole realm of everyday experience unconsidered, the province of the craftsman rather than the artist. Yet it was to attract a great deal of aesthetic attention after his death: the onset of industrialization (which Hegel did not foresee) would inevitably require that people who call themselves artists intervene in the shaping of the everyday. But "art," in Hegel's sense, could not include the making of such things as furniture or jewelry, which were far too material- and purpose-bound to engage true aesthetic attention. It was Goethe, though a whole generation older than Hegel, who foresaw the threat that mechanical reproduction posed for the visual arts and attempted to develop a conception of the artist as a supercraftsman to resist such tendencies.[7]

Hegel's vision of architecture had a minor echo in Victor Hugo's much-quoted meditation (inserted as a postscript into his *Nôtre-Dame de Paris*), *Ceci tuera cela*: architecture—whose acme is the stone book of the cathedral—shall be made obsolete by the more urgent and more accessible book of printed paper.[8] Unlike Hegel, Hugo, who certainly could only have heard of Hegel's views by word of mouth, did not see this as a stage in an ineluctable process, even if to him, as to Hegel, any experience of architecture was primarily cognitive. Architecture might therefore be retrieved, and it could take fire again from a new kind of embellishment of structure, which would return that cognitive element by relying on the letters of the alphabet and on natural motifs. That is what his friend, Henri Labrouste, was to attempt in the Bibliothèque Sainte-Geneviève, built between 1843 and 1850.

The absence of a style for the age, for the nineteenth century, then came to be considered one of the symptoms of the morbid condition of art (and of architecture in particular). There had to be a style to compare with those of a past time, whose characteristics were being worked out by various art historians—or so some architects and historians, not necessarily

Hegelians, thought. The call to formulate such a style was first explicitly made by the Karls-ruhe architect Heinrich Hübsch in 1828.[9] It was echoed by Augustus Pugin and Owen Jones, by Eugène Viollet-le-Duc and Gottfried Semper—and by many others. The establishment of *Kunstgeschichte* as an academic discipline was to be one of Hegel's legacies. Of course, to many of the new historians it seemed that the mark of any style was a repertory of ornament, and they therefore suggested that a wholly new one could be derived (as the different ornamental styles of the past seemed to have been) freely from the forms of nature, so that it would be clear of any historical reference.

The duly devised and much-heralded "style" appeared in the last decade of the nineteenth century and went by many optimistic names: *Art Nouveau, Stile Liberty, Yachting-style*. It was the aim of many of its practitioner-artists and publicists to subject the whole of life to aesthetic control by making every small object of use into a work of art. Many who considered themselves "artists" busied themselves with just those artifacts that had been considered too banal to have merited attention from their like in the past, and had certainly not been discussed by any aesthetic philosophers: not just furniture or jewelry, but door and window fittings, embroidery, tableware. Such objects, as I have already pointed out, were increasingly made by machine and manufactured in quantity. They would therefore have to be designed by architects and artists—the profession of industrial designer had not been thought of yet. Since they were the products of named authors, they were inevitably objects of "aesthetic" or at least "design" attention such as would not be lavished on the productions of mere craftsmen. The fragmentation that the conflict between such self-consciously shaped objects implied required a new style to control them and to bring them into harmony. It implied further—or so it seemed to some of these designers—that a new style should take control over the whole physical environment.

Ancillary to this process was the belief, which was fairly commonly held, that industry would offer society many of the solutions to those problems that had long beset it. The new organization of industrial labor and the steady rise in the quality of industrial products would bring about—or at any rate, might help to bring about—the ethical-aesthetic reform that would favor the stability as well as the prosperity of society.

This rather banal notion that art should include the making of any object capable of aesthetic attention consequently required adjustment. Georg Simmel, fascinated as he was by the outward presentation of the "self," had occasion to consider this very matter in *Sociology*, his treatise on societation. It comes in the section on secret societies, in which he finds himself considering *Schmuck* (which can variously be translated as "jewelry," "finery," "ornament"). This finery, he says, if it is to be generally appreciated, must have "style" beyond any inherent stone or metal worth, since it makes the form generally accessible.

In the *Kunstwerk* [the work of art proper], the more powerful the personal charge and the subjective vitality expressed in it, the more it is prized, since this [kind of work] invokes the unique response of the viewer—who is alone in the world with a work of art, as it were. We require a generalized and "typic" form in what we call *Kunstgewerbe*, the craftsman's products—which for its everyday usefulness is addressed to a generality. No unique personality, but a historical and social attitude should be expressed in it. . . . It is the greatest error to consider that individual finery should be a particular work of art because it adorns an individual. The contrary is true: since it serves the individual, it should have no specific character response, nor should the [chairs] on which we sit or the tableware we handle.[10]

To the continuous environment, which requires "type" objects, personal neutrality, "style," he opposes the work of art proper, which makes its own and separate world and therefore requires the viewer to enter into its life. Notwithstanding, the word *Kunst* governed both categories in German.

Simmel's distinction did not help to bridge, or even to rationalize, the business of invention in design. His concern with personal adornment (and even tattooing) and with the anonymity of elegance was paralleled by his Viennese near-contemporary, Adolf Loos, who approached the distinction from a slightly different point of view, though with similar premises. Loos held that most of the buildings the architect designed had to have the same characteristics as the crockery and jewelry that Simmel had considered; most buildings have no business to try and be "works of art" anyway, since art is reserved for the high emotional charge of the monument and the tomb. Therefore ornament "invented" for machine production (or indeed invented by an "artist") to make everyday objects "artistic," he held to be a regressive barbarism, and he therefore abhorred Art Nouveau and all its works.

Loos' attitude, stated allusively but clearly in his most famous essay, "Ornament and Crime," is often misunderstood, perhaps because of the vehemence of his language. In fact, it was not ornament itself that was criminal; delighting in ornamental objects was fine, provided such ornament was produced by people who could be considered more or less "primitive": contemporary craftsmen such as shoemakers or peasant weavers, or the artists and artisans of another age and civilization. Their ornaments were valid, since their invention was the by-product of pleasure in the unpremeditated gesture of the working hand. The quasi-bourgeois concert-going craftsmen of modern urban society such as the silversmiths and saddlers of Loos' Vienna should eschew such invention however, as should English tailors, and even more so the architect, though he was at liberty—as was the painter or sculptor—to "quote" ornaments from the past in order to establish historical legitimacy through such "typified" themes. But the architect should never think of designing or inventing them. There was therefore nothing inconsistent about Loos adapting an out-of-scale Doric column as a model for a Chicago skyscraper—for a monument. It was a type object if ever there was one. His reason for choosing the Doric order requires more comment and context than I could provide for it at the end of the first chapter of this book, when I introduced my entire project.

Loos and his circle had never been entirely convinced by the optimism, the belief in the promise of a well-ordered and well-equipped society that was cherished by many "enlightened" people at the turn of the century—a notion that had in fact been battered by the horrors of World War I and its aftermath. When all those ideas about the artistic control of industry had also been defeated by the pragmatic manufacturers (and their public), the problem Simmel once had approached gingerly had to be reformulated quite radically. This was done by the leading German thinker of the time, Martin Heidegger, but without any explicit reference to Simmel's distinction. What seemed crucial to him was to establish that the whole mode of being of a work of art should be beyond and apart from any context of use. This can be read as an echo of Kant's perception of the beauty in an object being revealed through the sense of its having a purpose, though separate from the idea that it is directed to any specific end.

Heidegger's insistence that the "work"—the work of art, that is—creates its "own world,"[11] that it demands its own context, gives the aesthetic experience a different coloration. Although he often professes himself an admirer of the craftsman, the handworker, like most philosophers who concern themselves with (for or against) aesthetics, he has little or nothing to say about the craftsman's product, *Kunstgewerbe*—and even less about any rationale of designing that might be separated from the inherently worthy process of hand labor. He concerns himself "aesthetically" only with *Kunstwerke,* and then only when they are what he terms "great art." What determines the quality of such works? "Das Sich-ins-Werk-Setzen der Wahrheit des Seienden" in them, Heidegger says, which might be translated—approximately, of course—as "the setting-to-work of the truth of being." This rather complex notion has been the subject of many commentaries. Whatever inner rules may govern the artist's operations in his or her work, such rules cannot concern the philosopher directly, since it is only confrontation with "the work," and his taking account of that experience, which can engage his attention. Great works, he readily concedes, have also to be beautiful, but their beauty may have nothing to do with an agreeable impression (such as craft objects might be expected to produce), since beauty—Heidegger says—is the mode in which truth has being as unconcealment, un-forgetting; his version of the Greek word *alētheia,* which has been crucial throughout his writings to his account of what might be meant by "truth." For that, Heidegger contended, personal confrontation is essential.[12]

The work of art, *das Werk,* therefore exists for itself; it does not offer itself to any current human use. Although it may originally have been produced as a useful artifact, a building, an altarpiece or something more humble—and as part of our artificial and manufactured context—*das Zeug* (in his terminology, the tool), if it is to be perceived as a work of art, has to be set in its own quite different world; this distinction is very important to Heidegger's thinking for other reasons.[13] He has often implied (and this theme has been taken over by his followers) that machine-made objects, being products of mass production for a mass market, are almost inevitably inauthentic—not even *Zeuge,* but *Gestell,* "tackle," a classification that would also cover most contemporary buildings, which are put together of industrially produced parts.[14]

Much of what I have just written draws on Heidegger's frequently discussed text "The Origin of the Work of Art," which was based on a lecture he gave in 1935 (now republished posthumously, with his notes.)[15] Inevitably, many works that he cites in that essay are literary; but insofar as he is concerned with painting, sculpture, and architecture, two exemplary objects are inspected and reviewed at some length (and they therefore return constantly in his commentators' many writings). The first of them is van Gogh's picture of a pair of shoes. Heidegger had remembered them as a peasant woman's well-worn and mud-caked ones; these essential *Zeuge* had been transformed by van Gogh's brush into a work of art that reveals an inner truth to the viewer, a truth that had been part of—indeed had been taken for granted as part of—a life of virtuous toil, and had therefore never attracted the attention of their owner. As it happens, there are several paintings of a pair (and even of several pairs) of shoes by van Gogh, though the one Heidegger might actually have seen probably shows the painter's own shoes: scuffed, but with highlights that speak of shoe polish. Whatever had been the truth painted into the picture by the artist, it does not seem to be the one "revealed" by the philosopher in his essay.[16] The claim of his commentators that he was in fact concerned with a "deeper" truth, with the *Zeugheit* of the *Zeug* in the picture, so that the details of appearance and context turn out to be irrelevant, seems to imply that the painting, the "work," was not in the end worthy of serious attention.

In the same passage Heidegger proposes another example, which, being nonrepresentational, nonmimetic, may allow the truth to become evident even more directly than was the case with the van Gogh canvas. It is the Greek temple on which he comments:

A building, a Greek temple, says nothing. It simply stands there, in the middle of a rugged rock valley. The building encloses the form of the god, and through this concealment allows it to stand forth through the open colonnade into the holy enclosure. . . . Standing there, the building rests on the rocky ground. This resting of the work draws up out of the rock the obscurity of that rock's bulky, yet spontaneous support.

It does appear that Heidegger meant "Doric" when he wrote "Greek"; and indeed his description is in line with all the vast literature about the German-Dorian connection. More importantly, he does mention one specific Doric temple elsewhere in the essay: the temple of Poseidon at Paestum. Now Paestum—or Poseidonia, as it was in antiquity—is neither rugged nor in a rocky valley, but stands on a fertile plain celebrated for its gardens and its twice-flowering roses. Four well-known ruined Doric temples are its most notable remains, though there are also an agora, a theater, a bouleterion, and all the appurtenances of a prosperous town. In fact it is difficult to think of a Greek temple—Doric, Ionic, or indeed Corinthian—sited in a rocky valley: rocky valleys are not what they are "about."

If Heidegger then has deliberately used these two examples in a way which seems unfaithful ("unmindful") to their makers' intention, I wonder if his own reading of them might not be "deconstructed" to suggest that he is using the "mimetic" work to signify a female, patient, earthy, toiling, setting-to-work-of-the-truth, while the masculine version, being unmimetic, touches the truth more directly. The nature of this *Zeugheit* is hard and stony; it conceals the god, but allows him to declare himself through its open front and the setting "in which disaster and blessing, victory and disgrace, endurance and decline acquire the form of destiny for humankind." [17]

Here—as elsewhere—Heidegger would not concern himself with specific features of the works of art, which therefore become mere instances of a theoretical position. He seems to have neglected Schopenhauer's wise injunction: "before a painting, everyone should stand as before a prince, waiting if and what it will say to him. As with a prince—do not speak first." [18]

Writing in 1934, a year or so before Heidegger's now canonic pronouncement, and very similarly, though more eclectically, Gottfried Benn made a strange collage of texts from Burckhardt, Nietzsche, and Taine that he called the "Doric World," in which he claimed that "when we . . . turn to the substance of Greek art, the Greek temple does not express anything, it is not comprehensible, the column is not natural, they [column and temple] do not assume any cultic or political intention." Benn's aim was more directly didactic, more banal—and therefore more obvious—than Heidegger's: to convince himself and his readers that the wholly abstract, wholly nonreferential Doric "style" represented an art created in parallel to naked violence, arbitrary power, racial pride, a homoerotic antifeminism; and to demonstrate that a vital art will only be created in such conditions. "The Dorians work at the stone, they leave it unpainted. Their statues are naked. Dorian is the skin, but tight over muscles, manly flesh. The body bronzed by the sun. . . ." Or again: "at the back of the Panhellenically conceived outline of the Greeks stands the grey column without a base, the temple of ashlar blocks, the men's camp on the right bank of the Eurotas." [19]

Of course, the pieces of Benn's collage do not add up. Sparta, evoked by his mention of the river Eurotas, had no major stone Doric temple; Doric columns, practically all of them, were not of grey, bare stone but much more often of marble or other light-colored limestone, and were in any case stuccoed and stained yellow, with their capitals, cornices, and statues even more garishly colored and inlaid. Moreover, the major stone monument on the Eurotas was the elaborately colored throne-altar of Apollo at Amyklae, which had those mixed Doric-Ionic columns I have already discussed.[20] But Heidegger's text presents analogous difficulties. Apollo at Delphi, on the slopes of Parnassus, is the only god whose temple could be described as "in a rocky valley"—at a pinch.[21] In any case, the temple of Delphic Apollo, like all its treasuries, or indeed any Doric temple—or other Greek temple for that matter—can hardly be considered "dumb" or "mute," being not only stained and inlaid, but bedizened with gilt bronze trophies, painted ornaments, and many inscriptions. "Buttonholing" or "garrulous" would seem a more appropriate description.

Many Doric temples were built over the rocky outcrops, if not in valleys, while some were set on quite different terrain, notably those of Paestum I mentioned earlier or that of Ephesian Artemis, which was sunk into the sumpy, alluvial seashore—a situation paralleled by other (especially Ionic) ones. Curiously enough it is in the temple of Artemis Ephesia that Heraclitus "the dark one," the ancient philosopher whom Heidegger most admires (and often comments on and whom he quotes a little later in the same essay on the "Origin of the Work of Art"), was a hereditary official. Such niggles are not mere pedantry: they show how evidence had to be manipulated for Heidegger, as for Benn (though for somewhat different reasons), to produce the "timeless" Doric temple as an exemplary "aesthetic" object. Yet in spite of all its gaps and faults, their view has been adopted by many of my own contemporaries.[22]

The Benn-Heidegger Doric temple is inexpressive; it carries no reference and is un-blemished by any historical development. It is not the building described either by archaeologists or by ancient writers; it is a ruin that hardly stands out from its natural setting, though it "creates its world," in Heidegger's terms. For him, as for Benn, that world is the Dorian-Aryan antiquity that had been fabricated by a group of German philologists a century earlier. Its main proponent, the fabulously learned and prolific Saxon scholar, Carl Otfried Müller, died young in 1840, on an expedition to Greece. His account of Greek tribes and cities had concentrated on the Dorians, Apollo's people, and I have already had occasion to refer to it in discussing the legends of the Dorian invasions. Müller's persuasive—if perverse—version of that legend had a powerful effect on Prussian and even German national identity; his book was quickly translated into English and that same image of the blond invaders from the north was assumed into British imperial ideology through the public school fostering of the classics. It also left its mark on a newly independent Greece. German philology and history ever since, in spite of Nietzsche's plea that it only showed half the picture, that the bright Apollonian world had to be complemented by the world of dark, mania-inspiring Dionysos, has been dominated by the Müllerian legend.

Equally inevitably, it had its architectural consequences. In spite of all Vitruvius had written, the Doric order was idealized as the quintessential one, the concentration of all that was best about Greek architecture. That is why Gaudí appropriated it as the fundamentally "Mediterranean" element in the Parc Güell, and why Adolf Loos thought that he could set it up for a canon of architectural rectitude. To Loos, the Doric column seemed a perfect object, not bound by scale, place or history—the "order of orders," which would overwhelm Chicago with its beauty and permanently change the configuration of the skyscraper as a building type. Nor is this belief in the Doric order of orders quite a thing of the past.[23]

By contrast, Hegel's contemporary and acquaintance Karl Friedrich Schinkel, who was Müller's senior by some fifteen years, could still see the Ionic column as in some sense the perfect one: he used it in his most famous buildings, the Schauspielhaus of 1818–1821 and the Altes Museum of 1823–1828, both in Berlin. It even framed the foreground in a large painted, idyllic vision of the creation of Greek architecture.[24] His one famous Doric building was a small guardhouse, quite Vitruvianly appropriate, the Neue Wache, by the Arsenal on Unter den Linden, built just before 1820. It is within this much-admired building that between 1930 and 1931 Mies van der Rohe designed his most Doric-grey and stark Monument to the Fallen of 1914–1918—though it was never executed.[25] This association sets Mies' endeavor within Schinkel's building, but also—by extension—beside the almost exactly contemporary literary Doric temples of Benn and Heidegger, since all have something of the quite unhistorical and dumb, the nonreferential character of building found in Mies van der Rohe's programmatic statements.

In spite of his insistence on the surrender of every formal consideration to the demands of production, even Mies was not quite absolved from the metaphor with which I have been concerned. In the manifesto of 1923 that I quoted earlier, he proclaimed that steel, concrete, and glass were the "materials of a skin-and-bone construction," and that was the form he primarily favored. Reduced to skin and bones, the body inevitably becomes a jejune thing.[26] Yet this reduced body of building has been perhaps the most influential "construction" of contemporary Western architecture. Its validity is buttressed by that parallel of Dorian mythology to which Heidegger and Benn both subscribed.

Heidegger remains one of the two or three dominant thinkers of the century;[27] Benn, one of the most influential German writers of his generation. The Graeco-Dorian mythology they have perpetuated retains its grip on the modern imagination. Anyone who today wishes to acquire some notion of how the Greeks went about making their buildings, or at least what they made of them, must come to terms with the position that Heidegger and Benn have staked out—and which has had a lasting effect on the historians and architects of their time. Their vision provides the most recent installment in the strange story of how these columns, these "orders" were understood; it may be deceptively clear and it may be distorted, but it retains its power.

Skin and bones

The Heidegger-Benn vision of the past—those dumb Doric temples wrought by blond, brutish, naked, sun-bronzed Dorians, which mutely draw strength out of the rock—is in turn closely related to Mies' assertions about the equally dumb buildings of his present (as well as of the future). Such interdependence was not suspected and has not been examined; and yet the two notions belong to the same time and the same intellectual climate. By an extension of the Miesian proclamation into an argument,[28] the *Sachlichkeit* he first professed half a century ago has been proposed as the only proper model for the builders of the late twentieth century. The isolation of the stylized, crystalline building from its surroundings, raised on a podium and on its tall columns (exalted in the work of Mies), emulates the stark temple of the Dorian myth on its rocky socle. This Miesian teaching, though it has no truck with veneration of the craftsman or contempt for the "machine age," has also been interpreted as representing an aesthetic of silence and absence that some have considered the only appropriate one for our alienated and technology-dominated time.[29]

I hope that I am not traducing a subtle argument in this brief summary, but it seems to run to the following conclusion: in the postcapitalist (and therefore alienated) society of the late twentieth century, any search for meaning that fabrication might "contain" or incorporate can only lead to frustration, since our mass society has lost any organic sense of the relation between the process of production and the producer—the builder. Also lost is the connection between that producer and the consumer, whether understood as a consumer of the tangible goods offered on the market, or of the intangible consolatory messages by which "good design" masks the inner contradictions of mass production (since both producer and consumer are ultimately powerless individuals manipulated by impersonal forces). The blasé attitude so well described by Georg Simmel, in which the particular character of different objects and all their qualities are annulled in their money value, so that any excitation they might provide has to be an added flavor, replaces—in the wholly market-oriented society—the constant and varied stimulation that had previously been provided by the environment in everyday circumstances.[30]

Seen from the vantage point of those observers who have understood the extent of inevitable late-capitalist alienation, it is clear that any talk of a "meaningful" architecture, of an urban planning that would claim to impose form on the city of mass production and mass communication, must be illusory; indeed it may be something rather worse, since it represents a monumental case of ideology as false consciousness. The banalities of industrial design are mere packaging: they do not deserve a critic's serious aesthetic attention and certainly not a philosopher's. Yet even the advocates of such arguments recognize (if only tacitly) that there cannot be any architecture without metaphor; they therefore conclude that a search for an architecture beyond (or distinct from) an art of building in the late twentieth century must be condemned to fringe activity—of which Louis Kahn's (inevitably frustrated) attempt to find a modern plastic expression for institutions in building is a symptomatic instance.[31]

Something about this formulation brings to mind the argument about the intellectual's betrayal, the *Trahison des Clercs* that Julien Benda, the paladin of reason and enemy of intuition—and of all things German—was denouncing in France about the time of Mies' first proclamations. Benda's ire was directed at those intellectuals who abjured the claims of rationality before the demands of "superior" interests, whether national, institutional, or organizational; but he also balked when he observed the claims of reason subsumed to those of historicism. In what seemed (to him) a caricature of Hegelian dialectic, he noted these "false clercs" surrendering to the urging of anonymous historical forces as if they were diving into a stream with which they wanted to swim—as if the truly enlightened historian and critic was a winner in a kind of downstream historical swimming contest.

However subtly it is elided and formulated, the abdication before historical forces, with its implied surrender of the individual's critical faculties, has dominated much twentieth-century building activity, and consequently much of the work of the historians and critics who live on the fringes of architecture and have seen their role as providing arguments to justify the industrial building process as the only true determinant of architects' formal endeavors. Against such resignation, I would maintain that no historicist construction can absolve the producer-builder, architect, or any other artist from the elementary and common human duty of acting by reason from choice and considered principle rather than from obedience to necessity.

Any insistence on efficiency as the only valid criterion for judging buildings and the emphasis on management, so strident in Miesian and post-Miesian pronouncements (some, such as those of Hannes Meyer, the second director of the Bauhaus, were extremely influential), might also be read as the architectural equivalent of Benda's *Trahison*. The insistence on architecture being merely the "aesthetic" supplement to building shows the way in which this argument has restructured Western thinking, and it has long ago led to a complete change in expectations about "beauty" in a work of art, which is not sought in response to an object beautiful-in-itself (that is, by virtue of certain identifiable qualities of surface and structure); instead, the qualification "beautiful" is allowed only to that which produces a certain kind of pleasure. It is a view that Martin Heidegger clearly found distasteful: he explicitly denied that the work (of art) might be considered a thing (his *Zeug*) that has been equipped with an aesthetic value as a kind of attachment.[32] Yet a version of that very belief has led some writers to posit the "decorated shed" as the only possible architecture of the late twentieth century—an argument that can serve as the "artistic" correlative and justification of a developer's design policy.

Sensation and production

Paradoxically, this "aesthetic reduction" reflects another preoccupation of Benda's: that aesthetic considerations were overwhelming intellectual ones among artists.[33] This may well be one of the reasons why the element of play has withered in architectural thinking. Human play, as I suggested earlier, is mostly play *at;* and "playing at" is a game. A game (whether it is hopscotch or football or chess) has patterns and rules, and must be played in and through them. The rules of a game are self-imposed by internal necessity and are rationally expressible—even for the most childish games like not stepping on flagstone joints when walking on a paved street—but they should also never be suborned to any exterior end: "professional sports" are an oxymoron, or at any rate, they are not games. When architecture is the servant not just of building technology, but of the real estate market, its element of play will (of course) vanish. Even if it is seen as the motor for "aesthetic emotion," its internal coherence will suffer.

The reader may have noticed that I have not considered aesthetic issues here in any case, because my concern is not primarily with *aisthēsis*—the way things are seen and perceived—but with *poiēsis*, the way in which they are made. The severance between the crafts of making and the apparatus of *aisthēsis* was only implicit to Alexander Baumgarten but had become apparent to Kant, whose most important writings on art are part of his *Critique of Judgment*. Hegel had influentially established aesthetics as the only true way of thinking about art, and his concentration on the spectator and his disinterested contemplation led to many conceptual difficulties, which Nietzsche knew exactly how to exploit:

> The illusion of disinterested pleasure and of the impersonal character and universality of aesthetic judgement arises only from the perspective of the spectator; but from the perspective of the producing artist we realize that value *appraisals* are necessarily value *positings*. The aesthetics of production unfolds the experience of the genius-artist who *creates values.*[34]

The same difficulties have (inevitably) also been raked over by neoscholastic writers, even though they have tended to elide some of the difficulties of the discussion in the postindustrial era a little glibly; the same difficulties led Heidegger—to take the most conspicuous example—to formulate his radical considerations about technology.

I have only mentioned briefly the means that are used to arouse "aesthetic emotion" in the viewer (and I produce this nineteenth-century commonplace deliberately, echoing Nikolaus Pevsner), concentrating on those elements, motifs, relationships that are devised or employed by a maker to constitute objects—cities, buildings, engines, household items—indeed the whole artificial world we inhabit. The way in which such objects are perceived, however, sanctions the place they take in the world, much as the effect of a rhetor's *ēthos* depended on the response he found in his audience—as I pointed out when discussing *ēthos* and character in the second chapter. Yet the literature of *poiēsis* has been regarded with suspicion since Hegel (following Kant) authoritatively dismissed all writings on the theory of artistic creation—from Aristotle's *Poetics* to the advice of Goethe—as based on an incomplete experience of masterpieces and fit only to dictate such rules as are needed at a time of decadence. In this many philosophers have followed him.

In the late twentieth century, aesthetics has wholly eaten up, has digested the theory of art; this has happened at a time when building has become one of the most energy- and capital-consuming of human activities, and the process of making has become almost exclusively industrialized. One consequence has been that the maker's craft can only be the subject of private speculation. I therefore want to direct attention to the importance of theories of making that are (*pace* Hegel) independent of (perhaps because they are prior to) a theory of perceiving what is made, what is artificial: and they require *poiēsis*.

The purpose of *poiēsis* is not just the making of certain (and very different) classes of objects, such as poems or houses, but (so Plato and Aristotle substantially agreed) it is the making of something out of nothing. However, because no human agent can make any one thing "out of nothing," it is the condition and limit of the business of making that it must always be an assemblage and an imitation. The artist's imitation is not making something that *looks* like something else, but rather something that has a *way of being* like something else. Houses cannot be required to *look* like people—that would clearly be absurd; instead, they need to "occupy a place in the world" analogous to the way that persons take their place in it. Nor is this merely a statement about the view of the philosophers of the Athenian Academy in antiquity; it has had echoes in much writing about art and design at the end of the twentieth century.

Making, imitating, loving

If what I say about the transformations of my base metaphor—the body and the column, the body and the building—is true, then there is also no sense in making our buildings look like those of the Greeks (though some have tried!). We therefore do not need to learn from them the exact shape and use of this or that of their elements, since our conceptions of the body and its working, our very differing views about the structure of the universe—never mind our building methods—make any attempt at reproducing the externals of Greek architecture vacuous. Of course, our minds work in ways that are not unlike theirs; we continue to use Greek terms in the vocabulary of our philosophy, and we conduct many of our debates within the limits that were first set by Greek thinkers. That is why we should attend to their beliefs about how buildings are projected and produced by that process by which living beings produce their own likeness; by love, as Plato has it in that passage in the *Symposium* that I discussed earlier, which requires that the two partners in any act of creation see each other as "good." It is the analogy that runs through Plato's recapitulation of the creation of the world in the *Timaeus:* how reason persuaded necessity into conceiving the world. Inevitably, there are three kinds of beings in this process: that which is born, that in which bearing takes place, and that which the born resembles. That model is the father, the matrix or womb is the mother, and that which is born, the child.[35]

What the demiurge, the craftsman (or architect) god operated on was the *chōros,* the precosmic space, the place and the "nurse" of all being. This creative act Plato advanced as the model of all right human action, whether it is the wise ruler acting in society or the craftsman impressing forms on material. The architect's disposing of his elements is therefore merely another instance of the workings of that model, and the articulation of the building could be interpreted, as by Vitruvius, as an evocation and fulfillment of that grand universal vision. The Creator playing at cosmic order taught architects how to play at building.

These crucial invocations of procreation are ignored by modern critics who worry retrospectively about the "purity" or "coherence" of the orders, and are disturbed by the fact that the Doric peristyle of the Athenian Parthenon embraces an Ionically molded wall, or that the columns of the Apollonian shrine at Amyklae are half Doric, half Ionic—a kind of hermaphrodite. They have mistaken the nature of these categories, since what they take for impurity or even confusion is in fact a form of *methexis,* the mix or "participation" that (according to Aristotle) was merely Plato's modification of what the Pythagoreans had called *mimēsis,* the imitation of heavenly forms in tangible objects.[36] The best architectural demonstration of this loving embrace was perhaps the temple of Apollo Epikurios at Bassae, where the Doric peristyle encloses a motherly Ionic, which in turn embraces the lonely daughter-column that appears as a unique, slender Corinthian. Vitruvius had been so emphatic about his girl-column, the Corinthian, being procreated by the man-column, Doric, and the woman-column, Ionic, just in order to prevent (or so it seems to me) such a misreading as is practiced by modern authors about the orders.[37]

Belief in the importance of mimetic action may be seen as an antique superstition. Kant had seen it as an incidental factor in the making of art: "everyone is agreed that Genius should be considered utterly opposed to the spirit of imitation. Since learning is nothing but imitation, it follows that even the greatest capability, a learning aptitude (capacity), can surely never be counted as genius". As far as the arts are concerned, Kant had previously made it quite clear that: "Genius is the talent (gift of nature) which gives rules to art. As this talent, being the inborn productive asset of the artist, belongs to nature, it is possible to restate the matter in this way: genius is the inborn cast of mind (*ingenium*) through which nature grants rules to art."[38] Rules, Kant maintained, are a mark of every art. The marks of genius are its originality, its capacity to produce prototypes, the unconscious formulation and unteachability of its method; nature gives to genius the rules for art, not for science. Genius does not act by imitating nature—on the contrary, nature (not any conscious thought or action) *dictates* rules to genius.

Hegel had taken the argument further; but it was Kant's relegating of mimesis to the margins of his aesthetic, dominated by the figure of the Genius and his essential originality, which was decisive for the future and which has been echoed by critics and philosophers throughout the nineteenth century. It had been of marginal importance in biologically oriented sociology, such as that of Walter Bagehot or Herbert Spencer. Toward the end of the century, however, imitation was erected into a structural principle of human association by the French jurist and "rational" sociologist, Gabriel Tarde, who is sometimes regarded as the father of social psychology and seems closer to Simmel than to his British contemporaries. His very rational and inductive method was wholly opposed to any historicist, linear account of social phenomena, and he maintained that social bonds proceeded by a dialectic (a word he would have eschewed, I suspect, for its Hegelian association) between imitation, which is the run of human relations, and invention, which produces the catastrophic change in the human condition that imitation diffuses and translates into custom, fashion, tradition. This formulation has influenced a generation of German and American sociologists and anthropologists, as well as many of his own countrymen.[39] He chose to explicate his notion through the example of the Greek and Egyptian arrangement of beam and column (for which he cites Georges Perrot): stone elements, which at first imitate wood construction, are then decorated with local plants (acanthus in Greece, lotus in Egypt); the massive stone pillar is next subdivided into base, shaft, capital; then new and invented ornaments are added to this form. Tarde denies any linear development (as, for instance, the procession from thick and stumpy to thin and etiolated columns in Greek architecture) as always being subject to too many counterexamples. Style is an independent social-mental phenomenon, which owes little either to industry or to physical necessity.[40]

The interplay between imitation and invention is the stuff of social history—even of history tout-court—in Tarde's thinking. His emphasis on imitation was inevitably contested, yet it has had its effect on the structure of modern psychology through Jean Piaget,[41] on anthropology (in the work of Arnold Gehlen and Helmut Plessner), and on ethology in general. It becomes a hidden factor in the linguistic theories of representation.[42] Imitation, which Kant had removed from its dominant place in the philosophy of art, had returned as a nuclear notion in philosophy. Of course, the instinctive imitation of sociologists and anthropologists is not quite the deliberate, even artful, mimesis of the critics and philosophers. That mimesis has been closely associated with play—and the Dutch historian, Johan Huizinga, has restored play as a social force much as Tarde had the idea of imitation.[43]

Aesthetic activity or play (linked as it is to imitation) is an end in itself. Against Kant, Schiller had already vindicated the value of play as an essential aesthetic, even human, occupation. It is the drive that reconciles the sensory appetites with the formal and nouminal ones—hence, his famous sentence, almost an aphorism: "Man plays only when he is man in the full sense of the word, and he is only wholly man when he is playing." This common ground has become the subject of much learned comment in recent literary criticism, and has led to imitation being again considered a fundamental factor in the making of a work of art; and indeed (though he professes to have forgotten his source), it is also echoed by Herbert Spencer when he associates play with the aesthetic potential of an evolving humanity.[44]

Mimesis was the title as well as the theme of Erich Auerbach's book, written over half a century ago; and that work became the stimulus for a renewed interest in the matter. René Girard (whom I mentioned earlier in connection with the origins of sacrifice) has given it a great deal of attention and for him the mimetic drive is almost coterminous with original sin, since all human desire is a desire to possess another, which desire, being specular, has to be mediated by some reflection—a model or image; desire will therefore, Girard maintains, always involve rivalry, hatred, violence.[45] It is the drama that was not revealed by the old philosophers (Plato, Aristotle and the rest) but that fully played out in the work of the great nineteenth-century novelists. Practically all Girard's examples are literary. References to matters visual are extremely rare.

At any rate, the vast output of Georg Lukács, certainly the most famous and learned apologist of Social Realism, revindicated the mimetic mode against the "classic" eighteenth-century systems of the "fine arts"[46] as being fundamental to an understanding of both architecture and music—and even of craft products (I have already quoted Georg Simmel, his teacher, about *Kunstgewerbe*) and film. In the course of this lengthy revindication, which claims to deal with the whole aesthetic realm—painting, sculpture, and architecture included—his account of art is concerned largely with theories and is drawn from handbooks and manuals rather than first-hand experience. The visual arts, he thought, went irremediably wrong after Cézanne and van Gogh.[47] And though he talks learnedly about Bramante and Michelangelo (out of Dvorak and Burckhardt), he only has occasion to mention one twentieth-century architect—Le Corbusier, who is condemned for an excessive faith in pure geometry. Again, the judgment is reached not through the examination of any of his buildings or even direct quotation, but at second hand: with Lukács, oddly enough, assenting to the damning reference of Hans Sedlmayr, whose politically right-wing views and Catholic integrism are the other end of the ideological spectrum from his own "critical" Stalinism. Sedlmayr and Lukács therefore condemned modernism in art for very different reasons: Sedlmayr looked back to a lost age of total world systems and of belief; Lukács, for his part, saw the grasp of essentials that characterized realist artists as a remedy against the danger of reification in an alienated age. Still, he, too, was nostalgic for a total worldview: on his home ground, the nineteenth-century novel, he was careful to distinguish inferior descriptive prose from the true artist's work, composed narration, and he allowed himself the daring conclusion that there is no narration without a world-hypothesis.[48]

Reification, the making a fetish of the work of art, has been the bugbear of the avant-garde (or what is left of it) since the 1960s: land-art, body-art, conceptual art have all been attempts to release the artist from the burden of commodity—even if the work produced by some look like packets of enigmatic power, like some exotic wrapped talisman. Earlier, in the 1940s and 1950s, Theodor Adorno had considered the art of the avant-garde a true remedy for the ills of the age; it was an art that offered experiential depth and technical advance, and allowed its practitioners to act within the contradictions of late capitalism, so as to oppose reification by work.[49] The negative dialectic that the dominant subject of Enlightenment rationality established with his or her environment was dissolved in a mimetic, a dialogic relationship, and it was therefore much more important to Adorno than it had been to Lukács.

For Lukács, the highest (and the most powerful) art was the narrative image of social reality that at the modern elevated state of culture corresponded to primitive magical (which to him was identical with religious) and humdrum representations of the exterior world, in a bid to control it.[50] For Adorno, in contrast, mimesis had become *the* model of desirable human social relation, since a prerational empathy allowed the interdependence of imitator and imitated in dialogue and play, not in hegemony or domination. Unlike Lukács, he was passionately involved in the controversies of recent music and literature. In an early book, done with Max Horkheimer, Adorno offered mimesis as a model for overcoming the shackles of instrumental reason, as a sort of positive analogue of intersubjectivity. Mimesis was not only a touchstone in discussing the work of the artist but the rationality of social relations in general.[51]

Like the theorists of the Romantic movement, Adorno also saw a strict parallel between music and architecture. Schönberg's technical reform of tonality was not merely parallel in intellectual history to Karl Kraus' battle against the journalistic cliché or Adolf Loos' condemnation of ornament: they had the same intention, exactly, and it is this attitude that Adorno called "Functionalist".[52] But he was aware that their modernity had to live a contradiction, since "the dark secret of art is the fetish-nature of commodities. Functionalism would break from the trap, and tears in vain at its chains as long as it remains in bondage to the ensnaring social fabric."[53]

The only weapon that an artist could wield against commodity fetishism was technical skill and invention; that could sometimes afford the public that experiential depth that offered the promise of a juster society—a society freed of fear. As it happened, Adorno saw one instance of that mimetic skill in a recent building. In his last major book, his *Aesthetic Theory,* he comments enthusiastically on the Philharmonic Hall in Berlin (which Hans Scharoun designed in the late 1950s but finished in 1963) and the words can be read as an index of his whole theory of art: the Philharmonie was, he said, "beautiful because, in order to create spatially ideal conditions for orchestral music, it assimilates itself to music without explicitly borrowing its elements. Its function is expressed in and through the building, yet it transcends mere purpose, though such transcending is not determined by the functional forms themselves." Referring back to some of the arguments I have set out earlier, he goes on: " 'New Objectivity' dismisses expression and mimesis as superfluous decorative adornments and as frivolously subjective trimmings. This condemnation is valid only in cases where construction is cluttered up with expression like a room is with furniture."[54] Adorno saw works of art as recollections, *anamnēses* of an archaic magical power that echoes "primitive" man's original and fearful dialogue with the other—the original objectification.

Anamnesis gave access to primitive power even in the time of the reason-bound Enlightenment. The excess of reason breeds fear. The desirable society is one in which fear has been banished. The language of the true artist can anticipate that which is beyond fear, and the primary mode of his language is mimesis, the mode that divides into two ways of relating to the world: the magical and the aesthetic, terms he applied very differently than did Lukács. Since for Adorno mimesis is the inevitable rationalization of and historical advance on the sympathetic magic of the first men and women, it was also a guarantee that art was both rational and cognitive: rational in the sense that it was always directed to some social and therefore intelligible end, and cognitive in that it was inevitably referential. Even if any artist working in late-capitalist society is performing alienated work, when he or she objectifies mimetic impulses, nature comes to be reflected in culture. Such works of art provide a model of how—in a fearless and liberated society—the ego could deal with its libidinous impulses without repressing them, and human labor could be organized without domination. Great art is that which summons such a form of mimesis. The work of the artist has to do with objectifying: it is social labor, which eschews any license, and by its very nature may contribute to social transformation.[55]

With Adorno's notion, the process by which architecture was removed from the group of the arts of imitation in the course of the eighteenth century is reversed, and it returns as the very type of mimetic operation. Though he is primarily concerned with music and literature, his notions seem to be equally workable in architecture and the other visual arts. And he further insists: if great art is to have the liberating power required of it, it has to offer the richness of experience and the technical virtuosity and innovation he so much admires.

That mimesis and rationality are the double spring of the artist's activity is a notion that subverts the majesty of Hegel's great system in which the arts were a less transparent vehicle of the truth than religion or philosophy because they could only convey such truths as the senses apprehended. Adorno, using the terminology of Kant and Hegel, attempts in his posthumous *Aesthetic Theory* (which strangely echoes Hegel's posthumous *Aesthetics*) to claim that for the arts which Hegel had denied them: namely, the capacity of communicating philosophical truths through sensory apprehension without using the vocabulary or the methods of philosophy. He could not offer his readers a triumphant vision of any ultimate accomplishment: Hegel, on the other hand, could point to the poetry of Goethe and the structure of the Prussian state as the two summits of human achievement.

Adorno therefore set a very different value on rational *poiēsis* and the modalities of mimesis. For all that, he did not concern himself overly with the technicalities by which mimesis had been achieved—or might be attained again. My primary concern in this book, on the other hand, has been to show how the mimetic artifice has been organized during a specific period in the past.

Such artifice, such techniques, have been so neglected that they will need to be reinvented for our own times if we are to master them again. Inevitably, this will be done spasmodically, by re-creation and reinvention, not by setting out a set of rules to be followed. One example of what can be done seems to be provided by Le Corbusier's *Modulor*. It was perhaps the most convincing attempt to set out a mimetic teaching, and it is at least partially successful, though not as Corbusier wished it—in becoming the basis for new industrial type measurements. Nevertheless, it is used by a number of younger architects as a standard in their building, in part because it relates imperial and metric scales to a common measure, and also because it introduces a canonic human figure and the proportional system based on it—an explicitly mimetic technical procedure—into the normal working of the architect's office.

The *Modulor* has so far been an isolated instance of a *technē* formulated for current conditions, and it has gained limited assent. For the bulk of building in the late twentieth century, it is barely relevant, since the production of income is generally considered building's main social function. Even the highly bureaucratized institutions of government and the law therefore adapt the current developers' manner to public office. Developers are the determinant patrons of architecture. Their views on architecture are therefore important, and one of the senior practitioners of their profession has offered an authoritative summary: Buildings, he says, are a commercial product "like laundry detergent. We need to know what kind of design is going to appeal to the consumers, who have become much more design-conscious, across the board. Design is a secondary added feature. . . . We are *not* dealing with the public at large: the *media* generate interest in the designs that we develop."⁵⁶

Much more than the products of painters or sculptors, never mind "conceptual" artists, buildings are viewed by their owners and patrons, who are their effective producers, as needing ornamental surface (which can intermittently and indifferently be neo-Greek, or neo-Gothic or postmodern, or even "modern"—but certainly mute) to be fed to the media so as to provide a distraction from the developers' real aim of producing "sheds," maximal rental space. The buildings of the late twentieth century have become raw commodity disguised in gift wrapping.

Developers are of course prepared to indulge the revived curiosity of the public about design, because it is only curiosity about the package. In fact many urban buildings are outsize phallic fetishes whose true form is disguised by their vast size: point blocks that repay the investors but in doing so suck up the energy of the city in their service ducts and elevator shafts and will eventually destroy its fabric. Are they in any sense of the word works of art? Insofar as they are a part of the sensible and visual environment, they are received as such by the public; hence the interest in design of which the developers are aware. Yet the problems and terrors against which artists of the avant-garde have struggled (if Adorno is to believed) are derided and travestied in their production.

"The ultimate barbarity is the plain reading."[57] It is not quite what Adorno said, but something like it underlies much of his work on aesthetics. "Plain reading" of a building is the univocal, antimetaphoric account of it that presents it as mute, as meaning no more than itself, the way it is built and used. The menace hidden in the discourse of a mute architecture is the ultimate barbarity that the developers' buildings may represent. Yet this notion of the mute architecture remains the dominant doctrine in architecture schools and of much architectural criticism. The mute Greek example has for too long and far too effectively been invoked to sanitize the building process and to present its products as incurably dumb.

Greek architecture spoke to another time, and in other circumstances, but I think we may also listen to its discourse. It will not invite us to mimicry, to repeating catalogues of details or features, but it does demonstrate the validity of such technical speculations about mimesis as the *Modulor*. Greek architecture shows how buildings should be conceived, how physical forms relate to the fabric of human groups—to societies and communities. Building is, after all, the group activity par excellence. What the Greeks really do show us is what we are entitled to expect of our buildings.[58] And since we are so entitled, we should protest (and that is what I am also doing) when we are given short change.

I therefore hope that I have presented Greek architecture as the most entrancing and forceful, the exemplary art of building, an architecture which still invites dialogue and touch, which requires physical contact across the millennia. It cannot teach us—history never can. But we may learn from it.

Notes

I Order in Building

1. Such disapproval persisted in spite of the recurrent "returns to order," such as the one in France and Italy in the early 1920s. The literature of postmodernism is far too copious to require separate reference.

2. The Marchese Giovanni Poleni (1739, pp. 8 ff.) was the first to establish this date: by inference, since the *editio princeps* carries no printer's name or date. Poleni supposes that the peace between the Pope and Ferrante of Naples of 11 August 1486 is the terminus post quem, and most bibliographers have followed; he thought the printer was Georg Herolt or Heroldt, one of the Germans established in Rome. However, A. W. Pollard (1908) names another prolific German printer, Eucherius Silber. The editor was Giovanni Sulpizio of Veroli, and the text of Vitruvius was followed by that of S. Julius Frontinus *On Aqueducts,* which Sulpizio edited with a more famous scholar, Pomponio Leto. For reasons he does not disclose, Auguste Choisy in his edition gives the date as 1488 (1909, vol. 2, p. iii.).

3. On Serlio, see Christof Thoenes (1989), though there has not been a monograph or a life for some time. See therefore, still, A. B. Amorini (1823) and L. Charvet (1869). Of his six or seven books the fourth was the first to be printed in Venice in 1537, though he had issued some engravings of orders in 1528. Five books were first published together in Venice in 1551, then in French (1551), Latin (1568), and German (1608). The Latin text was the basis of the popular Dutch version (Amsterdam, 1606), on which in turn the English (London, 1611) translation was based. A list of the editions is given by J. B. Bury in C. Thoenes (1989, pp. 100 f.).

4. L. B. Alberti *De Re Aedificatoria* VII.8 (1966, vol. 2, pp. 584 ff.; trans. 1988, p. 209). It could, Alberti suggested, therefore be considered a more "complete" kind of column than any of the Greek ones. The columns on the facade of the Tempio Malatestiano in Rimini may have been an attempt at working out this "genus."

5. These were the most important: but there were other capitals of this order on the Arch of the Argentarii, in the circular church of Santa Costanza (known as the temple of Bacchus), and in the central bay of the Tepidarium of the Baths of Diocletian (the Church of Santa Maria degli Angeli). And indeed, Serlio sees it as completing his sequence:

> which the old Romanes, peraduenture, being not able to goe beyond the inuention of the Greekes, finders of the Dorica after the example of men, and of the Ionica, resembled to women, and of the Corinthia, after the form of maydes, of the Ionica and the Corinthia made a composition, piecing the volute of the Ionica, with the Echino in the Capitall Corinthia. . . .

The seventh book, unpublished until the twentieth century, was printed from two manuscripts; one in the Avery Library at Columbia University, the other in the Bayerische Staatsbibliothek in Munich. See A. K. Placzek, J. Ackerman, and M. N. Rosenfeld (1978) and M. Rosci (1966). On this text see M. Carpo (1992).

6. The word *ordines* was probably first applied to the four genera of columns in Raphael's circle. See his letter to Fabio Calvo (dated 15 August 1514) thanking him for the Vitruvius translation manuscript, and offering to design a frontispiece for it: "e ve farò el frontespitio de hordine doricho." But when the orders are ennumerated, in the "anonymous" *Letter to Pope Leo X,* they are indeed five. However, the fifth order, the Attic, is in fact a square pillar ("ha le colonne facte a quattro facce"), as in Grapaldi's book, following Pliny the Elder. See V. Golzio (1971, pp. 35, 91 f.) and Raffaele Sanzio (1956, pp. 39, 63 f.). But see also Raffaele Sanzio (1984, pp. 38 ff.). For the letter to Leo X, see the critical edition by Renato Bonelli in A. Bruschi (1978, pp. 459 ff., esp. 483 f.). The text of its last redaction, now in Munich (and annotated by Raphael), has now been edited again by I. D. Rowland (1994, pp. 100 ff.), with reference to the practice of its principal scribe, Angelo Colocci—an apostolic secretary and bibliophile, whose part in the Raphael circle had not been appreciated.

In the letter, the word *ordini* rather than *modi* is used for the three ways of drawing (plan, elevation, perspective). Bonelli suggests that the section on setting out the five orders, for which the word *ordini* is also used, is written out in another hand. On the use of the word *ordo* for the "five orders," see Rowland (pp. 83, 97 f.). A recent discussion of the orders in fifteenth-century architecture is Arnaldo Bruschi's "L'Antico e la Riscoperta degli Ordini Architettonici nella prima metà del Quattrocento: Storia e Problemi" (1988) in S. D. Squarzina (1989, pp. 410 ff.), which provides an ample bibliography. Raphael's letter may also provide the first use of the architrave, though Leonardo also uses it about the same time (ed. J. P. Richter, 1970, vol. 2, p. 473). The *Gr. Diz.* quotes Bernardo Tasso as the first user of *ordine* in this sense: but his usage is later than Raphael's and the context shows that here the word still has the old sense of "story." It is interesting that Cesariano uses the words *ordini stilobati* and *parastatici* to distinguish piers from pilasters in the top-heavy and patently Gothic Milan cathedral.

The invention of noncanonic or even unclassical orders from the sixteenth century to the twentieth deserves a book on its own.

7. C. A. Daviler (1720, vol. 2, p. 736). The word *ordo* (like the Greeek *orthos*, upright, straight, just) is related to the Indo-European root **rta* from which words like *ars* and *ritus* also spring: see E. Benveniste (1969, vol. 2, pp. 100 ff.). Vitruvius assimilates *ordinatio* to the Greek *taxis*, as one of the three main operations of the architect (I.2.i; the others are *dispositio* and *distributio*) and he explains it in I.2.ii as "considering the minor dimensions of the work one by one, and relating them to the major proportion for symmetry. It is derived from quantity, which the Greeks call *posotēs*. Quantity is the calculation of modules from the work itself and the result of a harmony between the single members and the whole work." The version of *taxis* as *ordinatio* is the usual one, as in Cicero's translation (in *De Univ.* from Pl. *Tim.* 30a) of *eis taxin auto ēgagen ek tēs ataxias* into *idque ex inordinato in ordinem adduxit*. *Posotēs* as *quantitas* also has such a Ciceronian precedent (in *Acad.* I). S. Ferri (1960, pp. 50 f.) has suggested that this definition is an echo of some Hellenistic treatise in which *symmetria* was attainable through *posotēs* and *eurythmia* through *poiotēs:* but Vitruvius, so Ferri thinks, was not able to appreciate the distinction, which has led to much confusion. There are many commentaries, particularly the very scholastic one by Daniele Barbaro ([1556] 1584, pp. 27 ff.).

The word *ordo* occurs a little later, at the end of I.2.ix: it signifies the due order of things that should not be disturbed by mixing Doric and Ionic detail, but it is used in a quite general, unspecific sense, much as Alberti uses *ordines columnarum* to signify "rows of columns." *Ordo* on its own can mean anything from universal order to a course of bricks, though never "order" in the sense of Doric or Ionic; Alberti, like Vitruvius, uses interchangeably *compartitio, mos, opus,* and *ratio,* as if there were no standard word; and indeed the *ordo* of I.2.ix is translated as "arrangement." Francesco Maria Grapaldi's psychophilology of everyday life, which often appears in architectural bibliographies because of its title *De Partibus Aedium* (*On the Parts of Buildings*), divides columns into three categories or *genera: teretes, quadrangulas, striatas*— "smooth," "square," and "fluted" (1501, p. 7); the square ones are for corners, as antae. As for their height and breadth, he refers the reader to Vitruvius' book III. In the third edition (Cologne, 1508, fo. vii r.—the foliation is unreliable), he adds further comments about the flutes as well as the proportion of the heights of columns to their diameters: Doric (6:1), Ionic and Corinthian (9:1), and Tuscan (7:1). He adds a fifth, Attic order: all this is not out of Vitruvius (who does describe an Attic base in III.5.ii) but an abbreviated quotation of Pln. *NH* XXXVI.vi.179.

8. The first five volumes were published in many fascicules; volumes VI to VIII were not so subdivided. Georges Perrot, who during the publication had become director of the Ecole Normale Supérieure as well as perpetual secretary to the Académie des Inscriptions, issued two more volumes in 1911 and 1914; he maintained the double authorship.

9. Schliemann's *Trojanische Alterthümer* (1874) had a brief text and a collection of some 200 very rough photographs; Mykene did not appear until 1877. C. Chipiez refers to it (1876, p. 156); he must have seen it on publication.

10. Auguste Choisy (1899). Choisy was not an architect by training; starting as a mathematician, he was attached to the Ecole des Ponts et Chaussées, in which he spent his whole career, though he published a great deal on building technology and the history of building techniques. C. P. J. Normand (1819); the first English edition appeared in 1829 (trans. Augustus Pugin), the first German edition in 1830 (trans. M. H. Jacobi).

11. Vignola's work was first published in 1562; at times during the seventeenth and eighteenth centuries there was a version or reprint almost every year. See M. W. Casotti (1960, no. 11, vol. 1, pp. 99 ff.). A fuller bibliography is provided by A. G. Spinelli (1908; *Bio-bibliografia dei due Vignola* was reprinted separately [Rome, 1968]). Charles Normand's *Vignole des Ouvriers* first appeared in 1821 to 1823; but the practice of publishing pocket editions had started in the eighteenth century.

12. First published in 1860 (there were several later editions; see pp. 157 ff.).

13. On David-Pierre Humbert de Superville (1770–1849), who was the director of the first print room for Leyden University, and the cosmological context of his ideas, see B. M. Stafford (1972a, pp. 51 ff.; 1972b, pp. 308 ff.; 1973, suppl. pp. 31 ff.). He had adopted a third Christian name, Giottino, from a nickname given him in Italy where he was a precocious and enthusiastic pre-Raphaelite; on which see G. Previtali et al. (1964). Humbert's *Essai sur les Signes Inconditionnels dans l'Art* was published in 1827 and reprinted by the Royal Netherlands Academy in 1857. In England there was George Field (1839); on him see D. Brett (1986, pp. 336 ff.). He stated principles that influenced many artists directly, while his younger friend D. R. Hay's *Science of Beauty* (1856) developed such ideas a great deal further. In Germany a major painter and sculptor, Gottfried Schadow, published a study of human proportion (1834; English ed., 1883), which stood at the beginning of many German "scientific" and experimental studies of proportion, of which the most important were those of Adolf Zeising (1854; and posthumously, 1884). With Gustav Fechner's experiments, the whole subject moved out of aesthetics into experimental psychology.

14. G. W. F. Hegel (1965, vol. 2, pp. 64 ff.). ("Über deren architektonische Schönheit und Zweckmässigkeit hinaus früher und später nichts mehr erfunden ist").

15. A. Schopenhauer (1972, vol. 1.1, pp. 252 ff.; vol. 1.2, pp. 467 ff.). Much more will be said about the Platonic Ideas and the form of artifacts and buildings in chapter 5, but it should be noted here that Schopenhauer's Platonic Ideas did not have to do with categories of manufactured objects (table, chair, house) but with the more abstract ones of weight and support, which he enumerated. Schopenhauer rebuts Plato specifically on this matter (1972, vol. 1.1, pp. 249 ff.; vol. 2.1, pp. 449 ff.). On Schopenhauer's removal of the Ideas from ontology to aesthetics, see G. Simmel (1907, pp. 110 ff.).

16. F. W. J. von Schelling (1856–1861, vol. 5, pp. 488 ff.). But see also G. Lukács (1963–1974, vol. 12, pp. 402 ff.). The idea of *enstarrte Musik* in Schlegel seems rather different from Goethe's *gefrorene Musik*, which he in his turn borrowed from Madame de Staël's *Corinne*.

17. J. Burckhardt (1879, p. 4.). On his visit to Paestum, see W. Kaegi (1947–1982, vol. 3, pp. 456 f., 503 f.).

18. On the French and German background of Burckhardt's ideas, particularly on his debt to Franz Kugler, see W. Kaegi (1947–1982, vol. 2, pp. 48 ff., 146 ff., 229 ff., 264 ff.) and K. Löwith (1981–1988, vol. 7, pp. 91–100). Even in his last lectures on Greek civilization (1898–1902), published posthumously in four volumes (1929–1933), Burckhardt spoke of the *ieros gamos* between the German and the Greek peoples (see vol. 8, p. 5); this is echoed in such notions as Martin Heidegger's often-repeated conviction that German is the only language into which Greek texts may properly be translated. The Prussian assumption of Greek characteristics is discussed in more detail in chapter 6, n. 4.

19. J. Ruskin (1851, vol. 1, p. 14.).

20. J. Ruskin (1851, vol. 1, p. 275); on the useless varieties of orders, see pp. 359 f.

21. J.-N.-L. Durand (1819, pp. 8 ff.). On Durand, see W. Szambien (1983); on the diffusion of his ideas outside France, see particularly pp. 74 ff. D. van Zanten, "Architectural Composition at the Ecole des Beaux-Arts," in A. Drexler (1977, p. 506 n. 60), remarks on the enthusiasm of German and British theorists for Durand's ideas, as compared to the "stony silence" of the French. Surely he is right in thinking that this is due in part to the fact that Durand was codifying what was self-evident to them and in part to the enormous popularity of his book.

22. On Quatremère's relation to the teaching of architecture, see L. Hautecoeur (1943–1957), vol. 6, pp. 146 ff.) and R. Schneider (1910, pp. 10 f., 65 ff.). For a summary account of his views on the origin of the orders, see Quatremère de Quincy (1832), s.v. "Architecture" and "Dorique"; both were condensed versions of the same articles in the first edition of the dictionary, when the first volume appeared in 1794 as part of C. J. Pancoucke's *Encyclopédie Méthodique*.

23. On Gottfried Semper's changing views on the primitive hut, see the discussion in W. Herrmann (1984, pp. 164 ff.).

24. Fra Carlo Lodoli was the most important and influential exponent of this view: see J. Rykwert (1980, pp. 296 ff.; 1982, pp. 114 ff.).

25. L. Hautecoeur (1943–1957, pp. 335 ff.).

26. E.-E. Viollet-le-Duc (1877–1881, vol. 1, p. 104); a similar account of the gradual faceting of the square column (but in wood, before the *Stoffwechsel* into stone) was already given by D. Ramée (1860, vol. 1, pp. 462 ff.).

27. Although he does refer to Viollet's views obliquely: C. Chipiez (1876, pp. 224 f.).

28. C. Chipiez (1876, p. 371).

29. J. Guadet (1909, vol. 1, pp. 327 ff.).

30. The first Englishman, Richard Lane, was a Manchester man who went on to practice in his home town, where he trained (among others) Alfred Waterhouse and also seems to have influenced the second Englishman to study in Paris, John Billington. He was admitted to the Ecole in 1826; on his return after an undistinguished stay there he produced a missionary tract for its methods, *The Architectural Director;* the first edition of 1831 was still very Durandian, but the revised version of 1834 owed more to Quatremère de Quincy. On both Lane and Billington, see H. M. Colvin (1978), s.vv., and R. S. Chafee (1972, pp. 200, 234 ff., 321).

31. On Joseph Gwilt (1784–1863), see H. M. Colvin (1978), s.v. The *Encyclopaedia* was published in 1842, 1845, 1854, 1859, and posthumously (edited by Wyatt Papworth, who continued to use the Durand material until the very last edition, although he supplemented it with an archaeological discussion of the orders written by Edward Creasy) in 1867, 1876, and 1889. The edition used here is that of 1859.

32. J. Gwilt (1859): on Greek architecture, pp. 57 ff.; on beauty, pp. 673 ff.; on the orders, pp. 680–746; on Durand and composition, pp. 772 ff. Strangely enough, Ruskin expressed a similarly contemptuous view of aesthetics several times (1903–1912, vol. 4, pp. 7 f., 42 ff.; vol. 20, pp. 207 ff).

33. A. Alison (1790). Beauty is an emotion; it is produced by evidence of (1) design, (2) fitness, and (3) utility. On the analogy of design, machine and animal beauty, see vol. 1, pp. 291 f. On the three kinds of beauty (*natural* expression of qualities "arising from the nature of bodies"; *relative,* "arising from the subject or the production of art" and *accidental,* from "chance association"), see vol. 1, pp. 318 ff. On proportion and associative emotion, see vol. 2, pp. 21 ff. On the orders, their differing character and their supremacy as an instance of fitness for purpose, see vol. 2, pp. 33, 137 f., 141–168. On Alison and Loudon, see G. Hersey (1965). On the persistence of Alison's influence, see Sir Thomas Dick Lauder in Sir Uvedale Price (1842, pp. 4 ff.). In the form he professed them, Gwilt's ideas seem a flat application of Alison's associationism.

34. The effects of this exhibition had, earlier in the century, been regarded as an unmitigated disaster for American architecture: "Like something awful" (F. L. Wright, 1943, p. 126).

35. On Hunt's rather undistinguished career at the Ecole (promotion of 1846, 1st class in 1851), see R. S. Chafee, "Hunt in Paris," in S. R. Stein (1986); P. R. Baker (1980, pp. 24, 36); and A. Drexler (1977, pp. 464 ff.). Baker has clearly shown how Hunt's Beaux-Arts primacy gave him the dominant position on the East Coast scene. Ware had been recommended to MIT and to Columbia by Hunt (Baker, 1980, pp. 100 ff., 459).

36. On Perrault's treatment of the orders, see A. Perez-Gomez (1983, pp. 71 ff.) and J.

Rykwert (1980, pp. 33 ff.). For a different view, see W. Herrmann (1973, pp. 95 ff.). Hunt presumably learned the Perraultian trick from Normand's *Vignole* (1821–1823; but see above, n. 11) or one of its later derivatives.

37. On Gaudí and Viollet-le-Duc, see T. Torii (1983, vol. 1, pp. 94 ff.).

38. J. J. Sweeney and J. L. Sert (1960, pp. 122 ff.); T. Torii (1983, vol. 1, pp. 247 ff.; vol. 2, pp. 300 ff.). Gaudí himself occupied one of the two houses built on the site, in 1906. It was of a type called a chalet in Barcelona and was designed by Gaudí's disciple and assistant, Francisco Berenguer. Gaudí lived there until his final move to the workshop on the Sagrada Familia site toward the end of his life.

39. The extraordinary finish of broken pottery has been identified as one of the sources of collage technique in modern painting. So C. Giedion Welcker (1955).

40. Gaudí quoted in E. Casanelles (1965, p. 212).

41. H. d'Espouy (1893–1905, vol. 1, pl. 35, for Cori; vol. 2, pls. 6 f., for Paestum). The plates are not numbered, nor always bound in the same order. Labrouste's restoration had been published earlier in L. Dassy, *Compte-Rendu sur la Restauration de Paestum par Henri Labrouste* (Paris, 1879); and the restorations had also been printed in *Restauration des Monuments Antiques par les Architectes Pensionnaires de l'Académie de France à Rome* (Paris, 1877). The slight inclination of Greek Doric columns was exaggerated for purposes of demonstration in G. Perrot and C. Chipiez (1882–1889, pl. 7). But see chapter 8, figs. 8.7, 8.8.

42. The various writers about Gaudí have been very reticent about his historical information, but a close disciple, Domingo Sugranes, maintained that his main source was Luigi Canina's giant *Architettura Antica* (two editions in different formats, 1830 and 1844). Canina also discussed and engraved both the temple at Cori (vol. 3, pl. 15) and the "Basilica" at Paestum, which (like Labrouste) he called *Portico a Pesto* and showed having hipped roofs and therefore no pediments. On

the "civic" character of that hipped roof, see L. Canina (1830, 1844) and N. Levine (1975, pp. 770 ff., 796 ff.). On its exhibition at the Villa Medici in 1827, see Levine (pp. 1–123, with n. 912). Gaudí's Doric nevertheless feels more like Piranesi's than like Canina's. Piranesi's relevant publications are the *Antichitá di Cora* (Rome, 1764; nos. 537–550 in Focillon's catalogue); and *Différentes Vues de quelques Restes de Trois Grands Edifices qui subsistent encore dans le milieu de l'Ancienne Ville de Pesto* (Rome, 1778; Focillon's nos. 583–599).

43. Asplund had won the competition for the layout of the whole cemetery in partnership with Sigurd Lewerentz in 1914. See H. Ahlberg (1950, pp. 41 f., 94 ff.) and M. Capobianco (1959, pp. 36 ff.). Lewerentz designed a rather (to my mind) less successful "Corinthian" crematorium chapel in the same cemetery: see J. Ahlin (1987, pp. 70 ff.). He, too, turned modern about 1930, and remained steadfast in this new approach.

44. On the Skandia Cinema see H. Ahlberg (1950, pp. 46 ff., 112 ff.).

45. Chicago Tribune (1923, pl. 196).

46. The reference is to the Rookery Building in Chicago by Burnham and Root.

47. This was an Egyptian column by Paul Gerhardt of Chicago, who submitted an equally inept obelisk project (Chicago Tribune, 1923, pls. 159, 160). Yet another scheme, by Matthew L. Freeman of the Mississippi Agricultural and Mechanical College (pl. 162), shows a twenty-story pedimented but otherwise rather bare building with a Doric column, some ten stories high over the center, perhaps containing an electrical display. Of the many projects that have a central feature with some form of electric advertising, that of Erich J. Patelski (pl. 165) has a ten-story column carrying a lit-up globe.

48. It was reprinted in the *Inland Architect and News Record* for May 1896, and with small changes in the *Craftsman* for July 1905; in January 1922 the *Western Architect* reprinted it with a note that Sullivan saw no need to revise it in any way. See L. H. Sullivan (1947, pp. 8 ff., 202 ff.). On the other hand, Sullivan was scathing about the plan to erect the largest Doric column in the world in memory of the founding fathers of the city of Detroit (pp. 58 ff). Sullivan had in fact learned of the importance of tripartition from William R. Ware, whose lecture at MIT he had attended in November 1872. His fragmentary but conclusive notes are in the Avery Library, Columbia University; I would like to thank Dr. J. A. Chuning for bringing them to my attention. Curiously enough, Montgomery Schuyler, the most important architectural critic of the time, considered the Union Trust Building in New York by George B. Post to be the first tripartite skyscraper. See Winston Weisman in E. Kaufmann (1970, pp. 115 ff.).

49. Sullivan to Loos, around October 1920; the letter seems to have disappeared, though Heinrich Kulka, Loos' disciple and biographer, remembered having seen it: the Viennese-Californian architect Rudolph M. Schindler had acted as a go-between. See B. Rukschcio and R. Schachel (1982, pp. 246 ff., 252 f.).

50. Chicago Tribune (1923), pp. 14 f.).

51. A. Loos, quoted in L. Münz and G. Künstler (1966, p. 176). The curious notion that there was something jocular about Loos' scheme is still perpetuated, as in T. Benton, S. Muthesius, and B. Wilkins (1975, p. 39). See a discussion of the issues in B. Gravagnuolo (1983, pp. 175 f.), though even he persists in ascribing an ironic intention to Loos and in considering the column a fragment torn from its context—analogous to Duchamp's celebrated urinal. This seems to be quite contrary to Loos' whole view of ornament and of the orders in particular. The Greeks, he thought, were still capable of inventing ornament; the Romans were sufficiently less primitive that they could only adapt it. This is quite different from Duchamp's view of the artist's power to isolate and reveal by his or her choice. In any case, classical ornament was an essential part of the architect's training; see A. Loos (1931, pp. 206 f.). M. Tafuri in G. Ciucci et al. (1973, pp. 429 ff.) has, Humpty Dumpty–wise, heaped more meaning on this project than it could carry. A rather more sober view is taken by M. Cacciari in his introduction to A. Loos (1982, pp. 32 f.) The most reliable account of the episode is in B. Rukschcio and R. Schachel (1982, pp. 273 ff, 562 ff.). Indeed, Loos tried to beard Colonel McCormick (the owner of the *Chicago Tribune*) on the Train Bleu in the winter of 1922 to persuade him to build the column in spite of the competition results (Rukschcio and Schachel, 1982, p. 281).

52. Originally published privately in Nice (1922); quoted in L. Münz and G. Künstler (1966, p. 177); B. Gravagnuolo (1983, p. 170); and B. Rukschcio and R. Schachel (1982, p. 556). The same form as the base of the Chicago column had already appeared, in miniature, as the tomb for the art historian Max Dvořak. The small black granite chamber was to have been decorated in mosaic by Oskar Kokoschka.

II Order in the Body

1. Le Corbusier's best-known pronouncement on the Doric order is in *Vers une Architecture* (1923, pp. 165–192). But in the earlier (and much rougher) account of his encounter with the Parthenon, his travel diary (1966, pp. 123 ff; 1989, vol. 1, pp. 120 ff., vol. 3, pp. 98 ff.) he gives a rather more detailed and more perplexed account of his fascination. Jean Petit (1970, p. 39) discusses the relation between the two texts.

2. Gen. 6.13 ff; Exod. 25.10–27.19 ff., 36.8 ff.; 1 Kings 6.2 ff. (and 2 Chron. 3.3 ff.); Ezek. 40.5 ff. In architecture this led to a wide application of what Erwin Panofsky has called "the principle of disjunction" (1960, pp. 83 ff.). The most popular exposition of the correspondence between the Body of Christ and the building antetypes in the Old Testament is in Aurel. Aug. *De Civ. Dei* XV.26 f. However, the analogy between the body and the church as an institution (if not quite as a building) is already made by St. Paul (2 Cor. 12) The detailed study of the parallel between Noah's ark, the Ark of the Covenant, and Christ's body was graphically represented in the great polyglot Bible edited for Plantin by Benito Arias Montano in 1569 to 1572. On these speculations and their immediate effect on architecture in the sixteenth and seventeenth century, see J. A. Ramirez (1991).

3. On the springs and immediate influence of "higher criticism," see A. Fliche and V. Martin (1934—; vol. 19, by A. Preclin and E. Jarry, pp. 710 ff.) as well as H. Jedin and J. Dolan (1980–1982, vol. 4, pp. 99 f., esp. n. 11), though the capital importance of Benedict Spinoza's *Tractatus Theologico-Philosophicus*, which first appeared in 1670, is often underrated in this context. See still Matthew Arnold, "Spinoza and the Bible" (1895, pp. 307 ff.), and on Richard Simon and Spinoza, J. Steinmann (1960, pp. 50 ff., 208 f., 373 ff.); but see also B. Willey (1934, pp. 76 ff.) and P. Hazard (1961, pp. 239 ff.). On the effect of Richard Simon's two *Critical Histories,* see P. Barrière (1961, pp. 225 f., 277, 281 ff.) and E. Cassirer (1951, pp. 154 ff.). On Simon's critical view of the reception of the text and the nature of tradition, see J. Le Brun (1981, pp. 185 ff.).

4. R. S. Neale (1981, pp. 183 ff.). William Whiston, the astronomer and Arian theologian, exhibited a model of the Jerusalem temple in Bath in 1726, as well as in Bristol and Tunbridge Wells between 1726 and 1728; John Wood had arrived in Bath in 1726. Whiston lectured at the exhibitions of the models of the Mosaic tabernacle and of the temple made by a "Mr Crosdale, a very skilful Workman[,] . . . the substance of which important lectures I added about the year 1728 or 1729 to a large scheme of that model, after I had compared it to Sir Isaac Newton's scheme . . . to which it almost entirely agreed." He is presumably referring to *A Description of the Tabernacle of Moses and of the Temple in Jerusalem in a very large Sheet* (London, 1737), which was reprinted in Whiston's edition of Josephus; see Whiston (1753, pp. 333 f.), as well as M. Farrell (1981, p. 324) and J. E. Force (1985, pp. 21, 129 f.). Although Villalpando's speculations were well-known throughout the seventeenth and eighteenth centuries—if only through Fréart de Chambray's book (1650)—few drew the kind of consequences that Wood elaborated. A representative of the central, "established" tradition was Henry Aldrich, Dean of Christchurch, Oxford, architect, prolific musician, and latitudinarian divine, who composed a popular treatise (left incomplete at his death; 1750, 1818) in which the Villalpandan order is reproduced out of Fréart (pl. 14), though there is no word about ancient plagiarism.

5. J. Wood (1741, pp. 71 f.). On John Wood, see further T. Mowl and B. Earnshaw (1988). As the title of their book indicates—*John Wood, Architect of Obsession*—the authors take an unfortunately patronizing view of their hero's preoccupations. Wood's conception is a special

case of the argument from design, which was much favored by latitudinarian theologians (see G. R. Cragg, 1966, pp. 70 ff.) and supported by many eighteenth-century biologists such as John Ray (particularly in *The Wisdom of God Manifested in the Works of the Creation*)—on Ray, see C. E. Raven (1942, pp. 374 ff.) and S. Hales (1969, pp. xii ff., 181 f.)—though the locus classicus may be Robert Boyle (1685 [but Latin 1684, German 1729], pp. 24 ff.). Similar arguments appear in Whiston (1725, pp. 98 ff., 117 ff., 184 ff., 233 f.—this last quoting Boyle). One of the most popular expositions of the argument was William Derham's *Physico-Theology,* based on his Boyle lectures of 1711 and 1712 and first published in London in 1713, though much reprinted (lastly, with author's life; London, 1798: the first edition is used here). The microcosmic argument is addressed in connection with the eye (vol. 1, pp. 139 ff.) and with man in general (vol. 2, pp. 166 ff.). Derham also wrote a continuation, *Astro-Theology* (1714); much later, in 1729 he preached a sermon in Bath, on which he based his *Christo-Theology* (1730).

Speculation about the Druids was very popular in the late seventeenth and early eighteenth centuries in Britain, and earlier also in France: they were inevitably associated with megalithic monuments, such as Stonehenge and Avebury in Britain, or the Carnac stones in Brittany; see, for instance, D. P. Walker (1972, pp. 73 ff.), J. Baltrusaitis (1967, pp. 102 ff.), and S. Piggott (1975, passim). Wood's immediate source might have been the new edition of William Camden's *Brittania* by Edmund Gibson (1695), which incorporated some of John Aubrey's views on the subject that he had set out in his unpublished manuscript *Monumenta Britannica,* to which Gibson had access. The "shabby firebrand," John Toland, who did seem to know Breton and Gaelic, had published "A Specimen of the Critical History of the Celtic Religion and Learning, Containing an Account of the Druids . . . with the History of Abaris the Hyperborean" in his posthumous *Collection of Several Pieces* (1726). It was reprinted several times, last by R. Huddleston (1814).

6. Newton's own view of the subject is given in the "Scholium Generale" to his *Principia* (on whose composition and publication see I. B. Cohen, 1971, pp. 240 ff.) and in the final chapters of his *Opticks.* There is a neglected correspondence of his with the Rev. John Harrington, Fellow of

Wadham College, Oxford; Newton thanks Harrington (30 May 1693; corrected 1698) for his

> demonstration of the harmonic ration from . . . the 47th of Euclid . . . all resulting from the given lines 3, 4 and 5. You observe that the multiples hereof furnish those ratios that afford pleasure to the eye in architectural design. . . . I am inclined to believe some general laws of the Creator prevailed with respect to the agreeable or unpleasing affections of all our senses; at least the supposition does not derogate from the wisdom and power of God, and seems highly consonant to the simplicity of the microcosm in general.

See Newton (1967, vol. 4, pp. 272 ff.), NN [Sir John Harington] (1769), and F. Webb, (1815, appendix).

Wood had read about King Bladud (the father of King Lear and reputedly a magician killed in a flying accident) in Geoffrey of Monmouth's *History of the Kings of Britain* II.10 (written ca. 1150; 1929, p. 261; but see also J. S. P. Tatlock, 1950, pp. 47, 360 f.). In Gibson's edition of Camden (1695), he appears as Bleyden Doith. Toland seems to have first made the identification between Geoffrey of Monmouth's Bladud and the Abaris whom Iamblichus mentioned in his *Life of Pythagoras*. It had been translated from Dacier's French version by Nicolas Rowe, later poet-laureate, and published in 1707. Toland's work may, of course, have been independent of Rowe's publication. The monotheism of the Druids is maintained by Gibson (1695, pp. lxxxiv, cxvi) on the basis of some comments of Origen. Wood's Druids are almost a direct inversion of the account of them given by Inigo Jones in his *Most Notable Antiquity of Great Britain* (1725, pp. 2 ff.).

7. The diameter of the Circus, 318 feet, is related by R. S. Neale (1981, pp. 196 f.) to Stonehenge, though in his survey Wood gives its diameter as 312 feet to the edge of the ditch and bank, 342 to the outer ditch. W. Ison (1948, pp. 150 f.) relates it to the dimension of the Queen's Square (though he himself gives its dimension as 300 by 350 feet). T. Mowl and B. Earnsham (1988, pp. 96 f., 187 ff.), on the basis of an unpublished correspondence between Wood and Lord Oxford in the British Museum, argue convincingly that the numbers and dimensions of the King's Circle fit in

both with Wood's idea of the second temple in Jerusalem and his reconstruction of the Stone Circle at Stanton Drew and that the relation between the Circle and the Crescent are an invocation of the Sun-temple and Moon-temple Wood had imagined were located at Stanton Drew.

8. His one surviving play, *Fontana di Trevi* (ed. C. D'Onofrio), was printed in Rome in 1963; on his theatrical activities, see I. Lavin (1980, vol. 1, pp. 146 ff.).

9. Roland Fréart, Sieur de Chambray (1650; rev. ed., 1702).

10. Paul Fréart, Sieur de Chantelou ([1885] 1981, pp. 18 f.). I have discussed this passage briefly elsewhere (J. Rykwert, 1980, pp. 30, 34). On Bernini's visit, see most recently C. Gould (1981). On the competition, see A. Blunt (1958a, vol. 1, pp. 189 ff.) and D. del Pesco (1984). But Lady Dilke (1888, pp. 57 ff.), M. Soriano (1972, pp. 124 ff.), and W. Herrmann (1973) may also be consulted. The other side is presented by C. Perrault (1909).

11. N. Poussin (1929, p. 65). The letter, dated 20 March 1642, is addressed to Chantelou "à la Cour" and answers one of Chantelou's, dated from Nîmes.

On the ideas summarized by Bernini in the general context of his theory, see I. Lavin (1980, vol. 1, pp. 9 ff.). And see also R. Wittkower and H. Brauer (1931, vol. 2, pl. 54a; vol. 1, pp. 100 ff.).

12. Chambray's protector was his cousin, the powerful Soublet des Noyers, director of buildings under Richelieu. Chambray became an influential critic and translator: of Palladio, in 1650; of Euclid's *Perspective*, in 1663; of Leonardo's *Treatise*, edited and translated, with some illustrations after Poussin, in 1651 (it was to be one of the bones of contention between Abraham Bosse and the Academy of Painting: see below, n. 62). He also wrote one original essay, *Idée de la Perfection de la Peinture* (1662). Both brothers were patrons and correspondents of Poussin, Chambray being the recipient of the very important letter of 1 March 1665 that is often taken to have been the last statement of Poussin's theoretical position about the correspondence of the senses: on which see P. Alfassa (1933, pp. 125 ff.) and A. Blunt

(1958a, vol. 1, pp. 353 ff., 371 ff.). An earlier letter to Chantelou (24 November 1647) has some preliminary comment on the modes: see N. Poussin (1929, pp. 238 ff., 309 ff.).

13. R. Fréart de Chambray ([1650] 1702, pp. 39 f.). He also considered it iconographically inappropriate to substitute angels or virtues for columns (as was often done) since vices are what should have been punished by being weighed down; but see E. Forssman (1956, pp. 144, 213 ff.). The English version of the *Parallèle* was done by John Evelyn and went through several editions after its first publication in 1664. On the various editions see J. Schlosser-Magnino (1956, p. 634) and D. Wiebenson et al. (1982, III-A-14). The modern authorities discussed for each order are Alberti, Palladio, Scamozzi, Pietro Cataneo, Daniele Barbaro (out of his Vitruvius commentary of 1556/7), and the much less known Zanini (Gianni) Gioseffe Viola whose book on the orders appeared in Padua in 1626, as well as two Frenchmen, Jean Bullant and Philibert de l'Orme.

14. On Shute, see J. Summerson (1970, pp. 46 ff., 51 ff.) and J. Rykwert (1980, p. 200 n. 18). On the order:deity parallel, see S. Serlio (1584, p. 126 r.).

15. On Hans Vredeman de Vries and his two sons, Paul and Samuel, see *TB*, s.vv.; on their notions of the orders as analogues for completeness, see J. A. Ramirez (1983, pp. 166 ff.).

16. Shute was, at any rate, following Serlio: see J. Poleni (1730, p. 47), quoting the 1600 edition of Serlio's *Opere* (p. 81). Guillaume Philandrier's edition first appeared in Strasbourg in 1543, and was reprinted within a few months in Rome and in Paris. He was generally known by the Latinized form of his name, Guillelmus Philander or Philandrus.

17. V. Scamozzi (1615, vol. 2, p. 37); on Scamozzi as architect, see F. Scolari (1837), F. Barbieri (1952), and R. K. Donin (1948).

18. Varro *L. L.* VIII.14; VIII.79. So also in Greek: "capital" or "coping" is *kephalidon* (Philo Byz. II.147); *kranion* (as coping: Xen. *Cyr.* III.3.lxviii; as capital, epigraph); or *kiokranon* or

epikranis (Pind. frr. 88, 10; Eur. *Iph. Taur.* 51). In Hebrew, *rosh* (1 Kings 7.21).

19. J.-F. Blondel (1771–1777). The edition was prepared by Pierre Patte. See J. Rykwert (1980, pp. 417, 474 nn. 26–29).

20. J.-F. Blondel (1771–1777, vol. 1, pp. 195 ff.), where the Vitruvian legends are uncritically set out.

21. J.-F. Blondel (1771–1777, vol. 1, pp. 258 ff.). It can be read as a reference to the opening lines of Horace's *De Arte Poetica*.

22. The reference seems obvious: to Jean Le Blond, known chiefly as a painter and engraver, but also as the author of a small book on the orders, *Deux Examples des Cinq Ordres de l'Architecture Antique* (1683), which is in fact a digest of Vignola through Chambray and one of the first "Vignole de Poche." It did have a small public, and Jean's son, Alexandre-Jean Baptiste Le Blond (himself a prolific writer and engraver, though he was to die young in Russia), seems to have persuaded the critic-publisher Jean Mariette to reprint the book with a new title page and dedication as *Parallèle des Cinq Ordres d'Architecture* (Paris, 1710; and again 1716); see B. Lossky (1936, passim). It is at any rate possible that Blondel may also have been acquainted with the opinions of a quite different writer-artist, Jean-Christophe Le Blon, an artist born in Frankfurt of Flemish parents and best known as the engraver who perfected the three-color printing process on Newtonian principles, which led to his establishing a London atelier. After the financial failure of his "Picture Office" and a second bankruptcy, he settled in Paris, where he died in 1741; see J. Byam Shaw (1967, pp. 21 ff.). A pupil of Carlo Maratta, he (almost incidentally) translated into English the French essay on *le beau idéal* that Laurence Ten Kate had prefaced to his translation of Richardson's *Essay in the Theory of Painting* (Amsterdam, 1728). (This preface is also quoted by William Hogarth in *The Analysis of Beauty,* though much less favorably; ([1753], p. 80.) The translation was important enough for a mention in Horace Walpole's brief life of Le Blon (1828, vol. 5, pp. 258 f.); but see also H. W. Singer (1901). Although it is doubtful that Blondel knew that text, he may well have come across Le Blon's book on the three-color process, which was printed in Paris by Gaultier de Montdorge in 1756, and in which he made some analogous observations on harmony. I suggest that Blondel may have telescoped the two Jeans with such similar surnames: at any rate, he may be reflecting a view rather like the one expressed by Ten Kate on the arts in general (1728, pp. xxx ff.), rather than any specific reference to architecture.

23. Diego de Sagredo (whom Blondel calls Sangrado) was court chaplain to Juana la Loca. *Medidas del Romano*, which can be variously translated as *The Methods of the Roman* or simply *Roman Measurements* (sometimes treated as a mere paraphrase of Vitruvius), appeared in Spanish in 1526 (in Toledo: reprints in 1549 and 1564), in Portuguese in 1541/2, in French in 1531 (again in 1539, 1542, 1550, and 1608), and was the first "Vitruvian" book in all three languages: see W. Stirling-Maxwell (1891, vol. 1, pp. 157 f.). See also further, his notes 46 f. Diego's ideas were seconded by the research for the exact dimension of the Roman foot, which were carried out at Salamanca and were the subject of one of a set of lectures at the university.

24. Pomponius Gauricus [Pomponio Gaurico] ([1504] 1969). Pomponius (ca. 1480–1530) had arrived as a very young man in Padua, where he wrote his treatise, inserting—after an interesting but not very innovative discussion of human proportion—an account of physiognomy based on a Latin version of a Jewish author of the fourth or fifth century AD, Adamantius of Alexandria. Gauricus' book was first published in Florence in 1504. In the same year Bartolomeus Cocles (pseudonym of Bartolomeo della Rocca) published his pseudo-Aristotelian book on physiognomy, which was the first work of this type to be illustrated with wood engravings. Pomponius' brother Lucas republished both books after his death. See Gaurico ([1504] 1969, pp. 118 f.).

25. On physiognomy and the fine arts in general, see the all-too-brief chapter in J. Baltrusaitis (1983, pp. 9 ff.) and A. Niceforo (1952). On physiognomy in antiquity, see PW, s.v., and E. C. Evans (1969). On medieval physiognomy, see E. Male (1925, pp. 331 ff.), E. de Bruyne (1946, vol. 1, pp. 287 ff.), and J. Białostocki (1961, pp. 52 ff.).

The great vogue for astro-physiognomic publications in the later fifteenth and sixteenth centuries is described rather chaotically in J. B. Delestre (1866, pp. 23 ff.). See also E. Battisti (1962, pp. 209 ff.). There were many incunabulum editions of Michael Scot (some listed by L. Thorndike, 1965; see below, n. 33), for instance, and the book continued to be popular well into the sixteenth century.

26. *De Humana Physiognomonica* first appeared in Naples in 1586, some years after it had been submitted to the censors. It has been suggested that della Porta's vindication of free will was intended to calm that authority. There were many reprints, of which the most relevant here are the two French ones of 1655 and 1665. Della Porta's is only the most important and popular of such books, many written by clergy. Their authors tend to associate physiognomy with chiromancy, various forms of astrology, and therapeutic simples—even introductions to logic.

27. Little is known about Thomas of Celano: but see F. J. E. Raby (1927, pp. 443 ff.); E. R. Curtius (1953, pp. 317 ff., 389 f.). The translation is by W. J. Irons (*English Hymnal*, no. 351).

28. "In no other age, since the classical days of Roman law, has so large a part of the sum total of intellectual endeavour been devoted to jurisprudence" (F. Pollock and F. W. Maitland, *History of English Law;* quoted in C. H. Haskins, 1927, p. 194).

Honorius of Autun, describing the Last Judgment in his *Elucidarium* (III.paras. 12–15; *PL* 172, pp. 1165 ff.) a century before Thomas of Celano, has brought to judgment "the books of prophets, apostles, . . . and the most perfect, which is the life of Christ." Honorius and Thomas of Celano are referring to scriptural passages in which judgment is made and books are opened: Jer. 17.1; Dan. 7.10; Ps. 56.9; and above all Apoc. 5.1 ff., 20.12 ff. Honorius' book was often translated and used as a source by artists. Christ in majesty is often shown holding a gospel book; in early Christian art, he is often seen as the lawgiver, handing a scroll to Peter. The transition from the majestic book holder to the universal judge is clear in the tympanum of St. Trophime at Arles, to take an early and splendid instance. While at Vézelay the Judge is already displaying His wounded hands, in the Byzantine tradition Christ is often shown holding the gospel book, or an inscribed book; in Byzantine Last Judgments he is usually "harrowing," as in the great Torcello mosaic. This book of judgment is one aspect of the "great book" studied by Hans Blumenberg (1983), though it should be noted that the *liber* in question is a new kind of artifact: the rare codex did appear in Imperial times, but the *volumen* scrolls were still being written in the seventh century.

29. On Raymond of Marseilles and his apology, see M.-T. d'Alverny (1967, pp. 36 ff.). Zodiacal astrology had been explicitly condemned by many of the earliest church fathers (Minucius Felix; Tertullian; Aurel. Aug. *De Civ. Dei* V, 3 ff.; Greg. Magn. *Hom. Ev.* II; *PL* 76, pp. 1110 ff.), as divination in general had several times been forbidden in the Old Testament (Jer. 27. 9, Ezek. 28.28). They often relied on arguments used against the Stoic physicists by the Hellenistic Skeptic, Carneades of Cyrene (see D. Amand, 1945), as well as the Councils of Saragossa in 380, Toledo in 400, and Braga in 561, whose tenth canon is explicit: "Si quis duodecim signa vel sidera quae mathematici observare solent, per singula animae vel corporis membra dissipata credunt et nominibus Patriarcharum adscripta dicunt, sicut Priscillianus dixit, Anathema sit." Priscillian, whose ideas were condemned, was the first heresiarch to be burned at the stake, in 385. He had attempted a biblical typology of the zodiac by assimilating the signs to the twelve patriarchs. The Council of Toledo fulminated against him, and the canons of Braga show that Priscillian's ideas persisted in spite of persecution. In fact Priscillian seems to have been quite orthodox even if fiercely ascetic. See E. C. Babut (1909) and B. Vollmann (1965). But see also below, chapter 3, n. 1; moreover, Isidore of Seville had clearly distinguished astronomy, the natural study of the stars in their movements, from astrology, the divinatory kind of stargazing: *Etym.* III.27. And indeed mantic (as against medical) astrology was hardly practiced in Europe in the second half of the first millennium AD; see F. Saxl, "The Revival of Late Antique Astrology" (1957, vol. 1, pp. 75 ff.) and F. Boll (1926, pp. 30 ff.). All these circumstances made Raymond of Marseilles' first reasoned Christian apology for judicial astrology so very courageous.

The condemnations did not stop there, however. Paul IV had put a number of astrological and chiromantic books on the Index in 1559. A further general condemnation by Sixtus V was to follow in 1585. Condemnations in other countries were common enough, as in Britain: 4 Elizabeth I, 39. Modified by 13 Anne, 26; and further by 17 George II, 5. The Vagrancy Act (1824), which superseded the earlier legislation, limits penalties to palmistry. Of course all divinatory practices had been suspect for the church, even if ultraorthodox theologians admitted the influence of the stars on the body, if not on the soul; see *Aq. Ep. Reg.* Dante states the orthodox view in *Purgatorio XVI.* 16.64 ff. through the mouth of a Venetian gentleman, Marco Lombardo. The most famous scholastic condemnation of astrology is by Nicolas Oresme (1952; the views of some contemporaries—Jean Gerson, Pierre d'Ailly—are discussed on pp. 39 ff.). Nicolas, however, also demonstrates the unscientific and untheological nature of astrological procedure in his *Ad pauca respicientes*, propositions 19 and 20 (1966, pp. 425 ff.), arguing both that planetary conjunctions are infinitely variable and that the procedure denies free will.

30. W. Wetherbee (1972); B. Silvestris (1876), and trans. by Wetherbee (1973). On the general question, see R. Javelet (1967, passim). St. Thomas does not use the image of the microcosm often: but see Aq. *Sum. Th.* pt 1, qu. 96, n. 2; *De Reg. Prin.* I.12.

31. John of Salisbury (1979, p. 61; 1848, vol. 3, 261 ff.; vol. 4, 1 ff.). The text, which exists in several manuscripts, was first printed before 1500; the *Polycraticus* was reprinted often and in many languages. John claims to base his teaching on the *Institution of Trajan* by Plutarch, a work otherwise unknown: see R. H. Barrow (1967, pp. 47 ff.) and H. Liebeschütz (1950, pp. 24 f., 43 ff.).

32. (William) Durandus of Mende I.14 (1843, pp. 24 f.); a similar curt reference appears in the *Mitrale* by Sicardus of Cremona (*PL* 213, pp. 35 f). This dry text is based on the more exalted ones of Peter of Celle (*PL*) and Honorius of Autun, the enigmatic (probably German) theologian, though in fact Durandus quotes Richard of St. Victor. But these later fathers echo St. Augustine (*Enarr. in Ps.* 39, 126), as well as Maximus the Confessor, on whom see H. U. von Balthazar (1947, pp. 246 ff.). But see also J. Sauer ([1924] 1964, pp. 98 ff., 291 f.).

33. L. Thorndike (1965, pp. 5 f., on the manuscripts; 87 ff., on the Hellenistic and Arabic sources of the physiognomy). On Michael the Scot as a translator of Aristotle, see F. Copleston, S. J. (1946–1975, vol. 2, pp. 206 ff.). On his place at the court of Frederic II, see E. Kantorowicz (1936, pp. 313 ff., 323 ff.), and, more recently, D. Abulafia (1988, pp. 254 ff.).

34. On the astrological sources of Michael Scot and Pietro d'Abano, see F. Boll (1926, pp. 38 ff, 55 ff.). On Pietro d'Abano's *Physiologus* and its relation to his astrology, see S. Ferrari (1900, pp. 190 ff., 413 f.; 1915, pp. 629 ff.), and E. Pascietto (1984, pp. 139 ff.). On his part in the programming of the astrological paintings by Giotto in the Palazzo della Ragione in Padua, see F. Saxl (1934, pp. 14 ff.). The paintings were defaced in the fire of 1420 and restored by Niccolo da Miretto (or Mireti: Vasari, 1878–1906, vol. 3, pp. 639 f.), using Pietro's *Astrolabium* as guide.

35. U. V. Chatelain (1905, pp. 321 ff., 379 ff.) and B. Teyssèdre (1957, pp. 35 ff.).

36. Though the design was also attributed to Bernini himself: probably first by Piganiol de la Force in his guide to Paris, perhaps because Marin's son, the Abbé de la Chambre, wrote a life of Bernini. The tomb was carved by Jean-Baptiste Tuby, a sculptor of Italian origin, best known as a collaborator of Coysevox (A. Blunt, 1953, p. 254 n. 103; H. Jouin, 1889, pp. 254 n. 1, p. 615 f.). It was one of the sculptures moved to the Musée des Monuments Français at the Revolution (no. 190; A. Lenoir, 1810, p. 257) and is now at Versailles.

37. On the crucial disagreement between Descartes and Cureau on the status of light (Cureau maintained it was a quality only, against Descartes' view that it was a substance), see A. I. Sabra (1981, pp. 137 ff.), which also involves Descartes' controversy with Pierre de Fermat. On the Cartesian plenum, see A. Kenny (1968, pp. 204 ff., commenting on Descartes' "Reply to the Sixth Objection") and B. Williams (1978, pp. 208 ff.). Of Cureau's works *Les Charactères des Passion*, in four volumes, originally appeared between 1648 and 1662; *L'Art de Connoistre les Hommes*, in three volumes in 1669. But there was to be a great deal more; in the volumes of 1659 (p. 424) the list includes 21 chapters on the "Passions" (pre-

sumably those published), 100 on the "Virtues and Vices," 52 on the "Temperaments," 29 on "Animal Comparative Anatomy," 50 on the "Beauty of Man and Woman," 60 on the "Influence of Climate," 20 on the "Changes Induced by Age, Way of Life, etc.," and an extra volume on "Dissimulation." This was to form a vast "Art of Knowing Men."

38. In particular, the doctrine of "love by affinity" was directly borrowed from Ficino's commentary on Plato's *Symposium;* indeed he deferred to Marsilio Ficino on this very point. The working of spirits in our bodies was, in Cureau's view, not only instrumental in causing passions but was required to explain such different (and otherwise incomprehensible) phenomena as the rainbow, the fertilizing floods of the Nile, and what he called love by affinity. He returned to these themes throughout his life– though he also went on to refine his conception of the digestive process, which (he thought) was dependent on the dissolving spirits in the stomach: see A. Darmon (1985, pp. 18, 23 ff.). But then Galen had already found spirits indispensable to explain the workings of the body.

One matter, too important and too extensive for this footnote, concerns the way in which the microcosmic metaphor was adapted to take in the Copernican view of the solar system. However, a relevant and conspicuous instance is provided by the dedication of William Harvey's book on the circulation of the blood to Charles I ([1628] 1978, pp. A2, v): "The heart of creatures is the foundation of life, the Prince of all, the Sun of their microcosm, on which all vegetation does depend. . . . Likewise the King is the foundation of his Kingdoms, and the Sun of his Microcosm, the Heart of his Commonwealth, from whence all power and mercy proceeds." The division of the face in Le Brun therefore refers to the change from the "heart as prince" to the "head as prince" metaphor.

39. On the "anthropological" opposition between Cureau and Descartes, see A. G. A. Balz (1951, pp. 42 ff.). But see also the "Eloge" of Cureau by Condorcet (1847–1849, vol. 2, pp. 3 f.).

40. "For my part, it is my opinion that the soul receives the impressions of the passions in the brain, and that it feels the effect of them in the heart. The external movements which I have observed strongly confirm me in this opinion"; Le

Brun, *Conférence,* trans. Jennifer Montagu, in Montagu (1959, p. 27).

On Le Brun's borrowings from Descartes, Cureau de la Chambre, and other authorities, see Montagu (1959, passim). She lists mostly the direct ones from Descartes; however, Le Brun's first biographer, Claude Nivelon, felt it necessary to defend him from an anonymous charge of plagiarizing Cureau; the biography has remained in manuscript (Bibl. Nat., Fond Français 12987). See A. Fontaine (1909, pp. 100 f.). Of two younger academicians, Michel Anguier (who also taught anatomy and the proportions of ancient statues after 1669) analyzed more particularly the effects of anger (1675) and Pierre Mosnier (or Monier) of love (1698). See B. Teyssèdre (1957, pp. 126, 160, 207). With the revival of interest in physiognomy, some drawings were lithographed by L.-J.-M. Morel d'Arleux, in his "Dissertation sur un Traité de Charles Le Brun Concernant le Rapport de la Physiognomie Humaine avec celle des Animaux," as an appendix to J.-L. Moreau de la Sarthe's monumental edition of Lavater's *Physiognomonie* (1806). See J. Montagu in *Charles Le Brun* (1963, nos. 130–137). It was known to Charles Darwin from its reprint of 1820: see C. Darwin (1873, p. 1).

41. On Cureau's explicit rejection of Cartesian teaching about the pineal gland in his "Système de l'Ame," see A. Darmon (1985, pp. 36 f.). He attempts to reconcile his doctrine of the spirits with Harvey's description of circulation in his *Art de Connoistre les Hommes* (1669, pp. 189 ff.); in the same book (pp. 326 ff.) he suggests an outline of a "Treatise on Simulation" that he never wrote, but that was to relate the language of the body precisely to the passions of the soul. But see Darmon (pp. 39, 43 ff.).

42. Descartes had a very clear idea of the distinction between man, the "spiritual automaton," and the animal mechanism. See H. Caton (1973, pp. 74 ff.). Caton relates Descartes' somewhat fantastic account of the workings of the pineal gland (1973, pp. 89 ff.). He also clearly distinguishes and attempts to reconcile Descartes' view of the body-mind relationship in the Sixth Meditation with the more complex notions of the *Treatise on the Passions,* in which passion includes all perception, everything that is not "made by the soul" but comes from outside, from volition. Yet "all that senses is also rational and all appetition

is also volition." The conflict between volition and reflection is seated in the pineal gland (pp. 182 ff., 194 ff.). Spinoza poured contempt on Descartes' conceptual dependance on the pineal gland in the preface to Book 5 of his *Ethics* ([1883] 1915–1919, vol. 2, pp. 244 ff.; though see also the end of Section 3, "General Definition of the Emotions," pp. 185 f.). The duality that forced Descartes' dependence on the pineal gland has remained a problem for philosophers; see R. Rorty (1980, pp. 54 ff., 95 ff.).

Cureau's theory of color was set out in *L'Iris* (i.e., *The Rainbow,* 1650), though its influence on Le Brun has not, as far as I know, been studied.

Descartes' clearest formulation of the idea that the mouth is ruled by the heart is in the "Traité de l'Homme" in *Oeuvres* (1958, pp. 844 ff.). Although written in 1630 to 1635, it was first printed in French posthumously, in 1664, though a Latin translation had appeared in Leyden in 1662. It was accessible much earlier in his *Passions de l'Ame,* which Elzevier had first printed in 1649 (paras. 34 ff.; 1958, pp. 712 ff.).

On the face as a dial, see J. Montagu (1959, p. 60; 1993, pp. 53 f.); she attributes the term to Sir Ernst Gombrich.

43. See M. Fumaroli (1981). Pierre Le Moyne's *Peintures Morales* of 1640–1643 are particularly interesting in this connection: see Fumaroli (1980, pp. 379 ff.).

44. That was the view taken by the scholar Giovanni Pierio Valeriano of Belluno, writing on the head and other parts of the body as hieroglyphic sources (1625, pp. 396 ff., 478 ff). On this problem, see P. Rossi (1960, pp. 208 ff.), M. Fumaroli (1981, pp. 238 f.), and D. P. Walker (1972, pp. 98 ff.). *Chirologia, or the Naturall Language of the Hand . . . whereunto is added Chironomia or the Art of Manuall Rhetoricke* was the title of a book by John Bulwer, who was concerned in the first place with deaf-and-dumb aids, but who also saw the philosophical implications of the subject; on Bulwer, see *DNB,* s.v., and A.-M. Lecoq (1981, pp. 265 ff.). On the influence of these discussions on the notion of *actio,* the language of gesture in the theatre, see A. Goodden (1986, esp. pp. 12 ff.). Interestingly enough, Charles Coypel, who had a considerable influence on these ideas in the eighteenth century, was the grandson of Charles Le Brun's assistant and successor, Noel.

45. "Apologie de Raimond de Sebonde," in M. de Montaigne (1965, L. II, cap. 12, p. 436); but see also M. Fumaroli (1980, pp. 237 f.). A. G. A. Balz (1951, pp. 52 ff.) remains the best account of Cureau's debt to Descartes. Cureau considered that animals could also think, like human beings, but did not have the gift of imagination and of generalization; on the specific problem of the soul influencing or directing the human body through motion, see pp. 62 f. Balz's account is based largely on Cureau's *Traité de la Connaissance des Animaux* (1664).

46. Casaubon was then teaching at Montpellier. "Ethical" was prefaced to the title by Diogenes Laertius. Various editions are listed in Theophrastus' *Characters* (1961, pp. 31 ff.). A useful account of his debt to Aristotle is given by E. Voegelin (1957–1974, vol. 3, pp. 362 ff.). On the early reading of Casaubon's edition, see B. Boyce (1947, pp. 44 ff.). The English texts were anthologized by H. Morley (1891). For the influence of La Bruyère on later literature, see J. M. Smeed (1985), particularly the appendix "Character and Physiognomy" (pp. 292 ff.). That Theophrastus also described "good" characters in a book now lost is clear from his own preface and from Eustatius' quotation of the contrast between the brave man and the coward (*In Il.* 93.21).

47. *Les Caractères de Théophraste traduits du Grec avec les Caractères ou les Moeurs de ce Siècle* (1688), reprinted yearly. On La Bruyère, see C.-A. Sainte-Beuve (n.d., vol. 1, pp. 389 ff.). For the influence of La Bruyère on later literature, particularly the novel, see J. M. Smeed (1985), which contains an appendix (pp. 292 ff) on the relation of character to physiognomy.

48. Shaftesbury (1732, vol. 3, pp. 380 ff.); B. Rand (1914). Theophrastus is only mentioned as a source for La Bruyère in the posthumous fragments (B. Rand, 1914, p. 99 n. 5). At the beginning of the unfinished essay "Plastics" (pp. 90 ff.), Shaftesbury defines three kinds of characters: (1) notes (i.e., signifiers); (2) signs, signa, or sigilla, which some would now call iconic signs; and (3) the "middle" sort, emblematic. These he distinguishes radically into the "true" (i.e., natural and simple) and the "false" (barbarous and mixed, enigmatic). A mixture of the first and second category (as in Egyptian hieroglyphics) is quite unacceptable. But see also D. Leatherbarrow (1983; 1976, pp. 332 ff.).

49. Arist. *Rhet.* II.12 ff. (1388b) but also III.6,vii f. (1408a). See A. Plebe (1961, pp. 15 ff.) on pre-Aristotelian *politropia*. On Aristotelian *ēthos* and *pathos*, see J. Wisse (1989, pp. 36 ff., 60 ff.), Plebe (1961, pp. 77 ff.), and also G. Kennedy (1963, pp. 92 ff., 135 ff.). Of course, pathos introduced the whole matter of passion (in the modern sense) into rhetoric, and with it the passivity of possession, on which see E. R. Dodds (1964, pp. 185 ff.). In this context, he quotes the enigmatic Heraclitean (DK fr. 119) *ēthos anthropou daimōn*, which has been translated either as "character is man's inner voice" or "man's character is his genius." For earlier interpretations, see G. Colli (1977—, vol. 3, p. 112); M. Marcovich (1967, pp. 500 ff.); also Dodds (1964, pp. 46 ff.). On the enigmatic use of the word by the pre-Socratic philosophers—as in the Heraclitean fragment—see M. Heidegger, *Heraklit* (1970—, vol. 55, pp. 215 ff.). Charles Kahn (1979, pp. 252 f., 260 ff.) shifts the weight of the interpretation in the fragment from *ēthos* to *daimōn*. In view of all this, perhaps the most telling (if very cumbersome) translation might run: "the way a man relates to the world round him, that is his destiny."

However, in Aristotle's own account, the two words also have a more elementary sense. In describing the virtues of epic poets, he points out how Homer is at once simple and pathetic, full of feeling, even if in the *Odyssey*, which abounds in recognition scenes, he is complex and ethical, concerned with identity: see *Poet.* 24 (1459b). The use of pathos and ethos in ancient literary theory was recently discussed by C. Gill (1984, pp. 149 ff.).

Accuracy and fidelity of ethical description came inevitably to be highly prized by Roman and Byzantine critics of the visual arts, who were mostly rhetors since criticism is verbal, and they were masters of words; consequently the estimation of painting (as well as its practice) was closely dependent on rhetorical notions.

Later discussion of this matter is colored by the fact that the Latin words *actio* and *pronunciatio* both translated the Greek *hupokrisia*, dividing the actor's and speaker's activities as between his gestural, body language and his elocution. It did not have any negative sense in Aristotle (*Eth. Nic.* 1118a, *Rhet.* 1386a), nor did it acquire its wholly negative sense until late Hellenistic times; but see G. Kennedy (1963, pp. 12, 283 ff.).

50. Arist. *De An.* I.1 (403a5 ff.), I.3 (407b), II.1 (411b27 ff.); *Part. An.* II.10 (656a).

51. *Ethos* (from which also come words like *ethnos*, meaning groups of people accustomed to live in the same place; or *ethismos*, habituation) might be more correctly transliterated *ehthos* or even *aythos*, since the *e* is long whether you take the word to mean a "character" or a "moral attitude"; the verb *ethō*, though, has a short *e*, and when thus written, *ethos* is used to mean "custom" or "habit." The Aristotelian use of the word has recently been discussed by W. F. R. Hardie (1968, p. 36 ff.) and E. Schuetrumpf (1970). The shift in the meaning of the word is very noticeable in the writings of later Hellenistic rhetoricians: see, for instance, Hermogenes of Tarsus' *Peri Ideon* II.2 (trans. Cecil W. Wooten, *On Types of Style*: the only part of his *Technē Rhetorikē* to be translated into English, 1987, pp. 70 f.) An ingenious old translator, Theodore Buckley, suggested "humor" as an English equivalent (1851, p. 149 n. 2), by reference to Ben Jonson's *Every Man out of His Humour*. In particular, he quotes Asper the Puritan's long speech on humors in the play's introduction (lines 86 ff.; Jonson, 1954, pp. 431 ff.; also vol. 9, pp. 391 f.) and refers to the status of humoral theory in English literature at the break of the sixteenth century. The adjustment of humors was called *krasis*—mixture, blending, combination— as in the title of Galen's book on the temperaments, *Peri kraseōn*. "Temper" has been preferred by other translators.

The ethical precision of *ekphrasis* has not really been examined. The word appears in a somewhat different context in the rhetorical *Technē* attributed to Dionysius of Halicarnassus (XI.88), and the procedure is recommended as an exercise in the *Progumnasmata*, attributed to Hermogenes of Tarsus, which became very popular with Byzantine writers. They in turn transmitted the skill to fourteenth- and fifteenth-century Italian rhetoricians; see K. Krumbacher (1897, pp. 454 ff.) and M. Baxendall (1971). Recently this term has been somewhat abused by extending it to any description of painting and sculpture. See R. A. Macdonald (1993, pp. 112 ff.).

Aristotle returns to the matter several times in the *Rhetoric*: the way his own ethos is part of his persuasive power (I.2.iv; 1356a); how the nature of the argument must be related to the character of the government of the city in which the speech is made (I.8.i; 1365b); how the nature of different "ethical" dispositions relates to age and their fortune (at some length, II.12 ff., 1388b). On the value of ethical narrative in reported speech (particularly in the epideictic style), see III.16 (1416b) on establishing character through telling anecdote.

52. Quin. VI.2.ii.

53. The word is used almost interchangeably with *sphragis*, a seal, in early patristic literature. Augustine's account of the doctrine does not use the word, however, but rather *signum dominicum* or *consecratio* (CSEL, vol. 52, p. 355; vol. 51, p. 79). The Pauline precedent is usually cited from Eph. 1.13 and 2 Cor. 1.21 ff.

On the parallel between the Greek *charaxi* and the Akkadian *hurassu*, see W. Burkert (1992, pp. 38, 175 n. 19). As he points out, this parallel is mentioned by LSJ but ignored by PC. Nevertheless the word *hurassu* is used to mean "writing" in the Gilgamesh epic (I.1.viii).

54. Le Brun's lectures were first given in the Academy on 6 October and 10 November 1668; the drawings were presented to Colbert on 28 March 1671. The first publication of the *Physiognomy*, with only some drawings engraved and an abbreviated text, was J. de Lorme's and E. Picart's (Amsterdam and Paris, 1698); see B. Teyssèdre (1957, pp. 628 ff.). See also A. Fontaine (1903, pp. 105 f., on the mutilated text of Nivelon, and pp. 105, 117 f.; 1909, pp. 68 ff.; 1914). For a criticism of Le Brun's approach, which he shared with many painters, see R. de Piles (1708, pp. 146 ff). See also B. Teyssèdre (1964, pp. 162 ff.) and N. Bryson (1981, pp. 44 ff.).

The application of equilateral triangles to animal faces is clearly shown on some of the Louvre drawings. They are further lettered and explained in the publication by Morel d'Arleux. Some of them were reprinted by L. Métivet (1917). But on all this, see J. Baltrusaitis (1983, pp. 25 f.).

55. There were at least three separate engraved versions of the lectures (by Hubert Tastelin, Sebastien Le Clerc, and Bernard Picart, all splendid engravers) and they went through many editions throughout the later seventeenth, the eighteenth, and into the nineteenth century. A list is given in J. Montagu (1959, pp. 289 ff.).

A passionate interest in physiognomy developed among both eighteenth-century scientists and artists; probably its most famous representative was Johann Kaspar (or Gaspard) Lavater's generally benevolent *La Physiognomonie, ou l'Art de Connaitre les Hommes* (here quoted from the 1845 ed. of H. Bacharach), which was first published in 1773. In fact, he hardly mentions Le Brun (letter to Count Thun, p. 258; on writing and physiognomy, see p. 303). On the obsessional concern of one artist of the time, Franz Xaver Messerschmidt (1736–1783), with physiognomy, see E. Kris (1953, pp. 128 ff.). J. J. Lequeu's equally obsessive interest in it, applied directly to architecture, has not really been studied: but see P. Duboy (1986, pp. 179, 208, 224, 323 ff.).

On the importance of "animal physiognomy" for the development of caricature, see E. Kris (1953, pp. 192 ff. [with E. Gombrich]); E. Gombrich (1965, pp. 136 ff.); and W. Hofmann (1957, pp. 38 ff.). Like some other, later physiologists, Charles Bell (see his *Essays on the Anatomy and Philosophy of Expressions in Painting,* 1806, with many editions throughout the nineteenth century) absorbed Le Brun through Peter Camper (see n. 57 below). But see J. A. Lemoine (1865, pp. 31 ff.). Charles Darwin (1873) again reviews the literature in his introduction: he knew Le Brun through Morel d'Arleux's lithographs in the appendix of J.-L. Moreau de la Sarthe's monumental Parisian 1806 edition of Lavater; he had access to an 1820 reprint; and he was wholly contemptuous of Camper.

By the time Bell wrote, phrenologists had deduced a rather literal diagnostic system from earlier empirical observations. Goethe had been fascinated by Lavater and even collaborated on his *Physiognomische Fragmente* (see his observations on the physiognomy, 1948–1964, vol. 17, pp. 439 ff.). And although he broke with him on religious grounds, yet in later life he was obviously attracted by the teachings of Franz Josef Gall, the founder of phrenology, whose lectures he heard in Weimar in 1805 and whom he obviously found both brilliant and attractive (vol. 11, pp. 716 f., 755 ff.). However, the physicist and satirist Georg Christoph Lichtenberg, five years Goethe's senior, though he toyed with the idea of physiognomy, at first rejected Lavater's doctrine contemptuously ("Ueber Physiognomik wider die Physiognomen zur Beförderung der Menschenliebe und Menschenkenntnis") in the *Göttingischer Taschenkalender* for 1778 (1967, pp. 371 ff.). A year later, he added a parody of Lavater on how to tell character from the tails of wigs (pp. 411 ff.). Physiognomical doctrines were given a new scientific credibility in the "anthropological criminology" of Cesare Lombroso (1836–1909) and his disciples, and they were much extended into a positive classification of "constitution, temperament, character" by E. Kretschmer (1931).

There is still much to be learned about this matter. The sources and influence of Winckelmann's essay on the proportions of the human face are still very little known, and the papers belonging to the Società Colombaria in Florence have just been published: see H.-W. Kruft (1972, pp. 165 ff.) and Max Kunze (1994).

56. J. Baltrusaitis (1983, pp. 34 ff.) and Max Kunze (1994).

57. The drawings were published by J. Baltrusaitis (1983, pp. 19 ff.). The method was taken up, in a rather elementary way—as he himself says—by Lavater (1806, vol. 2, pp. 34 ff.), but developed into a full-blown "science" by the Dutch anatomist Peter Camper, who acknowledges Le Brun's priority in the field: see Baltrusaitis (pp. 37 ff.).

58. Gérard (or Gaspard) Désargues (or Des Argues) published very little on his own, and some of his publications have not survived. What is known was put together by the historian of mathematics, Noël Poudra (1864), who has vindicated the high place attributed to Désargues by some of his contemporaries and a few later mathematicians. See W. M. Ivins (1946, pp. 103 ff.).

59. G. Janneau (1965, pp. 144 ff., 166 f.) and H. Jouin (1889, pp. 142 ff.), very much from Le Brun's point of view. But see Bosse's own account in *Le Peintre Converty aux Precises et Universelles Regles de Son Art* (1667, pp. a v. f.); A. Blum (1924, pp. 46 ff.).

60. On the place of the Désargues-Bosse methods in the history of perspective constructions, see L. Vagnetti (1979, pp. 354 ff., 389 ff.).

61. On the precedents and the development of seventeenth-century anamorphisms, see J. Baltrusaitis (1984); he discusses the quarrel of

Bosse with the Academy in this context (pp. 69 ff.) and relates it to Descartes' view on visual perception (pp. 64 ff.). The Fourth Meditation is its most concerted statement. On Descartes' distrust of percepta, see B. Williams (1978, pp. 207 ff., 214 ff.); A. Kenny (1968, pp. 24 ff., 218 ff.).

62. Bosse himself publicized his quarrel with the Academy in three pamphlets of 1660, 1661, and 1667. But see also B. Teyssèdre (1957, pp. 53 ff.).

63. The two quarrels with the Academy were due first to the proposal that Fréart de Chambray's translation of Leonardo's *Trattato*, with its illustrations attributed to Poussin, be adopted as the standard perspective text in preference to Désargues-Bosse. Bosse thereupon wrote to Poussin, who replied—in a letter Bosse published after Poussin's death—with a condemnation of the illustrations in which (the letter said) he only drew the figures. Some scholars have considered the letter spurious. See B. Teyssèdre (1957, pp. 129 ff.) and J. Białostocki, "Poussin et Le 'Traité de la Peinture' de Léonard," in A. Chastel (1960, vol. 1, pp. 133 ff.). For the sarcastic remarks about Blondel and the entasis compass, see A. Bosse (1664, fol. 33). Bosse's principal aim was the reduction of the orders to a measurable rule.

64. See A. Bosse (1664, fol. 1).

65. J. Rykwert (1980, pp. 33 ff.); A. Perez-Gomez (1983, pp. 30 ff.).

66. C. Perrault (1696–1700, vol. 1, p. 91). Perrault had also taken Marin's side in another dispute—about whether animals could think and form general ideas; see A. Picon (1988, pp. 82 ff.).

67. *Livre d'Architecture contenant les Principes Généraux de cet Art* . . . (1745). The enormous vogue for Horatian hexameter treatises on the visual arts begins with Charles Alphonse Du Fresnoy's *De Arte Graphica,* first published in 1668, and in England popularized through Dryden's translation (1695). See J. Schlosser-Magnino (1956, p. 635) and L. Lipking (1970, pp. 38 ff.). Both Boffrand and Du Fresnoy published parallel texts, in Latin and French.

68. Robert Morris' ideas are summed up in his *Lectures on Architecture* (1736; 2nd ed. 1759). Of the dictionaries, Sulzer's (which is one of the longest), refers the term to painting, sculpture, and poetry only: J. G. Sulzer (1792), s.v. "Charakter."

69. J.-F. Blondel (1771–1777, vol. 1, p. 104).

70. The French version translates the title of Sagredo's book as *Raison d'Architecture Antique.* On Diego de (or da) Sagredo, see above, n. 23; W. Stirling Maxwell (1891, vol. 1, pp. 157 f.); N. Llewellyn (1977, pp. 292 ff.); and J. A. Ramirez (1983, pp. 132 ff.). A facsimile of the 1549 Toledo edition with commentary was published in 1986 (edited by Fernando Marias y Agustín Bustamante).

71. On Leon Picardo, see W. Stirling-Maxwell (1891); he seems better known for his friendship with Count Salvatierra than as a painter. However, he was court painter to the Constable of Castille at Burgos from 1514 to 1530, and two paintings by him, an Annunciation and a Purification of the Virgin, are in the Prado (nos. 2171, 2172).

72. P. Gauricus ([1504] 1969, pp. 76 f., pp. 94 ff.).

73. Diego de Sagredo (1526, pp. a v, ff.). He attributes this formula to Varro as an ancient authority and his acquaintance Felipe de Borgoña, otherwise Vigarny (on whom see W. Stirling-Maxwell (1891, vol. 1, pp. 145 f.).

74. *The Art of the Old Masters* of Cennino Cennini (1899, pp. 64 ff.).

75. Dionysius of Fourna (1974, p. 12). This *hermeneia* seems to have been compiled ca. 1730, drawing on much earlier material, though the proportional canon may—as is true of other matters in it—have also depended on Western sources. On other, unprinted canonic material, see P. Hetherington (1974, p. 113 n. 25).

76. H. Corbin (1973, pp. 237 ff.); but much Neopythagorean speculation inspired the

Brethren of Purity, *Ikhwan al-Safa*, a Shi'ite eso-teric group known through a collection of fifty-two "epistles," the *Rasa'il* (probably tenth century), also called *Kitab Ikhwan al-Safa*. Of the transla-tion by Susanne Diwald, only vol. 3 has so far ap-peared (1975). See H. Corbin (1973, pp. 94 ff., 140 ff.), Y. Marquet (1973, pp. 206 ff., 241 ff.), and Marquet in *Encyclopaedia of Islam*, s.v. "Ikh-wan-al-Safa."

There are other echoes of the doctrines in ear-lier Ismaili teaching about the heavenly Adam: see S. H. Nasr (1978, passim, but esp. pp. 25, 66 ff.)—he also has something to say about the micro-cosmic doctrines of Al-Biruni (pp. 149 ff.) and of Ibn-Sina (pp. 251 ff.). See also below, chapter 3. It may have been merely coincidental that one of the most illustrious thinkers in the Neopythagorean Is-lamic tradition, Ibn al-Sid of Badajoz, spent many years in Toledo four centuries before Diego was born there.

77. Several versions were in circulation at the time. For a list of the surviving copies, see G. Scaglia (1992, passim) and Francesco di Giorgio Martini (1967, vol. 1, pp. xxvii ff.). On supple-mentary material and their dating, as well as the authorship of the drawings and their various hands, see G. Scaglia (1970, 4 pp. 439 ff.) and C. H. Ericsson (1980, pp. 40 ff.). But see also R. J. Betts (1972, pp. 62 ff.; 1977, pp. 3 ff.).

78. Francesco di Giorgio Martini (1967, vol. 2, p. 390). I quote the Sienese-Magliabec-chiano version and have followed the editors in as-suming the separate integrity of the two groups of manuscripts: the earlier being the Turin and Lauren-ziana versions, the later the one quoted here. The illustration is copied (with an open mouth) in the New York Public Library manuscript. See R. J. Betts (1977, pp. 12 ff.). The reference to the cyma as "throat" is presumably a gloss on the pun "cima" (top, summit) and "the top of the head."

79. See G. Scaglia (1970, pp. 440 f.).

80. Francesco di Giorgio, Turin 15 (1967, vol. 1, fig. 25). This is about the most com-mon of all the canonic rules. On it, and much else of the material discussed here, see E. Panofsky, "The History of Human Proportions as a Reflec-tion of the History of Styles" (1974, pp. 55 ff.).

81. Only in the Turin/Laurenziana ver-sion (Turin 21 v.; Laurenziana 21 r.; 1967, vol. 1, pp. 90 f.). The face is here taken from the chin to the hair line, so that in fact the whole height is $9\frac{1}{3}$ faces. Francesco quotes other canons (Vitruvius' 10 and an unnamed author's 7) and says that they are acceptable alternatives (1967, vol. 1, p. 46). Fran-cesco himself seems to work with two canonic divi-sions of the figure: into 7 heads, as in vol. 1, p. 68, and into $9\frac{1}{3}$ faces as here (what is being specified is the face as one-ninth + one-third of the body). On proportion in Francesco's practice, see L. Lowic (1982, pp. 151 ff.) and H. A. Millon (1958, pp. 257 ff.), though Millon seems to have assumed that the face and the head form the same unit; but cf. also G. Hellmann, in [L. Bruhns] (1961, pp. 157–167).

82. It opens the section on church de-sign in the Magliabecchiano manuscript only, 38 v. (Francesco di Giorgio, 1967, vol. 2, pp. 394 f.). The body is divided into seven heights from the head and into four widths by two superimposed circles, one whose radius is from the top of the head to the center of the chest, the other from the knees to the bottom of the feet. The resulting over-lap is of course 1 module, and the grid 28 modules.

83. Magliabecchiano 42 r. (1967, vol. 2, pp. 402 f.); not in the Sienese codex. But cf. the Turin (16 v.) and Laurenziana (15 v.) version (as in 1967, vol. 1, pp. 68 f.).

84. This is implied by Magliabecchiano 22 r. (1967, vol. 2, pp. 348 f.) and Magliabec-chiano 40 v. ff. (1967, vol. 2, pp. 399 ff.). The 4 × 7 model is provided in Magliabecchiano fol. 41 (1967, vol. 2, pp. 400 f.). This diagram and there-fore presumably the whole manuscript were known to Philibert de l'Orme, who reproduces it, with the added complication of specified wall thicknesses, in his *Architecture* (1648, p. 235), though he does not give his source. This matter was entirely mis-conceived by A. Blunt (1958b, pp. 129 ff.). The problem of the commensurability of the side and the diagonal of the square was much debated in the fourteenth and fifteenth centuries: see V. Zoubov (1968, pp. 150 ff.).

85. Vitr. III.1.v. The problem of the per-fect number was formulated by Theon of Smyrna, in his mathematical commentaries on Plato, and the formula was generally known from Euclid (book IX, theor. 34, prop. 36). Plato discusses the "perfect" or "nuptial" number in *Rep.* VIII.546b ff.

86. Magliabecchiano 42 v. (1967, vol. 2, p. 403).

87. Turin 3 r. (1967, vol. 1, p. 4). The notion of the head "looking down" on the body appears in Pomponius Gauricus ([1504] 1969, p. 137), though Pomponius may have taken the image from Lactantius' *de Opificio Dei* VIII.11 and IX.2 rather than from Francesco. The more usual physiognomic image is of the eyes as gates or doors rather than windows of the soul.

88. Magliabecchiano 28, r. and v.; (1967, vol. 2, pp. 362 ff.). The passage is discussed by L. Lowic (1983, pp. 360 ff.).

89. Antonio Averlino detto Il Filarete, *Trattato di Architettura* (sometimes also called *Libro Architettonico;* (1972). The date is discussed on pp. xii ff.

90. Ibid. (pp. 14 ff., 23 f., 104).

91. John Donne, "Holy Sonnets," 5, line 1. The image reappears often in Donne's verse, most elaborately in the "Hymn to God my God on my Sicknesse" (1978, p. 50; see notes, p. 108).

92. "[C]ivitas philosophorum sententia maxima quaedam est domus, et contra domus ipsa minima quaedam est civitas": L. B. Alberti (1486), 14 r. (I.ix; 1966, p. 65). This is restated in a modified form in V.i., ii (1966, pp. 337 f.) The parallel between body and building is repeated too often to warrant separate reference.

III The Body and the World

1. Folio 14r. The miniatures of the uncompleted (perhaps because of the Duke's death in 1416) manuscript have been published several times. See J. Longnon and R. Cazelles (1969, no. 14); H. Bober (1948, pp. 1 ff.). On the Limbourg Brothers and the Duke's patronage, see M. Meiss, "The Limbourgs and Their Contemporaries" (1974, vol. 6, pp. 144 f., 308 ff., 421 ff.). The order in which the astrological signs are related to the

bodily parts is exactly the same as that given by M. Manilius in his *Astronomica* (II.453 ff.) probably written in the reign of Tiberius. They are also listed in the four corners according to the temperament and the humors. The *vesica piscis* enclosing the figures is calibrated by the signs as well as months and days. As for the gender of the figure, it has been suggested (J. Longnon and R. Cazelles) that the blond frontal figure represents a feminine temperament—female it plainly is not—while the dark-haired back is masculine.

A rather crabbed account of astrological medicine is given by R. Eisler (1946, pp. 94 ff., 246 ff.). On the pervasiveness of the zodiacal man, see H. Bober (1948, pp. 2 ff.). The correspondence between the human body and the zodiac, known as *melothesis,* was highly developed in late antiquity. On its origin in Egyptian iatromathematics, see A.-J. Festugière (1944–1954, vol. 1, pp. 125 ff.). However, Raymond of Marseilles offered the first reasoned Christian apology for judicial astrology, on which see above, chapter 2, n. 29.

The androgynous character of the zodiacal, microcosmic, or first man in the Judeo-Christian tradition was warranted by Gen. 1.27: "male and female created He them" (in His image). See L. Ginzberg (1913, vol. 1, pp. 88 ff.). On the androgyne as an archetype, see M. Eliade (1965, pp. 98 ff.). On the ancient, particularly the Iranian, background of this idea, see R. Reitzenstein and H. H. Schaeder (1926, pp. 210 ff.); on the Manichaean reading of Iranian and Jewish traditions, pp. 240 ff. But see also Festugière (vol. 4, pp. 178 ff.). The best known of such myths is Aristophanes' contribution to Plato's *Symposium* (189d ff.), which in a folksy way echoes Empedocles' ideas (DK frr. 60, 62), though of course Plato has to have three kinds of double-backed creatures: male/male, female/female, and "hermaphrodite."

A rather different idea of the monstrous and brutish androgyne beginnings of humanity is given by Lucretius (V.837 ff.). In the Zoroastrian Bunduhisn, the first couple Matro and Matroyao (also called Mashye and Mashyane) are a double child-devouring androgyne (XV.2 f.; E. W. West 1880–1897, vol. 1, p. 53). See R. C. Zaehner (1961, pp. 130 f.). The androgyne is a familiar figure in Indian tradition, and the notion of a microcosm is closely connected to that of a universal first androgyne, on which see M. Falk (1986, pp. 35 ff.); the Creator Atman is androgyne in the Brhad-aranyaka Upanishad I.4.iii (S. Radhakrishnan, 1953, p. 164). On the Siva-Sakti unity in the androgyne Ardhanari-svara, see S. Kramrisch (1981, pp. 199 ff.). Other comparative material has been studied by W. Doniger (1980, esp. pp. 283 ff.) But see also below, n. 16.

On the relation between antique and fifteenth-century astrology, see F. Saxl, "The Revival of Late Antique Astrology" (1957, vol. 1, pp. 73 ff.), and on the astrological medicine of the fifteenth and sixteenth centuries, see A. G. Debus (1978, pp. 121 ff., 133 f.).

2. "Sed et homo ipse, qui a sapientibus microcosmos id est minor mundus appellatur, iisdem per omnia qualitatibus habet temperatum corpus, imitantibus minirum singulis eius, quibus constat humoribus, modum temporum quibus maxime pollet": Bede, "De Temporum Ratione," xxxv (1688, vol. 1, pp. 114 f., and *PL* 90, p. 457). The transliteration from Greek seems to have had no currency in classical or even silver Latin (Roman writers preferred *minor mundus, brevis mundus*) before Bede. It seems to have been coined by Isidore of Seville a generation earlier. He uses it with reference to musical harmony: "Sed haec ratio quemadmodum in mundo est ex volubilitate circulorum, ita et in microcosmo in tantum praeter vocem valet, ut sine ipsius perfectione etiam homo symphoniis carens non consistat" (*Etym.* III.23). Isidore uses both the commoner Latin terms as well: see J. Fontaine (1959, pp. 376 ff., 423 ff., but esp. 662 ff.), and F. Rico (1970, pp. 40 ff.)

3. The Hon. Robert Boyle (1661, and many later editions, passim). The skepticism was about the four-element combination of the universe. Boyle, who was once described as the "son of the Earl of Cork and the father of chemistry" (often misquoted "father of chemistry and uncle to Earl of Cork": so *DNB*, S.V.; but see R. E. W. Maddison, 1951–1955, p. 197) wanted to seek the substance of matter in quite different principles, which he called "essences," of salt, sulphur, and mercury. This led him to recognize both the primary difference between mixture and compound and the importance of "fire-analysis." See A. G. Debus (1977), pp. 484 ff.

4. On the formation of the doctrine of the four elements, see M. R. Wright (1981, pp. 22 ff.); W. K. C. Guthrie (1962–1978, vol. 2, pp. 138 ff.); E. Bignone (1963, pp. 521 ff.); and S.

Sambursky (1959, pp. 2 f.). On the introduction of the *pneuma* as a fifth element, see Sambursky (1959, pp. 4 f.). On Empedocles' originality, see G. de Santillana (1961, pp. 108 ff.) and S. Sambursky (1956, pp. 16 ff.). The fifth element is sometimes identified with Aion, the world breath, the world soul; images of Aion with the world serpent and the zodiac (of which the most famous one is the statue in the Vatican Museum) are discussed in their different ways by C. G. Jung in *Aion* (1953–1979, vol. 9.2) and A.-J. Festugière (1944–1954, vol. 4, pp. 176 ff.). As it was also stated by Sextus Empiricus (*Adv. Math.* VII.115, 120 f.), Empedocles' teaching involves the four elements and the two forces of love and strife, which Hellenistic writers called six *kritēria tēs alētheias*, "warrants of truth."

Empedocles' main surviving texts on this matter are reported by Simplicius in his commentary on Aristotle's *Physics* (DK frr. 21.17–64. G. S. Kirk, J. E. Raven, and M. Schofield, 1983, pp. 280 ff.). The main editions are by J. Bollack (1965) and M. R. Wright (1981). Aristotle gives this teaching its "ethical" character in *Met.* I.4 (985a); III.4.xxvi (1001a); XII.10.vii (1075b). For a survey of the anti-Aristotelian literature of interpretation, see R. A. Prier (1976, pp. 123 ff.). It has long been pointed out that certain fragments of Heraclitus presuppose the four-element system: for example DK 12.126, out of Tzetzes scholia on the *Iliad* (see M. Marcovich, 1967, pp. 220 ff.). Others allude to the eternal strife: for example, DK 12.53, 80 (see Kirk, Raven, and Schofield, 1983, pp. 193 f.; Marcovich, 1967, pp. 132 ff.; and P. Wheelwright, 1959, pp. 34 ff.).

On the Islamic zodiac, see S. H. Nasr (1978, pp. 151 ff.). On the acceptance of the zodiac in India and in China, see F. Boll (1926, pp. 57, 85, 97, 150, 155); R. Berthelot (1949, pp. 96 ff.); and A. Rey (1942, pp. 341 ff, 359 ff.). On the Chinese conception of astrobiology, see P. Wheatley (1971, pp. 414 ff.).

5. Ambrosius, *Hexameron* (commentary on the six days of the Creation) III.4.xviii (*PL* 14, p. 164b): "sic sibi per iis jugales qualitates singula miscentur elementa . . . atque ita sibi per hunc circuitum et chorum quemdam concordiae societatisque conveniunt. Unde et Graece *stoicheia* [in Greek in the text] dicuntur, quae Latine elementa dicimus, quod sibi conveniant et concinant." Ambrose follows here (as in many other things) the earlier *Homilia V in Hexameron* of St. Basil of Cesarea (IV.9.xcii a, *PG* 29, pp. 19 ff.); see J. Fontaine (1959, p. 656 n. 4). This passage is in fact a free version of Empedocles' teaching in his *Peri Phusios* (*Book on Nature*) now known best through the summary quotations of Simplicius. Although he wrote in the sixth century AD, Simplicius is regarded as an accurate (if Neoplatonically tinted) reporter of earlier Greek philosophy. Ambrose presumably knew it through other, earlier sources, such as a lost work of Posidonius.

Implicit in Empedocles' scheme was a world body, the spherical *Sphairos kukloterēs,* complete—prior to differentiation by strife—though there is a problem in introducing the element of time into such an image. But see D. O'Brien (1969, pp. 4 n. 2; 28 f.). On the relation between Empedocles' cosmic man and other microcosmic ideas of the pre-Socratic philosophers, as well as their redaction by Plato, see A. Olerud (1951, pp. 43 ff.).

On the importance of this quadripartition as an interpretative element, see A. C. Esmeijer (1978); with particular reference to the human figure as microcosm, see pp. 50, 100 ff. On the operations of chance and necessity in this and other elemental schemes, see J. Barnes (1979, pp. 119 ff.). Inevitably, this was not the view of all the fathers. Gregory of Nyssa, Basil's younger brother, took a sceptical view of some aspects of microcosmic teaching in *De Opificio Hominis* (*PG* 44, pp. 132 ff., 177 ff.; trans. and ed. J. Laplace and J. Danielou, 1943, pp. 6 f., 90 ff., 148 ff.). On the early church fathers (particularly Tertullian, Athenagoras, and Clement of Alexandria), see M. Spanneut (1957, pp. 171, 390 f., 411 ff.). Unfortunately, Spanneut is dismissive about Tertullian's ideas.

The four and five elements appear in other cosmogonies: so the *ogdoad* of the Rig-Veda is made up of the four elements + ether + the sun and moon + the other stars. See S. Kramrisch (1981, pp. 108 f.).

6. The hypothesis of Paleolithic tally records of star movements was persuasively argued by A. Marshack (1972); he advances no strong hypothesis about the relation between the animals associated with tallies, though the supposition that the animals which almost exclusively occupy Paleolithic artists were obsessively grouped in some sort of classificatory scheme, probably gendered, had already been made by A. Leroi-Gourhan and Annette Laming-Emperaire. See A. Leroi-Gourhan

(1965), pp. 80 ff. For a recent review of the evidence, see E. Hadingham (1983).

7. Rig-Veda X.90, the Purusa-Sukta, is one of the best-known and most commented Vedic hymns, though it is considered rather late by C. F. Geldner (1951, pp. 286 ff.). It describes the creation of the world and of society from the sacrificed human body of Purusa, the primal man and world Involucrum. The word also signifies the measurement of 7½ feet, which can be taken from the patron of the sacrifice standing with his arms raised; see F. Staal (1983, vol. 1, pp. 96 ff.). See also A. K. Coomaraswamy (1933, pp. 69 ff.); J. Gonda (1975, pp. 137 ff., 330); M. Eliade (1969, pp. 41 f.); and R. Panniker (1977, 72 ff.), among others. I adapt from Panniker's version, verses 6 ff., capitalizing THE MAN for Purusa.

6 Using THE MAN as their oblation,
   The Gods performed sacrifice.
   Spring served for their clarified butter,
   Summer for fuel, Autumn for the offering[.]
      . . . .

11 When they divided up THE MAN,
   Into how many parts did they divide him?
   What did his mouth become, what his arms?
   What are his legs called? or his feet?

12 His mouth became the Brahmin; his two
      arms—
   Rajanya the Warrior-princes; his legs—
   Vaisya, the common people who ply their
      trade.
   Sudra the lowly serfs were born from his feet.

13 The moon was born from his mind; the Sun
   shone out of his eyes.
   From his mouth came Indra and Agni,
   and from his breath the wind was born.

14 From his navel issued the Air,
   from his head unrolled the sky,
   the earth was his feet, his ears the four
   directions, so the whole universe was
      organized.

15 Seven were the posts round the enclosure,
   thrice seven the fuel sticks were counted,
   when the Gods who performed the sacrifice
   bound THE MAN as their victim.

On its recitation and a commentary on it, see F. Staal (pp. 112 ff., 415 ff.). On the connection between the Purusa hymn and the Bundihisn, see R. C. Zaehner (1955, pp. 136 ff.). J. G. Frazer suggested in "The Scapegoat" (1911–1915, vol. 9, pp. 40 f.) that cosmogonies were related to the human body through human sacrifice without reference to the Rig-Veda. In a different context A. Olerud (1951) related the notion of microcosm to creation "out of" the body and therefore sacrifice, particularly in relation to Iranian cosmogony, on which see R. C. Zaehner (1955, pp. 134 f; 1961, p. 257 ff.). On the problem of the cosmogony out of the Creator's body, see H. W. Bailey (1943, pp. 121 f.), and G. Widengren (1965, p. 9). The date of the Pahlavi texts, particularly of the earlier material they incorporate, is heavily contested, and I merely report the references. It will be evident that the body of a sacrificial animal may be "quartered" in exactly the same way as a human body.

In any case, even in recent performances of sacrifices, the wooden altar enclosure and shelter are ceremonially burned after the sacrifice. See Staal (pp. 689 ff.).

8. I owe the last suggestion to Charles Correa. The literature of mandala and building is enormous. See, however, principally S. Kramrisch (1946, vol. 1, pp. 7 ff., 21 f.; vol. 2, pp. 357 ff.); P. K. Acharya (1981, pp. 37 f.). Purusa is sometimes the perfect primal man, or the spirit of the place, or a hunchback dwarf. Prajapati, the Lord of all creatures, is both sacrificer and victim. It is not clear how Prajapati as efficient cause of all things is different from Purusa as the type of sacrificial victim. Purusa, the hunchback, fell and was smashed to pieces; Purusa, the victim, was cut up by the gods (as in the Purusa-Sukta, above, n. 7). Since the fall or dismemberment of Purusa is the primal scene of Indian action, all Indian building—beginning with the five-layer fire altar—could be described as the attempt to put Purusa together again. On Kim-purusa ("what man?"), variously interpreted as mock-man, deformed man, depraved man, and so forth, see F. Staal (1983, vol. 2, pp. 62 ff.).

9. On the zodiac in classical antiquity, see F. Cumont in *DS*, s.v.; and more recently H. Gundel and R. Böler in PW, 2 Reihe, vol. 19; see also H. Gundel in *EAA*, s.v. "Tierkreis." On the formation of a zodiac of twelve houses of 30° each in Asia Minor, see O. E. Neugebauer (1957, pp. 102 ff., 188, 207) and R. Berthelot (1949, pp. 24 ff.), but also R. Guénon (1962, pp. 120 ff.). On the decans, see S. Schott in W. Gundel (1936, pp.

1 ff.); E. A. Wallis Budge (1934, pp. 245 ff.); and O. E. Neugebauer (1957, pp. 82 ff.). Egyptian astrology was based on this system, and in the Ptolemaic period the Babylonian zodiac, imported by the Greeks, seems to have been conflated with the native one. Much of the later complexity of calculation was based on the shifts between the basically solar Egyptian system of computation and the lunar Mesopotamian method. The hypothesis about the early relation between the observations and building was already made by Sir Norman Lockyer in ([1894] 1964).

The first clear statement of the "classical" zodiacal system is attributed to the great fourth-century astronomer Eudoxus of Cnidos, who had been a pupil of Plato, on whose doctrines the Hellenistic poet Aratus wrote a very popular verse account known in the West through a translation by Cicero of about half the poem: see in Callimachus (ed. A. W. and G. R. Mair, 1977); fragments are quoted in the English version by T. L. Heath (1932, pp. 112 ff.). A slightly different account is given by Simplicius, who was quoting out of the *Elementa Astronomiae* of Geminus (see Heath, pp. 123 ff.). Geminus would probably have been a near contemporary of Julius Caesar.

10. The astrological year was fixed sometime in the Hellenistic period, perhaps before 300 BC, though the "houses" were identified with planet and constellation names that were already old at the time. When it was fixed, the point of the spring equinox was in Aries. Because of the retrograde movement of the equinoxes (known as the "precession of equinoxes") it has by now moved into Pisces. This precession of "about 1° every century"—in fact, every 72 years, or 25,920 years for the whole cycle—was discovered by Hipparchus in 127 BC, according to Cl. Ptolemy *Syntaxis* VII.3 (quoted by T. L. Heath, 1932, pp. 142 ff.). See G. de Santillana and H. von Dechend (1969, pp. 142 ff., 186 ff.). The custom of casting personal horoscopes seems also to date from the Hellenistic period, before which stargazers made predictions about public events based on more general astronomical observations. See A.-J. Festugière (1944–1954, vol. 1, p. 75 n. 4) and O. E. Neugebauer (1957, pp. 102 f., 170).

11. That is, of the complete cycle. A number of the zodiacal animals and monsters appear on Babylonian boundary stones known as *ku-*

durru, though their precise relation to the constellations or the zodiacal cycle remains to be ascertained. See W. J. Hinke (1911, pp. vii ff.); L. W. King (1912, pp. xiv f.); F. Steinmetzer ([1922] 1968, pp. 14–15); and A. Bouché-Leclercq (1899, pp. 76 ff., 311 ff.). Cf. H. Bober (1948, p. 8) and O. Neugebauer (1957, pp. 67, 102 f., 140 f., 170).

12. On the decans, see above, n. 6. We are not well informed on the nature of the animals. But see C. Lévi-Strauss (1964–1971, vol. 4, pp. 102 f.), for a different context.

13. As for the humors, they derive their label from the notion that the "temper" of the body could be detected through the lack, or excess, of one of the four "liquid or fluent parts of the body comprehended in it, and either born with us or . . . adventitious and acquisitive": so Robert Burton (Democritus Junior) in *The Anatomy of Melancholy* ([1628] 1854, pp. 92 f.). The four liquids were blood, choler, phlegm, and melancholy; and of course the secondary sense of humor as meaning "caprice" signifies an excess of one humor over the others. Burton quotes as his authorities Crato and Laurentius "out of Hippocrates."

14. At the Nativity: Matt. 2.2 ff. At the Crucifixion: Matt. 17.45 ff., Luke 23.44 ff.

15. See E. Garin (1942, vol. 2, p. 192) and also C. Trinkaus (1970, p. 507 ff.); though in fact the whole of sections 6 and 7 of chapter 5 of Pico's *Heptaplus* might be quoted in support. But of course Pico, like his friend Savonarola, was a convinced enemy of astrology, whose postulates he considered (as did many of its earlier Christian enemies) to be an implicit denial of free will: see G. Saitta (1961, vol. 1, pp. 628 ff.) and E. Garin (1976, pp. 87 ff.). On other interpretative diagrams and figures (such as the "tree of wisdom," "tree of virtue," "tower of wisdom," "ladder of virtue," and "shield of faith" or "donkey's bridge" as medieval thinking aids, but also as structures of mystical visions), see M. Evans (1980, pp. 32 ff.). Unfortunately, H. Bober's material about this matter has remained unpublished, except for two papers: "An Illustrated Mediaeval School-Book of Bede's *De Natura Rerum*" (1958, pp. 65 ff.), and "In Principio, Creation before Time" in M. Meiss (1961,

vol. 1, pp. 13 ff.). See also M. J. Reeves and B. H. Hirsch-Reich (1972, pp. 73 ff.).

16. On this notion in general, see G. P. Conger (1922) and F. Saxl (1957, vol. 1, pp. 58 ff.). On the Western tradition, see R. Allers (1944, pp. 319 ff.). On the early medieval treatment of the ancient theme, see M.-T. d'Alverny (1973, pp. 69 ff.). Later medieval material is treated by M.-D. Chenu (1952, pp. 39 ff.) and M. M. Davy (1977, pp. 39 ff.). Allers has classified six different microcosmic approaches current in Western thinking, some of which (inevitably) overlap:

1. Man contains within himself all the world elements.
2. The structural view, in which his composition mirrors the rule and order of the whole world. This in turn has two "modes": cosmocentric, in which man understands himself to be organized on the same principles as the order of the world; and anthropocentric (as in Plato), in which the understanding of the world is read out of the human figure and soul.
3. The idea is extended from the individual to society, to cover (a) the organization of society as a world model and (b) the aesthetization of the notion by making human life into a work of art, as in Romantic thinking.
4. The symbolic approach, in which the microcosm corresponds to—or is a symbol of—the ordered universe.
5. The epistemic view, in which man contains the universe by knowing it.
6. The metaphoric view, in which the human body as a microcosm is simply one special case of every organized thing being a "small world."

Although not very precise, Allers' categorization is the best available. As for the word itself, it may have been coined by Democritus of Abdera (DK vol. 2, p. 72, fr. 34: *en toi anthrōpoi mikroi kosmoi onti,* on which see G. de Santillana, 1961, p. 155 ff.; though the fragment is recorded only by the fifth-century Armenian philosopher, David of Nerken) whom legend makes a year (or sometimes ten years) older than Socrates. The first writer to use the word properly was Aristotle: *Phys.* VIII.2 (252b.26), but cf. also *De Anima,* III.8. Though he does not actually name it, the idea was powerfully stated by Plato in the *Timaeus* (33b, 41d f., 69a ff.; see A. Olerud, 1951, particularly on the Eastern antecedents of the idea). That the head is another, separate degree of microcosm is a notion that returns several times in the *Timaeus* (44d ff., 69d f., 72d). The use of the word *kosmos* for world order was attributed to Pythagoras by some older writers: Aet. I.3.8 (in DK 45B, 15), Iamblichus (*V. Pyth.* 162, ed. A. Nauck, 1884, pp. 118 f., which is taken up by Firm. Mat. III.proem.4, and is discussed at length by Philo Jud. *De Opif. Mundi* 53 ff.). See however W. Burkert (1972b, p. 77 nn. 151 f.). On the other hand, Philo attempted to wrestle with the idea that man was the "image" of God and the anthropomorphism this implied (145 ff.). It was also a powerful strain in the Hermetic tradition. Although the actual word microcosm does not appear in the *Poimander*, the idea is clearly expressed there (I.12 ff., VIII.5—*ho anthrōpos kat'eikona tou kosmou genomenos;* see *Corp. Herm.,* ed. A. D. Nock and A.-J. Festugière, 1945, vol. 1, pp. 11 f., 89). On its treatment in recent Catholic theology, see C. Korwin-Krasinski (1956, pp. 206 ff.).

On Peter Abelard's allegorical identification of the Platonic world soul with the Holy Spirit, see T. Gregory (1955, passim); A. V. Murray (1967, pp. 94 ff., 110 ff.); and D. E. Luscombe (1969, pp. 123 ff., 188 f., 237 f.). To St. Bernard, Satan is a parody of the microcosm: "Tract. de Erroribus Abelardi" in *PL* 182, p. 1055. This obverse microcosm seems an echo of the second, evil world soul of certain Gnostics and its relation to various dualistic (particularly later Zoroastrian), conceptions—on which see H. Borst (1958, pp. 497 ff.); M.-D. Chenu (1955, pp. 75 ff.); L. Grill (1961, pp. 230 ff.); R. Klibansky (1961, pp. 1 ff.); and L. Ott, "Die Platonische Weltseele," in K. Flasche (1965, pp. 307 ff.).

On the relevance of this idea to medieval architecture, see O. von Simson (1956, p. 36 n.), quoting Peter of Celle. (See Peter of Celle in *PL* 202, p. 610; also Durandus of Mende, *Ratio Divinorum Officiorum* I.i.14, and the *Clavis Melitonis*—the "Key" of Melito of Sardis, in J. B. Pitra, [1885–1887] 1967, vol. 3, p. 184—as well as Hildegard of Bingen, passim.) Durandus' explicit text, which is an aside in his very popular manual of ecclesiology (trans. C. Barthélemy, 1854, vol. 1, pp. 19 f; but see also 1843, pp. 24 f.), seems to refer to an

idea wholly self-evident to him, though he does gloss it with a comment on the three orders of salvation that make up human society.

Painted representations of the idea seem to have been rare. Nevertheless, as A. C. Esmeijer insists, the study of the influence of the Vitruvian figure on medieval artists—particularly sculptors—is a neglected field (1978, pp. 100, 171 n. 18; also in J. Bruyn, 1973, p. 9 n. 26), and this figure seems alluded to in certain manuscript illuminations (F. Saxl, 1957, vol. 1, pp. 61 f.).

On the king's presenting the image of God through his "other" body, see E. Kantorowicz (1957, passim, but esp. pp. 500 f.). On the coronation robe as the ritual imposing of the microcosm on a king or emperor, see R. Eisler (1910, vol. 1, pp. 3 ff.) and F. Saxl (1957, vol. 1, pp. 87 ff.). For sixteenth- and seventeenth-century material, see S. K. Heninger (1974, passim, esp. the bibliography offered p. 199 n. 62). For the medieval background, see H. Liebeschütz (1950); on the Lullian tradition, see J. N. Hillgarth (1971, pp. 223 ff., 423 ff.). On the vast literature of the subject in the fifteenth and sixteenth centuries, see E. Cassirer (1953–1957, vol. 2, p. 109 and passim); P. O. Kristeller (1976, pp. 174 ff.).

17. L. Thorndike (1923–1958, vol. 4, pp. 132 ff., 374, 446 ff.). On early zodiacal pharmacopoeia and other forms of therapy, see A.-J. Festugière (1944–1954, vol. 1, pp. 139 ff.). In France, for instance, barber-surgeons were required (by a patent of Charles VII of 1427) to display a zodiacal man chart in their shops; see F. Saxl (1957, vol. 1, p. 67).

18. On the microcosm in Moslem thought, see S. H. Nasr (1978, passim, but esp. pp. 75 ff, 149 ff.), and H. Corbin (1960b, pp. 163 ff.). On the implications for Moslem art, see A. Papadopoulos (1976, pp. 40 ff., 93 f., 239 ff.). On the Ikhwan al-Safa, see above, chapter 2, n. 76. F. Dieterici ([1865–1878] 1969, vol. 3 [1865, *Die Propaedeutik*], pp. 136 ff.), gives an account of the microcosm according to the mathematician al-Muharrir, who also provides a canon of the proportions of letter forms.

Al-Muharrir's canon states that God made man's length commensurate with his breadth and with the hollow of his torso. The length of the upper and lower arms corresponds to the two parts of the legs, the neck to the backbone, the head to the rump. The curve of the face corresponds to the width of the breast, the form of the eyes to the mouth, the length of the nose to the width of half a face, the ears to the cheeks, the fingers to the toes, the length of the intestines to that of the veins, the stomach to the liver, the heart to one lung, the form of the liver to that of the kidneys, the width of the throat to the size of a lung.

The length and thickness of sinews correspond to the bones, the length of the flanks in their turn to the hollow of the chest, the length and thickness of veins to the thickness of the body section; only the Creator, God himself, can know these correspondences, which recur throughout the human body exactly. If the seed falls into the womb without any harmful admixtures or bad celestial influences, then the fully and well-developed newborn child will be 8 insteps high: 2 insteps from the base of the knees to the soles of the feet, 2 more to the waist, 2 to the base of the breast, and 2 to the top of the head. If the hands are spread as the wings of a bird in flight, they will also be 8 spans wide, half for the trunk where it meets the neck, and a quarter at the elbows. If both hands are stretched over the head and a compass set at the navel, it will circumscribe the toes with a circle 10 spans in diameter, a quarter more than the height. The length of the face from chin to hairline is 1¼ spans, the same as the distance between the ears; the nose is ¼ of a span long; the eye is ⅛ of a span long; the forehead is ⅓ of the face; the width of the mouth equals the nose. The sole of the foot is 1¼ of the instep—which makes the sole ⅙ of the whole height.

The hand from its root tip of the middle finger is one span; the small finger equals the thumb; the intermediate fingers are ⅛ of a span longer. The width of the breast is 2½ spans, and the distance between the nipples one span. There are 2 spans between the shoulders, one from the base of the breast to that of the neck. The internal organs, bones, and sinews follow the same rule. What is true of man is also true of animals, and judicious artists make their forms correspond to each other analogously and so imitate the work of the Creator. Indeed philosophy may be defined as the becoming more like God as far as it is possible to man. These same relations persist in the stars and their movements as well as in music. The passage ends with an invocation of thrice-holy Hermes as the Prophet Idris and of the soul of Pythagoras.

The essential difference between the several Western canons and that of the Ikhwan is the adoption of the instep, ¾ of the foot (i.e., the foot minus the toes), as the module.

19. Vincent of Beauvais ([1624] 1964, vol. 1, col. 1994; XXVIII.2). In spite of the title page, Vincent was never a bishop.

20. *Placides et Timéo* 214 (1980, p. 93). Probably composed in the middle of the thirteenth century, it was a fairly popular book: seven fourteenth- and fifteenth-century manuscripts survive, and three more have been catalogued (in 1373) but are lost. Several printed editions appeared in the sixteenth century, between 1504 and 1540.

Very few ancient authors are named: Aristotle, Hippocrates, Homer, Lucan, Ovid (and his friend Macer Floridus, to whom a later book on plants is falsely attributed), Plato, Porphyry, Virgil. The Cato of Rome is the author of the popular medieval distichs.

21. Although Restoro mentions fewer ancient authors than does *Placides et Timéo* (if more Arab ones), this passage seems to echo Vitr. III. praef.ii. and 1.ii.

22. Much material on this was collected by Hans Flasche in (1949, pp. 81 ff.). But see above, pp. 61 ff.

23. Much as anatomical information remains wedded to received ideas, backed by a textual tradition that was a guide to the practice of dissection but was not tested against it until the late fifteenth century. See A. Carlino (1988, pp. 33–51).

24. Aurel. Aug. *De Civ. Dei* XV.26. This is reasserted emphatically by Eugippius the Abbot, an African writing late in the sixth century (1885; *CSEL* vol. 8, p. 215 f.); to the Augustinian meditation on the identity of measurements between the ark and Christ's body, he adds the reflection that the Savior's resurrected Body must have retained exactly the same dimensions it had during His earthly life.

25. *PL* 107, pp. 150 ff., 167 f.

26. *PL* 197. The work of greatest interest is the *Liber Divinorum Operum Simplicis*

Hominis. The account of the microcosm begins with a grandiose statement of the human figure standing in the middle of the cosmic circle, his stretched hands touching its edge (pp. 761 ff.). The human figure was important to Hildegard as the summary and measure of all creation. She sets this out in section 43 of the fifth vision in the second book of the *Liber Divinorum Operum* (PL 197, 945c). See B. Widmer (1955, pp. 46 ff.).

A detailed canon, consonant with the "golden section" according to the precepts of the Beuron School, has been extracted from eight later passages by I. Herwegen (1909, pp. 443–446). The cosmic man is described in pp. 813 ff. The canonic passages are in pp. 815, chap. XVII (triple division of the face); 816, chap. XVIII (lips are the same size; the distance between the ears, from ear opening to the shoulder and to the base of the throat are the same); 819, chap. XXII (the forehead is divided into seven spaces to correspond to the planets); 831, chap. XXXII (eyes are equal in size and their movements are coordinated); 843, chap. LIII (equal parts from the top of the head to the base of the throat, from there to the umbilicus, and again to the *locum egestionis*); 844, chap. LV (from the shoulder to the elbow, from elbow to the end of the middle finger, is equal to the distance from the ankle to the big toe); 845, chap. LVI (the space between the thigh bones equals that between the umbilicus and the place of evacuation, the crotch); 869, chap. XCII (from the knees to the ankles equals the distance from the crotch—or the thighs—to the knees); these measurements are amply and elaborately moralized by Hildegard. They are also splendidly illustrated in Lucca, Munic. Libr. Cod. 1942, illuminated (perhaps?) at Rupertsberg ca. 1230, which is the basis of the Migne text, a reprinting of the edition of Giandomenico Mansi. Cf. K. Liebeschütz (1930, p. 86 ff.) and H. Schipperges (1961, pp. 14 ff.).

27. Agrippa von Nettesheim (1550, vol. 1, pp. 187 ff.). See E. Panofsky (1940, p. 113 with n. 1; figs. 105, 106, 107).

28. On Bernard Silvester (also Bernard of Tours), see C. H. Haskins ([1927] 1967, pp. 104 ff., 375 f.); E. R. Curtius (1953, pp. 106 ff., and passim); and B. Stock (1972).

29. In spite of persistent inquiries, the Biblioteca Civica in Lucca (to which it was con-

signed after secularization) could not tell me anything about its provenance. I am grateful to Prof. Albert Derolez, who told me that though it may have been brought to Italy by one of Cardinal Mansi's German ancestors, to judge by the *postille* in its margins, it was in Italy by the fourteenth century.

30. The vast figure of the Creator in the northern cloister of the Campo Santo in Pisa, making a circular universe which coincides with his body, is now attributed to Pietro di Puccio of Orvieto—though Vasari had given it to the Florentine Buonamico Buffalmacco, as did A. da Morrona (1821, pp. 61 f.) and L. Lanzi (1822, vol. 1, pp. 35 f.). But see G. Vasari's *Vite* (ed. G. Milanesi, 1878 1906, vol. 1, p. 513; ed. R. Bettarini and P. Barocchi, 1966–1976, text vol. 2, 172 f., commentary vol. 2, A, pp. 490 f.); see also A. Letalle ([1925?], pp. 101 ff.) and A. Caleca et al. (1979, pp. 86 f.).

The figure of the godhead surrounded by angels in the nearby *Sacrifice of Noah* by the same master seems based on a projected regular hexagon.

It is not usually observed that this Pisan world-man-god is the opening panel of a sequence on human creation and redemption mostly out of Genesis and Exodus (it ends with the story of Solomon and the Queen of Sheba) over the whole south wall of the Campo; it was continued by Benozzo Gozzoli nearly a century later (1468–1484), for which the grateful Pisans buried him in the Camposanto.

31. She is also known as Herrade of Landsberg. Her principal work, the *Hortus Deliciarum,* was destroyed in the 1870 campaign in Strasbourg and has now been reconstructed: see Herrad of Hohebourg (1979, vol. 1, pp. 96, 221). There are many less popular speculations on the anthropomorphism of the world and its parts, such as Opicinus de Canistris' meditations on the world map, which remained unknown until the twentieth century. See R. Salomon ([1936] 1969).

32. See K. Menninger (1969, pp. 12 f., 34 ff., 117 f.).

33. In fact the Book of Creation, apparently a pre-Talmudic text, bases its speculation on the completeness of the number 32, which is made up of 22 letters of the Hebrew alphabet and the 10 numbers. L. Goldschmidt's (1894) remains the best text available, but see also M. Lambert (1891). On the date and editions of the text, see G. Scholem ([1962] 1987, pp. 20 ff.); on its use by the early Cabbalists, see pp. 40 ff. and passim. On the Book of Creation and the golem, see G. Scholem (1960, pp. 165 ff.).

34. There was much meditation on this in medieval thinking. The tetragram ADAM was often examined for inspiration: Aurel. Aug. *Enarr. in Ps.* (ps. 95, v. 15.); *In Joann. Evan.* (IX.14, X.12). Augustine may have drawn it from the Sybilline Book III.24 f. See A. Kurfess (1951, p. 72 with n., p. 287). This presumably Jewish source speaks of the "tetragrammaton Adam," although the name Adam (*Ad'm*) is in fact a trigram in Hebrew. I know of no study about the iconography of the confrontation between the new and the old Adam: however, see F. Saxl (1957, vol. 1, pp. 61 ff.) on the gospels of Uta of Ratisbon. The tradition of the Vitruvian old Adam being reflected in the new is taken up by William Blake (1804, pl. 76), on which see D. Bindman (1977, pp. 179 f.). Blake's perfect man is in fact the giant Albion, and the figure on this plate is the rear view of the better known figure of "Albion Arose" with its deliberate echo of Scamozzi's proportion man. See M. D. Paley (1984, pp. 113 ff.).

35. The cult of the "measure of Christ" seems to have originated in Constantinople, where one of the three gold crosses (in the arch of Constantine, the Forum Artopolium, and the Forum Filadelfium) may have displayed such a measure. One of the earliest Western documents of the cult is a Florentine thirteenth-century manuscript (Laurenziana Plut. XXV.3, dated 1293), on which see A. M. Bandini (1774–1778, vol. 1, col. 748), and G. Uzielli (1899, pp. 9 f, 29). The other MSS are Riccardi 1294. and Riccardi 1763. There was also an early undated printed devotional sheet of the measurement of Christ. A stone tablet (on which, it is said, the soldiers played dice for the robe at the Crucifixion) supported on four quattrocento columns has long been known as the *mensura Christi;* it had been set up in the so-called Council Chamber of the Lateran palace, from where it was moved to the cloister of the church by Sixtus V; presumably *mensura* refers to the columns, that is, the underside of the tablet. See G. Rohault de Fleury (1877, pl. 51) and also C. Ginzburg (1982, pp. 72 f.). It is

still described in Murray's Guide of 1899, but not in the Rome CIT guide (Rome, 1965). There were also columns from the Jerusalem praetorium, which were regarded as the measure of Christ since they were thought to be the columns of the Flagellation to which His body was tied. In spite of the enormous popular devotion to the *mensura Christi,* particularly in Tuscany (see G. Uzielli, 1904, vol. 12, pp. 192 ff.), to date I have not found any connection between this devotion, which seems to concentrate on the body as 3 *braccie* (each measuring 2 feet) high, and the devotion to the footprints of Christ, which according to tradition He left in the paving stones when He turned St. Peter back at "Domine, Quo Vadis?" The footprints are preserved in St. Sebastiano on the Appian Way as well as in a seventeenth-century copy in the Quo Vadis chapel. The various manuscripts and documents do in fact provide varying dimensions. That such cults were widely spread is familiar material. The devotion to and imitation of the form of the Holy Sepulcher in Jerusalem was discussed by R. Krautheimer (1942, pp. 1 ff.).

36. F. Petrarch *Dial.* 93 (1649, pp. 590 ff.).

37. Innocentius PP III (Lotario Conti) (*PL* 207, pp. 701 ff.); F. Petrarch (1649, pp. 16 f., and passim). The theme is constantly reiterated in fifteenth-century letters. See G. Toffanin (1943, pp. 273 ff.); E. Garin (1984, pp. 94 ff.); and C. Trinkaus (1970, passim). On the theological debate round the theme, see J. W. O'Malley (1979, passim).

38. L. Coelius (Caecilius?) Firmianus Lactantius (*PL* 6).

39. The argument from design is implicit in St. Augustine's praise of Plato and his followers, *De Civit. Dei* VIII.6 and further in XI.4 and Sermon 241. It was formalized by St. Anselm in *Cur Deus Homo.* On its role in scholastic thinking, see E. Gilson (1944, pp. 73 ff.).

40. Cic. *De Nat. Deo.* 53 ff.

41. Nicholas Chrebs (Krebs, *Cancer*— the crayfish, which he adopted as his emblem) of Cusa or Cues (a small town just over the Moselle from Bernkastel) was born in 1401; he died as

Cardinal and Prince-Bishop of Tyrolean Brixen (now better known as Bressanone in the Italian Alps) in Todi in 1464. *De Docta Ignorantia*, II.3, III.3. See also *De Doctrina Christiana*, in P. M. Watts (1982, passim, and particularly pp. 96, 109 f., 206). But see also G. P. Conger (1922, pp. 54 f.). A glimpse of Cusanus, as seen by the Florentines in the next generation, is in V. de Bisticci (1859, p. 169).

42. *De Venatione Sapientiae* XII.32, quoted by H. Blumenberg (1976, p. 39). "Infinitum ad finitum proportionem non esse" (*De Doc. Ign.* I.3.i). *Proportio* is translated as "gradation" in the only English version (by G. M. Heron, 1954, p. 11), which conveys the sense but weakens the association. Cusanus is here glossing Arist. *Met.* I.1 (980a).

43. See P. M. Watts (1982, pp. 25 ff.).

44. Nicholas of Cusa (H. Blumenberg, 1983b, p. 525 f.). Protagoras of Abdera in Pl. *Tht.* 152a, 178b, and *Crat.* 383a, as well as Arist. *Met.* 1053a, 1062b, *Eth. Nic.* 1113a. See also Sext. Emp. *Adv. Math.* VII.60. On this much quoted and commented saying, see W. K. C. Guthrie (1962–1978, vol. 1, pp. 401 ff.) and H. Cherniss (1935, pp. 33, 55 ff.). On Nicholas' reading of this aphorism as one of his four basic assumptions in his late work, *De Beryllo*, see H. Blumenberg (1976, p. 83 f.) and P. M. Watts (1982, pp. 174, 180, 185 f.). On the place of this aphorism in quattrocento thinking generally, see C. Trinkaus, "Protagoras in the Renaissance: An Explanation," in E. P. Mahoney (1976, pp. 190–214). In his book *De Globo* (*On the Globe Game*), Nicholas of Cusa again appeals to Protagoras so as to formulate a wholly anthropocentric theory of time as a mere measue of motion, according to which the schematic division into years, days, hours is a creation of the rational soul, without which there is no time. The rational soul, though unable to measure motion without the aid of time, stands outside it, is independent of it. See H. Blumenberg ([1975] 1987, pp. 485 f.).

45. H. Blumenberg (1976, pp. 63, 85 ff.). This is a diagram, and therefore inevitably a betrayal of Nicholas' conception that God is the enfolding, the enveloping, the *complicatio* of His creation, while the creation is in turn the unfolding, the developing, the *explicatio* of God. See E. Van-

steenberghe (1920, pp. 310 ff.) and G. Santinello (1972, pp. 43 ff.) on the relation between the two concepts of complication and explication to another key Cusan concept, contraction: the world as a contraction of the Creator. See P. M. Watts (1982, pp. 66 ff.), relying on *De Docta Ignorantia* II.3; cf. G. M. Heron (1954, p. 77).

46. Antoninus (1740, pt. 1, p. 57):

Anaxagoras [*sic!*] qui interrogatus quid esset homo, ait mensura omnium rerum.

This section "de quidditate homini et ejus excellentia" begins on p. 54. Antoninus quotes "Method. 10" as his source, presumably Methodius of Olympia. This passage is quoted and attributed correctly around the same time by L. B. Alberti (1972, vol. 1, p. 132; 1969, p. 134). Protagoras was certainly known at the time through a translation of Diogenes Laertius' *Lives of the Philosophers* made in the 1430s by Ambrogio Traversari, Prior of St. Marco. See C. Trinkaus in C. H. Clough (1976, pp. 190 ff., esp. 194 f.). St. Antoninus may have conflated Alberti's quotation of Protagoras with another one Alberti attributed to Anaxagoras: "God created man to contemplate the heavens and the works of God, which is confirmed by his erect posture" (1972, vol. 1, p. 132). That saying is also quoted in his *Della Pittura* (1950, p. 69 f.; see 1972, vol. 3, p. 34), which he quotes out of Lact. Firm. *De Falsa Sapientia* (*Inst.* III.9.ii). On another confusion, Pythagoras/Protagoras, sometimes made by both Alberti's and Nicholas of Cusa's scribes, see C. Trinkaus (1976, p. 199 n. 18).

47. P. M. Watts (1982, p. 179). Quoted almost exactly by Pico della Mirandola, *Heptaplus* VII.7.

48. P. M. Watts (1982, pp. 184 ff.). On Nicholas' attempt to reconcile Plato and Aristotle through his doctrine of the distinction between forms (deduced as in Plato) and species (induced intellectual constructs, as in Aristotle), see G. Santinello (1972, pp. 46 ff.).

49. The effort to refer the empirical to the ideal led him to deduce the imperfect circularity of planets and their orbits: Cusa, *De Doc. Ign.* II.12. The break in method is well described by E.

Cassirer (1935, pp. 47 ff.), who sees it as a "ful-fillment" of Socrates' words on the unreliability of astronomical observation in Pl. *Rep.* VII.529 f. Cf. G. Santinello (1972, pp. 50 ff.).

50. Nicholas was involved in the negotiations before and during the Council of Florence; he was made a cardinal *in pectore* by Eugenius IV, and although not known as one, was sufficiently a public figure to be voted for at the conclave that elected his friend Nicholas V (L. von Pastor, 1923–1940, vol. 2, p. 11). St. Antoninus was also voted for on that occasion. Alberti was a member of the papal court at the time; both men were close friends of the mathematician Paolo Toscanelli and of a humanist bishop, Giovanni Andrea de'Bussi. It would have been almost impossible for them *not* to have met. A fifteenth-century manuscript of Alberti's *Elementae Picturae* in the Hospice of St. Nicholas at Cues (cod. 112) was presumably part of the founder's library: see J. Rykwert and A. Engel (1994, pp. 425 ff.) and E. Hempel (1957, pp. 14 f). Alberti, Toscanelli, and Cusanus shared a common passion for map making: see G. Santinello (1972, p. 101). The so-called Map of Cusanus is reproduced by E. Meffert (1982, pp. 72 f).

The circumstantial evidence for Leonardo's dependence on Nicholas' teaching was set out by M. N. Duhem (1906–1909, vol. 2, pp. 99 ff., but esp. pp. 101 ff., 149 ff., 180 f.). It was dismissed as "congetture troppo fragili" by E. Garin ("Cusano e i Platonici Italiani del quattrocento" in G. Flores d'Arcais (1962, p. 78). Leonardo's probable familiarity with Nicholas' *De Transformationibus Geometricis* is however suggested by M. Kemp (1980, pp. 250 f.). Garin (pp. 80 ff.) also dismisses Vespasiano da Bisticci's suggestion that Nicholas was "dotto in Greco"; and yet his friend Lorenzo Valla (whom Garin considers one of those humanists who prided themselves on their Greek) also calls Nicholas "doctissimus vir et graecarum litterarum peritus" (*Collatio Novi Testamenti*, quoted in S. I. Camporeale, 1972, p. 360). Pius II, who was the third pope to be a friend of Nicholas, suggested in his *Commentaries* (ed. L. Totaro, 1984, p. 2498) that he was "suae gentis nimis amans"; nevertheless Nicholas was a great public figure, who preached in Italian as well as in Latin and German and had a great following as a preacher in Rome: see E. Vansteenberghe (1920, pp. 158 ff.) and J. W. O'Malley (1979, pp. 92 ff., 142). His teach-

ing would therefore have been familiar to many who had no knowledge of his writings. But see G. Santinello (1958, passim); E. Hempel (1957); and W. Tatarkiewicz (1967, vol. 3, pp. 76 ff.). Tatarkiewicz seems rather dismissive of Nicholas' influence on even the immediately succeeding generations in Italy: it is certainly true that the only artists he mentioned by name were Rogier van der Weyden and Hans Memling. Useful illustrations of his possessions and patronage are in E. Meffert (1982). Although most of it could be described as "International Gothic," his tomb in his cardinaltial church, San Pietro in Vincoli, is in fine late quattrocento style, done by Andrea Bregna of Milan, the leading Roman tomb sculptor.

51. The canonic preoccupation is implicit in the famous dispute between Brunelleschi and Donatello about the figure of the crucified Christ: the story of Donatello breaking the eggs in astonishment at his friend's masterpiece is told by Vasari twice, in his life of Brunelleschi (ed. Milanesi, 1878–1906, vol. 2, pp. 333 f.) and of Donatello (vol. 2, pp. 398 f.). Both figures represent a man who is three Florentine *braccie* (i.e., 5 ft. 10 in.) high, rather than the *mensura Christi*, which is 6 palestine feet (i.e., 5 ft. 11 in.); see O. Cassazza (1978, pp. 209 ff.). See however also L. Steinberg (1983, pp. 131 ff.).

52. The thirty-three parchment leaves (Paris, Bibliothèque Nationale MS fr. 19.093; perhaps less than half have survived) were first published—in lithographic redrawing with a commentary by the restorer-architect and collaborator Viollet-le-Duc—by J. A. B. Lassus in 1858. It was a posthumous publication. The manuscript had belonged to the Félibien family since the fifteenth century and was widely known. An English translation of Lassus' edition by Robert Willis (as "Wilars de Honecort") appeared in the next year in London. There have been several facsimiles published since, which are discussed by C. F. Barnes (1982); it is clear that there is no agreement among the various authorities either about the purpose of the book or about its sources.

The standard edition remains that by Hans Robert Hahnloser (1935, reprinted with supplementary material in 1972). Hahnloser established convincingly that the drawings and texts were by three different hands; a more accessible edition was prepared by Alain Erlande-Brandenburg (1968). The geometric schemas have been discussed by R. Bechmann (1991, pp. 224 ff., 305 ff.).

Hahnloser also published another pattern book, an early thirteenth-century, rather Byzantine-style album in the Library at Wolfenbüttel (1929). D. J. A. Ross (1962, pp. 119 ff.) considers that two parchment pages bound as a bifolium were just such a fragmentary pattern book.

For the Villard drawings of the human canon, see R. Hahnloser (1935, pp. xxxv, xxxvi, xxxvii, xxxviii). Villard himself calls these pages "mode de portraicture/materia picturaturae," on which see Hahnloser (1935, pp. 84 ff., 275 f.) and R. Bechmann (1991, pp. 305 ff.).

53. One such has recently come to light: a sculpture workshop notes on human proportion. See B. Bischoff (1984, pp. 233 ff.). It uses the Vitruvian "module" of the face, but though it begins by dividing the work into 10 units, it in fact allows the artist latitude between 8 and 10½ modules.

54. Folio 36 v., part 2, Cod. Lat. 197, Bayrische Staatsbibliothek, Munich. See Mariano Taccola's *De Ingeneis* (1984a), vol. 1, p. 67; vol. 2, p. 72). F. C. Prager and G. Scaglia (1972, pp. 46, 167 ff.) see this drawing, with its garnish of compass, plumb line, and square, as linked to medieval traditions rather than initiating the new Vitruvian speculations, though in fact the circle in which Taccola's man is inscribed is—Vitruvianly—eight heads in diameter, and these eight units are clearly marked as compass points in the drawing. Two of these units (a quarter of the figure) are its breadth; the ratio between hand and head is 8:10, i.e., the Vitruvian face dimension. The feet of the figure do not touch the edge of the circle, however, and the figure seems to have been drawn as seven of its own feet high. Of Taccola's other manuscripts, *Liber Tertius de Ingeneis* (Florence Palat. 766) was edited by James H. Beck (1969); *De Rebus Militaribus* (MS Paris, Bib. Nat. Lat. 7239—a "presentation" book) was edited by Eberhard Knobloch (1984b).

55. Francesco's drawing in Cod. Saluzz. 148, Torino, folio 6 v. (ca. 1480, probably before 1486). Francesco lent another manuscript of his treatise (now Ashburnham 361, in the Laurenziana, Florence) to Leonardo in Milan about 1490,

and it is annotated in Leonardo's hand, though the attribution of some of the drawings in it to Leonardo has not been upheld; see C. Maltese in F. di Giorgio Martini (1967, p. xxi n. 1, p. xxiv n. 3). The standing man has his center within the square at the crotch; the circle is circumscribed rather lackadaisically and does not correspond to the Vitruvian prescription.

56. Venice, Accademia no. 228, dated variously 1476 or some time between 1485 and 1490. The text on the drawing has been transcribed by J. P. Richter (1970, vol. 1, p. 255 f., no. 343), and glossed in C. Pedretti (1977, vol. 1, pp. 244 ff.). Richter had grouped the Accademia drawings as his numbers 308–388 (of which 315—Venice, Accademia 236—may be a portrait of Leonardo's father, Piero; so at any rate R. S. Stites et al., 1970, pp. 115 ff.). To these must be added the drawings of the skull at Windsor (19057 r., 19058 r. and v.; K. Clark and C. Pedretti, 1969, vol. 3, p. 24) in which Leonardo investigates the canon of the bone structure. Leonardo's enterprise involved the reconciling of empirical observation with the antique. In this he confessed that he had not succeeded: Platino (the Milanese poet Piattino Piatti), who claimed his friendship, wrote a Latin epitaph in which the self-deprecating artist is made to say, "Sum Florentinus Leonardus Vincia proles; Defuit una mihi symmetria prisca: peregi Quod potui; veniam da mihi posteritas" (quoted in G. Bossi, 1811, p. 14.; but see also C. Pedretti, 1977, vol. 2, pp. 9 f.). On the relation between Villard's, Leonardo's, and Dürer's canons, see V. Mortet (1910, pp. 367 ff.).

57. Though in fact, as drawn by him, the equilateral triangle (eight of his palms to a side) does not have its center at the umbilicus, but five inches (on Leonardo's scale) below the center of the circle. On this drawing, see G. Hersey (1976, pp. 97 ff.) and M. Kemp (1980, pp. 113 ff.). Mine are rather tentative observations on the drawing, working from an engraving. See however G. Favaro (1917, pp. 167 ff.; 1918, pp. 109 ff.), as well as E. Panofsky (1974, pp. 92 ff.).

58. Windsor 19057 r. and v., 1908 r.

59. For Leon Battista Alberti's approach to empirical measurement, see his *De Statua* 12:

"satis constare certum est, ab vivo etiam exemplari cum dimensiones tum etiam finitiones captari adnotarique posse percommode." In fact, the canonic dimensions Alberti provides are "perfect" or canonic in the Vitruvian way (1972, pp. 132 ff.), though on the procedure of compiling such a canon he alluded to the story of Zeuxis deriving the proportions of the perfect female body from the five most beautiful girls of Croton, which is told by Pliny (*NH* XXXV.xxxvi.64) and Cicero (*de Inv.* II.1.i–iii).

On Pomponio Gaurico, see above, chapter 2, p. 35 and n. 24.

60. The most complete evidence of Leonardo's concern with the business of proportion in movement is the presumed copy of Leonardo's manuscript on the subject, the Codex Huygens, now in the Pierpont Morgan Library, New York, and attributed by C. Pedretti (1977, vol. 1, pp. 69 f.), who also reproduces the text) to a Milanese sixteenth-century painter, Girolamo Figino. On the basis of convincing documentary evidence, it now seems given to another painter from the same circle, Carlo Urbino; see S. Marinelli (1981, pp. 214ff.). It was engraved as Leonardo's about 1720 by Edward Cooper, who had restored and mounted the drawings for Constantine Huygens, a Dutch scientist and collector (and brother of the more famous Christian); he had bought the drawings in 1690. These engravings are reproduced by Pedretti (vol. 1, p. 53). The codex was published by E. Panofsky (1940), who there attributed it tentatively to Aurelio Luini, or another Milanese painter (unrelated to Girolamo), Ambrogio Figino. Two sheets of drawings now in the collection of Christ Church, Oxford, are said by Pedretti (vol. 1, pp. 244 ff.) to have been part of the original codex, though to judge on the evidence of photographs, they would seem to be by a different hand. It seems permissible to speculate that John Wood of Bath may have known the Cooper engravings, which were reprinted by Edward Oakley, an architectural publicist, architect, and masonic functionary, in his *Magazine of Architecture, Perspective and Sculpture in Five Parts* (Westminster, 1730; reprinted 1732, 1736). This work was largely based—for the parts on perspective and the orders—on Abraham Bosse's book and includes a reengraved section of Gilles Audran's measurements of ancient statues.

61. A. Dürer (1528). It was actually published a few months after Dürer's death. The manuscript seems to have been ready in 1523, but (as W. L. Strauss argues, 1972, p. v n.) Dürer may have hesitated about publication until he had examined Alberti's book on the subject (perhaps the *De Statua,* though no printed text was yet available), which had come into his hand after he had finished his own text. That Dürer had worked on the problem is well known; when very young he was fundamentally influenced in this matter by Jacopo de' Barbari:

> Jdoch so ich keinen find, der do etwas beschriben hett van menschlicher mas zw machen, dan einen man, Jacobus genent, van Venedig geporn ein liblicher moler. Der wies mir man und weib, dÿ er aws der mas gemacht het . . . und wen ichs hett so wolt ich ims zw eren in trug pringen gemeinem nutz zw gut. Aber ich was zw derselben zeit noch jung und hett nie fan solchem ding gehört. . . . Dan mir wolt diser forgemelt Jacobus seinen grunt nit klerlich an zeigen das merckett ich woll an jm. Doch nam ich mein eygen ding für mÿch und las den Fitrufium, der beschreibt ein wenig van der glidmas eines mans.

So much himself in a draft of the dedication to Wilibald Pirckheimer of the body-proportion book, now in the British Museum. See W. M. Conway (1889, pp. 165, 253 f.); H. Rupprich (1956–1969, vol. 1, pp. 101 ff.; vol. 2, pp. 31 ff.). The early Apollo and Diana drawing (BM L 233) and the manifesto-like Adam-and-Eve engraving (B.1., 1504, for which there are a number of constructional drawings: for Eve, Vienna, Albertina W.421/2, and London, BM L 242; for Adam, Vienna, Albertina W 423/4, and Dresden, folios 110a, 135a) and the two slightly later panels (*The Fall of Man* of 1507 in the Prado) are the most obvious fruit of Jacopo's teaching; on whom see L. Servolini (1944, pp. 87 ff.). Dürer was convinced that Barbari had held something back, that there may even have been some drawings or writings known to Barbari. He knew, of course, that there had been canonic writings in antiquity, and he felt that the early church fathers had done wrong in destroying them as well as the statues instead of allowing an *interpretatio Christiana* (W. M. Conway, 1889, pp. 178, 203). Dürer stated his intention in the measuring exercise in the so-called aesthetic excursus of the 1528 book on proportion (folios T1–T4); see H. Rupprich (1956–1969, vol. 3, p. 290 ff., with preparatory drafts). Both Leonardo and Dürer took the musical analogy of their proportional researches for granted (Rupprich, pp. 298 f.; ed. J. P. Richter, 1970, vol. 1, pp. 76 f., 243). Dürer's precocious topology may be seen clearly in his transformation of the male head within a square, divided into seven vertical sections, while the horizontal ones follow the main articulations of the face. Dresden Sketchbook 94a (Rupprich, vol. 2, p. 476); of course, there is much discussion of this procedure in Dürer's other papers (Rupprich, ibid., pp. 397 ff).

Lomazzo's proportions are set out in his *Trattato della Pittura* (in 1974b, vol. 2, pp. 38 ff.). On motion, see pp. 238 f., 255 ff. On the relation of color to humors, see pp. 269 ff.; on gesture, see pp. 388 ff. Advice on optics recurs throughout the book.

62. G. P. Lomazzo (1974b, vol. 2, pp. 44 ff.): *Trattato dell'Arte della Pittura, Scoltura e Architettura*, book 1, chapters 6–19 for the human bodies, 20–22 for the horses, and 22–29 for the orders; there follow two chapters of general principles. For Lomazzo's sources, see the preface by R. P. Ciardi (vol. 1, pp. xxix ff.).

63. G. P. Lomazzo (1974b, vol. 2, pp. 288 f.).

64. A. Condivi (1976, pp. 99 f.). On Michelangelo's criticism of Dürer's anatomy, see D. Summers (1981, pp. 380 ff.).

65. On Colombo, a disciple but later a critic and enemy of Vesalius, see *DBI*, s.v. On his association with Michelangelo, see A. Hyatt Mayor (1984, pp. 68 ff.). He seems to have treated Michelangelo (for the stone and kidney complaints: letter to Giorgio Vasari, 22 May 1557 [see E. H. Ramsden, 1963]), though he in fact predeceased Michelangelo, in 1559. It has been suggested that one of the bearded figures (to the right of the cadaver) on the title page of Colombo's *De Re Anatomica* was intended to be a portrait of Michelangelo; so at any rate A. Carlino (1994, p. 61).

66. On this statue, see M. Weinberger (1967, pp. 185, 200 ff.). On the problems of its composition, see D. Summers (1981, pp. 85, 395 f.). Summers suggests that both the reed and the cross held by Christ are indications of the module. On the problem of the nudity and the Incarnate Christ, see L. Steinberg (1983, pp. 18 ff., 130 ff.).

67. On the relation between anatomy and architecture, see the much quoted, undated (1550 to Cardinal Marcello Cervini? 1560 to Cardinal Rodolfo Pio da Carpi?) letter in which, after a few Vitruvian phrases, Michelangelo sums up: "chi non e stato o non e buon maestro di figura e massimo di notomia non se ne puo intendere" (E. H. Ramsden, 1963, vol. 2, pp. 129, 290 ff.; Milanesi, no. 490). Vincenzo Danti and Francisco da Hollanda both bear witness to this obsession of Michelangelo's. See D. Summers (1979, pp. 9 f., 46 f.).

68. Letter to Cosimo I de'Medici of 17 April 1548. See A. Carlino (1988, pp. 33–50).

69. On Vesalius' complaint, see "Epistola Rationem . . . Radicis Chynae" in his *Opera Omnia Anatomica et Chyrurgica* (1725, vol. 2, p. 680); see also A. Hyatt Mayor (1984, p. 102).

70. On Titian's part in the Vesalian enterprise, see W. M. Ivins in S. W. Lambert et al. (1952, pp. 65 ff., 85 ff.) and C. D. O'Malley (1964, pp. 124 ff.).

71. D. Summers (1981, p. 432), quoting T. Koch (1972, pp. 65–80, esp. p. 72).

72. In the dedicatory letter to Pope Paul III. Copernicus (1976, p. 26) quotes Philolaos, Heraclides of Pontus, and Ecphantus as having held Pythagorean views out of Plut. *Plac. Phil.* III.13. For an account of Copernicus' intellectual background and the nature of his immediate achievement, see H. Blumenberg ([1975] 1987, pp. 206 ff. and passim).

C. Singer (1925, p. 122) links Copernicus and Vesalius; but see G. Canguilhem (1970, pp. 27 ff.).

73. See A. Picon (1988, pp. 75 ff., 151 ff.).

‸
‸

‸

‸

‸

‸

‸

‸
‸

‸

‸

‸

‸

424

74. There were very many editions of Dürer's book:

First German edition	30 October 1528
Second German edition	Arnheim, 1603
First Latin trans. (by J. Camerarius, in two parts)	Nürnberg, 1532, 1534
Second Latin trans.	Paris, 1535
Third Latin trans.	Paris, 1557
French trans.	Paris, 1557
Second edition	Arnheim, 1613
Italian trans.	Venice, 1591
Second edition	Venice, 1594
Spanish trans.	Madrid, 1599 (?)
Dutch trans.	Arnheim, 1622
Second edition	Arnheim, 1662

The Spanish translator, Luiz da Costa, may also have prepared a Portuguese translation, which has remained in manuscript. See W. M. Conway (1889, pp. 229 f.).

On Dürer's influence in France, particularly on Poussin and Bosse, see G. Kauffmann (1960, pp. 22 ff.). Inevitably, Dürer was used in evidence against Le Brun.

75. An account of the precedents for the Vitruvian canon is given by S. Ferri in his edition of Vitruvius (1960, pp. 94 ff.). But see chapter 4.

IV Gender and Column

1. See above, chapter 1.

2. F. Zoellner (1987) seems to have assembled a great mass of material to revise the view of the microcosmic man that was advanced first by Aby Warburg himself and developed by Rudolf Wittkower. However, his own collection of material suggests that one of the two hypotheses, which both scholars invoked—that the recurrence of the theme is due either to its ahistorical, perhaps even archetypical validity, or that we are dealing with a phenomenon of "hidden" and broken textual transmission—can account for it. Since he does not himself advance any other hypothesis to explain the phenomenon, the case rests with the older scholars.

On the various quotations from Vitruvius in medieval literature, see H. Koch (1951, passim).

3. Vitr. III.1.i f. With Ferri (1960) and Fensterbusch (1964), I follow Rose's textual correction: "manus pansa" for "manus palma." Other variants do not affect the translation. The insertion of "a medio pectore" ("from the middle of the breast" before "to the crown of the head") made by Bernardo Galiani in 1758 was accepted by most editors (Rose, 1899; Granger, 1931–1934, 1955–1956; Ferri; and Fensterbusch—though not Stratico/Poleni, 1830, or Marini, 1836).

A great deal of trouble results from the minor inconsistencies of Vitruvius' canonic proportions: if the height of the full head is $1\frac{1}{4}$ (the face being $\frac{1}{10}$ of the body, the head $\frac{1}{8}$), then the fathom, of eight full heads, is equal to six (Polykleitan seven?) feet; nevertheless it is also equal to ten faces, and the face is equal to the hand; the discrepancy seems to be of $\frac{1}{16}$ of a fathom, and this has occupied many of the commentators. The fathom and not any of its parts is the basic unit of measure, as in the temple the length of the stylobate is subdivided to find the module; *symmetria* seems primarily a method of division.

At the beginning of the quotation—"non potest aedis ulla sine symmetria atque proportione rationem habere compositionis"—I have simply transliterated the words *symmetria* and *proportio.* Earlier in the same passage Vitruvius defines *symmetria* as the result or effect of *proportio,* a word he has taken as an accurate translation of the Greek *analogia.* Cicero claims (*De Univ.:* "Graece analogia, Latine—audendum est enim quoniam haec primum a nobis nouantur, comparatio proportioue dici potest") to have coined the term in translating passages from the *Timaeus,* but in fact Varro (who, though Cicero's senior by ten years, dedicated the book to him) used the word currently and defined it etymologically, without reference to Plato in *L. L.* X.35. Euclid defines the word much more closely (V.4; V.8, 9; VII.20, 21, on *analogous* numbers) to mean a four-term series of dimensions or numbers ($a:b = c:d$, reducible to three terms as $a:b = b:c$; Speusippos distinguished between the geometrical relations of *analogic* numbers and the arithmetical relation of *anacolouthic* ones, though this distinction is not generally adopted). On this matter see P.-H. Michel (1948, pp. 141 ff.).

On the fortune of *pro portio, pro ratio,* see S. Ferri (1960, pp. 93 f.). Euclid however is very strict in defining the word as applied to mean proportionals, particularly in the context of geometrical proportion.

In fact, although Pliny regrets that "non habet Latinum nomen symmetria" (*NH* XXXIV.xix.65), Vitruvius seems to have coined the adjective *commensus* and used it often (I.3.ii, III.1.iii; cf. III.2.ix, VI.1.vii) though frequency of use suggests that he preferred the Greek word. The verb *commensurare* and its derived noun, *commensuratio*, and adjective, *commensurabilis*, do not appear until Boethius' *Topic. Aristot. Interp.* III.1 and *De Mus.* I.31, or Aurel. Aug. *Enarr. in Ps.* See V. Mortet (1909, pp. 46 ff.).

4. Vitr. III.1.v ff. *Uncia, oungkia*, is said to be a word of Sicul origin. There is a curious gloss on this passage in St. Augustine's *Civ. Dei* (XV.26; on this passage and its quotation by Eugippius, see above, chapter 3, n. 24), though St. Augustine maintains, without giving any source for this information, that the human body is six times its width, and ten times its thickness: this he relates to the dimensions of the ark of Noah. This assertion may of course be inferred from the scriptural text, rather than derived from any artistic canon. It does correspond to the breadth of the Vitruvian man of Leonardo, taken at its narrowest.

A long and repetitive passage at this point has been removed from Vitruvius by some editors (notably Rose, 1899). The numerology is swollen out by many references to both Greek and Latin coinage, which confirm but do not alter that numerology, claiming that the coinage is also derived from the body. Aristoxenus of Tarentum, a quarrelsome and embittered pupil of Aristotle (his major disappointment was that he did not succeed his master as head of the Athenian Lykeion), maintained that Pythagoras came to be interested in mathematics because of their practical commercial application: see F. Wehrli (1944–1959, vol. 1, ii, fr. 23, p. 14, 49 ff.). There is a quasi-legendary connection with the coinage of Croton, and all this is reflected in what may be called a vulgar Marxist view of his role there; see G. Thomson ([1941] 1946, pp. 212 ff.). A quite different Marxist interpretation of Pythagoras, which links him with aristocratic reaction and considers the mystical founder of the brotherhood as quite unconnected with the scientific activities of his later followers, is in S. Luria (1963, pp. 34 ff.). It is difficult to give fixed rules for the dimensions and their subdivisions. The fathom, for instance, which is usually counted as six feet, is given as ten feet by Pliny, so the divisions

and subdivisions are conventional rather than normative.

5. Vitr. III.1.iii. The human body in this passage becomes the resolution of an archetypal geometrical problem, that of the squaring of the circle. This is suggested again in the treatment of the theater (V.6.i., V.7.1). Inevitably, it attracted some later commentators and is one of the themes of Leonardo's drawing discussed above (see chapter 3, n. 56). On this see H. K. Lücke, (1991, pp. 61–84).

There is, however, contradictory evidence about such beliefs; "Hippocrates" seems again to countenance them (*Peri ebdomadon*, ed. H. Roger and E. Littré, 1932–1934, vol. 8, p. 666: "definitio autem superiora partum, et inferiora corporum umbilicus"), yet Varro, in his etymology of *umbilicus*, comments on a verse fragment (attributed to Ennius by one editor):

O Sancte Apollo, qui
umbilicum certum terrarum optines
Unde superstitiosa
primum saeva evasit vox fera[.]

Cicero completes the fragment in *De Div.* II.115, though he has "obsides" for "optines": "Holy Apollo, who rules the immovable world's center, from which the harsh voice comes forth proclaiming your oracles." See Varro *L. L.* VII.17 (ed. R. G. Kent, 1938, vol. 1, p. 285), who denies explicitly that the umbilicus is the center of the body any more than the Delphic omphalos is the center of the world; he refers the doctrine to a Pythagorean belief in an *antichthon*, an anti-earth, which turns round the fire at the world center counter to the movement of the earthly sphere (on which see W. Burkert, 1972a, pp. 231 f., 347 ff.). A line drawn through the Delphic umbilicus is no more the axis of heaven and earth ("ut media coeli et terrae linea ducatur infra umbilicum") than a horizontal line drawn through the navel delimits the distinction between male and female or the place of generation.

The Vitruvian passage is invoked by Diego de Sagredo (1526 pp. a v., r.) to justify his canon of 9 ⅓ heads. On the dependence of this canon (of the *modernos autenticos*) on medieval workshop practice, see R. Klein in P. Gauricus ([1504] 1969, pp. 76 ff.). However, see also R. Hahnloser (1935, pp. 59 f., 86 ff.). One of the inconsistencies that puzzled later commentators is the indication by Vitruvius of the umbilicus as the body's center, while in the man-in-the-square it is clearly the base of the penis. This was already noticed by Lorenzo Ghiberti in his *Commentarii* III.45 (1947, p. 214), and in a sense resolved in the Leonardo drawing. It is discussed by all the Vitruvius commentators. Most remarkable is the solution of Cesare Cesariano, who makes the generative member one of his square units—which also equals half the face and half the distance between the base of the pubis and the umbilicus (¹⁄₃₀ of the large square, 21¹⁄₃ of the small one), thus reconciling the two centers. The drawing attributed to Pietro Paolo Sagazone (Vitruvius, 1521, fol. 49 v.), nobleman of Como, has other peculiarities, such as feet 5¹⁄₃ of the figure's height. This problem had occurred among medieval draftsmen in another form. R. Klein and A. Chastel point out (in Gauricus [1504] 1969, p. 78 n. 9) that Villard d'Honnecourt had drawn his stationary figure with the center at the crotch, while a walking one has it at the umbilicus.

6. As in the particular cases of Alberti, Leonardo, and Dürer, discussed in chapter 3.

7. In the Ashmolean Museum, Oxford (Michelis, 83). See most recently E. Fernie (1981, pp. 254 ff.), and also H. Ben-Menahem and N. S. Hecht (1985, pp. 139 ff.). In the Oxford plaque, however, the relation of the fathom to the visible footprint seems to be 1:7 (Vitruvius, 1:6); to the whole head the Lisyppeian (?) 1:8, the trunk at the armpits 1:4 (both of these agree with Vitruvius); and to the face 1:11 (Vitruvius 1:10). On the other hand, the orthogonal height of the slab, 26.15 cm, is in a ratio to the fathom of 1:8, while it cuts the side of the regular octagon (of which the pedimental triangle is a part) at 3:4; this may, of course, represent some as yet unidentified canon. There is a much more damaged and less known metrical slab recently found as part of the wall of the Chapel of St. Demetrios at Salamis (I. Dekolakou-Sideris in *AJA* 94, 1990 pp. 445 ff.), but with comparable indications: the recent suggestion—that the fathom of the Oxford slab is intended to measure 100 *dactyloi,* fingers, and that its peculiarities owe something to this unusual dimension, whereas a six-foot fathom of course measures 96 *dactyloi*—has recently been suggested: E. Berger, B. Müller-Huber, and L. Thommen (1992, chapter 4); but a comparative study of the two slabs remains to be done.

8. The explicit references to such a book are late: Plutarch (*On Virtue, On Hearing, On the Chicken and the Egg*), Philo of Byzantium; Galen (*On the Views of Plato and Hippocrates, On Temperaments*), and Lucian (*On Dancing, On the Death of Peregrinus,* where it is proverbially used: "this canon of Polykleitos!" meaning "you think he is perfect, don't you?"). There can be little doubt that such a book existed and that it was attributed to him, whoever may have been its real author. It was one of several such writings by artists (Pheidias, the best-known of them, seems to have left nothing in writing, though he was known to have an interest in such matters as optical correction), some of whom were certainly literate—since the sixth century, at any rate, which, after all, is only two centuries after the devising of the Greek alphabet. On this see H. Philipp, "Zu Polyklets Schrift 'Kanon,'" in *Polyklet* 1990, pp. 135 ff.; 1968; pp. 42 ff.).

However, Vitruvius' is the earliest surviving text to record any numbers; when compared with the measurements of the Doryphoros, they show some agreements and some discrepancies. So for instance, the foot is one-sixth and the face one-tenth of the whole height, the face is square (within 2 mm), and the three parts are equal to within 3 mm of the average.

9. *Kanōn* meant something straight, something to measure with or by; it was spelled earlier with omega and later with omicron. It also meant the tongue or beam of a balance, a loom rod, or the crossbar of a lyre and had many other subsidiary meanings—hence also a set of rules, a standard. On the relation of the measuring rod to a reed and the possible Semitic origin of the word from *kaneh,* "a reed," "measuring rod," or "beam of a balance" (as in Isa. 46.6), see W. Burkert (1992, pp. 38, 176 n. 21). The word *metron* as a near synonym is already familiar to Homer (*Il.* XII.422), who describes the closely fought battle before the wall of the Greek camp as being like a quarrel between two peasants about a common field. But in the next book *kanones* are the rods that keep Idomeneus' Cretan figure-of-eight shield rigid (*Il.* XIII.407).

In later logic, particularly Epicurean logic, the word came to mean a test of reality, truth, or both; see Diog. Laert. X.29 ff. and E. Zeller (1931, pp. 232 ff.). Cf. Cic. *Acad.* II.30 and n. 17, below; but see also E. Boisacq (1938), s.v.

10. Etymologists offer no general explanation of Greek words like *phythongos, klangos,* which, however, they relate to the Lithuanian *sphengti,* Northern Slavonic *zviek, dzwiek,* all meaning noise or sound. According to the *Euclidian Section of the Canon* (A. Barbera, 1991, p. 115), a *phthongos* is always the result of a percussion, *pligis.*

11. In the bog of Pythagorean speculation, in which many better scholars than I have lost their way, I propose to take Walter Burkert (1972b) as guide.

The accidental discovery of harmony was first described by Nichomachus of Gerasa (*Encheiridon* 6), probably from an older—but unrecorded—source. It was taken up in a number of later texts: Iamb. *V. Pyth.* 115; Macr. in *S.S.* II.1; Cens. X, XI. The details of the legend have been analysed by Giovanni Comotti, "Pitagora, Ippaso, Laso e il Metodo Sperimentale" in R. W. Wallace and B. MacLachlan (1991, pp. 20 ff.). The study relates how, after Pythagoras had heard harmonious strokes of the hammers, he noted that the hammers were of different sizes; he then attached corresponding weights to thongs of twisted sheep or ox gut and obtained the same harmonies. He further found that the same arithmetical relations persisted in the heights of columns of air (as in pan pipes) and quantities of water in vessels—which could be struck to produce the notes, and so on. In fact the number of harmonics was increased from four to six, which, being the sum of its factors $(1 + 2 + 3 = 1 \times 2 \times 3)$, is a number as "perfect" as ten, if not more so. These consonances that, as Vitruvius put it, the voice can modulate had both Greek and Latin names: *epitritos, diatessaron,* a fourth; *hemiolios, hemiolius* or *diapente,* a fifth; *duplaris, diapason,* an octave; further there was *triplaris, diapason kai diapente,* ⅓; *quadruplus, disdiapason,* ¼ or double octave. These six intervals were called *symphoniae* as Vitruvius (V.iv.7) rightly says, though he does not give the full technical terminology, according to which the octave was also called *diplasios;* the octave plus a fifth, *triplasios;* and the double octave, *tetraplasios.* The tone was *tonus* or *epogdoos.* These exhaust the possible or at least permissible relations of the *tetraktus.*

There was no room on this scale for thirds and sixths. Even in the most explicit statement of elementary Pythagorean music theory, the so-called *Sectio Canonis,* the production of intervals by cross-multiplying simple ratios is not entirely consequent; 8:3, an octave and a fourth, for instance, is not mentioned. But see A. Barker (1981, pp. 1 ff., esp. pp. 4, 9). Although that interval was generally recognized as a *symphonia,* and indeed sounded like one, it was considered illicit because it could not be got out of the *tetraktus.* In fact Ptolemy (*Harm.* I.6 f.) is the only theorist who attempted a solution to this problem, on which see A. Barker (1984, vol. 2 pp. 284 ff.) and A. Barker, "Reason and Perception in Ptolemy's *Harmonics,*" in R. W. Wallace and B. MacLachlan (1991, pp. 104 ff.). Aristoxenus is invoked by Vitruvius to introduce his account of musical theory in V.4, in connection with the acoustics of the theater. As for the harmonic series, it is one in which the reciprocals of the terms are in arithmetic proportion; the simplest instance is 1, ½, ⅓, ¼, and so on. The rule was formulated by the Pythagorean philosopher and friend of Plato, Archytas of Tarentum (DK 35 A2), though it did not have any of the musical implications it has for a modern reader.

12. The belief that the number ten "was something perfect, and contained within itself the whole nature of number" as Aristotle reported (*Met.* I.5.ii ff., 985b), seems to have been one of the fundamental tenets of the Pythagoreans. Indeed when they swore by their founder Pythagoras, whose name they did not utter, they simply referred to him as the one "who handed to us the *tetraktus* [the holy fourfold], source and root of everlasting nature" (*Ver. Aur.* 47 f.; see Iamb. *V. Pyth.* 18, where the *tetraktus* is identified with the Delphic oracle). Other "sympathies" are $6 + 10 = 4^2$; and $10^2 = 1^3 + 2^3 + 3^3 + 4^3 = 100$. See J. Lomas (1974, pp. 74 f.). Much of this reasoning is really only comprehensible in the figural terms in which the Greeks carried out much of their arithmetical speculation.

The terminology of the Cartesian (three-dimensional) universe is an elementary barrier to our understanding the appeal of this idea, since the Pythagorean world was not three dimensioned, but four pointed—1, a point; 2, a line; 3, a triangle; 4, a tetrahedron—even if these notions underlie the fundamental definition of point, line, and so forth at the opening of Euclid's book I. On the *tetraktus* and harmony, see Iamb. *V. Pyth.* 150; *Ver. Aur.* 47 f. Some commentators have even suggested that the immortal soul of Pythagorean theology is the numerical harmony of the *tetraktus:* as Sext. Emp.

Adv. Math. IV.3 ff. See A. Delatte (1915, pp. 249 ff.); W. K. C. Guthrie (1977, vol. 1, pp. 224 f.); W. Burkert (1972b, pp. 72 f., 186 ff.). All this inevitably recalled the four-element world of the Stoics, on which see chapter 3, n. 4, above.

On the perfect number in the Euclidian sense, see book IX, theor. 34, prop. 36. The Euclidian perfect numbers only occur when factors equal their sum: 1, 6, 28, 496, 8128, and so on. On this passage see also P. H. Scholfield (1958, pp. 29 ff.), and for an elementary treatment of the arithmetical issues, L. Hogben (1936, pp. 190 ff.). On the *teleion* in the theory of art, see J. J. Pollitt (1974, pp. 262 f.). My surmise is that Vitruvius is referring to some "pebble" figure rather than to a strict number theory. He does not mention another class of "friendly" numbers: those whose integers multiply to make the other number, as 10 and 24 ($1 + 2 + 3 + 4$; $1 \times 2 \times 3 \times 4$). The importance of such a relationship is that it reconciles a decimal and a duodecimal system of reckoning.

The *interpretatio Christiana* of this reasoning was provided by St. Augustine when he commented on the whole of the universe being created in six days: see *In Gen.* IV.2, "De Senarii Numero perfectione"; *PL* 34, pp. 190 ff.

13. The chord instruments were tuned according to one of seven "modes" or *harmoniai*. Their nature is discussed most accessibly in A. Barker (1984, vol. 1, pp. 163 ff.) and M. L. West (1992, pp. 177 ff., 233 ff.) Plato sets out his ideas on them in *Rep.* III.399a and more ambiguously in *Leg.* II.664b. Aristotle is altogether more latitudinarian on the matter—as may be expected (*Pol.* VIII.3, 1339b)—and seems to owe much of what he says on the subject to the enigmatic Sophist, Damon, an adviser and friend of Pericles (Plut. *V. Per.* 4, 153), on whom see A. Barker (1984, vol. 1, pp. 168 ff.) and P. A. Stadter (1989, pp. 68 ff.). On Damon's possible relation to the Pythagoreans of his time, see Diod. Sic. X.4.iii. He may well have been the author of Plato's moralizing interpretation of the modes. On their origin and relation to oriental tunings, see C. Sachs (1943, pp. 211 ff.). Plato's relatively tolerant attitude to the Phrygian mode is considered by Barker to be related to the sobering change in the nature of Dionysian worship at the time.

Later theorists divided the modes into the autochthonous (Hypodorian and Dorian: solemn, severe, proud, solid, joyous) and the Asiatic or barbarian (Phrygian and Lydian: Bacchic, *pathetikon,* "enthusiastic"). On these epithets and their implications, see J. Combarieu (1953, vol. 1, pp. 84 ff.). Tuning of instruments was referred to the Dorian mode, if tradition is to be believed; H. Helmholtz (1885, pp. 262 ff.).

14. W. Burkert (1972b, p. 371).

15. E. Frank (1923, p. 11; on which see, however, J. Lomas (1974, p. 75), W. Burkert (1972b, p. 373). But Burkert goes on to emphasize that it is the recognition of a direct relation between a theory of proportion and the quantifiable, vibrational nature of sound that makes the "Pythagorean" discovery so original. For all that, tuning and even the building of instruments went on being guided by the ear and approximation, as Burkert recognizes. And indeed, the warning of philosophers against attempting any great virtuosity—such as Aristotle at the end of the *Politics*—has something of the suspicion of a theorist for the improviser.

16. Numerical speculations are a commonplace of Greek musical writings. The earliest of them to survive may be the passage, perhaps quoted via Aristotle, in the treatise *On Music* attributed to Plutarch (1131a ff.). It is reproduced and commented on by A. Barker (1984, vol. 2, pp. 230 ff.), though such teaching was almost certainly stated explicitly by Philolaos of Croton (DK 44 B 6), who is presumably reporting the earlier teaching of his group, though writing in the late fifth century. Much of it revolves around the basic numerical statement of the octave as two fourths (4:3) separated by a tone (8:9), which can be stated as four numbers (another tetrachord), 12:9:8:6.

On the conservative nature of this "musical" thought, see H.-I. Marrou (1949, pp. 204 ff.). Here Vitruvius is at any rate typical in discussing numerical harmony in Pythagorean terms, while later in the book, when he reviews music in the theatre (V.3.vi ff.) he would do so in neo-Aristotelian and Aristoxenian terms: see F. Wehrli (1944–1959, vol. 1, p. ii).

The physical objection to the Pythagorean teaching on harmony—that the numerical equivalents of the musical scales do not work out, that the octave is smaller than the sixtuple tone—is irredeemable. This "comma" may arise in the tuning of the monochord; it is described both by Ptolemy and in the *Sectio Canonis*. This is also considered by A. Barker (1984, vol. 2, pp. 199 f.). But cf. H. Helmholtz (1885, p. 228 f.). On the concept of the comma, the "leftover," see also J. Lohmann (1970, pp. 67 ff., 113). So, for instance, twelve fifths do not (as they numerically should) work out to seven octaves: the difference, known as the Pythagorean comma, 531441:524288, was called *diabolus in musica* during the latter Middle Ages (S. Sadie, 1980, s.v. "Tritone interval"). The simpler comma was known as the comma of Didymus (after the grammarian who set out the formula of the difference between the whole major and minor tones about 20 BC, or perhaps after a more musical Didymus—known from Plutarch and Ptolemy—who may have lived at the time of Nero), though perhaps his assertion of a sensory appreciation of musical intention against the speculations of the "mathematicians" or "physicists" was even more important: see A. Barker (1978, pp. 9 ff.). Aristoxenus is the first "musician" mentioned by Sextus Empiricus (*Adv. Math.* VI.1 ff.) in his much more aggressive statement of the "aesthetic" position.

On the conflict between the two harmonies and their rival resolutions in late Pythagorean thinking, see F. R. Levin (1975, pp. 42 ff.). On the common ground of Aristoxenus and the Pythagoreans, see A. Barker (1981, pp. 9, 16), who also gives the basic differences as being on two points: Pythagoreans (1) considered relations of length equivalent to musical intervals and to numerical ratios and (2) gave reason (numerical calculation) priority over sensation; Aristoxenus and his disciples (1) considered intervals as distances on a linear continuum and (2) considered perception as the final arbiter of harmony—which last point, as Barker says, is hard to pin down (1981, p. 3).

17. Isidore of Seville was the first Christian writer to speculate on this parallel. On these speculations, see J. W. McKinnon (1978, pp. 1 ff.). Of the nine medieval theorists that Abbot Martin Gerbert of St. Blasien collected ([1794] 1931), all mention Pythagoras, while only three (Isidore, Aurelianus Reomensis, and Johannes Cottonius or Afflighemensis) mention Jubal or Tubal.

18. The first to assert explicitly (and experimentally!) that the business with the hammers and anvils could not have been true was the Parisian seventeenth-century Minim Friar, Marin Mersenne, the friend and agent of Descartes, who

devoted much of his working energy to musical theory (1634, pp. 166 ff.); see also W. Burkert (1972a, p. 375 n. 23). He was, as might be expected, too radical. Some variants of the story relate how Pythagoras then worked with strings on which the harmonically related weights were suspended. Had the story told that these weights were related not directly but according to their roots, the results would have been correct. The mistake had been suspected by Ptolemy (*Harm.* I.8), as has been pointed out by A. Barker (1984, vol. 2, p. 291)—though he did not provide the answer. I owe all this information to Professor Barker.

There was, of course, music magic and smithing much nearer home. The Idaean Dactyls (on whom see H. Jeanmaire, 1939, pp. 438 ff.), who were said to have come to Crete from Phrygia with Minos, were smith-dancers and musicians, and Orpheus was reputed their disciple or initiate. See W. Burkert (1972a, pp. 171 n. 34, 376 f.).

In fact Pythagoras' oriental precedents or heritage seem to have been much richer than is usually acknowledged. Plutarch, in his commentary on Plato's *Timaeus* ("De Anima Procreatione" in *Opera*, vol. 2, p. 1028d), has the Chaldeans equate the four Pythagorean harmonies with the four seasons; his ancient biographers quote older writers as claiming that Pythagoras was either Syrian or Persian, and that he wandered long in Syria and Egypt before settling on Samos. Indeed there are Greek traditions, wholly mistaken ones as it happens, that while the nobler string instruments were autochthonous, the wilder wind ones were oriental. See H. G. Farmer, "Music of Ancient Mesopotamia," in E. Wellesz (1957, pp. 250 ff.). On the background, see M. L. West (1971, pp. 228 ff.).

19. Vitr. III.1.v–viii. See W. Burkert (1972b, p. 290 n. 64). *Antiqui, palaioi*—"the ancients"—is a common reference to authority and does not usually mean a specific group, whereas *mathematici* is a common way of referring to the Pythagoreans.

The term *teleion* was also important as a critical as well as a numerical term, meaning "perfectly finished," "flawless." J. J. Pollitt (1974, pp. 262 f., 302 f., 419 f.) sees it as the equivalent of the Latin *absolutus, consummatus, perfectus.*

20. Pl. *Rep.* VII.531b ff. On the monochord (or *pandoura*), see A. Barker (1984, vol. 1, pp. 268 ff.). See, however, C.-E. Ruelle (1897, pp. 309 ff.).

21. Pl. *Rep.* VII.530e, 531a. See A. Bélis (1986, pp. 95 f.). Even monochords could be used for this, since—in spite of their name—they often had two, or even three chords, one or two scaled and the other unscaled, for comparison.

22. Very roughly the "Pythagoreans" against the "Aristoxenians"—or to be less anachronistic, the "Canonists" against "Musicians," as Ptolemy has it; but see n. 11, above.

23. See below, chapter 8.

24. See D. P. Walker (1979, passim).

25. The *Thesaurus Linguae Latinae* lists nine separate meanings. The word is sometimes derived from the Babylonian-Assyrian *qanu,* "pipe," or the Western Semitic *k'n'h,* "reed," from which we also get our "canon"—though the Spanish *canon,* "psaltery," is the transliteration of the Arabic *quanun,* which may be a simple homonym. In Hebrew it acquired overtones similar to the Greek: measuring rod (Ezek. 40.3 ff.) and balance beam (Isa. 46.6) or even a measure, as of grain. The word appears in the New Testament both for a rule (as in Gal. 6, 16 and to signify the measurable against the immeasurable: 2 Cor. 10, 13 ff.

The simple early distinction of Herodotus and Thucydides was elaborated in the Alexandrian library into a list of four heroic, three iambic, four elegiac, and nine lyric poets; nine historians; and ten orators: there was also an alternative Pergamene list. These were, of course, the *classics.* On the background, see R. R. Bolgar (1963, pp. 19 ff.) and A. E. Douglas (1956, pp. 30 ff.). See particularly Cicero (*Brut.* 36, *Or.* 1 ff.) and Quintilian (I.4.iii, X.1.liv), who were well aware of such a process of redaction.

26. Another measure is the distance from the second to the third joint, Sanskrit *angula,* though this is also the width of a finger and therefore is sometimes calculated as a module of the statue: a fourth part of the fist. In other documents, it is the top phalange of the thumb. The measurements of the donor of the building to be used as a module are taken in some documents to be his

whole height with his hands raised (S. Kramrisch, 1946, vol. 1, pp. 25 f, 134, esp. n. 16.; F. Staal 1983, vol. 1, pp. 195 ff.). In fact the regulations about the proportions of statues are much more explicit and unambiguous than those for the proportioning of buildings. Some have inevitably claimed that the canonic Indian system is a product of post-Alexandrian Greek influence. It seems to me, however, rooted in the conception of the plan as mandala, a conception that has no Mediterranean, let alone Greek, precedent or equivalent. The Buddhist iconic proportions were also adopted in China.

Some of these texts from Sinhalese sources are given by A. K. Coomaraswamy (1956, 124 ff., 150 ff.), who is very informative about the use of a stringed frame as a device for proportioning figures. For the use of such a frame in the Near East, see Isa. 45.13. For the general references, see B. Rowland (1967, pp. 164 ff.).

27. Aul. Gel. I.1. The lost book by Plutarch about Herakles supposedly dealt (among other things) with Pythagorean musing on human stature that is not mentioned elsewhere. Pausanias reports that Herakles was four cubits and one foot high: nine feet, rather than the normal six. *Stadium* was of course also the Latinized form of the Greek measure, *stadion* (100 orgyai or 600 feet), a furlong or 125 paces, a pace being five feet. As it happens, the stadium at Olympia was 192.3 m long (1 foot = 0.32 m), though the hippodrome was probably more like 300 meters, the measure that would correspond to Aulus' 600 Herculean feet. The stadium at Delphi was nevertheless only just shorter: 177.5 m (1 foot = 0.296 m). But see W. Dörpfeld (1935, vol. 1, pp. 40 f.). Aristophanes manages to caricature both the notion of the Heraklean foot and perhaps also the idea of man as the measure of all things by suggesting that the flea's jump can only be measured by a flea's foot: *Nub.* 144 ff., 830; see the allusion in Xen. *Symp.* VI.8.

On the *interpretatio Christiana* of this passage, see above, chapter 3, n. 35.

28. Pind. *Ol.* X.44 ff. The word for "measured out," *stathmato,* suggests measuring out by a standard.

29. Ageladas is said to have worked almost exclusively in bronze, ca. 520 BC and after, as did Polykleitos, whose dates are not securely known (though it is assumed that he worked 450–410 BC). Myron was the third of Ageladas' famous disciples.

The principal texts by and about Polykleitos were collected first by J. A. Overbeck (1868, pp. 929 ff.), and in DK 28 A and B, vol. 1, pp. 391 ff., which also establishes the quotations from the canon. These were critically collected by R. Bianchi Bandinelli (1938). The texts have recently been studied by J. J. Pollitt (1974, pp. 14 ff.), while the copies of the sculptures have been surveyed, with a view to establishing the dimensional and numerical canon, by H. von Steuben (1973). The impossibility of such an enterprise, in view of the variations between copies and the general unreliability of copyists, was asserted by T. Lorenz (1972), who also lists further variations and copies. Since these two books on the sculptures were published, some new fragmentary casts—perhaps of the original statues—were found at Baiae. See W.-H. Schuchhardt (1974, pp. 631 ff.) and A. F. Stewart (1978, pp. 122 ff.). The main sources for his life are Xen. *Mem.* III.10 and Pln. *NH* XXXIV.xix.55 (cf. Ael. *Nat. An.* XIV.8). But see P. F. Arias (1964): pp. 41 ff. for the texts, p. 52 for the two signed sculpture bases that were first published by E. Loewy (1885, pp. 42 f., 71). Pliny (*NH* XXXIV.xix.55) says that he came from Sikyon, 35 miles north of Argos. Some confusion is caused by the existence of two other sculptors called Polykleitos, also of Argos, who seem in fact to have been related to the great Polykleitos: see PW, s.v.

The passage about the *sophia* of Pheidias and Polykleitos occurs in Arist. *Eth. Nic.* VI.7 (1141a19). J. J. Pollitt (1974, pp. 92 ff.) has suggested (in my view rightly) that the *sophia* of which Aristotle speaks here conforms with his description of *sophia* elsewhere in *Met.* (I.1.xvii ff., 982a) and as a combination of *nous* and *epistemē*. Therefore, he is clearly alluding to their importance as theorists, as intellectuals, and not just to their manual skills, even if *sophia* in earlier texts was sometimes used to signify manual dexterity only, and thus, as Pollitt has pointed out, cannot be regarded as meaning "theory" or "doctrine" exclusively. On *sophia* and skill, see also W. Tatarkiewicz (1963, pp. 231 ff.), commenting on Phil. *Icon.* I.1.iii. Much Hellenistic speculation around the notion of *sophia* as the knowledge of things human and Divine on the one hand, and as the *technē peri ton bion*—"the art of life"—is discussed in M. Isnardi

Parente (1966, pp. 337 ff.), since of course *sophia* is the object of the philosopher's love: the formulation of the word, of the very notion even, is attributed to Pythagoras by Heraclides Ponticus, a pupil of Plato. On the more probable Platonic origin of the word, however, see W. Burkert (1960, pp. 59 ff.; 1972a, p. 65 f.). About the origin of *sophia* in its craft context, see B. Snell (1973, pp. 178 f.).

The names of Pheidias and Polykleitos as the two wise sculptors continued to be known through the Middle Ages: see Aq. *In Eth. Nic.* VI, lectio V, 1180, or Dante *Purgatorio* X.29 ff. On the ideological background, see also L. Stefanini (1944, pp. 84 ff.).

30. On Polykleitos as a sculptor, see most recently P. E. Arias (1964, pp. 15 ff.); M. Robertson (1975, vol. 1, pp. 328 ff.); and B. S. Ridgway (1981, pp. 201 ff., 244 f.), though J. Overbeck (1881, pp. 385 ff.) still provides an interesting survey of some questions that are by no means closed. A marble copy, said to be finer than the Naples statue, is at the time of writing on offer to an unnamed U.S. Museum.

The reliability of the copies was first discussed by C. Anti (1931). See G. M. A. Richter (1962, pp. 52 ff.) and M. Bieber (1977, pp. 6 f.). On Plato and the canon, see P.-M. Schuhl (1933, pp. 51 f.). The fidelity in the copying of curling hair is particularly striking (as H. von Steuben, 1973, has insisted); it was usually chased on the cold bronze. But see also S. Casson (1933, pp. 164 ff.).

31. Chapter 2, p. 33 f. with n. 17. The list of all surviving copies has been compiled by D. Kreikenbom (1990).

32. First so described by A. de Ridder and W. Deonna (1924, pp. 222 ff.). Its application to the Doryphoros by B. S. Ridgway (1981, pp. 202 ff.), does not seem to throw any light on the problem. She has relied on the solution of the canonic measurements that invokes a knotted rope, with the joints related as $1:\sqrt{2}$, which does not, prima facie, seem at all plausible: but see R. Tobin (1975, pp. 307 ff.). While the canon may well have been "operated through" the sort of knotted rope that Tobin proposes, his $\sqrt{2}$ dimensions seem to fit neither the living body nor the statues he examines at all comfortably. A more plausible suggestion about such procedure is offered in A. Stewart (1978, pp. 122 ff., esp. pp. 129 f.).

33. Pln. *NH* XXXIV.xix.56 f; J. J. Pollitt (1974, pp. 263 f.). Unfortunately the manuscripts seems to be corrupt at this point. "Proprium eius est uno crure ut insisterent signa excogitasse, quadrata tamen esse ea ait Varro et paene ad unum exemplum"; or "quadratam tamen excogitasse [esse] eam ait Varro. . . . Proprium eius est uno crure ut insisterent signa et Paene ad unum exemplum." See S. Ferri (1946, pp. 79 ff; 1940, pp. 117 ff.). It is clear that "squareness" as well as "standing on one foot" are regarded as Polykleitos' characteristics. *Quadratus,* everyone is agreed, is Latin for the Greek *tetragonos,* which could simply mean "rectangular," though it also had explicit moral overtones: as in the quotation from Simonides in Pl. *Prot.* 339a, where the word means "four square" (or, extrapolating, "with all four elements balanced"), "morally perfect." It was also used in the somewhat dated slang sense of "square" as stiff or old-fashioned. See H. Philipp in *Polyklet* (1990, p. 142); against her rhetorical interpretation of the term, T. Lorenz (1972, pp. 60 ff.) argues for restricting the term to mean "equidimensional." Other texts are discussed by A. Stewart (1978, p. 131) and G. V. Leftwich, "The Canon of Polykleitos: Tradition and Content" (Princeton, 1986, typescript).

34. Galen *De Plac. Hipp. Plat.* (ed. I. Mueller, 1874, pp. 425 ff.); *De Usu Partium* 28A, 3 II.441H. See, most recently, J. J. Pollitt (1974, pp. 14 ff.). In fact *carpos* also means "fist," while *metacarpion* is used for the bone structure of the hand. Unlike "finger," "foot," or "ell," these are names of body parts that were not used for standard dimensions. Galen may therefore be taken to be writing here about the internal dimensional organization of the body, without any reference to measurement standards. But see H. Philipp in *Polyklet* (1990, p. 139); A. Stewart (1978, p. 131).

On Chrysippus and the unity of the person, see J. M. Rist (1969, pp. 256 ff.). The other mention is brief and not referred back to Hippocrates: in Hip. Aer. I.9, Galen writes of "the much-praised statue known as the Canon of Polykleitos, which has that name because of the perfect symmetry of the members to each other."

35. E. W. Marsden (1971, pp. 50 ff., 106); also Philo Byz. *Mech. Synt.* IV.1. This is one of the few books mentioned in Vitruvius' "bibliography" (VII.praef.xiv ff.) to have survived. *Para mi-*

krōn has caused commentators the most problems. So, for instance, Walther Kranz adds an alternative version, "by the smallest steps," to Hermann Diels (DK 1906, vol. 1, p. 229; 1934, vol. 1, pp. 392 f.) "when every smallest detail matters," though various other versions have been suggested, including "approximately." They are discussed by J. J. Pollitt (1974, pp. 86 f. n. 6) and by A. Stewart (1978, p. 126), who prefers "with each little step." The other attractive interpretation, "by something small [or 'by a je ne sais quoi,' to use a helpful if anachronistic phrase] *beyond* many numbers" seems to be ruled out by the sense of the context. As for *to eu,* it has been taken to color the passage Pythagorically by both Diels and Kranz. This was argued persuasively by J. E. Raven (1951, pp. 147 ff.). See Pl. *Tim.* 68e and Arist. *Met.* XIV.5 (1092b), which is a sceptical discussion of "the good which arises from number."

36. Plut. *De Aud.* 45, 13. The word *kairos* is a technical term here; it has been understood as a kind of "extracanonic" grace by some writers: but see H. Philipp in *Polyklet* (1990, pp. 140 f.) and A. Stewart (1978, p. 126). Philipp indeed suggests that it may mean something similar to *suntaxis* as used by Philo of Byzantium.

37. This saying is a compound, made up by Diels (DK 28 B, 1) from two separate quotations in Plutarch: *De Prof. Virt.* 17 (86a, which contains an elaborate allusion to a poem of Pindar now surviving only in a fragment, Schroeder 194, and in which "Polykleitos" is used in its common meaning, the "far famed") and *Quaest. Con.* II.3 (636c, "Which came first, the chicken or the egg?"). It seems as if Plutarch's two slightly differing quotations were made from memory and that the saying had passed into common parlance. However, what Polykleitos meant by "at the nail" is puzzling. Kranz (DK, vol. 1, p. 295) translates: "Wenn man mit der Tohnbearbeitung zur Nageldicke gekommen ist." In Latin, at any rate, the phrase "in unguem" (Virg. *Geor.* II.277) or "ad unguem" (Hor. *Sat.* I.v.32 f.: "ad unguem factus homo," i.e., "the perfect gentleman") meant "to perfection"; commenting on the Virgilian verse, Servius explains that it is a term taken from sculptors ("a marmoriis"), who will test joints with a nail. This *could* make sense here: but it could also refer to the way clay gets *under* the nail in the final detailing of a figure. T. Visser-Choitz, "Zu Polyklets Canon," in

M. Schmidt (1988, pp. 127 ff.) suggests that this passage is concerned with the refinements of bronze-casting technique, the clay being that of the outer mold "getting under" the nail of the wax model prepared for cire-perdue casting.

38. Vitr. VII.praef.xii ff. Of those mentioned, Nexaris, Theocydes, Leonidas, and Melampus Sarnacus seem otherwise unknown; the others all reappear in Pliny's *Natural History*: Demophilus of Himera (XXXV.xxxvi.61) was the master of the great painter, Zeuxis, and therefore a near contemporary of Polykleitos; Pollis (or Polis) was a painter (XXXIV.xix.91); while Silanion (XXXIV.xix.81) was probably the Silanion of Athens, who according to Pausanias (VI.4.v; VI.14.iv, xi) made three famous athlete figures at Olympia that were as well-known as his bronze portrait of Plato. There are some attempted emendations. Granger's "Melampus of Sarnaka" (Vitr. 1931–1934, 1955–1956, pp. 74 f.) is no less of a cypher than the two names separately, while Marini (1836) suggests tentatively "Melanthius"—an artist known to Pliny and Pausanias, whose inferior Apelles confessed himself (Pln. *NH* XXXV.xxxvi.80) and who is reckoned among theorists by Quintilian (XII.10); according to Diogenes Laertius (*in Pol.* IV.18) he wrote a book on painting. For Plato's presumed views on the renewal of the canon (*Soph.* 334), see P.-M. Schuhl (1933, pp. 6 ff.).

39. On Euphranor, see Pln. *NH* XXXIV.xix.50, XXXV.xl.128; Dio Chrys. XXXVII.43.

40. Cic. *Brut.* 275; however, in another quotation, Pliny (*NH* XXXIV.xix.61) reports him as having answered Eupompos, a fellow artist who had asked him which older master he followed, by pointing at a crowd of people to signify that he followed only nature. These two comments were not at all in contradiction, of course.

41. On Lysippus' canon, see M. Collignon (1905, pp. 101 ff.). Euphranor was well-known, although again few works survive even as copies: he himself is quoted as having said that whereas Parrhasios' Theseus was fed on roses, his own looked fed on beef (Pln. *NH* XXXV.xl.129, Plut. *De Gl. Ath.*). Elsewhere (*NH* XXXV.xl.128 ff.) Pliny, who admired his book learning and versatility, criticizes him for having made the trunks of his figures too slim and exaggerating the size of other members—and that exaggeration was presumably the novelty of Euphranor's canon. It made his heroes really look the part—"Hic primus videtur expressisse dignitates heroum"—and seems also to have become the main authority on the canon of proportion: "et usurpasse symmetriam," he "made symmetry his own," "took over the subject of symmetry." He does seem to have made them look particularly sinewy (*artipous*, Dio Chrys. XXXVII.43). See J. J. Pollitt (1974, pp. 358, 368 f., 378 ff.) and M. Robertson (1975, vol. 1, 412, 433 ff, 493 f.). Pliny identifies a number of his works, most now lost, but the lower part of the cult statue of Apollo Patroos from his temple on the Athenian agora has survived (now in the Agora Museum, S 2154, where there is also a 1:10 complete bronze model of it, S 877). There are also two copies (in the Villa Torlonia, Museo Capitolino) of a statue of Leto, the Great Mother of the Gods, with her two children, Apollo and Artemis, which Pliny (*NH* XXXIV.xix.77) had seen in the Temple of Concord and which also appears on the coins of Ephesus and Stectorium. See, still, J. A. Overbeck (1881, vol. 2, pp. 86 f.). S. Ferri, counting the number of artists in this Vitruvian passage (1960, pp. 253 ff.), notes that there were four Mausoleum sculptors, nine "less noble ones," and twelve "mechanics," and suggests that Vitruvius is in fact setting down some artistic canon, on the model of the literary Alexandrian and Pergamene ones.

42. See, however, J. Charbonneaux (1971, p. 145) on Spartan statuettes. The small ivory nude woman wearing a polos from the Kerameikos grave (National Museum, Athens) is another early exception. Twenty-six small nude female bronzes are listed by T. J. Scanlon in W. J. Raschke (1988, pp. 203 ff.). And there are a number of small orientalizing female nudes. One curious exception is a couple in the museum at Thessaloniki: a marble girl and a boy from the mid-fifth century—the boy with rather prominent genitals under his chiton—both draped in the same way.

The Spartans alone held races among naked girls, or allowed boys and girls to wrestle naked, provoking censorious remarks about the consequent inconstancy of Spartan girls, who in any case had a bad role model in their mythical queen, Helen. But even in Sparta, athletic nudity was limited to virgins, and such fights and races have been interpreted as a prenuptial ritual. Outside Sparta, partly clothed girls (in "Amazon costume") raced at Brauron and at the Olympian Hera festival (Paus. V.16.ii ff.). Otherwise female nudity was as rare in Greek life as it was in Greek art.

However, a canon of the female body is reconstructed by H. von Steuben (1973, p. 56 ff.) using the restored copy of the Amazon in the Palazzo dei Conservatori (signed "Sosikles"). There is no literary document until sometime in the fourth century, though there is a compilation, *Anecdota Graeca et Graecolatina*, which relies on three earlier authors (Polemo, Loxus, and Aristotle: in V. Rose, 1864, pp. 106 ff.), on which see V. Mortet (1909, pp. 68 ff.) and E. C. Evans (1969, pp. 14 ff.); even there, the differences are described in general terms without dimensions being given. Of the two statues that Praxiteles made for the temple of Aphrodite on Kos—one clothed, the other naked—the Koans preferred the draped one (Pln. *NH* XXXVI.iv.20–22). The nude statue was later bought by the Cnidians for their sanctuary of Aphrodite and caused a sensation. It was reputed to be the first cult statue representing the naked female body. See A. Furtwängler (1964, pp. 386 f., esp. 387 n. 1).

Another view of the subject is offered by S. B. Pomeroy (1976, pp. 36, 141 ff.). Cennino Cennini in his *Libro* ([ca. 1450] 1899, para. 70) refused to consider the proportions of the female body, since it is, unlike the male, fatally imperfect: "[le misure] della femmina lasci stare, perché non ha nessuna perfetta misura."

43. On this competition, see Pln. *NH* XXXIV.xix.53. Polykleitos is said to have won because the sculptors were asked to name whose statue should be second; though of course that is the type of win established by Themistocles after the battle of Salamis (Her. VIII.123, Plut. *V. Them.*)

In fact there were not five competitors but four: Kresilas of Kydon (in Crete) becomes two persons in the manuscript. See A. Furtwängler's (1964) doubts about some of Pliny's account, which was revindicated by B. S. Ridgway (1974, pp. 1 ff.).

44. D. Kreikenbom (1990) lists 67 copies and emulations of the Doryphoros but excludes the Amazon. However, R. Bol in *Polyklet* (1990, pp. 213 ff.) has attempted a reconstruction and at-

tribution of the various types. The Amazon and the Doryphoros herms (Inv. 4885) in the Museo Nazionale, Naples, are in fact signed by Apollonios, who may be the same sculptor who signed a copy of the Mantua Apollo in the Ny Carlsberg Glyptothek (no. 59).

45. Vitr. IV.1.iii. Although brief, this passage poses a number of problems: the colonization of Asia Minor gives rise to a distinct proportional system when the sequence of traditional craft transmission is broken. That may be the implication of this text. Hellen, the eponymous hero of the Hellenes, is variously the son of Zeus or of Deucalion and Pyrrha; his wife bore a Thessalian eponym: Pythiade or Othrys, Orstis. His sons Dorus, Xuthos (who in turn fathered Achaios and Ion), and Aiolos provide the eponyms for the main Greek dialects.

In fact the Panionian deity was Poseidon, and his sanctuary, dedicated to Poseidon of Helicon, overlooked the Samian strait northwest of Priene. It belonged to Melia near Ephesus, a town that was put down by an Ionian league ca. 700 BC. The league appropriated the sanctuary, which was called the Panionion from that time and was still functioning in the fourth century AD.

For the purpose of illustrating the canon, Apollo is a much more suitable or typical deity than Poseidon; that the pairing of Apollo and Artemis informed the legend, and/or that Vitruvius may be thinking of the ancient temple of Apollo at Klaros (Strb. XIV.642, Paus. VII.3.1), which was one of the great Ionian oracles, may account for his mistaking the dedication of the sanctuary.

46. The "accidental" work of art has a long and as yet unchronicled genealogy. But see H. W. Janson (1974, pp. 55 ff.) and J. Baltrusiatis (1983, pp. 55 ff.): though it is not clear from the text whether Vitruvius' *fortuito* refers to luck or accident. In fact this seems to me a neglected problem: translators are content to render Vitruvius' account that the temple was built "eius generis fortuito formae" as "by chance" or "accidentally"; yet since the account must have come from a Greek informant or source, *fortuito* could only have been the Latin for *tuchē, kata tuchēn*. As Aristotle's translators have repeatedly pointed out, such renderings are inadequate: "fortunately," "luckily," or even "on purpose" would all be more appropriate to the sense here. But see on the distinction between *tuchē* and *to automaton* in Arist. *Phys.* II.3–6 (195a ff.); cf. *Met.* XI.9 (1065b ff.). An interesting meditation on this notion is found in John Selden, *De Dis Syris syntagmata* (1617).

On the great shrine of Argive Hera, see C. Waldstein (1902–1905, vol. 1, pp. 105 ff.).

47. Vitr. IV.1.iii–vi. The paternity of Ion is scandalous: Xuthos, his supposed father, married Kreusa, the daughter of the Athenian king Erechtheos; she (according to Euripides' *Ion,* prologue) had exposed her illegitimate first child, whom the god had adopted through a Delphic priestess and thus by implication acknowledged paternity (Pl. *Euthy.* 302c, d). There were variations on this myth in Herodotus (V.66, VIII.44.), Velleius Paterculus (I.iv), Strabo (VIII.383), and Pausanias (VII.1.ii). On his brother Achaios, see Apol. I.50 and Her. II.98. Pausanias saw Ion's tomb in Thorikos, Attica (I.31.ii). This Ion, the eponymous founder of the Ionian deme, may have been no relation.

48. Vitr. IV.1.vii. Gisela M. A. Richter (1968) provides the most recent survey of these statues (in collaboration with Irma A. Richter). For earlier material, see W. Deonna (1909).

At this point it might be useful to establish that both boy and girl are represented by a footprint, *vestigium,* and not the foot itself in relation to the height. The text therefore reads as if it referred to some event, perhaps a ritual occasion, when the module was taken from a real person. The relation of Dorians and Ionians will be considered in greater detail in chapter 6.

49. A. Thumb (1932–1959, vol. 1, pp. 14 ff.): on Dorian literary usage, see vol. 1, pp. 217 ff.; on Ionian, vol. 2, pp. 202 ff. Strabo (VIII. 1.ii) tried to organize the multiplicity of dialects into two main groups: Ionic-Attic and Aeolian-Doric. Although the Ionic-Attic is accepted by most grammarians, in fact most of them prefer to divide the dialects into three groups: see Thumb (vol. 1, pp. 47 f.). In Corinth, a variety of Doric was spoken.

50. On the use of "as spoken" and jocular dialect forms, see A. Thumb (1932–1959, vol. 1, pp. 124 ff.).

v The Literary Commonplace

1. C. Lévi-Strauss in M. Mauss (1950, pp. xlvii); my assertion may be read as an extension of I. A. Richards' "romantic" theory of metaphor (1936, pp. 89 ff.), as elaborated not by him but by G. Lakoff and M. Johnson (1980, passim). There is a vast older literature on the subject of metaphor, most of which is listed by W. A. Shibles (1971a), and there is further specific discussion in A. Ortony's (1979) book. The problem of metaphor has occupied a number of philosophers recently: see M. Merleau-Ponty (1960, pp. 155 ff.) and J. Derrida (1967, pp. 412 f., 425 ff.); cf. P. Ricoeur (1975, esp. pp. 356 ff.; 1962–1968, vol. 2.2, pp. 228 ff.).

Metaphor has never looked quite the same since E. Husserl's distinction between what is signified, *Bedeutung,* and what is manifested, *Kundgabe.* See H. Spiegelberg (1965, vol. 1, pp. 104 ff.) on its influence—on English critical thought, for instance, even if indirectly (see C. K. Ogden and I. A. Richards, 1930, pp. 260 ff.; most recently, for a specifically Husserlian and very sympathetic reading of English romantic poets, see S. R. Levin, 1988), and on Russian, particularly "Formalist," thinking much more vitally (V. Erlich, 1965, pp. 62 ff., 180 ff.). See also A. Warren and R. Wellek (1949, pp. 190 ff.) and C. Perelman and L. Olbrechts-Tyteca (1969, pp. 191 f., 398 ff.). For a psychological description of the metaphorical process, see J. Piaget (1967, pp. 115 ff.) and E. Neumann (1954, passim). It is the view implied by Aristotle in the fourth chapter of the *Poetics,* in which metaphor makes its first systematic appearance. On the a priori social nature of reason, see J. Piaget (1965, pp. 143 ff.).

2. In part, this is what Hegel meant when he described the self-reflexive conscious man as a sick animal ([1931] 1969, pp. 213 f.) It refers the reader inevitably to the concept of alienation, so important for Hegel, and later transformed by Feuerbach, Marx, and their followers. See C. Taylor (1975, pp. 177 ff., 381 ff.). On the pre-Hegelian fortunes of alienation, see L. Kolakowski (1978, vol. 1, pp. 16 f., 25 ff., 38); in Hegel's other followers, pp. 88 ff., 155; in Marx, passim, but esp. pp. 133 f., 177 f., 265 f. The dialectic of in-itself and for-itself in the bodily subject is discussed by M. Dufrenne (1966, pp. 145 ff.).

On animal "language," see H. and M. Frings (1975) and D. R. Griffin (1976, pp. 15 ff.). On bee language, see K. von Frisch (1967). The distinction between animal communication and human language is stated tersely and categorically by N. Chomsky (1988, pp. 38 ff.). There is a vast literature on the origins of language to which G. W. Hewes (1975) has provided a summary guide. Elsewhere (in R. W. Wescott, 1974) he has usefully caricatured such theories as belonging to one or more of twelve categories: (1) interjectional, or pooh-pooh; (2) animal imitation, or bow-wow; (3) object imitative, or dong-dong; (4) work song, or yo-he-ho; (5) mouth movemental, or ta-ta; (6) babble; (7) instinctivist; (8) conventionalist; (9) contract; (10) God-given; (11) chance; and (12) gestural. A more articulated summary of language origin theories is offered in the anthology edited by J. Gessinger and W. von Rahden (1989). On the relation between the biological and cultural problems of language origins, see J. Wind et al. (1992). Still, whichever of Hewes' twelve categories is given primacy, the real issue is between evolutionary/learning and innate/generative theory. See S. Kanngiesser in J. Gessinger and W. von Rahden (1989, vol. 2, pp. 438 ff.); and M. Piattelli-Palmarini ([1989?], esp. pp. 68 ff.); M. Piattelli-Palmarini (1979, esp. pp. 406 ff.).

3. E. Fink (1969, pp. 16 ff., 71 ff.) and H.-G. Gadamer (1975, pp. 97 ff., esp. p. 100 n. 2). Gadamer acknowledges his debt to J. Huizinga (1949, pp. 101 ff., esp. 104). Kostas Axelos (1969, 1964 and Gustav Bally (1945, pp. 52 ff.) discuss the distinction between human and animal play first explicitly made by F. J. J. Buytendijk (1976). A useful discussion of the earlier literature of games and play appears in A. C. Haddon (1898, pp. 219 ff.). Play is often discussed in ethological literature; but see—in relation to curiosity and experience—K. Lorenz (1965, vol. 1, pp. 104 ff., 182 f.; vol. 2, pp. 180 ff., 236 ff.).

4. Quoted by W. Köhler (1947, pp. 126 f, 234).

5. The seer Teiresias foretold Narcissus' mother that he would live a long time if he never saw his reflection in water. He did look and fell in love with his image. The reason for his drowning in that water are variously explained in Greek and Latin literature: Phil. *Icon.* I.23, Paus. IX.31, Ovid *Met.* III.339 ff. His excessive self-love and consequent spurning of the nymph Echo transformed her

in a disembodied voice, according to one version (Moschus VI, Longus III.23, Dio Chrys. VI.204 r). The flower (if Sophocles is to be believed in *Oed. Col.* 681 f.) grew abundantly at Colonus.

The complex seems to have been identified and named by Havelock Ellis and Rémy de Gourmont, as well as the psychiatrist Paul Näcke. Freud appropriated it (with acknowledgment) and described it in his "Case of President Schreber" (1953–1974, vol. 12, pp. 9 ff.) and "Introduction to Narcissism" (vol. 14, pp. 73 ff.), and since then much literature has been devoted to the role of reflection in the constitution of the ego; see especially G. Stuart (1955). On the mirror image and its role in the construction of the ego, see J. Lacan (1966, pp. 88 ff.); he returned to the subject a number of times in the *Séminaire* (1975, no. 11, pp. 71 f., 131 ff., 188 ff.). The complex has had a particularly bad press lately: see C. Lasch (1979, esp. pp. 33 ff., 170 ff.).

In any case, it is the scent of the flower *Narcissus poeticus* that is supposed to have the intoxicating properties, beguiling Kore so that she could be carried off by Hades: so Hom. *Hym. Dem.* 8 ff. An extract from the flowers was supposed to have antispasmodic properties. The superstition reported by Artemidorus (*Onei.* II.7) that dreaming of your reflection in water is a presage of death may be related to this legend; on this and related beliefs, see J. G. Frazer (1911–1915, vol. 3, pp. 92 ff.). And of course, since mirrors have been made of glass, the breaking of it (and your image with it) has been taken as a token of bad luck. Narcissus may even have had a female twin, Narcissa—which opens a whole series of other possibilities; on the cult, see K. Chirassi (1968, pp. 143 ff.).

6. The "superstitions" relating man to shadow were gathered by J. G. Frazer (1911–1915, vol. 3, pp. 77 ff.). The fate of Peter Schlemiel, who actually sold his shadow in Adelbert von Chamisso's novel of that name, is its best-known literary representation.

Perhaps because of the presence of funerary deposits, archaeologists sometimes take the realization of death to be the original deposit of meaning in language: see D. Schmandt-Besserat in J. Wind et al. (1992, pp. 225 ff.).

7. The difference between an instinctual or encoded *model for* and the reflective *model of* is clearly shown by C. Geertz (1973, pp. 89 ff.).

On gesture and body language, see G. Cocchiara (1932, pp. 15 ff.), who quotes the distinction between *instinctual* and *conventional* gesture made by G. Malley (1881). J. Bremmer and H. Roodenburg (1992) have attempted a fragmentary history and bibliography of gesture.

8. On the contrast between "gradualist" and "catastrophic" theories of language origins, see E. H. Lenneberg (1967, pp. 227 ff.). But this problem has occupied biologists and paleontologists as well as philosophers: see C. F. Hockett and R. Ascher (1964, pp. 20 ff.). An account of recent experimental evidence is given in R. E. Leakey (1981, pp. 127 ff.). See also P. L. Berger and T. Luckmann (1967, pp. 34 ff.). But such a view was already outlined by Jean-Jacques Rousseau in his *Essay on the Origin of Language,* and may indeed be traced back to Cordemoy's theories. Ancillary to the problem of origins (see the bibliography above, n. 2) is a problem of whether the first humans to speak had actually something to say before they developed the capacity to speak: see A. Marshack in J. Wind et al. (1992, 421 ff.).

9. H.-G. Gadamer (1989, pp. 138 ff., 151 ff.). On presentation and re-presentation, see pp. 108 ff.

10. Some anthropologists and biologists wish to relate the development of speech—not language—to the appearance of the subspecies *sapiens sapiens;* see G. S. Krantz (1980). Others deny such relationship, for instance, B. Arensburg et al. (1989, pp. 758–760). An extended discussion of the anatomical and accoustic problems is found in P. Lieberman (1975, pp. 121 ff.).

11. Louis de Bougainville was the first European to note the word as *tattou* (1771); in English the word was fresh coined as *tattow* by Captain Cook in 1769, on his visit to Tahiti (1955, vol. 1, passim, but esp. pp. 278 f., 585 ff.). In many Polynesian languages *ta* signifies a "blow" or "knock," from which tattooing, which usually involves a point and a mallet to pierce the skin, is obtained by duplication.

In spite of the scriptural prohibition of body scarifying in Leviticus, certainly Christian groups (notably the Copts and the Armenians) practice it, particularly as a souvenir of pilgrimages. See J. Carswell (1958).

Tattooing is treated as one of the major arts of Polynesia (with house building and boat building) by M. Mead (1928, pp. 71 ff.). The subject is treated very perfunctorily in older anthropological works: Sir John Lubbock (1870, pp. 46 ff.) is largely descriptive, and E. B. Taylor (1903) is concerned to relate it to etiological myths. While tattooing is often described in ethnographical studies, few general treatments have been attempted. Perhaps the very first was a curious positivist essay by Wilhelm Joest (1887), while a much more refined and extensive treatment was W. D. Hambly (1925); for a very brief discussion, see V. Ebin (1979) and A. Virel (1980). See most recently C. P. Jones (1987, pp. 139 ff.).

Cesare Lombroso, and some of his contemporaries and followers, confined the study of the modern tattoo to the province of criminal psychopathology: see J.-A. Lacassagne (1881, 1908) and H. Prinzhorn (1922, 1926). It is certainly true that tattooing has been adopted by many subcultures and deviant groups as a membership token: see C. R. Sanders (1989, pp. 18 ff.). On its connection with body art, see Bryce Bannatyne Gallery (1992). The medical aspects have recently been discussed by C. S. Zwerling, F. H. Christensen, and N. F. Goldstein (1986, pp. 179 ff.); the danger both of infection and of allergic reaction from tattooing seems surprising small.

The problem of the tattoo and body painting is reposed by C. Lévi-Strauss (1958, pp. 282 ff.). There are many ethnographic studies of the tattoos of individual regions, for instance, A. Hamilton ([1900] 1961, esp. pp. 308 ff.); K. von Steinen (1925); and J. C. Faris (1973). On the Caduvei of the Upper Amazon, see C. Lévi-Strauss (1955, pp. 183 ff.), though they had already been recorded by G. Boggiani (1895, pp. 105 ff.). On the Maori, see D. R. Simmons (1986). That these markings also involved separation and reintegration in the community is a separate and extremely important issue. On the general business of the deformation of organs, see R. D. Guthrie (1976).

12. M. Mauss (1968, vol. 2, pp. 409 ff.); on circumcision, immortality, and sociability, in a review of J. G. Frazer (1911–1915), vol. 1, pp. 141 ff.; on tattooing in a review of W. Wundt (1908), vol. 2, pp. 195 ff. See also E. Durkheim ([1912] 1961, pp. 136 ff., 262 ff.), and J. G. Frazer (1911–1915, vol. 5, pp. 278 ff., vol. 11, 255 ff.). The stenciled hands of European cave painting sometimes lack a phalange, but this is inconclusive evidence of mutilation. See A. Leroi-Gourhan (1965, pp. 109 ff.).

Many of the older ethnographers separated circumcision from other body markings: So C. de Pauw (1792, vol. 2, pp. 149 ff.). And indeed the circumcision of infants may diminish the traumatic effect, and so sets it apart from the more usual puberty rites. Like everyone else, I am here indebted to A. van Gennep ([1909] 1960).

13. H. Melville ([1851] 1983, p. 1307). Melville had occasion to see many tattooed Polynesians. See H. P. Vincent (1965, pp. 6, 756ff., 371 f.) and R. S. Forsythe (1935, pp. 99 ff.; 1936, pp. 1 ff.; 1937, pp. 344 ff.).

14. E. Grosse (1902, p. 41) quotes Captain Cook on the inhabitants of Tierra del Fuego: "They are content to be naked, but ambitious to be fine." The pages that follow the quotation are among the few to treat the subject generally. Flattened foreheads were induced in children in certain Polynesian islands (New Hebrides) or of American Indians of the Northwest (Kwakiutl and Chinook). See B. Lincoln (1981, pp. 34 ff.).

15. See E. Strauss (1963, pp. 207, 365 f.). On the Australian ceremonies, see B. Spencer (1914, pp. 166 ff.); B. Spencer and F. J. Gillen ([1899] 1938, pp. 133 f.; 1927, vol. 1, pp. 207 ff.). More recently subincision has been performed with a sharpened reed (conversation with author, R. Guidieri).

16. Wolfgang Porzig, *Das Wunder der Sprache* (1950), quoted by K. Lorenz (1965, pp. 229 f.).

17. See E. Strauss (1966, pp. 137 ff.); G. von Kaschnitz-Weinberg (1944–1961, vol. 1, pp. 12 ff.); and S. Giedion (1964, vol. 2, pp. 440 f.). On posture and the ego's relation to the *Umwelt*, see K. Koffka ([1935] 1963, pp. 389 ff., 514 ff.).

18. Denise Schmandt-Besserat (in J. Wind et al., 1992, p. 225) has suggested that the oldest remnants of a "visible language" were concerned with representing ideas, not things.

19. In this context, see J. Lacan (1966, pp. 697 f.), who compares Ernest Jones' theory of symbol to an artful building:

Cet édifice . . . pour métaphorique qu'il soit . . . est bien fait pour nous rapeller ce qui distingue l'architecture du batiment: soit une puissance logique qui ordonne l'architecture au-delà de ce que le batiment supporte de possible utilisation. Aussi bien nul batiment, sauf à se réduire à la barraque, ne peut-il se passer de cet ordre qui l'apparente au discours. Cette logique ne s'harmonise àl'efficacité qu'à la dominer, et leur discord n'est pas, dans l'art de la construction un fait seulement éventuel [ ].

He thus contrasts Jones' theoretical preoccupations with the professional empiricism of his contemporaries, as architecture contrasts to mere building. It is an interesting comment on the late-twentieth-century status of architecture that this passage is discussed with some disapproval by a French critic, Denis Hollier (1974, pp. 68 f.).

20. An interesting exception was Gottfried Schadow's book *Polyklet* (1834; it was republished in English, under the auspices of the Department for Trade and Industry, in 1883), which attempted to recalculate the Dürerian differentials on living bodies, though Schadow thought himself to be applying the principles of Leonardo and Gérard Lairesse. He also compiled a series of national physiognomies, which he considered an extension of Peter Camper's work. See Dr. *[sic!]* Johann Gottfried Schadow (1849, pp. 251 f.). David R. Hay of Edinburgh, who is remembered chiefly as an interior decorator (Sir Walter Scott's Abbotsford, the Scottish National Gallery), published a number of books: he also continued Camper's speculations on the geometry of the skull and the face in art and nature in *On the Science of those Proportions by which the Human Head and Countenance as represented in Works of Ancient Greek Art are distinguished from those of Ordinary Nature* (1849). His books were received with the greatest acclaim, but their influence is difficult to estimate. In the second half of the century, there were a number of studies culminating in Sir D'Arcy Wentworth Thompson (1917), which, however, applied to "organic" nature as a whole and did not privilege the human body.

21. "Unsere Sprache ist eine Verkörperung alter Mythen. Und der Ritus der Alten Mythen war eine Sprache." This should perhaps rank as an obiter dictum since Wittgenstein omitted the handwritten sentence from his 1931 typewritten fair copy of the "Bemerkungen über Frazers *Golden Bough*" (1967). See R. Rhees (1979, pp. 35 ff.). Later in the essay (pp. 56 f.), Rhees exemplifies Wittgenstein's thinking by specific buildings (a Norman church, a Gothic cathedral) as giving "visible form to religious thought." Although, as Rhees points out, the differences are self-evident, yet the sense in which architecture as well as ritual can be described as a language could be extended and developed.

A discussion of the relation between metaphor and hypothesis in relation to myth is found in E. R. MacCormac (1976, pp. 102 ff.).

22. G. Bataille ([1929] 1971, vol. 1, pp. 171 ff.). On this, and the whole question of architecture and building in Bataille, see D. Hollier (1974, passim, but esp. pp. 66 ff.). The whole dictionary only had twenty entries (by different hands), in any case: see Hollier (pp. 28 f.).

23. J. Verdenius ([1957] 1972); W. Weidlé (1963, pp. 249 ff.); H. Blumenberg, (1957a, pp. 266 ff.); H. Koller (1954); and E. Grassi (1962, pp. 76 ff., 120 ff.).

A survey of recent thinking on the subject is offered in R. Bogue (1991). The constitution of the concept in the work of one modern philosopher is discussed by J. Früchtl (1986). F. Tomberg (1968), who uses Wilhelm Pinder's combative—and as it seems to me, mistaken—statement that "neither a cathedral nor a symphony represent reality" to justify Georg Lukács' realism, has to reject Koller's carefully documented and argued distinction in order to do so. But see also more general works: M. Heidegger (1979–1984, vol. 1, pp. 166 ff.); M. Dufrenne (1967, pp. 225 ff.); S. K. Langer (1953, pp. 46 f., 71 ff., 352; 1957, pp. 94 ff.); and E. Auerbach ([1946] 1964, passim). The fundamental irrelevance of mimesis to architecture was influentially asserted by M. Dessoir (1970, pp. 366 ff.). Nevertheless a number of twentieth-century architects have been developing extreme applications of the idea: perhaps most insistently and obsessively the Cuban-French architect Riccardo Porro.

I am aware that I have not yet referred to another important aspect of body imitation in building, what I might call the psychosomatic kind, by which the spectator identifies, empathizes with, parts of the building through the experience of aspects of his or her own body—as lips and eyelids with the moldings that outline doors and windows. This is, however, very much a spectator and a post facto identification, while I am concerned with it from the outlook of the maker and the theorist. Adrian Stokes has dealt with this kind of imitation-identification in much of his writing; more recently, R. Wollheim (1987, pp. 305 ff.) has examined Venetian late-fifteenth- and sixteenth-century painting from that point of view. More surprisingly he has, by a brilliant ellipsis, pointed out how the metaphor can still dominate the perception of a painting even when there is no direct representation of a human figure.

24. See C. Bragdon (1918, pp. 188 ff.; 1922, pp. 64 ff.; 1925, pp. 114 ff.); M. C. Ghyka (1931, vol. 1, pp. 99 ff., vol. 2, pp. 127 ff.; 1932, pp. 19 ff.); and even Le Corbusier (1954, pp. 18 ff.).

25. Pl. *Symp.* 201d ff. According to Socrates, she was a prophetess or diviner of a shrine at Mantinea in Arcadia, though he did not specify which. A guide to them is provided by Paus. VIII.9.

26. The specific mime plays of late antiquity were maskless comedies; the Delian hymn to Apollo (1. 163) praises the temple girls who *mimeisth' isasin*, mimicked the sound and babble of all foreign tongues. Jane Harrison already protested against its being abstracted from its sacral context for the purpose of "imitation" (1913, p. 46). The Greek *hupokritēs* derives from **krit*, which has to do with judging and speaking; it is in any case a rather later word. In tragedy the *hupokritēs* was an impersonator of a great figure who entered into dialogue with the chorus as a whole, or with the chorus leader, the *exarchos*; his art or skill was also assimilated to the rhetor's *actio*. See M. Bieber (1961, pp. 18, 80, 147 f.) and H. -G. Gadamer (1989, pp. 113 ff.). H. Koller (1954) has argued persuasively—and with a wealth of example—the intimate connection between mime, music, and dance, and hence the coloring of mimesis by this precedent. The Roman grammarians considered the word *histrio* (the *hupokritēs* of the Roman theater), like *persona*, the mask that was his accoutrement, to be of Etruscan origin. On the transformation of the *mimos* into the mime, even pantomime, of modern theater, see H. Wiemken (1972). On the fraught relation between dance and drama in Indian theoretical writings, see M. Bose (1991, pp. 256 ff.).

For Aristotle's usage, see *Phys.* II.2 (194a21), on which see further G. Vattimo (1961, pp. 23 ff.) and P. Somville (1975, pp. 45 ff.). On the *mimēsis* of *ēthos* in Aristotle, see also above, chapter 2, n. 49. Buildings appear early in this connection, since Herodotus notes that the columns of the tomb of the Saistan Pharaoh Apries (589–570) were of stone, carved to imitate palm leaves.

27. In a work that is often read as paean to homosexual love, it is notable that *poiēsis* is honored as an analogue of human generation and reproduction: but see G. Vlastos (1973, pp. 38 ff.). On this and the figure of Diotima as a fictionalized Aspasia, see D. E. Anderson (1993, pp. 70 ff.) and S. Rosen (1987, p. 191 f., 223 ff., 243 ff.).

The word Plato uses for generating, childbearing, here is *tiktein,* which is close enough to *technē* to have led older philologists (such as Georg Curtius) to associate the two words and their roots **tek, *tik,* with **tih* (I am using "i" to translate upsilon) in a common concept of "begetting," "producing." This connection is discounted by recent etymologists, such as Chantraine: but see LSJ, s.v. *tiktō.*

28. The last sentence paraphrases H. Blumenberg (1957a, 272).

29. Pl. *Symp.* 205b; see M. Heidegger (1962, pp. 64 ff.; 1970N, vol. 54, p. 190). This account is inevitably summary: an extended discussion of the idea in Plato, underlying certain internal contradictions implicit in Plato's use of the word is given by W. J. Verdenius ([1957] 1972). But see also P. -M. Schuhl (1933, pp. 42 ff.).

30. Arist. *De Mundo* V. 396b. He is commenting on how nature reconciles opposites, much as artists achieve harmony by combining opposites, and quotes Heraclitus "the Obscure" *(ho skoteinos)* in support: which quotation becomes DK B12.

31. Arist. *Phys.* II. 8(199a); modified trans. R. Hope, 1961, p. 37). The strict parallel between *kata phusin* and *kata technēn* has been the subject of many commentaries. On this passage, and those from Plato's *Symposium* (above, n. 25),

see H. Blumenberg (1957a, pp. 267 f.) and G. Agamben (1970, pp. 85 ff.).

32. As Arist. *Poet.* 2 (1448a ff.), echoing Pl. *Rep.* X. 603c. See G. Finsler (1900, pp. 40 ff.). On this and the following pages in this chapter, see more generally H. Koller (1954) and G. Sörbom (1966). It is however worth mentioning here that by the Hellenistic period mimesitic doctrines of imitation were not of imitation from nature but from "old masters": so the very fragmentary *Peri mimēseos* by Dionysius of Halicarnassus is largely concerned with the imitation of "the ancients," in his case Thucydides and Demosthenes. The second chapter of Quintilian's tenth book is entirely devoted to the subject, but makes no mention at all of the imitation of nature. Sir Joshua Reynolds, in the sixth of his *Academic Discourses* (1778), deliberately limits the topic of imitation to the matter of copying pictures by masters. On the fate of mimesis in eighteenth-century artistic theory, see A. Rey, "Mimesis, Poétique et Iconisme: pour une relecture d'Aristote," in P. Bouissac, M. Herzfeld, and R. Posner (1986, pp. 17 ff.).

33. But much of the *Sophist* and the *Ion* are also concerned with it. For other loci, see J. G. Verdenius ([1957] 1972). It does appear that in the later dialogues, particularly in the *Laws,* Plato took a rather less censorious view of *mimēsis:* see P. Friedländer (1958–1969, vol. 1, pp. 116 ff.; vol. 3, pp. 406 f.). But even in the *Statesman (Pol.* 300c) Plato admitted that laws should be *mimēmata tēs alētheias,* "imitations of true being."

34. Pind. fr. 194 (Sandys); see J. Svenbro (1984, pp. 157 f.). The argument about the change of the poets' status is, of course, Svenbro's.

35. Pl. *Soph.* 219a ff. On this puzzling figure and his intentions, see most recently S. Rosen (1983, esp. pp. 17 ff.). On the *mimēta tou ontos,* see pp. 309 ff. But see also W. K. C. Guthrie (1962–1978), vol. 5, pp. 135 ff.).

36. Pl. *Soph.* 264e ff.; see G. Finsler (1900, pp. 22 ff.). The primary distinction between production, poetics, and the various *technai* of acquisition is not relevant here. On the different Platonic and Aristotelian uses of the word, see A. B. Neschke (1980, pp. 82 ff.); also F. M. Cornford ([1935] 1963, pp. 323 ff.).

37. Pl. *Rep*. III. 397f. The "Apollonian" instruments, the kithara and the lyre, are the only ones permitted—even if the shepherd is to be allowed to play his panpipes, the syrinx. The instruments of Marsyas are excluded from the city. The whole passage is discussed by A. Barker (1984, vol. 2, pp. 128 ff., 163 ff.). Pythagoras' lyre, the reader may remember (see above, chapter 4, n. 13), was tuned in the Doric mode. It is a strange reflection on the theme that the first literary mention of mimesis (Pind. *Pyth*. XII. 11 ff.) tells how Pallas Athena invented the reed flute (aulos) to imitate the wailing of the Gorgon mixed with the hiss of her serpent hair; see Barker (pp. 14 f, 57 f. On the aulos players' bad habit of hogging the virtuoso's position, see H. Koller (1954, pp. 74 ff.).

38. Pl. *Rep*. X. 595a ff. On this passage see M. Heidegger (1979–1984, vol. 1, pp. 173 ff.).

39. Aristotle already confessed himself bemused by this terminological ambiguity in Plato in *Met*. I.6.iii (987b). But then Plato also considered other forms of imitation. In the sense of man either being or making himself the likeness of a god, he uses *omoiousis* (*Parm*. 132e), which serves him for that which subsists between sensible and intelligible order (*Rep*. VI.510a); the likeness of the created kosmos to the demiurge required a different term, *paraplēsis* (*Tim*. 18c). These words, and allied concepts in Plato, are briefly discussed by C. Rutenber (1946, pp. 19 ff.).

40. Xen. *Mem*. III.10. The artists were Parrhasios, one of the trio of great Greek artists (with Zeuxis and Apelles); the sculptor Kleiton (not otherwise recorded, though some think the name a contraction of Polykleitos); and the bronze armor maker Pistias. For all of them, he discussed what it was they imitated and how imitation of what was bad in itself could then turn out to be good as an imitation, a discussion that Xenophon actually thought useful to artists. Zeuxis, Polykleitos, and Pheidias turn up in the Platonic dialogues as exemplary figures. But see B. Schweitzer (1953, pp. 22 ff., 77).

41. Pl. *Phil*. 55d ff. This was probably written a little later than the *Sophist*. It is not clear how much significance should be attached to the fact that Plato/Socrates is concerned with carpentry and house building, not with stone architecture.

The whole argument is virtually reversed by Schopenhauer in the passage I quoted in chapter 1 (see n. 15).

42. Arist. *Poet*. 25 (1460b ff.; 1907, pp. 96 ff., 121 ff.; 1909, pp. 80, 324 ff.). The whole chapter, which deals with the relation between the possible and the poetic, is in the nature of an appendix: its crucial importance is emphasized by D. de Montmollin (1951, pp. 99 ff.), who regards it as the key to his interpretation of the text as resulting from two redactions, while the inconclusive G. F. Else (1957, p. 632) sees the passage as relatively independent of the rest of the text, albeit "thorny." A linguistic analysis of the concept of *mimēsis* in relation to *technē* and *katharsis* is found in A. B. Neschke (1980, pp. 76 ff., 118 ff.). Aristotle discusses problems presented by the poet's craft, as well as by other forms of *eikonopoiēsis*, such as painting and sculpting, though there is no specific mention of architecture here. And indeed it is the idea of purification to which the spectator can aspire in identifying with the tragic hero that separates Plato's notion of *mimēsis* from Aristotle's.

43. H. Blumenberg (1957b, pp. 270 ff.). On Nicholas of Cusa, see chapter 3, n. 41.

44. The Kantian terms are *Lust* for pleasure, *Genusz* for appropriation; see I. Kant (1922–1923, vol. 5, p. 379). Kant's view is rebutted by Nietzsche (1975–1980, vol. 6.2, pp. 305 ff.). See also P. Lacoue-Labarthe in S. Agacinski et al. (1975, pp. 194 ff.). J. Derrida has pointed out (ibid., pp. 68 ff.) that this question is closely tied, in Kant's mind, with the problem of the artist's social responsibility, but that Kant then seemed to admit architecture *(Baukunst)* rather reluctantly among the fine arts (vol. 5, p. 398).

45. An ancillary problem is that the Greek *technē*, not *poiēsis*, becomes the Latin *ars*, the scholastic "recta ratio factibilium" (Aq. *Sum. Th*. I.ii, q. 57 a. 3 ad. 3). The grammarians derived the Latin *ars* from *aretē*, more usually translated as *virtus* (see Donatus' commentary on Terence), or alternatively from **aro*, "I plough." Varro enumerates three *artes*: medicine, shoemaking, and music. Cicero distinguished *artes liberales* from the *sordidae* (*De Off*. I.42), that is, those that involve private gain and incur human hate (tax gathering, usury), all mechanical trades, and those that involve mere gratification of the senses (cooking, scent making). Medicine, architecture, and the teaching of the liberal arts he considered suitable to the gentleman. The distinction was already operative in the Hellenistic period. The word *architektōn* was applied to all inventive, "superior" craftsmen (as it was still by Pappus of Alexandria in the fourth century AD). See M. S. Cohen and I. E. Drabkin (1948, pp. 183 ff.) and G. E. R. Lloyd (1970, pp. 91 ff., 111 f.).

46. "Die Natur war schön wenn sie zugleich als Kunst aussah; und die Kunst kann nur schön gennant werden, wenn wir uns bewusst sind sie sei Kunst, und sie uns doch als Natur aussieht . . . " (I. Kant, 1922–1923, vol. 5, p. 381). This is of course an inversion of the well-known ancient principle that the skill of the artist is to hide his art, chiefly coined for the rhetorician: Arist. *Rhet.* III.2 (1404b), III.7 (1408b); Long. XXII.1; *ad Her.* IV.7.x; and constantly in Cicero and Quintilian. Inevitably, too, since the unteachability and originality of genius are conditions of the production of great works of art, the whole doctrine of imitation, however important as an aspect of representation, must be seen as secondary (Kant, vol. 5, pp. 383 ff.).

47. The exact location of the sanctuary is uncertain, though it seems to have been somewhere in the eastern Crimea. The Taurians were also called *Tauroskuthoi*: see PW, s.v. "Tauris." Aulis was an important anchorage on the Eubean strait, just northeast of Thebes, opposite Chalkis. Pausanias (IX.19.v) visited the temple of Artemis, which was still a shrine in his day. The episode is the subject of Euripides' *Iphigenia in Aulis,* is described by Aeschylus (*Ag.* 185 ff.), and is referred to by Hesiod (*Op. D.* 650 ff.), as well as in the *Cypria.* On the connection between the three Artemisian sanctuaries at Aulis, Halai, and Brauron, see M. B. Berg Hollinshead (1979) and H. Lloyd-Jones (1983, pp. 87 ff.). Iphigenia has sometimes been identified with Hekate (Paus. I.43.i), but there was also a cult of Artemis Iphigenia (Paus. II.35.i, VII.26.v). The cult statue at Halys was said to be a xoanon brought by Orestes, Iphigenia's brother, from Taurus (Paus. III.16.vii). Pausanias also quotes Herodotus as having said that the Taurians sacrificed castaways to Iphigenia herself. Iphigenia was revered moreover as the founder of the cult at Brauron. On the sacrifice of Iphigenia, and Greek

human sacrifice generally, see A. Henrichs (1981, pp. 195 ff.).

On the skene building of the Greek stage as a representation of heroic temple and palace, see R. Padel in J. J. Winkler and F. Zeitlin (1990, pp. 345 ff.).

48. Aesch. *Ag.* 1438 ff.

49. Aesch. *Ag.* 896 ff. (for commentary, see 1950, vol. 2, pp. 403 ff; 1966, vol. 2, pp. 72 f.).

50. Eur. *Iph. Taur.* 68 ff. It is not clear from the text whether the skulls were hung under the altar edge or up on the cornice of the building: "*thrichōmata / thringkois*" seems to associate linguistically the edge of the bloodstain and the cornice of the temple. Ovid seems to be echoing these lines in *Pont.* III.ii.53 f.: "Araque, quae fuerat natura candida saxi. / Decolor adfuso tincta cruore rubet." In the *Bacchae* (1214 ff.), Euripides again refers to the placing of skulls at the cornice, when Agave, still possessed by Dionysos, returns from her ecstatic expedition with her son's head, which she thinks is that of a lion cub. She calls Pentheos (this very son) to bring a stout ladder, set it against the wall, and nail the head "up in the triglyphs" *(krata trigluphois),* a much more precise instruction than is warranted by the modern usage of displaying hunting trophies.

51. Eur. *Iph. Taur.* 42 ff.

52. Some 200 examples, including atlantes and caryatids, have been collected by F. Schaller-Harl (1973). Earlier material, much of it non-Greek, was collected by E. Wurz (1906).

53. Vitr. I.1.v.; and see P. Fleury (1969–1992, ad loc.). The male and female figures are mentioned in this passage to show how necessary the architect will find historical scholarship if he is to explain the iconography of his building to his client. For this and the following section on caryatids, see the survey by F. Schaller-Harl (1973).

54. "*Epi ton kionon*": Paus. III.11.iii (1896–1910, vol. 1.2, pp. 687, 768 f.; 1913, vol. 1, p. 149, vol. 3, p. 328.); see also K. Lange (1885, pp. 105 ff.). More recently, see R. E. Martin (1951, p. 495), and J. J. Coulton (1976, pp. 39 ff.). For this and much of what follows, see A. Schmidt-

Colinet (1977) and Eva M. Schmidt (1982). But see Athen. VI.241d f. A number of porticos with male "supporters" attached to columns were built in imperial times: see Cornelius C. Vermeule III, "Figural Pillars: From Asia Minor to Corinth and Rome," in M. del Chiaro (1986, pp. 71 ff.). On the whole, the Greeks did not mix the genders of their supporters; the Romans did.

55. Although the temple seems to have suffered a serious collapse in 1401 due to neglect and pillage, the ruins continued to be known as the Palazzo dei Giganti, and the telamones figured on the arms of the town. Much stone was taken out of the ruins to build the Porto Empedocle in 1749 to 1763. The ruins were first inspected and drawn "archaeologically" by C. R. Cockerell (1860); see J. J. Haus (1814). The stones of two figures were assembled by a local antiquarian: see R. Politi (1826, p. 38); D. Lo Faso Pietrasanta, Duca di Serradifalco (1834, vol. 3, pp. 52 ff.); and W. Wilkins (1807, pp. 30 ff., pls. 14 ff.). The full account and reconstruction are in R. Koldewey and O. Puchstein (1899, vol. 1, pp. 153 ff., vol. 2, pls. 21 f.). This account is criticized by W. B. Dinsmoor (1950, pp. 101 ff.) and A. W. Lawrence (1983, pp. 179 ff.). But see also P. Marconi (1935) and B. Pace (1935–1938, vol. 2, pp. 191 ff.).

The term *telamon* is used by Vitruvius (VI.7.v f.) for male supporters in gardens and banqueting rooms as an alternative to *atlantes*. But though he provides a little note on Atlas, he says nothing about Telamon, who was usually considered the father of Ajax and was a companion of Herakles in a number of adventures, particularly his fight with the Amazons. He has been identified with the world-carrier Atlas by some; the names are conflated as *Atlas-telamon* in some late inscriptions.

56. See however Diod. Sic. XIII.82.i f. Although he says nothing at all about the telamones, he does mention the sculptures in the pediments.

The only close copy of these statues is a bearded, over-life-size Hellenistic Atlas in the Reggio Calabria Museum (from Monte Scaglioso). See Eva M. Schmidt (1982, p. 116).

57. On the common metaphor giant/barbarian, see PW, s.v. "Gigantes; *EAA,* s.v. "Gigantomachia"; and F. Vian (1952, pp. 184 ff.).

Other barbarians were the crouching sileni who supported the skene of the theater of Dionysos in Athens—placed there, it would seem, by the Archon Phaidros (who held that position intermittently, 296–281); or even later, the tritons and titans of the porch of the Odeion in the Athenian agora (H. A. Thompson and R. E. Wycherley, 1972, p. 113), which was added when the building was reconstructed in the archonate of Dionysios in the middle of the second century AD after the collapse of the roof. The original building was ordered by Marcus Agrippa after 15 BC; it also had a series of male and female herms, of which some damaged heads survive as part of the stage structure. But these, like the "slaves" in Trajan's forum, were done long after Vitruvius' death.

58. See F. E. Winter (1976, pp. 143 ff.). However a screen, about half the height of the columns, seems to have closed in the peristyle of the somewhat earlier (ca. 540?) temple F (temple of Athena?) at Selinunte, also in Sicily: G. Fougères and J. Hulot (1910, pp. 246 ff.).

59. Paus. I.26.v ff.; see M. de G. Verrall and J. E. Harrison (1890, pp. 484 ff., 506 ff.) and C. Picard (1931, vol. 1, pp. 30 ff.). Kristian Jeppesen (1987, summing up his earlier essays on the subject) suggests plausibly that the shrine of Athena Polias—the Old Temple—was not the Erechtheion but the nearby, so-called House of the Arrhephoroi by the north wall.

60. See R. Demangel (1932, p. 267 f.). But then the caryatid, as the kore type generally, seems to have had an oriental origin: see B. Ridgway (1977, pp. 113 f., 203 f.). The sarcophagus of the pleureuses is no. 368 in the Istanbul Museum; it is of Pentelic marble, done perhaps by an Atticizing Syrian, perhaps by a Greek artist about the middle of the fourth century. Most of the Nereid monument is in the British Museum. But see P. Coupel and P. Demargne (1969); W. R. Lethaby (1915, pp. 208 ff.); and A. H. Smith (1900, pp. 1–46).

61. But the second-century (AD) grammarians—the Greek Diomede, commenting on Theocritus' *Bucolics,* and the Latin Marcus Valerius Probus, commenting on Virgil's *Ecl.* II.8—both carry a different story: that of the terrorized flight of the priestess-maids because of the war, and their being replaced at the feast by rough shepherds. Probus' calling the peasant song *Astrabikon* introduces yet another complication; but see M. P. Nilsson ([1906] 1957, pp. 196 ff.).

62. Athen. VI.241d, quoting a play by the middle comedian Alexis, which would put the remark back to ca. 350 BC; though perhaps it is fair to add that his book is almost a continuous aside.

63. G. E. Mylonas (1961, pp. 11, 158 ff.). Of the two surviving fragmentary statues, one is in the museum in Eleusis and the other in the Fitzwilliam Museum, Cambridge (no. 81). The casket decorated with reliefs may have been a metal-covered box; it rests not directly on the hair but on a cushioning pad on top of the head. Unlike the Erechtheion korai, the Eleusinian women seem not to have been symmetrical.

The eighteenth-century villagers, who were reluctant to release one of the statues, knew it as St. Demetra and believed her to guarantee the fertility of their fields. In antiquity the statues were popular as models: two highly carved and symmetrical (headless, armless, and probably attached, not freestanding) basket- or casket-supporting caryatids from Mylasa are in Istanbul (294, 294a; see G. Mendel, 1912–1914, vol. 1, pp. 281 ff.). Four reduced and much restored *kanēphoroi* from the villa of Herodes Atticus on the Via Appia were made into a porch—of a shrine of Ephesian Diana—in the Villa Albani: Peter Boll (1990, pp. 90 ff.; nos. 178–180). Two more (nos. 224, 225) are also in the Villa Albani, now Torlonia, and seem to have been copies of the Eleusinian ones. Other, more or less fragmentary and restored caryatids, are in the Vatican Museum (Braccio Nuovo, nos. 2270, 2296), in the British Museum (no. 1746), and so on.

64. The Pantheon caryatids are only known from Pliny's description (*NH* XXXVI. iv.38); he praises them fulsomely and attributes them to the otherwise unknown Diogenes of Athens. Since plaster casts of the Erechtheion caryatids seem to have existed in Rome—they were presumably the basis of those in the Augustan Forum and in Hadrian's Villa—it is tempting to attribute them to this otherwise unknown Greek sculptor.

The fragments of the Forum of Augustus caryatids (found in the 1930s) were in fact restored by Gismondi with the help of such plaster casts: see Erika E. Schmidt (1973, pp. 7 ff.). The inscription "C(aius) VIB(ius) RUF(inus)" on one base frag-

ment has been thought to be the artist's signature. In the Forum of Augustus the caryatids seem to have formed a kind of huge frieze, alternating with large circular shields. Schmidt also lists the three other full copies (Florence, Mantua, Vatican) and six heads, as well as an account of such statues that have vanished since the eighteenth century.

The four Parian marble caryatids of Hadrian's Villa (found in April 1952) stood beside the Canopus and were full-size copies of the Erechtheion ones, with the echinus and abacus. They were flanked by two squatter sileni carrying baskets of fruit on their heads, which made the six statues of equal size. See Erika E. Schmidt (1973, pp. 19 ff.) and S. Aurigemma (1961, pp. 113 ff.).

Yet another type of caryatid can only be partially reconstructed. The National Museum in Athens has two that seem to be an asymmetrical pair: one is dressed in a long himation over a chiton (no. 1641), the other in a very long peplos, wound like a sari (no. 1642). Eva M. Schmidt (1982, pp. 98 f.) associates these headless but two-armed statues with a much damaged caryatid head in the National Museum storehouse; if she is right then the two korai from the Roman Villa of Herodes Atticus on the Via Appia (Vatican Braccio Nuovo no. 2270, BM no. 1746) are copies of that type.

65. Such caryatids, with a basket supported by one hand, possibly copies of archaizing Attic statues, were also found at Tralles (G. Mendel, 1914, Istanbul Museum nos. 541, 1189; M. Collignon, (1903, pp. 5 ff.) and Cherchel (or Iol-Caesarea) in Algeria (no. 103); see C. C. Vermeule (1977, p. 25 n. 34) and S. Gsell (1952, pp. 70, 91).

66. The jewelry and the hand gestures follow those of the Roman copies, since all the Athenian statues have lost both hands. What exactly the caryatids carried on their heads is more of a problem. Certainly some of the satyrs (from Pompeius' Theater in the Capitoline Museum, no. 5.23, or the ones from Hadrian's Villa) carry baskets. The Delphic caryatids seem to carry cylindrical boxes, the Istanbul and the Via Appia ones have vases. But so many Greek ceremonies involved carrying baskets or offering-boxes on the head that such a reference would have come to mind quite naturally. See J. Schelp (1975, pp. 15 ff.).

67. For a general account of the figures and their location, see G. P. Stevens et al. (1927, pp. 111 f., 232 ff.). C. Picard ([1931?], pp. 30 ff.) connects the porch with the monument and cult of Kekrops, while J. A. Bundgaard's (1976) view that the porch was built as a strut for some of the branches of the venerable and overgrowing sacred olive tree that grew just outside (and perhaps also within) the temple has some force; but neither suggestion explains the special character of the porch. See most recently H. Lauter (1976) for a survey of the sculptures.

68. The Lycian prince was given his name by his father, who was an admirer of the Athenian ruler. Pericles of Limyra, who may have been the last dynastic Lycian ruler, was an emulator of all things Greek and enemy of the Persians: see J. Borchhardt (1967; 1976) and J. Zahle (1979); see also W. A. P. Childs (1981, pp. 73 ff.). The heroon at Limyra was first found in 1966.

69. The Erechtheion figures were already in place (even if not yet roofed-in) according to the building report of 408/9 BC inscribed on marble slabs (*IG* 1³. 474; trans. G. P. Stevens and J. M. Paton in Stevens et al. 1927, p. 290).

The hypothesis that the Delphic caryatids are in fact much later than had been supposed (the Cnidan treasury is usually dated 550–545, the Siphnyan ca. 525 BC) was advanced by E. D. Francis and M. Vickers (1983, pp. 48 ff.), who wish not only to update the sculptures but to give the two treasuries to some other, unnamed cities. It has never been quite clear which of the caryatids belong to the Siphnyan and which to the Cnidan building. See C. Picard and P. Frotier de la Coste Messelière (1928), and P. Frotier de la Coste Messelière (1943, pp. 321 ff.).

If Pausanias is to be believed, the small town of Sellasia, a neighbor of Carya, was destroyed by the Achaeans and its inhabitants taken into slavery after the defeat of Kleomenes of Sparta in 222. Perhaps Vitruvius incorporated that story into his legend as well.

70. For the buried korai, see P. Kavvadias and G. Kawerau (1907, pp. 23 ff.). The drawings were republished by J. A. Bundgaard (1974); but see also A. Boetticher (1888, pp. 78 ff.) and also G. Dickins and S. M. Casson (1912–1922, vol. 1, pp. 5 ff., 30 ff.). Most of the surviving figures are in the Acropolis Museum (nos. 582–696); some of these figures may indeed have been shown car-

rying a basket on their heads. On the kore type generally, see L. A. Schneider (1975, pp. 31 ff.).

71. W. H. Plommer (1979) suggests, based on the inscription quoted by G. P. Stevens and J. M. Paton (see above, n. 69), that they were not called "caryatids" in antiquity. On the political background, see G. L. Huxley (1967, pp. 29 ff.); but see also H. Lauter (1976, pp. 40 ff.) and M. Vickers (1985, pp. 3 ff.).

72. Paus. I.27.iii (1913, vol. 1, p. 40; vol. 2, pp. 344 f.) describes them as being maintained by the state and changed after the yearly performance of their rite. A number of copies of the statues exist in Istanbul (Archaeological Museum, from Tralles), Cherchel, Mantua, Leningrad, and Vienna. A replica of the Athenian caryatids is in Hadrian's Villa at Tivoli, now in the Metropolitan Museum, New York (*Bull. of Metropolitan Museum*, 1963, 1160 f.); see M. Robertson (1975, vol. 1, p. 347). The type of devout offering bearer who carries his or her offering in a basket steadied by one hand on the head, while the other hand carries an offering jug or dish, was in any case already very ancient: see C. H. W. Johns (1908).

Grouped caryatids with their elbows on either side of the head, their arms bent to support a beam on the neck and arms, are the four damaged life-size, bare-breasted "maenads" in the museums of Taranto and Lecce that come from a tomb near Vaste; better preserved, of this type, are two miniature (under 50 cm) limestone figures in the Musée de l'Art et de l'Histoire, Geneva. At Monte Iato in Sicily, two male and two female figures a little over life-size were found that alternately support the cornice on forearms and head. The females were in long gowns, the bearded males in garland and skin aprons. They have been interpreted as maenads and satyrs, an appropriate Bacchic association for a theater. See H. Bloesch and H. P. Isler (1976, vol. 1, pp. 15 ff.).

73. On the cult, see L. Preller (1860, pp. 234, 236 n. 2); and M. P. Nilsson (1967, vol. 1, pp. 161, 486, 491). See W. Burkert (1983, vol. 8) on the cult (pp. 128, 222) and on Dionysos (who fathered the walnut-nymph Carya—at Carya, pp. 320, 326); but see also S. Wide (1893, pp. 102 f., 108). There was also a cult of Artemis Caryatis connected with the village and the dance, known to Servius (ad *Ecl.* VIII. 29) and Statius (*Theb.* IV.

225). According to Athenaeus (X.392 f.), Pratinas wrote a play called *Karuatides* of which only the title survives. The dances are also mentioned by Pausanias (III.10.vii, and IV.16.ix) and Lucian (*Salt.* 10). Pliny the Elder used the word as synonymous with *thyas* (*NH* XXXVI.iv.23). Hesychius and Photius (s.vv.) called the feast of the goddess "Karyteia."

74. For a detailed discussion of the figures, C. Picard and P. Frotier de la Coste Messelière (1928, pp. 1 ff., 57 ff). On the frieze, see R. Demangel (1932, pp. 260 ff.). Vitruvius, who elsewhere is insistent on the difference between the Doric frieze with its mutules and the Ionian with its dentils, here curiously says that the architect should "insuper mutulos at coronas conloca(re) (I.1.v). That this emphatically Ionic architrave should be described in such Doric terms is—so Roy W. Lewis has suggested (Philadelphia, 1992, typescript)—due to Vitruvius' describing the situation of the caryatids in the Augustan Forum, where they almost form a giant Doric frieze.

One of the curious and unexplained features of the caryatids is that the structure of the capital—a curved circular echinus supporting a square abacus—is more Doric than Ionic, though their elaborate hairdos, the artfully pleated dresses, and the sandals are the features that Vitruvius incorporates in his Ionic column.

75. The custom was common among the megalith builders of southern and western Europe. Stones of this kind appear in Iberia (Folha da Baradas, Casainhos, Moncarapacho); France-Brittany (Locmariacquer, Carnac); the Marne Valley tombs, Provence (a group at Orgon-Sénas); and southwest Russia. They have been curiously "theologized" by O. G. S. Crawford (1957); but see H. Müller-Karpe (1966–1989, vol. 2.1, pp. 287 ff., 548 ff., 592 f., 763 ff.), as well as S. Piggott (1965, pp. 59 ff.) and G. Clark (1977, pp. 125 ff.).

76. R. Fréart de Chambray seems to have had some inkling of this in his rather generalized condemnation of the figured supporters ([1650] 1702, p. 39).

VI The Rule and the Song

1. There were variously said to be about seventy or over one hundred of these children. Several tribes claimed descent from them; the *Iliad* tells of Heraklidae in Tiryns, in Rhodes, and on Kos, though it was also the name of a Spartan clan. The material has been gathered by L. Preller (1964, vol. 2, pp. 648 ff.) and N. G. L. Hammond (*CAH,* vol. 2.2, pp. 678 ff.). On the identity of the Dorians and the Heraklidae, see A. Jardé (1923, pp. 127 ff.). The story of the invasion—from the northeast rather than northwest, as modern archaeologists have preferred—was told by Herodotus (I.56 ff.). On the archaeological evidence, see A. M. Snodgrass (1971), who finds that it does not justify the account of an external invasion. But see also V. R. d'A. Desborough (*CAH,* vol. 2.2, pp. 686 ff.) and O. Murray (1980, pp. 17 ff., 170 f.). On the political use of the Dorian-Heraklidean mythology, see A. M. Snodgrass (1980, pp. 115 f.). On the place of the Heraklidae in Greek hero-cults, see L. R. Farnell (1921, pp. 109 f.).

The best known of the Heraklidae, "noble and great" Tlepolemos, who had settled in Rhodes and led the Rhodian troops at Troy, was killed by the Lycian prince, Sarpedon (Hom. *Il.* V.628 ff.).

2. Thuc. I.12 (trans. Thomas Hobbes). The date was calculated back 80 years after the Trojan War, and 329 years before the first Olympiad (i.e., 1104), by Apollodorus of Athens in the second century; see R. H. Drews (1988, pp. 204 f.). Drews further discusses the various twentieth-century theories about the Dorian invasion and the return of the Heraklidae.

3. On the dialect literature, see C. D. Buck (1928, pp. 12 ff.), A. Thumb (1932–1959, vol. 1, pp. 70 ff.), and J. Chadwick (*CAH,* vol. 2.2, pp. 805 ff, esp. pp. 811 f.).

The name *doriei* is often related by modern lexicographers to *doru,* "tree" (particularly oak), "ship timber," "shaft," "mast," "spear." The ancients were usually content to derive the name from their eponymous hero.

4. The many problems about Dorian origins were inextricably confused by Peloponnesian war propaganda. An extremely skeptical view of the "ethnicity" of the nations was taken by E. Will (1956); see, however, J. Alty (1982, pp. 1 ff.). By Hellenistic times, the Ionian-Dorian boundary was allegedly set up by Theseus (the Ionian Herakles?; Plut. *V. Thes.* 27) while the earlier accounts were summarized by Thucydides (I.3 ff.) and survived in the histories of eponymous heroes of the several cities and tribes. Many of them agree that Hellenus or Hellen, eponymous ancestor of the Hellenes, had three sons, Dorus, Aiolus, and Xuthos. Xuthos in turn had two sons, Achaios and Ion, who are the hero-fathers of the four Greek nations. The Hellenes figure in the catalogue of boats: Hom. *Il.* II.684 (as does the term *Panhellenas*, "of all the Greeks": II.530). The Hellenes open Thucydides' *History* (I.1). Hellen was known to Hesiod (frr. 1, 4) but the cult, though recorded, was not very important: see A. Brelich (1958, pp. 144 f.). The hypothetical genealogy is given by C. Kerenyi (1959, table C), but the difficulty of arriving at consistent genealogies of heroes is well-known. A variant (followed by Vitruvius) is given by Euripides (*Ion* 1575 ff.).

The racial myth of the blond Heraklidae-Dorians from the north conquering the dark-haired, olive-skinned "Achaeans" was virtually created by C. O. Müller (1844) with a sense of a high, even a holy mission (vol. 2, pp. v, ix ff.). Müller's mythological Dorians were virtually monotheistic, since they only worshipped Apollo (his cult receives 327 pages, all the other gods *and* Herakles only 117 pages) with virtually bloodless rites and produced the "higher" Pythagorean philosophy (vol. 2, pp. 47 ff., 250 ff.; vol. 3, pp. 382 ff.). He discusses the highly idealized Dorian view of themselves (vol. 2, pp. 392 ff.) and the coming of the cult of Apollo from the northern Hyperboreans (vol. 2, pp. 275 ff.). Müller later condensed his view of the Dorians (1852, pp. 19 ff., 25 ff.). He was one of the last exponents of the concept of Greek civilization that Nietzsche attacked in *The Birth of Tragedy.* However, Müller's curious picture of the Dorians seems to have been held by those philosophers in the twentieth century who believed that the German language was uniquely privileged as a vehicle for translation from the Greek. On the background and implications of Müller's "Prusso-Dorianism," see A. Momigliano (1980, pp. 48 ff.); a forthright and detailed, if more highly colored, account of it is found in M. Bernal (1987–1991, vol. 1, 308 ff., 329 f.; vol. 2, 72 f., 177).

That the Spartans were particularly associated with the kouros-like display of male nudity is witnessed by a number of writers; it fitted all too well with their self-image. Thucydides (I.6) reports that they were the first to do so. The very word *gumna-sion* is obviously derived from *gumnos*, "naked," and suggests homosexuality—or pederasty—as well as exercises. See Stephen L. Glass, "The Greek Gymnasium," in J. Raschke (1988, pp. 158 ff.), who has also collected the appropriate texts. Spartan girls also exercised naked, or nearly so, which led to accusations about their low moral standards and inconstancy (Helen was, after all, queen of Sparta), on which see Thomas F. Scanlon, "Virgineum Gymnasium," in Raschke (1988, pp. 189 ff.). Even in Sparta, however, athletic nudity was limited to virgins, and it has been regarded as a nuptial ritual. Outside Sparta partially naked girls ("in Amazon costume": presumably with one breast bare) raced at Brauron and at the Olympian Heraia festival (Paus. V.16.ii ff.). Otherwise, female nudity was as rare in Greek life as it was in art. As in other matters, Greek writers contradictorily extol the virtue and fidelity of Spartan women and rationalized adultery there "for the sake of healthy children"; E. N. Tigerstedt (1965–1974, vol. 1, pp. 166 f.; vol. 2, pp. 21 f., 234).

Pausanias (I.44.i) says that he saw the tomb of Orsippos, whose epitaph claimed that he was the first (in 720) to run the Olympic race naked at Megara; but the Spartan tradition seems the more ancient (Thuc. I.6; Dion. Hal. VII.72 names one Acanthus of Sparta as having done so in the 15th Olympiad, the one Orsippos won). On the ritual nudity of the *kouros* and the games, see J. Drees (1962, pp. 126 ff.) and F. M. Cornford ([1935] 1963), who relates the custom of naked racing to the legend of the Cretan *kouroi*—on which, more extensively, see H. Jeanmaire (1939, pp. 413 ff.). For the main literary evidence in PW, s.v. "Sparta" (II.3.ii, esp. 1373 ff.).

5. The Olympic games had been started by Herakles in heroic times (Pind. *Ol.* II, X), either as a commemoration of the hero, Pelops (who gave his name to the Peloponnese, and whose heroon was a tumulus between the temples of Hera and Zeus; so K. Meuli, 1975, vol. 1, pp. 881 ff.), or as a festival of the Olympian high god (L. Weniger, 1904)—who may indeed have had Pelops as his avatar (L. Drees, 1962, pp. 30 ff.). Drees indeed sees the substitution of celestial Zeus for the fertility-king *pharmakos*, Pelops, in terms of the takeover of the pre-Hellenic fertility cult by the northern Dorian sky-god festival. Recent scholarship has tended to dismiss the pre-Hellenic antiquity of the

Pelopeion and the tradition of games on the Olympic site earlier than the eighth century: see Colin Renfrew and Alfred Mallwitz in W. J. Raschke (1988, pp. 13 ff., 81 ff.). However, the sculptures on the east pediment of the great temple of Zeus showed the race between Pelops and Oenomaos; and indeed some excavators have revindicated the Early Helladic (II: before 2100) date under the Pelopeion. So *BSA Arch. Rep.* 36 (1989–1990, p. 30); 37 (1990–1991, p. 31).

The calendric games were reinstituted by King Iphitus of Elis on the order of the Delphic oracle in 884. They were refounded in 776 (V. Vanoyeke, 1992, pp. 72 ff.); Paus. V.4.v ff., V.8.v ff.). The original prize in the races was an apple, until Apollo of Delphi ordered the crowning with wild olive wreaths, which was practiced from the seventh Olympiad onwards (Phlegon of Tralles, in *FHG*, vol. 257, fr. I.2). The wild olives in Olympia were reportedly descended from those that the Hyperboreans had given Herakles, which he had planted in Elis (Pind. *Ol.* III.11–35). In the other Panhellenic games, the crowns were of bay (Delphi), where a palm leaf and an apple were also sometimes given; of pine (Isthmus); and of celery/parsley (*selinon*, at Nemea). See J. Fontenrose in Raschke (1988, pp. 136 f.).

6. On the formation of the polis and the estimated population of eighth- and seventh-century Greece, see A. M. Snodgrass (1980, pp. 21 ff.) and *CAH* vol. 3.3 (pp. 286 ff.).

On the Roman addition of a base to the Doric order, and the Roman Doric generally, still see W. J. Anderson, R. Phené Spiers, and T. Ashby (1927, pp. 43 ff.), D. S. Robertson (1945, pp. 207 ff.), and G. von Kaschnitz-Weinberg (1961–1963, vol. 1, pp. 43 ff.; vol. 3, pp. 74 ff.). See also G. Worsley (1986, pp. 331 ff.).

7. A. W. Lawrence (1983, pp. 239 ff., 337 ff.); W. B. Dinsmoor (1950, pp. 117 ff.).

8. On varieties of the pre-Hellenic sanctuary, see specifically B. Rutkowski (1986); more generally, see J. Chadwick (1976, pp. 141 ff.); F. Schachermeyr (1976, pp. 99 ff.; 1964, pp. 158 ff.); E. Vermeule ([1964] 1972, pp. 282 ff.); C. G. Starr (1962, pp. 172 ff.); T. B. L. Webster (1958, pp. 106 f., 139 f., 212); H. L. Lorimer (1950, pp. 439 ff.); M. P. Nilsson (1950, pp. 77 ff., 110 ff., 117 ff.); and H. R. Hall (1915, pp. 145 ff.). On the

siting of the Greek temples within the city; at the edge of the city, in a "suburban" situation; or outside inhabited territory, in an isolated sanctuary, see F. de Polignac (1984, pp. 30 ff.).

9. Vitr. IV.1.v. Later sixteenth-century grammarians even wanted to derive the Latin *fanum* from the Greek *naos* (accusative, *naon*, transposed to *anon*, and with harsh breathing . . .). However, most scholars now relate the word to the still puzzling *fas* ("in accord with divine law," "fortunate," "morally right": *ius ac fas* means "right according to civil as well as moral law"), which relates to other Italic languages: *fasnom* in Old Latin, *fesna* in Umbrian, *fiisma* in Oscan. See G. Devoto (1954–1962, vol. 1, pp. 359 f.); E. Benveniste (1969, vol. 2, pp. 134 ff). They further relate *fas* to *fascia*, "a bundle"; also *fascinus*, "a phallic charm," "a bewitchment." However (by way of *fanum*), the word gives *fanaticus*, "a temple servant"—a word that has now changed meaning completely.

On *templum* and *temenos*, see J. Rykwert (1976, pp. 45 ff.). In a sense, *fanum* is almost the best translation for *fetish* in its older sense, on which see W. Pietz (1985–1989, no. 16, pp. 105–124).

10. Their rotted remains were later brought up to be used in making piglet-shaped images for luck that were buried in fields for fertility. On the two words and the ritual use of such pits, see Kevin Clinton, "Sacrifice in the Eleusinian Mysteries," in R. Hägg, N. Marinatos, and G. C. Nordquist (1988, pp. 72 ff.); W. Burkert (1984, pp. 350 ff.); H. Jeanmaire (1939, pp. 266 ff., 305 ff.); and J. E. Harrison ([1903] 1977, pp. 120 ff.). The main ancient sources are Xen. *Hel.* V.2.xxix; Plut. *Quaest. Gr.* 31.; Luc. *Dial. Het.* II.1 (and scholia; this was first published by E. Rohde, 1870, pp. 548 ff., and became capital for the interpretation of the other texts); Athen. VII.37 ff.; Clem. Alex. *Protr.* II.

11. Hes. *Theog.* 534 ff. (ed. R. Merkelbach and M. L. West, 1966, pp. 131 f., 316 ff.), the passage that Lucian mocks in his *Prom.* 3; though Aeschylus makes it one of Prometheus' general religious regulations (*Prom. Vinc.* 496 ff.). In fact the bones that were burned in that altar after the sacrifice were the thigh bones (*mera*, on which below, p. 147), pelvis, and tail, which were recomposed on the altar slab as they were in the beast before being burned. The pact was supposed to have taken place during an assembly of men and gods at "Mekone," the place of poppies—one of the ancient names, it is said (Str. VIII.7.xxv; Steph. Byz., s.v. "Sikyon"), for Sikyon or perhaps a place near it. For this deceit, as well as the more famous theft of fire (Hes. *Op. D.* 45 ff.), Prometheus was chained to the rock. See M. P. Nilsson ([1955] 1967, vol. 1, pp. 27 f., 143) and W. Burkert (1983, vol. 2, pp. 252 ff., 190 f.; this condenses his earlier statement, 1972a, pp. 10 ff., 42.4). A detailed criticism appears in G. S. Kirk (1981).

On the place of sacrifice in the mythologeme, see C. Kerenyi (1963, pp. 42 ff.); see G. S. Kirk (1974, pp. 137 ff.) for a positive interpretation of the myth. M. Detienne and J. -P. Vernant (1979, pp. 15 f., 37 ff.) have an extended analysis of this passage and the ritual practices that it guarantees, and they account also for the elaborate ritual of butchering by which the beast was cut up (pp. 23 f., 154 f.). See also L. M. Nield (1969, pp. 160 ff.). Although the notion of sacrifice as it was set out by H. Hubert and M. Mauss (1899; in M. Mauss, 1968, vol. 1, pp. 193 ff.) is still worth reading, the phenomenon has been much discussed recently: see R. G. Hamerton-Kelly (1987, passim). On the symmetry of the two betrayals and the percipience of Zeus, see H. Lloyd-Jones (1971, pp. 33 f., 82).

12. Prometheus and Hephaestus are sometimes assimilated to each other. But Prometheus (forethought) also had a brother, Epimetheus (afterthought), who in fact, according to Plato (*Prot.* 320d ff.), ran out of valuable qualities when he made man—hence the problem.

The poets called him *ho purphoros theos, titan Promētheus*: "the torch-bearing god, Prometheus the titan" (Soph. *Oed. Col.* 55 f.). His lineage is very complicated and the legends about him (as collected by Ludwig Preller) were very confusing. As the bringer of fire, he also figures—with Hephaestus—as one of the deities who taught humans the arts, as Pliny still witnesses (*NH* VII.lvi.199, 209).

13. The Haghia Triada sarcophagus (Archaeological Museum, Heraklion) seems to be showing the worship of a cult statue, or at least a kind of herm in front of rather than inside a building. A very explicit reading in that sense is found in B. Rutkowski (1986, pp. 101 ff.). On the origin of the cult statue, see this chapter, p. 143, and chapter 7. But see also M. Robertson (1975, vol. 1, pp. 11, 35 ff.) and H. L. Lorimer (1950, pp. 442 ff.).

T. B. L. Webster dismisses the Homeric temples as "modern" fictions invented for those particular scenes (1958, p. 221); though elsewhere (p. 242) he speaks of the rite of *supplicatio* that was practiced there as being "very old, and . . . known only to Mycenaean poets" (but see also p. 212). The word *nuos* or *naos* occurs several times in the Homeric epics: in Ilion itself (*Il.* VI.269, 297 ff.) Queen Hecuba offers a robe to Pallas (or rather, to the Palladium) in a temple served by Theano, the only Trojan priestess mentioned in the story. The offering of a robe is a common enough cult practice in Greece—witness the Panatheneia. Chryses the priest (*Il,* I.35 ff.) reminds the god in his prayers that he *erepsa* the temple: this is usually translated as "roofed," though *erepsei, erefsei,* may have been more specifically, "to thatch"; so H. L. Lorimer (1950, p. 441 n. 1). If Chryses means that he thatched the temple himself, he may be thinking not of a permanent shrine but of some temporary hut, like the one built for the Delphic *septentrion,* on which see Plut. *De Def. Or.* 15 and *Quaest. Gr.* 12; Str. IX.422. In *Od.* VIII.80 ff., Agamemnon consulted the oracle about the end of the war. The stone threshold or porch stands in the poem for the whole temple, which is also mentioned in *Il.* IX.404 f. A temple is also vowed to Apollo by Odysseus' crew before they kill his flocks in *Od.* XII.346, but since the god had taken his revenge, the vow was broken and the temple not built. The founder of Phaecia was said to have built temples as part of his first foundation (*Od.* VI.5 f.). On the implications of all these texts, see G. S. Kirk (1962, pp. 185 f.).

14. J. Chadwick (1958, p. 114). Named gods also held land, though it is not quite clear through whom they did this. But see also A. J. B. Wace and F. H. Stubbings ([1962] 1974, pp. 454 ff.) and T. B. L. Webster (1958, pp. 105 ff.).

15. On the etymology of *wanax, wanass,* and *basileus,* see E. Benveniste (1969, vol. 2, pp. 23 ff.) and R. H. Drews (1983). Drews starts (pp. 3 ff.; see also pp. 108 ff.) with a critical report of Juri Andreev's thesis (1979, pp. 361 ff.) that the *basileus* was a village chieftain, who became a member of an aristocracy when his village was absorbed into a polis as the result of the synoikismos movement of the eighth century. As Drews observes (pp. 101 ff.), the plural of *anax, anaktes*, is virtually unknown to epic, while the plural *basileis* is common.

16. It may well be that the name *Nestor* was not personal, but also covered a kingly function. See D. Frame (1978, pp. 81 ff.).

17. Her. VIII.137 ff.; Apol. II.8.ii ff.; Diod. Sic. IV.58.i. The Heraklidae of Lydia, the second dynasty, descend from the Assyrian and Hittite Herakles, Sandos or Sandon (see M. C. Astour, 1965, pp. 65 f.). It is notable that their military "baton of command" is the double-edged ax that Herakles had taken from the Amazon Hippolyta and given to Omphale. See Nicolas of Damascus (*FHG*, vol. 3, pp. 382 ff.); Plut. *Quaest. Gr.* 45; and G. Radet ([1893] 1967, pp. 64 ff.).

18. The chief archon was the eponymous one, who gave his name (as the consul did in Rome) to the year; the archon basileus was second. W. Burkert (1972a, p. 94; though cf. p. 387) thinks that the priestly duties of kingship would not descend through a state officer even if he carried the *basileus* title (his wife was known as *basilissa*), and that instead they were mostly connected with the new cult of Dionysos, not with Athena Polias and Erechtheus. However this seems based on a rather narrow interpretation of his duties, on which see also A. Mommsen (1898, pp. 103 ff., 218 ff., 379, 399). On the archaic relation between *basileus* and *monarchos*, see R. H. Drews (1983, pp. 114 ff.).

19. On the *wanax* and his temenos, see A. J. B. Wace and F. H. Stubbings ([1962] 1974, pp. 454 f.); on the divinity of kingship, see A. M. Hocart (1927, pp. 9 ff.) and J. G. Frazer (1911–1915, vol. 4, 190 ff.). On the Egyptian and Mesopotamian divine kings, see H. Frankfort (1948b), particularly on their cult (pp. 291 ff.). That many kings achieved heroic and therefore quasi-divine status is familiar stuff: see A. Brelich (1958: pp. 83, 178, on the king-hero; pp. 99 f., 315 ff., on the worship of the living hero).

20. This personage appears in Homer as a Lycian prince who fought with the Trojans; he killed Tlepolemos and was in turn killed by Patroclus (*Il.* II.876 ff., V.479, XII.292, XVI.419 ff.). He is sometimes the son of Laodemia, daughter of Bellerophon, or the brother of Minos and Rhadamantes (and therefore the son of Europa and Zeus the Bull) who fled to Lycia after the brothers quarreled (Apol. III.1.ii; Appian *Bel. Civ.* IV.78; Her. I.173). See PW, s.v.; also A. B. Cook (1914–1940, vol. 1, 464 ff.); E. Rohde (1925, pp. 151 f.). He was also worshipped as an oracular divinity, and the name *Zrppdoni* occurs on a Xanthan inscription.

21. See PW, s.v. "Minotauros." The custom of wearing animal masks, though not usual on the Greek or the Roman stage, was very common in Egyptian religious ceremony. The changeling, the half-animal, half-human figure, has no determinable relation to the masked figure. On the other hand, the relation between wearer and mask has been discussed by R. Merz (1978, pp. 60 ff., 83, 266 ff.) and E. Canetti (1960, pp. 428 ff.). But see also R. L. Grimes (1982, pp. 75 ff.); A. D. Napier (1986, pp. 204 ff.); and M. Eliade (1964, pp. 93 f., 165 ff.). On Egyptian ritual masks, see E. Otto (1958, pp. 22 f.); on the persistence of Egyptian mask-wearing into the classical world, see R. E. Witt (1971, pp. 97 f., 208). On the mask in Greek, particularly Dionysian ritual, see A. D. Napier (1986, pp. 50 ff.); H. Jeanmaire (1951, pp. 11 ff.). On the Gorgoneion as a mask, see A. D. Napier (1986, pp. 92 ff., 168 ff.); E. A. S. Butterworth (1966, pp. 60, 151 f., 163 ff.). More generally, see M. Bieber (1961, pp. 22 ff. and passim); PW, s.v. "Maske."

The Minoans are well-known to have had Egyptian connections: Cretan envoys carrying Minoan products appear in Egyptian painting from the Eighteenth Dynasty onward; Egyptian scarabs of a much earlier period have been found on Crete. The Cretans were certainly familiar to the Egyptians, probably as *Keftiu*, and in some way related to the "People of the Sea": on which see H. R. Hall, "Keftiu," in S. Casson (1927, pp. 31 ff.) and F. Schachermeyr (1964, pp. 109 ff.). The Minoans certainly had a cult of the beetle analogous to that of the scarabeus, and used a very Egyptian-looking sistrum for cultic purposes. The Egyptians worshipped a bull in at least three sanctuaries, and in one of them the cult figure was a bull-headed man: Ur-mer, one of the sacred bulls of Heliopolis. More surprisingly, another feature of the Minotaur legend also appears in one version of the central one of Osiris and Isis: Isis reassembles the dispersed members of Osiris except for the generative one in an artificial bull, by which she is nevertheless impregnated. See Plut. *De Is. et Os.* 14 ff., Diod. Sic. I.21.lv, for this inevitable variant of the generating legend of a people for whom cattle were the main source of livelihood, though neither in Egypt nor for that matter in Asia Minor does it assume the importance that it has for the Cretans, as the Egyptians themselves recognized.

22. Of the thirty-two pre-500 BC temenos excavations tabulated by B. Bergquist (1967, pp. 62 ff.), over twenty are rectangular, and orientated east-west; the two superimposed ones are at Larissa and Samos. These and the others are accounted for by Bergquist for topographical reasons. The only square one seems to have been the Delion at Paros.

23. On tree and stone as a commonplace in Greek ritual places, see S. Mayassis (1966, passim). Some material on stone outcrops as the bones of Mother Earth appears in M. Eliade (1977, pp. 35, 164).

24. PC, s.v. *"meiromai."* Also see M. Detienne and J.-P. Vernant (1979, pp. 23 ff.). The distinction between the *moira*, by which the victim's body is hieratically divided, and the *isodomia* of later sacrifices in the polis, when the meat is divided into parts of equal weight to be distributed by drawing lots, is made by Detienne. On *moira* and *themis* in the Homeric poems, see A. W. H. Adkins (1972, p. 1 ff.). In later Greek, *themis* had a sense similar to *moira*, though it could also be translated as "goods"—both literal and metaphoric. On the substances used and the various methods of sacrifice, see M. P. Nilsson ([1955] 1967, vol. 1, pp. 132 ff.).

25. Pausanias (V.13.viii ff.) describes it as being 22 ft. high, 125 ft. in circumference at the base; the ash of the thigh bones burned in the sacrifices were mixed yearly with the lime-rich waters of the Alphaeus by the sanctuary *mantes*, "diviners," into a plaster. On the dimensions, see E. N. Gardiner (1925, pp. 193 ff.). W. Dörpfeld (1935) supposes this altar to have been the original one of

Zeus and Hera, founded in the fifteenth century BC, and corresponding to the Idean Cave in the Kronos Hill opposite (about 50 m away; see vol. 1, pp. 70 f.). For the reconstruction of the shape according to Pausanias, see C. G. Yavis (1949, pp. 210 ff.).

26. "Impromptu hearths," Pausanias has the Attic people call them. It is not clear whether the Pergamene altar was the famous one now reassembled in Berlin. On this M. Kunze and K. Volker (1985, pp. 25 ff.).

27. In cremation the bones were considered the eternal and therefore incorruptible part of the body, so that the pyre was extinguished before they became shapeless, and the remains of the skeleton were sought out and "read," or assembled into some ossuary; on the analogy between sacrifice and cremation, see W. Burkert (1972a, pp. 64 ff.).

28. E. Rohde (1925, pp. 18 ff., 166 ff.). It is a difference taken up in apostolic and patristic theology, and warranted by St. Paul's distinction between *sarx* and *sōma* in 1 Cor. 15.45 ff. See C. A. von Peursen (1966, pp. 88 ff.) and J. A. T. Robinson (1963, pp. 28 f.). But see R. Rorty (1980, 43 f.).

29. These remarks reflect a position in the debate on the nature of sacrifice, which I would locate somewhere between René Girard's psycho-structural account of the ritual-mimic/substitute-victim complex and Walter Burkert's evolutionary/formative conception of interspecific aggression and the ritualized hunt of Paleolithic men. The debate is set out and mediated in R. Hamerton-Kelly's (1987) book to which both protagonists contribute papers and discussions. My own view, as I suggested in the previous chapter, inclines to Adolf Jensen's (1963) conception of ritual as a formalization of what he calls "ontic shock."

30. Fragments of several such temple buildings survive at Asine in the Argolid; on that and other such remains, see R. Hägg (1992, pp. 9 ff.). On Crete, temples A and B at Prinias and the one at Dreros are examples; in Sicily, the so-called Megaron Temple at Selinus: the one at Korakou outside Corinth may even go back to the Bronze Age. As for the so-called temple G at Eleusis—perhaps part of a megaron complex—see P. Darcque (1981, pp. 593 ff.).

31. On these proto-Doric temples and the structure and use of their "fireplaces" (or perhaps, more accurately, "fire pits"), see M. P. Nilsson (1950, pp. 454 ff.); M. Guarducci (1937a, pp. 159 ff.); and C. Picard (1930). On the possible fire pits in Minoan palaces and villas (Mallia), see B. Rutkowski (1986, pp. 13, but also 120 ff. on the more usual Cretan open-air altars). On the distinction between Minoan and Mycenaean cult-places, see J. C. van Leuven, "Problems and Methods in Prehellenic Naology," and Colin Renfrew, "Questions of Minoan and Mycenaean Cult," in R. Hägg and N. Marinatos (1981, pp. 11 ff., 27 ff.).

On the keeping of "immortal" sacrificial remains (animal horns and skulls and bones generally), as well as those of animals killed in the hunt, see K. Meuli (1975, vol. 2, pp. 957 ff.). But see also n. 27 above.

32. Such processions (the Panathenaia and the Eleusinian in Athens are simply the most obvious examples) were very much a feature of classical Greek religion, as they were of the cities in the Near East: see W. Andrae (1941).

33. Such as the Lefkandi heroon; on which see chapter 7, pp. 198 ff.

34. About the earliest known hearth-centered megaron prefaced by a porch with two in antis columns is in Dimeni in Thessaly, a walled village that has for conventional purposes given its name to the civilization of the late Neolithic "invaders" who are sometimes named as the first Greek-speakers in Hellas. Though the autochthonous Neolithic Sesklo also has a hearth-centered megaron, which was considered to be slightly later, recent reexamination of the sites has upset this tidy arrangement. See R. Treuil (1983, pp. 297 ff.) and a summary of the evidence in *CAH* vol. 1.1, pp. 575 ff.; on the Dimeni megaron and its implications, see pp. 605 f. By the middle Bronze Age, hearths become relatively commonplace, as in the "palace" of Phylakopi on the island of Melos and even as far as Kültepe in Central Anatolia; but see R. Naumann (1955, pp. 316 ff., 417 ff.). There is a remarkable hearth, presumably a ceremonial one, with a hollow in the shape of a double ax in the House BG (early Helladic) at Lerna in the Argolid.

The fireplace, *eschara*, retains the dual meaning of "hearth" and "brazier" in early Greek literature: as "hearth," *Od.* VI.52, 305; more or less synonymous with *hestia*, "campfire," *Il.* X.418; "brazier," Aristoph. *Achrn.* 888. But it also retains the meaning "sacrificial fire" or even "fire altar"—but one that is hollowed out, as against *bomos*, a built-up altar: so Aesch. *Pers.* 205, *Eum.* 108.; though the two words in combination, *bomioi escharai* (Eur. *Phoen.* 274), can be taken to mean "built-up fire altar." And, of course, it had a secondary meaning, "sanctuary for suppliants" (*Od.* XXIII.71), while in Vitruvius it has the technical meaning of "gridded base" (for a war machine: X.11.ix, 14.i). It survives in the French and English *eschar*, a burn mark or scab. The shape of the brazier was much the same as that of the offering table: a flat top supported on three legs—a tripod, in fact.

35. See M. Eliade (1977, pp. 72 ff.).

36. See, e.g., N. Marinatos (1985, pp. 118 f.).

37. S. P. Morris (1992, pp. 229 ff.).

38. A. J. B. Wace and F. H. Stubbings ([1962] 1974, pp. 368 ff., 494 ff.).

39. The question has recently been reopened by Frank Brommer (1987, p. 20), who suggests that the upturning of tree trunks—which in the case of Minoan architecture, following Arthur Evans (*PM* vol. 1, pp. 343 f.; vol. 2, pp. 145, 321; vol. 4.1, p. 970), he believes to have been of cypress—would have been a natural way of preventing rot at the base, a procedure still followed by woodwrights in the Black Forest. But the issue is a bigger one, since these columns, as Brommer rightly sees, were tree trunks turned upside down: a major constructive-iconographic theme that has been treated by A. Coomaraswamy (1977, vol. 1, pp. 376 ff.) and has been found by many ethnographers and anthropologists in various parts of the world, though almost never in a structural position, as in Minoan building. In any case, the puzzle remains: why did the first Greeks revert to the right-way-up tree trunk as a column?

Exceptionally, the "Wrestler" rhyton from Haghia Triada (in the Heraklion Museum) shows the block-headed columns narrowing upward.

40. See the scanty material from buildings and representations from wall paintings and furniture collected by B. Wesenberg (1971, pp. 3 ff.). Wesenberg in fact speaks often about a "normal" Minoan-Mycenaean column, which is the kind Arthur Evans found painted in the "Spectator" fresco and another fresco fragment from the palace at Knossos (both ca. 1450, now in the Heraklion Museum). On the other hand, the "pedestal lamp" in *rosso antico* (before 1450) as well as the "gryphon-and-column" painted relief from the East Hall of the palace (ca. 1600; both also in the Heraklion Museum) seem to refer to a column with a capital ringed by leaves: but see also n. 49 below. Another stone lamp (also Late Minoan I, ca. 1450) from the north side of the central court of the palace has a similar "capital" but an uncolumnar trilobate composite shaft. The column fragments that survive in situ, such as the two in the anteroom of the north wing of the second palace at Phaestos (Middle Minoan or Late Minoan I), do have the sloping shafts and disk bases. Their upper parts do not survive.

41. Strabo, who had not visited Mycenae, seemed to think that nothing was left of it (VIII.6.x); however it was known to travelers and was one of the monuments that ensured the identification of the site. Pausanias saw the "Gate with the Lions over it" (II.16.v). Philip Hunt, Lord Elgin's agent and chaplain, inspected it covetously, but was deterred (as many before him must have been) by its weight and distance from the sea (W. St. Clair, 1967, pp. 101 f.). The site around the gate was carefully observed by Schliemann, who first noted, for instance, that the heads that had vanished must have been attached to the relief by dowels (for which there are holes in both lions' necks) and may have been of a lighter and perhaps more precious material: see K. Schuchhardt (1891, pp. 138 ff.). The problem of the date of the sculpture being later than that of some of the polygonal walls was already raised by Schliemann's assistant, F. Adler (see L. Deuel, 1977, pp. 230 ff.), and the matter was linked to the relation between the gate and the tomb circle within it; there was a suggestion that there may be a tomb under the gate itself, though nothing has yet been found under the stone threshold. See most recently G. E. Mylonas (1957, pp. 7 ff., 13, 18, 22 f., 34 ff.). Mylonas, at any rate, thinks it contemporary with the Treasury of Atreus, ca. 1250, and considers "Atreus" the

builder of both (pp. 88 f.); on the finds under the threshold, see pp. 114 ff. The material of the doorjambs, the beam, and the sill are of conglomerate, more elaborately hammer-dressed than the surrounding walls. The relief is in pale gray, hard limestone. See A. J. B. Wace (1949, pp. 51 ff.).

42. It is used as an allover pattern to represent the sea on a number of Cycladic "frying pans," as the strange hollow vessels of dark earthenware, which were probably used as water mirrors, are usually called.

The fragments of the facade of the Treasury of Atreus were presented to the British Museum by the Earl of Elgin and the Marquess of Sligo (nos. A 51–57). The greater part are in greenish-gray limestone, and there are two fragments of a bull relief in gypsum and a band in red marble. They are described, and their acquisition discussed, by F. N. Pryce (1928, pp. 14 ff.). Several restorations of the facade have been made, though I do not find any of them wholly convincing: it seems to me that the fragments of relief must certainly, like the one in the Lion Gate, have belonged to a sculpture that appeared in the relieving triangle over the beam.

43. The ones on the Lion Gate columns seem to have been common both in Mycenae and in Minoan Crete: a thin flat base under the column; at the top of the shaft, a separating fillet; above it, one thick hollow molding. The hollow molding is filled with a ring of leaves as often as not. Next a thicker swelling molding—usually half-cylinder in section—supports a square abacus on top. A number of fillets and other minor moldings sometimes act as spacers between these elements.

44. The earliest to survive are the Hittite guardians of the gates at Alaça Hüyük and Boghazköy, of the fourteenth or thirteenth century (either in situ or at the Ankara Museum). The surviving guardians from Nimrud and Khorsabad are at the British Museum; they are, of course, much later than the Lion Gate (Nimrud, ca. 850; Khorsabad, ca. 730) and were emulated both by Cyrus' sculptors at Pasargadae and by Xerxes' at Persepolis in the next century. But it would be fair to assume that many such figures have not survived, and that the theme is traditional.

The sphinx, the Egyptian man-headed lion from whom they seem to derive, is much earlier of course: it appears fully formed in Fourth Dynasty

Egypt, in the middle of the third millennium. On the Greek word *sphinx* and its probable Egyptian origin, see I. E. S. Edwards (1967, pp. 141 ff.). Even earlier monstrous creatures (such as the scorpion-man) are represented in both Mesopotamian and Egyptian art. The commonest Mesopotamian monsters are the man-faced, bearded bull and the winged bull, which are shown on some of the earliest cylinder seals, in reliefs, and on small sculptures. They are conflated into a human-faced, bearded, winged bull about the middle of the second millennium. The unbearded winged sphinxes that guard the gates at Alaça Hüyük look rather Egyptian. See the evidence collected by H. Demisch (1977, pp. 16 ff., 42 ff.). However, long before that, the bearded hero grips two man-faced bearded bulls (in shell mosaic) on the sounding box of the lyre of Queen Shub-Ad of Ur (in the British Museum; ca. 3500–3200?); a gaming board from the same cemetery has (with the exception of the two end and opening panels) panels of goats and bulls on either side of a tree alternating with two lions heraldically attacking a deer and goat; see C. L. Woolley (1935, pp. 76 ff.).

On the Sumerian door guardians, see E. D. Van Buren (1945, pp. 4, 46 ff.). The most famous of such guardian columns were Jachin and Boaz, the keepers of the door of the sanctuary in the Jerusalem temple: see 1 Kings 7.15 and 2 Chron. 3.17. But see E. Vermeule ([1964] 1972, pp. 214 ff.).

45. In a sense, glyptic art is older than metal; baked clay seals were found at Çatal Hüyük (J. Mellaart, 1967, pp. 220 ff.). They were probably used as fabric dye-stamps and are dated to the end of the seventh millennium. In the fifth millennium, they also appear in India: at Mundigak, in south Afghanistan. See B. and R. Allchin (1968, pp. 106 f., pl. 7b). The first Sumerian cylinder seals are probably of the second half of the fourth millennium, the Uruk period.

The heraldic theme appears almost at once: see H. Frankfort (1939, pp. 24 ff.), who also discusses Egyptian seals (pp. 291 ff.), Aegean ones (300 ff.), and the recurrence of the heraldic motif (pp. 308 ff.); but see also H. Frankfort (1954, pp. 103 ff.). In Egypt the heraldic theme appears about the same time, in a tomb painting at Hierakonpolis (now in the Cairo Museum) and in a near-contemporary ivory relief on the handle of a flint knife from Gebel-el-Arak (now in the Louvre). On the importance of those two figurations in Egyp-

tian prehistory, see W. B. Emery (1961, pp. 38 ff.). Although the theme is not common in Egyptian art, it does appear, for instance, on the side of the pharaonic throne to represent the union of Upper and Lower Egypt: this is particularly striking on the statue of Sesostris I (1971–1928) in the Cairo Museum.

46. The heraldic *theme* as a protector of the door first appears in Sumeria: the grand beaten copper plaque (now in the British Museum) of the lion-headed eagle, the sky-god Imdugud, holding two deer in his talons was over the door of the First Dynasty temple at Al'Ubaid (C. L. Woolley, 1935, pp. 72 ff.). It is repeated on a statue base from the Abu temple at Eshnunna/Tell Asmar, and in the somewhat later silver vase of Entemena of Lagash (in the Louvre), presumably representing the local storm-god Ningirsu, now holding lions. The bird of prey holding two animals is interlaced with the hero holding two lions on a Serpentine vase from Kafajeh: these two last are in the Baghdad Museum. This device was much loved in Lagash. It appears multiplied and interlocked on mace heads from Kish (H. Frankfort, 1954, pp. 31 f.); on cylinder seals (Woolley, 1935, p. 124, pl. 69k); on reliefs such as the "Vulture Stele" in which King Ennatum celebrates a military victory and the god holds enemies in a net by a symbol (by carved handle?) of the lion-seizing heraldic eagle, or the "Tablet of the Scribe/High Priest Dudu" from Telloh/Lagash (both in the Louvre), which in turn served as a stand for a votive mace, of the kind that were carved with the same scene. They were, it seems, used for swearing oaths: all of which alludes (however remotely) to the use of the double ax in Crete (Frankfort, 1954, pp. 32 f.). On the votive maces, see E. D. Van Buren (1945, pp. 166 ff.). On the formal origins of the heraldic theme in Mesopotamia, see H. A. Groenewengen-Frankfort (1951, pp. 154 ff.); on the implications of the heraldic theme in the Aegean, see M. P. Nilsson (1950, pp. 355 ff.). On the sacred tree, see H. Danthine (1937, pp. 69 ff., 153 ff.) and Van Buren (pp. 23 ff.).

47. The lions were presumably taken from art rather than nature. The last native lion in Greece must have been the one Herakles killed at Nemea, which is a mere seven miles from Mycenae. A. J. B. Wace (1949, p. 53—following Arthur Evans, 1901, pp. 99–204) suggests an identification between Cybele, often associated with lions (W. Burkert, 1979, p. 102 f.), and the column; the identification *Kubeles-pelekes,* made by Suidas, associates the column here with the double ax. G. E. Mylonas (1957, pp. 27 ff.) rejects this and considers the column with its entablature a representation of the palace, and through it the Atreid dynasty, which would make the relief strictly narrative-heraldic. This does not seem to me satisfactory either, as the theme is very common in Mycenaean glyptic art (particularly in ring bezels), in which it goes through a number of transformations; this suggests an apotropaic character. The identification of the column with Cybele presents another difficulty: in the heraldic position the male god/hero is more usually accompanied by carnivores, the goddess by herbivores.

48. The openings over the stone doors are always made by cutting the stones at an oblique angle to the wall surface, to make a kind of relieving arch for the heavy beam, even though the arch is not a true one. The principle had of course also been managed by the Egyptians, in the burial chambers of the pyramids, notably that of Cheops. However, the insertion of a plaque of a highly finished (and therefore denser and heavier) stone would seem to argue against this explanation.

The best survey of the models is provided by H. Drerup (1969, pp. 69 ff.), who calls them *Hausmodelle,* though without committing himself to the nature of the buildings. The actual opening in the steep gable is sometimes square (from the Argive Heraion). Some are square ended, others apsed. Unlike the many Italian house models (which were almost all cremation urns), the few Greek ones seem to be votive offerings—*anathēmata.* On Crete some twenty examples have been catalogued (by R. Hägg, 1990, pp. 95 ff.). The one well-preserved and highly decorated circular model in Crete came from Archanes (now in the Heraklion Museum, sigma gamma 376, probably late Protogeometric). Robin Hägg tentatively suggests a Syrian origin of the form, citing the Ras Shamra model urns (now in the Louvre).

As for the roofs, in some of them (from the Samian Heraion, for instance, whose roof is partly hipped) the thickness of the thatch and the twisted straw ridging are clearly marked. Indeed S. Marinatos, *"Aetos,"* in D. Zakithinos et al. (1965, pp. 19 ff.) suggests that the term, as used in building terminology for gable or ridge piece, assimilated the thatch finishing of such roofs to the pinions of birds, especially omen-birds of prey.

The only flat-roofed (terracotta) model comes from Chania Tekke near Knossos. It is the only one, too, that has a clearly marked chimney (as did the circular model from Archanes). Like practically all of these models, this also has a row of small windows under the roofline.

49. The very word *baetyl* seems to have been a Semitic loanword (from *bet-el,* or perhaps *bet-eloah*—"god-house"); the alternative derivation from *baety,* a shepherd's skin-coat, through the goatskin that Kronos swallowed for Zeus, is proposed by M. Mayer (WHR, s.v. "Knossos"), but it is not generally accepted.

Jacob called the place where he dreamed sleeping with his head on a stone *Bet-El* (Gen. 28.12 ff., 35.1 ff.), and that became the name of a town on the site, which had previously been called *Luz,* "hazel (or almond) tree." There was also a prominent oak in the place, so that the original Bethel seems to have been a holy place with standing stone and tree. Sir Arthur Evans (1901) pointed out the important association of tree and pillar (pp. 7 ff.) but unfortunately also spoke of the baetyl as a kind of column; again he referred to the Semitic cult of Asheroth as being associated with baetyls. E. Goblet d'Alviella ([1891] 1983, pp. 170 ff.), however, had already pointed out the pre-Exodic Canaanite nature of the cult. M. P. Nilsson (1950, pp. 244 ff.) repeats Evans' baetyl/ashera association, but they seem to me to be related only in that they are both cult objects and nonstructural. The baetyl *(mazzebah)* is always "found" and may act as a fetish, monument, or tombstone. This is shown in Jacob's annointing of the stone on which he rested his head during the vision of the angelic ladder (Gen. 28.18) where on another occasion God would order him to change his name to Israel (Gen. 35.14). The same words for setting up a stele are used a few sentences later for Jacob's monument to Rebecca, his favorite wife (Gen. 35.20).

These were Jewish holy places and remained places of pilgrimage; but there were a number of other holy stones *(Hamanim, Margemah, Eben, Gal, Gilgal)* that were regarded with antipagan horror by the prophets and the Talmudic rabbis. Many such holy places have been recorded and excavated: the best preserved may be the one at Hazor. The ashera may be a growing tree, perhaps specially decorated, but it may also be a constructed object, composite and human-made. Its nature is illustrated in the Mishnaic prohibitions against using any leaves of wood that had been part of an ashera, or even its fruit. See *A. Zar.* 3.7, *Orl.* 17 f., and *Sukkah* 3.1 ff., about palm, willow, and myrtle branches, which may not be used for the construction of a tabernacle, and citron, which may not be used for carrying in the procession, if they had been part of an ashera. The most important passage, however, is *Nez. Ab Nis.* 2.7, in which *ashera* is defined as "any tree that is an idol," though R. Simeon is quoted with approval when he "authorized" the use of a tree under which a heap of stones with an idol in it was being worshipped— since only the idol was at fault. The earlier passages refer to cultic constructions, not to growing trees. But see also W. O. E. Oesterley and T. M. Robinson (1930, pp. 9 f., 43 f., 58 ff., 175 ff.); M. -J. Lagrange (1905, pp. 120 ff., 171 f., 204 ff.); and P. Torge (1902, pp. 29–33). See also J. Rykwert (1994).

50. Though W. O. E. Oesterley and T. M. Robinson (1930, p. 177) think that they were two different deities. It is curious that the Hebrew language knows both a feminine *(asheroth)* and a masculine *(asherim)* plural for the noun. The *djed* column is discussed later in this chapter.

51. See H. Danthine (1937, vol. 1, pp. 22 ff., on the date palm in Mesopotamian agriculture; on the palm and the sacred tree, see pp. 28 ff.

52. B. Wesenberg (1971, pp. 23 ff.).

53. A. Evans (1921–1936, vol. 1, pp. 221 f., fig. 166); M. P. Nilsson (1950, pp. 86 ff.).

54. H. -G. Buchholz and V. Karageorghis (1973, p. 109 n. 1303); M. P. Nilsson (1950, pp. 86 f.); A. Evans (1921–1936, vol. 1, pp. 221 f., fig. 166).

55. The state of current debate is represented by Margareta Lindgren, Antonis Zois, Peter Warren, and Stefan Hiller in R. Hägg and N. Marinatos, *AIARS* 35 (1987, pp. 39–74).

56. See B. Bergquist (1988, pp. 21 ff.); M. P. Nilsson ([1955] 1967, vol. 1, pp. 401 ff.); and P. Demargne (1967, pp. 137 f.).

57. Previous record in E. F. Jomard (1809–1822, vol. 4, pp. 334 ff.). They had also been seen by James Burton, who published only some inscriptions; see F. Champollion (1833, pp. 72 ff.); cf. G. Perrot and C. Chipiez (1882–1889, vol. 1, pp. 252 ff.), but see also C. Chipiez (1876, pp. 43 ff.). The standard account is that of P. E. Newberry and G. Willoughby Fraser (1893–1900). Most recently, see A. Badawy (1954–1968, vol. 1, pp. 173 ff.; vol. 2, pp. 292 ff.); W. Stevenson Smith (1958, pp. 315, 320 ff.); W. C. Hayes (1953–1959, vol. 1, p. 174); and K. Michałowski (1969, pp. 518 ff.). On the development of the columns and their place in Egyptian architecture generally, see S. Clarke and R. Engelbach (1930, pp. 136 ff.).

58. On the real date, see P. E. Newberry and G. Willoughby Fraser (1893, vol. 1, pp. 2 ff.). All seem to be of the Twelfth Dynasty, ca. 2500 BC. The group that Champollion took for Greek, which are labeled *aamu*, are in tomb no. 3 (vol. 1, p. 69). On Beni-Hasan and other rock-cut tombs, see G. Jéquier (1924, pp. 99 f.); J. Vandier (1954–, pp. 293 ff.).

59. So D. Ramée (1860, vol. 1, pp. 462 ff.); Carl Boetticher (1874, vol. 1, pp. 2 ff.); and most memorably, E. E. Viollet-le-Duc (1858, vol. 1, pp. 35 ff.). However, since the tomb chambers were certainly not visited during their patron-denizens' lifetimes, the problem of lighting them does not seem to have been of any real interest to the ancient Egyptians. Champollion's discovery and surmise about the Egyptian origins of truly lithic Doric architecture, not based on wooden prototypes, echoes a model that had been pervasive in the eighteenth century; see J. Rykwert (1980, pp. 322 ff.).

60. C. Chipiez's problems with the dating of the Egyptian proto-Doric columns are due largely to his being unable to place the columns at Khalabaseh (the Rock Temple of Rameses II at Beit-el-Wali near Kalabashah, Talmis in antiquity; see A. Badawy, 1954–1968, vol. 2, pp. 298 f.). They have four hieroglyphed flanges, dividing the twenty flutes (as in the Doric order described by Vitruvius) into groups of five.

61. So J. Cerny (1952, pp. 26, 30 f.). Various other explanations have been advanced; for example, that the leaves are curled shavings of a tree trunk (H. Kees, 1941, pp. 96 ff.), though Kees confesses about the *djed* that "über sein eigentliches Wesen wissen wir so gut wie Garnichts." Other suggestions have been that the original form was a bundle of poles, at the top of which new corn sheaves were tied: so W. Helck in *Or* 23 (1954), pp. 408 f., though the usually green coloring of the capital would not seem to confirm this. Others have proposed that it be read as a tree trunk with lopped branches; see A. Moret (1927, pp. 80 ff, 130 ff.; it should be noted that Moret calls it *Zed*); or again as a pillar or post around which "vegetable matter"—leaves and whatnot—were tied in rings: so Z. Y. Saad (1947, p. 27, pl. 14b).

It seems already to have been both a ritual object—a kind of fetish—and the focus of a notion, a protohieroglyph, in the predynastic period. On similar notions, see W. S. Arnett (1982, esp. pp. 7 ff.).

Old Kingdom texts already call it "*Holy djed, which belongs to the Lord of Heaven*" (H. Goedicke, 1955, pp. 31 ff.). In some temples it was raised once a year (on the 30 of Khoiak: E. Chassinat, 1966–1968), in others more often. The association with Osiris may have been the result of his absorbing (as Ptah did in other rites and myths) the necropolis god, Sokar. As a hieroglyph, the *djed* form signified stability, permanence; it is often personified, given eyes and arms. It is also a common amulet and used on coffin lids (usually at the foot of the coffin), in the west walls of tombs, and, with the Ankh-sign and Was sign ("the knot of Isis"), as a monogram. But see also B. van de Walle (1954, pp. 283 ff.) and E. Diroton (1957, pp. 151 ff.), as well as H. Schäfer in *MDAIK* (1933, p. 3, pl. 2, pp. 1–17).

62. E. A. E. Reymond (1969, pp. 27 ff., 94 ff., 106 f., 273 ff.). See A. Badawy (1954, 1968, vol. 1, p. 86; 1951, pp. 1 ff; 1957, pp. 51 ff.); A. Erman (1934, pp. 42, 183); H. Frankfort (1933); G. Jéquier in J. Vandier (1924, vol. 6, pp. 25 ff.); K. Sethe (1930, pp. 15 f., para. 19; 80 f., para. 96); W. M. Müller (1918); E. A. Wallis Budge (1911, vol. 1, pp. 37, 46 ff.; vol. 2, pp. 63 f, 176, 199). But see also L. Borchardt (1937) on the relation between *Dd* and the "Ionic" column-symbol of the god *Dw3*.

On the west wall of the Osiris shrine of Seti I at Abydos, there is a vivid relief of the ceremony; R. David (1981, p. 134); there are also vast *djed* reliefs (p. 131).

63. On this identification, see J. B. Hurry (1926); on his building activities and his father, the "architect" Kanofer, see pp. 4, 9 ff. In the fifth century BC a master builder who was King Darius' "minister of works" in Egypt left an inscription in a quarry at Wadi-Hammamat tracing his pedigree through a number of master builders down to Imhotep and Kanofer; transcribed by H. Brugsch (1877, pp. 753 ff.). See also K. Sethe (1902, vol. 2, pp. 95 ff.) and S. Giedion (1964, vol. 2, pp. 269 ff.). The actual cult of the divinized vizier (whose name means "he who comes in peace") is only documented in the sixth century BC, though Sethe (pp. 23 f.) does not question the historicity of Imhotep as vizier of Zoser. This has been confirmed by the decipherment of a statue base from the Zoser enclosure at Sakkara (in the Cairo Museum, no. 49889; only the feet survive of the statue itself), which seems to have been a near-contemporary statue of him, and enumerates his titles. On this and other more recent evidence, see D. Wildung (1977, pp. 12 ff.). On the cult of the helper-architects, see L. Kakosy (1968, pp. 109 ff.).

64. Diod. Sic. I.97.vi; but cf. I.61 ff. On archaic Greek figures that "imitate" Egyptian art, see W. Deonna (1930, vol. 2, pp. 216 ff.). S. P. Morris (1992, p. 238 ff.) tends to dismiss Diodorus' Egyptian musings, since she is primarily interested in relating Greek to Near Eastern arts and crafts.

65. On Egyptian nude statues, both male and female, see J. Vandier (1954–, vol. 3, pp. 248, 252, 499). Their extreme rarity focuses attention on such exceptions as the *ka* figure of Pharaoh Hor (Twelfth Dynasty). See H. Frankfort (1948b, pp. 61 ff.). Some of the earliest of such statues wear a belt: notably the giant kouros of the Naxians at Delos. M. Robertson (1975, vol. 1, pp. 43 ff.) suggests that the early sculptors found the surface of the male nude unacceptably articulated.

66. Diod. Sic. I.98.v ff. tells the legend about the sons of Rhoecos, Tlecles and Theodoros. Rhoecos was famous for a wooden cult-statue of Pythian Apollo who appears in other legends as the architect of the first Heraion at Samos, as well as a bronze caster. According to Diodorus, his two heroes used Egyptian grid measurements to make two halves of a statue (cut down the middle from head to crotch), one in Samos, the other in Ephesus, and then joined them together. Diodorus specifically mentions a canon of 21¼ modules, presumably measuring the figure from the soles of the foot to the hairline, as in the new canon that was established in the metrological reforms of the Twenty-sixth Dynasty in the eighth century BC; on this canon, see E. Iversen (1975, pp. 17 ff., 75 ff.).

67. See Pl. *Leg.* II. 656d ff. about sculpture and music; but see more generally *Tim.* 21e ff.

68. This issue was sharply, sometimes even acrimoniously, debated in the second half of the eighteenth century (see J. Rykwert, 1980, pp. 297 ff.), and went on being discussed in France and Germany in the first half of the nineteenth. The transformation of materials was not asserted as a universal principle of architecture until Gottfried Semper (1878, vol. 1, pp. 213 ff.).

69. On the Heb-Sed festival, see E. Frankfort (1948a, pp. 1 ff.). On its connection with the beginnings of Egyptian double kingship and the Egyptian New Year, see K. Sethe (1930, pp. 181 f.). On the Heb-Sed field in Zoser's pyramid, see J. -P. Lauer (1976, pp. 93 ff.).

70. Schliemann was the first proponent of this interpretation (see H. Schliemann, 1886, pp. 47 f., 229). But his excavation was rather destructive, and subsequent opinion divided. M. Nilsson (1950, pp. 475 ff.; [1955] 1967, vol. 1, p. 346) was its active proponent, but it was dismissed by G. Rodenwaldt (1912, pp. 137 ff.). A summary of the positions can be found in H. Drerup (1969, pp. 17 f.). The Hera statue stood in the Argive Heraion on a votive column beside the Polykleitan image, where both were seen by Pausanias (II.17.viii). He called it "the oldest" Hera statue.

A separate problem for the opponents of the Hera shrine on the megaron is provided by the Doric capital that Schliemann found there, and that was in the Museum at Nauplion (K. Schuchhardt, 1891, p. 133; B. Wesenberg (1971, pp. 50 ff.). Earlier literature is reviewed in H. Sulze in *AA* (1936, pp. 14 ff.), who maintained that this and the similar capital from Agrigento (as Wesenberg argued) were already stone capitals for wooden columns, carrying a wooden beam.

VII The Hero as a Column

1. "Ita quod non potest in veritate fieri, id non putaverunt in imaginibus factum posse certam rationem habere"; Vitr. IV.2.i, v. Perhaps it might just be worth being clumsy and translate *imagines* by "representations."

2. Ramps have survived at the temple of Zeus at Olympia and Aphaia at Aegina. The steps of the *krēpis* were not always of the same height, but usually they were somewhere between one and two feet tall; the usable stairs were sometimes cuttings in the bigger steps rather than blocks positioned on them.

3. The Greek Doric was revived for such public buildings as the Propylaion of the agora—the gate of Athena Archegetes financed by Julius Caesar. See M. de G. Verrall and J. E. Harrison (1890, pp. 197 ff.).

4. Helpfully, Vitruvius provides a bibliography in VII.*praef*.11 ff.

5. On the sources of Vitruvian terminology, see (still) S. Ferri (1960, pp. 2, 4 f.).

6. See PW, s.v. "*krēpis*"; see also R. E. Martin (1965, p. 335 n. 2).

7. Hesy., s.v. "*krēpis*" 4074 (ed. K. Latte 1953–1966, vol. 2, p. 529). On the variant *krēpidōma* (inner foundations) and *krēpidia* (paving blocks), see R. E. Martin (1965, p. 335 n. 2).

8. The word refers to a putting right, correction, leveling: it could almost be translated as "datum level." Vitruvius' rather odd term *stereobates* (III.4.i: its unique occurrence), which has caused commentators some trouble, might just be a transliteration from Greek, even if the text here may well be dealing with a Roman and not a Greek constructional device; but see W. Alzinger (1972–1973, pp. 95 ff.).

9. On Agesilaos' quip, see Plut. *Apoth. Lac.* 210d; 227c offers a similar anecdote about Leotychidas I: On Lycurgos' *rhetrae* and his motivation, see n. 16.

10. The tomb was first named after Mrs. Schliemann, who had conducted the excavations. In contrast, the Treasury of Atreus and the Lion Gate were already known by those names in antiquity; witness Paus. II.16.v, who did, of course, record that all the Atreidae were buried around the city, already ruined in the time of Thucydides (I.ix.2–5), at the end of the fifth century. On the excavation, see L. Deuel (1977, pp. 227 f.) and K. Schuchhardt (1891, pp. 148 ff.). On the tomb and the fragment, see G. E. Mylonas (1957, pp. 91 ff.); the fragment is illustrated on pl. 30. The "treasuries" and beehive tombs have been surveyed by O. Pelon (1976). The tomb of Clytamnestra is his 1 H (pp. 166 ff.); he discusses the relative date of the treasuries (pp. 385 ff.) and the decoration (pp. 425 ff.).

11. This fragment has proved unique so far: see A. Evans (1921–1936, vol. 1, p. 344, vol. 2, pp. 520 ff.; 1913, pp. 76 f.; cf. C. W. Blegen and M. Rawson (1967, vol. 1, p. 40). Evans found "negative" traces of convexly "fluted" columns in the "Lustral Area" of the "Little Palace." His restoration counts 28 flutes and refers the form to the clustered reeds of Egyptian architecture (vol. 2, 519 ff., figs. 323 f.). Fresco fragments from Knossos and Pylos also seem to imply fluted columns, but they are not clear enough to deduce the columns' plan or method of construction.

12. C. W. Blegen and M. Rawson (1966, vol. 1, pp. 39 f., 61, 80, 191 f., 250 f.); C. W. Blegen in D. Zakithinos et al. (1965, vol. 1, pp. 122 f.). Where a whole ring is preserved, the flutes seem to number 32, 44, or 64.

It could even be argued that the whole Mycenaean-Minoan column was inverted, since the capital was made of three moldings, not unlike the *torus-scotia-torus* of the Attic base, while the molded base may be taken as an anticipation of the Doric echinus.

13. Darius in his inscription claimed that the fluted stone columns were the work of Greek masons (R. Ghirshman, 1964, p. 215), though of course he says nothing about the less important wood and plaster ones. They were found and reconstructed by R. C. Haines: see Erich F. Schmidt (1939, pp. 53 f.).

14. Hom. *Od.* XXIII.187 ff. (Chapman lines 276 ff.). On a possible punning misunderstanding of this passage, see J. T. Kakridis (1971, pp. 151 ff.). V. Bérard (1925, vol. 3, p. 157) makes the olive tree one of the four legs of the bed. Homer had written that only a god could move the bed.

15. In fact, modern Greek carpenters and shipwrights in many rural communities work very skillfully using an ax alone. They can shape quite complicated symmetrical and three-dimensional curves, such as boat prows, relying on the eye and the handheld ax, without the aid of templates. It is therefore at least conceivable that quite complex structural members could have been roughed out in timber, or even finished with an ax.

16. On Lycurgos' laws regarding carpentry, see Plut. *Es. Car.* 2 (997c-d); but cf. *V. Lyc.* 13 (47c). Plutarch adds that Lycurgos had no quarrel with adzes or chisels but thought the rough workmanship would discourage corrupting luxury.

17. See G. Germain (1954, pp. 211 ff.) on the rooted bed as a magical sign of stability in marriage and of fidelity.

18. R. E. Martin (1965, pp. 39 f.). It is worth noting that, as Martin observes, *"le type n'évolue pas."* The attachment of Greek carpenters to the ax is odd in view of the antiquity of the ax-adze, a common form even in polished stone throughout Europe: in the Aegean, see H. -G. Buchholz and V. Karageorghis (1973, pp. 46 ff.); in early Bronze Age Europe, see H. Müller-Karpe (1966–1989, vol. 3.3, pl. 384B, Crete; 455A, Hungary; 475C, G, H, Czechoslovakia; 675, Romania; 699, north Iran). On the general problem of the distribution of axes in southern Europe and the Aegean, see J. Bouzek (1985, pp. 41 ff.). However, the tool is already displaced in Minoan Crete by the double ax, which is inherited by the Mycenaeans and the heroes. It often appears as an attribute of heroes doing violence: for the list of such images, see A. Orlandos (1966–1968, vol. 1, p. 30 n. 1). But the double ax is also an attribute of workers and specifically of Hephaestus (who used it to crack open the skull of Zeus for the birth of Athena), as on the BM red figure *pelekus* from Vulci (BM E 410). Sometimes it is called *distomos*, "two mouthed," "two edged."

19. Red figure lekythos from Capua (Louvre G 210); see A. Orlandos (1966–1969, vol. 1, pp. 27 f.) and R. E. Martin (1965, p. 40, and pl. 4, fig. 3). The other figures are Syleos (and his daughter Xenodike—Xenodoke in Tzetz. *Chil.* II.429 ff.), though the legends have him kill Syleos with a hoe or spade; Apol. II.40.iii ff., Diod. Sic. IV.31. The subject is often shown on Attic vases (usually with Herakles wielding a double ax: PW, s.v. "Syleos").

20. Eur. *Krit.* (ed. Nauck, fr. 472).

21. *Hupo,* "under"; *trachēlos,* "the neck or throat." Curiously there is no member of the order called *trachēlos, collum,* or *cervix,* or even *fauces* for that matter, though such terms were fairly common as applied to clay or metal pots.

Hypotrachelium (Latin spelling: Vitr. III.3.xii, 5.xii; IV.3.iv, 7.iii) or *hypotrachelion* (transliteration from Greek) occurs only in relation to the column or to human anatomy; in the human body, it is the part of the neck just above the collar bone.

22. *Anta* is both a Greek (as in *anta antein,* "in front of") and a Latin word. *Parastas* is anything that stands beside or in front of—such as doorposts, for instance. Although *pastades* is sometimes used as a contraction for *parastades,* it in fact has a different origin. For the origin of the Greek words, see PC, s.v.; for the origin and the arrangement, R. E. Martin (1965, pp. 470 ff.).

23. Varro *L.L.* VIII.14, 79. Varro provides alternative spellings; *capitellum, capitulum.* In IV.3.vi, Vitruvius also calls the top molding of the triglyph *capitulum.*

24. They are not synonymous: *kranon* means "the skull," "the top of the head"; *kēphalon,* "the head," also "the person" is used in building for "a coping." Euripides uses *epikranon* for the column capital from which hair had sprung in Iphigenia's dream (*Iph. Taur.* 51); but see above, chapter 5, n. 50.

25. Leon Battista Alberti (VIII.8) introduced Latin variants into this terminology: *collum* for hypotrachelion; *lanx* (platter, dish) for the echinus; and for the abacus, *operculum* (lid). There was even a proverb, reported by St. Jerome (Hier. *In Ep. Eph.* 15): *dignum patella operculum,* the dish is worthy of the cover." However, only the first of Alberti's terms has been commonly accepted.

26. Vitruvius does not give any specific term for the space between the hypotrachelion and the annulets; although it is very much part of the column, the capital is always reckoned from the hypotrachelion *up.*

27. The great variety of annulets can only be registered on large-scale sections. No coordinated survey of Doric capitals has been done recently.

The annulets are the only member of the order for which Vitruvius gives only the Latin (and no Greek) term. And yet they are about the most archaic bit of modenature, the most persistent witness to the fictile nature of this molding. "Annulus," a medieval misspelling of "anulus," slipped the second *n* into the English term.

28. These examples come nearest to the moldings of the Mycenaean capitals and perhaps even the Minoan ones, which seem to have been very similar.

29. E. D. Van Buren (1923, 1926).

30. Hip. *De Foem. St.* III.24. Demosthenes uses *echinos* almost technically for the pottery or metal vessel in which the documents of an action were to be deposited and sealed: to be considered, a punning commentator said, as untouchable as a sea urchin or a hedgehog.

31. On the echinus, see Bernardo Baldo (in Vitruvius, 1649, part 2, p. 42). In Hor. *Sat.* I.vi.117, the *echinus vilis,* the "common bowl," is part of the setting for the poet's simple supper. The Greek for sea urchin was simply transliterated in Latin, as Varro pointed out (*L.L.* V.77) when discussing the various native or adapted names for sea animals. It was also a fairly common name, and an eponymous hero called Echinus had a number of towns (on which see Steph. Byz.) named after him. There was an island of the name in the northwest corner of the Gulf of Kalydon. In fact, PW produces seventeen different uses of the word (the architectural part of the article is by R. Delbrück); see also R. Hampe (1938, pp. 359 ff.).

Carl Boetticher based his interpretation of the echinus on the notion that it is identical with the cyma (on which see below, n. 54) and that the cyma was a molding that signified the resolution between horizontal and vertical function, a sort of marble analogue of dialectical synthesis (1874, vol. 1, pp. 38 ff., 63 ff., 70 ff.); but in order to argue this consistently, he had to adopt a "higher critical" view that the term *echinus* was corrupt, since he calls the equivalent member of the Ionic capital *cymatium*. His use of the term depends on his closely reasoned derivation of the cymatium from the drooping and spiky (hence echinus) leaves that almost always (according to him) decorate Ionic cymatia in relief, while on the Doric echinus they are painted (pp. 70 ff., 187 f.). Since Boetticher was the first historian of architecture to develop the study of parallels between stone architectural detail and pottery shapes (pp. 128 ff.), this curious byway is worth noting: its weakness had already been pointed out by Josef Durm (1910, pp. 254 ff.). Inevitably, very few examples of the echinus have survived: they were eminently breakable. But see B. Schweitzer (1971, pp. 37 ff.).

32. Varro (*L.L.* IX.46) uses *abacus* to mean "sideboard," but does not consider the etymology; the word is very common, and its diminutive, *abaculus*, is used for counters and beads in gaming. The Greek *abax* (diminutives *abakion*, *abakiskos*) has been derived by some philologists (taking their cue from Sext. Emp. *Adv. Math.* IX.282, who insists on the materiality of the lines drawn on the abacus against the immateriality of geometrical lines, or from Pers. *Sat.* I.131) from the West Semitic *abᵃq*, "dust"—that is, from the dust or sand on which early mathematicians wrote. But *abᵃq* is "blowing dust"; the Hebrew for "settled dust" is *afᵉr;* this derivation therefore *reste indémontrable* (as PC, s.v.).

33. So already L. B. Alberti (VII.8) in the middle of the fifteenth century. Although this form of the column was based on Roman precedent (the Basilica of Aemilius on the Roman Forum and the Tablinium provided obvious examples), Alberti's recipe was dismissed harshly by R. Fréart de Chambray (1650, pl. 13). It is very remarkable that the only surviving Greek Doric colonnade in Rome, the twenty Hymettus marble columns in San Pietro in Vincoli (a church well-known as the home of Michelangelo's tomb of Julius II), remained unnoted and unrecorded until the late nineteenth century: see R. Krautheimer et al. (1937–1977, vol. 3,

pp. 178 ff.; 1941, pp. 353 ff.); also G. Matthiae (1969).

34. This convention seems to have been carried through into the Ionic order (painted capitals in the Agora Museum, Athens: A 1595 from Sounion, A 2973 from an unknown building), where the fret on the flat bed and egg-and-dart on the echinus are painted under the volutes. The unique (now destroyed) surviving Corinthian capital from the temple of Apollo at Bassae also had a fret painted on the abacus (C. R. Cockerell, 1860, pt. 2, pl. xv). The placing of a fret over an egg-and-dart pattern is too common to require specific reference. A great deal has already been written on the origins of these two patterns. A glossary and index of the patterns is provided by J. N. Coldstream (1968, pp. 395 ff.). For a bibliography of the controversies around the patterns, see C. G. Starr (1962, p. 144 n. 9) and N. Himmelmann-Wildschütz (1968).

35. Varro (*R.R.* III.5) used *epistilium* as did Pliny (*NH* XXXV.xlix.172), and it is defined by Festus ("trabs, quae super columnas ponitur; 58, p. 72). In Greek the word is found in inscriptions (*IG* I².372; *CIG* 2751) and occasionally in Hellenistic writers (Plut. *V. Per.* 13).

36. Both *epistilium* and *architrave* would have been offensive to Alberti, who simply uses the word *trabs* (VII.9). But Filarete and Leonardo both use *architrave*, as does Castiglione. By Palladio's time, it is quite familiar.

37. The binding of hair with a woollen fillet was very common, but the word was also used for ceremonial dress. *Tainia* is merely the commonest word for the many forms of head crowning the Greeks went in for—from *mitra* down. When drunken Alcibiades arrives at Agathon's Symposium (Pl. *Symp.* 212e f.), he is wearing an ivy wreath, *stephanos*, wound with *tainia* and makes much play with the ivy and the ribbons. A. Krug (1968) recognizes fourteen different types of such bands. Woollen fillets, which were also called *tainia*, were used in many forms of rite (like burials) that required the tying of knots. The application of the word to the fillet that ran all the way around the building was therefore a kind of visual metonymy.

38. B. Baldo in Vitruvius (ed. de Laet, 1649, vol. 2, p. 144). In Italian the word had a wide medieval use as referring to ornamental borders: for instance Dante (*Inferno* VIII.46 f.) on the vain Filippo Argenti:

Quei fu al mondo persona orgogliosa
Bontà non è che sua memoria fregi.

Or see Petrarch (Sonnet Rime 263):

Il bel tesoro
Di castità par ch'ella adorni e fregi

which he, like Dante, rhymes with *dispregi!*

Again, Alberti avoids the term; *fregio* may well have been brought into building terminology by Grapaldi, as *architrave* has been.

39. Vitr. III.5.x. In Hellenistic Greek the word was commonly used as a synonym for *Kuklos,* to mean the zodiac, and the two words are sometimes coupled, *zōphoros kuklos,* as by Aristotle (*De Mundo* 392a) and in the *Corp. Herm.* XIII.12. See A. D. Nock and A. J. Festugière (1945–1954, vol. 2, p. 206), although the word also meant "life bearing," "life giving."

40. Palladio translates *regula,* "a rule" in both senses of the word, as *listello,* "a lath or slat."

41. The earlier guttae are almost always cylindrical pegs, the later become wedges. Alberti's attempt to replace guttae by the Latin *claviculi* (VII.9) does not seem to have been adopted by anyone else. In some self-coloring buildings (as the temple of Apollo at Bassae), the guttae are of a different material from the rest of the cornice, white marble against the blue-green limestone.

42. Vitr. IV.2.ii.

43. Vitr. IV.2.iii.

44. A. von Gerkan (1959, p. 386) holds "das Metope als femininum nicht ein substantiviertes Adjektivwort ist, sondern ein Hauptwort, folglich ein Loch zwischen anderen Teilen," which sets up, following Vitruvius all too literally, the possible interpretation of *metope* as a "hole-in-between-other-things" as against the common "what-is-between-the-holes," while it neglects the etiology of the notion (as in H. Kähler, 1949, pp. 13 ff.). This etymological twist is worth quoting as the only possible (as far as I can see) solution to the problem that the word must otherwise pose, which I suggest, following S. Ferri (1960).

This passage has provided material for much comment. See R. Demangel (1931, pp. 117 ff.; 1946, pp. 132 ff.).

45. Vitr. IV.2.iv; but cf.X.4.ii, where *columbaria* are openings in water-drawing machines. But see Vitruvius (ed. C. Fensterbusch, 1964, pp. 138 f.).

46. The François Vase (Athenian, found at Chiusi) is signed by the potter, Ergotimos, and the painter Klitias; on it, see M. Robertson (1975, vol. 1, pp. 124 ff.) and J. D. Beazley (1951, pp. 26 ff.); Athenian Black Figure Vases 76 I §29. For the Artemision on Corfu, see A. W. Lawrence (1983, pp. 113 ff.) and W. B. Dinsmoor (1950, pp. 73 ff.). The main publication is by G. Rodenwaldt (1940).

47. Eur. *Iph. Taur.* 112 f. This text has already been cited by Winckelmann to support the idea of the metope as an opening ([1762] 1964, pp. 24 f.). In *Or.* 1371, Euripides has the Phrygian coward escape "over the cedar porch, between the Dorian triglyphs." Again the alliteration almost makes him stammer.

In *Iph. Taur.* 119, *hopoi* is a preposition, "whence"; although this may be intended to emphasize a double entendre, I do not think it can quite be used (as I. Beyer has done, 1972, p. 204) to support the idea that there were other triglyphs beside Doric ones.

Metōpon, in the sense brow, forehead, front— of anything, including a temple—is assimilated to this notion by G. Hersey (1988, pp. 32 f.); it derives from *ōps,* usually spelled with omega and psi in the nominative, and therefore seems to me to be of no help here, though also related to **op* (any hole or opening). In any case, "brow" or "front" do not help with the meaning of the technical term.

48. Vitruvius uses the word not only for this projecting bit of the triglyph, but also for the whole face of it, as a synonym for *femur* (IV.3.5), and for the plasterer's *kanōn* (VII.3.5), a straight edge or even an iron bar (V.10.iii). This is one of the reasons I find George Hersey's (1988, pp. 31 f.) identification of the *meros* with sacrificial, thrice cloven thighs dripping "sacred liquids" not wholly convincing: you cannot have thighs in the head.

49. Both were discussed by J. de Laet in "De Verborum Significatione" in his Vitruvius (1649; s.vv. "*metoche,*" "*metope,*" "*triglyphus*"); see, however, S. Ferri (1960, pp. 162 ff.). Festus was clear that *trabs* could never be a solid beam; it can only mean "duo ligna compacta" (ed. W. M. Lindsay, 1913, p. 504). Vitruvius explains the *laxationes* in IV.7.iii, when he considers the "Etruscan disposition" (to which I will devote my chapter 11).

50. For these various alternatives, see P. Zancani-Montuoro (1940, pp. 49 ff.), who bases herself however on the almost unique instances of triglyphs narrowing upward in the Heraion at Foce del Sele near Paestum, which she excavated. The Mycenaean derivation is asserted by W. Dörpfeld (1935, vol. 1, pp. 199 ff.), though this frieze seems to have been used for orthostats rather than cornices. More plausibly perhaps, E. Lorenzen calculates the triglyphs in terms of unworked, split logs (1959, pp. 241–262). H. Robert (1970) sees the Doric cornice as a formalized version of the storage attic in a proto-Germanic barn building. This rather implausible thesis (shades of Strzygowski!) is the basis of a very much more learned and elaborate version by I. Beyer (1972, pp. 197 ff.), which gives his restoration of the temple at Thermon (see below, n. 74); however, it requires him to have an immoderately high epistyle—and produces other problems.

51. H. Kähler (1949, pp. 28 ff.) has even suggested that there were wooden painted *metopai* before terracotta ones, though so far no remains of such metopes have come to light.

52. *Mutuli* in Latin commonly are "brackets," projecting bits of wood, as in Varro *R.R.* III.5 or Colum. *de R. R.* VIII.9, where the nature of the wooden bracket is made clear. Dictionaries ascribe an Etruscan root or derive it from *mitulus,* a homonym of the Greek word meaning "lopped off," "broken off," "mutilated," which transforms normally into the Italian *modiglione,* from which is derived the English *modillion.* The spaces in between the modillions are called *viae* (ways? Vitr. IV.3.vi), for which there seems to be no standard translation.

The inclination is rarely more than 17.5°. More often than not, it is the same inclination as the angle of the roof, since the mutules represent the underside of the roof boards.

53. As coping: Eur. *Or.* 1570, 1620; *Phoen.* 1158, 1180. As hem or fringe of a garment (analogous to frieze): Aristoph. *Fr.* 762. It also appears on inscriptions. Stephen of Byzantium (s.v. "*monogissa*") says the word is of Carian origin.

The most complete account of Greek moldings was given by L. T. Shoe (1936, 1952).

54. In fact the sloping member of the pediment usually repeats the corona and takes up the cyma, which run along the long sides of the building. But see B. Wesenberg (1972, pp. 1 ff.) and E. Wistrand (1942, pp. 191–225). But see also PW, s.v. "*kymation,*" and *EAA,* s.v. "Cornice."

The cyma is not in fact a single profile, but a whole group of moldings. A. W. Lawrence (1983, pp. 132 ff.) has divided cymas into two groups: (1) gutters, such as cyma proper, cyma reversa, sima, and cavetto, and (2) bed molds, such as square corbel, ovolo, and bird's beak; cyma reversa may also act as a bed mold. While this smacks a little of Carl Boetticher's division between *Werkformen* and *Kunstformen,* the distinction of type seems valid.

55. Vitruvius suggested, typically, that only the lions' heads over the columns should discharge water, while the intermediate ones should be spoutless masks (III.5.xv). The connection with Plut. *Symp.* IV, quaest. 5 ("On why the Jews do not eat pork"; ed. G. Xylander, 1599, 11, 669e), cited by some sixteenth- and seventeenth-century commentators, seems obscure.

56. Vitruvius (II.8.xviii) uses the word *corona* for any coping or other molding used to throw water off the wall surface. Since some of these moldings have a sharp profile called "the beak," some rather curious etymology associates the Italian derivative *cornice* with *cornacchia,* "a raven," through *corniccione.* In fact, since the sixteenth century *cornice* has come to mean any kind of framing. The English word is simply the Italian anglicized. The original meaning of *corona*, "wreath" as well as "crown," has remained unchanged in Italian, while the building term has undergone transformation.

57. E. Akurgal (1978, pp. 119 f.); R. Hampe and E. Simon (1981, pp. 56 f.).

58. A quite different model, but also of painted terracotta, comes from Sala Consilina in Lucania. It has no porch but a large side window: notable are the horned acroteria at either end of the roof and the bird statues on the ridge. It seems to belong to yet another building tradition, though it is classed with the Greek ones by H. Drerup (1969, p. 69).

Yet another painted terracotta model, from Ithaka, has not yet been published by its excavator; but see I. Beyer (1976, pl. 25).

59. J. Boardman in *Brit. Sch. Ann.* 62 (1967, pp. 66 ff.). And see below, chapter 8, n. 14.

60. On the importance and the development of circular buildings, see F. Seiler (1986, passim).

61. Paus. X.5.v ff. Sweet bay or laurel is of course *daphnē* and Daphne was the girl who was changed into a bay tree while evading the amorous god. The honey-and-wax association (all considerations of cire-perdue apart) had a powerful connection with Delphi, where the Sybil was called "a bee"; Pind. *Pyth.* IV.60 f. See also C. Sourvinou-Inwood (1991, pp. 194 ff.) and R. Triomphe (1989, pp. 257 ff.).

62. The best-known source of the story is in Ovid *Met.* I.452 ff.; but there are differing versions in Paus. X.7.viii, Non. XLII.387 ff. For other references, see PW, s.v. "Daphne."

The worship of Apollo Daphnephoros, "bearer of the bay," seems to have been primarily Boeotian. Pausanias was given summary information about it at Thebes (IX.10.iv), but it is described in much more detail by Photius (Schol. ad Clem. Alex. *Protr.* II.27). It involved a "made-up" tree: an olivewood trunk to which bay branches were tied was decorated with ribbons and flowers and crowned with a metal ball from which several others are sus-

pended to represent, most probably, the heaven tree. See M. P. Nilsson ([1955] 1967, vol. 1, pp. 125 f.); J. E. Harrison ([1912] 1963, pp. 473 ff.).

63. Olen was one of the legendary founders of the oracle (Paus X.5.viii ff.). Pausanias also knows a Lycian Olen, "the author of the oldest Greek hymns" (IX.27.ii; his hymns to Achaeia, V.7.viii; to Hera, II.13.iii; to Eilaithyia, I.18.v, VIII.21.iii, IX.27.ii), who gave the Delians their songs: the two may have been the same person (see PW, s.v.; M. P. Nilsson, 1967, vol. 1, p. 548).

Apollo's hut of wax and feathers appears as an example of divine frugality in Phil. Jun. *V. Apol. Thy.* VI.10. But there is another connection with Apollonian prophecy: the Homeric *Hymn to Hermes* (552 ff.) tells of the three sisters, *Triai*, who lived under Mt. Parnassus and were taught prophecy by Apollo; they will speak the truth only if they are fed honey.

64. Feather fern is the common name of *Aspidium filix-mas* or *Pteris aquilina*, both of which have tough stalks, were sometimes used for thatching, and had medicinal uses. Feather grass, *Stipa pennata*, is not unlike Esparto grass, a common thatching material that is also used for rope making: either of the plants may have figured in the legend, whose play on the different meanings of *pteron* from "feather" to "wing of a building" is not always easy to catch.

65. Hephaestus is, after all, seen as an experienced builder in the *Iliad* (I.607 ff.; XIV.166 ff., 338 ff; XX.9 ff.).

66. Paus. X.5.xii, quoting a lost *Paean* by Pindar (ed. A. Puech, 1922–1923, fr. 12; vol. 4, pp. 134 ff.; Schroeder fr. 61.). This paean was based on a legend with which Pausanias may have become familiar independently many centuries later. The fragment was found—with others—in one of the Oxyrrhincus papyri and published in 1922; it also speaks of "walls of bronze, and columns all of bronze." Presumably the song of the "charmers" was metaphoric rather than mechanical or miraculous: but see C. Sourvinou-Inwood (1991, pp. 201 f.).

67. In much the same way, Pope Urban VIII had the bronze (or perhaps bronze-clad) beams of the Roman Pantheon replaced with a timber truss and melted them down for metal to cast Bernini's *baldacchino* for St. Peter's.

68. But the god himself (according to the Hom. *Hym. Pyth.* 294 ff.) laid the foundation. The two architects appear jointly in a fragment of Pindar (ed. A. Puech, 1922–1923, fr. 2; vol. 4, p. 87) as having been given advice by the arrow-shooting god. Whether this advice was the same as the gift of death with which the god rewarded them (Plut. *Consol. ad Apoll.* 14, Cic. *Tusc.* I.47), the fragment does not make clear.

Trophonios was killed by savage bees and was probably the same Trophonios (son of Apollo! or of Zeus) who was the patron of a very famous and much-discussed oracle at Lebadia in Boeotia (Paus. IX.39.iii ff.; J. G. Frazer, 1913, vol. 5, pp. 198 ff.), which had, in fact, been discovered by bees. See C. Sourvinou-Inwood (1991, pp. 192 ff., 204 ff.).

69. The first reconstruction by Paul Auberson, one of the excavators, was published by C. Bérard (1970, pp. 59 f.), and republished by P. Auberson and K. Schefold (1972, pp. 118 f.). It was criticized sharply by H. Drerup (1986, pp. 608 f.), and less radically by J. J. Coulton (1988, pp. 59 ff.). The hairpin-plan dwarf wall with postholes on either side suggests that the rest of the building was not wattle and daub, as Auberson stipulated, but of sun-dried brick; however, Coulton rejects Drerup's notion that posts and wall may have been of different periods. On the other hand, the building may have been a temporary structure, built for special occasions to be removed or destroyed after one use—and the dwarf wall may have provided the edging of the floor rather than the underpinning of the walls.

70. Published by M. R. Popham, P. G. Calligas, and L. H. Sackett (1990–1993, vol. 2). I am grateful to Mr. Popham for having generously shown me a large-scale survey drawing of the excavation before publication.

71. M. Popham and L. H. Sackett (1968), pp. 34 ff. for a summary.

72. The fundamental publication is by S. Marinatos (1936, pp. 214 ff., 257 ff.); he dated it to the first half of the eighth century. On the relation between the shrine and the agora, see R. E. Martin (1951, pp. 107, 180); see p. 226 on the relation to the other Geometric Cretan agorai. That one of the earliest Dreran inscriptions was bilingual in Greek and Eteocretan witnesses to the persistence of the "Minoan" population in the Crete of the time.

G. M. A. Richter (1970) dates the bronzes a century later, as does P. Demargne (1964, pp. 350 ff.). A slightly earlier terracotta triad was found in the shrine of Apollo at Amyklae near Sparta; it is now in the National Museum in Athens, while the Dreros bronzes are in the Heraklion Museum (nos. 2445–2447). See E. Walter-Karydi (1980, pp. 8 f.) and I. Beyer (1976, pp. 154 ff.), who date the bronzes to the first half of the eighth century.

73. I. Beyer (1976, pp. 17 f.).

74. I. Beyer (1976, pp. 24 ff.); but see H. Drerup (1969, pp. 187 ff.). Drerup's reconstruction with oblique struts was criticized by J. J. Coulton (1988, pp. 63 ff.).

75. No single monograph is available. The best documentation is still the excavation report: G. Soteriadis (1901, pp. 165 ff.; K. A. Rhomaios (1915, pp. 225 ff.). For a reconstruction, see G. Kawerau and G. Soteriadis (1902–1908) and I. Beyer (1972, pp. 197 ff.). A recent discussion of the earlier remains appears in A. Mazarakis-Ainian (1985, pp. 12 ff.), and B. Wesenberg (1982, pp. 149–157).

76. For which attack Philip was reproved by Polyb. V.8–18, XXVIII.4 ff.

77. But see a putative reconstruction of the wooden, double-pitched temple by J. J. Coulton (1977, pp. 35 ff.).

78. R. Hampe and E. Simon (1981, pp. 53 f.); A. Snodgrass (1971, pp. 421 ff.); and P. Demargne (1964, pp. 315 f.).

79. P. A. Clayton and M. J. Price (1989, passim). The lists of the seven wonders were collated by M. L. Madonna (1976, pp. 25 ff.); the statue of Olympian Zeus figures in most of them.

80. Several such tablets, some perhaps from columns, survive, notably in the National Museum in Athens. Some columns carried several dedications; one column, numbered S 2 by Dörpfeld, had as many as ten (1935, vol. 1, pp. 170 f.).

81. According to W. Dörpfeld (1935, vol. 1, pp. 137 ff.), the seventh-century temple was the third one on the site; and the first had two separate phases. He saw this first temple as a kind of megaron with a clerestory and considered the range of columns inside the cella as a successor of the supports of that clerestory. The discrepancy between the columns and capitals and the relation between that and Pausanias' two timber columns (as well as Oenomaos' column) were already pointed out by Adolf Boetticher (1886, pp. 193 ff.).

82. See chapter 4, n. 27, and chapter 6, n. 5. Aul. Gel. I.1 quotes a lost work of Plutarch, who claimed to have recorded Pythagoras' teaching that Herakles' height—as well as his strength and virtue—exceeded other men as the Olympic stadium exceeded others in Greece. There seems to have been a women's race over two hundred feet (the foot measured on the temple of Hera) and a corresponding men's one, also of two hundred feet (measured from the temple of Zeus): see H. M. Lee in W. J. Raschke (1988, pp. 113 ff.), who extrapolates from Paus. V.16.ii.

83. Only the head and some fragments of the statue remain. It may in fact be later than the temple building, and its identification as the cult statue from the temple has inevitably been disputed: see L. Drees (1963, pp. 131 f. 176 nn. 20 ff.).

It is nevertheless curious to record that at the beginning of the games, the judges and competitors took the oath in the *bouleutērion,* the "council house," which consisted of two hairpin-plan buildings with a central colonnade, parallel to each other: the first probably mid-sixth century, the second somewhat later. The oath was taken before a statue of Zeus Horkios, which was housed somewhere in that group of buildings.

84. On the relation between the eleventh- and the seventh-century temples, see W. Dörpfeld (1935, vol. 1, pp. 125 ff.); E. N. Gardiner (1925, pp. 207 ff.); and K. Schefold (1967, pp.

237 ff.). The cella is a *hekatompedon,* a "hundred footer." The only true peculiarity of the plan are the cross-walls from the alternate columns to the cella wall, which act as internal buttresses.

85. The main sources of the legend are Pind. *Ol.* I.36 ff.; Apol. *Epit.* II.3 ff.; Hyg. *Fab.* 83. For his crime, Tantalos still suffers in the underworld, and gives his name to forms of frustration. He has two avatars, a Lydian hero and an early king of Argos, as well as two tombs, one at Argos, the other on Mount Sipylus near Magnesia. See Paus. II.22.iii (ed. J. G. Frazer, 1913, vol. 1, p. 104; vol. 3, pp. 202 ff.; ed. H. Hitzig and H. Bluemner, 1896–1910, vol. 2, pp. 436, 588); Hom. *Od.* XI.582; Eur. *Or.* 5. He also had the dubious fame of having been Clytamnestra's first husband: Paus. loc. cit. On the antiquity of the games, see chapter 6, n. 5, and C. Renfrew in W. J. Raschke (1988, esp. pp. 21 ff.).

86. Paus. V.20.vii; J. G. Frazer (1913, vol. 1, pp. 267 ff.; vol. 3, pp. 620 ff.). Ludwig Drees (1968, p. 28) thinks Pausanias got it quite wrong, and that none of the pre-Hellenic remains on the site could have been the foundations of a Mycenaean palace. This leads him to argue that the pillar had merely been a phallus, connected with the races. However, Wilhelm Dörpfeld (1935, vol. 1, pp. 32 ff., 71, 93) had already pointed out that the text suggests that the pillar had been brought from elsewhere to the position in which Pausanias saw it.

87. Paus. VIII.10.i ff. On Trophonios and Agamedes, see above, n. 68.

88. P. Courbin (1980a, pp. 11 ff., 26 ff., 43 ff.; 1973). Original publication by Théophile Homolle and Maurice Holleaux (1910, fasc. 33), for the older building; on its date and its relation to the colossus, see pp. 29 f. On the colossus itself, which was four times life-size, of one piece with the base, and of Naxian marble, see also G. M. A. Richter (1970, pp. 51 ff. no. 15) and W. Deonna (1930, vol. 1, pp. 112 f., 426, 465). Like many archaic statues, the colossus had a bronze necklace (with a bulla?) and a belt (and an apron; gilt?) riveted onto it.

89. Pind. *Ol.* XIII.21 f.

90. *Etym. Magn.,* s.v. But see Paus. III.17.iv. On this passage it is still worth reading Solomon Reinach, "Aetos Prometheus" (1908, vol. 3, pp. 68 ff.), who rejects such an interpretation and thinks the phrase is concerned with bird-acroteria, not with the assimilation of shapes of which the *Etymologicon* spoke.

91. T. N. Howe (1985, pp. 370 ff.); A. W. Lawrence (1983, pp. 143 ff.); H. S. Robinson, "Temple Hill, Corinth," in U. Jantzen (1976, pp. 239 ff.); W. B. Dinsmoor (1950, pp. 89 ff.); and H. N. Fowler and R. Stillwell (1932, pp. 115 ff.).

92. Pl. *NH* XXXV.xliii.152; he talks also of Butades' invention of *prostypa* and *ectypa,* which may well mean high and low relief. It is not clear whether the *fastigium* of which Pliny here speaks is the pediment with its sculptures, or merely the acroteria.

93. Vitr. IV.3.i. He quotes as his authorities Arcesius, who appears later in his book as the designer of a temple of Asklepios in Tralles; Pytheios, a fourth-century architect; and the most famous of all, Hermogenes, who actually had all the material ready for a Doric temple but changed his mind at the last moment and built a temple to "Liber Pater"—presumably Vitruvius means the temple of Dionysos at Teos (Sigaçik) near Smyrna/Izmir. On this temple, see P. Hermann (1965, pp. 29 ff.); Y. Béquignon and A. Laumonier (1925, pp. 281 ff.); and R. W. Chandler, N. Revett, and W. Pars (1769, pp. 1 ff.). On Arcesius' name the codices disagree—Arcesius, Arkesios, Argelius, Terchesius—though nothing is known about any of them. On Pytheios and Hermogenes of Alabanda (Vitruvius' favorite), see PW, s.v., and W. B. Dinsmoor (1950, pp. 273 ff.).

VIII The Known and the Seen

1. The most famous of the irregular temple plans was that of the Athens Erechtheion, though a number of efforts have been made it to present it as a fragment of an incomplete symmetrical building.

2. Vitruvius recommends roofless or hypaethral temples—*sub divo*—for sky gods: Jupiter Thunder, the sky, the sun, the moon (I.2.v).

Elsewhere (III.2.viii), he demands that they should be decastyle, while the only temple of that type to which he refers is the Corinthian one of Olympic Zeus in Athens, which (as he also points out) was in fact an octastyle. The Didymaion outside Miletus was an Ionic hypaethral decastyle (though without the internal colonnade Vitruvius demands), while the Artemision at Ephesus and the Heraion at Samos were both Ionic octastyle. Some authors have also interpreted the Doric temples of Aphaia at Aegina and Apollo Epikurios at Bassae as hypaethral, but the evidence is inadequate.

Pierced tiles, *karēmides opaiai,* were described by Pollux in his *Onomasticon* II.54; see R. E. Martin (1965, pp. 78 f.) and A. Orlandos (1966–1969, vol. 1, pp. 107 f.). A number have been found in excavations at Priene, Olynthus, and Corinth. Neither these examples nor Pollux's dictionary entry make it clear whether these marble or ceramic tiles, pierced by large rectangular or circular holes and almost always framed by a raised edge, were intended for lighting, for ventilation, or as an outlet for smoke.

3. However, a number of students suggested that the pattern emerged (as Pind. *Ol.* XIII may be read to have implied) in Corinth, and that there could have been a single "inventor"; so, for instance, R. M. Cook (1973, p. 17).

4. J. J. Coulton (1977, pp. 37 ff.) However, a quite different view of the structure at Thermon was taken by Immo Beyer (1972), who considers the roof structure to have been a kind of amalgam of queen-post and king-post timber trusses, which also had the direct support of the central range of columns. These columns therefore had to be much higher (though otherwise of the same design) than those of the peripteron. His elaborately worked-out reconstruction, which also provided for a storage attic behind the triglyph-and-metope frieze, has not been welcomed by scholars.

5. On Artemis Orthia, see R. M. Dawkins et al. (1929, passim).

6. W. B. Dinsmoor, (1947, pp. 109 ff.); W. Judeich (1905); T. Wiegand (1904, pp. 1 ff.); M. de G. Verrall and J. E. Harrison (1890, pp. 464 ff.). But see Hesy., s.v. "*Hekatompedon naos.*"

7. W. Doŕpfeld (1935, vol. 1, pp. 147 ff.); A. Bammer in P. Gros (1983, pp. 276 ff.).

8. The foot standard varied from city to city. A number of attempts have been made to arrive at some ordered reading of dimensions: see, for instance, H. Bankel (1991, pp. 151–163) on the use of both the Doric (32.7 cm) and the Ionic (29.4 cm) foot in Attica, and more specifically on the Athens Acropolis; however, the Erechtheion standard seems to have been an ell of 49.0287 cm.

9. See, most recently, K. Schefold (1967, pp. 236 ff.). The Olympian Heraion may have been modeled on the Argive one. In any case, the change of direction in the Ephesian Artemision (see above, n. 7) was contemporary with the radical changes in the organization of the temple space.

10. Many results of these excavations are exhibited in the Archeological Museum in Thessaloniki. Unfortunately publication has been sporadic and fragmentary: but see W. A. Heurtley (1939, pp. 5 ff.).

11. It was also one of the earliest European buildings (Early Helladic II) to use baked instead of sun-dried bricks. The concentric walls within may have supported floors, while the whole was probably domed. See Klaus Kilian, "The Circular Building at Tiryns," and Daniel J. Pullen, "A House of Tiles at Zyguries?" in R. Hägg and D. Konsola (1986, pp. 65ff., 79 ff.). See also E. Vermeule ([1964] 1972, p. 35 f.); A. W. Lawrence (1983, pp. 17, 31 f.). The structural problems in the earliest Cretan circular tombs are discussed briefly in J. O. S. Pendlebury (1963, pp. 64 f.).

On tholoi in general, see F. Seiler (1986) and Georges Roux "Trésors, Temples, Tholos" in G. Roux (1984, pp. 153 ff.). The idea that the building had a flat or even a unified roof (as Kilian seems to suggest) seems to me constructionally improbable.

12. At the Mallia "villa" there is a twin range of four circular "granaries" (each about five meters in diameter and some with a central post); these represent a group similar to that of the little soapstone model of seven circular granaries arranged as a court, with a porch door (or shrine?), from the island of Melos, now in the National Museum, Athens. Granary models became relatively common in Archaic Greece: in fact the squashed-cone individual containers, found in tombs, were an awkward shape to interpret. They had been called beehives and spinning tops. The issue seems to have been settled by the model identified as a granary bench found in a grave in the Athens Kerameikos (now in the Agora Museum); see H. Drerup (1969) for that and other granary models.

13. E. Vermeule ([1964] 1972, pp. 120 ff.).

14. Some eighteen models of different size and finish survive and have been catalogued by R. Hägg (1990, pp. 95 ff.), who suggests that the type is an oriental import and quotes as a parallel hut urns with similarly barred doors from Ras Shamra (now in the Louvre). The hut from Archanes, which is most carefully preserved, was in the Giamalakis Collection and is now in the Heraklion Museum. It is a mere 22 cm high: see B. Schweitzer (1971, p. 220 pl. 238), who considers it very early (eleventh century). No building quite like these models has survived. They seem to be treasuries or granaries (the figures on the roof of the Archanes model being thieves?) while the statue of the goddess in her shrine is the guardian of the contents, to be worshipped, as was the Phrygian Mother, by being either revealed by disbarring the door or inserted for cult purposes.

There are a number of large Cretan *underground* circular chambers, notably at the palace at Knossos, which presumably served for storage, though nothing definite is known about their use.

15. See G. Roux (1984, pp. 166 ff.).

16. The distinction between the *basileus*, the hereditary king who relied on the support of an aristocracy, and the *tyrannos*, who seems—certainly by the fifth century—to be regarded as a despot raised against aristocratic abuse by the trading and moneyed classes, was not at all as clear when the word was first introduced. Its earliest recorded appearance is in a poem by Archilochus (fr. 19), in connection with the Lydian Gyges, who was a regicide as well as a tyrant. The word is conjecturally derived from Phrygian or Lydian, and is an Eastern importation, like *basileus*. On the terminology as well as the phenomenon, see A. Andrewes (1956, pp. 20 ff.).

17. On *pteron,* "the feathery," see above, chapter 7, n. 64.

18. I have in mind something like the pacing out of the Olympian stadium by Herakles, which I described in chapter 4, p. 104.

19. The original publication by Auguste Choisy (1883–1884) is still worth examining. Although it has not been replaced by any single collection, many documents have been published separately.

20. J. J. Coulton (1977, pp. 57 ff.).

21. Vitr. III.4.3 for the Ionic; IV.4.1 f. for the Doric and the cella. Though Vitruvius gives the modification necessary for different sizes in the case of the Doric temple, practically all exceed the length specified by him, especially those of Magna Graecia. Of the Ionic examples, the temple of Athena at Priene comes closest: but even that exceeds double the width by one column diameter, though the intercolumniations are doubled.

22. Pl. *NH* XXXVI.lvi.178, in his description of the archaic orders; the temples in which this seems to have been true are Athena at Assos and Apollo at Corinth. J. J. Coulton (1977, pp. 65 f., with n. 44).

23. Vitr. III.3.vii, IV.3.iii. All these dimensions have been the subject of much textual and paleographic wrangling. Vitruvius himself did not help by calling the full diameter a module (perhaps used only as a unit of measure rather than a technical term) in the first passage.

Fra Giocondo (Vitruvius, 1511, 1513) already suggested that the first set of dimensions—in diameters—refers to the eustyle discussed in the text. It works out as follows:

tetrastyle: $4D + (2 \times 2.25D) + 3D = 11\frac{1}{2} D$
hexastyle: $6D + (4 \times 2.25D) + 3D = 18 D$
octastyle: $8D + (6 \times 2.25D) + 3D = 24\frac{1}{2} D$

as presented by Pierre Gros (in Vitruvius, 1990, vol. 3, p. 109). Gros points out that if the prescriptions of III.5.i are also followed, an *ekphora,* a projection or overhang of the base of half a diameter *(sextantem)* beyond the shaft should be allowed for, since Vitruvius takes these dimensions beyond

the *crepidines et projecturas spirarum.* In the fourth book, the second set of dimensions is concerned with the establishment of the *embater,* the half-diameter module: the word is presumably related to *embateia* (entering, marching; perhaps to be considered a "first step"); note the rather dismissive comment on this passage in S. Ferri (1960, p. 109). It may be worth pointing out, as Gros has half-suggested (pp. 111 ff.), that the side elevation of the Vitruvian Hermogenes tetrastyle will also measure 27 units.

Nevertheless, no calculation on the lines of the eustyle equations given above is workable for the Doric column. Since it has no base and therefore no *ekphora,* the arithmetic would look as follows:

tetrastyle: $27 - 8$ (or four column-diameters) $= 19$; divided by 3 equals 6 with one module over
hexastyle: $42 - 12$ (or six column-diameters) $= 30$; divided by 5 equals 6 exactly.

The disparity between the two formulae, both of which seem to be for a diastyle, a three-diameter arrangement, has never been adequately explained. S. Ferri (1960, p. 161) corrects the second numeral, 42, to 32; this would make the hexastyle a sistyle, a two-diameter intercolumniation.

24. Specified with the others in III.3.vi–viii; no example of it, says Vitruvius there, could be offered in Rome, but in Asia there was the monopteral temple of Liber Pater in Teos. Vitruvius, who attributed a monograph on it and another one on the pseudo-dipteral temple of Artemis in Magnesia on the Meander to Hermogenes, admired them very much.

25. C. J. Moe (1945, pp. 23 ff.). Twenty-seven is Plato's seventh (and final) number of the world soul (*Tim.* 35a ff.); in the Middle Ages, following Bede's commentary on Genesis, 27 was taken to stand for the cube and stability; while 42, not otherwise interesting to ancient numerology, was read by Bede and Honorius of Autun into various passages of the Old Testament. See H. Meyer (1975, pp. 154 f., 161).

26. Vitr. III.3.xii, which relates the hypotrachelion radius to that at the foot of the column; on the other hand, it adjusts the exact proportion to the absolute height of the column, in the Ionic mode specifically. See also III.5.viii. However, while the first passage gives five variants between 15 feet and 50 feet, the second, which adjusts the epistylium (architrave) to that of the stylobate, has only four variants between 12 and 30 feet.

27. Vitr. III.5.xii. On the curious etymology of the English word "pediment" or "periment," see the note in the *Oxford English Dictionary*, s.v. It has no classical precedent, in spite of its Latin look. In Latin, the word *tympanum* is also used for the leaf of a door: so Vitr. IV.6.iv. The word seems to be onomatopaeic— "drum," "kettledrum," even "drumstick"—and suggests that whatever filled that opening in remote early days might have been stretched. The common form of execution, *apotumpanimos*, was assumed to be a form of bastinado, whereas it was a cruel form of death by exposure—against a wooden pole or plank. See L. Gernet (1981, pp. 252 ff.). This introduces a curious secondary meaning of the word, which may indeed affect the way in which the architectural element was read in antiquity.

28. E. Lapalus (1932, esp. pp. 325 ff.); T. N. Hope (1985, pp. 188 f.).

29. J. J. Coulton (1977, pp. 77 ff., 84 f., 157 ff.). See also K. Lehmann and P. W. Lehmann, (1969, vol. 3.1, pp. 199 ff., esp. n. 132), who consider that the Phrygians were the originators of the truss as early as the eighth century, following R. S. Young's (1957, pp. 322 ff.) description of large Phrygian rooms with elaborate burned timber roofs. However, A. T. Hodge (1960, pp. 38 ff.) makes the interesting suggestion that roof trusses were known to the Sicilians long before the mainland Greeks, and that they learned about them from the Carthaginians. This is dismissed by Coulton and the Lehmanns, leaving the problem of the much wider spans in Sicilian temples than those of mainland Greece unsolved.

30. Vitr. III.3.xi f.

31. W. H. Goodyear (1912). Goodyear began his studies of the matter by considering Byzantine, Gothic, and Italian sixteenth-century buildings; he thought that the correction of optical illusions by curvature and inclination was a universal art that had somehow become atrophied and lost after 1600. See W. H. Goodyear (1902a, 1904). On the whole subject, see W. Lepik-Kopaczyńska (1959, pp. 69 ff.). The matter has been partially discussed more recently, as in J. J. Coulton (1977, pp. 60 ff.); A. W. Lawrence (1983, pp. 222 ff.); R. E. Martin (1965, pp. 352 ff.); and W. B. Dinsmoor (1950, pp. 78 ff., 86 ff.). The most detailed discussion is in D. S. Robertson (1945, pp. 106 ff.); see also A. W. Baker (1918, p. 1 ff.) and G. P. Stevens (1934, pp. 532 ff.; 1943, pp. 135 ff.). Stevens developed a theory of *scamilli impares* in which the "steps" are odd-numbered units rising over a regular horizontal grid to produce a parabolic curve.

32. Pl. *Soph.* 235d ff. (trans. F. M. Cornford, slightly modified). Plato gives an account of the optical process, with particular reference to reflection, in *Tim.* 45b ff. But he returns to the problem of deformation in the visual arts in *Phil.* 41e f. (distance confuses our vision of measurements); *Rep.* X.602c ff. (contrast between misleading appearance and measurable reality: color with convexity or concavity, specular deformations and their corrections); and *Crit.* 107c (acceptance of sketchy representations of landscape, very critical attitude to representations of the human body). On other optical fictions, see *Rep.* VII.523b ff. (judgment of sensory data can only be comparative); *Tht.* 208e; *Parm.* 165c-d (paintings look fragmentary close-up, assume their proper form from a distance); *Leg.* II.663c (puzzle pictures, which look wrong from one point, right from another). Socrates makes ironic remarks about the reliability of artistic representations to Parrhasios in Xen. *Mem.* III.10.i. Lucian (*Pro Ikon.* 12) makes reference to it. Lucian also shows Pheidias adjusting the statues according to overheard popular criticism (*Pro Ikon.* 14), but this seems to have nothing to do with optical adjustment, *pace* Lepik-Kopaczyńska (1959, p. 75); although Paus V.11.ix attaches much importance to the dimensions of the Statue of Zeus at Olympia, he does not give them, but criticizes others who have not given them correctly (the god five cubits higher than the throne, the *nikē* in his hand six feet high). There is another description in Str. VIII.353.iv. Vitruvius (VII.praef.11) attributed the invention of stage perspective to Agatharchos of Samos, for a play of Aeschylus, and suggests that Democritus and Anaxagoras were prompted by his account to write about the subject.

The terms *eikastikē* and *phantastikē* are discussed by J. J. Pollitt (1972) and Myles Burnyeat (typescript).

33. See above, chapter 4, n. 20.

34. The essential text is Pl. *Symp.* 207a ff.

35. See M. S. Cohen and I. E. Drabkin (1948, pp. 257 ff., 543 ff.). For the bibliography of Greek works on perspective, see Vitr. VII.praef.12. But see also J. White (1956, pp. 43 ff.).

On the variation of letter sizes in inscriptions, see H. Usener (1892, pp. 414–456). However, the contradiction between the apparent sizes of close and distant cylinders (known as the column illusion) has much occupied theorists recently; see M. H. Pirenne (1970, pp. 116 ff.). Many commentators, from Dürer onward, have applied Vitruvius' teaching about diminution to inscriptions.

36. John Tzetzes VIII.333 ff. (ed. P. A. M. Leone, 1968, pp. 312 f.) and *Epistulae* 77. On his sources, see H. Spelthahn (1904) and C. C. Harder (1886), who is very nit-picking but unhelpful about this legend. On John Tzetzes and his brother Isaac, see K. Krumbacher (1897, pp. 526 ff.). Tzetzes pleaded poverty as a reason for selling his library, which dispensed him from quoting his sources. Alkamenes is well-known to ancient writers (Pausanias, Cicero, Pliny, Valerius Maximus) and historians of art, and one signature of his is well attested. A. Stewart (1990, vol. 1, p. 26) calls this story purely anecdotal.

37. Perhaps the curving was first observed by the astronomer-architect Francis Cranmer Penrose, in 1837. He did not publish his findings until much later (1888); he claims that he had noted similar deformations in Egyptian architecture in 1833 (pp. 22 ff.). At the same time they seem to have been noted by J. Hoffer and E. Schaubert, who published their observations in the Wiener Bauzeitung of 1838. While Hoffer saw the corrections as an attempt to make the stark proportions of the Doric order live and breathe, Penrose, like the slightly older John Pennethorne, saw the corrections as an attempt to present the buildings to a definite point of view and a predetermined path. Pennethorne published his *Geometry and Optics of the Ancient Architecture* toward the end of his life, in 1878. But see also G. Hauck (1879).

The matter is later confused because of the use of the term "fatter minerva" to mean a coarser, rougher way of thinking or procedure: on the *pinguis Minerva*, see Cic. *De Amic.* V; *crassa*, Hor. *Sat.* II.ii.3 and Colum. I.praef.32, XI.1.xxxii; *tenuis*, Virg. *Aen.* VIII.409, Serv. ad loc., and Macr. *Sat.* II.24. On the proverb, which was also used by John of Salisbury (*Polycraticus* II.22 in *PL* 199, pp. 449), see A. Otto (1890, pp. 224 f.); M. C. Sutphen (1902, 248); and R. Häussler (1968, pp. 110–187). On the importance of perspectival and projective illusion for the creation of the "tragic interior" in the Greek theater, see R. Padel in J. J. Winkler and F. I. Zeitlin (1990, pp. 336 ff.), and "Agatharchus and the Date of Scene-Painting" (typescript, 1993).

38. So, at any rate, W. H. Goodyear (1912, pp. 35 ff., 151). At present no agreement among the experts about horizontal curvatures can be reported.

39. Perhaps the most spectacular concave curvature is that of the stylobate of the Apollo temple at Delphi; since the temple stands on a slope, one face adjoins an escarpment cut in the rock, and it is this that has a pronounced concave curve, which is not discussed in the excavation report.

The exact measurements of many of the curves—particularly the plan curves in the pediments of some temples—remain to be accurately published.

40. The exact meaning of the Greek word is not easily established. The framing provided by the jointing edges on a rectangular stone has been assimilated to a door- *(thera)* frame: so J. J. Coulton (1977, pp. 46 f.). But, of course, it was practiced on column drums as well. The procedures are described in detail by R. E. Martin (1965, pp. 191 ff., 297 ff.). Greek masons brought the stones onto the building site with projecting surfaces and bosses (and/or with clay coating) and left all the finishing to site work, working downward, inversely to the builders.

Perhaps the most elaborate and helpful of the surviving specifications is that of the new paving at the temple of Zeus at Lebadia (*IG* 7.3073). See A. Choisy (1883–1884, pp. 173 ff.); and it is discussed by R. E. Martin (1965, pp. 189 ff.).

41. *Rhuthmos* is often assimilated to the modern meaning of the word; philologically, it remains a puzzle; the conventional derivation is from *reō*, "I flow," which implies a succession in time, repetition, and an even beat; but another, from **eri-*, "pull," "draw," has been suggested. A summary of the argument is given in W. Jaeger ([1933] 1973, p. 126).

On the changes of sense in Greek usage, see J. J. Pollitt (1974, pp. 218 ff.) and J. Svenbro (1984, pp. 134 f.). See also T. Georgiades (1956, pp. 29 ff.). There seems little doubt that the earlier texts (Archilochus, fr. 66; Theo. 1. 966) maintained the sense of repetition (perhaps the up and down of life), as well as disposition in general. Georgiades (1956, pp. 32 f.) insists on the static and additive rhythm (in the modern sense) of Greek verse enunciation. Plato seems to have used it almost technically (*Rep.* III.400a; *Crat.* 424c; *Leg.* II.665a.) and indeed distinguished it (as applying to dance figures) from pitch and harmony, which he applied to the blending of voices; the union of voice and dance is *choreia*.

The word also had a less definite meaning—something quite straightforward, like "shape" or "outline"—and that is how Diodorus Siculus (I.97.vi) used it when he said that ancient Egyptian sculpture had the same *rhuthmos* as the statues made by Daedalus. That is also how it appears in Aristotle, who devotes chapter 8 of his third book of *Poetics* to the subject. Elsewhere (*Met.* I.5, 985b16; VIII.2, 1042b14; though in *Poet.* 4, 1448b he also bluntly makes meter a part of rhythm), he distinguishes *rhismos* (using the Ionic spelling here), *diathigē*, and *tropē* as aspects of phenomena. He is quoting Democritus and older scholiasts who even suggested that these words had a special meaning in the dialect of Abdera, Democritus' home town. Pollitt translates them respectively as "shape," "contact," and "inclination"; but they might alternately be rendered as "figure," "order," and "position" as an older translator did. The changing implications of the word are of course determinant for the interpretation of the puzzling Vitruvian *eurythmia*.

Pythagoras of Rhegium (and/or Samos; only one person may be meant) is mentioned by Diogenes Laertius (VIII.46) as the first who aimed at achieving *rhuthmos* and *summetria* in his sculptures, and is therefore considered the first theoretician-critic among Greek sculptors. He is also mentioned by Pliny (*NH* XXXIV.xix.59), while Pausanias saw some statues by him in Olympia (VI.18.i), though none have been securely identified.

42. Pl. *Polit.* 258e ff., *Charm.* 163a ff., *Gorg.* 450d ff.; though it is almost scholastically set out by Aristotle, as a distinction between *poiētikē*, and *praxis: Eth. Nic.* VI.4 (1140a ff.).

43. On the teaching of music and theoretical concerns, see I. Henderson, "Ancient Greek Music," in E. Wellesz (1957, pp. 336 f., 377 ff.); H.-I. Marrou (1955, pp. 44 f., 49 f., 74, 189 f., 194 ff.). The contrast between thought music and heard, felt music remained a problem for St. Augustine: H.-I. Marrou (1949, pp. 197 ff.). On this problem in the philosophy of late antiquity, see H. J. Blumenthal (1982, pp. 1 ff.).

44. Philo Byz. *Mech. Syn.* IV.4: "have the appearance of being well-shaped"; J. J. Pollitt (1974, pp. 170 ff.).

45. Vitr. III.3.xiii. On the extrapolation of the Didymean construction, see L. Haselberger and H. Seybold (1991).

46. This suggestion, first made by S. Ferri (1960, pp. 50 ff.), though disregarded by later commentators, seems to me to have some force.

47. Philo. Byz. *Mech. Syn.* 56.

48. Vitr. VI.2.ii.

49. F. C. Penrose (1888, pp. 35 ff., 106 f.).

50. Lucr. IV.386.

51. Lucr. IV.513 ff. It is worth quoting the original in full, if only for its vehemence:

Denique ut in fabrica, si pravast regula prima,
Normaque si fallax rectis regionibus exit
Et libella aliqua si ex parti claudicat hilum
Omnia mendose fieri atque obstipa necesse est
Prava cubantia prona supina atque absona
tecta
Iam ruere ut quaedam videantur velle,
ruantque
Prodita iudiciis fallacibus omnia primis,

Sic igitur ratio tibi rerum prava necesset
Falsaque sit falsis quaecumque ab sensibus
ortast.

This violent condemnation is hardly surprising: see
Diog. Laert. X.32 ff. It is worth adding perhaps
that the *regula* in the first line of the quotation is
(if Cic. *De Fin.* I.19.1xiii is to be believed) a translation of the Greek *kanōn,* Epicurus' first rule of
truth, which Lucretius announced earlier in the
same book (IV.478 f.):

Invenies primis ab sensibus esse creatam
notitiem veri neque sensus posse refelli.

What is not entirely explicit, however, is the relationship of *sensa* to the geometry of the object. Presumably implicit in this passage is the conviction
that what the *sensa* report is the truth about the
perceived object and that the way in which the report is arrived at (optical refinements, deformations, etc.) is irrelevant to the philosopher. As
earlier commentators have noted, this passage echoes Plato's words in *Leg.* VII (793c) that unwritten
laws and customs are the supports of the written
and enacted ones, and that if they are disregarded,
society will collapse, as would a building whose
supports had given way: the whole elaborate structure would tumble down.

52. Vitr. I.2.iii.

53. Vitr. I.1.xv. The camber of the top
step of the Parthenon, for instance, is 0.8%. On
optical refinements in stoa buildings, see J. J. Coulton (1976, pp. 48, 57, 59, 110 ff.). It is not recorded after 300 BC.

54. Designed in 1919–1920; [E. Lut-
yens] *Gallery,* (1981, no. 289). In fact Lutyens is
there said to have calculated the dimensions according to Jay Hambidge's "Dynamic Symmetry"
which was then being published in his magazine,
The Diagonal. The first cenotaph was a temporary
structure in wood and canvas, and the main problem that Lutyens set himself was the calculation of
the curves and of the sloping diagonals.

55. Vitr. VI.2.ii. There is, of course, no
agreement about what kind of blue *caeruleus* was,
and opinions vary from ultramarine to dark Prussian blue. There is no doubt however that even in

Vitruvius' time, the stuccoed triglyphs were still often painted some kind of blue.

56. Pln. *NH* XXXIV.xxi.99. But Pliny's
later books provide much information about all
pigments: metal, stone, earth, vegetable. Modern
authors have treated the subject only incidentally.
However, see L. V. Solon (1924).

57. See chapter 7, p. 193.

58. These columns were in the first portico of Octavia in Rome.

59. These and other stones are discussed
by Pln. *NH* XXXVI.v.44 ff.

60. Vitr. VII.3.vii.

61. The word *ganōsis* is used by Plutarch to mean "varnishing," "refreshing of color":
Quaest. Rom. 287b.

62. The temple of "Minerva" at Elis: so
Pln. *NH* XXXVI.lv.177.

63. Pl. *Rep.* IV.420c-d.

64. This seems to have become proverbial; it is described by Critias (with many other inscriptions there) as being "the salutation that the
god offers his worshipper" (*Charm.* 164d ff.) and
was therefore presumably visible from (or at) the
entrance. It is referred to explicitly in *Leg.* XI
(923a), *Phaedr.* 230a, *Phil.* 48c, *Prot.* 343b; and
alluded to as if it were a proverb in *Phil.* 45a, *Hipp.
maj.* 290e, and *Epis.* VII.341b. Plutarch also has
much to say about this inscription and the *mēden
agan,* "avoid extremes": *De E Delph.* 385d, 392a.
He also mentions the succession of *E*s in three materials: first wooden, then bronze, and finally the
Roman one of gold (385a).

65. The temple at Cori south of Rome
(usually known as the temple of Hercules) was
quoted as the most extreme example of the tendency. On the inversions allegedly practiced by the
architect at Cori (perhaps in emulation of the front
of the Hera temple at Paestum and temple E at Selinus), see G. Giovannoni (1908, pp. 109 ff.), modified by R. Delbrück (1907–1912, vol. 2, pp. 21 ff.).
These speculations were dismissed by Armin von

Gerkan (1925, pp. 167 ff.), who maintained that the "curvatures" resulted from earth movements. However, his measurements, arrived at by methods marginally more reliable than Giovannoni's or Delbrück's, have not been checked.

66. Vitr. III.3.ix.

IX The Mask, the Horns, and the Eyes

1. Vitr. IV.8.vi: "Non enim omnibus diis isdem rationibus aedes sunt faciundae, quod alius alia varietate sacrorum religionum habet effectus." Though free, I hope my translation is faithful.

2. Vitr. I.ii.5 ff. *Decor*: it is worth noting here that it need not be man-made. *Naturalis decor* is the result of an appropriate choice of site. *Thematismos* is an irregular derivative from the name of Themis, the goddess of justice, who stands for obedience to divine laws more generally. **The*, the root of her name, associates her with Gea as the "well-established one." She signifies both stability and faith to any oath.

3. Vitr. IV.1.vi, I.2.v.

4. Vitr. IV.1.vii.

5. On Liber Pater, see below, n. 54.

6. Paus. III.18.ix. J. G. Frazer (1913, vol. 3, pp. 322, 351 ff.) thought that these artists may have been sent over by King Croesus (which presumably meant that Bathycles worked in the middle of the sixth century) from whom the Spartans wished to buy gold for another shrine of Apollo at Thornax; but Croesus made them a present of it. See Paus. III.10.viii, Her. I.69. On Bathycles, see W. B. Dinsmoor (1950, pp. 141 f).

7. As in the Erechtheion, in the temple of Nike Apteros, or even the half-Doric Propylaea—all on the Athenian Acropolis.

8. The implications of the different terms for the module in the Doric and Ionic genus, the Greek *embatēr*, a "footstep," are discussed by B. Wesenberg (1983, pp. 143 f., 159 ff.).

9. The art of working on the lathe was an invention attributed by tradition to the sixth-century sculptor-architect Theodore of Samos, who was to be involved in the building of the Samian Heraion as well as the Ephesian Artemision, and whom I shall have occasion to mention later. Lathe marks have been found on the unfinished column drums of the Artemis-Cybele temple at Sardis, as well as at the temple at Naucratis.

The craft character of tor-words was already noted by the French humanist Claudius Salmasius (Claude Saumaise, 1588–1653), in his commentary on Solinus and Pliny, *Exercitationes Plinianae* (1689, p. 1044); this is quoted by Simone Stratico's edition of Vitruvius (1830, vol. 4.2, pp. 157 ff.).

As for the detail of the bases, many early examples not only were quite different from Vitruvius' two rules but also had the torus fluted. This was particularly the case in the early examples in Athens such as the Nike temple or the Propylaea of the Acropolis or the temple on the Ilissos, where it was sharply separated from the scotia underneath it. But see U. Schadler (1991, pp. 91 ff.), who also discusses the relation between column, anta, and wall bases. On the implications of the design of the bases on an Ionian architect's working methods and alliances, see J. Boardman (1959, pp. 170 ff.).

10. "Truncoque toto strias uti stolarum rugas matronali more dimiserunt" (Vitr. IV.1.vii).

On Greek chisels (*gluphanoi* or *smili* or even *skarpelloi*), see A. Orlandos (1966–1969, vol. 1, pp. 43 f.) and H. Blümner (1891, pp. 210 ff.).

11. In fact, there is a wind-rose drawing in the BM MS Harl. 2767, which is the earliest; some rather fanciful capitals appear in the slightly later Sélestat MS 1153. There are rough diagrams and miniatures in others, but none of them have Vitruvius' authority and there is no trace of an enthesis diagram.

12. The more coarsely swelling Aeolic spiral was centered on one diagonal of the oculus only, not on cross-diagonals. Many early Ionic capitals have spirals drawn without any visible oculus—as the ones in the Croesan Artemision at Ephesus.

An attempt to classify the early Ionic capitals and give an outline of their development from the early votive columns was made by W. Kirchhoff (1988). On the Sicilian examples, see D. Théodorescu (1974).

13. Vitr. III.5.v. There is a vast literature on the drawing of the spiral. See, most recently, B. Lehnhof, "Das Ionische Normalkapitell vom Typus 1:2:3 und die Angaben Vitruvs zum Ionischen Kapitell," in H. Knell and B. Wesenberg (1984, pp. 97 ff.). The reduction of the recipe to these very simple proportions may well be a later simplification of archaic rules and details.

14. The many oculus implants have disappeared. In the Erechtheion, they were probably of glass; in the Epidaurus Asklepeion, of metal.

15. Also a head pillow, like the French *oreiller*, though later applied to any pillow or cushion. This curve is very like the one taken up by the normal Greek cushion draped over the headboard of a *klinē*, which was usually in the form of the "Aeolic" capital.

16. Alternatively, the relief of the spiral may be convex on the outside and concave on the inside, as in the capitals from the archaic temple of Therme, now in the Archeological Museum, Thessaloniki.

17. See chapter 7, p. 219.

18. J. J. Coulton (1977, pp. 126 ff.); G. Roux (1961, pp. 49 ff.). The most detailed survey is provisionally presented by F. A. Cooper (1970) though C. R. Cockerell (1860), the report of the first excavations, should still be consulted; see pp. 57 f. on the order, its corner character, and the curious abacus.

As for the temple on the Ilissos, it was measured by J. Stuart and N. Revett (1762–1816, vol. 1, pl. 7; vol. 2, pl. 2; for discussion see vol. 1, pp. 72 ff.). It was mostly destroyed during the Turkish building of city walls in 1778. It was probably dedicated to Artemis Agrotera, on which see W. Judeich (1905, pp. 370 f., but also p. 355).

19. For a discussion of these and related terms, see above, chapter 7, pp. 183 ff. The beam is here to the capital as 16:18, or 8:9—one tone.

20. This thesis was first worked out by R. Demangel (1932). While Demangel insisted on the elaborate figuration of top-gutter moldings, such as that of the Ephesian Artemision, he did not

perhaps take sufficient account of the top members of Lycian tombs, which though later (fourth century) than the first Ionic friezes seem deliberately archaizing stone versions of the wood-and-terracotta shrines, with elaborate roof moldings and even figured ridges. This argument has been taken up more recently by Å. Åkerstrom (1973, pp. 319 ff.), who recognizes two types of tall figured tile: with a lower lip, used as a cyma; with an upper lip, used as a surrogate orthostat. Fragments of both kinds of tile have been found at Gordion.

21. For a discussion of these terms, see above, chapter 6, pp. 148.

22. The river Helliconus flows under Mount Olympus in Boeotia, on the coast of the Gulf of Corinth, south of Delphi. The town, Helike, was on the north coast of Achaia (where the shrine of Poseidon had been "swallowed up" by the sea, probably in 373 BC: see Paus. VII.24.vi; and J. G. Frazer, 1913, vol. 4, p. 165; Ovid *Met.* XV.293 ff.). It is not sure that the Poseidon cult of the same name, already known to Homer (*Il.* II.506, XX.404; Hom. *Hym.* XXII.3), has any definite connection with the three localities. See, however, L. R. Farnell (1896–1909, vol. 4, pp. 28 f.). In Roman times, the cult is associated with Apollo and the Muses: so A. B. Cook (1914–1940, vol. 1, pp. 130 ff.) and M. P. Nilsson (1955, vol. 1, pp. 446 ff.). On the site proper, now known as Otomatik Tepe (because of the machine-gun emplacement there during the Greek-Turkish war) near Gözelamli, see G. E. Bean (1979, pp. 178 ff.). The shrine does not seem to have had a temple, but an altar and cave—and perhaps a stadium and theater. The site has suffered much and no excavation report has ever been published, though there is some material in G. Kleiner et al., *JDAI*, suppl. 23 (1967); in *JHS, Arch. Rep.* (1964–1965), pp. 50 ff.); and in C. Emlyn-Jones (1980, pp. 17 f., 21, 32). There were traditions about Panionian councils meeting there, as in the Scholia in Pl. *Tht.* 153c; see M. Moggi (1976, pp. 40 f.). In the fifth century the conflicts around the site made celebration difficult and they were moved to Ephesia, near Ephesus. They were revived there on Alexander's orders. For a dedication to Pan-Ionian Apollo, see, however, *CIA* 3.175 and *CIG* 2.32(2).4995. But the cult has also been discussed by F. Schachermeyr (1950, pp. 45, 159) and by I. Malkin (1987, pp.

119 ff.). But see also C. Roebuck (1959, pp. 7 ff., 30).

23. The relation between the two legendary building histories was pointed out by C. Picard (1922, p. 15 n. 2). A double dedication to both deities was unusual, though there was a temple dedicated to both Artemis—or rather Diana—and Apollo on the Palatine in Rome, and the ancient temple on Lesbos (near modern Klopedi) seems to have one. For all that, it was common enough to find cult statues of one in temples dedicated to the other, as there was a shrine of Artemis at Claros.

24. G. E. Bean (1978, pp. 60 ff.); H. Metzger (1966, pp. 103 ff.); 1967, pp. 113 ff.); and H. Metzger et al. (1979).

25. Hes. *Theog.* 497 ff.; Paus. X.24.vi. There was also a tradition preserved in Mantinea (where there was a famous shrine of Poseidon the Horse that for Poseidon, Rhea had given Uranus a foal to swallow (Paus. VIII. 8.ii). In Arcadia the actual location of the stone episode was supposed to be a cave in Mt. St. Elias close to the village of Nemnitsa, near the ancient Methydrium.

According to another account, the stone was the omphalos in the Delphic sanctuary; it marked the center of the world, where the two eagles sent by Zeus from either end of the world had met. See A. B. Cook (1914–1940, vol. 3, pp. 921, 937).

26. See chapter 7, p. 193.

27. It may well be that the royal dynasty of Ephesus were called "Androklids" (see PW, s.v.), although Androklos' status as oikist was not confirmed by the usual honors: see I. Malkin (1987, pp. 251 ff.). On Androklos himself, see Pherekydes in *FHG* vol. 3, fr. 155; on Neleos (who was in turn a son of Poseidon), see Hellanikos in *FHG* vol. 4, fr. 125. Following Hellanikos, Strabo has Kodros, father of the Neleid Melanthos, as king of Pylos. Melanthos was dispossessed and expelled by the "sons of Herakles." On the Mycenaean version of the legend of Kodros and the late intervention of Athens in Attic affairs, see M. P. Nilsson (1932, pp. 152 ff.) and F. Schachermeyr (1983, pp. 306 ff.). The most famous servant of the temple, Heraclitus, may have been one of the Kodridai.

Kodridai may well have been a title, "rich in

fame," which was mistaken for a patronymic and led to the insertion of a Kodros into the Athenian king list. At any rate, by the fifth century he had a temenos in Athens, mentioned but not described by Pausanias (I.19.v); also mentioned in inscriptions (*IG* Suppl. 2.53/a). See M. de G. Verrall and J. E. Harrison (1890, pp. 228 f.). *Kodridai* is a word that Aristotle applies (*Ath. Pol.* III.3) to the Athenians in general. On the Ionian Kodridai and the king list, see *CAH*, vol. 3.3, p. 364.

The identification of Koressos remains disputed: Kuru Dag and Aysoklu Dag have also been proposed. See A. Bammer (1972, p. 43 n. 16); but see already J. G. Frazer in Paus. VII.5.x (1913, vol. 4, p. 129). The account of Ionian settlement in E. Meyer (1937, vol. 3, pp. 397 ff.) is still worth reading. On the eastern colonies generally, see also *CAH*, vol. 3.3, pp. 207 ff. On Kodros, see PW, s.v.

28. The tradition about the oracle is mentioned rather vaguely by Athen. VIII.62. A Pythion on the shore might be considered a tribute to Delphic Apollo and the fulfillment of an oracle, though C. Picard (1922, p. 117) thinks a closer oracle (Claros, Didyma) would have been more likely.

The expulsion of the natives by Androklos is described in Str. XIV.1.xx f. But see F. Schachermeyr (1983, pp. 307 f.).

On Claros, see K. Buresch (1889); C. Picard (1922, pp. 6 ff.); and H. W. Parke (1985, pp. 112 ff.). On the excavations, see L. Robert, in *AnSt* 1 (1951), pp. 17 ff.; 2 (1952), pp. 17 ff.; 4 (1954), pp. 15 ff.; 5 (1955), 16 ff.; 6 (1956), 23 ff.; 8 (1958), 28 ff.; 10 (1960), pp. 21 ff.; also "Claros," in C. Delvoye and G. Roux (1967, vol. 2, pp. 305 ff.).

An extensive account of the foundation legend is given in the *Epigoni* (*TGF*, fr. 4), according to which Claros was founded by Manto, the daughter of the great Theban seer, Teiresias, or perhaps rather by Mopsos—the son she had with Rhoikos, a Cretan or Mycenaean whom she met at Delphi. Mopsos is the name of a *ktistēs* in other Ionian cities, such as Perge. Like Manto, he is also generally associated with the gift of prophecy, though the name seems to have been a princely one. Legends put his career both before and after the Trojan War; there may have been several persons of that name, or indeed office. Recent excavations have shown (almost too neatly) that the temple of Apollo was a Doric one.

The tempting identification of Ephesus with

Apsa or Apasa, the capital of the Arzawa who so often appear in Hittite inscriptions, is frustrated by A. Bammer (1988, pp. 131 ff.), who wants it situated at Ilicatepe, on the coast, some 25 km south. Perhaps at an earlier time the capital of the region—if it was the Arzawa of Hittite royal records—was the palace city, near modern Beycesultan, which was burned about 1500 BC. See H. Seton Lloyd and J. Mellaart (1962–1972, vol. 2, pp. 3 ff.).

The Ahhiyawa of the Hittite documents were identified as the Achaeans (Mycenaean Greeks) by the early decipherers of Hittite records such as B. Hrozny (1963, pp. 222 ff.), though this identification was inevitably disputed later. Recent learned opinion has on the whole tended to restore it (a useful summary of the current literature is given by G. Bunnens in D. Musti (1986, pp. 230, 247 n. 23; see also *CAH*, vol. 2.1, pp. 261 f.), as well as the idea that Apasa/Apsa may be considered a Mycenaean town in contact with the Hittites—which the Ephesian finds have tended to confirm.

29. There is a vast amount of literary and archaeological evidence about what the Greeks thought the Amazons looked like and how they dressed: see PW, s.v., and D. von Bothmer (1957). They were certainly thought of as sexually desirable.

Theseus married (or raped) an Amazon variously called Antiope or Hippolyta, and the Athenians sacrificed to them on the day before the Theseia: see L. R. Farnell (1921, p. 339); F. Pfister (1910–1912, pp. 127, 452). The infamous story of Achilles and Penthesilea is probably depicted in the unlettered Munich cup (2688/j.370). There were graves of Amazons in Athens, Megara, Chalkis, Chaironea, and in Thessaly. In Lakonia, the limit of their advance into Greece was commemorated by a temple to Artemis Astrateia (the unwarlike?), south of Sparta, where Pausanias (III.25.ii f.) also locates a cult of Amazonian Apollo. A "scientific reconstruction" of the myth was first attempted by the Swiss mythographer, Johann Jakob Bachofen in his *Versuch über Gräbersymbolik der Alten* (1856) and his *Mutterrecht und Religion* (1861). A curious anthropological account of the myth is given by Emanuel Kanter (1926). A feminist critique of the myth is given by A. W. Kleinbaum (1983, pp. 180 ff.). It is worth noting that J. Garstang (1930, pp. 86 f.) maintained that one of the guardian figures on the gate at Hattusas represented a female

warrior and dated her weapons—on the basis of Syrian finds—to the late fourteenth century BC. The myth seems to be too rich and important to be a mere mistake about the gender of long-haired and robed Asiatic (Hittite?) warriors. The idea that it has some connection with the legends about patriarchy ousting matriarchy has been suggested by recent authors: see W. B. Tyrrell (1984, pp. 26–30, 44–49) and M. Zografou (1986, pp. 36–38). The Amazons were, as Angelo Brelich has pointed out (1958, pp. 325 f.), one of those ambiguously "savage" mythical collectives, mythemes (like satyrs, centaurs, Telchines, etc.), against whom the heroes carried on their civilizing strife.

30. On Samorna/Smyrna, see C. Picard (1922, pp. 63 f., 432 ff.). While Ephesus was part of the Ionian confederacy, Smyrna belonged to the Aeolic one. On the foundation legends, see M. B. Sakellariou (1958, pp. 223 ff.). Recent archaeological finds (J. M. Cook, 1958–1959, pp. 1 ff.) have made Old Smyrna one of the best-documented early Greek settlements on the Asiatic coast. Alyattes of Lydia besieged and sacked the city, when its population was incorporated into that of Colophon; that was about the year 600, though Smyrna seems to have been "Ionized" by 688 (or so Paus. VII.5.i).

Samorna also gave her name to the quarter of Ephesus nearest the temple, as well as to the great city. See Athen. VIII.62, Str. XIV.14.xxi, though it seems as if Ephesus was regarded as the older town, and Smyrna to have taken its name from the Ephesyne quarter.

31. There is mention of the Amazons in Hom. *Il.* III.188 f. But the evidence about their presence at Troy is dispersed. See PW Suppl., s.v. "Penthesilea."

32. On the cult and its statue, see A. Galvano (1967). There were many versions of the statue, discussed by H. Thiersch (1935). A handy (though incomplete) index of the replicas in S. Reinach (1897–1898, vol. 1, pp. 298–300, 302; vol. 2, pp. 321 f.). Thiersch (1935) lists some forty stone figures and fragments and eleven bronzes, as well as many coins and seals. Since Thiersch wrote, two large statues have been found in Ephesus proper; the largest one (nearly 3 m high) of the first century AD is from the Prytaneion. It wears a high polos-hat crowned with arcades. A smaller figure (about

1.6 m) and somewhat later is very much like the one in the Tripoli museum, which probably comes from Leptis Magna (Thiersch, no. 29, pp. 38 ff.). There may have been different types of the cult image, though they all seem to share certain characteristics, discussed by Thiersch. Whatever the earlier fate of the image, it was diffused on coins from about the time of Mithridates V of Pontus. It is virtually impossible to determine the relation between the stone replicas and the original idol. Of the silver images from which many jewellers in Ephesus made a good income about the beginning of the Christian era (see Acts 19.23 ff.), no trace remains.

It is also impossible to reconstruct the nature and material makeup of the image. Pliny maintains that it was very ancient, had survived all the changes of the temple, and was made of vinewood (though others maintained it was of ebony), according to an ocular witness, who also named its maker, the artist Endoeos, of whom nothing else is known. This witness, the consul Mucianus, added that it was drilled with holes through which it was fed spikenard oil, to prevent cracking (*NH* XVI.lxxxix.213 f.).

In the later word lists (Hesy. and *Etym. Magn.*) the word *xoanon* could refer to wooden, stone, or even ivory figures; although *xoanon* seems derived from *xuein*—to "scrape," "strip," or "polish"—in later Greek philology, as practiced by antiquarians like Pausanias and Plutarch, it assumed the sense of "wooden and barely shaped," and therefore "archaic." Inevitably, it did not apply to metal or pottery agalmata; the restrictive use of the word for "primitive" statues seems to have been fairly late. Earlier Greek texts seem to use the two terms almost interchangeably with each other and with other words: *andriantes, aphidrumata, bretta, eidola, kolossai*. See A. A. Donahue (1988, pp. 9 ff.) and J. Papadopoulos (1980), though the issue had already been raised by H. Blümner (1891, vol. 2, pp. 176 ff.).

That xoana and daidala often wore clothing of real cloth—wool or linen—is common knowledge enough, but the matter has not been adequately studied: see Donahue (pp. 140 ff.).

33. Heraclid. Pont. in Steph. Byz., s.v. "Ephesos"; *Etym. Magn.*, s.v. "Ephesos"; Eustat. *In Dion. Per.* V.826. Valerius Maximus was shown the tomb of Hippo on the coast near Erythrae (VI. 1.xiv). Androklos was buried in the city, "between the Olympeion and the Magnesian gate; his monument was an armed man" (Paus. VII.2.viii). Strabo situates "the spring Hypelaeos near the Atheneion" (XIV.1.xxi).

34. The acropolis is now the site of the Byzantine castle of Selçuk, and the church of St. John Theologian (built on Justinian's order to overlook the Artemision) now stands there.

35. Its remains were wrecked by an earthquake a century later, and it was then looted for spolia to build the churches and mansions of the new Christian city.

The moving of the town was the result of negotiation and pressure: Lysimachos as its *ktistēs* also renamed the city Arsinoe, after his second wife, who was later Queen Arsinoë II, having married her brother Ptolemy II. See C. Picard (1922, pp. 635 ff.); PW, s.v. "Ephesos." Strabo (XIV.1.xxi ff.) describes the early shape of the city, and Lysimachos' forcing migration not only from Ephesus but also from the nearby towns, Teos and Lebedos.

36. Hom. *Il.* XXI.470; see Hier. *In Ep. Eph. Praef.* who there also quotes the words of M. M. Felix from his *Octavius* XXI: "mammis multis et uberibus extructa." Felix is enumerating the contradictory characteristics of Artemis—Huntress, many breasted, three-headed.

37. See R. Fleischer (1973, pp. 88 ff., 170 ff., 280 ff., 345 ff.) and H. Thiersch (1936, pp. 63 ff.).

38. The site is described by G. Bean (1979, pp. 137 f.).

39. It was divided into horizontal stripes, one of which carried heads of sun and moon, another the three graces, a third a nereid on a triton. The figure had similarly extended arms and many necklaces, from one of which a crescent moon hung between her breasts; she also wore a polos-hat. Stephen of Byzantium (s.v. "Aphrodisias") believed that the city of Aphrodisias had originally been called Ninoes after its eponymous founder Ninos, the namesake of the husband of Semiramis and founder of Niniveh.

On Nin, "The Lady" in Sumerian, in relation both to Aphrodite and Harmonia, see M. C. Astour (1965, pp. 159 ff.). Ninos, the son of Bel (who is presumably a Hellenized Ninurta, son of Enlil), was inserted into the genealogy of the Lydian kings at Sardis by Herodotus (I.7.iii). See C. Talamo (1979, pp. 40 f.). Whatever her legendary connections, Semiramis was the Hellenized version of Semmyrammat (whose real husband had been King Shamshi-Adad V of Assyria) and who, after her husband's death in 810, became regent for her son Adad-Nirari III until 805.

Aphrodisias had in fact been settled since Neolithic times, though it does not appear in literature until the Hellenistic period. Nin is one of the soubriquets of Astarte, and Stephen may well be recording the name of an Assyrian trading post or colony whose remains have not yet been found. All the images of this group have most recently been surveyed and the various views about them discussed by R. Fleischer (1973).

40. On the Zeus Stratos of Labranda, see P. Merlat (1960, p. 52 f., 57 ff.). Merlat also discusses the association of Jupiter Stratos and the Heliopolitan Zeus of Baalbek. On the Lydian sanctuary, which has been excavated since 1948, see A. Westholm, (1978, pp. 543 ff.). The temple of Zeus there was Ionic, while the treasuries or "Androns" of Mausolos and his brother Idrieos (d. 344) had Ionic columns carrying Doric entablatures; on Mausolos' pious building, see S. Hornblower (1982, pp. 345).

On the Heliopolitan Zeus, see R. Fleischer (1973, pp. 326 ff.); on Zeus Labrandeus and Lepsinos (of Euromos) and the god of Amyzon—known only from coins—see pp. 310 ff.

41. The fortified headdress of Cybele-Kubaba already appears on some of the goddesses at Yazilikaya. The embroidered polos appears on the orientalizing ivory female figure from the Dipylon, ca. 750 (Athens, National Museum). Another, even more elaborately mitered wooden figure in the Museum at Vathy on Samos (R. Fleischer, 1973, p. 202).

42. On the cult of the Great Mother at Pessinus, and on its political and economic importance, see B. Virgilio (1981, pp. 60 ff.). The archigallos always assumed the name "Attis." The heaven-sent stone image was removed to Rome, threatened by the Carthaginians in 205 in obedience to an oracle. A similar cult statue seems to have been worshipped in Perge, of which a great

many images have survived, especially on coins and gems: see R. Fleischer (1973, pp. 233 ff.).

The term *diopetus* was generally applied to images that reputedly fell from heaven, such as the Taurian Artemis (Eur. *Iph. Taur.* 977) as well as the image of Ephesus (Acts 19.35).

43. At Ephesus the highest functionary, who had the foreign (Persian?) title *megabuzēs*, was an eunuch (like the archigallos at Pessinus, and many priests of Cybele); he was said to "come from far away" and in ceremonies may indeed have worn an ependytes, though all we know about his vestments is that they were "made of purple and gold." In later Hellenistic times he was replaced by a senior priestess, yet for many centuries he was virtually the only major priest of a Greek cult to be emasculated (a practice that the Greeks on the whole found offensive). And he presided over a female college variously known as *melissai*, "bees," *korai*, "girls," or *parthenoi*, "virgins." There were other groups of temple servants, but the megabyzes and his "bees" seems to have been central to the cult.

Maha-bahu, Sanskrit "of the long hand," and Old Persian *Bagha-buksa* are plausible suggested origins of the word. But see C. Picard (1922, pp. 163 ff., 222 ff.). Hesychius makes it a Persian royal title. Xenophon was a friend of the megabyzes who came to Greece as theoros for the Olympic Games of 389 BC, with whom he had deposited a substantial sum of money in the keeping of the goddess. When the megabyzes paid it back, Xenophon used it to build a shrine to her at Scillus (near Olympia) where he lived, and which, like the Ephesian shrine, had a river called Selinus running through it (so Xen. *An.* V.3). Apelles and Zeuxis are both supposed to have painted portraits of a Megabyzes at the time of Alexander the Great, but it is not certain when the priesthood lapsed or passed to a woman. Several distinguished Persians were called *Megabazēs, Magabazēs,* a corruption of the Persian *Bagabuxsa,* "god saved; but see PW, s.vv.

The Phrygian galli were treated with some contempt by the Greeks and the Romans after them. Apart from the Ephesian, the only respected evirated priest is at Delphi: Labys, to whom the saying *gnōthi sauton* has been attributed (C. Picard, 1922, p. 223, quoting P. Perdrizel, *REG* 11, [1898], pp. 245–249; 12 [1899], p. 40); PW, s.v. "castratus."

The bee is a common attribute of the goddess. Honey is both a symbol of sexual pleasure and a guarantor of purity, a preservative from corruption. Bees also swarm like the souls of the dead (which are called *melissai* as are the priestesses of Artemis and the moon goddess herself). Porphyry points this out (*De Antr. Nymph.* 16 ff.), commenting on the association of bees with the carcasses of large animals, particularly bulls—the bull also being holy to the goddess because of his crescent-shaped horns. Strength that decays and gives birth to sweetness is the point of Samson's riddle to the Philistines, whose bees are generated from a lion carcass (Judg. 14). A parallel account, in which the bees swarm from four putrifying bull carcasses, appears in Virgil's *Georg.* IV.317 ff. There are many references in Hellenistic literature to such beliefs, though they seem even implicit in Democritus (DK 68 B, 29a).

The bee appears on the coins of Ephesus, while the term *melissai* for her priestesses appears only once. They seem to have had a leader, who had a fawn skin for an ependytes:

at the head of the *parthenoi* walked Anthia, . . . the most beautiful of the virgins, blond hair, partly braided, but most free and woven by the wind; eyes lively and brilliant as those of a young girl, yet intimidating, as those of a chaste virgin; her chiton purple, belted at the waist falling to her knees and over her arms. A fawn's skin round her, a quiver over her shoulder, she carried a bow and some spears, hounds followed her. . . . "Here comes the goddess," the crowd cried, . . . or "the image the goddess made of herself in her own likeness! (Xenophon of Ephesus, *Ephesiaca* II.6 f.)

This description tallies with the dress of the two archaic ivory female figures found in the "foundation deposit" of temple D by J. G. Wood. See D. G. Hogarth (1908, vol. 1, pp. 173 ff.).

As for the "breasts," they are called that in the oldest sources available to us, the two Christian writers quoted in n. 36. The various hypotheses were last discussed at some length (but inconclusively) by R. Fleischer (1973, pp. 74 ff.) and by W. Burkert (1979, pp. 130 ff.). That they are the severed breasts of the Amazons, or natural breasts at all, is unlikely, since none have any trace of nipples; the other anatomical explanation, that they are the severed scrota of sacrificed bulls, is also unlikely, since they are egglike objects, while scrota are always twinned: *doppiette* or *coppiette* is what roast

ram's scrota offered for sale in Italy are still called (see G. Seiterle, 1979). Whatever the breasts represented, they are certainly "artificial" by the time they come to be shown in art: an interesting confirmation is the Cybele in the Kunsthistorisches Museum in Vienna, who wears her "breastplate" almost as a belt.

44. On the mask and the face, see above. Presumably, if the xoanon really was boardlike, the feet might also just have been a pair of shoes attached to it.

R. Fleischer (1973, pp. 76, 180 f.) suggests that the dark face and hands in the bronze and marble copies of Ephesian Artemis in fact show the effect of the constant oiling of the image.

45. Servius Tullius, a legendary king of Rome (575–538), already emulated the Ionian confederacy around the Ephesian temple in his Latin federation centered on the temple of Diana on the Aventine, while the Phoceans took her worship westward to their colony of Massalia (Marseilles) at the beginning of the sixth century, where Artemis became Stella Maris. And there were shrines of Ephesian Artemis in several Greek and Roman cities: Corinth, Epidaurus, Syracuse, Cyrene, and Carthage.

On the "missionary" cult, see C. Picard (1922, pp. xix ff. 374 f., 682 f.); he also offers a summary of sometimes tenuous numismatic evidence for the expansion of the cult (p. xxix n. 8). The emigration to Massalia is usually dated ca. 540 BC. There were several such cults, such as those of Delian and of Pythian Apollo.

46. See E. Will (1955, pp. 212 ff.).

47. On the nature of Lenaea, see L. R. Farnell (1896–1909, vol. 5, pp. 208–214); C. Kerenyi (1976, pp. 282 ff.). It was one of the four principal celebrations of Dionysos in Attica: Lenaea (the 12th of Gamelion, in January/February) and Anthesteria (the 12th of Anthesterion, in April, on the day after the new wine was "uncorked"; the vase paintings may well have had more to do with this celebration) in Attic-Ionic lands; the Doric-Aeolic Agronia and the more general Dionysiaca, which were country celebrations, as well as the Catagoghia, which were introduced to Athens in the sixth century. See W. Burkert (1983, pp. 163 f.); M. P. Nilsson ([1955] 1967, vol. 1, p. 829). On the Greek tree cults, see still C. Boetticher (1856, pp. 104 ff., 226 ff.). On the difficulty of "interpreting" Dionysos, and the varieties of interpretation, see P. McGinty (1978).

48. The Lenaean vases were first recorded by Theodor Panofka (1843) and first seen as a thematic group by Carl Boetticher (1856, pp. 103 f., 226 ff.). A more thoroughgoing study of twenty-seven such vases by August Frickenhaus (1912) was published by the 22nd Winckelmannfest in Berlin. Additional material appears in A. Pickard-Cambridge (1968, pp. 30 ff.).

49. That Dionysos is always "the stranger" says more about his character within Greek religion than about his historical importation. His essentially Greek nature was vindicated when his name was read on some Linear B tablets from Pylos (di-wo-nu- so-jo: PY Xa 102 + Doc. p. 127). See Jaan Puhvel, "Eleuther and Oinoatis: Dionysiac data from Mycenaean Greece," in E. L. Bennett (1964, pp. 161 ff.) and more recently in a text from Cretan Chania: see *Kadmos* 31 (1992), pp. 75 f.; see also C. Kerenyi (1976, pp. 68 ff.) and W. Burkert (1983, pp. 68 f., 239).

50. Dionysos is often counted as one of the twelve Olympians; see WHR, s.v. "Zwölfgötter." On the name Dionysos, see W. Burkert (1983, pp. 162 f.). Bacchos is particularly the god who sends ecstasy and the god of the initiate who sometimes takes the god's name if he is male; the female initiates were called bacchae (H. Jeanmaire, 1951, pp. 58 ff.). On Dionysos and the "wonderland" of Nyssa, see W. F. Otto (1933, pp. 58 ff.). M. Schultze (1876, p. 230) had suggested a curious neologism referring the second part of the name to the Semitic root *Nes, Nis,* a "stake" or "pole" (as in Num. 21. 8 ff., where it supports the brazen serpent). On the defeat of the Amalekites, Moses builds an altar to Jehovah-Nisi ("God is my support"? Exod. 17.15). On all this, see M. C. Astour (1965, p. 190 n. 4).

On the possible relation between Pentheos (from *pentheō,* "to mourn, wail") and Bacchos, from the Semitic—variously *baku* (Akkadian), *b'ky* (Ugaritic), or *baka* (Hebrew)—see M. C. Astour (1965, pp. 174 ff.).

51. See above, n. 47.

52. Athen. III.78c, quoting the Naxian historian Aglosthenes. Naxos was, of course, the island on which Dionysos "took over" Ariadne. See also A. B. Cook (1914, vol. 2, pp. 1108 ff., figs. 944–946). On Dionysos Melichios and Dionysos Baccheos, see C. Kerenyi (1976, pp. 123 ff.).

53. Paus. II.2.v f. As J. G. Frazer points out in his 1913 commentary, Eur. *Bac.* 1064 f. and Phil. *Icon.* I.17 make it a pine; Theocritus XXVI.11, a mastic. Plutarch (*Quaest. Con.* V.3) makes the pine Dionysos' special tree. There were two of them in the temple of Ephesian Artemis in Corinth, both gilt except for the face, which was painted red. One was called Lisias, the other Baccheos. Frazer's copious comparative material about the red color and blood does not seem strictly relevant here; moreover, as he points out himself, Pausanias also saw other statues of painted Dionysos: at Egira (VII.27.iv), near Pellene in Achaia, and another, growing out of a bronze base of ivy and bay leaves; the wooden parts of both were stained with cinnabar (VIII.39.iv).

See also Plut. *Symp.* V.3.i and Hesy., s.v. "Endendros."

54. Persecution of bacchoi in 186 BC and in the following years (as decadent) is told by Livy (XXXIX.8 ff.). But see H. Jeanmaire (1951, pp. 454 ff., 502 f.). On the distinction between Liber Pater (with his consort Libera) and Bacchos-Dionysos, see A. Grenier (1948, pp. 99, 122). For a full bibliography, see J. Bayet (1943, pp. 339 ff.). On later developments of the "mystery," see J.-M. Palisser, "Lieu Sacré et Lien Associatif dans le Dionyisme Romain," in O. de Cazanove (1986, pp. 199 ff.).

55. The Great Mother had a male paredros in many of her forms: Adonis-Tammuz-Dumuzi has become the most famous of them. But although there is an obvious relation between Dionysos and the "dying god," he does not appear in Greek religion as a paredros, but on the contrary has a number of paredrai, such as Ariadne or Erigone.

56. On phalloi in Dionysiac procession, see C. Kerenyi (1976, pp. 71 ff., 285 ff.); H. Jeanmaire (1951, pp. 40 ff.). That this seems uncommonly like a lingam of Shiva has not really been pointed out.

57. On Apollo, see above, chapter 7, p. 193. His historical genealogy (as against the mythical one) is uncertain. He is *kouros,* the "young man"—but he may also be young in the sense of a "recent arrival" in Greek mythology (see W. Burkert, 1983, pp. 212 ff.) That he has foreign antecedents is suggested by the fact that in the *Iliad* he is represented as an enemy of the Greeks. During the Persian wars, Delphic Apollo sided with the Persians. He "comes out of" Lycia—or at any rate so Simonides (in *PLG,* p. 519, 55a) and, much later, Serv. ad *Aen.* IV.143. In the *Iliad,* however, the paean is his particular song, and although paeans were also sung to other gods, he is the one worshipped as Apollo Paean. Paiawon seems to have been an independent deity in Crete, as perhaps Phoebos was in northern Greece. Phoebe is the mother of Leto, sometimes identified with the moon and with Artemis. The sun and moon were variously represented by Apollo and Artemis, Phoebos and Phoebe, Helios and Semele—who in her turn is most frequently (as against Demeter, Pasiphae, even Ariadne) considered Dionysos' mother. In the killing of Pytho, Helios and Apollo are allies (Hom. *Hym. Pyth.* 370 ff.), though in fact the explicit identification of Apollo as the sun god may not antedate the fifth century. A giant statue of Apollo stood in the Ephesian temenos, old enough for its base to be cracked with age in Augustan times ("Nostra vero memoria . . . a vetustate diffracta"; Vitr. X.2.xiii).

No name like his appears in the Linear B fragments (though see W. Burkert, 1983, p. 78 n. 53); all this may point to an oriental origin, as does the abundance of the Anatolian oracles of Apollo (of which Claros and Miletus have been mentioned here); attempts to connect him with Hittite deities have not been successful. He is sometimes also identified with the Semitic Resep. It is worth noting that in at least one fragment (Aesch. fr. 187 ex Macr. *Sat.* I. xviii.6) he appears as a double of Dionysos: *o kisseus Apollōn, o Baccheis, o Mantis*— "ivy-crowned Apollo, reveller, seer").

58. Historically rather than formally: Hellenistic critics, such as Lucian or Strabo, preferred citing that of Artemis Lycophryna at Magnesia, which had been designed by Hermogenes, to the Ephesian "Archaic" one, as a type of Artemision.

59. The architects are also named by Pln. *NH* XXXVI.xxi, 95. But see F. Miltner (1958, pp. 3 ff.); J. M. Cook (1962, pp. 101 f.). On its destruction, see Trebellius Pollio, "Gallieni Duo," in *Hist. Aug. Scr.;* Jordanes *De Or.* XX; E. Gibbon ([1776–1788] 1912–1921, vol. 2, p. 207). Before the temple stood a vast enclosed sacrificial altar, remarkable in that the officiating priest faced neither temple door nor the setting sun, but looked south, at right angles to the axis of the temple. On the altar and its direction, see A. Bammer (1984, pp. 130 ff.; 1972, pp. 40 f.).

Nicholas Purcell tells me that the fragments excavated by Wood have recently been reinvestigated to reveal the unsuspected presence of a long hairpin-plan temple at the pre-Croesus level, with the usual central colonnade, as at Samos.

60. The main source for the story of Croesus is Her. I.56 ff., 71 ff.; Pind. *Pyth.* I.94; Bacch. 31.3; and Nic. Dam. 65.68 in *FHG* vol. 3, pp. 406 ff. The occasion for the rebuilding may have been the ravages of the Cimmerian invasion (ca. 655)—of which little archaeological evidence has so far appeared—or the mantic order to rebuild after the sacrilegious break of the right of asylum by the tyrant Pythagoras (ca. 670–650). The bibliography is in Vitr. VII.praef.12. See A. Bammer (1984, pp. 74, 223 f.).

It is notable that the word *tyrant* was first applied to Croesus' ancestor Gyges (who died in the Cimmerian invasion of 655) by the poet Archilochos (fr. 25); on this usage, see G. Radet ([1893] 1967, pp. 146 ff.) On the Ephesian dedication Croesus calls himself *basileos,* however: fragments of his dedications from at least three separate columns survive (BM frr. B 16, B 32, B 136).

On Croesus' self-immolation after his defeat as a typic "apotheosis" on a pyre, see M. Delcourt (1965, pp. 68 ff.); see also J. G. Frazer (1911–1915, vol. 4.1, pp. 174 ff.). That it might have seemed sacrilegious to Cyrus in view of his Mazdean allegiance is suggested by G. Radet (1967, pp. 249 ff.). On the succession of Archaic shrines and Croesus' temple, see A. Bammer (1984, pp. 165 ff.).

61. Vitr. VII. praef. 12, 16.

62. Pln. *NH* XXXVI.xxi.95.

63. On the Heraion at Samos, see E. Buschor (1930, pp. 1 ff.).

64. An inscription in the British Museum (fr. B 17) was read as a dedication of Agesilaos by F. N. Pryce (1928, vol. 1.1, pp. 39 f.). But see also W. Schaber (1982, pp. 20, 130 n. 66), who thinks it a kind of preparatory "graffito" scratched on a roughed-out torus in preparation for a finished inscription on the astragal, which may not have been executed. On Agesilaos, see also above, chapter 7, p. 175. He most certainly visited Ephesus more than once and spent some time there; it was at Ephesus that he announced his plan to conquer Asia, so as to avenge the Persian invasions. See P. Cartledge (1987, pp. 152, 213 f.).

Central columns are used on the almost equally vast Olympeion at Agrigentum (see above, chapter 10, p. 347) though the layout and the context is quite different. On the significance and the exact number of columns, see A. Bammer (1984, pp. 217 ff.).

65. See W. R. Lethaby (1908, pp. 12 ff.).

66. Both the eight-petaled rosette capitals and the spiral volute capitals are restored in the British Museum: see W. R. Lethaby (1908, pp. 1 ff.); A. W. Lawrence (1983, pp. 161 ff.).

67. See W. B. Dinsmoor (1950, pp. 121 f., 326). The column, probably votive and of uncertain date, with spiraling flutes, is described by T. Wiegand (1904, p. 177).

68. A particularly rich one is known from Cyrene in North Africa. An eyeless volute capital of the same type was found in Athens, and it seems also to have served as the base for a statue. The whole group has been surveyed recently by P. P. Betancourt (1977), pp. 106 ff.

As for sphinx ex-votos, it is not clear why Dia Naxos, Dionysos' particular island, should especially have offered them.

69. Two such capitals, one called a "lyre" capital, are in the Metropolitan Museum, New York (nos. 11.185 cd, 17.230.6). The first has considerable traces of color and has been restored. On the sphinx, see above, p. 162.

70. Pausanias (VI.19.ii), who is the chief source for the treasuries, is a late witness, and though usually very interested in the date of the buildings he visited, he judged the relation between legend and monument by criteria rather different from ours. The Sikyonian was the most westerly of the Olympian treasuries, and therefore comes first in Pausanias' account since he worked from west to east. Many fragments have been found, including the dedicatory inscription. Both the materials and the masons' marks confirm that the stone was quarried in Sikyon and that it was done sometime shortly before 450 BC, though the metal clamps used suggested a slightly later date to its excavator, Wilhelm Dörpfeld. See Dörpfeld (1883, pp. 67 ff.); A. Boetticher, (1888, pp. 220 ff.). The treasury has recently been partially rebuilt. See K. Herrmann (1980, pp. 351 ff.).

71. The facade is made up of paired half-columns carved into the rock face. The column shafts are not fluted, but they do narrow upward. Their height is just over six times the lower diameter (almost exactly seven times the higher one). The lower end of the shaft is edged by a narrow fillet and stands on a roughly shaped torus, under which is a square-cut base. By contrast, the capitals seem surprisingly refined: over the spirals is a thin abacus, and a shallow cross-beam is shown to carry a cornice divided into three main members, crowned by a castellated crenellation.

The tomb (and another, rougher, but apparently "Median" one) was first measured and published by E. E. Herzfeld (1941, pp. 207 ff.). But see J. Boardman (1959, p. 215 n. 2) and E. Porada (1965, pp. 138 f.).

72. P. P. Betancourt (1977) is careful not to call these capitals and columns an "order" (p. 4). Very little is known about its entablature—but it seems clear (and important) that in some Syrian examples as well as in the Cypriot ones the beam would have lain at right angles to the capital, and not over the "bracket" on which the volutes were inscribed. It has even been suggested that such columns were not intended to support any timber roof, though it is difficult to understand how a peripteral building—such as the Athena temple in Old Smyrna, which was destroyed by Alyattes of Sardis about the year 600—could have been configured without any roof or entablature, as the

Turkish excavators seem to suggest: but see F. Schachermeyr (1983, pp. 319 f.) and E. Akurgal (1962). The use and arrangement of the Neandrian columns has been reexamined by Stefan Alterkamp in *AMS* 3 (1991), pp. 45 ff.

The Aeolian dialect was spoken in Boeotia and Thessaly, and tinged some Ionic poetry; and indeed Fritz Schachermeyr (1980, pp. 420 ff.) suggests that evidence of earlier, Mycenaean/Aeolic-speaking colonization was suppressed by the Ionians. However, even in later times Aeolis patronized the northern range of Greek colonies on the Aegean, where many of these capitals were found; and the name has also been applied to the similar capitals from the eastern Mediterranean, which were found somewhat later.

73. The abacus is even represented on stelae such as the one from Golgoi in the Metropolitan Museum, as well as on the Tamassos "pilasters"; the Salomonic pilasters from Ramat Rachel or Jerusalem have a plain block over the palmette.

74. As on the red figure vase showing Telephos in Mycenae (Boston 98931) or the killing of the Minotaur on a krater in the National Museum, Athens (Acropolis 735). There are other, unidentified scenes showing such columns, as a skyphos in Münster (Wilhelms-Universität Museum 45), or a plaque in the National Museum, Athens (Acropolis 2549).

Another plaque in the same collection (no. 2547) shows an interesting, apparently deliberate, isomorphism between the capital and the head of an owl held by Athena.

75. A general survey of these capitals was provided by P. Betancourt (1977); but see also Y. Shiloh (1977, passim).

76. On the ritual use of garlands, see most recently K. Baus (1940, pp. 7 ff.).

77. W. B. Dinsmoor (1950, pp. 290 f.). They appear uniquely on the protomes of a Delian treasury in the third century, too late to have any great interest here—too late even to be considered a product of Persian influence.

78. See p. 186.

79. See p. 303. Two ibexes on a glass-paste vase from Hassanlu in the archaeological museum in Teheran are thus positioned around a "made-up" tree crowned with a palmette; in a private collection in Berne, there is an ivory plaque from Ziwiye with an analogous composition, both are ninth to eighth century.

Earlier this composition was common on cylinder seals (from Susa, ca. 3000, in the Louvre) and on ceremonial staffs and horse trappings from Luristan. It survives into the Christian era, as in the Sasanian stucco relief in the Louvre.

80. Particularly the Lurs and the Kurds; but also around Isfahan and Shiraz in Iran. See E. Porada (1965, p. 75) and H. Field (1936, p. 223).

The seventeenth-century relief at Kurangun in the Bakhtiari Mountains is in situ; the fragments of the stele of Untashgal, perhaps later (fifteenth century?), are in the Louvre (Acropole Sb 14); see E. Carter and M. W. Stolper (1984, pp. 154 f., 185 ff.). On a representation of the Elamite ziggurat in Susa from Assurbanipal's palace at Niniveh (BM Niniveh Gallery, Slab 25), two ox skulls appear to be embedded in the corners of the crowning temple.

Horns are very obvious in Elamite buildings, and they may indeed have been true animal horns that were walled into them, though it is difficult to deduce anything definite from the reliefs and no archaeological evidence is available. The ancient Ziggurat at Susa may have been restored in the energetic building campaign of King Shilhak-Inshushinak (1150–1120). See *CAH,* vol. 3.2, pp. 488 ff.; W. Hinz (1972, pp. 55 ff.).

An early Hittite (eighteenth century?) ceremonial vessel from Büyükkale at Boghazköy, in the form of the model of a tower, incorporates acroteria-corner horns as well as a number of horned heads, one of which is the spout (Ankara Museum 144.7.64).

81. As a much-quoted inscription of Darius at Persepolis explicitly says (E. Porada, 1965, p. 156); though he also used Egyptian masons, and their influence has also been noted in Persian buildings, particularly of Darius' time. On Pasargadae, see C. Nylander (1970, pp. 103 ff.).

82. So for instance, Strabo, 15.733. But this form of criticism of the Persian barbarians runs through Greek literature, from Queen Atossa's dream of the two women in Aesch. *Pers.* 176 ff.

83. The traditions about Cyrus' death and burial are recorded by Herodotus (I.127), Strabo (XI.514 f.), Aelian (*VH* IV.1), Arrian (*An.* VI.29.iv ff.), and Eusebius (*Chron.* I.29). Arrian's description has been taken to mean that Cyrus's corpse was mummified; the tomb, which was known locally as that of King Solomon's mother (Meshed-i-Madr-i-Suleiman), has long stood empty and shows no trace of burial. Like many other buildings at Pasargadae, it was probably begun in the third quarter of the sixth century; Cyrus had defeated Croesus of Lydia in 546 (see n. 60, above).

It has been strongly argued (most recently by C. Nylander, 1970, passim, but esp. pp. 91 ff.) that it is typically Ionian, though its proportions and outline are those of a miniature ziggurat; at the summit is the double-pitched block of a tomb chamber. That he should have been buried as an Egyptian mummy in a Greek-built miniature ziggurat is in itself a tribute to Cyrus' syncretistic policy.

84. See W. Kleiss (1989, pp. 4 ff.).

85. An early investigator of the Boghazköy archives, Ernst Forrer, counted eight of them, but his calculations were later modified.

86. Or 1275; the date of the battle of Khadesh has recently been discussed by L. de Meyer in D. Musti (1986, pp. 210 ff.).

87. The vast surviving Egyptian hypostyles were in any case quite different in character, the columns being very close to one another, often barely a diameter apart; all too little is known about palatial hypostyles, whose columns would have been wooden.

88. It is clearly translated into the Akkadian *sar* in the word lists and bilingual inscriptions (C. Rüster and E. Neu, 1989, pp. 146 f.). The cuneiform character does not look like the hieroglyph, however.

89. *CAH,* vol. 2.2, pp. 417 ff.

90. The palace at level V was burned ca. 1750, perhaps by the (first?) Hittite Great King Labarnash.

91. H. Seton Lloyd and J. Mellaart (1965, vol. 2, pp. 3 ff.).

92. On the (by now) notorious Achijawa question, see F. Schachermeyr (1986, pp. 81 ff.); Lord William Taylour (1983, pp. 157 ff.); *CAH,* vol. 2.1, pp. 678, 812, vol. 2.2, pp. 186, 261; I. Singer (1983, pp. 205 ff.); and Guy Bunnens in D. Musti (1990, pp. 227 ff.).

On the marriage of Rameses II and the daughter (perhaps even two daughters?) of Khattushilish III, who became his third "official" wife and was known in Egypt as Maetnefrure, see J. G. MacQueen (1986, p. 50) and *CAH,* vol. 2.2, pp. 226 ff.

93. The temple at Musasir, which Sargon II plundered in 714 during his campaign against Rusa I of Urartu, was shown on a relief unfortunately lost in Tigris mud, though a drawing of it by Flandrin was published by P.-E. Botta (1849–1850); it seems to have had a tentiform, pyramidal roof supported on pillars decorated with bronze shields on which snarling dogs were depicted. Sargon's relief is the most explicit representation of an Urartian building. On either side of its doors, beside the tree spears, were two great cauldrons on tripods. See T. B. Forbes (1983, pp. 46, 81 ff.). Forbes (whose book is the only survey of Urartian building) lists other "square" temples and has surveys of the main Urartian palaces, including the one at Amin-Berd/Erebani, near Yerevan: it contained a hypostyle hall, some 20 by 23 meters, comparable to the main hall at Büyükkale, 32 by 32 meters. Unfortunately the Achemenid remains on that site have not yet been clearly distinguished from the Urartian. This is also the case with the gridded settlement at Zernaki Tepe, at the northeastern end of Lake Van, on which see Forbes (pp. 121 ff.); C. Nylander (1966, pp. 141 ff.); and H. F. B. Lynch (1965, vol. 2, p. 29). Lynch has it as a foundation of King Arghisti II (713–685), though it is the unique Urartian grid-settlement, and no systematic excavation has yet taken place. The most impressive figural work is the colossal relief of the storm god, Teisheba, from Adilcevaz, just north of Lake Van; see B. B. Piotrovskii (who has nothing to say about Zernaki: 1967, pp. 64 ff.).

94. He committed suicide by drinking bull's blood, according to Str. I.61. Herodotus (I.14) saw Midas' judgment throne—reputedly the first barbarian gift to the oracle—in the Corinthian treasury at Delphi, where it was exhibited with the gold and silver treasure presented by Gyges of Lydia, the ancestor of Croesus.

95. *CAH,* vol. 2.2, pp. 265 ff. The language they spoke and wrote was Aramaic, the lingua franca of the northern Semites.

96. The discoverer of Tel-Halaf, Max von Oppenheim, who had abandoned a diplomatic career to take charge of the excavations, exhibited the statue-columns in his private museum in Berlin, which was destroyed by bombs during World War II. See H. Schmidt, R. Naumann, A. Moortgat, and B. Hrovda (1943–1962); W. Orthmann (1971, pp. 119 ff., 178 ff.); and R. Naumann (1955, pp. 362 ff.).

On possible Neo-Hittite influence in Cyprus, see V. Karageorghis (1973, pp. 362 ff.).

97. Sir Charles Leonard Woolley (1953, pp. 101 ff.).

98. The parallel was first pointed out by E. Akurgal (1955, pp. 87 ff.).

99. See S. I. Rudenko (1953, pp. 294 ff., 349 ff.). The origin of the Pasaryk carpet has been much debated, and the analogies of the patterns to those of the Assyrian stone "carpets" used as paving in gates pointed out (there is one each in the British Museum—from Niniveh—and the Louvre). See Crawford H. Greenewalt and Lawrence J. Majewski, "Lydian Textiles," in K. de Vries (1980, pp. 133 ff.).

100. The spectacular "patchwork" mosaic from Gordion—which its excavator claimed as the first mosaic ever (it is certainly the earliest floor mosaic discovered so far)—looks as if it were a durable replacement of mats randomly strewn over a floor. But see R. Young (1973, pp. 15 f.). By the time this technique had been adapted in Macedonia, pebble mosaic had become an independent pictorial device.

101. The vaulted house seems to have been formalized into a hieroglyphic Luvian charac-ter, which also appears in the Phaestos disk. See J. Best and F. Woudhuizen (1988, pp. 37 f., 59) and M. Mellink (1964–1965, pp. 1–7).

102. See D. Asheri (1983, pp. 31 ff.).

103. This is notable in the house tomb near the Athenian agora and the "Eastern Basilica" tomb at Xanthos. See P. Demargne, "Tombes-Maisons, Tombes Rupestres et Sarcophages," *Fouilles de Xanthos* 5 (1974), pp. 20 ff.

104. J. G. MacQueen (1988, pp. 22 ff., 101 ff.).

105. They are first recorded in the pre-Hittite buildings at Alaça Hüyük, north of Hattusas/Boghazköy, and in the palace of level VII, considered to have been that of Yarim-Lin, protégé of the Pharaoh Tuthmosis III.

106. The literature on the subject has already been discussed. But see also N. Perrot (1937).

107. On the association of Semitic *asseroth* and *asserim* with the Egyptian *Dd,* see M. J. Lagrange (1905, pp. 176 ff.) and E. A. Wallis Budge (1934, pp. 184 f.).

108. The original excavation report, by Max Ohnefalsch-Richter, although revised about 1910, was never published; but it was summarized together with his new finds by H.-G. Buchholz (1973, pp. 297 ff.).

The architectural arrangement of timber-hut-like underground chambers, whose doors are flanked by relief volute columns on stelae, are the so-called royal tombs V and XI (Buchholz provides a concordance of the different numberings, p. 324). They are not unlike the chambers in Archaic Lycian and Lydian tombs; Buchholz even calls them imitations. The two tombs are part of a "royal" burial ground.

109. See generally *CAH,* vol. 3.1, pp. 511 ff.; V. Karageorghis (1982, pp. 63 ff.); also L. R. Palmer (1965, pp. 156 ff.). As late as the fourth century, syllabary inscriptions with Greek and Phoenician transliterations were being executed in Cyprus. Elsewhere, Karageorghis (1973, pp. 363 ff.) even suggests that similarities between Anatolian and Cypriot tombs were so striking that the Cypriot ones could only have been the work of itinerant Lydian craftsmen.

110. *CAH,* vol. 1.1, pp. 542 ff. At any rate, no Paleolithic or Mesolithic remains had been found at the time of writing. But Neolithic settlement seems to have begun about 6000; it may well have had two separate stages, though evidence about the origin of the settlements is inconclusive, as it is about the fallow period between them. The island's great prosperity begins when the Cypriot population passes into a Chalcolithic phase in the fourth millennium and the exploitation of the rich metal mines begins.

111. When Cambyses of Persia conquered Egypt, he also got Cyprus. Alexander the Great returned it to the Greek koiné.

112. See Hes. *Theog.* 194 ff., who accounts for Aphrodite's name, *ounek en'aphrō thrēphthi,* "for she rose out of the sea foam." The rocky beach, now called Petra to Romou, was shown as the actual place of her rising. Hesiod's account of the birth of Aphrodite has been compared to the account of the castration of Anu by Kumarbi in the Hurrite account of the struggle for the kingship of heaven. Kumarbi did not cut off Anu's genitals, however, but bit them off and swallowed them: he spat out semen mixed with his spittle, from which other deities were born. But like Kronos, he tried to eat his son born of this impregnation, Teshub, and was given a stone instead. See H. A. Hoffner (1990, pp. 40 f.). On this and Hesiod's other oriental echoes, see W. Burkert (1992, pp. 5 f.); P. Walcot (1966, pp. 2 f., 111). The birth of a "bearded" Aphrodite from severed genitals was illustrated on a plaque found at Perachora: R. J. H. Jenkins (1940, pp. 231 f., no. 183).

There were many accounts of her birth, of course; in the *Iliad* she is the daughter of Zeus and Dione (with whom she is also sometimes identified). The *Odyssey* recounts her illicit liaison with Ares and the birth of her daughter, Harmonia, who was to marry the Phoenician Kadmos, the founder of Thebes. Pausanias (IX.16.iii) saw a triple xoanon (made of the stern timbers of Kadmos' ship) of the goddess in Thebes. He was told that it was consecrated by Harmonia to signify three different kinds of love. Herodotus, for his part, saw her temple in Askelon (I.106), which he says was the first ever consecrated to the goddess.

The alternative site of her emergence from the sea was Cytherea (Kythira), an island between western Crete and Monmenvassia, which seems to have had a large Phoenician colony associated with fishing for the purple snails that are plentiful there (*Murex trunculus* and *Murex brandaris,* which secrete the "Tyrian" purple: Paus. III.21.vi). The name of the island has been derived from the Western Semitic *K't'r, K't'r't,* "crown," by V. Bérard (1903, vol. 1, pp. 297 ff.) and M. C. Astour (1965, p. 143). Herodotus (I.105) had already called one of its bays *Phoinikos,* and Pausanias noted (I.15.vii) that the main temple of the goddess was a Phoenician foundation. Xenophon (*Hel.* IV.8.vii) records another place, Phoenix. Phoenix was a familiar name of the father or brother of Europa (Her. IV.147.iv) and the eponymous father of the Phoenician nation. The Aphrodite of Cytherea was known also as Urania, and her warlike image was associated with Semitic cults.

Her other famous shrines were at Corinth and on Mount Eryx in Sicily. Identified with the old Latin divinity, Venus (originally perhaps the patroness of vegetable gardens), she is established in the Roman pantheon: as Venus Genetrix the divine ancestor—through her son Aeneas—of the *gens Iulia,* she became the patroness of the Imperial cult.

113. Agapenor, son of Ancaeus the Argonaut, appears in the *Iliad* (II.609 ff.) as the Arcadian leader. The story of his settlement in Paphos and the building of the sanctuary was told by Pausanias (VIII.5.ii ff.); while J. G. Frazer (1913, vol. 4, p. 193) suggested that the legend is a by-product of the similarity of dialects. It is also referred to by Apollodorus (*Epit.* VI.15). Kinyras is a more complex figure than Agapenor, and seems to belong to an earlier age: he is variously the son of Apollo and Paphos, of the Assyrian King Thias, or of Phoenix. His incestuous amours with his daughter, Smyrna (who was changed into a myrrh tree), generated Adonis. This is the subject of much speculation in ancient literature and most famously in Ovid (*Met.* X.476 ff.). It was also commemorated in a lost poem, usually known as *Zmyrna,* by Gaius Helvius Cinna, the magistrate mistakenly lynched ("I am Cinna the poet!") after Caesar's funeral. Kinyras' name is derived by some scholars from the Western Semitic *K'n'r,* "Kinnor," the deified harp—a deity in Ugaritic mythology; see M. C. Astour (1965, pp. 137 ff., 191, 308).

The Kinyran was probably the only priest-king dynasty in Cyprus, but Kinyras (or merely a namesake?) is also the reputed founder of the temple of Aphrodite in Byblos, according to Lucian (*De Dea Syr.* 9). Tacitus (*Hist.* II.3, *Ann.* III.62), who knew the dynasty, was also told of an even more ancient founder of Paphos, Aeris: his son, Amathor Amathos, was the eponymous founder of the shrine of Aphrodite and Adonis, only a little eastward from Paphos on the southern coast. The name (like the cult, of course) is connected with Phoenicia; the breathing on the initial *A* of the name is hard, "Hamath." Steph. Byz., s.v. "Amathos," calls the cult one of Adonis Osiris.

114. Paphos and Kition were the only major settlements in Cyprus to be marginally affected by the natural cataclysm of 1075, which seems to have had repercussions in all Cyprus and in southern Anatolia. *CAH,* vol. 2.2, pp. 315 ff.; vol. 3.1, pp. 511 f.

On this site, see F.-G. Maier (1985, passim). General Cesnola had, in fact, identified the site and its ruins more than a century ago (1914, pp. 204 ff.). However, see the account of the excavations in F.-G. Maier and V. Karageorghis (1984, pp. 15 ff.).

Tacitus describes the conical baetyl in the course of his account (*Hist.* II.3) of Titus' visit to Paphos, though he confesses himself puzzled by the shape: "simulacrum deae non effigie humana, continuus orbis latiore initio tenuem in ambitum metae modo exsurgens; [s]et ratio in obscuro."

He notes that no blood sacrifice was offered there, though *exspicitinium* from the innards of billy goats (the goddess did not care for female victims in any case) had, according to legend, been introduced from Cilicia.

The baetyl described by Tacitus has now been identified as a conical block of nonnative red granite found as part of a wall in Paleopaphos (personal communication from Nicolas Purcell). Another candidate, the grey-green conical stone 1.22 m high, now in the Cyprus Museum in Nicosia, is illustrated by F.-G. Maier and V. Karageorghis (1984, pp. 99 ff.).

115. Although from the *Odyssey* (VII.362) onward it appears in Greek literature as a Greek shrine.

116. On the problem of the masonry, see V. Tatton-Brown (1989, p. 20) and V. Karageorghis (1976, p. 171).

117. 1 Kings 16.31 ff.–2 Kings 10; but cf. 2 Chron. 21.5 ff.

118. The other common type of temple involved an oblong shrine surrounded by small chambers, and probably most commonly covered by a barrel vault. On the type in general, see W. Andrae (1930). The oldest temples, such as the White Temple at Warka, seem to have been moved to the foot of the ziggurats; a particularly grandiose one stood at the foot of the ziggurat at Tell-el-Rimah.

Another archaic sanctuary with a covered portico has recently been excavated at nearby Bamboulia by a French mission. See A. Ciubet, "Le Sanctuaire Chypro-Archaique de Kition-Bamboulia," in G. Roux (1984, pp. 107 ff.).

119. About 4 to 4.5 m, depending which cubit was being used (Str. III.170). Steph. Byz., s.v. "Gadeira," an almost exact Greek transliteration of the West Semitic word for "enclosure," "walled place." While the Phoenician foundation of the town was never in doubt, it took the Roman side after the Second Punic War, and the awesome shrine (including some "archaic" timber buildings) is described as very splendid by Silius Italicus at the beginning of book III of his *Punica* (the longest Latin epic as well as the worst, someone remarked), though no special reference is made to the Herculean columns.

A small (28 by 30 cm) but early (seventh century) limestone "Aeolic" capital of the horns-and-triangle type was found in Cadiz and is now in the museum there (exp. 780).

120. That emerald column has raised many scholarly eyebrows. Herodotus had the temple dedicated to Herakles, of course (2.44). M.-J. Lagrange (1905, pp. 213 ff.) thought that they might in any case have been ex-votos rather than entrance pylons, which he discusses with a wealth of parallel examples.

121. 1 Kings 7.13 ff.; cf. 2 Chron. 3.15 ff., 4.11 ff. The scriptural texts seem to disagree about the dimensions of the columns (18 cubits for the shaft, 5 for the capital in 1 Kings; 35 for the shaft, 5 for the capital in 2 Chronicles), and this has caused much speculation: see R. de Vaux (1961, pp. 314 f.); W. O. E. Oesterley and T. M. Robinson (1930, pp. 42 f.). There was also much speculation about the exact meaning of the two names of the columns. In some rabbinic traditions they were taken to represent the sun and the moon. So at any rate holds the Talmudic rabbi Pinhas ben Yair (in R. Patai, 1947, pp. 108 ff.). The craftsman who cast them was called Hiram (or Ahiram), like the contemporary king of Tyre; he was the son of a Jewish mother and a Phoenician father (though a rabbinic tradition reads the two texts to say that both sides were Jewish) from whom he learned his craft: presumably his work looked Phoenician. The columns were probably replaced after the Exile. Ezek. 40.49.

Hiram the king was said by Menander of Ephesus (as reported by Ios. *In Ap.* I.118; R. L'Estrange, 1709, vol. 2, p. 1224) to have put a single gold column in the shrine of Olympic Zeus in Tyre—presumably Baal-Shamim.

On the Phoenician workmen and their influence, see a review of the literature by T. A. Busink (1970–1980, vol. 1, pp. 261 ff.). On the entrance columns of the shrine and the columns Jachin and Boaz, see vol. 1, pp. 173 f., 299 ff.; vol. 2, pp. 752 f.

The parallel to the twin obelisks that flanked the entrances of Egyptian shrines has often been made. In the much-quoted building inscription (Cairo 34025) near the Colossi of Memnon at Thebes, Amenhotep III (ca. 1415–1375) boasts, "all its doorways [are decorated] with gold. Two great obelisks have been erected, one on each side, so that my father [Amon-Re] may appear between them, while I am in his retinue" (J. B. Pritchard, 1955, p. 376).

122. A full survey of the thirty-four surviving capitals found in Israel is offered by Y. Shiloh (1977).

123. On Astarte and Ta'anit, see D. Harden (1962, pp. 86 ff.) and S. Moscati (1979, pp. 161 ff.).

124. A. Parrot, M. H. Chéhab, and S. Moscati (1975, pp. 217 f.). There are two very similar such stelae in fact in the Museo Nazionale, Cagliari (86032, and one without inventory number).

125. Fifth century, of uncertain origin (Louvre AO 27197). A much later (second century) Doric stele from Sulcis is in the Museo Nazionale, Cagliari (54386).

126. Mostly from Lilibea and Motya, many of them painted. On the whole the stelae do not display individuated "orders" as pilasters. Sometimes they are just framed by half-cylinders, with token capitals and bases.

127. B. Soyez (1977, pp. 31 ff.); *Antiquités Syriennes* 3 (1965), pl. 18.

128. A. Lézine ([1960?], pp. 43 ff.). Lézine (pp. 63 ff.) documents the relatively late appearance of the Doric column—at the end of the sixth century—which he associates with the importation of the cult of Demeter and Kore from Sicily. On the Monte Sirai aedicule, see A. Parrot, M. H. Chéhab, and S. Moscati (1975, pp. 226 ff.); on Tharros, see p. 217.

129. It was found by Melchior de Vogüé, and is now in the Louvre. See PC, vol. 3, p. 277, and S. Moscati (1988, pp. 162 f.), where the capitals are called "Hathoric." Although the principal cult at Idalion was of Melquart, this model, with its women at windows and pigeonholes, inevitably suggests that of Astarte. The human-headed bird was also associated with one of the Egyptian ideas of the soul, as a Ba-bird.

130. BM 90.000. It is also reproduced in a clay cover for the relief, 90.002. In fact it records the restoration and renewal of the statue of Shamash in his temple at Sippur. The tablet speaks of the image and temple being founded 300 years earlier and damaged by sacrilegious enemies. The iconography is conventional: the king is led to the god by his high priest. Unusual are the ornaments, such as the sun disk before the statue, which seems to be operated by ropes pulled by a divine figure perched on the canopy. The canopy and ornaments are also "ancient." H. Frankfort (1954, pp. 106 f.) thought the costume and regalia to be made deliberately archaic, like third-millennium ones, while the temple ornaments were more "modern." The volute capital is almost identical with the base of the column, and the stand of the sun disk repeats the same shape. See L. W. King (1912, pp. 120 ff.); Frankfort (1963, pp. 106 f.). This proto-Ionic form has already been noted by O. Puchstein (1907, pp. 30 ff.) and again by F. von Luschan (1912, pp. 31 ff.).

131. So. W. Andrae (1930, pp. 41 f.). Assur-nasir-pal's canopy is in an encampment on a relief from room B, NW Palace, Nimrud (in the British Museum, WA 1245–1248, dated ca. 850–860).

132. Notably on the bronze gates of King Shalmaneser III from Tel Balawat in the British Museum (WA, no number, ca. 850–840). Double "Aeolic" capitals on a "summer palace" in a relief in the British Museum (WA, 124939) which was made for Assur-bani-pal about 650, though probably representing the palace built for Sennacherib half a century earlier.

133. In the Vorderasiatisches Museum, Berlin.

134. One, from Sargon's palace at Khorsabad, is in the Louvre: G. Perrot and C. Chipiez (1882–1889, vol. 2, pp. 250 f.). An even larger one, from the North Palace of Niniveh, is in the British Museum (WA 124962). A much earlier, fourteenth-century Mycenaean ceiling panel from a tomb in Orchomenos is surprisingly similar: it may well represent a textile, though the detail of the ornament, which was almost certainly colored, is that of many Mycenaean wall paintings. It has remained in situ in the side chamber of the so-called Treasury of Minias, which Pausanias (IX.36.v, 38.ii) described. See also E. Vermeule ([1964] 1972, pp. 120 ff., 21 f.); R. Hampe and E. Simon (1981, p. 36, with pl. 51).

135. Such double spirals appear at the joint between leg and seat of chairs, but even more commonly under the headpiece of beds. A notable early example of a chair with such ornaments is the throne of the pomegranate goddess on the west side of the Harpy Tomb from Xanthos (BM B 287; F. N. Pryce, 1928, pp. 122 ff.). In sixth-century vase painting such chairs are fairly common: see H. S. Barker (1966, pp. 264, 268 ff.) and G. M. A.

Richter (1966, pp. 59 ff.). In a stool (a diphros), this ornament usually hides the mortise end of a transverse strut, where it is tenoned into the plank that serves as a leg; in beds and couches (klinai, etc.) this transverse piece is the low headboard, over which cushions are commonly plumped. Two terracotta sarcophagi (in the Louvre and in the Villa Giulia Museum in Rome) are notable extraterritorial examples of such klinai: particularly interesting, because the headboard seems to be flat, unlike the Greek examples, so that the volutes are made to support an abacus. Very impressive is the double image of such a klinē (one black on red, the other red on black) by the Andokides painter (Munich 2301). On the Phrygian "stone textiles," see above, p. 283).

136. See T. Howard-Carter (1983, pp. 283 f.).

137. One, from an unknown site in the Transjordan, is in the Rockefeller Museum, Jerusalem; the other, from Tel-el Farah, was published by R. de Vaux, *Rev. Bib.* 62 (1955), pl. 13; and A. Jirku (1960, pl. 87, p. 249); see Y. Shiloh (1977, p. 32). There are two other models in the Louvre and one in the Israel Museum from the Mt. Nebo area (no. 82.24.15).

Burial in a house-shaped urn was already common in the Neolithic period. The Jerusalem Museum has a large collection of such urns, from one Chalcolithic (fourth millennium) cave site near Tel-Aviv: some of them are animated with schematic faces and rumps indicated at either end.

138. The two models, both from Kotchti in the center of the island, are now in the Cyprus Museum, Nicosia; from the end of the early Bronze Age, ca. 1900. See V. Tatton-Brown (1979, p. 28; 1970/IV-28/1, 1970/IV-30/1).

139. Such prepared skulls were found at Enkomi and Kition: V. Karagheorgis (1976, pp. 171 f.). Literature on the use of animal skulls as masks is unfortunately very scarce. On the Minotaur as a ritually masked figure, see A. B. Cook (1914–1940, vol. 1, pp. 492 ff.; vol. 3, pp. 1087 ff.). Two small figures of men shown actually putting on bull-head masks (ca. 625) survive from a large group found in the Apollo sanctuary at Kourion, about 20 km east of Paphos.

140. The most striking literary evidence is Hermes' use of a horned skull to construct a lyre, whose sound box was made of a tortoise shell, and with sides made of bull's horns: Hom. *Hym. Herm.* 49 f. See also PW, svv. "Lyra," "Schildkröte"; B. P. Aign (1963), pp. 379 ff.); J. R. Jannot (1979, pp. 469 ff.); P. Courbin (1980, pp. 93 ff.); and H. Roberts (1980–1981, pp. 303 ff.). A tortoiseshell sound box for a lyre is known both from a figure-shaped unguent vase (BM 5114) and an actual tortoiseshell that had been so used (BM 38171). See R. D. Anderson (1976, pp. 3 f., 70 ff.).

Bull's heads, adorned with rosettes on the forehead or with an upright double ax between the horns, are a common ornament in Minoan as well as Mycenaean art. A chlorite one (gilded horns, no rosette) is from the palace at Zakros: see N. Platon (1971, pp. 161 ff.). A silver one, with gilded horns and a large gold rosette on the forehead, was recovered from grave circle IV at Mycenae (under life-size, after 1500, in the National Museum, Athens). A steatite one is from the "Little Palace" at Knossos (Iraklion Museum; *PM*, vol. 2, fig. 330). Two such rhytons appear in Egyptian Eighteenth Dynasty tomb paintings, as gifts of Cretan envoys ("from Keftiu"): of User-Amun and Re-Khmere, dated about 1480–1420; on which see F. Schachermeyr (1964, pp. 112 ff.).

Offerings of horns, presumably of sacrificed beasts, were found in several Greek shrines: notably under three bronze statues at Dreros and at Samos. That Gilgamesh, when he had killed the Bull of Heaven, cut off his horns "plated with lapis lazuli two fingers thick, and thirty pounds in weight . . . which he carried into his palace and hung on the wall," may be no more than an early instance of a hunting trophy.

141. This is true of Hittite and Phoenician deities. In Cyprus there are two well-known statues: the god (or worshipper) from Enkomi near Famagusta (1949/V-20/6), and another statuette, of a god in a horned cap, which stands on an ingot (no inventory number) and is known as "the Horned God" and "the Ingot God" (both dated to ca. 1200); see N. K. Sandars (1987, pp. 129 ff.) and V. Tatton-Brown (1979, p. 42). In these Cypriot examples, as in divine headgear for the most part, the horns are a bull's. It has been suggested that the first of the statues at any rate represents the Arcado-Cypriot Apollo Kereatas; some bulls'

horns have also been found there. The sanctuary is built of well-jointed ashlar masonry, not unlike the temples at Kition and Paphos.

142. Virg. *Aen.* IX. 626 ff.:

Ipse tibi ad tua templa feram sollemnia dona
Et statuam ante aras aurata fronte iuvencum
Candentem. . . .

But see also Serv. ad loc.

143. Molded bovine skulls (some of very large specimens of *Bos primigenius*) are a common ornament of the Neolithic buildings (shrines?) in Çatal Hüyük. See J. Mellaart (1967, pp. 77 ff.); modeled bucrania were about the most conspicuous and commonly repeated ornaments in the settlement "shrines," but did not seem to be used by the later Neolithic or Chalcolithic dwellers at Beycesultan and Haçilar. A single (miniature) bovine head, containing no bone fragments, was found in the temple of Ba'alat Gibal at Byblos (Beirut, Musée National). In Egypt, a whole bench of some three hundred bull and/or cow heads, on which real horns were mounted, surrounded the rather gruesome mastaba of Uadjit, perhaps the fourth pharaoh of the First Dynasty (Manetho's Uenephes? Horus Djed, the Serpent King?), at Sakkara (tomb 3504); on which see W. B. Emery et al. (1949–1958), vol. 2, pp. 8 f., 102 ff.). He also had a tomb at Abydos (Z) like other pharaohs of the First Dynasty, who were also buried with sacrificed servants, though the bull bench seems to be very unusual. See *CAH*, vol. 1.2, p. 55; W. B. Emery (1961, pp. 71 ff.). An even earlier painted ox skull "from the grave of a pre-dynastic chieftain" is in the British Museum (59262). But bull and ox skulls, usually painted, are a common enough grave-deposit in later mummy tombs, such as the one exhibited in the Metropolitan Museum, New York.

144. His Spartan sanctuary was described by Paus. III.13.iii ff. On him as a column, see Eumelos fragment in Paus. IV.33.iii. Karneios was celebrated in an important yearly festival, the Karneia, on which see W. Burkert (1983, pp. 234 ff.). On the cult, see C. Leroy (1965, pp. 371 ff.).

145. M. Gimbutas (1991, pp. 244 ff.).

146. See R. S. Ellis (1968, pp. 46 ff.) and E. D. Van Buren (1931, esp. pp. 2, 77 ff.). The nailing of a peg may well have been used in Mesopotamia as a token of agreement generally. Perhaps the calendar nailing of the temple Jupiter Optimus Maximus on the Roman Capitol has a Mesopotamian precedent?

147. See C. L. Woolley and M. E. L. Mallowan (1976, pls. 85–87; p. 180). The animal-headed demon, Pazuzu, was also represented for such amulets. Some scholars have suggested that Humbaba hides behind the Greekified Kombabos, of whose self-castration and that of Stratonice (as the foundation myth of the Galli) Lucian tells in *De Dea Syr.* 19 ff.

148. The masks seem to be of two periods: Early Assyrian (ca. 1800) and Nuzi (ca. 1600); the first masks were reused in the Middle Assyrian floor. They are in the Baghdad Museum (IM 69731, 73922). See D. Oates (1967, pp. 76 ff.). Theresa Howard-Carter has suggested (1983, pp. 64 ff.) that the masks were placed on one side as capitals, and the reliefs of bull-men found with them on the other. See also S. Dalley (1984, pp. 184 ff.).

149. Some of the earliest composed columns have survived in the first "palace" (level XII, 2700–2350) at Alalakh. In the Louvre there is a contemporary (Early Dynastic Sumerian) group of quadrant bricks, which constituted hollow columns.

150. A corner section of the wall is in the Pergamon Museum in Berlin. The female figure pouring out vessels held about the level of the breasts is familiar in Mesopotamian art: a nearly life-size stone goddess of this type is in the Aleppo Museum (no. 1659); a channel at the back of the statue allowed real water to flow out of the vase. It comes from the Zimrilim palace, Mari, ca. 1800–1750.

A strange postscript to this notion was the fortifying bath that the Byzantine emperor would take once a year in water poured out from a standing stone "icon" (presumably a stone relief) of the Blessed Virgin in the Blacherna church. The bath was probably destroyed when the church was burned in A.D. 1070, and the ceremony was discontinued. See A. Berger (1982, pp. 81 ff.).

151. As that of Ptah-iru-kah at Sakkara.

152. The alternation is common. An early example is the base of a statue of Imhotep from the Zoser enclosure at Sakkara, now in the Cairo Museum (no. JE 49889). The repetition is almost obsessive on the gilt walls of the shrine "in the form of a Heb-Sed pavilion" from the tomb of Tutankhamun (Cairo Museum, Tomb of Tutankhamun, 739).

153. E. D. Van Buren (1945), pp. 44 ff.). On the development of the cuneiform star sign (C 3), see pp. 82 ff. See also W. Andrae (1930, pp. 48 ff.) on the notion of the bundle, both Egyptian and Sumerian. On the building process, see E. Heinrich (1934, pp. 13 ff.). The alabaster vase is in the Iraq Museum, Baghdad, no. 19606, while the trough is in the British Museum, no. 120000. Both are dated to the Uruk 4 period, ca. 3100–3000 BC. They are discussed by H. Groenewengen-Frankfort (1951, pp. 150 ff.). Representations of the bundle on seals are reviewed by H. Frankfort ([1939] 1965, pp. 115 ff.).

154. The method is described by W. Andrae (1930, pp. 47–62) and E. Heinrich (1934, pp. 13 ff.). The terminology is confused, in that the more expensive and elaborate (as against the double-pitch roofed) srefe huts are usually used as mudhifs, while smaller ones are often divided between mudhif and harem, or domestic quarters.

155. A somewhat impressionistic account of srefe huts is given by W. Thesiger (1967, pp. 26 ff., 48, 71, 205 ff.). On their place in Marsh Arab society, see R. A. Fernea (1959, pp. 89 ff.).

156. Gottfried Semper had already observed this a century and a half ago (1878, vol. 1, pp. 73 ff.). Particularly striking are the images on either side of the throne of Sesostris I (1971–1928; painted limestone, Cairo Museum 301) of Horus and Seth on one side, and two Nile gods on the other, binding papyrus and bindweed plants to the hieroglyph *sema*, "unite," "join." Semper had seen a rather imperfect representation of the two Nile gods, though the plant forms (without the deities) already appear on the sides of the throne of the diorite statue of King Chephren (builder of the second-largest pyramid, ca. 2600; the diorite statue, over life-size, is in the Cairo Museum, 138) and are often repeated.

157. They already seem to have ceremonially accompanied "King Scorpion," an Upper Egyptian pharaoh, perhaps before Narmer, on his commemorative macehead (in the Ashmolean Museum, Oxford).

158. The nome sign was later written as a face on a stick surmounted by a sistrum. Sistra with Hathor faces appear in many reliefs: for instance, several times in the temple of Seti I at Abydos. And indeed there are some forty more or less complete ones in the British Museum, all with Hathor heads. See R. D. Anderson (1976, pp. 40 ff.).

There are two types of sistra: one with a spring protruding from the head, the other with a shrine on top of the head and with two sprigs on either side of the shrine. Certainly, both types seem to be equally popular. Moreover Hathor heads appear on other musical instruments, such as ivory hand-clappers.

A "theosophical" interpretation of the sistrum is provided by Plut. *De Is. et Os.* 63: its noise is apotropaic. "Typhon," whom Plutarch identifies with Seth, signifies death and stasis, while the noise of the sistrum indicates the life-giving motion of the four elements, which are identified with the four bars of the sistrum, its coil signifies the lunar sphere and its rising out of the head (as Plutarch thought) of Isis or Nephtys both generation and ending. But see *LÄ,* s.v.

Hathor's figure incorporated an even more archaic deity, Bat (on whom *LÄ,* s.v.), who may—rather than Hathor—be the deity on the Narmer Palette; her sign were two cow heads back-to-back, with huge sprawling horns, which become formalized in later images into the spirals on either side of the shrine over Hathor's head. See *LÄ,* s.v. "Hathor-Kapitell," and E. Staehelin (1978, pp. 76 ff.).

159. Her status as archaic mother-goddess was suggested by R. T. Rundle Clark (1978, pp. 87 ff.). Her hieroglyphic name was *Hethor,* "House of Horus"; alternately, *Het-Hert,* "House of the Face": E. A. Wallis Budge (1934, pp. 59 ff., 228 ff.). She is often identified with other goddesses, especially Isis, and is often called the consort of the chief god (of Ra, for instance). She was worshipped in Syria, at Byblos. She is often "multiplied": there were seven Hathors at Denderah and twelve in some texts; in the ritual of mummification she is the special patroness of the face.

Mykerinus had himself shown between Hathor and the patronesses of various nomes for his temple, about half life-size. Four of these green schist triads survive, one in the Boston Museum of Fine Arts (09.200) and three in the Cairo Museum (the one I mentioned is 180).

160. Cairo Museum 3055. It is described in all the standard works on Egyptian art. "Narmer" is usually assumed to be the Horus-name of Menes, the founder (ca. 3100) of the First Dynasty and uniter of Upper and Lower Egypt, which union the palette commemorates. On the other hand, no one has seriously suggested that the cow heads are anybody's but Hathor's. An even earlier image of her may be the predynastic relief of a cow's head surrounded by three stars (32124), on which see E. A. Wallis Budge ([1904] 1969, vol. 1, p. 78 f.).

161. H. Frankfort (1948a, pp. 11 ff.).

162. On which see "The Deliverance of Mankind," in A. Erman (1966, pp. 48 f.). Her beer had been stained red to simulate blood; see also "The Contendings of Horus and Seth," trans. Edward F. Wente, Jr., in W. K. Simpson (1972, pp. 112 f.).

163. See the inscriptions in the temple, which was begun under Ptolemy VIII Evergetes II about 115 BC; Cleopatra VII (Caesar's and Marc Anthony's Cleopatra) and Cesarion are all represented in inscriptions and reliefs; the upper edge of the cornice is inscribed to Aphrodite by Tiberius (perhaps in AD 34). The outerworks were not finished till the reign of Trajan.

Other inscriptions in the temple record Cheops' foundation, Pepi I's restoration of the shrine after he had found an ancient drawing for it on a hide, and the contributions of many other pharaohs. But there are tombs from the First Dynasty onward in the neighborhood. The main cult action was the yearly journey of Hathor to the shrine of her son-consort Horus at Edfu.

164. The obverse of this was, of course, ceremonial baldness (head shaving, epilation), which was practiced in certain Egyptian and Mesopotamian cults; on Mesopotamian beards, which certainly involved curling and oiling (sometimes artificial additions and even covering in gold leaf), see G. Contenau (1954, pp. 66, 121, 280 f.).

In the various images of Tutankhamun, the insignia character of the beard is obvious: his mummy shows him to have been beardless, while the gold coffin shows him with the beard; the canopic images are beardless. The Pharaoh Zoser is the first shown wearing such a beard when running his Heb-Sed race in a relief of the southern tomb in his pyramid at Sakkara. And on her sphinxes, the she-Pharaoh Hatshepsut wears the ceremonial beard.

Wigs were even more common: the life-size limestone statues of the Memphite high priest, Ranufer (end of the Fifth Dynasty, ca. 2425; Cairo Museum 224, 225), show him in everyday dress with short hair, wearing a ceremonial apron and a thick wig. By then there was an office holder at court whose title was "Keeper of the Diadem and Inspector of the Royal Wig Makers": he was called Hetep-ka and was buried in the mastaba (no. 3509 at Sakkara).

The head of a wooden statue (probably of the time of Sesostris I; Cairo Museum 4232) has a hairpiece made separately and of a different wood, covered with paste and gold-leaf fragments. Even Queen Hatshepsut is sometimes shown with a beard, which (as in the case of beardless male pharaohs) seems to have been attached on a tape, then tied under the wig or crown. But very many hairpieces, like the strange, asymmetrical hairdo of the god Khons (probably a representation of Tutankhamun; Cairo Museum 462), were meant explicitly to be read as wigs.

Separate wigs have also survived. One for a more or less life-size head from Ebla (Tell Mardish), composed of several pieces of hard limestone, has wooden peg and bitumen attachments, and was presumably part of a basically wooden statue; a smaller (probably painted one) also came from the same site (Aleppo, TM 76 G 433a; 77 G 115+155+175+184 a-c; 78 G 178+221 for the larger head; 76 G 830 for the smaller). Of ancient Greek examples, the most spectacular is perhaps the bronze wig for the (over-life-size) stone head of the Apollo Aleus temple at Ciró in Calabria. An Etruscan bronze wig from Chianciano is in the Etruscan Museum in Florence. Another, smaller,

Roman bronze wig (for Jupiter? for Asklepios?) in the Antiquarium in Bonn. Perhaps the most impressive of all such objects, however, is the golden (and therefore presumably ceremonial rather than defensive?) helmet-wig of Mes-kalam-dug, namesake of the king from Ur (Iraq Museum, U. 1000).

There are two stone objects with hair curled Hathor-wise in the Heraklion Museum, from the palace at Knossos (*PM,* vol. 3, pp. 419 ff.), but their purpose and the nature of the statue to which they were fitted (terracotta? wood?) has not been investigated. On the whole problem see PW, s.v. "akrolithon," and *EAA,* s.v. "acrolito." In this connection, E. D. Van Buren's view (1945, pp. 106 ff.) that the archaic Sumero-Akkadian character *XX* represents swaddling bands rather than a wig might be reexamined.

165. The two large heads are in the Louvre. These capitals are known as the sistrum type, since this is a common instrument that consists of a loop of metal with beads threaded on rods spanning the loop; it is held by a handle and shaken rythmically. The joint between the handle and the loop is usually such a Hathor head, with a little shrine for a crown, as in the capitals; many such instruments survive. Two further heads are from Amathos (Limassol, District Museum 853; sixth–fifth century) and Paphos (Louvre AM 2755); the seventh- or sixth-century stela is from an unknown site (Cesnola collection in the Metropolitan Museum, New York, 1414).

166. H.-G. Buchholz (1973, p. 336, figs. 34, 35).

167. The funerary chapel (also called the "small" temple: before 1237) is dedicated to Hathor. See E. Naville (1894, vol. 3, pl. 168); M. Werbrouck (1949, pls. 42 ff.). The tomb of Senmut is no. 51.

X The Corinthian Virgin

1. Vitr. IV. 1. i: "praeter capitula omnes symmetrias habent uti ionicae."

2. The diameter is taken on the underside of the shaft (as in the other columns), and the abacus dimension conforms to it, not to the top of the shaft.

3. Vitr. IV. I. xi. The sides of the abacus are to curve inward by one-ninth of the sides of the square; he says nothing about the cropping of the corners. At Bassae and at Epidaurus—and in several other earlier examples, particularly where the abacus is much thicker in relation to the capital than Vitruvius demands—this is not done; but it may be seen in other early examples, as at the temple of Athena Alea at Tegea and the monument of Lysikrates in Athens. On these, see below, nn. 67, 85. Many of these matters are dealt with by B. Wesenberg (1983, pp. 110 f. n. 486) and H. Lauter-Bufe (1987, pp. 71 ff.).

4. The exceptions are very few: the so-called Italic temple of Peace at Paestum, built after the entry of Paestum into the Latin league in 273 (M. Napoli, 1969, pp. 60 ff.; C. Sestieri, 1958, pp. 23 f.), and the temple of Augustus at Philae in Egypt are two explicit examples. See W. B. Dinsmoor (1950, p. 279). On "mixed" orders generally, see R. Demangel (1932, pp. 299 ff.).

5. Vitr. IV. 1. iii: "Ita . . . e generibus duobus . . . tertium genus in operibus est procreatum." The word *procreare* means "beget," "generate"; it is not often used figuratively.

6. Vitr. IV. 1. ix. Callimachus' name was not unusual, the most famous Callimachus being the Alexandrian poet who formulated the canon of literary classics; however, the artist here mentioned by Vitruvius was generally acknowledged as a master. The source material on him was already discussed by J. Overbeck (1881, vol. 1, pp. 381 ff.). The evidence of Pliny (*NH* XXXIV. xix.92) and of Dion. Hal. (*Isoc.* 113) suggests that he worked about 430–400. Pausanias had seen a statue by him of the "bridal" Hera at Plateia (IX. 2. vii), as well as the golden lamp with a palm tree (as a smoke escape) in the Erechtheion (I.26. vi). The ancient critical view of Callimachus was that he was a pernickety and excessively self-critical artist. In fact Pliny and Pausanias have his nickname as *catatexitechnus,* "who put his art in the crucible." Pliny thinks that his *saltantes Lacaenae* (caryatids?) were spoiled by his pedantry, and Pausanias that although he was superior for his *sophia,* yet he was not in the first rank of the *technitai.* The signature *Kallimachos epoiei* appears on an Archaic-style relief published by A. Furtwängler (1893, p. 202 ff.).

7. S. Stratico (Vitruvius, 1830, vol. 2.1, p. 137): "Sunt qui hanc a Vitruvio relatam narrationem pro fabula habent; S. Ferri (Vitruvius, 1960, ad loc.): "storiella più degna di un epigramma dell 'Antologia' anziché di un trattato di Architettura"; J. Durm (1910, p. 343): "anmutiger Mythos"; C. Chipiez (1876, p. 303): "légende gracieuse et touchante"; and J. J. Coulton (1977, p. 128), who finds it "unthinkable that a Greek architect should introduce a new capital . . . simply because he was attracted by the nature and novelty of the shape," and sees formal innovation as the true impulse for Callimachus' reaction.

8. E. Will (1955, pp. 231 f.) and Charles K. Williams II, "Corinth and the Cult of Aphrodite," in M. A. Del Chiaro and W. R. Biers (1986, pp. 12 ff.); see also T. J. Dunbabin (1948a, pp. 59 ff.).

9. Pind fr. 122 (in Athen. XIII. 573e; Sandys fr. 122) which tells of the vow Xenophon of Corinth made to consecrate a hundred courtesans to the goddess if he won the Olympic crown (Puech thinks the poet counted a hundred members or limbs by *hekatonguion,* which would halve the number of "the hospitable girls," as Pindar called them). But see also Pind. *Ol.* X.105, which celebrates the double victory in the stadion and the pentathlon; the Xenophonic offering and the reputation of the Corinthian hetairai are discussed by J. B. Salmon (1984, pp. 397 ff.). The proverbial "not everyone can go to Corinth" meant "not everyone can pay the girls"; thus Aristoph. *Fr.* 902 (which Horace translated in *Ep.* I. xvii.36) and Aul. Gel. I.8. The epithet is mentioned in a Locrian connection in Pind. *Pyth.* IV.316. On the origins of Greek Aphrodite, see above, chapter 9, n. 112.

10. On the relevance of Homeric-type cremations in "lower-class" funerals and in Iron Age Greece generally, see I. Morris (1987, pp. 46 f.). On the relation of cremation to inhumation in Greek funerary practice, see R. Garland (1985), pp. 34 ff.); but see still E. Rohde (1925, pp. 165 ff.). Although many dead were buried in enclosed family plots—*peribola*—separate burial is a frequent mark of sorrowing parents mourning a favorite child. On the importance of the family plot, see Lynn Goldstein in R. Chapman, I. Kinnes, and K. Randsborg (1981, pp. 53 ff.).

11. W. W. Tarn (1952, p. 318). Demetrios was a pupil of Theophrastus and a peripatetic. Of this decree on Athenian sculpture, see Tarn (p. 318) and A. J. B. Wace, review of Hans Möbius in *JHS* 50 (1930), p. 154. There is one element that Demetrios' law does not mention and that is often found and shown on vase paintings: the *trapeza,* a flat slab, bench, or table, on which mourners would be shown sitting by the tomb.

Burial regulations in many cities had a sumptuary character, forbidding, as Solon had done, the excesses that grief (or misplaced family pride) might dictate to the rich. On Solon's sumptuary law, see Arist. *Ath. Pol.* 1–13, Plut. *V. Sol.* XII.4–5, and Cic. *De Leg.* II.64 f. Cicero also recalls a later decree forbidding the placing of herms and the use of plaster or of anything that would require more work than could be done by ten men in three days. In Plato's city (*Leg.* XII.950b), the limit is set as five men working for five days.

12. The practice of using such vases starts about 950 in Athens, and 1000 in Lefkandi; see J. N. Coldstream (1977, pp. 56 f.) and B. Schweitzer (1971, pp. 37 ff.). On the problem of the ritually pierced vessel as a grave marker, see also A. B. Cook (1914–1940), vol. 3, pp. 373 ff.). On the practice in the context of beliefs about the underworld, see E. Rohde (1925, pp. 19 f., 162 ff.).

13. So Festus, s.v., and Serv. ad *Ecl.* II.45 (the word occurs in 1. 46). *Quasillus* is a diminutive of *quasus.* But in fact the Greek word is used not only by Virgil but also by Pliny the Elder, Ovid, and Juvenal. Was Vitruvius transliterating from a Greek source or had the Greek word passed into common Latin use? No definite answer is available.

The most common basket shown in grave offerings on lekythoi was in fact a *kaneion* or *kaniskion* with a wide, flat bottom, not unlike a liknon. But see R. Garland (1985, p. 108).

A funerary offering, the kind Vitruvius seems to be describing here, in the form of a square basket filled with black-figure pocula, is in the National Museum in Athens.

14. *Anthē te plekta,* "wreathed flowers and the fruitful olive" (Aesch. *Pers.* 618), are called for by Atossa when summoning the ghost of Darius: it is the first literary mention of the custom. On the nature of the plants used for the wreaths at the prothesis, as well as at the tomb, there is no rule, though selinon is also used for victors' crowns. See Pind. *Ol.* XIII.33 and Plut. *Symp.* 676c, d, though he also makes it a plant of bad augury, because of its funerary association. But see also below, n. 20.

15. Hom. *Od.* XI.538, 572; XV.13. There were a number of varieties of asphodel: according to Hesiod (*Op. D.* 41), it was a kind of basic food. The Pelasgians were said to have found its roots nourishing. Pythagoras recommended it as a food according to Pln. *NH* XXI.lxviii.190, who appreciated both seeds and roots; he also regarded it as an antidote to certain poisons.

16. R. Lattimore (1942, pp. 129 ff.) has suggested that the Greek dead were being rewarded for maintaining the fertility of the earth. But see (contra) S. C. Humphreys (1983, pp. 159 ff.). While Lattimore's suggestion may be too instrumental, a connection between fertility and death is hardly exceptionable.

17. As in the Vitruvius editions of Claude Perrault (1675, 1684) and Bernardo Galiani (1758). The account of the tomb remained a puzzle to commentators, particularly in view of the acanthus growing on top of it. That this monument could only have been earth heaped over the grave was first suggested *textually* by Joseph Ortiz y Sanz in his translation of Vitruvius (1787).

Daniele Barbaro (Vitruvius, 1584, pp. 164 f.) refuses to be drawn into a learned commentary quoting Pliny, Pausanias, and Strabo, since he believes that this passage "a maggior cura mi strigne [*sic*] che narrar le historie, descriver luoghi & dipinger herbe." He prefers to use the occasion to explain how nature provided the key to all the artists' operation in forming different kinds of beautiful bodies.

Francesco di Giorgio, who seems to have read a defective Vitruvius manuscript, mistook the text to say that the Corinthian maid was actually buried upright inside the basket—which was therefore both coffin and flowerpot. On the implication of the mistake, see J. Rykwert (1981b, pp. 78 ff.).

18. Sixth-century lekythoi were on the whole bulbous, with a short neck; the broken outline with a longer neck is common in the fifth century, and the white-ground ones are all of this latter kind. An even squatter type, related to the aryballos but with a flat bottom (Hesychius thought that *aruballos* was merely a Doric word for *lekuthos*), also appears about the same time and is a common cosmetic container. They all have a handle to one side only. See D. C. Kurtz (1975, pp. 77 ff.); also J. D. Beazley (1938).

On their discontinuation, see D. C. Kurtz (1975, pp. 73 f.); she discusses the relevance of Arist. *Eccl.* 538 ff. But see also W. Riezler (1914) and A. Fairbanks (1907). Of the lekythoi Fairbanks knew, about half the early ones and nearly all the later ones represent graveside scenes: while they were mostly used as oil or unguent bottles, some were solid dummies (pp. 349 ff.).

19. Such lekythoi exist in many collections: for instance, Athens, National Museum (1956, 14517); Chicago Art Institute (07.18); New York, Metropolitan Museum (06.1021.135, 06.1169); and Zurich Univ. (2518).

20. The enormous variety of plants that the ancients called "acanthus" was clearly sorted out by J. Yates (1846, pp. 1 ff.). But see further H. O. Lenz (1859, pp. 545 ff., 735 f.) and J. Billerbeck (1824, pp. 164 f.). Since the name derives from *akaina*, "spine," "spike" (even "fishbond"), and *to anthos* "flower," it is applied to all sorts of spiky and flowering plants: thistles and acacias as well as the Brancursine or Acantheae, a family of the order Acanthaceae that includes many more tropical species. It is a herbaceous plant or shrub; the usual English term is bear's foot or bear's breech. However, both plants are discussed by Diosc. III.19; Pln. *NH* XXII. xxxiv. 76; and Colum. IX.4. But see also Theoph. *De C. Pl.*, who uses the word for various forms of acacia and thistle: IV.2.i ff., 10.vi; IX.1.ii, 18.i. See also Virg. *Ecl.* III.45 and *Geor.* IV.124, and Serv. ad loc.

On the other hand Pausanias (II.10.iv ff.) tells of an Aphrodite at Sikyon in whose shrine (served by virgins) was a golden statue wearing a polos, holding a poppy in one hand and an apple in the other. In this shrine *paiderōs*, *"boys' love,"* "which had leaves smaller than *drus* [oak?] and larger than the *phēgos* [an unidentified species of oak], not unlike the *prinos* [holly oak]" grew plentifully. If Pausanias is to be believed, this plant grew nowhere else in Greece; branches of it were burned with the thighs of the animals sacrificed there. But the description of the plant seems vague, and very unlike

acanthus, yet Pliny (*NH* XXII.xxxiv.76) identifies *paideros* as *Acanthus mollis;* though elsewhere (*NH* XIX.liv.170) he says it is the same as *caerefolium,* "chervil." See, on all this, J. G. Frazer (1913, vol. 3, p. 68); and J. Murr (1890, pp. 207 ff.). *A. spinosus* is really a kind of thistle, and belongs to the order of Astracae, like the artichoke. *Onopordon acanthium* is one of the plants on which the Scottish thistle was modeled.

The names Akanthis, Akanthillis, Akanthos have nine entries in PW: four place names (the most important being a colony of Andros in Thrace, in Chalkidike) and five personal names; some are associated with the thistles as a sign of infertility. Diod. Sic. I.19 describes an Akanthon in Egypt where 360 priests daily brought Nile water to pour through a pierced altar.

The myth of Ocnus was represented by a man who platted a straw rope, which was being constantly undone by people behind him. In one Greek version, Ocnus' rope was eaten by a donkey as he platted it. He is represented thus in the Cnidan Lesche at Delphi (Paus. X.29.i; but see below, n. 29). See also Plut. *De Tranq. An.* 473.x.; Apul. *Met.* VI.18. On the connection of all this to myths of the underworld and rebirth, see J. G. Frazer ad loc. (1913, vol. 5, pp. 376 ff., listing other representation of this legend, all in "underworld connections"). But see also J. J. Bachofen (1989, pp. 522 ff.

21. Pln. *Ep.* V.6: "lubricus et flexuosus"; "mollis et paene liquidus."

22. *EAA*, s.v. "Palmette." But see H. Möbius (1968, pp. 10ff.). An even earlier attempt to trace its origin, interesting if not sufficiently articulated, is found in W. H. Goodyear's two papers in the *Architectural Record* (1894a, pp. 263 ff.; 1894b, pp. 88 ff.). A review of more recent arguments is given in J. Rykwert (1994b).

23. It is already recognizable as a configuration on proto-Elamite cylinder seals of the third millennium (N. Perrot, 1937, pp. 64 ff.) and on the edge of the robe of the Kassite king Marduk Nadin Ahze as he appears on his Kudurru-stone (ca. 1100; BM 90, 841). It is elaborately and constantly repeated on reliefs of the artificial trees.

24. They appear early on Mittanian seals (H. Frankfort, 1939, fig. 63). On his orthostat, Bar-Rekub of Zinçirli (ca. 740–720) holds a palmette scepter. Palmettes figure on the helmet of Sarduri II and on wall paintings and ivories from Kamir Blur, as well as on bronze ornaments from Zakim and the Melgunov scabbard: B. B. Piotrovskii (1967, figs. 18f, 27a, 28, 29b).

25. Palmette-and-acanthus (or selinon?) motifs, alternating with lion-head gargoyles, ran along the crowning cyma-moldings of temple cornices of Poseidon at Isthmia, Apollo at Delphi, and—of course—Athena Alea at Tegea, the Tholos, the Asklepeion, the North Propylaea, and the Artemision at Epidauros (in the last, lions' heads were replaced by hounds).

26. As in the Attic stele (ca. 540–530) of Megakles in the Metropolitan Museum, New York (11.185).

27. An Attic fourth-century palmette from the stele of Timotheos and his son Nikon (stele in National Museum, Athens) is in the Metropolitan Museum, New York (07.206.107), as well as even more elaborate acroterion (20.198.0). I have undertaken a detailed study of the palmette elsewhere: see J. Rykwert (1994b).

28. Some of these ridge tiles are exhibited in the Nemea museum: S. G. Miller (1990, pp. 58 ff.).

29. At the Delphic hestia-altar, Greek cities and sanctuaries would seek new fire if their altar had been defiled in any way; that is the sense of the legend of Euchidas of Platea who died of exhaustion from bringing the Delphic fire back to his home town after the battle of Platea (in 479), because the fire altars of Greece were considered sullied by barbarian presence. See Plut. V. Arist. XX.331d; M. Delcourt (1965, pp. 105 f.).

The most extensive account of the murder of Neoptolemos is given by Eur. Androm. 1085–1157. On the probable political coloring of this account, see G. Roux and J. Pouilloux (1963, pp. 102 ff.). Pausanias mentions the shrine in several connections. See I.4.iv, IV.17.iv, where the killing of Neoptolemos at an altar (as he had killed Priam) is called proverbially "the vengeance of Neoptolemos"—what he had done to another is done to him; on the Delphic shrine and the cult, see X.24.iv.

But see also J. Fontenrose (1960) and P. Vidal-Naquet (1986, pp. 304 ff.).

Priam's killing by Neoptolemos was often alluded to: so Eur. Hec. 23 ff. It was probably told in lost works by both Arctinus in the Iliupersis and by Stesichorus, though the first explicit account occurs in Virg. Aen. II.499 ff. and variations on it, Serv. ad v. 506. The burial under the temple threshold is cited in the scholia on Eur. Or. 1648, its removal by Menelaus in the scholia on Pind. Nem. VII.62. The sacrifices to Neoptolemos are described in some detail at the beginning of book III of Heliodor. Æthiopica.

30. So the legend preserved in scholia on Pind. Nem. VII.62 and on Eur. Or. 1655. By his marriage to Hermione, Neoptolemos was Menelaus' son-in-law—though Hermione had also married Orestes, and the quarrel over Hermione is one of the reasons given for his murder. Others are the old enmity between Achilles and Agamemnon, or even between Achilles and Apollo himself, and Neoptolemos' seeking revenge at Delphi from the god for his father's death—indeed he may have gone there to consult the oracle about it (H. W. Parke and D. E W. Wormell 1956, vol. 1, pp. 315 ff.; vol. 2, pp. 78 f.). But the matter of the legend seems connected with rights that the Delphians claimed over meat from sacrifices at the shrine. The name of Neoptolemos' murderer, Machaireos, is associated with machaira, the sacrificial (or just "meat") knife: Pind. Nem. VII.33 ff., Pae. VI.83 ff.; Apol. Epit. VI.14. The Trojan exploits of Neoptolemos, though not the killing of Priam, were celebrated in Polygnotos' paintings in the Lesche (which were among the most famous, as well as the largest, paintings of the ancient world): Polygnotos painted the Lesche well before 467, if an epigram attributed to Simonides (who died that year) is genuine.

The vast literature is reviewed by R. B. Kebric (1983; on Neoptolemos, see pp. 22 f.). But see Plut. De Def. Or. 47 (436b); Paus. X.25.i ff.; Arist. Pol. VIII.5.xxi (1340a), Poet. 6(1450a). The Lesche stood just above Neoptolemos' shrine, and was—or so Pausanias thought—related to it. Whatever body was buried or commemorated in the little temenos, excavations have shown it to have been occupied in Mycenaean times. The outsize Bronze Age amphora found on the site, as well as bone fragments and carbonized organic matter, suggest a cult place, perhaps a bothros (*FD* 2.2, "La Région Nord du Sanctuaire," pp. 49 ff.). See also B. Bergquist (1967, pp. 30 ff.). It is flanked on either side by a Thessalian votive group (of Daochos of Pharsalia) and on the other by the stoa of Attalos of Pergamon. Pergamos, the eponym-founder of the town, was the son of Neoptolemos. The nearby treasury (XVII) is variously identified as that of Corinth or Akanthon. During the Pythian games, the Thracian Aenianes also sacrificed to Neoptolemos as their mythical king; Paus. X.24.iv, vi.

Neoptolemos/Pyrrhos is one of a small number of double-named heroes such as Paris/Alexandros. The kings of Epirus regarded him as their ancestor and used both names. Neoptolemos I was the father of Alexander of the Molossians (whose son became Neoptolemos II) and of Olympias, mother of Alexander the Great. See PW, s.v. "Neopto-lemos," and M. Delcourt (1965, p. 33 n. 1). The variants of the Neoptolemos legend were gathered and compared by C. Vellay (1957, pp. 123 f., 303, 379 f., 449 ff.).

31. Pausanias, who was much impressed by the story of Neoptolemos and his presence at Delphi (I.4.iv, IV.17.iv), says nothing about it when he describes the temenos (X.24.v), which has suggested to some commentators that it was thrown down in the earthquake of 373—though this is disputed: see *FD* II.2 (pp. 61 ff.). The fragmentary base carried the signature PAN, which may well be a fragment of the signature of a well-known contractor, Pancrates of Argos, who worked ca. 340. On its discovery and decipherment, see *FD*, vol. 2, pp. 60 ff.; vol. 3.5, pp. 340 ff; vol. 4, pp. 32 ff.; and "Les Danseuses de Delphes," in G. Roux and J. Pouilloux (1963, pp. 123 ff.).

32. The procession is described in Heliod. *Aethiopica* III.

33. Quite how many dancers there were at Delphi, or how they related to Thyia, the authorities do not quite know, but see J. E. Harrison (1912, pp. 401 ff., 523 ff.). The word meant a wine festival in Elea (Paus. VI.26.i). *Thuein* was an ecstatic dance (C. Kerenyi suggests an erotic ecstasy: 1976, pp. 182 ff.). Pausanias (X.6.iv) has Thyia be a daughter of the autochthonous Castalius, eponym of the main spring in the shrine; she also had

a son, Delphos, by Apollo. Pausanias also knows of an Athenian sorority of Thyades who would come to Delphi to join the local sisterhood in their orgies, but who also had shows of their dances in various towns on the way there—which suggests that they were more disciplined than some at any rate would have them. See H. Jeanmaire (1951, pp. 78 ff.).

Some commentators wanted the women of the sculpture to be the Graces (who, however, usually turn toward each other), or the Charities. However, the back-to-back attitude is more characteristic of chthonic figures, as of the Hekate groups, for instance.

The discovery of Dionysos in the Parnassan cave by the Thyades is described by Plut. *De Is. et Os.* 35 (365a); but see C. Kerenyi (1976, pp. 44 f.), who also accounts for other divine children cradled in a liknon.

On the alternating lordship (between Apollo and Dionysos at Delphi), the *trietēris,* see C. Kerenyi (1976, pp. 198 ff.).

34. This is the tripod that was taken to Constantinople by Constantine the Great to stand on the spina of his hippodrome. The triple-coiled snake that supported the bowl still remains on the place where Constantine had it set.

35. The story provided one of the commoner themes in Greek art. Pausanias (X.13.viii) mentions the legend; he did not notice it in the gable of the Siphnian treasury, although he recorded it on a relief in the temple at Lycosura (VIII.37.i). The legend is told by Apol. II.6, 2, Plut. *De E Delph.* 387; Vet. Schol. in Pind. *Ol.* IX.29(43); Cic. *De Nat. Deo.* III.16.xlii; Hyg. *Fab.* 32; Serv. ad *Aen.* VIII.300.

36. Inside the tripod lebes, under the holmos, were the bones and teeth of Pytho the serpent, whom Apollo killed when he took over Delphi, while his skin was wrapped round the tripod itself. So Hyg. *Fab.* cxl. 5; Serv. ad *Aen.* III.92, 360. Suidas, s.v. "Pytho," says it contained *tas mantikas . . . psiphous*—"the divining pebbles" (grains, counters?), which gave the Pythia her answers: see P. Amandry (1950, pp. 259 ff.); C. Kerenyi (1976, p. 228 ff.). The omphalos was also considered Dionysos' tomb.

That Pytho and Dionysos were in some way

identical was already suggested by several ancient writers: see J. Fontenrose (1960, pp. 376 f.). In any case, Dionysos' lordship of Delphi seems to have alternated every second year with that of Apollo.

37. Pausanias records the omphalos in two places: by the Pythian tripod inside the main temple (X.16.iii) and between the Lesche and the temenos.

38. On the form of the Greek lebes-stand, see PW, s.v. "Dreifuss." Carl Otfried Müller was the first to provide a "scientific" account of the subject (1820). The lebes on a tripod stand seems to have been an Ugaritic or at any rate a Levantine invention; the stand was made of thin rods, not of plates, as in Greece. It became common in the Ugaritic form in Asia Minor, and was so taken over by the Cretans and the Etruscans, who sometimes had bronze lebes on iron tripod stands. The plate tripod seems to originate in the Peloponnese. See B. Schweitzer (1971, pp. 164 ff.). Some of the earliest tripods in Greece are the Geometric pithoi "with looped feet": Athens, National Museum no. 2451 (281) and Argos Museum, c 209. See Schweitzer (pp. 29, 61, 172). Terracotta, rod, and plate tripods all seem to appear in Geometric Greece about the middle of the ninth century.

On the early development of the tripod stand as a votive object and its economic significance, see A. Snodgrass (1980, pp. 52 ff., 104 ff.).

39. This lebes has been identified as a perirranterion of the kind Herodotus (I.51) saw at Delphi. See O. Picard (1991, p. 31). Here and there such Hekate figures survive: the Metropolitan Museum, New York, has a small marble group, first century (AD) with figures back-to-back around a plant or column.

40. Medea is said (*Nostoi* fr. 6) to have performed the same operation on Jason and Aison: Ovid *Met.* VII.163 ff. She was a witch and the sister of Circe; both are poisoners as well as givers of immortality. Medea's magic herbs have interested a number of writers; see J. Murr (1890, pp. 207 ff.).

41. The lebes is dated ca. 340; now in the National Museum, Athens, no. 3619.

42. Diod. Sic. XVII.27.

43. The Hellenistic bronze called Corinthian had a characteristic red tinge, which has been attributed to the ochre in the spring Pirene (on which see F. Glaser, 1983, pp. 76 ff.; B. H. Hill, 1966b), 1.6, pp. 1 ff.), which was used for tempering the bronze (Paus. II.3.iii). But Pausanias adds that the Corinthians had no bronze of their own, referring presumably to the constituent ores. On the early Corinthian bronze exports, see J. B. Salmon (1984, pp. 118 ff.).

W. B. Dinsmoor (1950, p. 157 n. 6) asserts the metallic origin of the acanthus column. Pliny, as Dinsmoor points out, called the porticus of Caneaus Octavius in Rome "Corinthian" because it had bronze acanthus leaves attached to the columns; in Palmyra, bronze leaves were fastened onto stone column shafts. Callimachus was, in any case, reputed to be as much a bronze caster as a stone carver; I have already mentioned his golden lamp in the Erechtheion.

44. Paus. X.24.v had seen this stone and was told that it was the one Rhea offered Kronos in lieu of Zeus, which Kronos vomited or spat out (the legend is told by Hes. *Theog.* 453 ff.). It was annointed daily. See also A. B. Cook (1914–1940, vol. 3, 929, 937).

45. *Pronoia* (of the forethought) or *pronaia* (of the temple porch)—it was not resolved in antiquity; see J. G. Frazer (1913, vol. 5, pp. 251 f.). In front of the temple and by the treasury stood a vast bronze statue of Athena.

46. Massalia was settled on the orders of the Delphic oracle, and had a shrine to Delphic Apollo: the story of the oracle is given in Str. IV.179; but see I. Malkin (1987, pp. 69 ff.).

The very similar arrangement of palmette leaves into a capital on a Minoan columnar "lamp stand" of *rosso antico* marble (ca. 1450, from the southeast house, Knossos; Heraklion Museum), whose shaft has spiral flutes alternating with ivy bands, may be coincidental. It may relate to Anatolian "Aeolic" palmettes rather than to its Massalian derivative at Delphi. These kinds of stone lamps were in fact exported to Anatolia and Syria as well as to mainland Greece. See M. S. Hood (1978, pp. 150 f.).

47. See above, chapter 5, pp. 137 ff.

48. The Romans called her Proserpine (by a curious transliteration), on which see St. Augustine (*De Civ. Dei* IV. 8), who produced his *Volksetymologie:* "goddess of the corn's first leaves and buds" (*sepes*). But the Romans also called her Coré.

49. On the *kukeōn* as (possibly) an alcoholic drink, see C. Kerenyi (1976, pp. 117 ff.).

50. The entwined serpents around Hermes' caduceus, according to some mythographers, represented the serpent Zeus and the serpent Persephone (Ath. gr. *Pro Christ.* 20). But then, as Athenagoras is quick to point out, Persephone was herself the daughter of Rhea (the mother, who after her seduction became Demeter the spouse) and her divine son.

The legend of Zagreos echoes—or mirrors—that of Athena; since Zeus feared that his son by Metis would become the world ruler by dethroning him as he had dethroned his father, he swallowed pregnant Metis and had her son turn into a daughter, Athena, who was born from his head.

51. This Orphic legend seems alluded to first in a Pindaric fragment (from Plato's *Meno* 81b; 133 Bergk, 127 Bowra). But the gory details are told at great length by many later authors, including Nonnos (VI, esp. 155 ff.; X.47 ff.; XXIV.43 ff.). and Firmicus Maternus (*De Error.* 6); the sources are reviewed by O. Kern (*Orphicorum Fragmenta,* frr. 209 ff.); see also PW, s.v. "Zagreos." The legend is considered the etiology of ritual substitution by J. E. Harrison (1903, pp. 489 ff.; 1912, 14 ff., 52 ff., 118 f.), who also has much to say on the omophagia as an initiation rite. For a further account of this epidode as an epic version of an initiation rite, see C. Kerenyi (1976, pp. 266 ff.).

Firmicus Maternus gives an extended account of the legend often alluded to contemptuously by church fathers: Athan. *Adv. Gr.* 9; Clem. Alex. *Protr.* II.15; Eus. *Prep. Ev.* II.5. But Isocrates (*Bus.* XI.38 f.) had already blamed Orpheus for telling cruel and shameful stories about the gods; the immorality of poetic mythology is a Platonic commonplace. The first to be explicit about it is Xenophanes (frr. 1, 11, 12; ed. J. M. Edmonds, 1931, vol. 1, pp. 192 f, 200 f.).

52. Aesch. fr. 124/228 calls him "son of the hospitaller" (sc. "of the dead"). On the passage from hunting to agriculture, and therefore viticulture, see also C. Kerenyi (1976, pp. 80 ff.), who refers his name to the Ionic word *zagrē,* a trap. But he also sees this legend as a survival of Cretan religious practice in Greek myth (pp. 114 ff.). But see K. Meuli (1975, p. 173).

53. Heraclitus fr. 15 (Clem. Alex. *Protr.* II.22.ii; cf. Iamb. *De Myst.* I.11 and Plut. *De Is. et Os.* 28): "Were it not for Dionysos that they paraded and hymned shameful things, their doings would just have been smut; but Hades is one with Dionysos, for whom they cavort and celebrate Bacchic rites." In Greek the pun plays on *asma,* "hymning"; *aidoioisin,* "shameful things," "genitals"; *anaidestata,* "lewdness," "shamelessness"; and *(H)aidis.* See K. Axelos (1962, pp. 139 ff.) and W. K. C. Guthrie (1962–1978, vol. 1, pp. 475 ff.). I have followed the usual interpretation; but see M. Marcovich (1967, pp. 250 ff.) who wants it to mean: "If they did not make processions in honor of Dionysos and sing hymns to the shameful parts, they would be acting impiously. But Hades and Dionysos are the same, however much they cavort and rave at their Bacchic rites." He would thus have it read as a rejection of the Bacchic rites, which only celebrate fertility.

Iamblichus explains this passage (referring presumably to its lost context) by saying that Heraclitus calls the mysteries *pharmaka,* "remedies," because they deliver souls from all the misfortunes of birth.

54. On Hades as the guardian of seed grain, see WHR, s.v. On Triptolemos' first initiates, Herakles and the Dioscuri, see Xen. *Hel.* VI.3.vi; or so Xenophon reports Callias the Eleusinian torchbearer saying in his speech to the Spartans.

55. Clem. Alex. *Protr.* II.12.

56. Pl. *Crito* 54d, *Phaedo* 69c, 81a; but the mysteries were by that time also seen as tokens of good-fellowship, primarily social bonding activities (*Ep.* VIII.333e).

57. On the shrine and its antiquity, see G. E. Mylonas (1961, pp. 29 ff., esp. on the Homeric *Hymn to Demeter* and its relation to the topography). But against his view, see P. Darcque (1981, pp. 593 ff.), who sees no evidence of a Mycenaean shrine in the remains of buildings under the Telesterion.

58. W. Burkert (1983, pp. 276 ff.).

59. Pl. *Phaedrus* 245c ff. But see the logical demonstration of the indestructible immortality of the soul in *phaedo* 105b ff.

60. Paus. VIII.41.vii. Although Pausanias (who offers the attribution to Ictinus as hearsay) thinks the god had received the temple as a vow protecting the Arcadians of Phigaleia from the plague of 430, there was almost certainly an older stone temple on the site; Ictinus' archaic plan may repeat that of the older temple, probably built about the year 500. But the site probably had a shrine to Apollo since the recapture of Phigaleia from the Spartans after 659. See F. A. Cooper (1970); A. W. Lawrence (1983, pp. 224, 227, 231 ff.); W. B. Dinsmoor (1950, pp. 154 ff.). The first full report on the ruins was published by C. R. Cockerell (1860).

The dedication was understood by Pausanias as being to "Apollo the helper" (sc. "against the plague") and therefore related the building to an epidemic outbreak in Phigaleia. However, following H. T. Wade-Gery, "The Rhianos-Hypothesis," in E. Badian (1966, pp. 289 ff.), Cooper suggests that the real meaning of the dedication pointed to the Arcadian *epikouroi* (auxiliaries or mercenaries), which would explain the number of military ex-votos on the site.

The nature of the cults on the mountainside is in any case very difficult to distinguish and to ascribe to a specific location. Beside the temenos of Apollo, the shrine seems to have enclosed a larger temple of Artemis and a smaller one of Aphrodite; there was also a Pan Bassas and a sacred spring. The whole mountainside was probably the background of the legend of Callisto the Bear, her son Orkos, and Arcadian origins. Other literature about the location is discussed by J. G. Frazer (1911–1915, vol. 4, pp. 394 ff). The Delphic oracle concerning the relics of Arcas reported by Pausanias is also discussed by H. W. Parke and D. E W. Wormell (1956, vol. 1, pp. 196 ff.).

The phrase that Pausanias uses in praise of Ictinus' work, *an tou lithou te es kallos kai tēs armonias heneka,* has been something of a poser for translators (see J. G. Frazer, 1913, vol. 4, p. 405), since it could refer to the beauty of the stone and the precision of its jointing; alternatively, it could refer to the perfect proportions of the building (e.g., Paus. II.27.v, praising the proportions and perhaps even the acoustics of the theater that Polykleitos designed at Epidauros). On the term *(h)armonia,* see J. J. Pollitt (1974, pp. 151 ff.).

61. The orientation toward Delphi and the archaic plan in emulation of the Delphian shrine have been suggested by G. Gruben (1984, pp. 122 f.). On the Archaic plan, see A. Mallwitz (1962, pp. 140–177).

62. Unlike the spur columns at the Olympian Heraion, which are centered on those of the outer peristyle, those at Bassae correspond to the centers of the intercolumniations. This curious feature of the plan has been explained as a second thought on the part of Ictinus' successors. The improbability of such an alteration is argued persuasively by F. A. Cooper (1970, pp. 130 ff.).

63. Dinsmoor maintains that the final two attached columns were also Corinthian, so that they formed almost an iconostasis.

64. This was, according to Pausanias, originally a bronze statue that had been moved, for reasons he does not give, to Megalopolis, where he saw it (VIII.30.i). It was replaced by an acrolithic statue, probably one whose foot and other fragments are in the British Museum (1815. 102042-49, 51), on which see Brian Madigan, "The Statue of Apollo at Bassae," in O. Palagia and W. Coulson (1993). The layout of the floor slabs will not support either idea definitely.

The short spur in the northwest corner of the adyton is sharply illuminated by the rising sun at certain times of the year because of the deliberate alignment of the edge of the eastward door and the peripteral columns. This has been studied by F. A. Cooper (1968, pp. 103 ff.).

65. The column is known through a very careful drawing by Cockerell's most important companion, Haller von Hallerstein. These, and the subsequent reconstructions, are reviewed by H. Bauer (1973, pp. 14 ff.). But see also G. Roux (1961, vol. 1, pp. 43 ff., 337, 356) and Karl Haller von Hallerstein in H. Bankel (1986, pp. 122 ff.).

The surviving pieces of the capital were apparently taken to the Danish consulate on the island of Zante/Zakynthos together with the frieze of the cella (which was sold to the British Museum); the capital seems to have remained on Zante. What was left of it was probably destroyed in the earthquake of 1953. Haller von Hallerstein's drawings in his excavation notes are to be found bound in the manuscript section of the Municipal Library in Strasbourg. He had already suggested (and this was echoed by many scholars, including Dinsmoor) that there were three columns, and that they formed a screen between the cella and the adyton.

66. On the "formal" reasoning, see J. J. Coulton (1977, pp. 127 ff.); A. W. Lawrence (1983, p. 234). W. B. Dinsmoor (1950, pp. 164 ff.) is more interested in the Egyptian origin of the form. He also suggests that the Ionic was used here to provide a continuous zophoron frieze. However, though he finds that the Ionic capitals, with their upward sweep of the astragal and the Attic cyma reversa between the volutes instead of the Asiatic ovolo, look "surprisingly modernistic and functional" (pp. 157, 184), he does not comment on the remarkable square block that issues between the volutes in lieu of an abacus, which gives the capitals a slightly Egyptian look.

67. The name is puzzling. Alea was an old town on the northern borders of Arcadia and the Argolid; *al-* is unfortunately too common a root to give any guidance to the nature of the goddess before she was amalgamated with Athena. At any rate, when the inhabitants of old Alea were invited to join the synoikismos of Mantinea, the town maintained its identity. At Tegea the cult was introduced by the hero Aleus, some three generations before the "Dorian Invasion"; whoever the goddess, she was certainly regarded as a powerful protectress of asylum. Pausanias is our chief informant about this (II.17.vii; III.5.vi, 19.vii; VIII.9.vi, 45.iv–vii, 47.i–iii), but see also Xen. *Hel.* VI.5. xxvii and Str. VIII.3.ii). The refoundation act of Tegea, which he dates to 479, is discussed by M. Moggi (1976, pp. 131 ff.).

A number of other, less important settlements bore the name (Steph. Byz. and PW, s.v.). It is not clear whether the Latin word *alea* for games of chance (and chance in general) had any connection with this goddess.

68. However, the intercolumniations reproduce faithfully not only the proportion (2⅓ diameters) but also the dimension (4¾ Doric feet) of Mnesicles' Doric intercolumniations in the Athenian Propylaea. Pausanias (VIII.45.iv) maintained that it was not only the most beautiful, but also the largest temple of the Peloponnese, although it was only half as big as the temple of Zeus at Olympia; but see W. B. Dinsmoor (1950, pp. 218 f.). On Skopas' commission and his procedure at Tegea, see A. F. Stewart (1977, pp. 80 ff.).

69. C. Dugas, J. Berchmans, and M. Clemmensen (1924). Because the pronaos was rather deeper than the opisthodomos, however, the symmetry of the exterior (the northward door corresponded to the intercolumniation between the seventh and the eighth column) did not correspond to the interior, where the door comes between the third and the fourth half-column; the arrangement is emphasized by antae in the corners of the building, making each side a pseudo-heptastyle in antis and also stressing the presence of the number seven in the building.

70. While the relation between the Aetolian legend of Calydon and of Boeotian Atlanta is not clear, Telephos was the son of Auge (daughter of Aleus) and Herakles. The story of his birth in a shrine of Athena, his exposure, his migration to Mysia, his encounter with Achilles, and his wound not healing until it was cured by Achilles applying rust from the wounding spear is told in the *Cypria*.

71. On this enigmatic lion, whose lair Pausanias (II.15.ii) was shown on his visit to Nemea, see—still—J. G. Frazer (1913, vol. 3, p. 88 f.).

72. The garlands of the victors in the games were of selinon: of the living plant at Nemea, of a dried one in the Isthmian games, as they were of bay at Olympia and Delphi.

73. Paus. II.16.iv.

74. Arist. *Ath. Pol.* 43.3; 44.1. That is where the prytaneia regularly sat. It was also called *skias,* the "shady place."

75. See chapter 8, n. 11.

76. Four of the six sides are equal, and two opposite each other are in a 4:3 ratio to the other sides. On the building, its function, and its measurements, see H. A. Thompson (1940, passim).

77. P. Auberson and K. Schefold (1972), pp. 123 ff.); F. Seiler (1986, pp. 36 ff.). Some authorities have thought this a fourth- (and even third-) century building; there is disagreement also about the "fence," which some have considered props for a roof.

78. H. Pomtow (1910, pp. 97 ff., 153 ff.; (1911, pp. 171 ff.) and F. Seiler (1986, pp. 40 ff.). On it and the newer Marmaria tholos, see also P. Frotier de La Coste-Messelière (1936, pp. 50 ff., 63 f., 74 ff., 79 ff.)

79. J. Charbonneaux (with Kaj Gottlob), "La Tholos" in *FD,* vol. 2, fasc. 4, pt. 2; F. Seiler (1986, pp. 56 ff.).

80. Theodorus of Phocea, who according to Vitruvius (VII. praef.12) wrote a monograph on the Delphic tholos, was presumably its designer. But see F. Seiler (1986, p. 56 n. 222).

81. F. Seiler (1986, pp. 73 ff.).

82. On the artists of that name, see above, chapter 4, n. 29.

83. On this and the earlier buildings, see F. Robert (1939).

84. The Philippeion was presumably built by Alexander. On a semicircular pedestal, concentric with the column ring, stood the chryselephantine statues of Philip, his father Amyntas and mother Eurydice, his ex-wife Olympia, and their son Alexander the Great, all by the sculptor Leochares. Fourteen meters in diameter, it has eighteen Ionic columns on the outside, nine Corinthian attached half-columns (and therefore a central one opposite the main door) on the inside (Paus. V. 20. ix–x. On Leochares, to whom the Apollo Belvedere is attributed, see A. Stewart (1990, pp. 180 f., 282 ff.); M. Robertson (1975, vol. 2, pp. 460 ff., 513 ff., 700 n. 11). He may also have been its architect. On the building, see F. Seiler (1986, pp. 89 ff.,

99 n. 413) on the identity of the architect; also A. Mallwitz (1972, pp. 128 ff.).

85. The term first appears in Vitr. VII. praef.12. The monument of Lysikrates, which was known in the sixteenth century (and perhaps long before) as the "Lantern of Demosthenes," continued under that name although the dedication had been read in the seventeenth century. It was first surveyed by James Stuart and Nicolas Revett (1762–1816, vol. 1, pp. 27 ff., pls. 1–26). But see F. Seiler (1986, pp. 135 ff.). Seiler lists a number of other analogous monuments: the one at Termessos is very well preserved; less remains of one at Delos and even less in Magnesia on the Meander—where the phrase *ho tholos* is applied to it in an inscription.

86. Paus. I.20.i ff.; Athen. XII.542 f., XIII.591; F. Seiler (1986, pp. 138 ff.); H. Bauer (1973, pp. 197 ff.); W. B. Dinsmoor (1950, pp. 263 ff.). But see M. de G. Verrall and J. E. Harrison (1890, pp. 243 ff.) and W. Judeich (1905, pp. 170 f., 274 ff.). Two further bases survive overlooking the theater. Theocritus' *Epigram* 12 was intended for inscription on such a base.

87. Arist. *Pol.* V.ll.ix (1313b).

88. Thuc. I.138 f.

89. G. Gruben (1984, pp. 230 ff.); A. W. Lawrence (1983, pp. 146, 275 ff.); W. B. Dinsmoor (1950, pp. 280 ff.).

90. Vitr. VII. præf.15, 17. Vitruvius also records the name of Peisistratos' architects of the original Doric temple on the site—Antistates, Callescheros, Antimachides, Porinos—which he presumably latinized following some inscription. Nothing else is known about them. Vitruvius also regrets that no writings by the "ingenious and learned" Cossutius have survived. A dedication, probably on a statue base in the shrine, was inscribed *Dekmos Kossoutios Popliou Romaios,* "Decimus Cossutius, son of Publius, a Roman" (*CIA* 3.561); it may well have been the base of the architect's commemorative monument. On the character of his work, see G. Roux (1961, vol. 1, pp. 373, 378).

91. J. G. Frazer (1913, vol. 2, p. 178) already suggested that these were probably monolithic columns from the cella, which may have been of more precious stone than the Parian marble ones of the exterior that were made up of drums, including their capitals.

92. This identification with the legendary first shrine was made by F. C. Penrose (1888, pp. 74 ff.), though it has inevitably been dismissed by later writers.

XI A Native Column?

1. Vitr. IV.7; a few references also in IV.8.

2. *Varicae, barycephalae, humiles, latae:* Vitr. III.3.v on araeostyle temples. On the atrium, see VI.3.iii ff. (as against atrium-less Greek houses, VI.7.i), though the main lines of house building are expounded "[ut] italico more et Graecorum institutis conformantur" (VI.7.vii). In the extensive list of earlier writers, Vitruvius quotes the (presumably) Etruscan Fufidius (also known from Cic. *Fam.* XIII.11.iii) of Arpinum and his friend Publius Septimius, as well as Varro; in the list of architects, he only names two: Cossutius, who worked in Athens, and Gaius Mucius, who built the temple of Honor and Virtue for Marius (III.2.v; VII. praef.17). Although in the next section he claims that *antiqui nostri* might be considered as fine architects as any of the Greeks, he was clearly unable to name many such, especially any who had written about the subject.

The apologia for Vitruvius as the exalter of Italic ways is presented by Henner von Hesberg, "Vitruv und die Italische Tradition," in H. Knell and B. Wesenberg (1984, pp. 123 ff.).

3. Vitr. III.3.v; cf. Pln. *NH* XXXV. xlv.158. Some of these sculptures were later transformed into bronze and the bronze was also often gilt. However, Pliny was very disapproving of the excessive use of gilding on statues, such as Nero's (Pln. *NH* XXXIV.xix.63 f.).

4. The only writer whose name has survived was one Volnius, whom Varro (*L.L.* V.9.lv) quotes as the author of *tragoedias Tuscas:* they may, of course (as R. Bloch, 1958, p. 140, suggests) have been emulations of Greek tragedies. The poet

Annianus, a friend of Aulus Gellius (XX.9), collected the "Tuscan" songs of his Faliscan estate; these improvised, rough, and ribald Fescennini were also known from the account of Livy (VII.2) and Horace (*Ep.* II.i.139 ff.), though by Horace's and certainly Annianus' time they would have been sung in some form of Latin. Neither Serv. ad *Aen.* VII.695, who considered them of Athenian origin, nor Val. Max. II.5 mention their being in Etruscan. On the other hand, Etruscan dances to the music of flutes—familiar from their depiction on tomb paintings—were, if Livy (loc. cit.) is to be believed, introduced in Rome in the year 364 to accompany a lectesternium held to ward off a plague. They were wordless and had no mimic gestures, which the Romans were to add. Livy adds that the Etruscan word *(h)ister*, which translated the Latin *ludio*, was later turned into the Roman word *histrio* for "actor," though Livy describes how theatrical representation proper was first devised in Rome a century later by a Greek-born slave-writer, Livius Andronicus. *Tusca Historia*, presumably largely town chronicles, survived into Imperial times. They were used by such late writers as Verrius Flaccus and the Emperor Claudius, whose books are also lost.

As for the origin of Etruscan writing, this is less of a puzzle; it is clearly of Greek derivation and was probably learned from the Greek settlers in Ischia and Cumae (see G. and L. Bonfante, 1983, pp. 7, 106 ff.; G. Colonna, 1976, pp. 9 ff.). By 650, objects inscribed with entire alphabets witness to the preexistence of Etruscan writing. The earliest such survival may be an ivory tablet (perhaps a writing tablet), from Marsiliana d'Albegna in the Museo Archeologico in Florence. It has a raised border on which the twenty-four letters are inscribed and may well have been used as a base for a wax writing surface; but see J. Heurgon ([1961] 1979, pp. 270 ff.).

5. On the appearance of scenes connected with the legend of the Seven against Thebes on funerary urns and the legend's relation to Roman-Etruscan lore, see J. P. Small (1981, pp. 92 ff., 165 ff.). The most spectacular representation of the legend in Etruscan art has recently been recognized in the columen-cover sculpture of the temple B at Pyrgi, on which see G. Colonna (1985, pp. 137 ff.) and the discussion later in this chapter.

There is little doubt that there were also Etruscan verse forms, even some kind of "bardic" tradition. The late-seventh-century epitaph of Aulus Feluske from Vetulonia (Florence, Museo Archeologico) seems to scan. Verse fragments have been identified by F.-H. Massa-Pairault (1985, pp. 38 ff.). But while meter and even syntax seem relatively clear, the meaning of the verses remains obscure.

Since the earliest times the Etruscans had been avid collectors of Greek ceramic ware of all kinds, as well as employers of Greek potters: see R. Bianchi Bandinelli and A. Giuliano (1973, pp. 150 ff.). While this helps to explain the transmission of the figurative themes, it does nothing to explain their enormous popularity.

6. Vitr. I.7.i ff.; see C. O. Thulin (1906–1909, pt. 3, pp. 41 ff.). It is however notable that while Vitruvius sets out Greek customs about temple siting as part of his general account of architecture, he relegates the Etruscan rules among town-planning regulations. Plutarch (*Quaest. Rom.* 47) had rationalized this rule, based (he says) on a ridiculous tradition that Romulus (who was a son of Mars) did not want jealous Vulcan's temple to be within the same walls.

Vitruvius continues the paragraph after an *item* that could either be read as conjunctive or disjunctive: "to Ceres also on a site outside the city; in a place where people never go except to perform sacrifice. That place must be maintained with veneration, chastely, and in a pious manner." Although Thulin (p. 44) excludes this sentence, it does read very much as part of the paragraph.

There seemed to be two translations of the ritual books current in Vitruvius' time: a prose one by Aulus Caecina, a friend and correspondent of Cicero, and a verse one by one Tarquitius Priscus: see J. Heurgon (1979, pp. 142 ff., 289 f.; 1953, pp. 402 ff.).

7. "Rasenna" was, according to Dionysius of Halicarnassus (I.30.iii), the eponymous hegemon or leader of the Etruscans, while Herodotus (I.94) has the Lydian Prince Tyrrhenus, the son of King Atys, lead a colony of them into Umbria. Atys elsewhere is also the father of Lydus, the eponymous founder of the Lydian nation. Many ancient and modern writers have struggled with this matter, which is still far from settled: see M. Pallottino (1975, pp. 81 ff.). *Rasna* occurs once in an Etrus-can inscription, on a boundary stone from Cortona.

The Greek *tursēnoi* already appear in Hes. *Theog.* 1010. Some recent linguists have suggested that the word derives from **tur, *tyr,* as in *turris,* meaning therefore something like "people of the castle, the citadel." *Turskum, tuscom, tuscer, tursce* are all forms that occur on the Iguvine tables, naming one of the alien peoples who have to be expelled during the ceremonies. On these terms see H. H. Scullard (1967, pp. 15 f., 34 ff.) and G. Devoto (1951, pp. 81 ff., 228 ff.). The rather extreme suggestion of Sergio Ferri that the *k* in Etruski was a mere suffix and that the Tuschi were the original immigrants into Italy from the Caucasus at the beginning of the Bronze Age who constituted the Italic substratum, whereas the Etrusi were Sea People–time immigrants from Anatolia who formed a directing elite (*Studi in Onore di Calderini e Paribeni,* 1956–1957, vol. 1, pp. 111 ff.), has not had much support from Etruscologists. In antiquity certainly Etruria was more often territorial: Vitruvius, at any rate, uses "Etruschi" to refer to territory or to ritual rules, and "Tuschi" when speaking of building types or techniques; in this he seems to have echoed general usage. *Tuscia* as a territorial term is a silver Latin coinage. The rather inconclusive evidence has been gathered and discussed by D. Briquel (1991, pp. 510 ff.). Briquel had earlier (1984) discussed the many versions of the pre-Hellenic Greek origin of the Etruscans and suggested that the account of their Lydian origins was a by-product of Lydian economic policy in the sixth and fifth centuries. As for the Latin usage, Cicero, for instance, details "Etruscan" divination (*De Div.* I.2, 18, 41, II.23; *De Nat. Deo.* II.3 f., where he also quotes his father having on one occasion—wrongly—rejected the view of Etruscan haruspices as those of "Tuscans and barbarians" and therefore having no authority over "Roman augury"); the Emperor Claudius appealed to the scientific respectability of Etruscan haruspicinium as a historic discipline in his drive to put down "foreign" (Pythagorean, Christian, Jewish, etc.) superstition. His speech on the subject is reported by Tacitus (*Ann.* XI.15).

8. It is this base, with a flight of steps at the front, that became the prototype of eighteenth- and nineteenth-century "classical" buildings.

9. Vitr. IV.7.ii; "ternae partes dextra ac sinistra cellis minoribus, sivi ibi aliae futurae sunt, dentur"; "the three parts to the left and to the right are for the smaller shrines, or other structures, if there are any." Fra Giocondo (1511) had corrected *aliae* to *alae*, "wings"—as the rooms on either side of the tablinium of a Roman house. This was accepted by many editors, even though it made the *sivi* redundant, as S. Ferri (Vitruvius, 1960, p. 175 n.) points out. Ferri's version is accepted by C. Fensterbusch (Vitruvius, 1964, p. 194), but rejected by A. Andrén (1940, pp. xl ff.) and Pierre Gros (Vitruvius, 1969–1992, vol. 4, pp. 181 f.). It does not seem to me that the question is finally resolved. Ferri's syntactic difficulty has not been removed by Gross, and the archaeological evidence about the different types of temple does not—it seems to me—point as firmly as some commentators would like to an alternative "type" of temple contemporary to the three-cella Vitruvian one. What A. Boëthius (1978, pp. 40 ff., 131 ff.) says about the two types of Etrusco-Italic temples, with and without alae, is made redundant if the textual "correction" is avoided. Boethius (pp. 38, 222 no. 9.) was familiar with Ferri's argument, but chose to ignore it.

Ferri's other suggestion, that the Etruscan temple derives from a north Balkan hut, half-enclosed (for people), half-open and fenced in (for animals), as against the peripteral Greek temple, half-megaron, half-hut, seems very far-fetched.

How Vitruvius distinguishes between *area,* "clearing," "bald patch," "threshing floor," "site," and *locus,* "place" (in every sense of that English word), is not altogether clear.

10. Too much has been made, I think, of Vitruvius' "reluctance" to use the terms *posticus* and *anticus* for the fore and aft parts of the temple, given that most of the buildings he described were presumably "inaugurated." In fact although he uses the word *posticus* without technical implications to refer to the back or aft part of a building, he seems not to have been interested in the technical augural sense of the word. The term *peripteros sine portico* was coined by Ferdinand Krohn in 1911 and given more importance by Ferdinando Castagnoli; they transposed the passage in Vitr. III.2.v to do so (*MDAIR* 62 [1955], pp. 139 ff.). Following Fensterbusch (Vitruvius, 1964, pp. 142 f.), S. Ferri (Vitruvius, 1960, pp. 100 ff.) and even P. Gros (Vitruvius, 1969–1992, vol. 4, p. 88), I will follow the usual reading, which limits the type to one temple in Rome, that of Honor and Virtue built under Marius by Gaius Mucius. On the modular arrangement of both Vitruvius' and the archaeologists' temples, see P. Barresi (1990, pp. 251 ff.). There is no doubt that the simple numerical relationships and plan shapes were the result of the builder's intention. The rectangle 3:4:5 (and its derivatives) is more frequent than the Vitruvian 5:6 formula.

11. Vitr. IV.8.v, vi: "de tuscanicis generibus sumentes columnarum dispositiones transferunt in corinthiorum et ionicorum operum ordinationes," though this may not always have been straightforward. However, exactly that was done, and with the consent of the haruspices in the case of Vespasian's rebuilding of the Capitoline temple.

Others, Vitruvius says a bit censoriously, move the walls of the shrine behind the outer line of columns and attach them to the columns of the pteroma, and thus take more of the area (of the podium) to enlarge the plan area of the cella. Maintaining the proportions and symmetries of the other parts of the building, they seem to produce a novel type of building called pseudoperipteral, of which the Corinthian Maison Carrée at Nimes or the Ionic temple of Fortuna Virilis in Rome are probably the best-known instances.

12. This is the formula Pliny gives (*NH* XXXVI.lvi.179) for the Artemis temple at Ephesus.

13. Pln. *NH* XXXVI.lvi.178 f. He begins by remarking that the closer they are, the thicker they will seem. But having enumerated the varieties, Pliny goes back to his lime plasters (lvii. 180 ff.). The unstated implication was that columns were always covered with a colored stucco.

14. Although Vitruvius is very explicit about the torus, the monuments are much more ambiguous: see L. Polacco (1952, pp. 36 n. 104, 58 ff.).

15. The description of the capital has also led to various interpretations: see P. Gros (Vitruvius, 1969–1992, vol. 4, pp. 188 ff.) and B. Wesenberg (1983, p. 27).

16. Now in the Palazzo dei Conservatori in Rome. This temple was reputedly consecrated by the King Servius Tullius (before the Capitoline one, therefore) and rededicated by Camillus: so Livy V.19.vi and Ovid *Fasti* VI.479 ff., and the archaeological results seem to confirm the traditional date. Unfortunately, there is no "normative" publication of the site and the remains: see F. Coarelli (1988, pp. 205 ff.). On the detail see A. Somella Mura (1977, pp. 62 ff.). Similar profiles were listed by L. T. Shoe (1965, pp. 26 ff., 131 ff.). The relation of such moldings to Greek leaf-capitals and the early Doric ones with a "throat" molding (as well as such later instances as the Massalian treasury at Delphi, on which see above, chapter 10, p. 331) and their relation to Mycenaean precedent are discussed by B. Wesenberg (1971, pp. 43 ff.).

17. Literally, "or they will heat and quickly rot." "Two fingers" is a precise measurement: 16 fingers to a foot of 0.296 m makes it 3.7 cm. As for the joints, Vitruvius does specify: "compactae subscudibus et securiclis." On these and the corresponding Greek terms, see Vitr. IV. 7.iv (ed. P. Gros, 1969–1992, vol. 4, pp. 190 f.).

18. G. Morolli (1985, pp. 86 ff.); L. Polacco (1952, pp. 66 ff.). On the notion of the "order," see above, chapter 1.

19. The older editors have assumed that the wall mentioned here is a dwarf wall carried by the beam (corresponding to a frieze), though it can equally be read as referring to the walls of the cella.

20. The term *antepagmenta* usually refers to moldings or carving fixed or nailed on: the nearest word in twentieth-century usage could in fact be "ornament." But see Cato *De A. C.* XIV. Festus (s.v.) thinks it means any of the things that *adpanguntur,* that "are fixed on." The fundamental collection of the remains was made by Arvid Andrén (1939–40; *Acta Instituti Romani Regni Sueciae* 6). Vitruvius has no word about such column sheaths as the one from the temple of Mater Matuta, mentioned earlier. It was clearly intended to clothe a cylindrical column, and both capital and base are molded into the ceramic cover; they show no joints.

21. Vitruvius does not comment on the word, which gave some trouble to grammarians and to later philologists. It seems at times to have been used (though not by Vitruvius) interchangeably with *columna*—"support," even "column." But it is obviously close to *culmen,* "summit," "high point," and in that sense occurs in Ennius and Varro. But the etymology of each word has remained a problem. The *Acta Fratrum Arvalium* use interchangeably "sub . . . culmime" and "sub . . . culmine." The Greek obsolete word (Theocritus XVIII.34) *keleontes* for the horizontal beam in an upright loom may have a relation. Etymologists relate it variously to *celsus,* "elevated," "high," "summit," or *collis,* "hill," "ascent," "mountain." Their relation to the Greek *kaleō* or the Latin *calo,* both meaning "I call," "I summon," is not clear. Although the Etruscans were famous for the bulk and straightness of their timbers, the *columen* of such a temple as the Capitoline could not have been made of a single timber.

22. From the temple of Diana (Villa Giulia no. 12642); R. A. Staccioli (1968, pp. 39 ff., cat. no. 30). Another model from the same site is lost. A similar, but much simpler arrangement is found in two other models: one from Orvieto (Staccioli no. 19; Museo Faina, Orvieto, nos. 857, 858) and another from the sanctuary of Mater Matuta at Satricum (Staccioli no. 39, pp. 48 f.; Villa Giulia no. 11614). On their relation to temple building, see A. Andrén (1960, pp. 21 ff.).

23. The tablets were found during the excavations of 1964; they have been dated to ca. 500. What was the office of the dedicator, Thefarie Velianas (*m'l'ch,* "king," in West Semitic) and where he ruled are still matters of dispute; but see A. J. Pfiffig (1965, passim). The temples and the tablets were discussed in M. Pallottino et al. (1981).

24. Another such a pediment was the second-century one from Telamon, reassembled in the Museo Archeologico in Florence.

The use of the Attic foot at Pyrgi and the stone columns made up of white tufa drums (with a capital of peperino) were among the many unexpected features of the two temples. See M. Pallottino and G. Colonna (1966, pp. 251 ff.).

25. This is not clear: one-third (presumably on either side of the double-pitched roof) seems a lot. The overhang appeared excessive to some of Vitruvius' editors, but the arrangement has turned out acceptable to archaeologists, and it assured a gentle slope to the roof. Granger translated here: "so that the slope of the roof should be as one to three." I prefer the sense of Ferri and Fensterbusch, but see the bibliography of this dispute in P. Gros (Vitruvius, 1969–1992, vol. 4, pp. 193 ff.).

26. These fragments are all in the Villa Giulia. The "Apollo of Veii" is (deservedly) about the best known of all Etruscan sculptures. On the style and the *facture* of those statues, see M. Pallottino, "La Scuola di Vulca" (in 1969, vol. 3, pp. 1003 ff.) and O. Brendel (1978, pp. 237 ff.).

27. Such as the bronze ash urn from Falerii or the ceramic one from Monte Abetone, both in the Villa Giulia, or another bronze one in the Metropolitan Museum, New York. Similar ornamental finials had already appeared on the Villanovan house urns.

28. F. Prayon (1975, pp. 56 ff., 149 ff.). On the huts at San Giovenale, see A. Boëthius et al. (1962, pp. 292 ff.). The Palatine hut was first published in *Monumenti Antichi* 41, pp. 45 ff. It is discussed by E. Gjerstad (1953–1973, vol. 3, pp. 48 ff.). On the thatched-roof hut as a prototype of a tomb, see A. Boëthius (1965, pp. 3 ff.). The two tombs at Cerveteri I have already mentioned: see also the "Campana 1" tomb at Monte Abetone or that of the "Animali Dipinti," also at Cerveteri.

29. But see P. Gros (Vitruvius, 1969–1992, vol. 4, pp. 192 f.).

30. See recent reviews of the material by C. de Simone, "Gli Etruschi a Roma," and G. Colonna, "Quali Etruschi a Roma," in *Gli Etruschi a Roma* (1981, pp. 93 ff., 159 ff.). See also M. Torelli (1987, pp. 68 ff.) and L. Banti (1973, pp. 13 ff.). On Lars Porsena, king of Clusium and perhaps a warlord of a real (and unusual) Etruscan confederacy, see Dion. Hal. V.21, Tac. *Hist.* III.72, Livy II.9, and also H. H. Scullard (1967, pp. 261 ff.). There is no doubt that he defeated Rome, and exacted a capitulation, if a relatively clement one (Livy II.12 ff.), which involved hostages and a gift of "an ivory throne, a golden crown, a scepter with an eagle, and a purple triumphal toga, the insignia of royalty" (Dion. Hal. V.35.i).

31. The *saeculum* is defined by a late writer, Censorinus *De Die Natali* 17: "spatium vitae humanae longissimum, partu et morte definitum." Censorinus further quotes the opinions of various writers who made it anything between 25 and 120 years. Varro's authority (set out in a lost book) supports the belief that a lost *Tusca Historia* allowed them ten saecula, while Plutarch (*Sulla* 7) reports a tradition of about eight. But see C. O. Thulin (1906–1909, pt. 3, pp. 63 ff.) W. Burkert (1992, pp. 48, 182 n. 9) considers that the first four saecula were 100 years each, after which they shortened progressively.

32. Hor. *Ep.* II.i.156 f. An epistle written specifically for Augustus, it is a kind of brief account of Latin letters.

33. Virg. *Aen.* X.145, XI.567; *Geor.* II.533; and Serv. ad loc. The reality of Etruscan rule was much more modest: they did control most of Tuscany, the Roman Campagna, much of Umbria, and also the eastern part of the Po Valley. Their attempts to establish themselves in the south, around Capua, were thwarted by the Greeks, who dominated much of the Italian coast. But see M. Sordi (1989).

34. Perhaps the most showy witness to this is Julius Caesar's dedication of a temple to Venus Genetrix in his own forum, as the mother of Aeneas and therefore ancestress of the Julian clan. But there are many other, more "historical" instances: Maecenas, called "eques Hetrusco de sanguine Regium" in a dedication (Prop. III.ix.1; but see also Hor. *Carm.* III.xxix.1, *Ep.* I.xiii), was very proud of his descent from the Ciluii of Arezzo.

There was a later elaboration of the legend (reported by Serv. ad *Aen.* III.167), according to which Troy itself was founded by an Italian colonist, Dardanus from Cortona, so that Aeneas could have been said to be returning home when he landed in Italy. See D. Briquel (1984, pp. 150 ff.). There were many other legends, connecting Homeric heroes (Ulysses, Diomedes) with Italian, particularly Etruscan places.

35. Livy V.1.vi; Serv. ad *Aen.* II.781; Clem. Alex. *Strom.* I.306; Is. Sev. *Etym.* XIV. 4.xxii. *Tusci* was even derived from *thus*, "incense," while Dion. Hal. (I.30.iii) derives it from *thuoscooi*, "sacrificing priests." G. Bonfante suggests that Etruscan deities were appealed to on "serious," ritual occasions, while the Greek names, sometimes of the same deities, are used when the occasion is merely "literary" (1986, p. 113).

Ancient literature on the religious superiority of the Etruscans was gathered by Thomas Dempster of Muresk (1723, vol. 1, pp. 56 ff.).

36. Although Vitruvius (V.11.i, VI.3.x) uses analogous terms to distinguish Greek from native ways: the *Italica consuetudo* did not include *palaestrae*; the Greek (Cyzicene) *oeci* were quite different from the *cavaedia* of the Tusci.

37. The texts relating to the Etruscan rite were collected by C. O. Thulin (1906–1909, passim).

38. The college was in charge of Greek but also other "imported religions," the *di novensides*, on which see G. Wissowa (1904, pp. 534 ff.). However, these importations do not seem to have produced conflicts with the other colleges. See W. Warde Fowler (1922, pp. 253 ff.).

According to tradition, the college was founded by Tarquinius Priscus as one of the four priestly colleges of the city, after his bargain with the Sybil (recounted by Dion. Hal. IV. 62; Varro apud Serv. ad Aen. VI.72; Lact. Firm. *Inst.* I.6.x ff.). The first known Sibylline books—those of the Cumaean Sibyl—were written on palm leaves (Varro apud Serv. ad *Aen* III.444). The Sybil offered her nine books to the king at a price he refused and then burned three of them; she offered six at the same price, and on the king's again refusing to pay, she burned three more. Tarquin bought the three remaining ones for the same price as she had asked for all nine. That the "original" Sybil was the Cumaean one suggests the way Greek documents arrived in Rome. Such books as existed were in fact burned on several occasions (notably in the Capitol fire of 83) and reassembled. A "definitive" textual revision was ordered by Augustus, and this official collection was finally burned at the time of Stilicho (Rutilius Namatianus II.52 ff.). A number of texts circulating under the title *Libri Sibillini* are mostly Hellenistic fabrications.

39. The rite was known as *evocatio*. It was practiced particularly before the final assault on a besieged town, when the god would be offered a temple and sacrifices in Rome. The carmen and

the rite were recorded by Macr. *Sat.* III.9; Pln. *NH* XXVIII.iv.18 gives a definition; Serv. ad *Aen.* II.244, 351 suggests that Rome had a secret name to prevent this rite being practiced on her, as does Plut. *Quaest. Rom.* 61; Liv V.21 tells of Camillus "calling out" Juno before the final onslaught on Veii. See G. Wissowa (1904, pp. 39, 321 f.). This rite, and an analogous one by which army commanders exposed themselves to certain death in battle as a sacrifice to the infernal powers, *devotio*, may indeed have been part of the original "Italic" religious practice.

40. The most extensive document of this Italic religion are the seven bronze (once nine; two were lost) Gubbio tablets inscribed with an elaborate "liturgy" for a priesthood of the town, which seems parallel to that of the Arval brothers in Rome. It is also a prime document for the development of the Umbrian language. The best edition is still Giacomo Devoto's (1937). Georg Wissowa's attempt to identify native gods as those called *di indigetes*, as distinguished from *di peregrini* or *novensides*, has not been generally accepted.

41. The Tarquins exemplify the insistently double heritage of the Romans. According to both Greek and Roman writers, the family was descended from a Corinthian merchant Demaratus, who married an Etruscan noblewoman of Tarquinia, where he had settled after the tyrant Cypsellus expelled the noble clan of Bacchiades (to which he belonged) from Corinth in the year 657. One of his sons, Lucumo, married the prophetess Tanaquil, changed his name to Lucius, and reigned as king in Rome from 616 to 578, in succession to Ancus Marcius. Versions of this legend appear in Livy I.34 and Dion. Hal. III.46 ff. Pliny (*NH* XXXV.xliii.152) even thought that Demaratus was responsible for bringing the art of cutting profile portraits, invented in Corinth by Butades or Debutades, to Italy. He goes on to make him the head of a school of artists (he names three: Eucheir, Diopos, Eugrammos); this passage is augmented by a quotation in Ath. gr. *Pro Christ.* 17. On this, see M. Torelli, (1977, pp. 305 ff.).

42. He had also forced Roman plebeians to do corvée work on his building: Livy, I.56; Cic. *In Verr.* II.5.xix (48). Varro (V.158) and, following him, a number of modern authorities (G. Wissowa, 1904, p. 126 n. 3) believe that an older shrine, the sacellum of the triad, was moved from the Quirinal by Tarquinius. E. Gjerstad (1967, pp. 9 f.) supposes that the title, *Optimus Maximus,* means not the best and greatest of all the gods, but the best Jupiter around. As for the other (and presumably earlier) triad, Jupiter, Mars, Quirinus, it was invoked by P. Decius Mus in his famous devotio before the battle of Minturnae in 340 (=414 AUC) according to Livy (VIII.9.vi), and its function in Roman religion is considered by Gjerstad (pp. 16 f.). But see Wissowa (pp. 151 ff.).

43. Livy X.23.xii, Pln. *NH* XXXV.xlv. 157. In some older books on ancient art (as Franciscus Junius, 1694, p. 219) he is identified as Turianus of Fregene, presumably owing to a scribal error in the Paris manuscript of Pliny (no. 6801). Vulca's statue was replaced in 296 with a chryselephantine one (by an otherwise unknown Apollonios), which was based on Pheidias' Olympic statue. So Chalcidius in Pl. *Tim.* 338c. On the painting of the face with *minium* (cinnabar, a mineral vermillion; it is an oxide of mercury, of which there are deposits in Tuscany), see Pln. *NH* XXXIII. xxxvi.111; Serv. ad *Ecl.* VI.62, X.27. Servius also discusses the importance of its regalia.

When he records the omen (XXVIII.iv.16) Pliny does not mention Vulca, nor does Plutarch when he refers to the same matter (*V. Publ.* 13). It has however been assumed—because of these and his other important commissions—that Vulca created a kind of studio and school in Rome, on which see M. Pallottino (1969, passim). The name of the artist who received the second commission was not recorded, and some have assumed that it went to the same workshop.

On the other early Italic artist-signatories, Novios Plautios and Vibis Pilipus, see F.-H. Massa-Pairault (1985, pp. 95 ff.).

44. Plut. *V. Publ.* 13. The consecrator was one of the first consuls of the new republic, M. Horatius Pulvillus (the other was Valerius, who had been a colleague of Brutus), whose name had been drawn by lot (Livy II.8, VII.3.viii; Tac. *Hist.* III.72). On the date and the significance of the consecrator, see E. Gjerstad (1967, pp. 168 ff.). Dion. Hal. V.35.iii suggests that his name was inscribed on the temple: but this has been doubted: see K. Hanell (1966, pp. 135 ff.).

45. So Livy I.38.vi: "iam praesagiente animo futuram olim amplitudinem loci," "his [Tarquinius Priscus'] spirit forecasting the future grandeur of the place."

46. The metric dimension given is 56.83 by 61.57, which—taking the Roman foot at 0.296 m—gives the dimension 192 by 208 or 12 by 13, which is squarer than Vitruvius' proportion: so R. A. Lanciani (1897, p. 298). The so-called Tempio del Belvedere at Orvieto comes closest to the Vitruvian formula. But on the poverty of the earlier Etruscan temples, see L. Polacco (1952, pp. 81 ff.). On the other hand, E. Gjerstad (1953–1973, vol. 3, pp. 162 ff.; vol. 4.2, pp. 388 ff.) gives it as 180 by 210 feet, or 6 by 7. His reworking of the absolute datings of the monarchy involves placing the building of the temple entirely in the reign of Tarquinius Superbus.

Of that platform, which Gjerstad thinks 12 feet high (though raised both in 83 BC and again in AD 69), some ten courses of dry-laid tufa are still visible in the garden of the old Palazzo Caffarelli, now part of the Palazzo dei Conservatori. When the Palazzo Caffarelli was being enlarged in the 1680s, some fourteen courses were removed from the platform.

47. A curious echo of this arrangement is the late (second-century) Tomba Ildebranda at Sovanna, whose stumpy pseudo-Corinthian columns surround the mass of the tomb on three sides. It has been suggested that the extra columns were an echo of the Greek peripteral temples in south Italy and Sicily: so A. Boëthius (1978, p. 50).

48. It is not clear under what form this god was worshipped, though "he" may well have been a baetyl: but see G. Piccaluga (1974, pp. 123 ff.). It was certainly *nefas* to worship him "non nisi sub divo," as Servius says (ad *Aen.* IX.448).

Ovid (*Fasti* II.669 ff.) found the hole small— "Exiguum templi tecta foramen habet"—but this opening in the roof and the reason for it was well known: see Festus 505 L; Livy I.55.iii–iv; Dion. Hal. III.69.v; Serv. ad *Aen.* IX.446.

A second aedicule, of Juventas associated with Mars, seems also to have been located close to the cella of Minerva; Juventas may have been (according to some accounts) another of the deities who refused to surrender her place to the Capitoline triad. But this may be a later interpolation into the legend. There is no doubt that Terminus was a more ancient and much more important deity, as is clear from Piccaluga's account (pp. 196 ff.). As Terminus had his calendar function, as the "terminator" of the calendar year, so Juventas was associated with the assumption of the toga virilis by Roman youths: Serv. ad *Aen.* IV.49, Dion. Hal. IV.15; see Piccaluga (p. 239 n. 23).

In fact Livy records the foundation of the first temple on the Capitol, inaugurated by Romulus to another Jupiter, Jupiter Feretrius (I.10.v f.).

49. M. Torelli (1976, pp. 13 ff.); Pietro Romanelli in *NS* 73 (1948), pp. 238 ff. The original shrine has been dated (also on the basis of terracotta finds), to the mid-sixth century.

50. F. E. Brown (1980, pp. 53 ff.; "The Capitolium," in F. E. Brown, E. H. Richardson, and L. Richardson 1960, pp. 50 ff.). The Capitoline temple of Cosa is dated ca. 160. The curious thing about the fabric of this temple is that the podium was not structural at all, but a stone facing around the brick walls of the cella, which went down to their foundations.

According to Aulus Gellius (II.10), who comments on the term *favissae Capitolinae,* there were several of them in the temple of Jupiter O.M. He quotes the letter of one of its restorers, Quintus Catulus, to M. Varro; the restorer wanted to lower the temple platform to increase the steps needed for ascent to the temple (to correspond to the greater height of the new fastigium) but could not do it because it had so many *favissae:* some were *cellae,* others *cisternae.* That last word almost always means an underground chamber for storing rainwater, but many of the chambers contained remains of the older temples and other sacred objects.

It is the use of area here which is odd: were the *favissae* sunk into the ground around the temple, as some commentators want, or were they part of the actual platform (as on the Capitol of Cosa or in the "Ara della Regina" at Veii) so that the ascent was a double-stepped one? That seems the implication of this gloss, at any rate.

51. Livy (VII.3) maintained that the same was done in the temple of the goddess Nortia at Volsinii by their chief judge, the Praetor Maximus, as a calendric rite. In the Capitolium it was done as part of the ceremonies of the Ides of September (13), when the face of the cult statue was colored before the *epulum.* The Roman *clavi* were fixed by a consul or praetor. Livy seems to imply that the nails were numbered, presumably *ab urbe condita.* But sometimes there was also a special magistrate for performing the rite, the *dictator clavi fingendi causa.* For reasons that are not clear (but presumably were related to crises) such *dictatores* were specially appointed in 363, 331, 313, and 263. See N. Turchi (1939, pp. 98 f.); and A. Momigliano, "Ricerche sulle Magistrature Romane" (1969, pp. 274 ff.). The rite of driving nails to avert bad fortune and disease is common in Africa—hence the hedgehoglike statues of human beings or animals familiar from ethnographic collections, but also known to the Romans: Pln. *NH* XXVIII.vi.63.

52. Tac. *Hist.* IV.53 gives an account of the complicated foundation rites on that occasion. The haruspices were consulted and told the officiating magistrate that the gods did not want the plan changed—"nolle deos mutari veterem formam"—but they did allow an increase in height. The haruspices were not always so conservative: Pliny the Younger took their advice about rebuilding the small dilapidated Ceres shrine on his estate; he was told to enlarge it. He made it a tetrastyle for which he ordered "prefabricated" column shafts from Rome, and added a "portico" (*Ep.* IX.39, to Mustius).

53. One of Sulla's great regrets was that he could not dedicate the Capitoline temple, on which he had spent much of his own fortune, and his death was caused by a stroke due to his learning of a fraud at the expense of the temple construction fund. See C. Lanzani (1936, pp. 330 f.). Some fragments of the marble columns were reused by seventeenth-century sculptors.

54. R. Lanciani (1897, pp. 298 ff.).

55. The dedication occurred during his second consulate. The consulates of Spurius Cassius are associated with the "secession" of the plebeians and the tightening of the Latin confederacy, directed against the Etruscans. He was later executed for seeking absolute power by being thrown from the Tarpeian rock.

56. Vitr. III.3.v. See F. Coarelli (1988, pp. 67 ff.); S. B. Platner (1929, 109 ff.). The temple of Hercules had been vowed by the dictator Aulus Tubertus Postumius.

57. Dion. Hal. VI.17.ii ff., 94.iii; H. Le Bonniec (1958, pp. 213 ff.). Cicero thought it "pulcherrimum et magnificentissimum" (*In Verr.* IV.108).

58. Damophilus' name has prompted some scholars to consider him a relation with Demophilus of Himera, master of one of the greatest Greek artists, Zeuxis—who in any case would have lived a generation or two later than the artists working in Rome. Still, both were almost certainly Greek (their names appear Dorian) and therefore possibly Sicilian.

Pliny (*NH* XXXV.xlv.154) writes as if he had actually read the epigram, though he does not quote it verbatim. An analogous distich-signature of Kimon—he did the right side of the door going in, Dionysius the right going out—appeared in the *Palatine Anthology* (IX.758); another epigram-signature of Kimon is quoted in XVI.84.

59. Vitr. III.3.v; Pln. *NH* XXXV.xlv. 154; H. Le Bonniec (1958, pp. 254 ff.). On the framing of such "primitive" fragments for exhibition in public or even private galleries, see M. Cristofani (1978, p. 5).

60. The Veian temple has been restored both as a *peripteron sine postico* (*NS* 73 [1948], p. 255) and as a three-chamber shrine with a distyle *in antis* portico (M. Torelli, 1975, pp. 18 ff.). A brief survey of the arguments by F. Prayon in L. Bonfante (1986, pp. 193 ff.).

61. The lowest columns of the Theater of Marcellus conform roughly to the description given by Vitruvius. But since they have no bases and carry a metope-and-triglyph cornice, they do not really qualify as "Tuscan," even if, despite variations, they were so identified in some modern works: so W. J. Anderson and R. P. Spiers (1927, p. 96). Older authorities (C. Fontana, 1694) tend to opt for Doric. The different labelings are discussed by S. Maffei (1731–1732, vol. 3, pp. 98 ff.).

62. Wood certainly needed renewing regularly. The tiles and revetments would have had to be refurbished every few years and completely replaced once or twice every century. They were often exhibited, reused, or buried in the same or in a neighboring temple.

63. They have been collected and catalogued by R. A. Staccioli (1968). As his title, *Modelli di Edifici Etrusco-Italici,* implies, he excluded any that must be considered hut urns. For these, see G. Bartoloni, F. Buranelli, V. d'Atri, and A. de Santis (1987); it lists 194 (mostly Villanovan) hut urns and 33 roof-shaped urn covers, but still does not include Etruscan material, such as the ash urn from Chiusi in the Museo Archeologico in Florence (see n. 72)

64. The evidence of the tombs about the architecture of the living has been collected by F. Prayon (1975), who is primarily concerned with the Cerveteri-type tomb, rock cut but topped with a tumulus. There were other types, not directly related to house and temple building, but built up with corbeled and pillar-sustained vaults—as the one from Montagnola, or another from Casal Marittimo (reassembled in the Museo Archeologico, Florence) or yet the Pozzo dell'Abbate at Vetulonia. Later tomb groups, whether rock cut (as at Norchia or San Giuliano) or built-up (as at Orvieto), were actually laid out as quasi-urban complexes.

65. For example, the tomb "of the Thatched Roof," also rock cut, at Caere, and usually dated to the early seventh century. At least one sixth-century tomb at Tarquinia, called del Cacciatore, is painted to represent a hunter's tent with trophies.

In some rock-cut tombs, flat roofs are also represented: they are mostly coffered by purlins and rafters laid at right angles to each other, and they have strong stereation at 45° to the coffers, suggesting a reed ceiling to be covered with clay or mud.

66. The best known of such models is the stone urn in the Altes Museum, Berlin. Although many earlier authorities dismissed this filiation, arguing from the absence of any developed Etruscan domestic architecture, it has recently become clear that by the early sixth century they did build quite elaborate quasi-domestic buildings (palaces? shrines?). These include the one at Poggio Civitate near Siena (*NS* 101 [1978], pp. 114 ff.)

and at Aquarossa (Complex A-C), which in turn are associated with the first Regia on the Roman Forum.

The even greater antiquity of the three-room plan has also become apparent in recent excavations: see F. Prayon (1975, pp. 152 ff, 179 f.; and in L. Bonfante, 1986, pp. 190 ff.). The insula-houses of Marzabotto already seem to have fully developed central plans, featuring a main room (perhaps open to the sky) with side alae.

67. Reputedly it was called Agylla and was a "Pelasgic" settlement: so Dion. Hal. I.16, III.193; Str. V.220, 226; Steph. Byz., s.v. "Agylla." Servius (ad *Aen.* VIII.597) traces the name back to an eponymous founder, Agella. The charming story that Caere was a corruption of the Greek *chaire,* the greeting of a (presumably Greek-speaking) local to the newly arrived Etruscans who were inquiring after the name of the place, was reported by Strabo, Stephen of Byzantium, and Servius.

On the Tomba dei Capitelli columns and the capital from Chiusi, see L. Polacco (1952, pp. 36 ff.).

68. Notably in the tomb "of the Doric Columns" in the Banditaccia necropolis at Caere, and the tomb "of Princess Margarethe of Danemark" at San Giuliano. But see F. Prayon (1975, pp. 43 f.).

69. The Aeolic capital was discussed in detail above, chapter 9, p. 269. A similar but isolated capital was found in Chiusi, and it is now in the Museo Archeologico in Florence.

70. Though the highly decorated tomb of the Volumni outside Perugia, still fully Etruscan, was prepared entirely for cremation urns about 150–100; C. Shaw (1939, pp. 58 ff.). At least one Perugian tomb, now the crypt of a church (and known as the Ipogeo di San Menno), has the main space covered by a barrel vault, whose voussoirs are well-shaped travertine blocks.

One curious aspect of this problem is the creation of many crowded colombaria, sometimes inside desecrated older rock-cut tombs. Many of them were indeed used to keep pigeons, especially in late Roman times. Were any of them intended for funerary use? The question remains unresolved; but see S. Quilici Gigli (1981, pp. 105 ff.).

71. On this whole issue, see A. Ciasca (1962, esp. pp. 27 ff.). There are sarcophagi with such capitals in the British Museum and Villa Giulia (from Tarquinia and Veii) and the Museo Guarnacci (Volterra). On its use in furniture, see S. Steingräber (1979).

72. Called a Palazzetto, now in the Museo Archeologico in Florence (no. 5539). It has other curious details: above the arch is a strip of vertical bars, which has been interpreted as the window of an upper story or a clerestory light of a two-level "hall." Such windows appear also on the long sides of the urn, framed by taller "Aeolic" pilasters. A lower "story" on the long sides is framed by Tuscan-style pilasters; at any rate, the two kinds of column seem to coexist quite easily within the one building. In fact the bars in the "windows" are held by strips that suggest shutter panels. The urn has been interpreted as the model of a temple, more specifically the Roman temple of Janus; see R. A. Staccioli (1984, pp. 91 ff.). The temple of Janus, though ancient, is known only from Neronian coins. But there may be other common sources for both constructions. The only similar urn is in the Villa Danzetto in Veliano near Perugia, and was published by A. Brunn and G. Körte ([1870–89] 1965, III, CII, 2).

73. Both are also 100 Roman feet high, and the shaft is very close (1:6.8) to the Vitruvian (1:7) rule for the Tuscan. The column of Marcus Aurelius has reliefs of not quite such high quality, has had most of the ornament on the base destroyed, and has a less pronounced entasis—and therefore looks a little squatter. In Constantinople, Constantine erected a porphyry column to himself (now called "the Burned Column"), which housed a number of relics in the base, rather like Trajan's (which had a shrine for the golden urn containing the emperor's ashes): it contained the original Palladium of Troy (which had been kept in Rome by the Vestals) and the baskets from the miracle of loaves and fishes. The Emperors Arcadius and Theodosius emulated him, even if Constantine's self-serving syncretism seems to have provided subject for jibes: see G. Fowden (1991, pp. 119 ff.).

Of other surviving Roman columns, that of Antoninus Pius was only half as high, though a monolith. It was excavated in the eighteenth century, and unfortunate attempts were made to set it up again; most of what remains is in the Vatican Museum.

The Corinthian one of the eastern Emperor Phocas in the Roman Forum was about a meter smaller; it was erected in AD 608, presumably from spolia.

The bronze statues of St. Peter and St. Paul by Tommaso della Porta and Leonardo da Sarzana replaced the (by then vanished) emperors in 1587 or 1588.

74. Pln. *NH* XXXIV.xii.27: "The use of columns is to raise [a statue] above other mortals." Pliny goes on to say that arches could serve the same purpose and that the Greeks were the first to put up honorific statues, though he implies that the columns were a Roman usage.

Ennius (fr. var. 2) wrote:

Quantam statuam faciet populus Romanus,
quantam columnam quae res tuas gestas
loquatur[.]

This is addressed to Scipio: "How great a statue will the Roman people make, how great a column shall speak of all the things you did[.]"

75. It stood with a group of relics, associated with the earliest history of the city, which had been miraculously moved there from the Palatine by Attius Navius: the fig tree under which the founding twins had been fed by the she-wolf, the bronze figure of the she-wolf and the children, and the whetstone that Attius Navius cut with a razor. On all this, see F. Coarelli (1985, vol. 2, pp. 28 ff.). Coarelli is noncommittal about the shape of the base: but see G. Becatti (1960, pp. 33 ff.). Pliny (*NH* XXXIV.xi.22) mentions the statue in the context of statues on columns.

76. They are not oblique bands, in any case; they have been described as toruslike swellings, and may, of course, represent horizontal narrative or figural bands.

77. Pln. NH XXXVI.xix.91 ff.; A. Boëthius (1978, p. 99; M. Cristofani (1978, pp. 6 f.); G. Becatti (1960, p. 36). On the tomb of Lars Porsena, see G. Morolli (1985, pp. 180 ff.) and F. Messerschmidt, (1942, pp. 53 ff.).

Varro's description—dimensions of 300 feet by 300 feet, with a labyrinth in the base, and the five pyramids crowned with bronze circles from which bells are hung—cannot be identified with any known ruin. Leandro Alberti in his guide to Italy (1550, p. 55) gives the description out of Pliny, insists on the folly of the enterprise, but says that no fragment of it remains. However, see H. H. Scullard (1967, pp. 155 ff.); the tomb between Albano and Ariccia, known variously as that "of the Horatii and Curatii" or "of Arruns," though much smaller, has been compared to Varro's description and restored accordingly.

78. The next column whose date is recorded was set up after C. Maenius' naval victory over the Volsci at Antium in 338; although dedicated for a naval victory, it was not rostral (Maenius decorated the Rostra, the curved sweep of the orator's tribune on the Forum, with the prows of the ships he had captured, whose brass and iron "beaks," or *rostra*, gave it its name, which later became the common term for any place of public speaking). The custom of decreeing rostral columns to naval victors began early in the third century. The first recorded one, to Caius Diulus, was decreed in 260 after a victory over the Carthaginians (Serv. ad *Geor.* III.29). Maenius' column was an important landmark in the Roman judiciary dispensation and had a view over the Forum.

79. The arch may well have been "restructured" after the siege and fire of AD 40 and carries the inscription *Augusta Perusia,* hence the name. Certainly its second inscription, *Colonia Vibia,* is much later. Its date and the date of the other gates remain uncertain: see C. Shaw (1939, pp. 22 ff.).

80. This was common enough in Etruscan sarcophagi: see the splendid one of Lartha Seianti, from Chiusi (Florence, Museo Archeologico).

81. Similar figurated capitals appear later in Greece, ca. 175 at the Asklepeion in Messene in the Peloponnese; with animals for volutes even later, on the inner propylon at Eleusis. They are not common until Roman Imperial times.

82. For all of these, but especially the "Tomba Ildebranda," see R. Bianchi Bandinelli (1929, pp. 89 ff.) who quotes a Punic example from Nora in Sardinia, probably even earlier, fifth century. G. Patroni (1904, p. 142). But see also M. Torelli (1975, pp. 168 ff.). For the Tarentine columns see H. Lauter (1986, pp. 268 ff.). The inner propylon at Eleusis was vowed by Appius Claudius

Pulcher during his consulship in the year 54 BC, but only finished after his death.

83. The building was begun as a hexastyle and finished (perhaps in Sulla's time) as a tetrastyle. It opened to the south, as did many Etruscan and Roman temples. This meant that its axis was at right angles to the east-west orientation of the famous Greek temples of the town.

An account of the building and its context is given by E. Greco and D. Théodorescu (1987, pp. 27 ff.), who also suggest that it may have been dedicated to Bona Mens; see also D. Théodorescu (1985, pp. 187 f.). The most extensive and detailed (though now superannuated in parts) account of the building itself is by Friedrich Krauss and Reinhard Herbig (1939). The context and recent excavations are reported briefly by J. G. Pedley (1990, pp. 114 ff.). On the orientation of Etruscan temples specifically, see F. Prayon (1991, pp. 1285 ff.).

84. Though it is worth noting that the restoration of Börje Blomé for Einar Gjerstad (1953–1973, vol. 3, pp. 178 ff.) shows "necked" capitals like those of the temple of Mater Matuta.

85. See M. Wilson-Jones (1994, passim).

86. Jacopo Barozzi da Vignola (1567; very many reprints). Vignola had been a surveyor of ancient monuments for the Accademia Vitruviana, which gives his words more weight. Palladio had appealed to the three amphitheaters for his reconstruction. But see J. S. Ackerman (1991, pp. 508 ff.). Philibert de l'Orme in his *Premier Tome,* which had appeared in 1567 (1648, pp. 134 ff.), echoed Vignola's comments about the absence of examples, but also made the Tuscan and the Albertian Italic the basis of his new French order of columns.

87. Dante *De Monarchia* II.3, *Inferno,* IV.85 ff. But see also M. Sordi (1989, pp. 17, 26).

88. See H. Galinsky (1932, passim).

89. G. Cipriani (1980, pp. 1 ff.); H. Baron (1966, pp. 54 ff., 414 ff.).

90. On the aspirate *H* in *Hetruria,* see Leandro Alberti (1550, p. 22), who interprets Ser-

vius (ad *Aen.* XI.598). The title was first used officially in the brief of 24 August 1569 by which Pius V raised Cosimo de' Medici from Duke of Florence to the title of Grand Duke: "Dilecto filio . . . Cosmo Medices Ethruriae . . . Magno Duci." It was, of course, the culmination of a decade of Medicean propaganda. Within months a French humanist, Marc Antoine Muret (quoted in G. Cipriani, 1980, pp. 107 ff.), was comparing Cosimo to Lars Porsena as a uniter of Etruria:

Me Ianus tenuit primus, Porsenna secundus
Tertius hetrusco Cosmus in orbe regit.

Another French scholar then resident in Italy, Guillaume Postel, was enthusiastic about such notions, which were taken up in France in the next generation by court poets such as Guy de la Boderie and Symphorian Champier: see D. P. Walker (1972, pp. 66 ff.). Walker points out that these ideas remained current well into the seventeenth century and had some appeal during the "Quarrel of the Rites," when an *interpretatio Christiana* of Chinese polytheism was attempted (pp. 214 ff.).

91. L. B. Alberti's version of the history of architecture appears in VI.3. The Italic column is described in VII.8:

the same design as the Corinthian for the vase, abacus, leaves, and flowers; but instead of stalks, they had handles, two complete modules in height, protruding under each one of the four corners of the abacus. The otherwise plain front of the capital took its ornament from the Ionic, with stalks sprouting into the volutes of the handles and the rim of the vase, like the echinus carved with eggs and lined below with a string of beads.

Variants on the capitals he described, says Alberti, do exist, but are "not approved by the learned."

92. Francesco di Giorgio S 15 r (ed. C. Maltesse and L. M. Degrassi, 1967, vol. 2, p. 61); C. Cesariano (Vitruvius, 1521, LXIII r.). Cesariano also includes Pliny's square-plan "Attic" pier.

93. In fact Montepulciano seems to have been founded about the seventh century AD. As for the statue, Vasari only saw it once, and believed it had come to grief. See G. Vasari (1878–1906, vol.

4, p. 522). Porsena remained famous for Pliny's account of his maze (as well as Livy's telling how Mutius Scevola burned his hand in defiance of him). He was therefore one of the few Etruscans whose memory endured: see G. Cipriani (1980, p. 23). But the maze continued to fascinate: see, for instance, the circumstantial account of it in P. Cluver (1624, pp. 567 ff.). The idea that Antonio de Sangallo's heavily Doric church of San Biagio at Montepulciano was intended to emulate the maze of Lars Porsena seems to me very far-fetched.

94. G. Villani (1559, vol. 1, pp. 4 ff.); R. Malespini (1718, pp. 1 ff.).

95. A version of the story appears in the fourteenth century in the *Dittamondo* of Fazio degli Uberti (1952, vol. 1, pp. 35 ff.), "Noé, che si può dire un altro Adamo"; it is told to the author by C. Iulius Solinus, who plays a part in the *Dittamondo* similar to Virgil's in the *Divine Comedy*.

96. It was built by Jean Boullant; see A. Blunt (1953, pp. 93, 105 n. 15) and Mrs. Mark Pattison (1879, vol. 1, pp. 95 f.). The palace is gone, but the column survives at the side of the old Halle du Blé. Much of the ornament was chiseled off at the time of the Revolution. The molded capital of the column is certainly modeled on the Roman ones, though some of the detail looks surprisingly—and presumably accidentally—like that of the capital from the temple of Mater Matuta on the Forum Boarium.

97. G. Morolli (1985, pp. 152 ff.); G. Morolli and C. Acidini Luchinat, "Il Linguaggio dell'Architettura," in *Firenze e la Toscana dei Medici* (1980, the volume *Il Potere e lo Spazio*, pp. 229 ff.).

98. This notion was taken up in other German order books, notably in Wendel Dietterlin's fantastic compilation of 1598, and in the much more workmanlike pattern book by the joiner Georg Caspar Erasmus, whose verse preface confusingly declares:

Die Tuscana erstlich wird von Tuscano so
genannt,
Von dem Tusci oder teutsche, hergestammt
aus
Griechenland[.]

See E. Forssman (1961, pp. 54 ff.; 1956, pp. 222 ff.).

99. C. Fontana (1694, pp. 187 ff.) describes the order as having a Tuscan base, an Ionic shaft and cornice, and a Doric capital. Such a description recalls Alberti's construction of his Italic order. But Fontana also identifies it with the lowest order of the Colosseum, as indeed, the whole eliptical plan of the *theatro* (Bernini's word) suggests. The many recent writers on the colonnade, while fascinated by the geometry, avoid the issue of the order and its choice. See D. del Pesco (1988); M. Birindelli (1980); R. Wittkower (1975, pp. 54 ff.); T. K. Kitao (1974, p. 24 ff.); H. Hibbard (1965, pp. 151 ff.); and H. Brauer and R. Wittkower (1931, vol. 1, pp. 69 ff., 56a ff., 161 ff.).

100. On its use for the Palazzo Massimi—a family who claimed venerable ancestry—by Baldassare Peruzzi (who used it often in any case), and on the early projects for the church of San Giovanni dei Fiorentini by Michelangelo, see J. S. Ackerman (1970, pp. 227 ff.). Ackerman also considers the "programmatic" use of the Tuscan by the Medici of Milan, particularly Giovanni Angelo (later Pope Pius IV) in the Palazzo Medici; in the tomb of his brother, the Marquis de Marignano, in Milan Cathedral (by Leone Leoni, with the help of Michelangelo); and in the unfinished Palazzo Medici as well as the Palazzo dei Giuresconsulti. All this has not been fully examined.

XII Order or Intercourse

1. The last few lines are a paraphrase of Hans Blumenberg's account of this phenomenon (1960, p. 10).

2. It could, some rhetoricians thought, be further confined to a single word, a verb or adjective deliberately applied out of place to imply a comparison that the reader or listener could supply, as when we speak of a ship ploughing the sea; so P. Fontanier (1977, pp. 99 ff.).

3. W. Benjamin (1972–1979, vol. 2, pp. 206 ff.); see also W. Menninghaus (1980, pp. 60 ff.).

4. Above, chapter 2, pp. 43 ff. Thomas Fuller, *The Holy State, The Profane State* 1642; rpt.; 1841), II.vii.73.

518

5. L. Mumford (1951); N. Pevsner ([1943] 1968, p. 15). On the two-term, one-word metaphor, see P. Fontanier (1977). Fontanier may have had the same relation to the European rhetorical tradition that Quatremère de Quincy had to the earlier *trattatisti*.

6. The term had been current in polemics and criticism at the beginning of the twentieth century. It is almost the key word of Hermann Muthesius' *Stilarchitektur und Baukunst* (1902); he praises early-nineteenth-century "classicism" for its *Sachlichkeit*. Even Heinrich Wölfflin (1921, p. 252) had spoken of the new linear German style at the beginning of the nineteenth century: "Die neue Linie kommt im Dienst einer neuen Sachlichkeit." This meant for Wölfflin no approximation, no seeking for general effects, but representation of the *Gestalt wie sie ist*.

However, the term became the label of a movement as a result of the exhibition called *Die Neue Sachlichkeit*, organized in the museum at Mannheim by G. F. Hartlaub in the spring of 1923 (see F. Schmalenbach 1940, pp. 161 ff.). Hartlaub saw the "movement" he was launching as consisting of two "wings": the "neo-classicists," such as Picasso, or Kay H. Nebel; and the "verists," such as Max Beckmann, Otto Dix. It is rather odd to find the little-known Nebel (1888–1953) classed here with Dix and Beckmann, never mind Picasso.

There was much talk about the "object," the "thing," at the at time. In 1922, Ilya Ehrenburg had launched the Russo-Franco-German periodical *Veshch/Object/Gegenstand*, in Berlin, and El Lissitzky, who was an editor of *Veshch*, also worked for "G". "G" stood for *Gestaltung*, another awkward word; indeed the review was called variously *Material zur elementaren / Zeitschrift für elementare Gestaltung*. *Gestalt* means both "form" and "shape," and *gestalten* is often translated as "shaping" or "designing"; however, Mies' credo sounded a different note from the Constructivist belief in the importance of the "designed" object—whether aeroplane or poem—as the organizing force of everyday life. His manifesto text accompanied the publication of his Bürohaus in "G", no. 1 (July 1923).

In all these manifestos and polemics may be heard some echo, however remote, not of *Verdinglichung*, "reification" (of which more later), but of Husserl's repeated call to approach the things themselves—which is the first step in the phenomenological method.

7. J. W. von Goethe (1948–1954, vol. 13, pp. 126 ff., 1124 f.). See also B. Schubert (1986, pp. 34 f., 50 ff.).

Of course Goethe (and later Morris) had to fight against the view set out by Kant (1922–1923, vol. 5, p. 378) that *Kunst*, fine art, which is free, had to be differentiated from *Handwerk*, which is paid for. Kant's *Handwerk* is therefore a *Lohnkunst*, or mercenary, paid-for art. Art and craftsmanship are therefore as different as play is from work.

Although Ruskin (and following him, Morris) developed analogous views in a very different climate, Goethe's influence—through Carlyle—whose doctrine of work as a sacred task runs through several of his writings (notably *Past and Present*) should not be discounted.

8. This first appeared in the eighth edition of 1832. Labrouste was asked by Hugo to read the proofs of this essay.

9. H. Huebsch, *In welchem Style sollen Wir bauen?* (1828).

10. G. Simmel (1992, pp. 416 ff.).

11. Although this may seem to echo Simmel's formulation, I suppose that Heidegger would deny any such association.

12. M. Heidegger (1975, pp. 102 ff.) also uses the word *alētheia* as a key to his interpretation of Heraclitus—even though the word appears only as an abstract noun in the surviving fragments (DK B 112). But see below, n. 22.

13. Heidegger, *Sein und Zeit*, sections 15 ff.

14. On *Gestalt* and *Gestell* see J. J. Kockelmans (1985, pp. 203 ff.). Heidegger himself outlines the distinction *Ding-Zeug-Werk* (1988, pp. 71 f.).

15. M. Heidegger (1988).

16. The details of van Gogh's shoe painting to which Heidegger referred had been studied in some detail by Meyer Schapiro in two essays of 1968 and 1994 (1994, pp. 138 ff.), who identified the paintings and the occasion when Heidegger might have seen any of them. His correspondence with Heidegger and Heidegger's partial retraction in a posthumous edition of the essay (M. Heidegger, 1988) do nothing to modify the original mismatch of the interpretation and the object. Nor do the elaborate commentaries of J. Derrida (1978, pp. 291 ff.), J. J. Kockelmans (1985, pp. 125 ff.), or G. Vattimo (1989, pp. 110 ff.).

17. M. Heidegger (1988), p. 37).

18. A. Schopenhauer (1972, vol. 2, p. 464). Sadly it recalls Octavio Paz' view of Jean-Paul Sartre's opinions about art: "I realized . . . that the object of his admiration was not the poems which Mallarmé actually wrote but his project of absolute poetry, the Book he never made. Despite what his philosophy declares, Sartre always preferred shadows to realities" (1986, p. 38).

19. G. Benn (1989, vol. 4.2, pp. 152, 135).

20. See above, chapter 9, pp. 237 f.

21. The remarkable thing about the site is the rock face of Parnassus above it; the valley of the Pleistos, which it overlooks, is, in any case, a carpet of olive woods stretching down to the Gulf of Itea.

22. Heidegger's repeated insistence that German was the only language into which Greek could adequately be translated is part of this attitude. The importance of truth as unconcealment (above, n. 12), translating the Greek *alētheia*, is fundamental to much of what he wrote; his etymology has, however, been completely rejected by P. Friedländer (1958–1969, vol. 1, pp. 221 ff.).

23. See above, chapter 1, pp. 16 ff., and chapter 6, n. 4.

24. *The Flowering of Greece*, painted in 1825, is now known only from a copy of 1836 by Wilhelm Ahlhorn in the Nationalgalerie, Berlin.

25. K. F. Schinkel (1982, pp. 29–30, 35). An earlier project (pl. 1) had plain piers; the

executed Doric one had victories articulating the cornice instead of triglyphs. At the time of the design, Unter den Linden ended at the west end into the Tiergarten with the "academically" correct (though Doric with bases) Brandenburg Gate, which Carl Gotthard Langhans designed 1788–1789. Mies' project is described in F. Schulze (1985, pp. 178 f.).

26. The anecdote was told by Mies himself to Peter Blake, who reported it in his *Four Great Makers of Modern Architecture* (1963, p. 101). On being asked why he did not consider himself influenced by the De Stijl group, specifically Theo van Doesburg, Mies answered: "Van Doesburg saw these drawings of the office building. I explained it to him and I said 'This is skin and bones architecture.' After that he called me an anatomical architect." Mies adds: "I liked van Doesburg, but not as though he knew very much about architecture."

27. It must be obvious to most of my readers that without Heidegger's formulation of a hermeneutic method this book could not have taken the shape it did.

28. It is a kind of neo-Marxian argument, which some who advance it would call neo-Nietzschean; it even smells a little of Lukács. But see M. Cacciari (1988, pp. 9 ff.).

29. See M. Cacciari (1993, pp. 199 ff.).

30. G. Simmel (1978, pp. 256 ff.).

31. Explicitly, in F. Dal Co and M. Tafuri (1976, pp. 408 ff.).

32. I again paraphrase: L. Ferry (1990, p. 15) and M. Heidegger (1988, p. 33).

33. J. Benda (1927, pp. 141 ff.; [1918] 1947, pp. 23 ff.). An interesting attack on Benda's exaggerations in this matter appears in C. Mauriac (1945).

34. J. Habermas (1990, p. 124), interpreting F. Nietzsche (1975–1980, 6.2, pp. 364 ff.). Though in fact, since Nietzsche is concerned here to exploit the weaknesses of Schopenhauer's sanitized and unsexed aesthetic as a version of Kantian "disinterest," and concerned with its corrupting influence on Wagner, the major point against Kant gets a little lost. On this passage, see the comments of G. Agamben (1970, pp. 9 ff.). For the neoscholastic commentaries, see J. Maritain (1932, pp. 162 ff.) and E. Gilson (1965, pp. 10 ff., 116 ff.).

35. Pl. *Symp.* 205b ff., *Tim.* 47e ff. On this passage and the concept of *mimēsis,* see M. Kardaun (1993, pp. 48 ff.).

36. I do not wish to gloss over the difficulty of this notoriously subtle notion. However, for all the problems it poses, most commentators agree that the relation that the Pythagoreans considered existed between numbers and the physical manifestations making up our world was analogous to Plato's conception of the way in which Forms were reflected or even expressed in phenomena. That Aristotle was here (*Met.* I.6.iii, 987b) taking his master to task does not concern me.

37. Here are Vitruvius' *expressa verba:* "Ita e generibus duobus . . . tertium genus in operibus est procreatum" (IV.1.iii). It may certainly be translated as "That is how the third gender in building was procreated out of the other two genders." I have already insisted on the almost exclusively sexual use of *procreare* above, chapter 5, n. 27.

38. I. Kant (1922–1923, vol. 5, pp. 382 f.). The paragraphs quoted are first #47, then #46. The second paragraph continues "Denn eine jede Kunst setzt Regeln voraus, and there follow the three principles I quote further. Kant may be deliberately echoing Diderot's (then) much-read article "Génie" in the Great Encyclopaedia: "Le goût est souvent séparé du génie. Le génie est un pur don de la nature . . ." (vol. 4, p. 582).

39. Gabriel (de) Tarde ([1895] 1979, pp. 15 ff.). While he attached no importance to the particule "de," his son Guillaume did: hence they are sometimes differently cited in bibliographies. Tarde carefully distinguished physical resemblance due to periodic recurrence or vibratory activity from biological (or as he put it, "corporeal") similarity or homology, which is genetically transmitted. All socially established similarity is due to various forms of imitation—custom, fashion, instruction; it may be naive or reflective.

40. G. de Tarde (1979, pp. 59 ff.). Although the Perrot and Chipiez *Histoire de l'Art* began publication in 1882, the material Tarde uses here seems drawn from C. Chipiez (1876).

41. J. Piaget (1926, pp. 11 [on *echolalia*], 41). On specularity and the development of Piagetian linguistics, see C. de Lemos in L. Camaioni and C. de Lemos (1985, pp. 23 ff.).

42. See Kent Kraft, "Mind's Mirrors," and Karl Eckhardt, "Concepts of Mimesis in French and German Anthropological theory," both in R. Bogue and M. Spariosu (1994, pp. 39 ff., 67 ff.). But see also S. Agacinski, "Découpages du *Tractatus,*" in J. Derrida (1978), and a discussion of her view in M. Spariosu (1982, pp. 54 ff.). See A. Gehlen (1993, 3. 1, pp. 467 ff.).; H. Plessner (1980, vol. 7, pp. 389 ff., 446 ff.; vol. 8, pp. 174 ff.).

43. J. Huizinga (1938; many editions).

44. Schiller's assertion is about the most quoted sentence from *Aesthetic Letters;* it is in letter 15. On Spencer's use of it, see M. Spariosu (1982, pp. 29 ff.).

45. R. Girard (1978, p. 93; 1977, p. 308); on which see P. Lacoue-Labarthe (1989, pp. 102 ff.).

46. So, for instance, in G. Lukács, *Die Eigenart des Ästhetischen* (1963–1974, vol. 12.2, pp. 330 ff.).

47. Though in fact Lukács (1963–1974, vol. 12.2, pp. 402 ff.) chooses Schelling as his adversary here. It is notable that in his earlier writing on art and aesthetics (see vols. 16, 17, which collect his papers written between 1912 and 1918) mimesis seems not come up at all, while in his last major work, *Die Eigenart des Ästhetischen,* it becomes the prime feature of the creative process (vol. 12. XII, 1, pp. 352–852; vol. 12.2, pp. 330–574) if only because of the space devoted to it. On Lukács as a moralist and a literary critic, see the fair if unfriendly estimate by S. Sontag (1967, pp. 82 ff.). Lukács would have resented my identification of his position with Social Realism, which is associated with Zhdanovite hangovers; however, it does not seem that his Critical Realism is different enough from it.

48. "Aber ohne Weltanschaung gibt es keine Komposition"; See Lukács, "Erzählen oder Beschreiben" in R. Brinkmann (1969, p. 69). This article was written in 1936, just before the Moscow trials.

On the problem of reification in this context, see L. Zuidervaart (1991, pp. 263 ff.). For Lukács' account of his own use of the term "reification" see, still, his *History and Class Consciousness* (1971, pp. 83 ff.). On the links between Lukács and Sedlmayr (who are discussed with Max Nordau and Oswald Spengler), see B. Wyss (1989, pp. 315 ff.).

49. It is characteristic that Le Corbusier is only mentioned once, by Lukács, and that no other twentieth-century architects appear in his book; his argument is almost entirely intertextual. A number of filmmakers (Chaplin, Eisenstein) are discussed, but no other visual artists show up. But then there are other exclusions: Bertolt Brecht had already made fun of Lukács' inability to get all of literature into his Stalinist portmanteau (quoted in B. Wyss, 1989, p. 328). Matisse and Picasso are mentioned twice: once to be dismissed for their blind-alley experiments, and once for Picasso's bitchy remark about Matisse's white-tiled Chapel for the Dominican Nuns at Vence—that "he liked it well enough, but which was the bathroom?" But then he seems to have no occasion to mention any Social Realist architect at all.

50. G. Lukács (1963, vol. 1, pp. 352 ff.).

51. On the more recent appeals to the concept of mimesis, see J. Habermas (1990, pp. 68 f., 225 ff.). Adorno's later ideas on this matter are discussed by C. Eichel (1993, pp. 35 f.). But see T. W. Adorno (1966, pp. 151 f., 178 ff.); T. W.

Adorno and M. Horkheimer in T. W. Adorno (1984b, pp. 204 ff.). Adorno and Horkheimer had warned their readers explicitly about confusing mimicry and mimesis.

Although the various classifications of the arts—of which the most popular was probably that of the Abbé Charles Batteux—counted architecture and rhetoric as the arts in which the imitative requirement was subsumed to utility, the idea of architecture as an art of imitation had its defenders even during the time of its eclipse; see notably Antoine-Chrysostome Quatremère de Quincy ([1823] 1980). On the "system of the arts" in the eighteenth century, see W. Tatarkiewicz (1980, pp. 60 ff.).

52. T. W. Adorno (1966, pp. 106 f.).

53. T. W. Adorno (1966, p. 124).

54. T. W. Adorno (1984a, p. 66, pp. 453 f.). This is a notable departure from his former identification with Loos, particularly in his view of ornament: see C. Eichel (1993, pp. 230 f.).

55. I have summarized (without caricature, I hope) some of the arguments set out by L. Zuidervaart (1991, pp. 101 ff.).

56. Ron Siskolne, vice president for design and development of Olympia and York, quoted by M. S. Larson (1993, p. 132).

57. What he actually wrote was "Das Barbarische ist das Buchstabliche" (T. W. Adorno, 1973, p. 97).

58. "[A]nyone who thinks that art can no longer be adequately grasped using Greek concepts is not thinking in a sufficiently Greek way": H.-G. Gadamer (1986b, p. 122).

Abbreviations and Ancient Texts

AA   *Archäologischer Anzeiger* (Berlin)

AAQ   *The Architectural Association Quarterly* (London)

AB   *The Art Bulletin* (Bulletin of the College Art Association of America, Boston)

AC   *L'Antiquité Classique* (Brussels)

Ad Her.   *Ad C. Herennium de Ratione Dicendi* (attributed to Cicero)
Ed. and trans. Harry Caplan, LCL

Ael.   Claudius Aelianus
VH   *De Varia Historia*
Nat. An.   *De Natura Animalium*
On Characteristics of Animals, trans. A. F. Scholfield, LCL

Aesch.   Aeschylus
Works, ed. Eduard Fraenkel, 3 vols. (Oxford, 1950)
Ed. G. Thomson and W. Headlam, 2 vols. (Prague, 1966)
Ag.   *Agamemnon*
Eum.   *Eumenides*
Fr.   *Fragmenta*
Pers.   *Persae*
The Persians, trans. J. Lembke and C. John Herington (New York, 1981)
Prom. Vinc.   *Prometheus Bound*

Aet.   Aetius

AH   *Art History* (London)

AJ   *The Antiquaries Journal* (London)

AJA   *American Journal of Archeology* (New York)

AJP   *American Journal of Philology* (Baltimore)

AK   *Antike Kunst* (Bern)

Ann. Rev. Anthr.   *Annual Review of Anthropology* (Palo Alto, Calif.)

AnSt   *Anatolian Studies* (British Institute of Archaeology at Ankara, London)

AntDenk   *Antike Denkmäler* (Berlin)

AntP   *Antike Plastik* (Berlin)

Apol.   Apollodorus (the Grammarian)
Trans. J. G. Frazer, LCL
Bibl.   *Bibliotheka* (*The Library*)
Epit.   *Epitome*

Apol. Ath.   "Apollodorus of Athens"
Chron.   *Chronika* (*Chronicles*)
Peri Th.   *Peri Theon* (*About the Gods*)

Appian   Appian of Alexandria
Bel.Civ.   *Bella Civilia*

APS   *Proceedings of the American Philosophical Society* (Philadelphia)

Apul.   Apuleius of Medaura
Met.   *Metamorphoseon Libri XI*
The Golden Ass, being the Metamorphoses of Lucius Apuleius, trans. W. Adlington (1566); rev. S. Gaselee, LCL
Ed. and trans. J. Arthur Hanson, LCL

AQ   *The Art Quarterly* (New York)

Aq.   St. Thomas Aquinas
Ep. Reg.   *Epistola de Substantiis Separatis ad Fratrem Reginaldum*
In Eth. Nic.   *In Ethica ad Nicomachum Expositio*
De Reg. Prin.   *De Regimine Principum*
Sum. Th.   *Summa Theologica*

ArchCl   *Archeologia Classica* (Rome)

ArchHom   *Archeologia Homerica* (Göttingen)

Arist.   Aristotle
De An.   *De Anima*
Trans. W. S. Hett, LCL
Ath. Pol.   *Athenion Politeia*
Eth. Nic.   *Nicomachean Ethics*
Trans. David Ross (Oxford, 1954)
Met.   *Metaphysics*

Part. An.   *De Partibus Animalium*
    Trans. A. L. Peck, LCL
Phys.   *Physics*
    Trans. R. Hope (Nebraska, 1961)
Poet.   *Poetics*
    S. H. Butcher (London, 1907)
    I. Bywater (Oxford, [1920] 1959)
    W. H. Fyfe, LCL
    Gerald F. Else (Leiden, 1957)
Pol.   *Politics*
Rhet.   *Rhetoric*
    Trans. J. H. Freese, LCL

Aristoph. Aristophanes
Achrn.   *Acharnians*
Fr.   *Fragmenta*
    Ed. A. Nauck (Halle, 1848)
Nub.   *Nubes* (*The Clouds*)

Arrian. Flavius Arrianus
An.   *Anabasis*

Artem. Artemidorus Daldianus
Onei.   *Oneirokritika*

Athan. Athanasios, Bishop of Alexandria
Adv. Gr.   *Oratio contra Gentes*

Athen. Athenaeus of Naucratis
Deip.   *Deipnosophistai*

Ath. gr. Athenagoras
    Works, ed. M. Dodds, G. Reith, and B. P. Pratten (Edinburgh, 1868)
Pro Christ.
    Apologia peri Christianon (*Plea for Christians*)

Aul. Gel Aulus Gellius, *Noctes Atticae*
    Trans. John C. Rolfe, LCL

Aurel. Aug.
    St. Augustine of Hippo
De Civ. Dei
    De Civitate Dei
    The City of God against the Pagans, trans. George E. MacCracken (Cambridge, Mass., 1957–1972)
Enarr. in Ps.
    Enarrationes in Psalmos
De Gen.   *De Genesi ad Litteram Imperfectus Liber*

AW   *Antike Welt* (Feldmeilen, Switzerland)

Bacch. Bacchylides
    In *Lyra Graeca,* ed. J. M. Edmonds, LCL

BCH   *Bulletin de Correspondance Héllenique* (Ecole Française d'Athènes, Athens and Paris)

Ber. Sächs. Akad. Wiss.
    Berichte der Sächsischen Akademie der Wissenschaften (Leipzig)

BIAO   *Bulletin, Institut Français d'Archéologie Orientale* (Cairo)

BICS   *Bulletin, Institute of Classical Studies* (London)

BM   British Museum, London

Boeth. Anicius Manlius Severinus Boethius
De Mus.   *De Musica*
    On Harmony, ed. Oscar Paul (Leipzig, 1872)
    Fundamentals of Music, trans. Calvin M. Bower (New Haven, 1989)
Topic. Aristot. Interp.
    Interpretatio Topicorum Aristotelis

Brit. Sch. Ann.
    Annual of the British School at Athens (London)

BSA Arch. Rep.
    Archaeological Reports of the British School at Athens (London)

CA   *Current Anthropology* (Chicago)

CAH   *Cambridge Ancient History,* ed. I. E. S. Edwards, C. J. Gadd, and N. G. L. Hammond, 3rd ed. (Cambridge, 1970–1991)

Cal. Callimachus, *Hymns and Epigrams,* trans. G. R. and A. M. Mair, LCL

Cato Marcus Porcius Cato
De A. C.   *De Agricultura*

C. Dio	Dio Cassius Cocceianus
Rom.	*Romaika Istoria*
	Roman History, trans. E. Cary, 9 vols., LCL
Cens.	Censorinus, *De Die Natali*
CIA	*Corpus Inscriptionum Atticarum* (Berlin, 1825–)
Cic.	Marcus Tullius Cicero
Acad.	*Academicae Quaestiones*
De Amic.	*de Amicitia*
Brut.	*Ad Brutum*
De Div.	*De Divinatione*
Fam.	*Epistulae ad Familiares*
De Fin.	*De Finibus*
De Inv.	*De Inventione*
	Trans. H. M. Hubbell, LCL
De Leg.	*De Legibus*
De Nat. Deo.	
	De Natura Deorum
	Ed. Harris Rackham, LCL
De Off.	*De Officiis*
Or.	*Orator*
Tusc.	*Tusculanae Disputationes*
De Univ.	*De Universo*
In Verr.	*Verrine Orations*
CIG	*Corpus Inscriptionum Graecarum* (Berlin, 1825–1877)
Clem. Alex.	
	St. Clement of Alexandria
	Works, trans. W. Wilson (Edinburgh, 1867)
Protr.	*Protreptikos pros Ellinas* (*Exhortation to the Greeks*)
	Trans. G. W. Butterworth, LCL
Strom.	*Stromata*
CM	*Classical Museum* (London)
CNR	Consiglio Nazionale delle Ricerche
CNRS	Centre National de la Recherche Scientifique
Colum.	Lucius Junius Moderatus Columella
De R. R.	*De Re Rustica*
	Trans. by Harrison Boyd Ash, E. S. Forster, and E. H. Heffner, 3 vols. LCL

Corp. Herm.	
	Hermes Trismegistus, *Corpus Hermeticum*
	Ed. and trans. A. D. Nock and A.-J. Festugière, 4 vols. (Paris, 1945)
CQ	*Classical Quarterly* (Oxford)
Crit. Arte	*Critica d'Arte* (Florence)
Cusa	Nicholas of Cusa
De Doc. Ign.	
	De Docta Ignorantia (*Of Learned Ignorance*)
	Ed. G. M. Heron (London, 1954)
CSEL	*Corpus Scriptorum Ecclesiasticorum Latinorum* (Vienna, 1866)
DBI	*Dizionario Biografico degli Italiani* (Rome, 1960–)
Dio Chrys.	
	Dio Chrysostomos
Or.	*Orationes*
Diod. Sic.	
	Diodorus Siculus, *Bibliotheca Historica*
	Trans. C. H. Oldfather et al., LCL
Dion. Hal.	
	Dionysius of Halicarnassus
Ant. Rom.	
	Antiquitates Romanae
Isoc.	*De Isocrate*
Diog. Laert.	
	Diogenes Laertius
In Pol.	*Life of Polemo*
Diosc.	Discorides Pedanius (of Anazarbus)
DK	*Fragmente der Vorsokratiker,* by H. Diels and W. Kranz, rev. 6th ed. (Berlin, 1985)
DNB	*Dictionary of National Biography,* ed. Leslie Stephen and Sidney Lee (London, 1885–1901); supplements 1 (1901), 2 (1970); new series (1901–)

DS   *Dictionnaire des Antiquités Grecques et Romaines,* ed. Charles Daremberg and Edmond Saglio (Paris, 1877–1919)

EA   *Ephemeris Archaiologike* (Athens)

EAA   *Enciclopedia dell'Arte Antica, Classica e Orientale* (Rome, 1958–1966)

Etym. Magn.
   Etymologicon Magnum (The "Great Dictionary")

Euc.   Euclid (of Megara)
   The Elements of Geometry, ed. I. Todhunter, intro. T. L. Heath, 2nd ed. (London, 1933)
   The Elements of Geometrie . . . of Euclid of Megara . . . translated into the English Toung by H. Billingsley . . . with a very fruitful preface made by M. J. Dee (London, 1570)

Eur.   Euripides
   Works, trans. Arthur S. Way, 4 vols., LCL
Androm.   *Andromache*
Bac.   *Bacchae*
Hec.   *Hecuba*
Ion   *Ion*
   Trans. H. D. (London, 1937)
Iph. Taur. *Iphigenia in Tauris*
Krit.   *The Cretans*
   In Papyri, ed. D. L. Page, LCL
Or.   *Orestes*
Phoen.   *Phoenissae*

Eus.   Eusebius of Caesarea
Chron.   *Chronica*
Prep. Ev.   *De Evangelica Praeparatione*
   Trans. K. Lake, 2 vols., LCL

Eustat.   Eustatius (of Salonica)
In Dion. Per
   Paraphrase of Dionysos Pereigetes
In Il.   *ad Iliadem*

FD   *Fouilles de Delphes,* Ecole Française d'Athènes (Paris, 1902–)

Fes.   Sextus Pompeius Festus, *De Verborum Significatione*

   Ed. W. M. Lindsay (Leipzig, 1913)
   Ed. C. O. Mueller (Hildesheim, 1975)

FHG   *Fragmenta Historicorum Graecorum* (Paris, 1841–1851)

Firm. Mat.
   Julius Firmicus Maternus
De Error.   *De Errore Profanarum Religionum*

Gal.   Galen of Pergamon
De Plac.   *De Placitis Hippocratis et Platonis*
   Ed. I. Mueller (Leipzig, 1874)

GR&BS   *Greek, Roman and Byzantine Studies* (Durham, N.C.)

Gr. Diz.   *Grande Dizionario della Lingua Italiana* (Turin, 1961–)

Greg. Magn.
   Gregorius Magnus Papa
Hom. Ev.   *Homilia in Evangelium*

Heliod.   Heliodorus of Emesa, *Æthiopica*
   Englished by T. Underdowne (1587); rev. ed. Charles Whibley (London, 1895)

Her.   Herodotus of Halicarnassus, *Histories*
   Trans. A. D. Godley, LCL
   Trans. A. de Selincourt (Harmondsworth, 1959)
   Trans. H. Carter (New York, 1958)

Heraclid. Pont.
   Heraclides Ponticus

Hes.   Hesiod *Theogonia; Opera et Dies,* ed. R. Merkelbach and M. L. West (Oxford, 1966)
   Trans. Hugh Evelyn-White, LCL
Op. D.   *Works and Days*
Theog.   *The Theogony*
   Trans. Richard Lattimore (Ann Arbor, 1959)

Hesy.   Hesychius of Alexandria, *Synagoge (Lexicon)*
   Ed. K. Latte, 2 vols. (Copenhagen, 1953–1966)

HF Hjalmar Frisk, *Griechisches Etymologisches Wörterbuch* (Heidelberg, 1960–)

Hier Hieronymus (St. Jerome)
In Ep. Eph.
 Commentary on St. Paul's Epistle to the Ephesians

Hip. Hippocrates
 Oeuvres Complètes, 4 vols., ed. H. Roger and E. Littré (Paris, 1932–1934)
Aer. *De Temporibus Aquis et Locis*
De Foem. St.
 De Foeminu Sterilitate (About the Sterility of Women)

Hist. Aug. Scr.
 Scriptores Historiae Augustae
 Trans. David Magie, LCL

HJ *Historisches Jahrbuch* (Munich)

Hom. Homer
Il. *Iliad*
Od. *Odyssey*
Hym. *Hymnes*
 Trans. J. Humbert (Paris, 1936)
 Ed. M. Cantilena (Rome, 1982)
Hym. Dem.
 Hymn to Demeter
Hym. Herm.
 Hymn to Hermes
Hym. Pyth.
 Hymn to Pythian Apollo

Hor. Quintus Horatius Flaccus
 Satires, with Epistles and Ars Poetica, trans. H. Rushton Fairclough, LCL
Ars Poet. *The Art of Poetry*
Carm. *Carmina*
Ep. *Epistles*
Sat. *Satires*

Hyg. Hyginus
Fab. *Fabulae*

HZ *Historische Zeitschrift* (Munich)

Iamb. Iamblichus of Chalcis
De Myst. *De Mysteriis*
V. Pyth. *De Vita Pythagorica*
 Ed. A. Nauck (St. Petersburg, 1884)

ICCA *International Congress of Classical Archeology*

IG *Inscriptiones Graecae* (Berlin, 1873)

IL *Inscriptiones Latinae* (Berlin, 1893)

Ios. Iosephus Flavius
 Works, ed. R. L'Estrange, 3 vols. (1709)
In Ap. *In Apionem*

Isoc. Isocrates
Bus. *Busiris*

Is. Sev. Isidore of Seville
Etym. *Etymologiae sive Origines*

Ist. For. *Istanbuler Forschungen* (Berlin)

JDAI *Jahrbuch des Deutschen Archäologischen Instituts* (Berlin)

JEA *Journal of Egyptian Archeology* (London)

JGH *Journal of Garden History* (London)

JHI *Journal of the History of Ideas* (Philadelphia)

JHS *Journal of Hellenic Studies* (London)

JOAIW *Jahreshefte des Oesterreichischen Archaeologischen Instituts in Wien*

Jordanes
De Or. *Getarum sive Gothorum Origine et Rebus Gestis*

JRA *Journal of Roman Archaeology* (Ann Arbor, Mich.)

JRS *Journal of Roman Studies* (London)

JSAH *Journal of the Society of Architectural Historians* (Philadelphia)

JWCI *Journal of the Warburg and Courtauld Institutes* (London)

JWI   *Journal of the Warburg Institute* (London; later *JWCI*)

LÄ   *Lexikon der Aegyptologie*, ed. W. Helck and E. Otto (Wiesbaden, 1972–1992)

Lact. Firm.
          L. Caecilius Lactantius Firmianus
Inst.   *Divinarum Institutionum adversus Gentes Libri Septem*
De Op. D.
          De Opificio Dei vel Formatione Hominis

LCL   Loeb Classical Library

Livy   Titus Livius, *Ab Urbe Condita*
          Ed. B. O. Foster, 14 vols., LCL

Long. Longinus Dionysius, *De Sublimitate*

LSJ   *A Greek-English Lexicon*, comp. Henry George Liddell and Robert Scott, rev. Sir Henry Stuart Jones, 9th ed. (Oxford, 1990)

Luc.   Lucian of Samosata
          Ed. A. M. Harmon and K. Kilburn, LCL
          Ed. M. D. McLeod, Oxford Classical Texts (Oxford, 1974)
          Ed. E. Chambry (1933)
De Dea Syr.
          De Dea Syria
Dial. Het.
          Hetairikoi Dialogoi
Pro Ikon. In Defence of Portraits
Prom.   *Prometheus or the Caucasus*
Salt.   *De Saltatione*

Lucr. Titus Lucretius Carus, *De Rerum Natura*
          Ed. and trans. H. A. J. Munro (Cambridge, 1886)
          Trans. W. H. D. Rouse (Cambridge, Mass., 1975)

MA   *Mediterranean Archaeology* (Sydney)

MAAR   *Memoirs. American Academy at Rome*

Macr. Ambrosius Theodorus Macrobius
Sat.   *Saturnalia*
In S. S. *In Somnium Scipionis*

MAL   *Monumenti Antichi della (Reale) Accademia dei Lincei (Rome)*

MDAI   *Mitteilungen des Deutschen Archäologischen Instituts*
MDAIA   Abt. Athen (Berlin)
MDAIK   Abt. Kairo (Mainz)
MDAIR   Abt. Rom

Mishnah *The Mishnah*, ed. Herbert Danby (Oxford, 1933)
Orl.   Tractate *Orlah*
Zar.   Tractate *Abodah Zarah*

M. Manilius
          Marcus Manilius, *Astronomica*
          Ed. Jacob van Wageningen (Leipzig, 1905)

M. M. Felix
          Marcus Minucius Felix, *Octavius*

MRS   Mediaeval and Renaissance Studies (London)

MQ   *The Musical Quarterly* (New York)

Nic. Dam.
          Nicholas of Damascus

NS   *Notizie degli Scavi di Antichità* (Rome)

Non.   Nonnos of Panopolis, *Dionysiaca*
          Trans. W. H. D. Rouse, LCL

OA   *Opuscula Atheniensia* (Acta Instituti Atheniensis Regni Sueciae, Lund)

Ovid   P. Ovidius Naso
          Works, ed. P. Burmann (Amsterdam, 1727)
Met.   *Metamorphoses*
          Trans. F. J. Miller, LCL
Pont.   *Ex Ponto*
          Trans. A. L. Wheeler, LCL

PAPA   *Proceedings of the American Philological Association* (Cleveland)

Paus. Pausanias, *Description of Greece*
          Ed. J. G. Frazer, 4 vols. (London, 1913)

Ed. W. H. Jones and R. Wycherley, 4 vols., LCL

Ed. H. Hitzig and H. Bluemner, 3 vols. in 6 (Leipzig, 1896–1910)

See also M. de G. Verrall and J. E. Harrison

PC   Pierre Chantraine, *Dictionnaire Etymologique de La Langue Grecque: Histoire des Mots* (Paris, 1968–1980)

Pers.   Aulus Persius Flaccus
Sat.   Satires

PG   *Patrologiae Cursus Completus,* accurante J.-P. Migne, Series Graeca (Paris, 1856–1861)

Phil.   Philostratus the Elder
Icon.   *Eikones (Imagines)*
Trans. Arthur Fairbanks, LCL

Phil. Jun. Flavius Philostratus
V. Apol. Thy.
Life of Apollonius of Tyana

Philo Byz.
Philo Byzantinus
Mech. Syn.
Mechanicae Syntaxis

Philo Jud.
Philo (of Alexandria or Judaeus)
Works, ed. Roger Arnaldez (Paris, 1961)
De Opif.   *De Opificio Mundi*

Phot.   Photius, Patriarch of Constantinople
Bibliotheca or *Myriobiblon*
Ed. Immanuel Bekker (Berlin, 1824–1825)
Trans. Nigel Wilson and Claudio Bevegni (Milan, 1992)

Pind.   Pindar
Works ed. Sir John Sandys, LCL
Ed. Aimé Puech, 4 vols. (Paris, 1922–1923)
Nem.   *The Nemean Odes*
Ol.   *The Olympian Odes*
Pae.   *Paeans*
Pyth.   *The Pythian Odes*

PL   *Patrologiae Cursus Completus,* accurante J.-P. Migne, Series Latina (Paris, 1841–1902)

Pl.   Plato
The Dialogues, ed. Edith Hamilton and Huntington Cairns, Bollingen Foundation (New York, 1961)
Charm.   *Charmides*
Crat.   *Cratylus*
Crit.   *Critias*
Ep.   *Epistles*
Euthy.   *Euthydemus*
Gorg.   *Gorgias*
Hipp. maj.
Hippias major
Leg.   *The Laws*
Trans. R. G. Bury, LCL
Parm.   *Parmenides*
Phaedr.   *Phaedrus*
Phil.   *Philebus*
Polit.   *Politicus*
Prot.   *Protagoras*
Rep.   *The Republic*
Trans. F. M. Cornford (Oxford, 1941)
Soph.   *The Sophist*
A. Diès (Paris, 1925)
Symp.   *The Symposium*
Tht.   *Theaetetus*
Tim.   *Timaeus*
Ed. R. G. Bury, LCL
Ed. A. Rivaud (Paris, 1925)
Trans. F. M. Cornford, ed. Oskar Piest (New York, 1937)

PLG   *Poetae Lyrici Graeci,* ed. T. Bergk, LCL

Pln.   Pliny the Elder (Caius Plinius Secundus)
NH   *Naturalis Historiae Libri XXXVII*
Ed. and trans. Jean Hardouin (Paris, 1685)
Trans. H. Rackham, 10 vols. LCL

Pln.   Pliny the Younger (Caius Plinius Caecilius Secundus)
Ep.   *Epistulae*

Plut.   Plutarch of Chaironea
Plutarchi Chaeronensis, Quae Exstant Omnia, cum Latina Interpretatione Hermanni Cruserii, Gulielmi Xylandri, et

Doctorum Virorum Notis (Frankfurt, 1599)

De An. Pr.

    De Animae Procreatione

Apoth. Lac.

    Apophthegmata Laconica

De Aud. *De Recta Ratione Audiendi*

Consol. ad Apoll.

    Consolatio ad Apollonium

De Def. Or.

    De Defectu Oraculorum

De E Delph.

    De E apud Delphos

Es. Car. *De Esu Carnium*

De Gl. Ath.

    De Gloria Atheniensium

De Is. et Os.

    De Iside et Osiride

Plac. Phil

    De Placitis Philosophorum

De Prof. Virt.

    De Profectibus in Virtute

Quaest. Con.

    Quaestionum Convivalium

Quaest. Gr.

    Quaestiones Graecae

Quaest. Rom.

    Quaestiones Romanae

Reg. et Imp. Apoth.

    Apophthegmata Regum et Imperatorum

Symp. *Symposion*

De Tranq. An

    De Tranquillitate Animae

De Trans. An.

    De Transitu Animae

V. *Vitae (Parallel Lives)*

Arist. *Aristides*

Lyc. *Lycurgus*

Per. *Pericles*

Publ. *Publicola*

Sol. *Solon*

Them. *Themistocles*

Thes. *Theseus*

PM Sir Arthur Evans, *The Palace of Minos at Knossos*, 4 vols. (London, 1921–1936)

Polyb. Polybius of Megalopolis, *The History*

Porph. Porphyry

De Antr. Nymph.

    De Antro Nympharum

PQ *Philological Quarterly* (Iowa City)

Prop. Sextus Aurelius Propertius

Ptol. Ptolemy (Claudius Ptolemaeus)

Harm. *Harmonics*

    Ed. J. Düring (Göteborg, 1930)

PW *Paulys Real Encyclopädie der Klassischen Altertumswissenschaft*, ed. A. F. von Pauly, rev. G. Wissowa et al. (Munich, 1894–1963)

Quin. M. Fabius Quintilianus, *Institutiones Oratoriae Libri XII*

    Trans. Harold Butler, 4 vols., LCL

RA *Revue de l'Art* (Paris)

REG *Revue des Etudes Grecques* (Association pour l'Encouragement des Études Grecques, Paris)

Rend. *Rendiconti* (Atti della Pontificia Accademia Romana di Archeologia, 3rd ser., Rome)

Rev. Arch.

    Revue Archéologique (Paris)

Rev. Bib. *Revue Biblique* (Ecole Pratique d'Etudes Bibliques, Jerusalem and Paris)

Rev. Hist. Sci.

    Revue d'Histoire des Sciences (Paris)

RhM *Rheinishes Museum für Philologie* (Frankfurt A/M)

Riv. Ist. Arch.

    Rivista del (Reale) Istituto Nazionale d'Archeologia e Storia dell'Arte (Rome)

SASAE Supplements: *Annales du Service des Antiquités de l'Egypte*

SE *Studi Etruschi* (Florence)

Serv.   M. Servius, *In Vergilium*

Sext. Emp.

Sextus Empiricus
Adv. Math.
Adversus Mathematicos

Soph.   Sophocles
Oed. Col.

Oedipus at Colonus
Ed. and trans. Hugh Lloyd-Jones, LCL

StA   *Studia Archaeologica* (Rome)

Stat.   Publius Papinius Statius
Theb.   *Thebaid*

Steph. Byz.

Stephen of Byzantium, *De Urbibus*

St MSR   *Studi e Materiali di Storia delle Religioni* (Bari)

Str.   Strabo, *Geographica (The Geography)*

Tac.   (P.) Cornelius Tacitus
Ann.   *Annales*
Hist.   *Historiae*

TAD   *Turk Arkeoloji Dergisi* (Ankara)

TB   *Allgemeines Lexikon der Bildenden Künstler von der Antike bis zur Gegenwart,* by Ulrich Thieme and Felix Becker (Leipzig, 1907–1950)

TGF   *Tragicorum Graecorum Fragmenta,* ed. A. Nauck (Leipzig, 1889)

Theo.   Theognis
Ed. J. M. Edmonds, in *Elegy and Iambus,* LCL

Theoph.   Theophrastus
Char.   *Characters*
Ed. J. M. Edmonds, LCL
De C. Pl.   *De Causis Plantarum*

Thuc.   Thucydides, *History of the Peloponnesian War*
Trans. Thomas Hobbes (1629)

Tzetz.   Tzetzes
Chil.   *Historiarum Variarum Chiliades*

Val. Max. Valerius Maximus, *Factorum ac Dictorum Memorabilium, Libri IX*

Varro   Marcus Terentius Varro
L.L.   *De Lingua Latina*
Trans. Roland G. Kent, 2 vols., LCL
Trans. Jacques Heurgon (Paris, 1978)
R.R.   *De Re Rustica*

Vel. Pat.   Gaius Velleius Paterculus, *Historia Romana*

Ver. Aur.   *The Golden Verses of Pythagoras*

Vet. Sch.   *Vetus Scholium*

Virg.   P. Virgilius Maro
Aen.   *Aeneid*
Ecl.   *Eclogae*
Geor.   *Georgica*

Vitr.   M. Pollio Vitruvius, *De Architectura Libri X*

Editions used:
Latin only:

De architectura libri decem, ed. Giovanni Giocondo [Iocundus] (Florence, 1513)
Ed. Joannes de Laet (Amsterdam, 1649)
Ed. Simone Stratico ex notis Ioannis Poleni, 8 vols. (Udine, 1830)
Ed. ab Aloisio Marinio [Luigi Marini], 4 vols. (Rome, 1836)
Ed. Valentinus Rose (Leipzig, 1899)

Latin and Italian, or Italian only:

De architectura libri dece tr. de latino in vulgare, ed. Cesare Cesariano (Como, 1521)
I Dieci Libri dell'Architettura, ed. Daniele Barbaro (Venice, [1556] 1584)
L'Architettura, ed. Berardo Galiani (Naples, 1758)
Vitruvio, ed. and trans. S. Ferri (Rome, 1960)

Latin and French, or French only:

Architecture, ou Art de bien bastir, ed. Jean Martin (Paris, 1547)

Les Dix Livres d'Architecture de Vitruve, trans. Claude Perrault, 2nd ed. (Paris, 1684)

Vitruve, ed. Auguste Choisy (Paris, 1909)

De l'Architecture/Vitruve, ed. and trans. Philippe Fleury, Pierre Gros, Louis Callebat, and Jean Soubiran, 10 vols. (Paris, 1969–)

Spanish only:

Los Diez Libros de Architectura de M. Vitruvio Polion, trans. Joseph Ortiz y Sanz (Madrid, 1787)

Latin and German:

Zehn Bücher über Architektur/Vitruv, trans. Curt Fensterbusch (Darmstadt, 1964)

Latin and English, or English only:

Ten Books on Architecture, trans. Morris Hicky Morgan (New York, [1914] 1960)

Ed. from the Harleian manuscript 2767 and trans. Frank Granger, 2 vols., LCL

Concordance:

Vitruve, De Architectura concordance: documentation bibliographique, lexicale et grammaticale, ed. L. Callebat, P. Bouet, P. Fleury, and M. Zuinghedau (Hildesheim and New York, 1984)

WA   *World Archaeology* (London)

WHR   W. H. Roscher, *Ausführliches Lexikon der Griechischen und Römischen Mythologie* (Leipzig, 1884–1937)

Xen.   Xenophon
An.   *Anabasis*
Cyr.   *Cyropaedia*
Hel.   *Hellenika*
Mem.   *Memorabilia*
Symp.   *Symposium*

Xeno.   Xenophanes
            Ed. J. M. Edmonds, LCL

Xenophon of Ephesus

Ephesiaca
Ed. and trans. G. Dalmeyda (Paris, 1926)

ZÄS   *Zeitschrift für Aegyptische Sprache und Altertumskunde* (Berlin)

Bibliography

Abrahams, Ethel and Lady Evans. 1964. *Ancient Greek Dress*. Chicago.

Abulafia, David. 1988. *Frederick II, A Medieval Emperor*. Oxford.

Académie des Inscriptions et Belles-Lettres. 1893. *Monuments et Mémoires*, Fondation Eugène Piot. Paris.

Acharya, Prasanna Kumar. 1981. *Indian Architecture According to Mānasāra-Silpasastra*. 2nd ed. New Delhi.

Ackerman, James S. 1970. *The Architecture of Michelangelo*. Harmondsworth.

———. 1991. *Distance Points: Essays in Theory and Renaissance Art and Architecture*. Cambridge, Mass.

Adam, Sheila. 1966. *The Technique of Greek Sculpture in the Archaic and Classical Periods*. London.

Adams, Robert M. 1981. *Heartland of Cities*. Chicago and London.

Adhémar, Jean. 1939. *Influences Antiques dans l'Art du Moyen Age Français*. London.

Adkins, A. W. H. 1972. Homeric Gods and the Values of Homeric Society." *JHS* 92.

———. 1973. "Arete, Techné, Democracy and Sophists: Protagoras 316b-328d." *JHS* 93.

Adorno, Theodor, W. 1966. *Negative Dialektik*. Frankfurt A/M.

———. 1984a. *Aesthetic Theory*. Trans. C. Lenhardt. London.

———. 1984b. *Gesammelte Schriften*. Vol. 3, *Dialektik der Aufklärung*. With Max Horkheimer. Frankfurt A/M.

Agacinski, Sylviane. 1992. *Volume: Philosophies et Politiques de l'Architecture*. Paris.

Agacinski, S., et al. 1975. *Mimesis des Articulations*. Paris.

Agamben, Giorgio. 1970. *L'Uomo senza Contenuto*. Milan.

Agrippa von Nettesheim, Heinrich Cornelius. 1550. *De Occulta Philosophia*, Cologne. Translated as *Three Books of Occult Philosophy and Magic*, book 1. Ed. Willis F. Whitehead (London, 1898; reprint, 1971).

———. 1550. *Opera in Duos Tomos . . . digesta*. Leyden.

Ahlberg, H. 1950. *Gunnar Asplund, Architect*. Stockholm.

Ahlin, Janne. 1987. *Sigurd Lewerentz, Architect, 1885–1975*. Cambridge, Mass.

Aign, B. P. 1963. *Die Geschichte der Musikinstrumente des Agäischen Raums*. Frankfurt A/M.

Åkerstrom, Åke. 1973. "Ionia and Anatolia: Ionia and the West." *ICCA* 10.

Akurgal, Ekrem. 1955. *Phrygische Kunst*. Ankara.

———. 1962. *The Art of the Hittites*. London.

———. 1978. *Ancient Civilizations and Ruins of Turkey*. Istanbul.

Alaux, Jean-Paul, 1933. *Académie de France à Rome, Ses Directeurs, Ses Pensionnaires*. 2 vols. Paris.

Alberti, Leandro. 1550. *Descrittione di Tutta Italia*. Bologna.

Alberti, Leon Battista. 1950. *Della Pittura*. Edizione Critica a cura di L. Mallè. Florence.

———. 1966. *De Re Aedificatoria*. Ed. G. Orlandi and P. Portoghesi. Milan.

———. 1969. *The Family in Renaissance Florence*. Trans. Renée Watkins. Columbia, S.C.

———. 1972. *On Painting and on Sculpture*. Trans. and ed. Cecil Grayson. London.

———. 1988. *On the Art of Building in Ten Books*. Trans. J. Rykwert, N. Leach, and R. Tavernor. Cambridge, Mass.

Aldrich, Henry 1818. *Elementa Architecturae: The Elements of Civil Architecture*. English by Rev. Philip Smyth. Oxford. First published 1750.

Alfassa, Paul. 1933. "L'Origine de la Lettre de Poussin." *Bulletin de la Société de l'Art Français*. Paris.

Alison, Archibald. 1790. *Essays on the Nature and Principles of Taste*. Edinburgh.

Allchin, B. and R. Allchin. 1968. *The Birth of Indian Civilization*. Harmondsworth.

Alleau, René. 1976. *La Science des Symboles*. Paris.

Allers, Rudolf. 1944. "Microcosmus from Anaximander to Paracelsus." *Traditio* 2.

Alty, John. 1982. "Dorians and Ionians." *JHS* 102.

Alverny, M.-T. d'. 1967. "Astrologues et théologiens au XIIe Siècle." *Mélanges Offerts à M.-D. Chenu*. Paris.

———. 1973. "Le Cosmos Symbolique du XIIe Siècle." *Settimane di Studio*. Centro Italiano di Studi sull'Alto Medioevo 20. Spoleto.

Alzinger, Wilhelm. 1972–1973. ". . . quae stereobates appellantur." *JOAIW* 50.

Amand, D. 1945. "Fatalisme et Liberté dans l'Antiquité Grecque." *Recueil des Travaux d'Histoire et de Philologie*. Ser. 3, fasc. 19. Louvain.

Amandry, Pierre. 1950. *La Mantique Apollinienne à Delphes*. Paris.

[Un Amateur]. 1843. *Nouveau Manuel Complet du Physionomiste des Dames*. Paris.

Ambrose, Bishop of Milan. 1896. *Some of the Principal Works of Saint Ambrose*. Trans. H. de Romestin, with the assistance of E. de Romestin and H. T. K. Duckworth. Oxford.

American School of Classical Studies in Athens. 1953–. *The Athenian Agora*. Princeton.

Amorini, A. B. 1823. *Elogio di Sebastiano Serlio, Architetto Bolognese*. Bologna.

Amyx, Darrell A. *See* Del Chiaro, Mario A.

Anderson, Daniel E. 1993. *The Masks of Dionysos: A Commentary on Plato's Symposium*. Albany.

Anderson, R. D. 1976. *Catalogue of Egyptian Antiquities in the British Museum (Medical Instruments)*. London.

Anderson, W. J., and R. Phené Spiers. 1927. *The Architecture of Ancient Rome*. Rev. Thomas Ashby. London.

Andrae, W. 1930. *Das Gotteshaus und die Urformen des Bauens im Alten Orient*. Berlin.

———. 1941. *Alte Feststrassen im Nahen Osten*. Leipzig.

Andreev, Juri. 1979. "Könige und Königsherrschaft in den Epen Homers." *Klio* 61.

Andrén, Arvid. 1940. *Architectural Terracottas from Etrusco-Italic Temples*. Leipzig-Lund.

———. 1959–1960. "Origine e Formazione dell'Architettura Templare Etrusco-Italica." *Rend.* 32.

Andrewes, A. 1956. *The Greek Tyrants*. London.

Andronikos, Manolis. 1968. "Totenkult." *Arch-Hom* 3.

Andronikos, Manolis, Manolis Chatzidakis, and Vassos Karageorghis. 1975. *The Greek Museums*. Trans. Kay Cicellis. London.

Anti, Carlo. 1931. *Policleto*. Rome.

Antoninus, Saint, Archbishop of Florence. 1740. *Summa Theologica*. Verona.

Aratus. *See* Mair, A. M., and G. W. Mair.

Architecture et Société de l'Archaïsme Grec à la Fin de la République Romaine. 1983. Ecole Française de Rome, C.N.R.S.

Arensburg, B., et al. (1989, 27 April). "A Middle Palaeolithic Human Hyoid Bone." *Nature* 338,

Arias, P. E. 1952. *Skopas*. Quaderni e Guide di Archeologia. Rome.

———. 1964. *Policleto*. Milan.

Arnett, William S. 1982. *The Predynastic Origin of Egyptian Hieroglyphs*. Washington, D.C.

Arnheim, Rudolph. 1956. *Art and Visual Perception*. London.

Arnold, Dieter. 1974–1981. *Der Tempel des Königs Mentuhotep von Deir-el-Bahari*. 3 vols. Mainz.

Arnold, Matthew. 1895. *Essays in Criticism*. London.

Asheri, David. 1983. *Fra Ellenismo e Iranismo*. Bologna.

Ashmole, Bernard. 1972. *Architect and Sculptor in Classical Greece*. New York.

Astour, Michael C. 1965. *Hellenosemitica*. Leiden.

Atkinson, T. D., et al. 1904. *Excavations at Phylakopi in Melos*. London.

Atzeni, Enrico, et al. 1985. *Ichnussa*. Milan.

Auberson, P. 1968. *Le Temple d'Apollon Daphnéphoros*. Berne.

Auberson, Paul, and Karl Schefold. 1972. *Führer durch Eretria*.

Audran, Gerard. 1785. *Les Proportions du Corps Humain*. Paris.

Auerbach, Erich. 1964. *Mimesis*. New York.

Aurigemma, Salvatore. 1961. *Villa Adriana*. Rome.

Austin, Michael M. 1970. *Greece and Egypt in the Archaic Age*. Cambridge.

Averlino, Antonio, detto Il Filarete. 1972. *Trattato di Architettura*. Ed. A. M. Finoli and L. Grassi. Milan.

Axelos, Kostas. 1962. *Heraclite et la Philosophie*. Paris.

———. 1964. *Vers la Pensée Planétaire*. Paris.

———. 1969. *Le Jeu du Monde*. Paris.

Babut, F. C. 1909. *Priscillian et le Priscillianisme*. Paris.

Bachofen, Johann Jakob. 1989. *Il Simbolismo Funerario degli Antichi*. Ed. A. Momigliano. Naples. Originally published as *Versuch über Gräbersymbolik der Alten*. (1859).

Badawy, A. 1954–1968. *A History of Egyptian Architecture*. 3 vols. Cairo (vol. 1) and Berkeley (vols. 2 and 3).

Badian, E., ed. 1966. *Ancient Society and Institutions: Studies Presented to Victor Ehrenberg*. Oxford.

Bailey, H. W. 1943. *Zoroastrian Problems*. Oxford.

Baker A. W. 1918. "The Subjective Factor in Greek Architectural Design." *AJA* 22.

Baker, Paul R. 1980. *Richard Morris Hunt*. Cambridge, Mass.

Bally, Gustav. 1945. *Vom Ursprung und von den Grenzen der Freiheit*. Basel.

Balthazar, Hans Urs von. 1947. *Liturgie Cosmique*. Paris.

Baltrusaitis, Jurgis. 1967. *La Quête d'Isis*. Paris.

———. 1983. *Aberrations*. Paris.

———. 1984. *Anamorphoses*. Paris.

Balz, Albert G. A. 1951. *Cartesian Studies*. New York.

Bammer, Anton. 1972. *Die Architektur des Jüngeren Artemision von Ephesos*. Wiesbaden.

———. 1984. *Das Heiligtum der Artemis von Ephesos*. Graz.

———. 1988. *Ephesos: Stadt am Fluss und Meer*. Graz.

Bandini, Angelo Maria. 1774–1778. *Catalogus Codicum Latinorum Bibliothecae Mediceae Laurentinanae*. Florence.

Bandmann, Günter. *See* Busch, Werner, et al.

Bankel, Hansgeorg, ed. 1986. *Haller von Hallerstein in Griechenland, 1810–1817*. Berlin.

———. 1991. "Akropolis-Fussmasse." *AA* 2.

Banti, Luisa. 1973. *Etruscan Cities and Their Culture*. Trans. Erika Bizzarri. Berkeley.

Barash, Moshe, and Lucy Freeman Sandler, eds. 1981. *Art the Ape of Nature: Studies in Honor of H. W. Janson*. New York.

Barbaro, Daniele. *See* Vitruvius.

Barbera, A. 1991. *The Euclidean Division of the Canon*. Lincoln, Neb.

Barbieri, Franco. 1952. *Vincenzo Scamozzi*. Verona.

Barkan, L. 1975. *Nature's Work of Art*. New Haven.

Barker, A. 1978. "Music and Perception: A Study in Aristoxenus." *JHS* 98.

———. 1981. "Methods and Aims in the Euclidean *Sectio Canonis*." *JHS* 101.

———. 1984. *Greek Musical Writings*. 2 vols. Cambridge.

Barker, Hollis S. 1966. *Furniture in the Ancient World*. New York.

Barnes, Carl F., Jr. 1982. *Villard de Honnecourt: The Artist and His Drawings*. Boston.

Barnes, Jonathan. 1979. *The Presocratic Philosophers*. London.

Barocchi, Paola. 1960. *Trattati d'Arte del Cinquecento*. Bari.

———, ed. 1977. *Scritti d'Arte del Cinquecento*. 3 vols. Milan and Naples.

Baron, Hans. 1966. *The Crisis of the Early Italian Renaissance*. Princeton.

Barresi, P. 1990. "Schemi Geometrici nei Templi dell'Italia Centrale." *Arch Cl.* 42.

Barrière, Pierre. 1961. *La Vie Intellectuelle en France du XVIe Siècle à l'Epoque Contemporaine*. Paris.

Barrow, R. H. 1967. *Plutarch and His Times*. London.

Barthélémy, J. J. 1789. *Voyage du Jeune Anacharsis en Grèce*. Paris.

Bartoloni, Gilda, et al. 1987. *Le Urne a Capanna Rinvenute in Italia*. Rome.

Bataille, Georges. 1971. *Oeuvres Complètes*. Ed. D. Hollier. Paris.

Battisti, Eugenio. 1962. *L'Antirinascimento*. Milan.

Battock, Gregory. 1977. *Why Art? Casual Notes on the Aesthetics of the Immediate Past*. New York.

Bauer, Heinrich. 1973. *Korinthische Kapitelle des 4 und 3 Jahrhunderts vor Christus*. Berlin.

Baumann, Hellmut. 1982. *Die Griechische Pflanzenwelt*. Munich.

Baumgarten, Alexander Gottlieb. 1750. *Aesthetica*. Bonn. (Reprinted Hildesheim 1961.)

Baus, Karl. 1940. *Der Kranz in Antike und Christentum*. Bonn.

Baxendall, Michael. 1971. *Giotto and the Orators*. Oxford.

Bayet, Jean. 1943. *Mélanges de Littérature Latine*. Paris.

Bean, George E. 1971. *Turkey beyond the Meander*. London.

———. 1978. *Lycian Turkey*. London.

———. 1979. *Aegean Turkey*. London.

Beazley, John Davidson. 1938. *Attic White Lekythoi*. Oxford.

———. 1947. *Etruscan Vase Painting*. Oxford.

———. 1951. *The Development of the Attic Black Figure*. Berkeley.

———. 1967. *Attic Red-Figured Vases in American Museums*. Rome.

———. 1971. *Paralipomena*. 2nd ed. Oxford.

Becatti, Giovanni. 1960. *La Colonna Coclide Istoriata*. Rome.

Bechmann, Roland. 1991. *Villard de Honnecourt: La Pensée Technique au XIIIe Siècle et sa Communication*. Paris.

Bede, the Venerable (Saint). 1688. *Opera Theologica, Moralia, Historica, Philosophica, Mathematica et Rhetorica Omnia*. 8 vols. Cologne.

Beit, Hedwig von. 1952. *Symbolik des Märchens*. 3 vols. Bern.

Bélis, Anne. 1986. *Aristoxène de Tarente et Aristote*. Paris.

Bell, Sir Charles. 1806. *Essays on the Anatomy of Expressions in Painting*. London. (Later editions add "as *Connected with the Fine Arts*.")

Benda, Julien. 1927. *La Trahison des Clercs*. Paris.

———. 1947. *Belphégor 1918*. Reprint, Paris.

Bendinelli, Goffredo. 1953. *Luigi Canina, Le Opere, i Tempi*. Alessandria.

Bengtson, Hermann, et al. 1969. *The Greeks and the Persians*. London.

Benjamin, Walter. 1972–1977. *Gesammelte Schriften*. Ed. Rolf Tiedemann and Hermann Schweppenhäuser. Frankfurt A/M.

Ben-Menahem, H. *See* N. S. Hecht.

Benn, Gottfried. 1989. *Sämtliche Werke*. Ed. Gerhard Schuster. 4 vols. Stuttgart.

Bennett, Emmet L., Jr., ed. 1964. *Mycenaean Studies: Third International Colloquium*. Madison, Wis.

Bennett, Florence Mary. 1967. *Religious Cults Associated with the Amazons*. New York.

Benton, S. 1938. "The Evolution of the Tripod-Lebes." *Brit. Sch. Athens Ann*. 35.

Benton, T., S. Muthesius, and B. Wilkins. 1975. *Europe, 1900–1914*. Milton Keynes.

Benveniste, Emile. 1966–1974. *Problèmes de Linguistique Générale*. 2 vols. Paris.

———. 1969. *Le Vocabulaire des Institutions Indo-Européennes*. 2 vols. Paris. Trans. Elizabeth Palmer as *Indo-European Language and Society* (Coral Gables, 1973).

Béquignon, Y., and A. Laumonier. 1925. "Fouilles de Téos." *BCH* 49.

Bérard, Claude. 1970. *L'Heroon à la Porte de l'Ouest*. Berne.

———. 1971. "Architecture Erétrienne et Mythologie Delphique." *AK* 14.1.

Bérard, Victor. 1903. *Les Phéniciens et l'Odyssée*. 2 vols. Paris.

———. 1925. *Introduction à l'Odyssée*. 3 vols. Paris.

Berg Hollinshead, M. B. 1979. "Legend, Cult, and Adventure at Three Sanctuaries of Artemis." Ph.D. diss., Bryn Mawr.

Berger, Albrecht. 1982. *Das Bad in der Byzantinischer Zeit*. Munich.

Berger, Ernst. *See* Schmidt, Margret.

Berger, Ernst, et al. 1992. *Der Entwurf des Künstlers Bildhauerkanon in der Antike und Neuzeit*. Basel.

Berger, Peter L., and Thomas Luckmann. 1967. *The Social Construction of Reality*. New York.

Bergquist, Birgitta. 1967. *The Archaic Greek Temenos*. London.

———. 1988. "The Archaeology of Sacrifice." *OA* 38.

Bergson, H. 1951. *The Philosophy of Poetry*. New York.

Bernal, Martin. 1987–1991. *Black Athena*. 2 vols. New Brunswick, N.J.

Bernini, Gian Lorenzo. 1963. *Fontana di Trevi*. Ed. C. D'Onofrio. Rome.

Berthelot, René. 1949. *La Pensée de l'Asie et l'Astrobiologie*. Paris.

Berthiaume, Guy. 1982. *Les Rôles du Mágeiros*. Coll. *Mnemosyne*. Leiden.

Berve, Helmut, ed. 1942. *Das Neue Bild der Antike*. 2 vols. Leipzig.

Beschi, L., et al. 1982. *Aparchai: Essays in Honor of Paolo Enrico Arias*. Pisa.

Best, Jan, and Fred. Woudhuizen. 1988. *Ancient Scripts from Crete and Cyprus*. Leiden.

Betancourt, Philip P. 1977. *The Aeolic Style in Architecture*. Princeton.

Betts, Richard J. 1972. Review of Francesco di Giorgio Martini, *Trattati*, ed. Corrado Maltese. *JSAH* 31.1.

———. 1977. "On the Chronology of Francesco di Giorgio's Treatises: New Evidence from an Unpublished Manuscript." *JSAH* 36.1.

Beyer, Immo. 1972. "Der Triglyphenfries von Thermos C." *AA* 87.

———. 1976. *Die Tempel von Dreros und Prinias A*. 2 vols. Freiburg.

Bhattacharyya, S. M. 1976. *The Alamkara-Section of the Agni-Purana*. Calcutta.

Biale, David. 1979. *Gershon Scholem: Kabbalah and Counter History*. Cambridge, Mass.

Białostocki, Jan. 1961. *Teoria i Twórczosc*. Poznan.

Bianchi Bandinelli, Ranuccio, 1929. *Sovana: Topografia ed Arte*. Florence.

———. 1938. *Policleto*. Quaderni per lo Studio dell'Archeologia. Florence.

———. 1973. *Les Etrusques et l'Italie avant Rome*. Trans. Jean-Charles et Evelyne Picard. Paris.

———. 1978. *Storia e Civiltà dei Greci*. Milan.

Bieber, Margarete. 1961. *The History of the Greek and Roman Theater*. Princeton.

———. 1977. *Ancient Copies: Contributions to the History of Greek and Roman Art*. New York.

Bignone, Ettore. 1963. *Empedocle*. Rome.

Billerbeck, Julius. 1824. *Flora Classica*. Leipzig.

Bindman, David. 1977. *Blake as an Artist*. Oxford.

Bingol, Orhan. 1980. *Das Ionische Normalkapitell*

in Hellenistischer und Römischer Zeit in Kleinasien. Tübingen.

Bintliff, John L. 1977. *Natural Environment and Human Settlement in Prehistoric Greece*. 2 vols. Oxford.

Birindelli, Massimo. 1980. *La Machina Heroica*. Rome.

Bischoff, Bernhard, ed. 1984. *Anecdota Novissima*. Stuttgart.

Bisticci, Vespasiano de. 1859. *Vite di Uomini Illustri del Secolo XV*. Florence.

Blacker, C. P. 1987. *Eugenics: Galton and After*. Westport. Conn.

Blake, Peter. 1963. *Four Great Makers of Modern Architecture*. New York.

Blake, William. 1804. *Jerusalem, The Emanation of the Giant Albion*. London.

Blanc, Charles. 1867. *Grammaire des Arts du Dessin*. 1860. Reprint, Paris.

Bleeker, C. J. 1967. *Egyptian Festivals*. Leiden.

——, ed. 1965. *Initiation*. Leiden.

——. 1973. *Hathor and Thoth*. Leiden.

Blegen, Carl William, and Marion Rawson. 1967. *The Palace of Nestor at Pylos in Western Messenia*. 3 vols. Princeton.

Bloch, Raymond. 1958. *The Etruscans*. Trans. Stuart Hood. New York.

Bloesch, Hansjörg, and Hans Peter Isler, eds. 1976–. *Studia Ietina*. 3 vols. to date. Zurich.

Blondel, Jacques-François. 1771–1777. *Cours d'Architecture ou Traité de la Décoration, Distribution et Construction des Batiments*. Ed. Pierre Patte. Paris.

Bluemel, Carl. 1969. *Greek Sculptors at Work*. 2nd ed. London.

Blum, André. 1924. *Abraham Bosse et la Société Française au Dix-Septième Siècle*. Paris.

Blum, Paul. 1960. *La Peau*. 2nd ed. Paris.

Blumenberg, Hans, ed. 1957a. "'Nachahmung der Natur': Zur Vorgeschichte der Idee des Schöpferischen Menschen." *Studium Generale*. Berlin.

——. 1957b. *Nikolaus von Cues: Die Kunst der Vermutung*. Bremen.

——. 1960. "Paradigmen zu einer Metaphorologie." *Archiv für Begriffsgeschichte* 6.

——. 1976. *Aspekte der Epochenschwelle: Cusaner und Nolaner*. Frankfurt A/M.

——. 1983a. *Die Lesbarkeit der Welt*. Frankfurt.

——. 1983b. *The Legitimacy of the Modern Age*. Cambridge, Mass.

——. 1987. *The Genesis of the Copernican World*. Trans. Robert M. Wallace. 1975. Reprint, Cambridge, Mass.

Blumenthal, H. J. 1982. "Proclus on Perception." *BICS* 29.

Blümner, Hugo. 1891. *Studien zur Geschichte der Metapher in Griechischen*. Leipzig.

Blunt, Anthony. 1953. *Art and Architecture in France, 1500–1700*.Harmondsworth.

——. 1958a. *Nicolas Poussin*. 2 vols. London.

——. 1958b. *Philibert de l'Orme*. London.

Boardman, J. 1959. "Chian and Early Ionic Architecture." *AJ* 197, vol. 39.

——. 1963a. "Artemis Orthia and Chronology." *BSA* 58.

——. 1963b. *Island Gems*. London.

——. 1980. *The Greeks Overseas: Their Early Colonies and Trade*. London.

Bober, H. 1948. "The Zodiacal Miniatures of the *Très Riches Heures* of the Duke of Berry—Its Sources and Meaning." *JWCI* 11.

——. 1958. "An Illustrated Mediaeval School-Book of Bede's *De Natura Rerum*." *Journal of the Walters Art Gallery* 19–20.

Boehlau, J., and K. Schefold. 1940–1942. *Larisa am Hermos: Die Ergebnisse der Ausgrabungen, 1902–1934*. Berlin.

Boehringer, Erich. *See* Gerkan, Armin von.

Boëthius, Axel. 1965. "La Tomba con Tetto Stramineo a Cerveteri." *Palladio* 15. 1–4.

——. 1978. *Etruscan and Early Roman Architecture*. Harmondsworth.

Boëthius, Axel, et al. 1962. *Etruscan Culture: Land and People*. Trans. Nils G. Sahlin. New York and Malmö.

Boetticher, Adolf. 1886. *Olympia: Das Fest und Seine Stätte*. 2nd ed. Berlin.

——. 1888. *Die Akropolis von Athen*. Berlin.

Boetticher, Carl. 1856. *Der Baumkultus der Hellenen*. Berlin.

——. 1874. *Die Tektonik der Hellenen*. 2 vols. Berlin.

Boffrand, Germain. 1745. *Livre d'Architecture contenant les Principes Généraux de cet Art . . . par le Sieur Boffrand*. Paris. (Reprinted, Farnborough, 1969.)

Boggiani, G. 1895. *I Caduvei*. Rome.

Bogue, Ronald, ed. 1991. *Mimesis in Contemporary Theory: An Interdisciplinary Approach*. Vol. 2, *Mimesis, Semiosis, and Power*. Philadelphia and Amsterdam.

Bogue, Ronald, and Mihai Spariosu, eds. 1994. *The Play of the Self*. Albany.

Boileau-Despréaux, Nicolas. 1722. *Oeuvres, avec des éclaircissements historiques*. Nouvelle édition revue, corrigée et augmentée. The Hague.

Bois, Page du. 1982. *Centaurs and Amazons: Women and the Pre-History of the Great Chain of Being*. Ann Arbor.

Boisacq, Emile. 1938. *Dictionnaire Etymologique de la Langue Grecque*. Paris.

Bol, Peter C., ed. 1989–1990. *Forschungen zur Villa Albani. Katalog der antiken Bildwerke*. Berlin.

Bolgar, Robert Ralph. 1963. *The Classical Heritage and Its Beneficiaries*. Cambridge.

———. 1976. *Classical Influences on European Culture*. A.D. *1500–1700*. Cambridge.

Boll, Franz. 1926. *Sternglaube und Sterndeutung*. Ed. C. Bezold and W. Gundel. Leipzig and Berlin.

Bollack, Jean. 1965. *Empedocle*. Paris.

Bommelaer, J. F., ed. 1985. *Le Dessin d'Architecture dans les Sociétés Antiques*. Actes du Colloque de Strasbourg, 26–28 Janvier 1984. Strasbourg.

Bonfante, Giuliano. 1986. "Religione e Mitologia in Etruria." *SE* 54.

Bonfante, Giuliano, and Larissa Bonfante. 1983. *The Etruscan Language*. New York.

Bonfante, Larissa. 1986. *Etruscan Life and Afterlife*. Detroit.

Borchardt, Ludwig, ed. 1937. *Beiträge zur Aegyptischen Bauforschung und Altertumskunde*, Cairo.

Borchhardt, J., ed. 1967. "Limyra: Sitz des Lykischen Dynasten Perikles." *Ist. For.* 17.

———. 1975. "Myra: Eine Lykische Metropole." *Ist. For.* 31.

———. 1976. "Die Bauskulptur des Heroons von Limyra: Das Grabmal des Lykischen Königs Perikles." *Ist. For.* 32.

Borelli, G. 1984. *I Motivi Profondi della Poesia Lucreziana*. Brussels.

Borgeaud, Philippe. 1988. *The Cult of Pan in Ancient Greece*. Chicago.

Borst, H. 1958. "Abelärd und Bernhard." *HZ* 3.14

Bosanquet, Bernard. 1934. *A History of Aesthetics*. New York.

Bose, Mandakaranta. 1991. *Movement and Mimesis*. Dordrecht.

Bosse, Abraham. 1664. *Traicté des Manières de dessiner les Ordres de l'Architecture en toutes leurs parties*. Paris.

———. 1667. *Le Peintre Converty aux Précises et Universelles Règles de son Art*. Paris.

———. 1973. *Traité des Pratiques Géométrales et Perspectives*. 1665. Reprint, Geneva.

Bossert, Helmuth. T. 1951. *Altsyrien*. Tübingen.

Bossi, G. 1811. *Delle opinioni di Leonardo da Vinci intorno alla Simmetria de' Corpi Umani*. Milan.

Bothmer, Dietrich von. 1957. *Amazons in Greek Art*. Oxford.

Botta, Paul-Emile. 1849–1850. *Monuments de Ninive, découverts et décrits par M. P. E. Botta; mesurés et dessinés par M. E. Flandin*. Paris.

Bottéro, Jean. 1987. *Mésopotamie: L'Ecriture, la Raison et les Dieux*. Paris.

Bottin, Francesco. 1982. *La Scienza degli Occamisti: La Scienza Tardomedievale dalle Origini del Paradigma Nominalista alla Rivoluzione Scientifica*. Rimini.

Bouché-Leclercq, Auguste. 1899. *L'Astrologie Grecque*. Paris.

Bougainville, Louis-Antoine de, comte. 1771. *Voyage autour du monde: par la frégate du Roi, La Boudeuse et la flûte, L'étoile: en 1766, 1767, 1768 et 1769*. Paris.

Bouissac, Paul, Michael Herzfeld, and Ronald Posner. 1986. *Iconicity: Essays on the Nature of Culture: Festschrift for Thomas A. Sebeok*. Tübingen.

Bouzek, Jan. 1985. *The Aegean, Anatolia, and Europe: Cultural Relations in the Second Millennium* B.C. Göteborg.

Boyce, Benjamin. 1947. *The Theophrastan Character in England to 1642*. Cambridge, Mass.

Boyle, Robert. 1661. *The Sceptical Chymist: or Chymico-Physical Doubts and Paradoxes*. London.

———. 1685. *Of the High Veneration Man's Intellect Owes to God*. London.

Bragdon, Claude. 1918. *Architecture and Democracy*. New York.

———. 1922. *The Beautiful Necessity*. New York.

———. 1925. *Old Lamps for New*. New York.

Brandenburg, Erich. 1915. *Ueber Felsarchitektur im Mittelmeergebiet*. Leipzig.

Branner, Robert. 1957, March. "Three Problems from the Villard de Honnecourt Manuscript." *AB* 39.

——. 1960. "Villard de Honnecourt, Archimedes, and Chartres." *JSAH* 19.3.

Brauer, Heinrich, and Rudolf Wittkower. 1931. *Die Zeichnungen des Gianlorenzo Bernini.* 2 vols. Berlin. (Reprinted New York, n.d.)

Brelich, Angelo. 1958. *Gli Eroi Greci.* Rome.

Bremmer, Jan, and Herman Roodenburg. 1992. *A Cultural History of Gesture.* Ithaca, N.Y.

Brendel, Otto. 1978. *Etruscan Art.* Harmondsworth.

Brett, David. 1986. "Aesthetical Science: George Field and the Science of Beauty." *AH* 9.

Brinkmann, Richard. 1969. *Begriffsbestimmung des literarischen Realismus.* Darmstadt.

Briquel, Dominique. 1984. *Les Pélasges en Italie.* Ecole Française de Rome.

——. 1991. *L'Origine Lydienne des Etrusques.* Ecole Française de Rome, Rome.

Brommer, Frank. 1987. "Zur Minoischen Säule." *AA* 1.

——. *See also* Höckmann, Ursula, and Antje Krug.

Brooke-Rose, Christine. 1958. *A Grammar of Metaphor.* London.

Brown, F. E. 1980. *Cosa, The Making of a Roman Town.* Ann Arbor.

Brown, F. E., E. H. Richardson, and L. Richardson. 1960. "Cosa II." *MAAR* 26.

Brugsch, H. 1877. *Geschichte Aegyptens.* Leipzig.

[Bruhns, Leo.] 1961. *Miscellanea Bibliothecae Hertzianae zu Ehren von Leo Bruhns.* Munich.

Bruit Zaidman, Louise, and Pauline Schmitt Pandel. 1992. *Religion in the Ancient Greek City.* Trans. Paul Cartledge. Cambridge.

Bruno, Vincent J. 1974. *The Parthenon.* New York.

Brunn, Enrico, and Gustav Körte. 1965. *I Rilievi delle Urne Etrusche 1870–1896.* Rome.

Brunschvigg, Leon. 1944. *Descartes et Pascal.* New York.

Bruschi, Arnaldo, ed. 1978. *Scritti Rinascimentali di Architettura.* Milan.

Bruyn, J., ed. 1973. *Album Amicorum: Festschrift van J. G. Van Gelder.* The Hague.

Bruyne, Edgar de. 1946. *Etudes d'Esthétique Médiévale.* 3 vols. Bruges. (Reprinted Geneva, 1975.)

Bryce Bannatyne Gallery. 1992. *Tattoo Art: Forever Yes.* Catalogue. Santa Monica.

Bryson, Norman. 1981. *Word and Image.* Cambridge.

Buchholz, Hans-Günter. 1973. "Tamassos, Zypern 1970–72." *AA* 3.

Buchholz, Hans-Günter, and Vassos Karageorghis. 1973. *Prehistoric Greece and Cyprus.* Trans. Francisca Garvie. London.

Buchthal, Hugo. 1979. *The "Musterbuch" of Wolfenbüttel and Its Position in the Art of the Thirteenth Century.* Vienna.

Buck, C. D. 1928. *Introduction to the Study of the Greek Dialects.* Boston.

Bulwer, John. 1644. *Chirologia, or the Naturall Language of the Hand . . . whereunto is added Chironomia or the Art of Manuall Rhetoricke.* London.

Bundgaard, J. A. 1957. *Mnesicles: A Greek Architect at Work.* Copenhagen.

——. 1974. *The Excavations of the Athenian Acropolis.* Copenhagen.

——. 1976. *The Parthenon and the Mycenaean City on the Heights.* Copenhagen.

Burckhardt, Jacob. 1929–1933. *Gesamtausgabe.* Ed. Jakob Oeri. Stuttgart and Basel.

——. 1879. *Der Cicerone.* Ed. Wilhelm von Bode. Stuttgart.

——. 1984. *Die Kunst der Betrachtung.* Ed. Henning Ritter. Cologne.

Buresch, Karl. 1889. *Klaros: Untersuchungen zum Orakelwesen des Späteren Altertums.* Leipzig.

Burkert, Walter. 1960. "Platon oder Pythagoras: zum Ursprung des Wortes Philosophia." *Hermes* 88.

——. 1972a. *Homo Necans.* Berlin.

——. 1972b. *Lore and Science in Ancient Pythagoreanism.* Cambridge, Mass.

——. 1979. *Structure and History in Greek Mythology and Ritual.* Berkeley.

——. 1983. *Anthropologie des religiösen Opfers: die Sakralisierung der Gewalt.* Munich.

——. 1984. *I Greci.* Milan. Originally published as *Griechische Religion* (Stuttgart, 1977).

——. 1987. "Die Antike Stadt als Festgemeinschaft." In *Stadt und Fest,* ed. Paul Hugger. Stuttgart.

——. 1992. *The Orientalizing Revolution.* Trans. Margaret Pinder. Cambridge, Mass.

Burl, Aubrey. 1976. *The Stone Circles of the British Isles.* New Haven.

Burnet, Joan. 1932. *Greek Philosophy, Thales to Plato.* London.

Burney, Charles. 1977. *The Ancient Near East.* Ithaca, N.Y.

Burton, Robert. 1628. *The Anatomy of Melancholy.* London.

Busch, Werner, Reiner Haussherr, and Eduard

Trier. 1978. *Kunst als Bedeutungsträger: Festschrift Bandmann*. Berlin.

Buschor, Ernst. 1930. "Die Heraion von Samos: Frühe Bauten." *MDAIA 55*.

Büsing, Hermann. 1984. "Optische Korrekturen und Propyläen-Fronten." *JDI 99*.

Büsing, Hermann, and Friedrich Hiller, eds. 1988. *Bathron: Beiträge zur Architektur und Verwandten Künste: Drerup Festschrift*. Saarbrucken.

Busink, T. A. 1970–1980. *Der Tempel von Jerusalem*. 2 vols. Leiden.

Butterworth, E. A. S. 1966. *Some Traces of the Pre-Olympian World in Greek Literature and Myth*, Berlin.

Buytendijk, F. J. J. 1976. *Wesen und Sinn des Spiels*. New York.

Byam Shaw, J. 1967. *Paintings by Old Masters at Christ Church, Oxford*. London.

Cacciari, Massimo. 1980. *Dallo Steinhof*. Milan.

———. 1988. "Mies's Classics." *RES 16*.

———. 1993. *Architecture and Nihilism*. New Haven.

———. 1994. *Geo-Filosofia dell'Europa*. Milan.

———. See also Loos, Adolf.

Caillois, Roger. 1950. *L'Homme et le Sacré*. Paris.

Caleca, Antonino, et al. 1979. *Pisa: Museo delle Sinopie del Camposanto Monumentale*. Catalogue. Pisa.

Camaioni, Luigia, and Claudia de Lemos, eds. 1985. *Questions on Social Explanation: Piagetian Themes Reconsidered*. Amsterdam and Philadelphia.

Camden, William. 1586. *Britannia: Sive Florentissimorum Regnorum Angliae, Scotiae, Hiberniae, et Insularum adiacentium ex Intima Antiquitate Chorographica Descriptio*. London. (Ed. Edmund Gibson, 1695.)

Camper, Petrus. 1962. *Optical Dissertation on Vision*. 1746. Trans. G. Ten Doesschate. Nieuwkoop.

Camporeale, Salvatore I. 1972. *Lorenzo Valla*. Florence.

Cancik, Hubert, Peter Eicher, B. Gladigow, and M. Greschat. 1983. *Die Religionen der Menschheit*. Storia delle Religioni. Milan.

Canetti, E. 1960. *Masse und Macht*. Hamburg.

Canguilhem, G. 1970. "L'Homme de Vésale dans le Monde de Copernic: 1543." *Etudes d'Histoire et de Philosophie de Science*. Paris.

Canina, Luigi. 1830. *L'Architettura Antica descritta e demostrata coi Monumenti*. Rome. (2nd ed. 1844)

Capobianco, Michele. 1959. *Asplund e il Suo Tempo*. Naples.

Carlino, Andrea. 1988. "The Book, the Body, the Scalpel." *RES 16*.

———. 1994. *La Fabbrica del Corpo*. Turin.

Carlyle, Thomas. 1924. *Past and Present*. 1843. Reprint, London.

Carpo, Mario. 1992. "The Architectural Principles of Temperate Classicism: Merchant Dwellings in Sebastiano Serlio's Sixth Book." *RES 22*.

Carritt, E. F. 1923. *The Theory of Beauty*. London.

Carruba, Onofrio. 1970. *Das Palaische Texte, Grammatik, Lexikon*. Wiesbaden.

Carswell, John. 1958. *Coptic Tattoo Designs*. Beirut.

Cartailhac, Emile. 1877. *L'Age de Pierre*. Paris.

Carter, Elizabeth, and Matthew W. Stolper. 1984. *Elan*. Berkeley.

Cartledge, Paul. 1987. *Agesilaos and the Crisis of Sparta*. London.

Caruchet, William, 1981. *Bas-Fonds du Crime et Tatouages*. Monaco.

Casanelles, E. 1965. *Nueva Vision de Gaudi*. Barcelona.

———. 1967. *Antonio Gaudi: A Reappraisal*. Greenwich, Conn.

Casotti, Maria Walcher. 1960. *Il Vignola*. 2 vols. Istituto di Storia dell'Arte Antica e Moderna 11. Trieste.

Cassazza, O. 1978. "Il Crocifisso Ligneo di Filippo Brunelleschi." *La Critica d'Arte 43*.

Cassirer, Ernst. 1935. *Individuo e Cosmo*. Florence.

———. 1951. *The Philosophy of the Enlightenment*. Trans. Fritz C. A. Koeller and James Pettegrove. Princeton.

———. 1953–1957. *The Philosophy of Symbolic Forms*. Trans. R. Mannheim. New Haven.

Casson, Stanley, ed. 1927. *Essays in Aegean Archaeology Presented to Sir Arthur Evans*. Oxford.

———. 1933. *The Technique of Early Greek Sculpture*. Oxford.

Caton, Hiram. 1973. *The Origin of Subjectivity*. New Haven and London.

Cazanove, Olivier de, ed. 1986. *L'Association Dionysiaque dans les Sociétés Anciennes*. Rome.

Cennini, Cennino. 1899. *The Art of the Old Masters as Told by Cennino Cennini in 1437*. Ed. C. J. Herringham. New York.

Cerny, Jaroslav. 1952. *Ancient Egyptian Religion.* London.

Cesnola, General Louis Palma di. 1914. *Cyprus, Its Ancient Cities, Tombs, and Temples.* New York.

Chadwick, J. 1958. *The Decipherment of Linear B.* Cambridge.

——. 1976. *The Mycenaean World.* Cambridge.

Chafee, R. S. 1972. "The Teaching of Architecture at the Ecole des Beaux-Arts and Its Influence in Britain and America." Ph.D. diss., University of London.

Chambray. *See* Fréart, Roland, Sieur de Chambray.

Champollion, F. 1833. *Lettres Ecrites d'Egypte et de Nubie entre 1828 et 1829.* Paris.

Chandler, Richard W., Nicholas Revett, and W. Pars. 1769. *Ionian Antiquities.* London.

Chantelou. *See* Fréart, Paul, Sieur de Chantelou.

Chapman, R., I. Kinnes, and K. Randsborg, eds. 1981. *The Archaeology of Death.* Cambridge.

Chapot, Victor. 1907. *La Colonne Torse et le Décor en Hélice dans l'Art Antique.* Paris.

Chapouthier, Fernand. 1935. *Les Dioscures au Service d'Une Déesse.* Paris.

Charbonneaux, J. 1971. *Archaic Greek Art.* New York.

Charles Le Brun 1619–1690: Peintre et Dessinateur. Exhibition Catalogue. 1963. Château de Versailles.

Charvet, Léon. 1869. *Sebastien Serlio.* Lyon.

Chassinat, Emile. 1966–1968. *Le Mystère d'Osiris au Mois de Khoiak.* Cairo.

Chastel, André, ed. 1960. *Nicholas Poussin.* 2 vols. CNRS. Paris.

Chatelain, E. *See* Mortet, V.

Chatelain, U. V. 1905. *Le Surintendant Nicolas Foucquet.* Paris.

Chaussier, M. 1838. *Nouveau Manuel du Physiognomiste et du Phrénologiste.* Paris.

Chenu, M.-D. 1952. "L'Homme et la Nature." *Archives d'Histoire Doctrinale* 19.

——. 1955. "Involucrum." *Archives d'Histoire Dogmatique et Littéraire du Moyen Age* 22.

——. *See also* d'Alverny, Marie-Thérèse.

Cherniss, Harold, 1935. *Aristotle's Criticism of Pre-Socratic Philosophy.* Baltimore.

Chicago Tribune. 1923. *The International Competition for a New Administration Building: Containing All the Designs Submitted in Response to the Chicago Tribune's $100,000 Offer Commemorating Its Seventy-Fifth Anniversary, June 10, 1922.* Chicago.

Childs, W. A. P. 1978. *The City-Reliefs of Lycia.* Princeton.

——. 1981. "Lycian Relations with Persians and Greeks in the Fifth and Fourth Centuries Reexamined." *AnSt* 31.

Chipiez, Charles. 1876. *Histoire Critique des Origines et de la Formation des Ordres Grecs.* Paris.

Chirassi, Ileana. 1968. *Elementi di Culture Precereali nei Miti e Riti Greci.* Incunabula Graeca 30. Rome.

Choisy, Auguste. 1883–1884. *Etudes Epigraphiques sur l'Architecture Grecque.* Paris.

——. 1899. *Histoire de L'Architecture.* Paris. (Reprinted Geneva, 1987.)

——. *See also* Vitruvius.

Chomsky, Noam. 1988. *Language and Problems of Knowledge: The Managua Lectures.* Cambridge, Mass.

Ciasca, Antonia. 1962. *Il Capitello detto Eolico in Etruria.* Florence.

Cipriani, Giovanni. 1980. *Il Mito Etrusco nel Rinascimento Fiorentino.* Florence.

Ciucci, G., et al. 1973. *La Città Americana.* Bari.

Clagett, Marshall. 1968. *Nicole Oresme and the Medieval Geometry of Qualities and Motions.* Madison, Wis. and London.

Clairmont, Christoph W. 1970. *Gravestone and Epigram.* Mainz.

Clark, Grahame. 1977. *World Prehistory.* Cambridge.

Clark, Kenneth. 1969. *The Drawings of Leonardo da Vinci in the Collection of Her Majesty the Queen at Windsor Castle.* 2 vols. Rev. Carlo Pedretti. London.

Clarke, Somers, and Reginald Engelbach. 1930. *Ancient Egyptian Masonry.* Oxford.

Clayton, Peter A., and Martin J. Price. 1989. *The Seven Wonders of the Ancient World.* New York.

Clough, Cecil H., ed. 1976. *Cultural Aspects of the Italian Renaissance: Essays in Honour of Paul Oskar Kristeller.* Manchester.

Cluver, Philip. 1624. *Italia Antiqua.* Leyden.

Coarelli, Filippo. 1985. *Il Foro Romano.* 2 vols. Rome.

——. 1988. *Il Foro Boario.* Rome.

Cocchiara, G. 1932. *Il Linguaggio del Gesto.* Turin.

Cockerell, C. R. 1830. *The Temple of Jupiter Olympius at Agrigentum.* London.

———. 1860. *The Temples of Jupiter Panhellenicus at Aegina and of Apollo Epicurus at Bassae*. London.

Cohen, I. Bernard. 1971. *Introduction to Newton's Principia*. Cambridge.

Cohen, M. S., and I. E. Drabkin. 1948. *A Source Book in Greek Science*. New York.

Coldstream, J. N. 1968. *Greek Geometric Pottery*. London.

———. 1977. *Geometric Greece*. New York.

Coleridge, Samuel Taylor. 1949. *The Philosophical Lectures Hitherto Unpublished*. Ed. Kathleen Coburn. New York.

Colli, Giorgio, ed. 1977–. *La Sapienza Greca*. 2 vols. to date. Milan.

Collignon, Maxime. 1903. "Sculptures Grecques Trouvées à Tralles." *Monuments Piot* 10.

———. 1905. *Lysippe*. Paris.

Colombo, Realdo. 1559. *De Re Anatomica*. Venice.

Colonna, Giovanni. 1966. "Elementi Architettonici in Pietra dal Santuario di Pyrgi." *Arch Cl.* 18.2.

———. 1976. "Il Sistema Alfabetico." *L'Etrusco Arcaico*. Florence.

———. ed. 1985. *Santuari d'Etruria*. Catalogue. Arezzo and Milan.

Colvin, H. M. 1978. *A Biographical Dictionary of British Architects*. London.

Combarieu, J. 1909. *La Musique et la Magie*. Paris.

———. 1953. *Histoire de la Musique*. 9th ed. Paris.

Combès, Joseph. 1960. *Le Dessein de la Sagesse Cartésienne*. Lyon.

Conche, Marcel. 1981. *Lucrèce et l'Expérience*. Paris.

Condivi, Ascanio. 1928. *Michelangelo: La Vita Raccolta dal suo Discepolo*. Ed. A. Maraini. Florence.

———. 1976. *Michelangelo*. Trans. and ed. H. Wohl. Baton Rouge.

Condorcet, Marie-Jean-Antoine-Nicolas de Caritat, Marquis de. 1847–1849. *Oeuvres*. Ed. A. Condorcet O'Connor and M. F. Arago. Paris.

Conger, George Perrigo. 1922. *Theories of Macrocosms and Microcosms in the History of Philosophy*. New York.

Contenau, Georges. 1928. *L'Art de L'Asie Occidentale Ancienne*. Paris and Brussels.

———. 1954. *Everyday Life in Babylon and Assyria*. London.

Conway, W. M. 1889. *The Literary Remains of Albrecht Dürer*. Cambridge.

Cook, A. B. 1914–1940. *Zeus: A Study in Ancient Religion*. 3 vols. Cambridge.

Cook, J. M. 1958–1959. "Old Smyrna." *Brit. Sch. Ann.* 53–54.

———. 1962. *The Greeks in Ionia and the East*. London.

Cook, Captain James. 1955. *The Journals*. 5 vols. Ed. J. C. Beaglehole. London.

Cook, R. M. 1973. *Greek Art: Its Development, Character, and Influence*. New York.

Coomaraswamy, A. K. 1933. *A New Approach to the Vedas*. London.

———. 1956. *Mediaeval Sinhalese Art*. New York.

———. 1977. *Selected Papers*. Ed. Roger Lipsey. Princeton.

Cooper, Frederick Alexander. 1968. "The Temple of Apollo at Bassae: New Observations on Its Plan and Its Orientation." *AJA* 72.

———. 1970. "The Temple of Apollo at Bassae. Ph.D. diss., University of Pennsylvania.

Copernicus, Nicholas. 1976. *On the Revolutions of the Heavenly Spheres (De Revolutionibus Orbium Celestium)*. Trans. A. M. Duncan. Newton Abbot, Devon.

Copleston, Frederic. 1946–1975. *A History of Philosophy*. 9 vols. London.

Corbin, Henry. 1960a. *Avicenna and the Visionary Recital*. Trans. W. K. Trask. London.

———. 1960b. *L'Imam caché et la Rénovation de L'Homme en Théologie Shiite*. Zurich.

———. 1973. *Storia della Filosofia Islamica*. Milan.

———. 1983. *Cyclical Time and Ismaili Gnosis*. London.

Cornford, F. M. 1963. *Plato's Theory of Knowledge*. 1935. Reprint, London.

———. 1969. *Thucydides Mythistoricus*. 1907. Reprint, New York.

Corpus Hermeticum. 1945. Ed. A. D. Nock and A. J. Festugière. Paris.

Coulton, J. J. 1974. "Lifting in Early Greek Architecture." *JHS* 94.

———. 1975. "Towards Understanding Greek Temple Design: General Considerations." *Brit. Sch. Athens Ann.* 70.

———. 1976. *The Architectural Development of the Greek Stoa*. Oxford.

———. 1977. *Greek Architects at Work*. London.

———. 1988. "Post Holes and Post Bases in Early Greek Architecture." *MA* 1.

Coupel, Pierre, and Pierre Demargne. 1969. "Le Monument des Néréides." *Fouilles de Xanthos* 3.

Courbin, Paul. 1973. "Le Colosse Naxien et le Palmier de Nicias." *BCH*, suppl. 1.

———. 1980a. *L'Oikos des Naxiens*. Paris.

———. 1980b. "Lyres d'Argos." *BCH*, suppl. 6, *Etudes Argiennes*.

Cragg, G. R. 1966. *The Church and the Age of Reason, 1648–1789*. Harmondsworth.

Crawford, O. G. S. 1957. *The Eye Goddess*. London.

Cristofani, Mauro. 1978. *L'Arte degli Etruschi*. Turin.

———. 1987. *Saggi di Storia Etrusca Arcaica*. Rome.

Cristofani, Mauro, and Paola Pelagatti, eds. 1985. *Il Commercio Etrusco Arcaico*. CNR. Rome.

Critchlow, Keith. 1976. *Islamic Patterns*. London.

Crossland, R. A., and A. Birchall, eds. 1973. *Bronze Age Migrations in the Aegean*. London.

Cumont, Franz. 1929. *Les Religions Orientales dans le Paganisme Romain*. 4th ed. Paris.

———. 1960. *Astrology and Religion among the Greeks and Romans*. 1912. Reprint, New York.

Cureau de la Chambre, Marin. 1648–1662. *Les Charactères des Passions*. 4 vols. Paris.

———. 1650. *L'Iris*. Paris.

———. 1664. *Traité de la Connaissance des Animaux*. Paris.

———. 1669. *L'Art de Connoistre les Hommes*. 3 vols. Amsterdam.

Curtius, Ernst Robert. 1953. *European Literature and the Latin Middle Ages*. New York.

Curtius, Ludwig. 1925. Review of *Phidias* by Hans Schrader. *Gnomon*. Berlin.

Cusa, Nicholas of. 1954. *Of Learned Ignorance*. Ed. G. M. Heron. London.

———. *See also* Blumenberg, Hans.

Czech, Hermann, and Wolfgang Mistelbauer. 1976. *Das Looshaus*. Vienna.

Dädalische Kunst auf Kreta in 7. Jahrhundert v. Chr. 1990. Exhibition Catalogue. Hamburg Museum für Kunst u. Gewerbe.

Dal Co, Francesco, and Manfredo Tafuri. 1976. *Architettura Contemporanea*. Milan.

Dalley, Stephanie. 1984. *Mari and Karana*. London.

———. trans. and ed. 1989. *Myths from Mesopotamia*. Oxford.

Dalley, Stephanie, and Norman Yoffee. 1991. *Old Babylonian Texts in the Ashmolean Museum*. Oxford.

Dani, Ahmad Hasan. 1981. *Prehistory and Protohistory of Eastern India*. Calcutta.

Danthine, H. 1937. *Le Palmier-Dattier et les Arbres Sacrés*. 2 vols. Paris.

Darmon, Albert. 1985. *Les Corps Immateriels: Esprits et Images dans l'Oeuvre de Marin Cureau de la Chambre*. Paris.

Darcque, Pascal, 1981. "Les Vestiges Mycéniens Découverts Sous le Télestérion d'Eleusis." *BCH* 105.

Darwin, Charles. 1873. *The Expression of the Emotions in Man and Animals*. London.

Dassy, L. 1877. *Compte-Rendu sur la Restauration des Monuments Antiques par les Architectes Pensionnaires de l'Académie de France à Rome*. Paris.

David, Rosalie. 1981. *A Guide to Religious Ritual at Abydos*. Warminster.

Daviler, C. A. 1720. *Cours d'Architecture*. 2 vols. Rev. Jean Le Blond and Jean Mariette. Paris.

Davis, Flora. 1978. *Eloquent Animals*. New York.

Davy, Marie-Magdeleine. 1977. *Initiation à la Symbolique Romane*. Paris.

Dawkins, R. M., et al. 1929. *The Sanctuary of Artemis Orthia*. London.

Debus, Allen G. 1965. *The English Paracelsians*. London.

———. 1977. *The Chemical Dream of the Renaissance*. New York.

———. 1978. *Man and Nature in the Renaissance*. Cambridge.

Dekolakou-Sideris, I. 1990. "A Metrological Relief from Salamis." *AJA* 94.

Delatte, A. 1915. *Etudes sur la Littérature Pythagoricienne*. Paris.

Delbrück, R. 1907–1912. *Hellenistische Bauten in Latium*. 2 vols. Berlin.

Del Chiaro, Mario A., and William R. Biers, eds. 1986. *Corinthiaca: Studies in Honor of Darrell A. Amyx*. Columbia, Mo.

Delcourt, Marie. 1965. *Pyrrhos et Pyrrha*. Paris.

Delestre, Jean Baptiste. 1866. *De la Physiognomonie*. Paris.

Della Porta, Giambattista. 1586. *De Humana Physiognomonica*. Naples. In Italian, *Della Fisionomia dell'Uomo*, trans. M. Cicognani (Parma 1988).

De l'Orme, Philibert. 1648. *Oeuvres*. Paris. (Reprinted Ridgewood, N.J., 1964.)

Delvoye, Charles, and Georges Roux. 1967. *La Civilisation Grecque de l'Antiquité à nos Jours*. 2 vols. Brussels.

Demangel, R. 1931. "Fenestrarum Imagines." *BCH* 55.

——. 1932. *La Frise Ionique*. Bibliothèque des Ecoles Françaises d'Athènes et de Rome. Paris.

——. 1946. "Fenestrarum Imagines, bis." *BCH* 70.

Demargne, Pierre, ed. 1947. *La Crète Dédalique*. Paris.

——. 1958–1989. *Fouilles de Xanthos*. Paris.

——. 1964. *The Origins of Aegean Art*. Trans. S. Gilbert and J. Emmons. New York.

Demisch, H. 1977. *Die Sphinx*. Stuttgart.

Dempster, Thomas, of Muresk. 1723. *De Etruria Regali*. 3 vols. Ed. Thomas Coke. Florence.

Deonna, Waldemar. 1909. *Les Apollons Archaiques*. Geneva.

——. 1914. *Les Lois et les Rythmes dans l'Art*. Paris.

——. 1930. *Dédale, ou la Statue de la Grèce Archaique*. 2 vols. Paris.

——. 1965. *Le Symbolisme de l'Oeil*. Paris.

Derham, William. 1713. *Physico-Theology, or a demonstration of the being and attributes of God from His works of Creation*. London.

——. 1714. *Astro-Theology or a demonstration of the being and attributes of God, from a survey of the Heavens*. London.

——. 1730. *Christo-Theology or a demonstration of the divine authority of the Christian religion*. London.

Derrida, J. 1967. *L'Ecriture et la Différence*. Paris.

——. 1978. *La Vérité en Peinture*. Paris.

Désargues, Gérard. 1648. *Manière Universelle "pour pratiquer la perspective."* Paris (reprinted 1987).

——. *See also* Bosse, Abraham.

Desborough, V. R. d'Arba. 1964. *The Last Mycenaeans and Their Successors*, Oxford.

Descartes, René. 1897–1909. 11 vols. *Oeuvres*. Ed. Charles Adam and Paul Tannery. Paris.

——. 1953. *Oeuvres et Lettres*. Ed. André Bridoux. Paris.

——. 1988. *Les Passions de l'Ame*. Ed. Jean-Maurice Monnoyer. Paris.

Desgodets, Antoine. 1682. *Les Edifices Antiques de Rome*. Paris.

Le Dessin d'Architecture dans les Sociétés Antiques. See Bommelaer, J.-F.

Dessoir, Max. 1970. *Aesthetics and Theory of Art*. Detroit.

Detienne, Marcel. 1967. *Les Maitres de Vérité dans la Grèce Archaique*. Paris.

——. 1977. *Dionysos mis à Mort*. Paris.

——. 1986. *Dionysos à Ciel Ouvert*. Paris.

Detienne, Marcel, and J.-P. Vernant. 1979. *La Cuisine du Sacrifice en Pays Grec*. Paris.

Deubner, Ludwig. 1932. *Attische Feste*. Berlin.

——. 1934. "Eine Neue Lenäenvase." *JDAI* 49.

Deuel, L. 1977. *The Memoirs of Heinrich Schliemann*. New York.

De Vinne, Theodore Low. 1910. *Notable Printers of Italy During the Fifteenth Century*. New York.

Devoto, Giacomo, ed. 1951. *Gli Antichi Italici*. Florence.

——. 1954–1962. *Tabulae Iguvinae*. 2 vols. Rome.

Dickins, Guy, and S. M. Casson. 1912–1921. *Catalogue of the Archaic Sculptures in the Acropolis Museum*. 2 vols. Cambridge.

Dieterici, Friedrich Heinrich. 1969. *Die Philosophie der Araber im Zehnten Jahrhundert nach Christi*. 2 vols. 1865–78. Reprint, Hildesheim.

Dieulafoy, Marcel. 1885. *L'Art Antique de la Perse*. Paris.

Dilke, Lady Emilia Frances Strong. 1888. *Art in the Modern State*. London.

——. *See also* Pattison, Mrs. Mark.

Dinsmoor, William Bell. 1947. "The Hekatompedon on the Athenian Akropolis." *AJA* 51.

——. 1950. *The Architecture of Ancient Greece*. London.

Dionysius of Fourna. 1974. *The Painter's Manual*. Trans. and ed. Paul Hetherington. London.

Diroton, Etienne. 1957. *Page d'Egyptologie*. Cairo.

Diwald, Susanne, ed. 1975. *Ihwan as-Safa: Arabische Philosophie und Wissenschaft in der Enzyklopädie*. Wiesbaden.

Dodds, E. R. 1964. *The Greeks and the Irrational*. Berkeley.

Döhmer, Klaus. 1976. *Im Welchem Style Sollen Wir Bauen?* Munich.

Donahue, A. A. 1988. *Xoana and the Origins of Greek Sculpture*. Atlanta.

Donaldson, Thomas Leverton. 1859. *Architectura Numismatica*. London.

Doniger [O'Flaherty], Wendy. 1980. *Women, Androgynes, and Other Mythical Beasts*. Chicago.

Donin, R. K. 1948. *Vincenzo Scamozzi und der Einfluss Venedigs auf die Salzburger Architektur*. Innsbruck.

Donne, John. 1978. *The Divine Poems*. Ed. Helen Gardner. Oxford.

Dörpfeld, Wilhelm. 1883. "Das Schatzhaus der Sikyoner." *MDAIA* 8.

——. 1935. *Alt-Olympia.* 2 vols. Berlin.

——. 1937. *Alt-Athen und Seine Agora.* Berlin.

Douglas, A. E. 1956. "Cicero, Quintilian, and the Canon of the Attic Orators." *Mnemosyne,* 4th ser., 9.

Drees, J. 1962. *Der Ursprung der Olympischen Spiele.* Schorndorf bei Stuttgart.

Drerup, Heinrich. 1969. "Griechische Baukunst in Geometrischer Zeit." *ArchHom* 2.

——. 1986. *Das Sogenannte Daphnephoreion in Eretria.* Studien zur Archäologie und Alten Geschichte. Saarbrücken.

——. *See also* Büsing, Hermann, and Friedrich Hiller.

Drews, Robert H. 1983. *Basileus: The Evidence for Kingship in Geometric Greece.* New Haven.

——. 1988. *The Coming of the Greeks: Indo-European Conquests in the Aegean and the Near East.* Princeton.

Drexler, A., ed. 1977. *The Architecture of the Ecole des Beaux-Arts.* New York.

Ducat, Jean. 1971. *Les Kouroi du Ption.* Paris.

Dufrenne, Mikel. 1966. *The Notion of the A Priori.* Trans. Edward S. Casey. Evanston, Ill.

——. 1967. *La Phénoménologie de l'Expérience Esthétique.* Paris.

Du Fresnoy, Charles Alphonse. 1668. *De Arte Graphica.* Paris. Trans. John Dryden (London, 1695).

Dugas, Charles, Jules Berchmans, and Mogens Clemmensen. 1924. *Le Sanctuaire d'Aléa Athéna à Tégée au IVᵉ Siècle.* Paris.

Dugas, Charles, and R. Flacelière. 1958. *Thésée: Images et Récits.* Paris.

Duhem, Pierre M. N. 1906–1909. *Etudes sur Léonard da Vinci.* 3 vols. Paris.

Dumézil, Georges. 1970. *Archaic Roman Religion.* 2 vols. Chicago and London.

Dunbabin, T. J. 1948a. "The Early History of Corinth." *JHS* 68.

——. 1948b. *The Western Greeks.* Oxford.

——. 1979. *The Greeks and Their Eastern Neighbors.* Westport, Conn.

Dunne, John S. 1974. *The City of the Gods.* London.

Dupont-Sommer, A. 1949. *Les Araméens.* Paris.

Durand, Jean-Nicolas-Louis. 1819. *Précis des Leçons d'Architecture données à l'Ecole Royale Polytechnique.* Paris.

Durandus [Durantis], Guilelmus, (sometime Bishop) of Mende. 1496. *Repertorium Aureum Gulielmi Durandi.* Venice. Trans. C. Barthélemy (Paris, 1854).

——. 1843. *The Symbolism of Churches and Church Ornaments: A Translation of the First Book of the Rationale Divinorum Officiorum.* Trans. John Mason Neale and Benjamin Webb. London.

——. 1859. *Rationale Divinorum Officiorum.* Ed. John Beleth. Naples.

Dürer, Albrecht. 1528. *Hierinn sind begriffen vier Bücher von Menschlicher Proportion.* Nürnberg.

Durkheim, Emile. 1961. *The Elementary Forms of Religious Life.* 1912. Trans. Joseph Ward Swain. New York.

Durm, Josef. 1910. *Die Baukunst der Griechen.* Leipzig.

Ebin, Victoria. 1979. *The Body Decorated.* London.

Eckstein, Felix. 1969. ΑΝΑΘΗΜΑΤΑ: *Studien zu den Weihgeschenken Strengen Stils im Heiligtum von Olympia.* Berlin.

Eco, Umberto. 1987. *Arte e Bellezza nell'Estetica Medievale.* Milan.

Ecole Française d'Athènes. 1991. *Guide de Delphes: Le Musée.* Paris.

Edwards, I. E. S. 1967. *The Pyramids of Egypt.* Harmondsworth.

Edwards, Ruth B. 1979. *Kadmos the Phoenician.* Amsterdam.

Ehrenberg, Victor. *See* Badian, E.

Eichel, Christine. 1993. *Vom Ermatten der Avantgarde zur Vernetzung der Künste.* Frankfurt A/M.

Einem, Herbert von. 1955. "Der Mainzer Kopf mit der Binde." *Arbeitsgemeinschaft für Forschung des Landes Nordrhein-Westfalen* 37.

Eisler, Robert. 1910. *Weltenmantel und Himmelszelt.* Munich.

——. 1946. *The Royal Art of Astrology.* London.

Eitrem, S. 1910–1920. *Beiträge zur Griechischen Religionsgeschichte.* Kristiania.

Elderkin, George Wicker. 1912. *Problems in Periclean Buildings.* Princeton.

——. 1924. *Kantharos: Studies in Dionysiac and Kindred Cults.* Princeton.

Eliade, Mircea. 1964. *Shamanism: Archaic Techniques of Ecstasy.* Trans. Willard R. Trask. New York.

——. 1965. *The Two and the One.* Trans. J. M. Cohen. London.

——. 1969. *Le Mythe de l'Eternel Retour*. Paris.

——. 1977. *Forgerons et Alchimistes*. Paris.

Elkin, A. P. 1938–1974. *The Australian Aborigines*. North Ryde, N. S. W.

Ellis, Richard E. 1968. *Foundation Deposits of Mesopotamia*. New Haven.

Else, G. F. *See* Aristotle.

Emery, Walter B. 1961. *Archaic Egypt*. Harmondsworth.

Emery, Walter B., et al. 1949–1958. *Great Tombs of the First Dynasty*. 3 vols. Oxford.

Emlyn-Jones, C. J. 1980. *The Ionians and Hellenism*. London.

Encyclopaedia of Islam. 1979. 2nd ed. Leyden and London.

Eppelsheimer, Rudolf. 1968. *Mimesis und Imitatio Christi*. Bern.

Ericsson, Christoffer H. 1980. *Roman Architecture Expressed in Sketches by Francesco di Giorgio Martini*. Commentationes Humanarum Litterarum 66. Helsinki.

Erlande-Brandenburg, Alain, ed. 1986. *Carnet de Villard de Honnecourt: d'après Le Manuscrit Conservé à la Bibliothèque Nationale de Paris*. Paris.

Erlich, Victor. 1965. *Russian Formalism*. 2nd ed. London and The Hague.

Erman, Adolf. 1907. *A Handbook of Egyptian Religion*. Trans. A. S. Griffith. London.

——. 1934. *Die Aegyptische Religion*. Berlin and Leipzig.

——, ed. 1966. *The Ancient Egyptians*. New York.

Esmeijer, Anna C. 1978. *Divina Quaternitas*. Amsterdam.

d'Espouy, Hector. 1893–1905. *Fragments d'Architecture Antique d'après les Relevés des Anciens Pensionnaires de l'Académie de France à Rome Publiés sous la Direction de H(ector) d'Espouy*. 2 vols. Ed. C. Schmid. Paris.

Eugippius, Abbot. 1885. *Eugipii Excerpta ex Operibus Scti Augustini*. Ed. Pius Knoell. Vienna

Evans, Sir Arthur. 1901. "The Mycenaean Tree and Pillar Cult." *JHS* 21. London.

——. 1921–1936. *The Palace of Minos at Knossos*. 4 vols. London.

——. *See also* Casson, Stanley.

Evans, Elizabeth C. 1969. *Physiognomics in the Ancient World*. Philadelphia.

Evans, M. 1980. "The Geometry of the Mind." *AAQ* 12.4.

Fairbanks, Arthur. 1907. *Athenian Lekythoi*. New York. Slightly revised, 1914.

Falk, Maryla. 1986. *Il Mito Psicologico nell'India Antica*. Milan.

Falkener, Edward. 1862. *Ephesus and the Temple of Diana*. London.

Faris, J. C. 1973. *Nuba Personal Art*. Toronto.

Farnell, L. R. 1896–1909. *The Cults of the Greek States*. 5 vols. Oxford.

——. 1921. *Greek Hero Cults and Ideas of Immortality*. Oxford.

Farrell, M. 1981. *William Whiston*. New York.

Fast, Julius. 1977. *Body Language*. New York.

Favaro, G. 1917. "Il Canone di Leonardo da Vinci sulle Proporzioni del Corpo Umano." *Atti del R. Istituto Veneto di Scienze, Lettere ed Arti* 77.

——. 1918. "Misure e Proporzioni del Corpo Umano secondo Leonardo," *Atti del R. Istituto Veneto di Scienze, Lettere ed Arti* 78.

Fehrle, Eugen. 1910. *Die Kultische Keuschheit im Altertum*. Giessen.

Fernea, Robert Alan. 1959. "Irrigation and Social Organization among El Shabana." Ph.D. diss. University of Chicago.

Fernie, Eric. 1981. "The Metrological Relief in Oxford." *AJ* 61.

Ferrari, S. 1900. *I Tempi, la Vita, le Dottrine di Pietro d'Abano*. Genoa.

——. 1915. "Per la Biografia e per gli Scritti di Pietro d'Abano." *Memorie. Classe di Scienza Morali, Storiche e Filologica, Reale Accademia dei Lincei*, 5th ser., 15.

Ferri, Silvio. 1940. "Nuovi Contributi Esegetici al canone' della Scultura Greca." *Riv. Ist. Arch.* 7.

——, ed. 1946. *Plinio il Vecchio, Storia delle Arti Antiche*. Rome.

Ferri, S. *See* Vitruvius.

Ferry, Luc. 1990. *Homo Aestheticus*. Genoa.

Festugière, A.-J. 1944–1954. *La Révélation d'Hermès Trismégiste*. 4 vols. Paris.

Field, George. 1839. *Outlines of Analogical Philosophy*. London.

Field, H. 1936. "Horns and Skulls on Buildings." *Antiquity* 10.

Filarete. *See* Averlino, Antonio.

Fink, Eugen. 1969. *Il Gioco come Simbolo del Mondo*. Trans. Nadia Antuono. Rome.

Finkelstein, Israel. 1988. *The Archaeology of the Israelite Settlements*. Jerusalem.

Finley, M. I. 1976. *The Olympic Games: The First 1000 Years*. London.

——. 1981. *Economy and Society in Ancient Greece.* New York.

Finsler, Georg. 1900. *Platon und die Aristotelische Poetik.* Leipzig.

Firenze e la Toscana dei Medici. 1980. 4 vols. Florence.

Flasche, Hans. 1949. "Similitudo Templi." *Deutsche Vierteljahrsschrift für Literaturwissenschaft und Geistesgeschichte* 23. Halle.

Flasch, K., ed. 1965. *Pariusia: Festschrift J. Kirchberger.* Frankfurt.

Flavio Biondo [Blondus]. 1559. *Works.* Basel.

Fleischer, Robert. 1973. *Artemis von Ephesos und Verwandte Kultstatuen aus Anatolien und Syrien.* Leiden.

Fliche, A., and V. Martin, eds. 1934–. *Histoire de l'Eglise depuis les origines jusqu'à nos jours.* 20 vols. to date. Paris.

Flores d'Arcais, G., ed. 1962. *Niccolo da Cusa.* Florence.

Fontaine, André. 1903. *Quid Senserit Carolus Le Brun de Arte Sua.* Paris.

——. 1909. *Les Doctrines d'Art en France.* Paris.

——. 1914. *Académiciens d'Autrefois.* Paris.

Fontaine, Jacques. 1959. *Isidore de Seville et la Culture Classique dans l'Espagne Wisigothique.* Paris.

Fontana, Carlo. 1694. *Templum Vaticanum et Ipsius Origo.* Rome.

Fontana, Vincenzo. eds. 1973. *Artisti e Committenti nella Roma del Quattrocento.* Rome.

Fontana, Vincenzo, and Paolo Morachiello, eds. 1975. *Vitruvio e Raffaello.* Rome.

Fontanier, Pierre. 1977. *Les Figures du Discours.* Paris.

Fontenrose, Joseph. 1960. *The Cult and Myth of Pyrros at Delphi.* Berkeley.

Forbes, R. J. 1965. *Studies in Ancient Technology.* Leiden.

Forbes, Thomas B. 1983. *Urartian Architecture.* Oxford.

Force, J. E. 1985. *William Whiston, Honest Newtonian.* Cambridge.

Forrest, W. G. 1968. *A History of Sparta.* London.

Forssman, Erik. 1956. *Säule und Ornament.* Stockholm.

——. 1961. *Dorisch, Ionisch, Korinthisch.* Stockholm.

Förster, P. R. 1893. *Scriptores, Physiognominici Graeci et Latini.* Leipzig.

Forsythe, R. S. 1935. "Herman Melville in Honolulu." *New England Quarterly* 8.

——. 1936. "Herman Melville in the Marquesas." *PQ* 15.

——. 1937. "Herman Melville in Tahiti." *PQ* 16.

Fougères, Gustave, and J. Hulot. 1910. *Selinonte: La Ville, l'Acropole et le Temple.* Paris.

Fowden, Garth. 1991. "Constantine's Porphyry Column: The Earliest Literary Allusion." *JRS* 81.

Fowler, Harold N., and Richard Stillwell. 1932. *Corinth: Excavations of the American School of Classical Studies at Athens.* Cambridge, Mass.

Fox, Michael V., ed. 1988. *Temple in Society.* Winona Lake, Wis.

Frame, Douglas. 1978. *The Myth of Return in Early Greek Epic.* New Haven.

Francesco di Giorgio. *See* Martini, Francesco di Giorgio.

Francis, E. D., and M. Vickers. 1983. "Signa Priscae Artis: Eretria and Siphnos." *JHS* 103.

Frank, E. 1923. *Plato und die sogenannten Pythagoreer: Ein Kapitel aus der Geschichte des Griechischen Geistes.* Halle.

Frankfort, Henri. 1933. *The Cenotaph of Seti I at Abydos.* London.

——. 1939. *Cylinder Seals.* London.

——. 1948a. *Ancient Egyptian Religion.* New York.

——. 1948b. *Kingship and the Gods.* Chicago.

——. 1954. *The Art and Architecture of the Ancient Orient.* Harmondsworth. Revised 1963 and 1970.

Frankl, Paul, and Panofsky. Erwin 1945, March. "The Secret of the Mediaeval Masons." *AB* 27.

Frazer, Sir James George. 1911–1915. *The Golden Bough.* 12 vols. London.

——. 1919. *Folk-Lore in the Old Testament.* 3 vols. London.

——. *See also* Pausanias.

Fréart, Paul, Sieur de Chantelou. 1981. *Journal de Voyage du Cavalier Bernin en France.* 1885. Ed. L. Lalanne. Paris.

Fréart, Roland, Sieur de Chambray. 1650. *Parallèle de l'Architecture Antique et de la Moderne.* Paris. Rev. ed. 1702. Trans. John Evelyn, *A Parallel of the Ancient Architecture with the Modern* (London, 1664).

——. 1662. *Idée de la Perfection de la Peinture. . . .* Le Mans.

Freud, Sigmund. 1953–1974. *Works.* 5 vols. Ed. Anna Freud with Alix Strachey and Alan Tyson. Trans. supervised by Joan Rivière. London.

Freyberg, Klaus S. 1990. *Stadtrömische Kapitelle aus der Zeit Domitian bis Alexander Severus.* Mainz.

Frickenhaus, August. 1912. *Lenäenvasen.* Berlin.

Friedländer, Paul. 1958–1969. *Plato.* 3 vols. Trans. Hans Meyerhoff. New York.

Friis Johansen, K. 1951. *The Attic Grave-Reliefs of the Classical Period.* Copenhagen.

Frings, Hubert, and Mable Frings. 1975. *Animal Communications.* Norman, Okla.

Frisch, Karl von. 1967. *The Dance Language and the Orientation of Bees.* Cambridge, Mass.

——. 1968. *Bees: Their Vision, Chemical Senses, and Language.* London.

——. 1974. *Animal Architecture.* New York and London.

Frobenius, Leo. 1933. *Kulturgeschichte Afrikas.* Vienna.

Frontinus, S. Julius. 1925. *The Stratagems and the Aqueducts of Rome.* Trans. Charles E. Bennett. London.

Frotier de la Coste Messelière, P. 1936. *Au Musée de Delphes.* Paris.

——. 1943. *Delphes.* Paris.

——. *See also* Picard, Charles.

Früchtl, Josef. 1986. *Mimesis: Konstellation eines Zentralbegriffs bei Adorno.* Würzburg.

Frye, Richard N. 1976. *The Heritage of Persia.* London.

Fubini, Enrico. 1964. *L'Estetica Musicale dal Settecento a Oggi.* Turin.

Fuller, Thomas. 1841. *The Holy State and the Profane State.* 1642. Ed. James Nichols. London.

Fumaroli, Marc. 1980. *L'Age de l'Eloquence: Rhétorique et "Res Literaria" de la Renaissance au Seuil de l'Epoque Classique.* Geneva.

——. 1981. "Le Corps Eloquent." *Dix-Septième Siècle.* Paris.

Furtwängler, Adolf. 1893. *Meisterwerke der Griechischen Plastik.* Berlin. Translated as *Masterpieces of Greek Sculpture: A Series of Essays on the History of Art,* ed. H. Oikonomides (Chicago, 1964).

Gadamer, Hans-Georg. 1986a. *The Idea of the Good in Platonic-Aristotelian Philosophy.* Trans. P. Christopher Smith. New Haven.

——. 1986b. *The Relevance of the Beautiful and Other Essays.* Cambridge.

——. 1989. *Truth and Method.* 2nd ed. Trans. J. Weinsheimer and Donald Marshall. New York.

Gage, J. 1981. "A Locus Classicus of Color Theory: The Fortunes of Apelles." *JWCI* 44.

Gaiser, Konrad. 1963. *Platons Ungeschriebene Lehre.* Stuttgart.

Galinsky, H. 1932. *Der Lucretia-Stoff in der Weltliteratur.* Breslau.

Gall, H. von. 1966. "Zu den 'Medischen' Felsgräbern in Nordwestiran und Iraqi Kurdistan." *AA* 81.

Galvano, Albino. 1967. *Artemis Ephesia.* Milan.

Gardiner, E. N. 1925. *Olympia: Its History and Remains.* Oxford.

Gardner, Percy. 1921. *The Principles of Greek Art.* New York.

Garin, Eugenio, ed. 1942. *De Hominis Dignitate, Heptaplus, de Ente et Uno, e Scritti Vari di Pico della Mirandola.* Florence.

——. 1976. *Lo Zodiaco della Vita.* Bari.

——. 1984. *L'Umanesimo Italiano.* Bari.

Garland, Robert. 1985. *The Greek Way of Death.* Ithaca.

Garstang, John. 1930. *The Hittite Empire.* New York.

Gauricus, Pomponius [Pomponio Gaurico]. 1969. *De Sculptura.* 1504. Ed. A Chastel and R. Klein. Geneva.

Gebauer, Gunter, and Christoph Wulf. 1992. *Mímesis: Kultur, Kunst, Gesellschaft.* Reinbek bei Hamburg.

Geertz, Clifford. 1973. *The Interpretation of Cultures.* New York.

Gehlen, Arnold. 1978–. *Gesamtausgabe* Frankfurt am Main.

Gelder, J. G. van. *See* Bruyn, J.

Geldner, C. F. 1951. *Der Rig-Veda.* Cambridge, Mass.

Gennep, A. van. 1960. *Les Rites de Passage.* 1909. Reprint, Paris.

Geoffrey of Monmouth. *See* Monmouth, Geoffrey of.

Georgiades. Thrasybulos. 1956. *Greek Music, Verse, and Dance.* Trans. Erwin Benedikt and Marie Louise Martinez. New York.

Gerbert, Martin, ed. 1931. *Scriptores Ecclesiastici de Musica Sacra . . . Typis San-Blasianis.* 1794. Reprint, Milan.

Gerkan, Armin von. 1925. "Die Krümmungen im Gebälk des Dorischen Tempels in Cori." *MDAIR* 40.

——. 1959. *Von Antiker Architektur und Topographie: Gesammelte Aufsätze.* Ed. Erich Boehringer. Stuttgart.

Germain, G. 1954. *Homère et la Mystique des Nombres*. Paris.

Gernet, Louis. 1981. *The Anthropology of Ancient Greece*. Baltimore and London.

Gerold, T. 1936. *Histoire de la Musique*. Paris.

Gessinger, Joachim, and Wolfert von Rahden, eds. 1989. *Theorien vom Ursprung der Sprache*. 2 vols. Berlin.

Ghiberti, Lorenzo. 1947. *I Commentari*. Ed. O. Morisani. Naples.

Ghirshman, Roman. 1962. *Iran: Parthians and Sassanians*. London.

——. 1964. *The Arts of Ancient Iran from Its Origins to the Time of Alexander the Great*. Trans. Stuart Gilbert and James Emmons. New York.

Ghyka, Matyla. C. 1931. *Le Nombre d'Or*. 2 vols. Paris.

——. 1932. *Essai sur le Rythme*. Paris.

Gibbon, Edward. 1912–1921. *The History of the Decline and Fall of the Roman Empire*. 1776–1788. 7 vols. Ed. J. B. Bury. New York.

Giedion, Sigfried. 1964. *The Eternal Present*. 2 vols. London.

Giedion Welcker, C. 1955. "Bildhafte Kachel-Kompositionen von Antonio Gaudi: Ein Vorspiel zu den 'Papiers Collés.'" *Werk* 42.4.

Giesen, Karl Joseph. 1929. *Dürers Proportionsstudien im Rahmen der allgemeinen Proportionsentwicklung*. Ph.D. dissertation, Friedrich Wilhelms Universität, Bonn.

Gill, Christopher. 1984. "The *ethos/pathos* Distinction in Rhetorical and Literary Criticism." *CQ* 34.

Gilson, Etienne. 1944. *L'Esprit de la Philosophie Médiévale*. Paris.

——. 1952. *La Philosophie au Moyen Age, des Origines Patristiques à la fin du XIVe Siècle*. Paris.

——. 1965. *The Arts of the Beautiful*. New York.

Gimbutas, Marija. 1989. *The Language of the Goddess*. San Francisco.

——. 1991. *The Civilization of the Goddess: The World of Old Europe*. San Francisco.

Ginouvès, René, and Roland Martin. 1985–. *Dictionnaire Méthodique de l'Architecture Grecque et Romaine*. 3 vols. to date. Rome.

Ginzberg, Louis. 1913. *The Legends of the Jews*. 7 vols. Philadelphia.

Ginzburg, Carlo. 1982. *Indagini su Piero*. Turin.

——. 1989. *Storia Notturna*. Turin.

Giovannoni, G. 1908. "La Curvatura delle Linee nel Tempio d'Ercole a Cori." *MDAIR* 23.

Girard, René. 1977.

——. 1978. *Des Choses Cachées depuis la Fondation du Monde*. Paris.

Gjerstad, Einar. 1953–1973. *Early Rome*. 6 vols. Lund.

Glaser, Franz. 1983. *Antike Brunnenbauten in Griechenland*. Vienna.

Gli Etruschi a Roma, Studies in Honour of Massimo Pallottino. 1981. Rome.

Goblet d'Alviella, Eugène. 1891. *La Migration des Symboles*. Paris.

Goedicke, Hans. 1955. "A Deification of a Private Person in the Old Kingdom." *JEA* 41.

Goethe, Johann Wolfgang von. 1948–1964. *Werke (Gedenkausgabe)*. 27 vols. Ed. Ernst Beutler. Zurich.

Goldman, Bernard. 1966. *The Sacred Portal*. Detroit.

Goldschmidt, L. 1894. *Sepher Jesirah, Das Buch der Schöpfung*. Frankfurt A/M.

Golzio, Karl-Heinz. 1983. *Der Tempel im Alten Mesopotamien und seine Parallelen in Indien*. Leiden.

Golzio, Vincenzo. 1971. *Raffaello nei Documenti: Nelle Testimonianze dei Contemporanei e nella Letteratura del Suo Secolo*. Rev. ed. [Westmead], England.

Gombrich, Sir Ernst. 1965. *Meditations on a Hobby Horse*. London.

——. 1968. *Art and Illusion*. London.

Gonda, Jan. 1975. *Vedic Literature*. Wiesbaden.

Goodden, Angelica. *"Actio" and Persuasion: Dramatic Performance in Eighteenth-Century France*. Oxford.

Goodenough, Erwin R. 1938. *The Politics of Philo Judaeus*. New Haven.

Goody, Jack, ed. 1968. *Literacy in Traditional Societies*. Cambridge.

Goodyear, W. H. 1894a, March. "The Lotiform Origin of the Greek Anthemion." *Architectural Record*.

——. 1894b, September. "The Acanthus Motive and of the Egg-and-Dart Moulding." *Architectural Record*.

——. 1896. *A Discovery of Greek Horizontal Curves in the Maison Carrée at Nimes*. Washington, D.C.

——. 1902a. *The Architectural Refinements of St. Mark's at Venice*. New York.

——. 1902b. *A Renaissance Leaning Façade at Genoa*. New York.

——. 1904. *Vertical Curves and Other Architectural Refinements in the Gothic Cathedrals of*

Northern France and in Early Byzantine Churches. New York.

———. 1912. *Greek Refinements.* New Haven.

Gordon, Cyrus H. 1966. *Ugarit and Minoan Crete.* New York.

Gould, Cecil. 1981. *Bernini in France: An Episode in Seventeenth-Century History.* London.

Gould, Stephen Jay. 1981. *The Mismeasure of Man.* New York.

Grapaldi, Francesco Maria. 1508. *De Partibus Aedium.* Parma.

Grassi, Ernesto. 1962. *Die Theorie des Schönen in der Antike.* Köln.

Gravagnuolo, B. 1983. *Adolf Loos.* New York.

Greco, Emanuele, and Dinu Théodorescu. 1987. *Poseidonia-Paestum.* Rome.

Greek Art of the Aegean Islands. 1979. Exhibition Catalogue, November 1979–February 1988. Metropolitan Museum of Art, New York.

Green, Rosalie, et al., eds. 1979. *Herrad of Hohebourg: Hortus Deliciarum.* 2 vols. London and Leiden.

Gregory of Nyssa. 1943. *De Opificio Hominis.* Trans. and ed. J. Laplace and J. Danielou. Paris.

Gregory, R. L. 1966. *Eye and Brain.* London.

Gregory, Tullio. 1955. *Anima Mundi.* Rome.

Grenier, A. 1948. *Les Religions Etrusques et Romaines.* Paris.

Griffin, Donald R. 1976. *The Question of Animal Awareness.* New York.

Grill, L. 1961. "Die neunzehn 'Capitula' Bernhards von Clairvaux gegen Abelärd." *HJ* 80.

Grimes, Ronald L. 1982. *Research in Ritual Studies.* Chicago.

Groenewegen-Frankfort, H. A. 1951. *Arrest and Movement.* London.

Gros, P., ed. 1983. *Architecture et Société.* Paris.

Grosse, E. 1902. *Les Débuts de l'Art.* Trans. E. Dirr. Paris.

Grottanelli, C., and N. F. Parise, eds. 1988. *Sacrificio e Società nel Mondo Antico.* Bari.

Grube, G. M. A. 1965. *Greek and Roman Critics.* London.

Gruben, Gottfried. 1984. *Die Tempel der Griechen.* Darmstadt.

———. 1988. "Fundamentierungsprobleme der ersten Archaischen Grossbauten." In H. Büsing and F. Hiller (1988).

Gruppe, Otto. 1906. *Griechische Mythologie und Religionsgeschichte.* 2 vols. Munich.

———. 1921. *Geschichte der Klassischen Mythologie und Religionsgeschichte.* Leipzig.

Gsell, Stéphane. 1952. *Cherchel: Antique Iol-Caesarea.* Algiers.

Guadet, Julien. 1909. *Eléments et Théorie de l'Architecture.* 4 vols. 3rd ed. Paris.

Guarducci, M. 1937a. "La 'Eschara' del Tempio Greco Archaico." *StMSR* 13.

———. 1937b. "Un Giudizio del Santuario di Alea a Mantinea." *StMSR* 13.

Guénon, René. 1962. *Symboles Fondamentaux de la Science Sacrée.* Paris.

Guilaine, Jean. 1976. *Premiers Bergers et Paysans de l'Occident Méditerranéen.* Paris and The Hague.

Gundel, Hans Georg. 1992. *Zodiakos: Tierkreisbilder im Altertum. Kosmische Bezüge und Jenseitsvorstellungen im Antiken Alltagsleben.* Mainz am Rhein.

Gundel, W. 1936. *Dekane und Dekansternbilder.* Hamburg.

Guralnick, Eleanor. 1978. "The Proportions of Kouroi." *AJA* 82.

Güterbock, Hans G. *See* Hoffner, H. A.

Guthrie, R. Dale. 1976. *Body Hot Spots.* New York.

Guthrie, W. K. C. 1962–1978. *A History of Greek Philosophy.* 6 vols. Cambridge.

Gwilt, Joseph. 1859. *An Encyclopaedia of Architecture.* London.

Habermas, Jürgen. 1990. *The Philosophical Discourse of Modernity.* Trans. Frederick G. Lawrence. Cambridge, Mass.

Haddon, Alfred C. 1898. *The Study of Man.* New York.

Hadingham, Evan. 1983. *Early Man and the Cosmos.* London.

Hägg, R. 1986. "Die Göttliche Epiphanie im Minoischen Ritual." *MDAI* 101.

———. 1990. "The Cretan Hut-Models." *OA* 18.

———. 1992. "Geometric Sanctuaries in the Argolid." *BCH* suppl. 22.

Hägg, R., and Dora Konsola, eds. 1986. *Early Helladic Architecture and Urbanization.* Göteborg.

Hägg, R., and N. Marinatos. 1981. *Sanctuaries and Cults in the Aegean Bronze Age.* Stockholm.

———, eds., 1987. *The Function of the Minoan Palaces.* Stockholm.

Hägg, R., N. Marinatos, and G. Nordquist, eds. 1988. *Early Greek Cult Practice.* Stockholm.

———, eds. 1993. *Greek Sanctuaries.* London.

Hahnloser, Hans R., ed. 1929. *Das Musterbuch*

von Wolfenbuttel: mit einem Fragment aus dem Nachlasse Fritz Ruckers. Vienna.

——. 1935. *Villard de Honnecourt: Kritische Gesamtausgabe des Bauhüttenbuches.* Vienna. (Reprint, Graz, 1972.)

Hales, Stephen. 1969. *Vegetable Staticks.* London.

Hall, H. R. 1915. *Aegean Archeology.* London.

Hambidge, Jay. 1919–1920. "Dynamic Symmetry." *The Diagonal* 1.

Hambly, W. D. 1925. *History of Tattooing and Its Significance.* London.

Hamerton-Kelly, Robert G., ed. 1987. *Violent Origins.* Palo Alto.

Hamilton, Augustus. 1961. *The Art and Workmanship of the Maori Race in New Zealand.* 1900. Reprint, London.

Hamilton, Victoria. 1982. *Narcissus and Oedipus.* London.

Hampe, Roland. 1938. "Ein Bronzenes Beschlagblech aus Olympia." *AA* 3–4.

——. 1960. *Ein Frühattischer Grabfund.* Mainz.

Hampe, Roland, and Erika Simon. 1981. *The Birth of Greek Art.* New York.

Hanell, Krister. 1966. "Probleme der Römischen Fasti." *Fondation Hardt* 13.

Hanfmann, G. M. A., and J. Waldbaum, eds. 1975. *A Survey of Sardis and the Major Monuments Outside the City Walls.* Cambridge, Mass.

Hanfmann, George. *See* Mitten, David Gordon.

Hani, Jean. 1962. *Le Symbolisme du Temple Chrétien.* Paris.

Haran, Menahem. 1978. *Temples and Temple Service in Ancient Israel: An Inquiry into the Character of Cult Phenomena and the Historical Setting of the Priestly School.* Oxford.

Harden, Donald. 1962. *The Phoenicians.* London.

Harder, Carl Christian. 1886. *De Joannis Tzetzae Historiarum Fontibus: Quaestiones Selectae.* Kiel.

Hardie, W. F. R. 1968. *Aristotle's Ethical Theory.* Oxford.

Harrison, J. E. 1903. *Prolegomena to the Study of Greek Religion.* Cambridge.

——. 1913. *Ancient Art and Ritual.* London.

——. 1963. *Themis: A Study of the Social Origins of Greek Religion.* 1912. Reprint, London.

——. *See also* Verrall, M. de G.

Harvey, William. 1978. *Exercitatio Anatomica de Motu Cordis et Sanguinis in Animalibus.* Trans. G. Keynes. 1928. Reprint, Birmingham, Ala. First published Frankfurt, 1628.

Haselberger, Lothar. 1985, December. "The Construction Plans for the Temple of Apollo at Didyma." *Scientific American.*

——. 1988. "Der Eustylos des Hermogenes." In *Hermogenes und die Hochhellenistische Architektur,* ed. W. Hoepfner and E.-L. Schwandner. Mainz.

Haselberger, Lothar, and Hans Seybold. 1991. "Seilkurve oder Ellipse." *AA* 1.

Haskins, C. H. 1927. *The Renaissance of the Twelfth Century.* Cambridge, Mass.

Hauck, Guido. 1879. *Die Subjektive Perspektive und die Horizontalen Kurvaturen des Dorischen Styls.* Stuttgart.

Haus, Josef Jakob. 1814. *Saggio sul Tempio e la Statua di Giove in Olimpio e sul Tempio dello stesso in Agrigento.* Palermo.

Häussler, Reinhard. 1968. *Nachträge zu A. Otto . . . Sprichwörter und Sprichwörtlichen Redensarten der Römer.* Hildesheim.

Hautecoeur, L. 1943–1957. *Histoire de l'Architecture Classique en France.* 7 vols. Paris.

Hawkes, Terence. 1972. *Metaphor.* London.

Hay, David R. 1849. *On the Science of those Proportions by which the Human Head and Countenance as Represented in Works of Ancient Greek Art are distinguished from those of Ordinary Nature.* Edinburgh and London.

——. 1856. *The Science of Beauty as Developed in Nature and Applied in Art.* Edinburgh and London.

Hayes, William C. 1953–1959. *The Scepter of Egypt.* 2 vols. New York.

Hazard, Paul. 1961. *La Crise de la Conscience Européenne, 1680–1715.* Paris.

Heath, Sir Thomas L. 1932. *Greek Astronomy.* London.

Hecht, Neil S. 1985. "A Modest Addendum to the Greek Metrological Relief in Oxford." *AJ* 65.

Hegel, G. W. F. 1965. *Die Aesthetik.* Ed. F. Bassenge after H. G. Hothos. Berlin and Weimar.

——. 1969. *Jenaer Realphilosophie.* 1931. Ed. J. Hoffmeister. Hamburg.

Heidegger, Martin. 1970–. *Gesamtausgabe.* Frankfurt A/M.

——. 1975. *Early Greek Thinking.* Trans. David F. Krell and Frank A. Capuzzi. New York.

——. 1979–1984. *Nietzsche.* 2 vols. Trans. David F. Krell. New York.

——. 1988. *Der Ursprung des Kunstwerkes.* Intro. H.-G. Gadamer.

Heinrich, Ernst. 1934. *Schilf und Lehm: Ein Beitrag zur Baugeschichte der Sumerer.* Berlin.

Helck, Wolfgang. 1972. *Lexikon der Aegyptologie.* Wiesbaden.

Hellström, Pontus, and Thomas Thieme. 1982. *Labraunda.* Stockholm.

Helmholtz, H. 1885. *On the Sensations of Tone.* Trans. A. J. Ellis. London.

Hempel, E. 1957. "Nicolas von Cues in seinen Beziehungen zur Bildenden Kunst." *Berichte über die Verhandlungen der Sächsischen Akademie der Wissenschaften zu Leipzig, Philologische-Historische Klasse,* Bd. 102.

Heninger, S. K., Jr. 1974. *Touches of Sweet Harmony: Pythagorean Cosmology and Renaissance Poetics.* San Marino, Calif.

Henrichs, A. 1970. "Pagan Ritual and the Alleged Crimes of the Early Christians." *Kyriakon: Festschrift Johannes Quasten.* Ed. Patrick Granfield and Josef A. Jungmann. Münster.

———. 1981. *Human Sacrifice in Greek Religion.* Fondation Hardt. Geneva.

———. 1987. *Die Götter Griechenlands: Ihr Bild im Wandel der Religionswissenschaft.* Bamberg.

Henry, Michel. 1965. *Philosophie et Phénoménologie du Corps.* Paris.

Hentze, Carl. 1932. *Mythes et Symboles Lunaires.* Antwerp.

Hermann, Peter. 1965. "Antiochos der Grosse und Teos." *Anatolia 9.*

Hermogenes of Tarsus. 1987. *On Types of Style.* Trans. Cecil W. Wooten. Chapel Hill, N.C.

Heron, G. M. *See* Cusa.

Herrad of Hohebourg. *See* Green, R., et al.

Herringham, C. J. *See* Cennini, Cennino.

Herrmann, Klaus. 1980. "Bericht über Restaurierungsarbeiten in Olympia." *AA.*

Herrmann, Wolfgang. 1973. *The Theory of Claude Perrault.* London.

———. 1984. *Gottfried Semper: In Search of Architecture.* Cambridge, Mass.

Hersey, George. 1965. *Loudon's Associationism.* New Haven.

———. 1976. *Pythagorean Palaces.* Ithaca and London.

———. 1988. *The Lost Meaning of Classical Architecture.* Cambridge, Mass.

Herwegen, Idelfons. 1909. "Ein Mittelalterlicher Kanon des Menschlichen Körpers." *Repertorium für Kunstwissenschaft 32.*

Herzfeld, Ernst E. 1941. *Iran in the Ancient East.* Oxford.

Heurgon, Jacques. 1953. "Tarquitius Priscus et l'Organisation de l'ordre des Haruspices sous l'Empereur Claude." *Latomus 12.*

———. 1979. *La Vie Quotidienne chez les Etrusques.* 1961. Reprint, Paris.

Heurtley, W. A. 1939. *Prehistoric Macedonia.* Cambridge.

Heuzey, Léon, and Jacques Heuzey. 1935. *L'Orient.* Paris.

Hewes, Gordon Winant, Ed. 1975. *Language Origins: A Bibliography.* The Hague.

Hibbard, Howard. 1965. *Bernini.* Harmondsworth.

Hill, Bert Hodge. 1966a. *Corinth.* Princeton.

———. 1966b. *The Temple of Zeus at Nemea.* Rev. and suppl. Charles Kaufman Williams. Princeton.

Hillgarth, J. N. 1971. *Ramon Lull and Lullism in Fourteenth-Century France.* Oxford.

Himmelmann-Wildschütz, Nikolaus. 1968. "Über einige gegenständliche Bedeutungsmöglichkeiten des frühgriechischen Ornaments." *Abhandlungen der geistes- und sozialwissenschaftlichen Klasse.* No. 7. Akademie der Wissenschaften und der Literatur. Wiesbaden.

Hinke, W. J. 1911. "Selected Babylonian Kudurru Inscriptions." *Semitic Study Series 14.*

Hinks, Roger. 1938. "The Master of Animals." *JWI 1.4.*

Hinz, Walter. 1972. *The Lost World of the Elam.* London.

Hitzig, Hermann, and Hugo Bluemner. *See* Pausanias.

Hocart, A. M. 1927. *Kingship.* Oxford.

Hockett, C. F., and R. Ascher. 1964. "The Human Revolution." *Current Anthropology 5.*

Höckmann, Ursula, and Antje Krug, eds. 1977. *Festschrift für Frank Brommer.* Mainz.

Hodge, A. Trevor. 1960. *The Woodwork of Greek Roofs.* Cambridge.

Hodge, Carleton T. 1975. *Ritual and Writing.* Ghent.

Hoenn, Karl. 1946. *Artemis Gestaltwandel einer Göttin.* Zurich.

Hoffner, H. A., trans. 1990. *Hittite Myths.* Atlanta, Ga.

Hoffner, H. A., and G. M. Beckman. 1986. *Kanišuwar: Essays in Honor of Hans G. Güterbock.* Chicago.

Hofmann, Werner. 1957. *Caricature from Leonardo to Picasso.* Trans. M. H. L. London.

Hogarth, David George, ed. 1908. *Excavations at*

Ephesus: The Archaic Artemisia. 2 vols. London.

Hogarth, William. 1753. *The Analysis of Beauty: Written with a View of Fixing the Fluctuating Ideas of Taste.* London. Ed. with an intro. by J. Burke (Oxford, 1955).

Hogben, Lancelot. 1936. *Mathematics for the Million.* London.

Hollander, John. 1961. *The Untuning of the Sky.* Princeton.

Hollier, Denis. 1974. *La Prise de la Concorde.* Paris.

Hölscher Fernande. 1972. *Die Bedeutung Archaischer Tierkampfbilder.* Würzburg.

Homolle, Théophile, and Maurice Holleaux. 1910. *Exploration Archéologique de Délos Faite par l'Ecole Française d'Athènes.* Paris. Later editions revised by Gustave Fougères.

Hood, M. Sinclair. 1978. *The Art in Prehistoric Greece.* Harmondsworth.

Hooker, J. T. 1976. *Mycenaean Greece.* London.

——. 1980. *The Ancient Spartans.* London.

Hope, T. N. 1985. *Pictures from Eighteenth-Century Greece.* Athens.

Hopper, R. J. 1976. *The Early Greeks.* London.

Hornblower, S. 1982. *Mausolus.* Oxford.

Howard-Carter, Theresa. 1983. "An Interpretation of the Second-Millennium Temple at Tell-el Rimah." *Iraq* 45.

Howe, Thomas Noble. 1985. *The Invention of the Doric Order.* Cambridge, Mass.

Hrozny, Bedrich. 1947. *Histoire de l'Asie Antérieure.* Paris.

——. 1963. *Ancient History of Western Asia, India, and Crete.* Prague.

Huart, Clément, and Louis Delaporte. 1943. *L'Iran Antique.* Paris.

Hubbard, B. A. F., and E. S. Karnofsky. 1982. *Plato's Protagoras.* London.

Huebsch, Heinrich. 1828. *In welchem Style sollen wir bauen?* Karlsruhe.

Huizinga, Johan. 1949. *Homo Ludens: A Study of the Play-Element in Culture.* 1938. Trans. R. F. C. Hull. London.

Hultsch, Friedrich. 1882. *Griechische und Römische Metrologie.* Berlin.

Humann, Carl, and Julius Kohte. 1904. *Magnesia am Maeander: Bericht ueber Ergebnisse der Ausgrabungen der Jahre 1891–1893.* Berlin.

Humbert de Superville, D.-P. 1827. *Essai sur les Signes Inconditionnels dans l'Art.* Leyden.

Humphreys, S. C. 1983. *The Family, Women, and Death.* London.

Huntley, H. E. 1970. *The Divine Proportion: A Study in Mathematical Beauty.* New York.

Hurry, J. B. 1926. *Imhotep: The Vizier and Physician of King Zoser.* Oxford.

Huxley, G. L. 1962. *Early Sparta.* London.

——. 1967. "The Medism of Caryae." *GR&BS* 8.

Hyatt Mayor, A. 1984. *Artists and Anatomists.* New York.

Ikhwan al Safa. *See* Diwald, Susanne.

Ingholt, Harald. 1954. *Parthian Sculptors from Hatra.* New Haven.

Invernizzi, Antonio. 1965. *I Frontoni del Tempio di Aphaia ad Egina.* Turin.

Isnardi Parente M. 1966. *Techné: Momenti del Pensiero Greco da Platone ad Epicuro.* Florence.

Ison, Walter. 1948. *The Georgian Buildings of Bath.* London.

Istituto di Studi Etruschi e Italici. 1976. *L'Etrusco Arcaico.* Colloquio 4–5 ottobre 1974. Florence.

Iversen, Erik. 1975. *Canon and Proportion in Egyptian Art.* Warminster.

Ivins, William M., Jr. 1946. *Art and Geometry.* Cambridge, Mass.

Jaeger, Werner. 1973. *Paideia.* 1933. Reprint, Berlin.

Jakobson, Roman, and Krystyna Pomorska. 1983. *Dialogues.* Cambridge.

Janneau, Guillaume. 1965. *La Peinture Française au XVIIe Siècle.* Geneva.

Jannot, J. R. 1979. "La Lyre et la Cithare." *AC* 48.

Janson, H. W. 1974. *Sixteen Studies.* New York.

——. *See also* Barasch, Moshe, and Lucy Freeman Sandler.

Jantzen, Ulf, ed. 1976. *Neue Forschungen in griechischen Heiligtumern.* Tübingen.

Jardé, Auguste. 1923. *La Formation du Peuple Grec.* Paris.

Javelet, Robert. 1967. *Image et Ressemblance au Douzième Siècle.* 2 vols. Paris.

Jeanmaire, Henri. 1939. *Couroi et Courètes.* Lille.

——. 1951. *Dionysos: Histoire du Culte de Bacchus.* Paris.

Jedin, Hubert, and John Dolan, eds. 1980–1982. *History of the Church.* 10 vols. New York.

Jeffery, Lilian Hamilton. 1961. *The Local Scripts of Archaic Greece.* Oxford.

——. 1976. *Archaic Greece: The City States 700–500 B.C.* London.

Jenkins, Romilly James Heald. 1940. *Dionysius Solomos.* Cambridge.

Jensen, Adolf. 1963. *Myth and Cult among Primitive Peoples.* Trans. Marianna Tax Choldrin and Wolfgang Weissleder. Chicago.

Jeppesen, Kristian. 1958. *Paradeigmata: Three Mid-Fourth-Century Main Works of Hellenic Architecture Reconsidered.* Aarhus.

——. 1987. *The Theory of the Alternative Erechtheion.* Aarhus.

Jéquier, Gustave. 1908. "Les Temples Primitifs et la Représentation des Types Archaiques." *BIAO.*

——. 1924. *Manuel d'Archéologie Egyptienne.* Paris.

——. *See also* Vandier, J.

Jirku, Anton. 1960. *Die Welt der Bibel.* Stuttgart.

Joest, Wilhelm. 1887. *Tätowieren, Narbenzeichnen und Körperbemalen.* Berlin.

[John of Salisbury] Joannis Saresberiensis. 1848. *Opera Omnia.* Ed. J. A. Giles. Oxford. (Reprinted Leipzig, 1969.)

——. 1979. *Policraticus: The Statesman's Book.* Ed. and trans. Murray F. Markland. New York.

Johns, C. H. W. 1908. *Ur-Enqur: A Bronze of the Fourth Millennium BC in the Library of J. Pierpont Morgan: A Treatise on the Canephorai.* New York.

Jomard, E. F. ed., 1809–1822. *Description de l'Egypte.* 9 vols. Paris.

Jones, C. P. 1987. "Tattooing and Branding in Graeco-Roman Antiquity." *JRS* 77.

Jones, Inigo. 1725. *The Most Notable Antiquity of Great Britain Vulgarly Called Stone-Heng.* 3rd ed. London.

Jonson, Ben. 1954. *Everyman out of His Humor.* Ed. C. Herford. Oxford.

Jouin, Henry. 1889. *Charles Le Brun et les Arts sous Louis XIV.* Paris.

Juan de Arfe. 1585. *De Varia Commensuración.* Seville.

Judeich, W. 1905. *Topographie von Athen.* Munich.

Jung, Carl Gustav. 1953–1983. *The Collected Works.* 21 vols. Ed. Herbert Read, Michael Fordham, and Gerhard Adler. London.

Jung, Helmut. 1982. *Thronende und Sitzende Götter.* Bonn.

Junius, Franciscus. 1694. *De Pictura Veterum.* Rotterdam.

Junker, Hermann. 1959. *Die Gesellschaftliche Stellung der Ägyptischen Künstler im Alten Reich.* Vienna.

Kaegi, Werner. 1947–1982. *Jakob Burckhardt: Eine Biographie.* 7 vols. Basel.

Kähler, Heinz. 1949. *Das Griechische Metopenbild.* Munich.

——. 1964. *Der Griechische Tempel.* Berlin.

Kahn, Charles H. 1960. *Anaximander and the Origins of Greek Cosmology.* New York.

——. 1979. *The Art and Thought of Heraclitus.* Cambridge.

Kakosy, L. 1968. "Imhotep and Amenhotep Son of Hapu, as Patrons of the Dead." *Acta Orientalia* (Academiae Scientiarum Hungaricae) 21.

Kakridis, Johannes T. 1971. *Homer Revisited.* Lund.

Kalpaxis, Thanassis E. 1986. *Hemiteles.* Mainz.

Kanellopoulos, Panagiotis, et al. 1980. *Stili: Tomos eis Mnimin: Festschrift Nikolau Kontoleontos.* Athens.

Kant, Immanuel. 1922–1923. *Werke.* 12 vols. Ed. E. Cassirer. Berlin.

Kanter, Emanuel. 1926. *The Amazons: A Marxian Study.* Chicago.

Kantorowicz, Ernst. 1936. *Kaiser Friedrich der Zweite.* Berlin.

——. 1957. *The King's Two Bodies.* Princeton.

Karageorghis, V. 1969. *Archaeologia Mundi—Cyprus.* Geneva.

——. 1973. "The Relations Between the Tomb-Architecture of Anatolia and Cyprus in the Archaic Period." *ICCA* 10.

——. 1976. *View from the Bronze Age: Mycenaean and Phoenician Discoveries at Kition.* New York.

——. 1982. *Cyprus: From the Stone Age to the Romans.* London.

——. 1988. *Blacks in Ancient Cypriot Art.* Houston.

Kardaun, Maria. 1993. *Der Mimesisbegriff in der Griechischen Antike.* Amsterdam.

Karo, George. 1970. *Greek Personality in Archaic Sculpture.* Westport, Conn.

Kaschnitz-Weinberg, Guido von. 1944–1961. *Die Grundlagen der Antiken Kunst.* Frankfurt am Main.

——. 1961–1963. *Römische Kunst.* 4 vols. Reinbek bei Hamburg.

Kauffmann, G. 1960. *Poussin-Studien.* Berlin.

Kaufmann, E., Jr., ed. 1970. *The Rise of an American Architecture.* New York.

Kavvadias, P., and G. Kawerau. 1907. *Anakaphaitis Akropoleos.* Athens.

Kawerau, G., and G. Soteriadis. 1902–1908. "Der Apollo-Tempel zu Thermos." *AntDenk* 2.

Kebric, Robert B. 1983. *The Paintings in the Cnidian Lesche at Delphi and Their Historical Context.* Leiden.

Kees, H. 1941. *Die Götterglaube im Alten Aegypten.* Leipzig.

Kemp, Martin. 1980. *Leonardo da Vinci: The Marvellous Works of Nature and Man.* London.

Kemp, P. 1935. *Healing Ritual: Studies in the Technique and Tradition of the Southern Slavs.* London.

Kennedy, George. 1963. *The Art of Persuasion in Greece.* Princeton.

Kenner, Hedwig. 1970. *Das Phänomen der Verkehrten Welt in der Griechisch-Römischen Antike.* Klagenfurt.

Kenny, Anthony. 1968. *Descartes: A Study of his Philosophy.* New York.

Kenyon, Kathleen. 1960. *Archaeology in the Holy Land.* 4th ed. London.

Kerenyi, C. 1959. *The Heroes of the Greeks.* Trans. H. J. Rose. London.

———. 1963. *Prometheus.* London.

———. 1976. *Dionysos: Archetypal Image of Indestructible Life.* Trans. Ralph Mannheim. Bollingen Series 65.2 Princeton.

Keuls, Eva C. 1978. *Plato and Greek Painting.* Leiden.

Kiechle, Franz. 1963. *Lakonien und Sparta.* Munich and Berlin.

King, Leonard William. 1912. *Babylonian Boundary Stones and Memorial Tablets in the British Museum.* London.

Kinzl, K. H., ed. 1977. *Greece and the Eastern Mediterranean in Ancient History and Prehistory: Festschrift Schachermeyr.* Berlin.

Kirchberger, J. *See* Flasch, K.

Kirchhoff, Werner. 1988. "Die Entwicklung des Ionischen Volutenkapitells im 6 und 5 Jhd. und Seine Entstehung." Ph.D. diss. Bonn.

Kirk, G. S. 1962. *The Songs of Homer.* Cambridge.

———. 1974. *The Nature of Greek Myths.* Harmondsworth.

———. 1981. *Some Methodological Pitfalls in the Study of Ancient Greek Sacrifice.* Fondation Hardt. Geneva.

Kirk, G. S., and J. E. Raven, 1983. *The Pre-Socratic Philosophers.* 2nd ed. With M. Schofield. Cambridge.

Kitao, Timothy K. 1974. *Circle and Oval in the Square of St. Peter's.* New York.

Kleinbaum, Abby Wettan. 1983. *The War Against the Amazons.* New York.

Kleiner, G., P. Hommel, and W. Müller-Wiener. 1967. "Panionion und Melie." *JDAI,* suppl. 23.

Kleiss, Wolfram. 1989. *Die Entwicklung von Palasten und Palastartigen Wohnbauten in Iran.* Vienna.

Klibansky, R. 1961. "Peter Abailard and Bernard of Clairvaux." *MRS* 5.

Knell, Heiner. 1979. *Perikleische Baukunst.* Darmstadt.

Knell, Heiner, and Burkhardt Wesenberg, eds. 1984. *Vitruv-Kolloquium.* Darmstadt.

Koch, Herbert. 1951. *Vom Nachleben des Vitruv.* Baden-Baden.

———. 1956. *Von Ionischer Baukunst.* Köln.

Koch, Tancred. 1972. "Die Schüler Vesals." *Anatomischer Anzeiger* 131.

Kockelmans, Joseph J. 1985. *Heidegger on Art and Art Works.* Dordrecht, Boston, Lancaster.

Koffka, K. 1963. *Principles of Gestalt Psychology.* Reprint, 1935. New York.

Köhler, W. 1930. *Gestalt Psychology.* London. *Rev. ed.* 1947.

Kolakowski, Leszek. 1978. Main Currents of Marxism. 3 vols. Oxford.

Koldewey, Robert, and Otto Puchstein. 1899. *Die Griechischen Tempel in Unteritalien und Sizilien.* 2 vols. Berlin.

Koller, H. 1954. *Die Mimesis in der Antike.* Bern.

Kontoleon, Nikolaos, et al. 1965. *Charisterion eis Anastasion K. Orlandon.* Athens.

Kopcke, Günter, and Mary B. Moore. 1979. *Studies in Classical Art and Archaeology.* New York.

Korwin-Krasinsky, C. 1956. "Die Schöpfung als 'Tempel' und 'Reich' des Gottmenschen Christus." *Enkainia, Gesammelte Arbeiten zum 800 jahr . . . der Abteikirche Maria Laach.* Düsseldorf.

Koyré, Alexander. 1971. *Descartes und die Scholastik.* Darmstadt.

Kramrisch, Stella. 1946. *The Hindu Temple.* 2 vols. Calcutta.

———. 1981. *The Presence of Siva.* Princeton.

Krantz, Grover S. 1980. "Sapienization and Speech." *CA* 21.

Krauss, Friedrich, and Reinhard Herbig. 1939. *Der Korinthisch-dorische Tempel am Forum von Paestum*. Berlin.

Krautheimer, Richard. 1942. "Introduction to an Iconography of Mediaeval Architecture." *JWCI* 5.

Krautheimer, Richard, et al. 1937–1977. *Corpus Basilicarum Christianarum Romae*. 5 vols. Rome.

——. 1941. "S. Pietro in Vincoli and the Tripartite Transept in the Early Christian Basilica." *APS* 84.

Kreikenbom, Detlev. 1990. *Bildwerke nach Polyklet*. Berlin.

Kretschmer, Ernst. 1931. *Körperbau und Charakter*. 10th ed. Berlin.

Krinsky, C. 1967. "Seventy-Eight Vitruvius Manuscripts." *JWCI* 30.

Kris, Ernst. 1953. *Psychoanalytic Explorations in Art*. London.

Krischen, F. 1938. *Die Griechische Stadt*. Berlin.

Kristeller, P. O. 1982. *Medieval Aspects of Learning: Three Essays*. Trans. and ed. E. P. Mahoney. Durham, N.C.

——. *See* Clough, Cecil H.

Kruft, Hanno-Walter. 1972. "Studies in Proportion by J. J. Winckelmann." *Burlington Magazine* 114.

Krug, Antje. 1968. *Binden in der Griechischen Kunst*. Hösel.

Krumbacher, Karl. 1897. *Geschichte der Byzantinischen Literatur*. 2nd ed. Munich.

Kulka, Heinrich. 1931. *Adolf Loos*. Vienna.

Kümmel, Hans Martin. 1967. *Ersatzrituale für den Hethitischen König*. Wiesbaden.

Kunze, M., and Kastner Volker. 1985. *Führer, Staatliche Museen zu Berlin: Antikensammlung*. Berlin.

Kurfess, Alfons, ed. and trans. 1951. *Sibyllinische Weissagungen*. Heimeran.

Kurtz, Donna Carol. 1975. *Athenian White Lekythoi: Patterns and Painters*. Oxford.

Kurtz, Donna Carol, and John Boardman. 1971. *Greek Burial Customs*. Ithaca, N.Y.

Labrouste, Henri. 1829. *Les Temples de Paestum*. Paris.

La Bruyère, Jean de. 1688. *Les Caractères de Théophraste traduits du Grec avec les Caractères ou les Moeurs de ce Siècle*. Paris.

Lacan, Jacques. 1966. *Ecrits*. Paris.

——. 1975. *Le Séminaire*. Texte établi par Jacques-Alain Miller. Paris.

Lacassagne, Jean-Alexandre. 1881. *Les Tatouages*. Paris.

——. 1908. *Le Criminel du Point de Vue Anthropologique*. Paris.

La Chambre, Marin Cureau de. *See* Cureau de la Chambre, Marin.

Lacoue-Labarthe, Philippe. 1989. *Typography: Mimesis, Philosophy, Politics*. Ed. C. Fynsk and L. M. Brooks. Cambridge, Mass.

Lagrange, M.-J. 1905. *Etudes sur les Religions Sémitiques*. Paris.

Lakoff, George, and Mark Johnson. 1980. *Metaphors We Live By*. Chicago.

Lambert, M., ed. 1891. *Commentaire sur le Séfer Yesira ou Livre de la Création par le Gaon Saadia de Fayyoum*. Paris.

Lambert, S. W., et al. 1952. *Three Vesalian Essays*. New York.

Lanciani, Rodolfo Amedeo. 1897. *The Ruins and Excavations of Ancient Rome*. Boston and New York.

Lange, Curt, and Max Hirmer. 1968. *Egypt: Architecture, Sculpture, Painting*. With contributions by Otto Eberhard and Christiane Desroches-Noblecourt. London and New York.

Lange, Konrad. 1885. *Haus und Halle*. Leipzig.

Langer, Suzanne K., ed. 1953. *Feeling and Form: A Theory of Art*. New York.

——. 1957. *Problems of Art*. New York.

Lanzani, Carolina. 1936. *Lucio Cornelio Silla Dittatore*. Milan.

Lanzi, L. 1822. *Storia Pittorica dell'Italia*. Florence.

Lapalus, Etienne. 1932. *Le Fronton Grec*. Paris.

Lappo-Danilewski, O. 1942. "Untersuchungen über den Innenraum der Archaisch-Griechischen Tempel." Ph.D. diss., Würzburg.

Larson, Magali Sarfatti. 1993. *Behind the Postmodern Facade*. Berkeley.

Lasch, Christopher. 1979. *The Culture of Narcissism*. New York.

Lattimore, Richmond. 1942. *Themes in Greek and Latin Epitaphs*. Urbana, Ill.

Lauder, Sir Thomas Dick. *See* Price, Sir Uvedale.

Lauer, J.-P. 1976. *Saqqara, The Royal Cemetary of Memphis*. London.

Lauter, H. 1976. "Die Koren des Erechtheion." *AntP* 16.

——. 1986. *Die Architektur des Hellenismus*. Darmstadt.

Lauter-Bufe, Heide. 1987. *Die Geschichte des Sikeliotisch Korinthischen Kapitells*. Mainz am Rhein.

Lavas, Georg P. 1974. *Altgriechisches Temenos: Baukörper und Raumbildung*. Zurich.

Lavater, Johann Kaspar. 1775–1778. *Physiognomische Fragmente zur Beförderung der Menschenkenntniss und Menschenliebe*. 4 vols. Leipzig and Winterthur.

———. 1806. *L'Art de Connaitre les Hommes par la Physiognomonie*. Ed. J.-L. Moreau de la Sarthe. Paris.

———. 1845. *La Physiognomonie, ou l'Art de Connaître les Hommes*. Trans. H. Bacharach. Paris.

Lavin, Irving. 1980. *Bernini and the Unity of the Visual Arts*. 2 vols. Oxford.

Lawrence, A. W. 1983. *Greek Architecture*. Harmondsworth.

Lazzaroni, Michele, and Antonio Muñoz. 1908. *Filarete, Scultore e Architetto del Secolo XV*. Rome.

Leakey, Richard E. 1981. *The Making of Mankind*. London.

Leatherbarrow, David. 1976. "Character, Geometry and Perspective: The Third Earl of Shaftesbury's Principles of Garden Design." *JGH* 4.4.

———. 1983. "On Shaftesbury's Second Characteristics." Ph.D. diss., University of Essex.

Lebessi, Angeliki, and Polymnia Muhly. 1990. "Aspects of Minoan Cult: Sacred Enclosures." *AA* 3. Berlin.

Le Blond, Jean. 1683. *Deux Examples des Cinq Ordres de l'Architecture Antique*. Paris.

———. 1710. *Parallèle des Cinq Ordres d'Architecture*. Paris.

Le Bonniec, Henri. 1958. *Le Culte de Cérès à Rome*. Paris.

Le Brun, J. 1981, April–June. "Sens et Portée du Retour aux Origines dans l'Oeuvre de Richard Simon." *Dix-Septième Siècle*, no. 131.

Lecoq, Anne-Marie. 1981, July–September. "Nature et Rhétorique." *Dix-Septième Siècle* no. 132.

Le Corbusier [Charles-Edouard Jeanneret]. [1923?] *Vers une Architecture*. Paris.

———. 1954. *Le Modulor*. London.

Leff, Gordon. 1975. *William of Ockham: The Metamorphosis of Scholastic Discourse*. Manchester and Totowa, N.J.

Leftwich, G. V. 1987. *Ancient Conceptions of the Body and the Canon of Polykleitos*. 2 vols. Princeton.

Lehmann, K. and P. W. Lehmann. 1969. *Samothrace*. 3 vols. Bollingen Series. Princeton.

———. *See also* Sandler, Lucy Freeman.

Lemoine, Albert. 1865. *De la Physiognomie et de la Parole*. Paris.

Le Moyne, Pierre. 1640–1643. *Les Peintures Morales*. Paris.

Lenneberg, Eric H. 1967. *The Biological Foundations of Language*. New York.

———. *See also* Miller, George A., and Elizabeth Lenneberg; Rieber, R. W.

Lenneberg, Eric H., and Elizabeth Lenneberg, eds. 1975. *Foundations of Language Development*. New York.

Lenoir, Alexandre. 1810. *Musée Impérial des Monuments Français*. Paris.

Lenz, Harald Othmar. 1859. *Botanik der Alten Griechen und Römer*. Gotha.

Leonardo da Vinci. 1970. *The Literary Works of Leonardo da Vinci*. 2 vols. Comp. and ed. Jean Paul Richter. London.

———. 1977. Reprint of Leonardo da Vinci (1970), with commentary by Carlo Pedretti. Oxford.

Leone, P. A. M., ed. 1968. *Ioannis Tzetzae Historiae. . . .* Naples.

Lepik-Kopaczyńska, W. 1959. "Die Optische Proportionen in der Antiken Kunst." *Klio* 37.

Leroi-Gourhan, A. 1965. *Préhistoire de l'Art Occidental*. Paris.

Le Roy, Christian. 1965. "Lakonika II: Nouvelles Antiquités de Kotronas." *BCH* 89.

Letalle, Abel. [1925?] *Les Fresques du Campo Santo de Pise*. Paris.

Lethaby, W. R. 1908. *Greek Buildings Represented by Fragments in the British Museum*. London.

———. 1915. "The Nereid Monument Reexamined." *JHS* 35.

Levi, Anthony S. J. 1964. *French Moralists: The Theory of the Passions, 1585–1649*. Oxford.

Lévi-Strauss, Claude. 1955. *Tristes Tropiques*. Paris.

———. 1958. *Anthropologie Structurale*. Paris.

———. 1964–1971. *Mythologiques*. 4 vols. Paris.

———. 1979. *La Voie des Masques*. 2nd ed. Paris.

Levin, F. R. 1975. *The Harmonics of Nicomachus and the Pythagorean Tradition*. University Park, Pa.

Levin, Samuel R. 1988. *Metaphoric Worlds: Conceptions of a Romantic Nature*. New Haven.

Levine, N. 1975. "Architectural Reasoning in the Age of Positivism: The Neo-Greek Idea of

Henri Labrouste's Bibliothèque Ste. Geneviève." Ph.D. diss., Yale University.

Lezine, Alexandre. [1960?]. *Architecture Punique.* Tunis.

Lichtenberg, Georg Christoph. 1967. *Werke.* 1778. Ed. Carl Brinitzer and Peter Plett. Hamburg.

Lichtenberg, Reinhold von. 1909. *Haus-Dorf-Stadt.* Leipzig.

Lieberman, Philip. 1975. *On the Origins of Language: An Introduction to the Evolution of Human Speech.* New York and London.

Liebeschütz, Hans. 1930. *Das Allegorische Weltbild der Heiligen Hildegard von Bingen.* Leipzig.

———. 1950. *Medieval Humanism in the Life and Writings of John of Salisbury.* London.

Lincoln, Bruce. 1981. *Emerging from the Chrysalis: Studies in Rituals of Women's Initiation.* Cambridge, Mass.

———. 1986. *Myth, Cosmos, and Society.* Cambridge, Mass.

Lipking, Lawrence. 1970. *The Ordering of the Arts in Eighteenth-Century England.* Princeton.

Llewellyn, Nigel. 1977. "Two Notes on Diego da Sagredo." *JWCI* 40.

Lloyd, G. E. R. 1970. *Later Greek Science.* London.

———. 1987. *The Revolutions of Wisdom.* Berkeley.

Lloyd-Jones, Hugh. 1971. *The Justice of Zeus.* Berkeley.

———. 1983. "Artemis and Iphigenia." *JHS* 103.

Lobeck, C. Augustus. 1829. *Aglaophamus, Sive de Theologiae Mysticae Graecorum Causis.* Königsberg.

Lockyer, J. Norman. 1964. *The Dawn of Astronomy.* 1894. Reprint, Cambridge, Mass.

Loewy, Emanuel. 1885. *Inschriften Griechischer Bildhauer.* Leipzig.

Lo Faso Pietrasanta, Domenico, Duca di Serradifalco. 1834. *Le Antichità della Sicilia.* Palermo.

Lohmann, J. 1970. *Die Komposition der Reden in der Ilias.* Berlin.

Lomas, Johannes. 1974. *Musiké und Logos.* Ed. A. Giannaràs. Stuttgart.

Lomazzo, Giovanni Paolo. 1970. *A Tracte Containing the Artes of Curios Paintinge, Carvinge, Buildinge written first in Italian by IO. Paul Lomatius. . . .* Englished by R[ichard] H[aydocke] [1598?]. Reprint, Farnborough, Hants.

———. 1974a. *Idea del Tempio della Pittura.* Trans. Robert Klein. Florence.

———. 1974b. *Scritti sulle Arti, a cura di Roberto Paolo Ciardi.* 2 vols. Florence.

Longnon, Jean, and Raymond Cazelles. 1969. *The Très Riches Heures of Jean, Duke of Berry.* New York.

Lonsdale, Steven H. 1993. *Dance and Ritual Play in Greek Religion.* Baltimore and London.

Loos, Adolf. 1931. *Trotzdem.* Innsbruck.

———. 1982. *Das Andere.* Ed. M. Cacciari. Milan.

Lorenz, Konrad. 1965. *Über Tierisches und Menschliches Verhalten.* 2 vols. Munich.

Lorenz, Thuri. 1972. *Polyklet.* Wiesbaden.

Lorenzen, Eivind. 1959. "De Dorike Traetemplers Bygningshistorie." *Arkitekten* 13.

———. 1964. *The Arsenal at Piraeus.* Copenhagen.

Lorimer, H. L. 1950. *Homer and the Monuments.* London.

Lossky, Boris. 1936. *J. B. A. Le Blond.* Prague.

Loudon, J. C. 1835. *An Encyclopaedia of Cottage, Farm, and Villa Architecture and Furniture; Containing Numerous Designs for Dwellings, Each Design Accompanied by Analytical and Critical Remarks.* New ed. London.

Lowic, Lawrence. 1982. "Francesco di Giorgio on the Design of Churches: The Uses and Significance of Mathematics in the *Trattato*." *Architectura* 12.

———. 1983. "The Meaning and Significance of the Human Analogy in Francesco di Giorgio's *Trattato*." *JSAH* 42.4.

Löwith, Karl. 1981–1988. *Sämtliche Schriften.* 9 vols. Ed. Klaus Stichweh and Marc de Launay. Stuttgart.

Lubbock, Sir John. 1870. *The Origin of Civilization and the Primitive Condition of Man.* London.

Lücke, Hans-Karl. 1991. "Mercurius Quadratus: Anmerkungen zur Anthropometrie bei Cesariano." *Mitt. des Kunsthistorischen Institutes in Florenz* 25.1.

Lukács, Georg. 1963–1974. *Werke.* 17 vols. Neuwied am Rhein and Berlin.

———. 1971. *History and Class Consciousness: Studies in Marxist Dialectic.* Trans. Rodney Livingstone. Cambridge, Mass.

Lullies, Reinhard, ed. 1954. *Neue Beiträge zur Klassischen Altertums Wissenschaft.* Stuttgart.

Luria, S. 1963. *Anfänge Griechischen Denkens.* Berlin.

Luschan, Felix von. 1912. *Entstehung und Herkunft der Ionischen Säule.* Leipzig.

Luscombe, D. E. 1969. *The School of Peter Abelard.* Cambridge.

[Lutyens, Sir Edwin]. 1981. *Lutyens.* Catalogue of Exhibition at the Hayward Gallery, November 1981–January 1982. Ed. Colin Amery, Margaret Richardson, and Gavin Stamp. London.

Lynch, H. F. B. 1965. *Armenia: Travels and Studies.* 2 vols. Beirut.

Macco, Michela di. 1971. *Il Colosseo.* Rome.

MacCormac, Earl R. 1976. *Metaphor and Myth in Science and Religion.* Durham, N.C.

Macdonald, Raymond A. 1993. "Ekphrasis, Paradigm Shift, and Revisionism in Art History." *RES* 24.

McDonald, William A. 1967. *Progress into the Past: The Rediscovery of Mycenaean Civilization.* New York and London.

——. 1976. *The Pantheon.* London.

McGinty, Park. 1978. *Interpretation and Dionysos: Method in the Study of a God.* The Hague.

Mach. Edmund von. 1903. *Greek Sculpture: Its Spirit and Principles.* Boston.

Mach, Ernst. 1914. *The Analysis of Sensations.* Chicago and London.

MacIntyre, Alasdair, ed. 1972. *Hegel: A Collection of Critical Essays.* New York.

——. 1984. *After Virtue.* Notre Dame.

McKinnon, James W. 1978. "Jubal vel Pythagoras, Quis Sit Inventor Musicae?" *Musical Quarterly* 64. 1.

MacQueen, J. G. 1986. *The Hittites.* London.

Maddison, R. E. W. 1951–1955. *Studies in the Life of Robert Boyle, F. R. S.* London.

Madonna, Maria Luisa. 1976. "Septem Mundi Miracula." *Psicon* 7.

Maeklenburg, Albert. 1914. *Darstellung und Beurteilung der Ästhetik Schopenhauers.* Borna-Leipzig.

Maffei, Scipione. 1731–1732. *Verona Illustrata.* 3 vols. Verona.

Magee, Bryan. 1983. *The Philosophy of Schopenhauer.* Oxford.

Maier, Anneliese. 1955. *Metaphysische Hintergründe der Spätscholastischen Naturphilosophie.* Rome.

Maier, Franz-Georg. 1985. *Alt-Paphos auf Cypern.* Mainz am Rhein.

Maier, F. G., and V. Karageorghis. 1984. *Paphos, History and Archaeology.* Nicosia.

Mair, A. W., and G. R. Mair. *See* Callimachus.

Male, Emile. 1925. *L'Art Religieux de la Fin du Moyen-Age en France.* Paris.

Malespini, Ricordano. 1718. *Istoria Fiorentina.* Florence.

Malkin, Irad. 1987. *Religion and Colonization in Ancient Greece.* Leiden.

Malley, G. 1881. "Language among North American Indians." *Bureau of American Ethnology Bulletin.* Washington, D.C.

Mallwitz, Alfred. 1962. "Cella und Adyton des Apollotempels in Bassai." *MDAIA* 77.

——. 1972. *Olympia und seine Bauten.* Munich.

——. 1981. "Kritisches zur Architektur Griechenlands im 8. und 7. Jahrhundert." *AA* 4.

Mansuelli, Guido. 1967. *The Art of Etruria and Early Rome.* New York.

Marcel, Pierre. n.d. *Charles Le Brun.* Paris.

Marconi, Pirro. 1935. *Studi Agrigentini.* Rome.

Marcovich, Miroslav. 1967. *Heraclitus.* Merida, Venezuela.

Mariette, Pierre Jean. 1854–1856. *Abecedario et Notes Inédites de cet Amateur* Ed. P. de Chennevières et A. de Montaiglon. Paris.

Marinatos, Nanno. 1985. *Art and Religion in Thera.* Athens.

Marinatos, Spiridon. 1936. "Le Temple Géométrique de Dréros." *BCH* 60.

Marinatos, Spiridon and Max Hirmer. 1960. *Crete and Mycenae.* New York.

Marinelli, Sergio. 1981. "The Author of the Codex Huygens." *JWCI* 44.

Maritain, Jacques. 1932. *Art and Scholasticism.* Trans. J. Scanlan. London.

Marquand, A. 1909. *Greek Architecture.* New York.

Marquet, Y. 1973. *La Philosophie des Ihwan as-safa.* Algiers.

Marrou, H.-I. 1949. *Saint-Augustin et la fin de la Culture Antique.* Paris.

——. 1955. *Histoire de l'Education dans l'Antiquité.* Paris. Translated as *A History of Education in Antiquity* (New York, 1956).

Marsden, E. W. 1971. *Greek and Roman Artillery.* Oxford.

Marshack, Alexander. 1972. *The Roots of Civilization.* New York.

Martin, Roland Emile. 1951. *Recherches sur l'Agora Grecque.* Paris.

——. 1965. *Manuel d'Architecture Grecque.* Paris.

——. 1987. *Architecture et Urbanisme.* Rome.

——. *See also* Ginouvès, René.

Martini, Francesco di Giorgio. 1967. *Trattati di*

Architettura, Ingegneria e Arte Militare. 2 vols. Ed. Corrado Maltese and Livia Maltese Degrassi. Milan.

——. 1985. *Il Vitruvio Magliabecchiano.* Ed. Gustina Scaglia. Florence.

Massa-Pairault, Françoise-Hélène. 1985. *Recherches sur l'Art et l'Artisanat Etrusco-Italiques à l'Epoque Hellenistique.* Rome.

Matheson, Sylvia A. 1972. *Persia: An Archaeological Guide.* London.

Matthiae, Guglielmo. 1969. *Ricerche Intorno a San Pietro in Vincoli.* Rome.

Mauriac, Claude. 1945. *La Trahison d'un Clerc.* Paris.

Maurmann, Barbara. 1976. *Die Himmelsrichtungen im Weltbild des Mittelalters.* Munich.

Mauss, Marcel. 1950. *Sociologie et Anthropologie.* Paris.

——. 1968. *Oeuvres.* 3 vols. Ed. Victor Karady. Paris.

Mayassis, S. 1966. *Architecture, Religion, Symbolisme.* 2 vols. Athens.

Mazar, Amihai. 1980–1985. *Excavations at Tell Qasile.* 2 vols. Jerusalem.

Mazarakis-Ainian, Alexandre. 1985. "Contribution à l'Etude de l'Architecture Religieuse Grecque des Ages Obscurs." *AC* 54.

Mead, Margaret. 1928. *An Inquiry into the Question of Cultural Stability in Polynesia.* New York.

Meffert, Ekkehard. 1982. *Nikolaus von Kues.* Stuttgart.

Meiss, Millard. 1961. "De Artibus Opuscula XL." In *Essays in Honor of Erwin Panofsky.* ed. Meiss. 2 vols. New York.

——. 1974. *French Painting at the Time of Jean de Berry.* 6 vols. London.

Mellaart, James. 1967. *Çatál Huyuk.* London.

Meller, Peter. 1963. "Physiognomical Theory in Renaissance Heroic Portraits." In *20th International Congress on the History of Art,* vol. 2.

Mellink, Machteld. 1964–1965. "Lydian Wooden Huts and Sign 24 on the Phaistos Disk." *Kadmos* 3.

Melville, Herman. 1983. *Moby-Dick.* 1851. Reprint, New York.

Mendel, Gustave, ed. 1912–1914. *Catalogue des Sculptures Grecques, Romaines et Byzantines: Musées Impériaux Ottomans.* 3 vols. Constantinople.

Menendez Pidal, Ramón, ed. 1947. *Historia de España.* Madrid.

Menninger, Karl. 1969. *Number Words and Number Symbols.* Cambridge, Mass.

Menninghaus, Winfried. 1980. *Walter Benjamins Theorie der Sprachmagie.* Frankfurt A/M.

Menut, Albert J., and Alexander J. Denomy. 1941–1943. "Maistre Nicole Oresme, Le Livre du Ciel et du Monde." *Mediaeval Studies* 3, 4, 5.

Mercklin, Eugen von. 1962. *Antike Figuralkapitelle.* Berlin.

Merlat, Pierre. 1960. *Jupiter Dolichenus.* Paris.

Merleau-Ponty, Maurice. 1960. *Signes.* Paris.

[Mersenne, Marin]. *Questions Harmoniques, dans lesquelles sont Contenues Plusieurs Remarquables pour la Physique, pour la Morale et pour les Autres Sciences,* Paris 1634 (Repr. Stuttgart 1972; Paris 1985).

Merz, Richard. 1978. *Die Numinose Mischgestalt.* Berlin.

Messerschmidt, F. 1942. "Das Grabmal des Porsenna." *Das Neue Bild der Antike* 2.

Métivet, L. 1917. *La Physionomie Humaine Comparée à la Physionomie des Animaux d'après les Dessins de Charles Le Brun.* Paris.

Metzger, Henri. 1963. *L'Acropole lycienne.* Vol. 2 of *Fouilles de Xanthos.* Paris.

——, ed. 1980. *Actes du Colloque sur la Lycie Antique.* Paris.

Metzger, Henri, et al. 1979. *La Stèle Trilingue du Létôon.* Vol. 6 of *Fouilles de Xanthos.* Paris.

Meuli, Karl. 1975. *Gesammelte Schriften.* 2 vols. Ed. Thomas Gelzer. Basel.

Meyer, Eduard. 1937. *Geschichte des Althertums.* 5 vols. Stuttgart.

Meyer, Heinz. 1975. *Die Zahlenallegorese im Mittelalter.* Munich.

Michałowski, Kazimierz. 1969. *Art of Ancient Egypt.* Trans. N. Guterman. New York.

Michel, Paul-Henri. 1948. "Les Mediétés." *Revue d'Histoire des Sciences* 2.

Miller, Andrew M. 1986. *From Delos to Delphi.* Leiden.

Miller, George A., and Elizabeth Lenneberg, eds. 1978. *Psychology and Biology of Language and Thought: Essays in Honor of Eric Lenneberg.* New York.

Miller, Norbert. 1978. *Archäologie des Traums.* Munich.

Miller, Stephen G., ed. 1990. *Nemea: A Guide to the Site and Museum.* Berkeley.

Millon, Henry A. 1958, September. "The Architectural Theory of Francesco di Giorgio." *AB* 40.

Miltner, Franz. 1958. *Ephesos: Stadt der Artemis und des Johannes*. Vienna.

Mirandola, Pico della. *See* Garin, E.

Mitten, David Gordon, John Griffiths Pedley, and Jane A. Scott, eds. 1971. *Studies Presented to George M. A. Hanfmann*. Cambridge, Mass.

Möbius, Hans. 1968. *Die Ornamente der Griechischen Grabstelen*. Munich.

Moe, C. J. 1945. *Numeri di Vitruvio*. Milan.

Moggi, Mauro. 1976. *I Sinecismi Interstatali Greci*. Pisa.

Momigliano, Arnaldo. 1969. *Quarto Contributo alla Storia degli Studi Classici e del Mondo Antico*. Rome.

——. 1980. *Sesto Contributo alla Storia degli Studi Classici e del Mondo Antico*. 2 vols. Rome.

Mommsen, August. 1898. *Feste der Stadt Athen im Altertum*. Leipzig.

Monmouth, Geoffrey of. 1929. *Historia Regnum Britanniae*. 1148. Ed. Acton Griscom and Trans. R. E. Jones. London.

Montagu, Jennifer. 1959. "Le Brun's Conférence sur l'Expression Générale et Particulière." Ph.D. diss., London University

——. 1994. "The Expression of the Passions: The Origin and Influence of Charles Le Brun's 'Conférence sur l'Expression Générale et Particulière.'" New Haven. Typescript.

Montagu, M. F. Ashley. 1968. *Culture: Man's Adaptive Dimension*. Oxford.

Montaigne, Michel de. 1965. *Oeuvres*. Ed. Pierre Villey. Lausanne.

Montmollin, D. de. 1951. *La Poétique d'Aristote: Texte Primitif et Additions Ultérieures*. Neuchatel.

Moran, William L., ed. and trans. 1992. *The Amarna Letters*. Baltimore.

Moravia, Sergio. 1972. "Philosophie et Médecine en France à la fin du XVIIIe Siècle." *Studies on Voltaire* 89.

Morenz, Siegfried. 1960. *Aegyptische Religion*. Stuttgart.

——. 1968. *Die Begegnung Europas mit Aegypten*. Berlin.

Moret, A. 1913. *Mystères Egyptiens*. Paris.

——. 1927. *The Nile and Egyptian Civilization*. New York.

Morley, Henry. 1891. *Character Writings of the Seventeenth-Century*. London.

——. 1915. *Anatomy in Long Clothes: An Essay on Andreas Vesalius*. Chicago.

Morolli, Gabriele. 1985. *Vetus Etruria*. Florence.

Morris, Desmond, ed. 1979. *Gestures: Their Origins and Distribution*. London.

Morris, Ian. 1987. *Burial and Ancient Society: The Rise of the Greek City-State*. Cambridge.

Morris, John. 1968. "The Dating of the Column of Marcus Aurelius." *JWCI* 15.

Morris, Robert. 1759. *Lectures on Architecture*. 2nd ed. London.

Morris, Sarah P. 1992. *Daidalos and the Origins of Greek Art*. Princeton.

Morrona, A. da. 1821. *Posa Antica e Moderna*. Pisa.

Mortet, V. 1909. "Recherches Critiques sur Vitruve et son Oeuvre." *RA* 4th ser., 13.

——, ed. 1910. *Mélanges Offerts à M. Emile Chatelain*. Paris.

Mortet, V., and Paul Deschamps, eds. 1929. *Recueil de Textes*. 2 vols. Paris.

Morton, James. 1841. *The Legend of St. Katherine of Alexandria*. London.

Moscati, Sabatino. 1979. *Il Mondo dei Fenici*. Milan.

——, ed. 1988. *The Phoenicians*. Milan.

Mossé, Claude. 1969. *La Tyrannie dans la Grèce Antique*. Paris.

Mowl, Tim, and Brian Earnshaw. 1988. *John Wood, Architect of Obsession*. London.

Muchau, Hermann. 1909. *Pfahlhausbau und Griechentempel*. Jena.

Mueller, Kurt, ed. 1930. *Tiryns: Die Ergebnisse der Ausgrabungen des Instituts*. 3 vols. Augsburg.

Mugler, Charles. 1964. *Dictionnaire Historique de la Terminologie Optique des Grecs*. Paris.

Mukerjee, R. 1959. *The Culture and Art of India*. New York.

Müller, Carl Otfried. 1820. *De Tripode Delphico*. Halle. Reprint in *Kunstarchaeologische Werke*, Berlin, 1873.

——. 1843. *Archaeologische Mittheilungen aus Griechenland*. Ed. Adolf Schöll. Frankfurt A/M.

——. 1844. *Geschichte Hellenischer Stämme und Städte*. 3 vols. 2nd ed. Breslau.

——. 1852. *Ancient Art and Its Remains*. Ed. F. G. Welcker and trans. J. Leitch. London.

Müller, Hermann. 1844. *Das Nordische Griechenthum und die Urgeschichtliche Bedeutung des Nordwestlichen Europas*. Mainz.

Müller, Wilhelm Max. 1916. *Egyptian Mythology*. London.

Müller-Karpe, Hermann. 1966–1989. *Handbuch der Vorgeschichte*. 4 vols. Munich.

Müller-Wiener, Wolfgang. 1988. *Griechisches Bauwesen in der Antike.* Munich.

Mumford, Lewis. 1951, November. "Function and Expression in Architecture." *Architectural Record.*

Münz, L., and G. Künstler. 1964. *Der Architekt Adolf Loos.* Munich.

———. 1966. *Adolf Loos, Pioneer of Modern Architecture.* London.

Murr, Josef. 1890. *Die Pflanzenwelt in der Griechischen Mythologie.* Innsbruck.

Murray, A. Victor. 1967. *Abelard and St. Bernard.* Manchester.

Murray, Gilbert. 1907. *The Rise of the Greek Epic.* Oxford.

Murray, Oswyn. 1980. *Early Greece.* Brighton.

Musti, Domenico, ed. 1986. *Le Origini dei Greci, Dori e Mondo Egeo.* Bari.

———. 1990. *Storia Greca.* 2nd rev. ed. Rome.

Muthesius, Hermann. 1994. *Style-Architecture and Building.* Intro. Stanford Anderson. Santa Monica, Calif. Trans. of *Stilarchitektur und Baukunst* (1902).

Muthmann, Friedrich. 1982. *Der Granatapfel.* Bern.

Mylonas, George Emmanuel. 1957. *Ancient Mycenae.* London.

———. 1961. *Eleusis and the Eleusinian Mysteries.* Princeton.

———. 1966. *Mycenae and the Mycenaean Age.* Princeton.

Myres, John L., ed. 1914. *Handbook of the Cesnola Collection of Antiquities from Cyprus.* New York.

Napier, A. D. 1986. *Masks: Transformation and Paradox.* Berkeley.

Napoli, Mario. 1969. *Il Museo di Paestum.* Naples.

Narcissus. 1982. Catalogue of Exhibition, January–February 1982. Rome.

Naredi-Rainer, P. von. 1984. *Architektur und Harmonie.* Cologne.

Nasr, Seyyed Hossein. 1978. *Islamic Cosmological Doctrines.* London.

Nauert, Charles G., Jr. 1965. *Agrippa and the Crisis of Renaissance Thought.* Urbana, Ill.

Naumann, Rudolf. 1955. *Architektur Kleinasiens, von Ihren Anfängen bis zum Ende der Hethitischen Zeit.* Tübingen.

Naville, Edouard. 1894. *The Temple of Deir-El-Bahari: Its Plan, Its Founders, and Its First Explorers.* 7 vols. London.

Neale, Ronald S. 1981. *Bath 1650–1850, a Social History; or a Valley of Pleasure Yet a Sink of Iniquity.* London.

Neschke, Ada B. 1980. *Die Poetik des Aristoteles.* Frankfurt A/M.

Neugebauer, Otto. 1957. *The Exact Sciences in Antiquity.* Providence.

Neumann, Erich. 1954. *The Origins and History of Consciousness.* Trans. R. F. C. Hull. New York.

———. 1955. *The Great Mother: An Analysis of the Archetype.* Bollingen Books 47. New York.

Neumeyer, Fritz. 1986. *Mies van der Rohe, Das Kunstlose Wort.* Berlin.

Newberry, Percy E., and G. Willoughby Fraser. 1893–1900. *Beni-Hasan.* 4 vols. Archaeological Survey of Egypt. London.

Newton, Charles Thomas. 1880. *Essays on Art and Archaeology.* London.

Newton, Sir Isaac. 1952. *Opticks, or a Treatise of the Reflections, Refractions, Inflections, and Colours of Light.* 1704–1730. Reprint, New York.

———. 1967. *The Correspondence.* 4 vols. ed. J. F. Scott. Cambridge.

Niceforo, A. 1952. *La Fisionomia nell'Arte e nella Scienza.* Florence.

Nichols, James H., Jr. 1976. *Epicurean Political Philosophy.* Ithaca and London.

Nield, Lawrence. 1969. "The Superstructure of the Greek Doric Temple." M. Litt. diss. Cambridge University.

Nietzsche, Friedrich. 1967–1972. *Werke.* Ed. Giorgio Colli and Mazzino Montanari. Berlin.

Nilsson, Martin P. 1932. *The Mycenaean Origin of Greek Mythology.* Cambridge.

———. 1950. *The Minoan-Mycenaean Religion and Its Survival in Greek Religion.* Lund.

———. 1957. *Griechische Feste von Religiöser Bedeutung.* 2 vols. 1906. Reprint, Stuttgart.

———. 1967. *Geschichte der Griechischen Religion.* 1955. Reprint, Munich.

NN [Sir John Harrington]. 1769. *Nugae Antiquae.* Bath.

Noack, Ferdinand. n.d. *Triumph und Triumphbogen.* Berlin.

Normand, C. P. J. 1819–1825. *Nouveau Parallèle des Ordres d'Architecture des Grecs, des Romains et des Auteurs Modernes.* Paris.

———. 1821–1823, 1839. *Le Vignole des Ouvriers;*

ou Méthode Facile pour Tracer les Cinq Ordres d'Architecture. Paris.

——. 1830. *Vergleichende Darstellung der Architectonischen Ordnungen der Griechen und Römer und der Neueren Baumeistern.* Trans. M. H. Jacobi. Potsdam.

——. 1951. *A New Parallel of the Orders of Architecture According to the Greeks and Romans and Modern Architects.* Trans. and two original plates by A. Pugin. 1829. Rev. and ed. R. A. Cordingley. London.

Nylander, Carl. 1966. "Remarks on the Urartian Acropolis at Zernaki Tepe." *Or. Suecana* 15.

——. 1970. *Ionians in Pasargadae.* Uppsala.

Oates, David. 1967. "Excavations at Tell al-Rimah." *Iraq* 29.

——. 1968. "The Excavations at Tell al-Rimah, 1967." *Iraq* 30.

O'Brien, Denis. 1969. *Empedocles' Cosmic Cycle.* Cambridge.

Oesterley, W. O. E., and T. M. Robinson. 1930. *Hebrew Religion: Its Origin and Development.* London.

Ogden, C. K., and I. A. Richards, 1930. *The Meaning of Meaning.* 3rd ed. London.

Ohnefalsch-Richter, Max. 1903. *Kypros: The Bible and Homer: Oriental Civilization, Art, and Religion in Ancient Times.* London.

Olerud, A. 1951. *L'Idée de Macrocosme et de Microcosme dans le Timée de Platon.* Uppsala.

O'Malley, C. D. 1952. *Leonardo da Vinci on the Human Body.* New York.

——. 1964. *Andreas Vesalius of Brussels, 1514–1564.* Berkeley.

O'Malley, C. D., and J. B. de C. M. Saunders. 1982. *The Anatomical Drawings of Andreas Vesalius.* New York.

O'Malley, John W. 1979. *Praise and Blame in Renaissance Rome.* Durham, N.C.

Onians, John. 1988. *Bearers of Meaning: The Classical Orders in Antiquity, the Middle Ages, and the Renaissance.* Princeton.

Oppenheim, A. Leo. 1964. *Ancient Mesopotamia.* Chicago.

Oppenheim, Max Adrian Simon, Freiherr von. 1899–1900. *Vom Mittelmeer zum Persischen Golf durch den Hauran, die Syrische Wuste und Mesopotamien.* Berlin.

——, ed. 1931. *Der Tell Halaf.* Leipzig.

Oresme, Nicolas, Bishop of Lisieux. 1952. *Livre de Divinacions.* Ed. and trans. G. W. Coopland. Liverpool.

——. 1966. *Ad pauca respicientes. . . .* Ed. and trans. E. Grant. Madison, Wis., and London.

——. *See also* Clagett, Marshall; Menut, Albert, and Alexander Denomy.

Orlandos, A. 1966–1968. *Les Matériaux de Construction et la Technique Architecturale des Anciens Grecs.* 2 vols. Paris.

Orsi, Paolo. 1933. *Templum Apollonis Alaei ad Crimisa Promontorium: A cura della "Società Magna Graecia."* Rome.

Orthmann, Winfried. 1971. *Untersuchungen zur Späthethitischen Kunst.* Bonn.

Ortony, Andrew, ed. 1979. *Metaphor and Thought.* Cambridge.

Otten, Heinrich, and Vladimir Souček. 1969. *Ein Althethitisches Ritual für das Königspaar.* Wiesbaden.

Otto, August. 1890. *Die Sprichwörter und Sprichwörtlichen Redensarten der Römer.* Leipzig.

Otto, Eberhard. 1958. *Das Verhältnis von Rite und Mythus im Aegyptischen Altertum.* Heidelberg.

Otto, Walter F. 1933. *Dionysos: Mythos und Kultus.* Frankfurt.

Overbeck, J. A. 1868. *Die Antiken Schriftquellen zur Geschichte der bildenden Künste bei den Griechen.* Leipzig. (Reprinted Hildesheim, 1959.)

——. 1881. *Geschichte der Griechischen Plastik.* Leipzig.

Pace, B. 1935–1938. *Arte e Civiltà della Sicilia Antica.* 4 vols. Milan.

Pacioli, Luca. 1509. *Diuina Proportione: Opera a Tutti Glingegni Perspicaci e Curiosi Necessaria.* Venice.

Padel, Ruth. 1993. "Agatharchus and the Date of Scene-Painting." London. Typescript.

Palagia, Olga, and William Coulson, eds. 1993. *Sculpture from Arcadia and Laconia.* Oxford.

Paley, Morton D. 1984. *The Continuing City: William Blake's Jerusalem.* Oxford.

Palladio, Andrea. 1642. *L'Architettura.* Venice.

Pallottino, Massimo. 1969. *Saggi di Antiquità.* Rome.

——. 1975. *Etruscologia.* 6th ed., rev. Milan.

Pallottino, Massimo, and G. Colonna. 1966. "Scavi nel Santuario Etrusco di Pyrgi." *ArchCl* 18.

Pallottino, Massimo, et al. 1981. *Die Göttin von*

Pyrgi. Colloquium held in Tübingen, 16–17 January 1979. Florence.

Palmer, Leonard R. 1965. *Mycenaeans and Minoans.* New York.

Panniker, Raimundo. 1977. *The Vedic Experience.* London.

Panofka, Theodor. 1843. *Dionysos und die Thyaden.* Berlin.

Panofsky, Erwin. 1940. *The Codex Huygens and Leonardo da Vinci's Art Theory: The Pierpont Morgan Library Codex M.A. 1139.* London.

———. 1960. *Renaissance and Renascences in Western Art.* Stockholm.

———. 1974. *Meaning in the Visual Arts.* Woodstock, N.Y.

———. *See also* Meiss, Millard.

Papadopoulos, Alexander. 1976. *Islam and Muslim Art.* Trans. R. E. Wolf. New York.

Papadopoulos, Jeannette. 1980. "Xoana e Sphyrelata: Testimonianza delle Fonti Scritte." *StA* 24.

Papini, Roberto. 1946. *Francesco di Giorgio Architetto.* Milan.

Parke, H. W. 1985. *The Oracles of Apollo in Asia Minor.* Dover, NH.

———. 1986. *Festivals of the Athenians.* London.

Parke, H. W., and D. E. W. Wormell. 1956. *The Delphic Oracle.* 2 vols. Oxford.

Parrot, André. 1961. *Nineveh and Babylon.* Trans. Stuart Gilbert and James Emmons. London.

Parrot, André, M. H. Chéhab, and Sabatino Moscati. 1975. *Les Phéniciens.* Paris.

Pascietto, E. 1984. *Pietro d'Abano, Medico e Filosofo.* Rome.

Pasteur, Georges. 1972. *Le Mimétisme.* Paris.

Pastor, L. von. 1923–1940. *The History of the Popes.* 10 vols. Ed. and trans. F. I. Antrobus and F. R. Kerr. London.

Patai, Raphael. 1947. *Man and Temple in Ancient Jewish Myth and Ritual.* New York.

Patroni, Giovanni. 1904. "Nora. Colonia Fenicia in Sardegna." *MAL* 14.

Patterson, Annabel M. 1970. *Hermogenes and the Renaissance.* Princeton.

Pattison, Mrs. Mark. 1879. *The Renaissance of Art in France.* 2 vols. London.

———. *See also* Dilke, Lady Emilia Frances.

Paul, Oscar, ed. 1872. *Boetius und die Griechische Harmonik.* Leipzig.

Pauw, Cornelius de. 1792. *Recherches Philosophiques sur les Américains.* 2 vols. Paris.

Payne, Humfry. 1931. *Necrocorinthia: A Study of Corinthian Art in the Archaic Period.* Oxford.

Paz, Octavio. 1986. *On Poets and Others.* New York.

Pedley, John Griffiths. 1990. *Paestum: Greeks and Romans in Southern Italy.* New York.

Pedoe, Dan. 1976. *Geometry and the Visual Arts.* New York.

Pedretti, C. *See* Leonardo da Vinci.

Pelon, Oliver. 1976. *Tholoi, Tumuli et Cercles Funéraires.* Paris.

Pendlebury, J. D. S. 1963. *The Archaeology of Crete: An Introduction.* New York.

Pennethorne, John. 1878. *The Geometry and Optics of the Ancient Architecture.* London.

Penrose, F. C. 1888. *An Investigation of the Principles of Athenian Architecture.* London.

The Peopling of Ancient Egypt and the Deciphering of Meroitic Script. 1978. Proceedings of symposium held in Cairo, 1974. Paris.

Perelman, C., and L. Olbrechts-Tyteca. 1969. *The New Rhetoric.* Trans. J. Wilkinson and P. Weaver. Notre Dame, Ind.

Perez-Gomez, Alberto. 1983. *Architecture and the Crisis of Modern Science.* Cambridge, Mass.

Perpeet, Wilhelm. 1977. *Ästhetik im Mittelalter.* Freiburg.

Perrault, Charles. 1696–1700. *Les Hommes Illustres qui ont paru en France pendant ce Siècle,* Paris.

———. 1909. *Mémoires de ma Vie.* 1669. Ed. Paul Bonnefon. Paris.

Perrot, Georges, and Charles Chipiez. 1882–1889. *Histoire de l'Art dans l'Antiquité.* 8 vols. Paris.

Perrot, Nell. 1937. *Les Représentations de l'Arbre Sacré sur les Monuments de Mésopotamie et d'Elam.* Paris.

Pesco, Daniela del. 1984. *Il Louvre di Bernini nella Francia di Luigi XIV.* Naples.

———. 1988. *Colonnato di San Pietro.* Rome.

Petit, Jean. 1970. *Le Corbusier lui-même.* Geneva.

[Petrarch], 1649. *Francisci Petrarchae De Remediis Utriusque Fortunae.* Rotterdam.

Petronotis, Arg. 1972. *Zum Problem der Bauzeichnungen bei den Griechen.* Athens.

Peursen, C. A. von. 1966. *Body, Soul, and Spirit.* Trans. H. Hoskins. Oxford.

Pevsner, Nikolaus. 1968. *An Outline of European Architecture.* 1943. Reprint, Harmondsworth.

Pfiffig, Ambros Josef. 1965. *Uni-Hera-Astarte.* Vienna.

Pfister, Friedrich. 1910–1912. *Der Reliquienkult im Altertum.* Giessen.

Pfühl, Ernst. 1923. *Malerei und Zeichnung der Griechen*. 3 vols. Munich.

Philip, J. A. 1966. *Pythagoras and Early Pythagoreanism*. Toronto.

Philipp, Hanna. 1968. *Tektonon Daidala: Der Bildende Künstler und sein Werk im Vorplatonischen Schrifttum*. Berlin.

Philippson, Paula. 1944. *Thessalische Mythologie*. Zurich.

Piaget, Jean. 1926. *The Language and Thought of the Child*. Trans. Marjorie and Ruth Gabain. London.

———. 1965. *Etudes Sociologiques*. Geneva.

———. 1967. *Biologie et Connaissance*. Paris.

Piattelli-Palmarini, Massimo, ed. 1979. *Théories du Langage: Théories de l'Apprentissage*. Paris.

———. [1989?] "Evolution, Selection, and Cognition: From 'Learning' to Parameter-Fixation in Biology and in the Study of Mind." *Center for Cognitive Science, MIT, 35*. Cambridge, Mass.

Picard, Charles. 1922. *Ephèse et Claros*. Paris.

———. 1930. *Origines du Polythéisme Hellenique*. Paris.

———. [1931?] *L. Acropole*. 2 vols. Paris.

Picard, Charles, and P. Frotier de la Coste Messelière. 1928. "Art Archaique: Les Trésors Ioniques." *FD 2, fasc. 2*.

Piccaluga, Giulia. 1974. *Terminus: I Segni di Confine nella Religione Romana*. Rome.

Piccolomini, Aeneas Silvius [Pius II]. 1984. *Commentaries*. Ed. L. Totaro. Milan.

Pickard-Cambridge, Arthur. 1968. *The Dramatic Festivals of Athens*. Oxford.

Picon, Antoine. 1988. *Claude Perrault ou la Curiosité d'un Classique*. Paris.

Piérart, Marcel, ed. 1992. "Polydipsion Argos: Argos de la Fin des Palais Mycéniens à la Constitution de l'Etat Classique." *BCH*, suppl. 22.

Pietz, William. 1985–1989. "The Problem of the Fetish." *RES 9, 13, 16, 17*.

Piggott, Stuart. 1965. *Ancient Europe from the Beginnings of Agriculture to Classical Antiquity*. Edinburgh.

———. 1975. *The Druids*. London.

Piles, Roger de. 1708. *Cours de Peinture par Principes*. Paris.

Piotrovskii, B. B. 1967. *Urartu: The Kingdom of Van and Its Age*. London.

Piranesi, G. B. 1764. *Le Antichità di Cora*. Rome.

———. 1778. *Différentes Vues de quelques Restes de Trois Grands Edifices qui subsistent encore dans l'Ancienne Ville de Pesto*. Rome.

Pirenne, M. H. 1970. *Optics, Painting, and Photography*. Cambridge.

Pitra, Jean Baptiste, ed. 1967. *Analecta novissima Spicilegii Solesmensis*. 1885–1887. Reprint, Farnborough, Hants.

Pius II, Pope. *See* Piccolomini, Aeneas Silvius.

Placides et Timéo ou Li Secrés as Philosophes. 1980. Ed. C. A. Thomasset. Paris.

Placzek, A. K., J. Ackerman, and M. N. Rosenfeld. 1978. *Sebastiano Serlio on Domestic Architecture*. New York.

Platner, Samuel Ball. 1929. *A Topographical Dictionary of Ancient Rome*. Completed and rev. Thomas Ashby. Oxford.

Platon, Nicholas. 1971. *Zakros: The Discovery of a Lost Palace of Ancient Crete*. New York.

Plebe, Armando. 1961. *Breve Storia della Retorica Antica*. Milan.

Plessner, Helmut. 1980. *Gesammelte Schriften*. 10 vols. Ed. Gunter Dux et al. Frankfurt A/M.

Plommer, W. Hugh. 1977. "Shadowy Megara." *JHS 97*.

Polacco, Luigi. 1952. *Tuscanicae Dispositiones*. Padua.

Poleni, Marchese Giovanni. 1739. *Exercitationes Vitruvianae*. Padua.

Polignac, François de. 1984. *La Naissance de la Cité Grecque*. Paris.

Politi, Raffaello. 1826. *Il Viaggiatore in Girgenti e il Cicerone di Piazza*. Girgenti/Agrigento.

Pollard, A. W. 1908. *Catalogue of Books Printed in the XVth Century, Now in the British Museum*. London.

Pollitt, J. J. 1972. *Art and Experience in Classical Greece*. Cambridge.

———. 1974. *The Ancient View of Greek Art*. New Haven and London.

Pollock, Frederick, and Frederic Maitland. 1896. *The History of English Law before the Time of Edward I*. Cambridge.

Polyklet: Der Bildhauer der Griechischen Klassik. 1990. Catalogue of Exhibition at the Museum Alter Plastik, October 1990–January 1991. Frankfurt A/M.

Pomeroy, Sarah B. 1976. *Goddesses, Whores, Wives, and Slaves*. London.

Pomtow, H. 1910–1911. "Das Alte Tholos und das Schatzhaus der Sikyoner zu Delphi." *Zeitschrift für Geschichte der Architektur 3, 4*.

Popham, M. 1982. The Hero of Lefkandi." *Antiquity 56*.

Popham, M., P. G. Calligas, and L. H. Sackett, eds. 1990–1993. *Lefkandi II: The Protogeometric Building at Toumba*. 2 vols. London.

Popham, M., and L. H. Sackett, eds. 1968. *Excavations at Lefkandi, 1964–66*. London.

Porada, Edith. 1965. *Ancient Iran: The Art of Pre-Islamic Times*. London.

Portmann, Adolf. 1969. *Le Forme Viventi*. Trans. Boris Porena. Milan.

Posner, Donald I. 1959, September. "Charles Le Brun's Triumph of Alexander." *AB* 41.

Poudra, Nöel. 1864. *Oeuvres de Désargues Réunies et Analisées*. Paris.

Pouilloux, Jean. *See* Roux, Georges.

Poulsen, Vagn. 1937. *Der Strenge Stil*. Copenhagen.

Poussin, Nicolas. 1929. *Correspondance*. Ed. P. du Colombier. Paris.

Prager, Frank C. and Gustina Scaglia. 1972. *Mariano Taccola and His Book "De Ingeneis."* Cambridge, Mass.

Prayon, Friedhelm. 1975. *Frühetruskische Grab- und Hausarchitektur*. Heidelberg.

——. 1984. "Zur Genese der Tuskanischen Säule." *Vitruv Colloquium, Schriften des Deutschen Archäologen-Verbandes* 8. Darmstadt.

——. 1987. *Phrygische Plastik*. Tübingen.

——. 1991. "*Deorum Sedes* sull'Orientamento dei Templi Etrusco-Italici." *ArchCl* 43.

Preller, Ludwig. 1964. *Griechische Mythologie*. 1860. New ed. 2 vols. Berlin.

Prestel, Jakob. 1901. *Des Marcus Vitruvius Pollio Basilika zu Fanum Fortunae*. Strassburg.

Previtali, G., et al. 1964. *Mostra dei Disegni di Humbert de Superville*. Catalogue. Florence.

Prince, Sir Uvedale. 1842. *On the Picturesque*. Edinburgh and London.

Prier, R. A. 1976. *Archaic Logic*. The Hague and Paris.

Prinzhorn, H. 1922. *Bildnerei der Geisteskranken*. Berlin.

——. 1926. *Bildnerei der Gefangenen*. Berlin.

Pritchard, James B. 1955. *Ancient Near Eastern Texts*. Princeton.

——, ed. 1969. *The Ancient Near East*. Princeton.

Prüchner, Helmut. 1992. "Ein Traum für Apollon." In *Kotinus: Festschrift für Erika Simon*. Mainz.

Pryce, F. N. 1928. *Catalogue of Sculpture in the Department of Greek and Roman Antiquities in the British Museum*. London.

Puchstein, Otto. 1907. *Die Ionische Säule*. Leipzig.

Quatremère de Quincy, Antoine-Chrysostome. 1832. *Dictionnaire Historique de l'Architecture*. 2 vols. Paris.

——. 1980. *De l'Imitation*. 1823. Reprint, Brussels.

Quilici Gigli, Stefania. 1981. "Colombari e Colombaie nell'Etruria Rupestre." *Riv. Ist. Arch.* 3rd ser., 4.

Raby, F. J. E. 1927. *A History of Christian-Latin Poetry from the Beginning to the Close of the Middle Ages*. Oxford.

——. 1934. *A History of Secular Latin Poetry in the Middle Ages*. Oxford.

Radet, Georges. 1967. *La Lydie et le Monde Grec au Temps des Mermnades (687–546)*. 1893. Reprint, Rome.

Radhakrishnan, S. 1953. *The Principal Upanishads*. London.

Raffaele Sanzio. 1956. *Tutti gli Scritti*. Ed. E. Camesasca. Milan.

——. 1984. *Il Pianto di Roma: Lettera a Leo X.* Ed. Piero Buscardi. Turin.

Rakob, Friedrich. 1979. "Das Groma-Nymphaeum im Legionslager von Lambaesis." *MDAI* 86.

Ramée, Daniel. 1860. *Histoire Générale de l'Architecture*. 2 vols. Paris.

Ramirez, Juan Antonio. 1983. *Edificios y Sueños*. Malaga and Salamanca.

——. 1991. *Dios Arquitecto: J. B. Villalpando y el Templo de Salomón*. Madrid.

Ramsay, Sir William M. 1969. *Asianic Elements in Greek Civilization*. New York.

Ramsden, E. H. 1963. *The Letters of Michelangelo*. 2 vols. London.

Rand, Benjamin, ed. 1900. *The Life, Unpublished Letters, and Philosophical Regimen of Anthony, Earl of Shaftesbury*. London.

——. 1914. *Second Characters*. Cambridge.

Raschke, Wendy J., ed. 1988. *The Archaeology of the Olympics*. Madison, Wis.

Raven, C. E. 1942. *John Ray, Naturalist, His Life and Works*. Cambridge.

Raven, J. E. 1951. "Polyclitus and Pythagoreanism." *CQ* 45.

——. *See also* Kirk, G. S.

Redfield, James M. 1975. *Nature and Culture in the Iliad*. Chicago.

Reeves. M. J., and B. H. Hirsch-Reich, 1972. *The Figurae of Joachim of Fiore*. Oxford.

Reinach, Salomon. 1897–1898. *Répertoire de la Statuaire Grecque et Romaine.* 3 vols. Paris.

———. 1908–1909. *Cultes, Mythes et Religions.* 3 vols. Paris.

Reinach, Théodore. 1926. *La Musique Grecque.* Paris.

Reitzenstein, R. 1904. *Poimandres: Studien zur Griechisch-Aegyptischen und Früchristlichen Literatur.* Leipzig.

Reitzenstein, R., and H. H. Schaeder, 1926. *Studien zum Antiken Synkretismus aus Iran und Griechenland.* Leipzig.

Renfrew, Colin. 1972. *The Emergence of Civilization: The Cyclades and the Aegean in the Third Millennium B.C.* London.

———. 1985. *The Archaeology of Cult: The Sanctuary at Phylakopi.* London.

Restoro d'Arezzo. 1976. *La Composizione del Mondo.* Ed. Alberto Morino. Florence.

Reuterswärd, Patrik. 1960. *Studien zur Polychromie der Plastik.* Stockholm.

Rey, Abel. 1942. *La Science Orientale avant les Grecs.* Paris.

Reymond, E. A. E. 1969. *The Mythical Origin of the Egyptian Temple.* Manchester.

Reynolds, Sir Joshua. 1975. *Discourses on Art.* 1778. Ed. Robert R. Wark. New Haven.

Rhees, Rush. 1979. "Wittgenstein über Sprache und Ritus." T. L. Wittgenstein (1960–1982, vol. 3).

Rhomaios, K. A. 1915. "Ek Tou Preistorikou Thermou." *Archaiologikon Deltion* 1.

Rice, Michael. 1991. *Egypt's Making: The Origins of Ancient Egypt, 5000–2000 B.C.* London.

Richard, Heinrich. 1970. *Vom Ursprung des Dorischen Tempels.* Bonn.

Richards, I. A. 1936. *The Philosophy of Rhetoric.* Oxford.

Richardson, J. 1725. *An Essay on the Theory of Painting.* London.

Richter, Gisela M. A. 1949. *Archaic Greek Art against Its Historical Background: A Survey.* New York.

———. 1954. *Catalogue of Greek Sculptures in the Metropolitan Museum of Art.* Cambridge, Mass.

———. 1962. "How Were the Roman Copies of Greek Portraits Made?" *MDAIR* 69.

———. 1966. *The Furniture of the Greeks, Etruscans, and Romans.* London.

———. 1968. *Korai: Archaic Greek Maidens.* London.

———. 1970. *Kouroi.* London.

Richter, J. P. *See* Leonardo da Vinci.

Rico, F. 1970. *El Pequeño Mundo del Hombre.* Madrid.

Ricoeur, Paul. 1962–1968. *Philosophie de la Volonté: Le Volontaire et l'Involontaire.* 2 vols. Paris.

———. 1975. *La Métaphore Vive.* Paris.

Ridder, A. de, and W. Deonna. 1924. *L'Art en Grèce.* Paris.

Ridgway, Brunilde S. 1974. "A Story of Five Amazons." *AJA* 78.

———. 1977. *The Archaic Style in Greek Sculpture.* Princeton.

———. 1981. *Fifth Century Styles in Greek Sculpture.* Princeton.

———. 1990. *Hellenistic Sculpture.* Madison, Wis.

Rieber, R. W., ed. 1976. *The Neuro-psychology of Language: Essays in Honor of Eric Lenneberg.* New York.

Riezler, W. 1914. *Weissgrundige Attische Lekythen.* Munich.

Rimmer, Joan. 1969. *Ancient Musical Instruments of Western Asia.* London.

Rist, J. M. 1969. *Stoic Philosophy.* Cambridge.

Robert, Fernand. 1939. *Thymelé: Recherches sur la Signification des Monuments Circulaires dans l'Architecture Religieuse de la Grèce.* Paris.

Robert, Louis. 1940. *Hellenica: Recueil d'Epigraphie, de Numismatique et d'Antiquités Grecques.* 5 vols. Limoges.

———. 1945. *Le Sanctuaire de Sinuri près de Mylasa.* Paris.

Robert, Louis and Jeanne Robert. 1954. *La Carie.* Paris.

———. 1983. *Fouilles d'Amyzon en Carie.* Paris.

Roberts, H. 1980–1981. "Reconstructing the Greek Tortoise-shell Lyre." *WA* 12.

Robertson, D. S. 1945. *A Handbook of Greek and Roman Architecture.* 2nd ed. Cambridge.

Robertson, M. 1975. *A History of Greek Art.* 2 vols. Cambridge.

Robinson, J. A. T. 1963. *The Body, a Study in Pauline Theology.* London.

Rodenwaldt, G. 1912. "Votivpinax aus Mykenai." *MDAIA* 37.

———. 1938. *Altdorische Bildwerke in Korfu.* Berlin.

———. ed. 1940. *Korkyra: Archaische Bauten und Bildwerke.* Berlin.

Roebuck, Carl. 1959. *Ionian Trade and Colonization*. New York.

Rohault de Fleury, G. 1877. *Le Latran au Moyen Age*. Abbeville.

Rohde, Erwin. 1870. "Unedirte Lucianscholien, die attischen Thesmophorien und Haloen betreffend." *RhM* n.f., 25.

———. 1925. *Psyche*. Trans. W. B. Hillis. London.

Romano, Elisa. 1990. *La Capanna e il Tempio: Vitruvio o dell'Architettura*. Palermo.

Rorty, Richard. 1980. *Philosophy and the Mirror of Nature*. Princeton.

Rosci, Marco. 1966. *Il Trattato di Architettura di Sebastiano Serlio*. Milan.

Rose, H. J. 1929. *A Handbook of Greek Mythology*. New York.

Rose, V., ed. 1864. *Anecdota Graeca et Graecolatina*. Berlin.

Rosen, Stanley. 1983. *Plato's Sophist: The Drama*. New Haven and London.

———. 1987. *Plato's Symposium*. New Haven and London.

Ross, David J. A. 1962. "A Late Twelfth-Century Artist's Pattern-Sheet." *JWCI* 25.

Rossi, Paolo. 1960. *Clavis Universalis: Arte Mnemoniche e Logica Combinatoria da Lullo a Leibnitz*. Milan and Naples.

Rouse, William H. D. 1902. *Greek Votive Offerings*. Cambridge.

Roussel, Denis. 1976. *Tribu et Cité*. Paris.

Roussel, Pierre. 1925. *Delos*. Paris.

Roux, Georges. 1949. *Le Problème des Argonautes*. Paris.

———. 1961. *L'Architecture de l'Argolide aux IVᵉ et IIIᵉ Siècles avant Jesus-Christ*. 2 vols. Paris.

———. 1979. *L'Amphictionie, Delphes et le Temple d'Apollon au IVᵉ Siècle*. Lyon.

———. ed. 1984. *Temples et Sanctuaires*. Lyon.

Roux, Georges, and Jean Pouilloux. 1963. *Enigmes à Delphes*. Paris.

Rowland, Benjamin. 1967. *The Art and Architecture of India: Buddhist, Hindu, Jain*. Harmondsworth.

Rowland, Ingrid D. 1994, March. "Raphael, Angelo Colocci, and the Genesis of the Architectural Orders." *AB* 76.

Rubin, Arnold, ed. 1988. *Marks of Civilization: Artistic Transformations of the Human Body*. Los Angeles.

Rubin, Ida E., ed. 1963. *Twentieth International Congress of the History of Art*. Princeton.

Rudenko, S. I. 1951. *Der Zweite Kurgan von Pasaryk*. Berlin.

———. 1953. *Kultura Naseleniia Gornogo Altaia*. Moscow and Leningrad.

Rudhardt, Jean. 1986. *Le Rôle d'Eros et d'Aphrodite dans les Cosmogonies Grecques*. Paris.

Ruelle, C.-E. 1897. "Le Monocorde, Instrument de Musique." *REG* 10.

Rügler, Axel. 1988. *Die Columnae Caelatae des Jüngeren Artemisions von Ephesos*. Tübingen.

Rukschcio, B., and R. Schachel. 1982. *Adolf Loos: Leben und Werke*. Vienna.

Rundle Clark, R. T. 1978. *Myth and Symbol in Ancient Egypt*. London.

Rupprich, H. 1956–1969. *Dürers Schriftlicher Nachlass*. 3 vols. Bonn.

Ruskin, John. 1851. *The Stones of Venice*. London.

———. 1903–1912. *The Complete Works*. 39 vols. ed. E. T. Cook and A. Wedderburn. London.

Russell, D. A. 1973. *Plutarch*. New York.

———. 1981. *Criticism in Antiquity*. Berkeley.

Rüster, Christel, and Erich Neu. 1989. *Hethitisches Zeichenlexikon*. Wiesbaden.

Rutenber, Culbert Gerow. 1946. *The Doctrine of the Imitation of God in Plato*. New York.

Rutkowski, B. 1986. *The Cult Places of the Aegean*. New Haven.

Rykwert, Joseph. 1976. *The Idea of a Town*. London and Princeton.

———. 1980. *The First Moderns*. Cambridge, Mass.

———. 1981a. *On Adam's House in Paradise*. Cambridge, Mass.

———. 1981b. "On an Egyptian (?) Misreading of Francesco di Giorgio's." *RES* 1.

———. 1982. "Lodoli on Function and Representation." In *The Necessity of Artifice*. London.

———. 1994a. "Greek Temples: The Polychromy." *Scroope* 6.

———. 1994b. "On the Palmette." *RES* 26.

Rykwert, Joseph, and Anne Engel, eds. 1994. *Leon Battista Alberti*. Catalogue of exhibition held at Palazzo del Tè, September–December 1994. Mantua.

Saad, Zaki Y. 1947. "Royal Excavations at Saqqara and Helwan (1941–45)." *SASAE* 3.

Saalman, Howard. 1959, March. "Early Renaissance Architectural Theory and Practice in Antonio Filarete's Trattato di Architettura." *AB* 41.

Sabra, A. I. 1981. *Theories of Light.* Cambridge.

Sachs, Curt. 1943. *The Rise of Music in the Ancient World.* London.

Sadie, Stanley, ed. 1980. *The New Grove Dictionary of Music and Musicians.* 20 vols. London.

Sagredo, Diego de. 1526. *Medidas del Romano.* Toledo.

———. 1549. *Medidas del Romano Agora Nueuamente Impressas y Añadidas de Muchas Pieças y Figuras Muy Necessarias alos Officiales que Quieren Seguir las Formaciones de las Basas, Colunas, Capiteles y Otras Pieças de los Edificios Antiguos.* Toledo. Edited by Fernando Marias y Agustín Bustamante. Madrid, 1986.

St. Clair, William. 1967. *Lord Elgin and the Marbles.* Oxford.

Sainte-Beuve, C.-A. n.d. *Portraits Littéraires.* Paris.

Saitta, Giuseppe. 1961. *L'Umanesimo.* 3 vols. Florence.

Sakellariou, M. B. 1958. *La Migration Grecque en Ionie.* Athens.

———. 1969. *Iraklion: Sammlung Metaxas.* Berlin.

Salmasius, Claudius [Claude Saumaise]. 1689. *Exercitationes Plinianae.* Utrecht.

Salmon, J. B. 1984. *Wealthy Korinth.* Oxford.

Salomon, Richard. 1969. *Opicinus de Canistris, Weltbild und Bekenntnisse eines avignonesischen Klerikers des 14. Jahrhunderts.* 1936. Reprint, Nendeln/Lichtenstein.

Sambursky, Shmuel. 1956. *The Physical World of the Greeks.* London.

———. 1959. *Physics of the Stoics.* London.

———. 1974. *Physical Thought from the Presocratics to the Quantum Physicists.* London.

Sandars, N. K. 1987. *The Sea Peoples.* London.

Sanders, Clinton R. 1989. *Customizing the Body.* Philadelphia.

Sandler, Lucy Freeman, ed. 1964. *Essays in Memory of Karl Lehmann.* Locust Valley, N.Y.

Santillana, G. de. 1961. *The Origins of Scientific Thought.* London.

Santillana, G. de, and H. von Dechend. 1969. *Hamlet's Mill.* Boston.

Santinello, G. 1958. *Il Pensiero di Niccolò Cusano nella sua Perspettiva Estetica.* Padua.

———. 1972. *Introduzione a Niccolò Cusano.* Bari.

Sauer, Josef. 1964. *Symbolik des Kirchengebäudes und seiner Ausstattung in der Auffassung des Mittelalters.* 1924. Reprint, Münster.

Saxl, Fritz. 1915–1927. *Verzeichnis Astrologischer und Mythologischer Illustrierter Handschriften des Lateinischen Mittelalters.* Heidelberg.

———. 1934. *La Fede Astrologica di Agostino Chigi.* Rome.

———. 1957. *Lectures.* 2 vols. Ed. G. Bing and F. Yates. London.

Scaglia, Gustina. 1970, December. Review of *Trattati,* by Francesco di Giorgio Martini. *AB 52.*

———. 1979. "A Translation of Vitruvius and Copies of Late Antique Drawings in Buonaccorso Ghiberti's *Zibaldone.*" *APS 69.*

———. 1980. "Autour de Francesco di Giorgio Martini." *RA 48.*

———. 1992. *Francesco di Giorgio: Checklist and History of Manuscripts and Drawings in Autographs and Copies.* Bethlehem, Pa.

———. *See also* Martini, Francesco di Giorgio; Taccola, Mariano.

Scamozzi, Vincenzo. 1615. *L'Idea della Architettura Universale.* Venice. (Reprinted Farnborough, Hants., 1964.)

Schaber, Wilfried. 1982. *Die Archaischen Tempel der Artemis von Ephesos.* Stiftland-Verlagwaldassen, Bayern.

Schachermeyr, Fritz. 1950. *Poseidon und die Entstehung des Griechischen Götterglaubens.* Bern.

———. 1964. *Die Minoische Kultur des Alten Kreta.* Stuttgart.

———. 1976. *Die Ägäische Frühzeit.* 2 vols. Vienna.

———. 1980. *Griechenland im Zeitalter der Wanderungen.* Vienna.

———. 1983. *Die Griechische Rückerinnerung im Lichte Neuer Forschungen.* Vienna.

———. 1984. *Griechische Frühgeschichte.* Vienna.

———. 1986. *Mykene und das Hethiterreich.* Vienna.

———. *See also* Kinzl, K. H.

Schadler, Ulrich. 1991. "Zur Entstehung der Attischen Basis und ihrer Verwendung im Kleinasiatischen Tempelbau." *Asia Minor Studies* 3.

Schadow, Gottfried. 1834. *Polyklet.* Berlin. English trans. John Sutcliffe (London, 1883).

———. 1849. *Kunst-Werke und Kunst-Ansichten.* Berlin.

Schäfer, Heinrich. 1932. "Djed-Pfeiler, Lebenszeichen, Osiris, Isis." In *Studies Presented to F. LL. Griffith.* London.

———. 1933. *Der Reliefschmuck der Berliner Tür aus der Stufenpyramide und der Königstitel Hrnb.* Deutsches Institut für Aegyptische Altertumskunde. Berlin.

Schäfer, J. 1983. "Bemerkungen zum Verhältnis

Mykenischer Kultbauten zu Tempelbauten in Kanaan." *AA*.

Schaller-Harl, Friederike. 1973. *Stützfiguren in der Griechischen Kunst*. Vienna.

Schapiro, Meyer. 1994. *Theory and Philosophy of Art: Style, Artist, and Society*. New York.

Schefold, Karl. 1960. *Meisterwerke Griechischer Kunst*. Basel and Stuttgart.

——. 1967. *Die Griechen und ihre Nachbarn*. Berlin.

Schefold, Karl, et al. 1968. *Eretria: Fouilles et Recherches*. Berne.

Schelling, Friedrich Wilhelm Joseph von. 1856–1861. *Werke*. 6 vols. Tübingen.

Schelp, Jochen. 1975. *Das Kanoun: Der Griechische Opferkorb*. Würzburg.

Schinkel, Karl Friedrich. 1982. *Collection of Architectural Designs*. Chicago.

Schipperges, H. 1961. *Das Menschenbild Hildegards von Bingen*. Leipzig.

Schliemann, H. 1874. *Trojanische Alterthümer*. Leipzig.

——. 1886. *Tiryns: The Prehistoric Palace of the Kings*. London.

Schlosser, Julius von. 1941. *Leben und Meinungen des Florentinischen Bildners Lorenzo Ghiberti*. Basel.

Schlosser-Magnino, Julius. 1956. *La Letteratura Artistica*. Florence.

Schmalenbach, Fritz. 1940, September. "The Term *Neue Sachlichkeit*." *AB* 22.

Schmaltz, Bernhard. 1983. *Griechische Grabreliefs*. Darmstadt.

Schmidt, Erich F. 1939. *The Treasury of Persepolis and Other Discoveries in the Homeland of the Achaemenians*. Chicago.

Schmidt, Erika E. 1973. "Die Kopien der Erechtheion Koren." *AntP* 13.

Schmidt, Eva Maria. 1982. *Geschichte der Karyatide: Funktion und Bedeutung*. Würzburg.

Schmidt, H., R. Naumann, A. Moortgat, and B. Hrovda, eds. 1943–1962. *Max Freiherr von Oppenheim, Tel Halaf*. 4 vols. Berlin.

Schmidt, Margret, ed. 1988. *Kanon: Festschrift Ernst Berger*. Basel.

Schmidt, Wilhelm. n.d. *Werden und Wirken der Völkerkunde*. Regensburg.

Schmidt-Colinet, A. 1977. "Antike Stützfiguren." Ph.D. diss., Cologne.

Schneider, Lambert A. 1975. *Zur Sozialen Bedeutung der Archaischen Korenstatuen*. Hamburg.

Schneider, René. 1910. *L'Esthétique Classique chez Quatremère de Quincy*. Paris.

Scholem, Gershon. 1960. *On the Kabbalah and Its Symbolism*. New York.

——. 1987. *Origins of the Kabbalah*. Princeton. Originally published as *Ursprung und Anfänge der Kabbala* (Berlin, 1962).

Scholfield, Peter Hugh. 1958. *The Theory of Proportion in Architecture*. Cambridge.

Schopenhauer, Arthur. 1972. *Sämtliche Werke*. 7 vols. Ed. Arthur Hübscher. 3rd ed. Wiesbaden.

Schrader, Hans. 1924. *Phidias*. Frankfurt A/M.

Schreiber, Hermann, and Georg Schreiber. 1957. *Vanished Cities*. New York.

Schubert, Bernhard. 1986. *Der Kunstler als Handwerker*. Königstein/Taunus.

Schuchhardt, K. 1891. *Schliemann's Excavations*. London.

Schuchhardt, W.-H. 1927. "Die Friese des Nereiden-Monumentes." *MDAI* 52.

——. 1974. "Antike Abgüsse Antiker Statuen." *AA* 89.

Schuetrumpf, Eckart. 1970. *Die Bedeutung des Wortes Ethos in der Politik des Aristoteles*. Munich.

Schuhl, P.-M. 1933. *Platon et l'Art de son Temps*. Paris.

——. 1934. *Essai sur la Formation de la Pensée Grecque*. Paris.

Schultze, Martin. 1876. *Handbuch der Ebräischen Mythologie: Sage und Glaube der Alten Ebräer*. Nordhausen.

Schulz, Dietrich. 1955. "Zum Kanon Polyklets." *Hermes* 83.

Schulze, Franz. 1985. *Mies van der Rohe: A Critical Biography*. New York.

Schulze, Werner. 1978. *Zahl Proportion Analogie*. Münster.

Schweitzer, B. 1953. *Platon und die Bildende Kunst der Griechen*. Tübingen.

——. 1971. *Greek Geometric Art*. London.

Scolari, F. 1837. *Della Vita e delle Opere di Vincenzo Scamozzi, Commentario*. Treviso.

Scullard, H. H. 1967. *The Etruscan Cities and Rome*. London.

Sebeok, Thomas A. 1968. *Animal Communication*. Bloomington, Ind.

——. *See also* Bouissac, Paul, Michael Herzfeld, and Ronald Posner.

Séchan, Louis. 1930. *La Danse Grecque*. Paris.

Seiler, Florian. 1986. *Die Griechische Tholos*. Mainz.

Seiterle, Gérard. 1979, September. "Artemis—Die Grosse Göttin von Ephesos." *AW* 3.

———. 1984. "Zum Ursprung der Griechischen Maske, der Tragödie und der Satyrn." *AK* 2.

Seltman, Charles. 1948. *Approach to Greek Art.* London and New York.

Selva, Giannantonio. 1814. *La Voluta Jonica.* Padua.

Semper, Gottfried. 1878. *Der Stil.* 2 vols. München.

Serlio, Sebastiano. 1584. *I Sette Libri dell'Architettura.* Venice. (Reprinted Bologna, 1978.)

———. 1982. *The Five Books of Architecture.* 1611. Reprint, New York.

Servolini, Luigi. 1944. *Jacopo de'Barbari.* Padua.

Sestieri, Pellegrino Claudio. 1958. *Paestum.* Rome.

Sethe, K., ed. 1902. *Untersuchungen zur Geschichte und Altertumskunde Aegyptens.* 15 vols. (vols. 12–15 ed. Hermann Kees). Leipzig.

———. 1930. "Urgeschichte und Aelteste Religion der Aegypter." *Abhandlungen für die Kunde des Morgenlandes* 18.

Seton Lloyd, Hugh. 1961. *The Art of the Ancient Near East.* New York.

Seton Lloyd, Hugh, and James Mellaart. 1962–1972. *Beycesultan.* 3 vols. London.

Settegast, Mary. 1986. *Plato Prehistorian.* Cambridge, Mass.

Seyrig, Henri. 1934–. *Antiquités Syriennes.* 6 vols. Paris.

Shaftesbury, Anthony Ashley Cooper. 1732. *Characteristicks of Men, Manners, Opinions, Times.* 3 vols. 5th ed. London.

———. *See also* Rand, Benjamin.

Shaw, Chandler. 1939. *Etruscan Perugia.* Baltimore.

Shelby, R. L. 1970. "The Education of Mediaeval Master Masons." *Speculum* 32.

———. 1977. *Gothic Design Techniques.* Carbondale, Ill.

Shibles, Warren A., ed. 1971a. *Metaphor: An Annotated Bibliography and History.* Whitewater, Wis.

———. 1971b. *Models of Ancient Greek Philosophy.* Southwick, Sussex.

Shiloh, Yigal. 1977. *The Proto-Aeolic Capital and Israelite Ashlar Masonry.* Jerusalem.

Shoe, Lucy T. 1936. *Profiles of Greek Mouldings.* Cambridge, Mass.

———. 1952. *Profiles of Western Greek Mouldings.* Rome.

———. 1965. *Etruscan and Republican Roman Mouldings.* Rome.

Shute, John. 1964. *The First and Chief Groundes of Architecture.* 1563. Reprint, London.

Sibylline Books. *See* Kurfess, Alfons.

Silvestris, Bernardus. 1876. *De Mundi Universitate.* Ed. C. S. Barach and J. Wrobel. Innsbruck.

Simmel, Georg. 1907. *Schopenhauer und Nietzsche: Ein Vortragszyklus.* Leipzig.

———. 1978. *The Philosophy of Money.* Trans. Tom Bottomore and David Frisby. London.

———. 1989. *Philosophie des Geldes.* Vol. 6 of Simmel 1989–.

———. 1989–. *Gesamtausgabe.* 24 vols. to date. Ed. Otthein Rammstedt. Frankfurt A/M.

———. 1992. *Soziologie.* Vol. 11 of Simmel 1989–.

Simmons, D. R. 1986. *Ta Moko: The Art of Maori Tattoo.* Aukland.

Simoni, Margherita Bergamini, ed. 1977. *Studi in Onore Di Filíppo Magi.* Perugia.

Simpson, R. Hope. 1965. *A Gazetteer and Atlas of Mycenaean Sites.* London.

Simpson, William Kelly, ed. 1972. *The Literature of Ancient Egypt.* New Haven.

Simson, Otto von. 1956. *The Gothic Cathedral.* London.

Singer, C. 1925. *The Evolution of Anatomy.* London.

Singer, Hans W. 1901. *Jakob Christoffel Le Blon.* Vienna.

Singer, I. 1983. "Western Anatolia in the Thirteenth century BC According to the Hittite Sources." *AnSt* 33.

Small, Jocelyn Penny. 1981. *Studies Related to the Theban Cycle on Late Etruscan Urns.* Rome.

Smeed, J. M. 1985. *The Theophrastan Character.* Oxford.

Smith, A. H., ed. 1900. *A Catalogue of Sculpture in the Department of Greek and Roman Antiquities, British Museum.* London.

Snell, Bruno. 1960. *The Discovery of the Mind.* 1953. Reprint, Cambridge, Mass.

———. 1967. *Scenes from Greek Drama.* Berkeley.

———. 1973. "Wie die Griechen Lernten, was Geistige Tätigkeit ist." *JHS* 93.

Snodgrass, A. M. 1971. *The Dark Age of Greece.* Edinburgh.

———. 1980. *Archaic Greece.* London.

Sokdowsky, B. von. 1887. *Die Musik des Griechischen Alterthums.* Leipzig.

Sokolowski, Franciszek. 1962. *Lois Sacrées des Cités Grecques.* Paris.

Solon, Leon Victor. 1924. *Polychromy: Architectural and Structural, Theory and Practice.* New York.

Sommella Mura, Anna. 1977. "La Decorazione Architettonica del Tempio Arcaico." *La Parola del Passato* 32.

Somville, P. 1975. *Essai sur la Poétique d'Aristote.* Paris.

Sontag, Susan. 1967. *Against Interpretation.* New York.

Sörbom, Göran. 1966. *Mimesis and Art.* Uppsala.

Sordi, Marta. 1989. *Il Mito Troiano e l'Eredità Etrusca di Roma.* Milan.

Soriano, Marc. 1972. *Le Dossier Perrault.* Paris.

Soteriadis, G. 1901, 1904, 1906. "Anaskephai en Thermon." *EA* 11, 14, 16.

Sourvinou-Inwood, Christiane. 1991. *"Reading" Greek Culture: Texts and Images, Rituals, and Myths.* Oxford.

Soyez, Brigitte. 1977. *Byblos et la Fète des Adonies.* Leiden.

Spanneut, M. 1957. *Le Stoicisme des Pères de l'Eglise.* Paris.

Spariosu, Mihai. 1982. *Literature, Mimesis, and Play.* Tübingen.

Spelthahn, Heinrich. 1904. *Studien zu den Chiliaden des Johannes Tzetzes.* Munich.

Spencer, B. 1904. *The Northern Tribes of Central Australia.* London.

Spencer, B., and F. J. Gillen. 1927. *The Arunta.* London.

———. 1938. *The Native Tribes of Central Australia.* 1899. Reprint, London.

Spiegelberg, Herbert. 1965. *The Phenomenological Movement.* 2 vols. The Hague.

Spinelli, A. G. 1908. *Memorie e Studi intorno a Jacopo Barozzi nel IV Centenario dalla Nascità.* Rome.

Spinoza, Benedict de. 1917–1919. *Works.* 2 vols. Ed. and trans. R. H. M. Elwes. 1883. Reprint, London.

Spuhler, James N. 1977. "Biology, Speech, and Language." *Ann. Rev. Anthr.* 6.

Sprague, Rosamond Kent, ed. 1972. *The Older Sophists.* Columbia, S.C.

Squarzina, S. D., ed. 1989. *Roma, Centro Ideale della Cultura dell'Antico nei Secoli XV e XVI.* Milan.

Staal, Fritz. 1983. *Agni: The Vedic Ritual of the Fire Altar.* 2 vols. Berkeley.

Staccioli, Romolo Augusto. 1968. *Modelli di Edifici Etrusco-Italici.* Florence.

———. 1984. "Il 'Sacello' di Giano Riprodotto in un 'Urna Cineraria Chiusina'?" *Colloqui del Sodalizio* 2nd ser.

Stadter, Philip A. 1989. *A Commentary on Plutarch's Pericles.* Chapel Hill, N.C.

Staehelin, Elisabeth. 1978. "Zur Hathor-Symbolik in der Ægyptischen Kleinkunst." *ZÄS* 90.

Stafford, Barbara M. 1972. "Les Deux Edifices: The New Areopagus and a Spiritual Trophy: Humbert de Superville's Vision of Utopia." *AQ* 35.

———. 1973. "Mummies, Herms, and Colossi: Easter Island and the Origin of Sculpture." *AQ* 36.

———. 1972. "Medusa, or the Physiognomy of the Earth." *JWCI* 35.

Starr, Chester G. 1962. *The Origins of Greek Civilization.* London.

Stefanini, Luigi. 1949. "Varietà. Ispirazione Pitagorica del 'Canone' di Policleto." *Giornale Critico della Filosofia Italiana,* 3rd ser., 1.

Stein, S. R., ed. 1986. *The Architecture of Richard Morris Hunt.* Chicago.

Steinberg, Leo. 1983. *The Sexuality of Christ.* New York.

Steinbrucker, Charlotte. 1915. *Lavaters Physiognomische Fragmente im Verhältnis zur Bildenden Kunst.* Berlin.

Steinen, K. von. 1925. *Die Marquesaner und ihre Kunst.* 2 vols. Berlin.

Steingräber, Stephan. 1979. *Etruskische Möbel.* Rome.

Steinmann, Jean. 1960. *Richard Simon et les Origines de l'Exégèse Biblique.* Bruges.

Steinmetzer, F. 1968. "Die Babylonischen Kudurru als Urkundenform," *Studien zur Geschichte und Kultur des Altertums.* Vol. 11. 1922. Reprint, New York.

Steuben, Hans von. 1973. *Der Kanon des Polyklet.* Tübingen.

Stevens, G. P. 1934. "Concerning the Curvature of the Steps of the Parthenon." *AJA* 38.

———. 1943. "The Curves of the North Stylobate of the Parthenon." *Hesperia* 12.

Stevens, G. P., et al. 1927. *The Erechtheion.* Ed. J. M. Paton. Cambridge, Mass.

Stevenson Smith, W. 1958. *The Art and Architecture of Ancient Egypt.* Harmondsworth.

Stewart, Andrew F. 1977. *Skopas of Paros.* Park Ridge, N.J.

———. 1978. "The Canon of Polykleitos: A Question of Evidence." *JHS* 98.

——. 1990. *Greek Sculpture: An Exploration*. 2 vols. New Haven.

Stirling Maxwell, Sir William. 1891. *Annals of the Artists in Spain*. 4 vols. London.

Stites, Raymond S., et al. 1970. *The Sublimations of Leonardo da Vinci*. Washington, D.C.

Stock, B. 1972. *Myth and Science in the Twelfth Century: A Study of Bernard Silvester*. Princeton.

——. 1983. *The Implications of Literacy*. Princeton.

Stokes, Adrian. 1978. *The Critical Writings*. 3 vols. Ed. Lawrence Gowing. London.

Stratton, Arthur James. 1931. *The Orders of Architecture: Greek, Roman, and Renaissance*. Philadelphia.

Strauss, Erwin. 1963. *The Primary World of the Senses*. New York.

——. 1966. *Phenomenological Psychology*. New York.

Strauss, Walter L. 1972. *Albrecht Dürer: The Human Figure*. New York.

Strzygowski, Josef. 1930. *Asiens Bildende Kunst*. Augsburg.

Stuart, Grace. 1955. *Narcissus: A Psychological Study of Self-Love*. New York.

Stuart, J., and N. Revett. 1762–1816. *The Antiquities of Athens Measured and Delineated*. 3 vols. London.

Stucchi, Sandro. 1955a. "La Decorazione Figurata del Tempio di Zeus ad Olympia." *Annuario della Scuola Archeologica di Atena* 30–32.

——. 1955b. "Nota Introduttiva sulle Correzioni Ottiche nell'Arte Greco fino a Mirone." *Annuario della Scuola Archeologica di Atene* 30–32.

——. 1982. "La Corrispondenza Metrica del Cosiddetto 'Piede Partenonico.'" *Atti dell'Accademia Nazionale dei Lincei*, 8th ser., 37.

Studi in Onore di Aristide Calderini e Roberto Paribeni. 3 vols. 1956–1957. Milan.

Sullivan, Louis H. 1947. *Kindergarten Chats*. New York.

Sulze, Heinrich. 1936. "Das Dorische Kapitell der Burg von Tiryns." *JDAI* 51.

Sulzer, Johann Georg. 1792. *Allgemeine Theorie der Schönen Künste*. Leipzig.

Summers, D. 1979. *The Sculpture of Vincenzo Danti*. New York.

——. 1981. *Michelangelo and the Language of Art*. Princeton.

Summerson, Sir John. 1963. *The Classical Language of Architecture*. Cambridge, Mass., and London.

——. 1970. *Architecture in Britain, 1530–1830*. Harmondsworth.

Sutphen, Morris Carter. 1902. *A Collection of Latin Proverbs*. Baltimore.

Svenbro, Jesper. 1984. *La Parola e il Marmo*. Turin.

——. 1988. *Phrasikleia: Anthropologie de la Lecture en Grèce Ancienne*. Paris.

Sweeney, J. J., and J. L. Sert. 1960. *Gaudi*. Stuttgart.

Szambien, Werner. 1983. *Jean-Nicolas-Louis Durand, De l'Imitation à la Norme*. Paris.

Taccola, Mariano. 1969. *Liber Tertius De Ingeneis*. 1433. Ed. James H. Beck. Milan.

——. 1984a. *De Ingeneis: Liber Primus Leonis, Liber Secundus Draconis*. Ed. Gustina Scaglia, Frank D. Prager, and Ulrich Montag. Wiesbaden.

——. 1984b. *De Rebus Militaribus (De Machinis)*. 1449. Trans. and ed. Eberhard Knobloch. Baden-Baden.

Tafuri, Manfredo. 1968. *Teorie e Storia dell'Architettura*. Bari.

——. 1969. "Per una Critica dell'Ideologia Architettonica." *Contropiano* 1.

——. 1973. *Progetto e Utopia*. Roma-Bari.

——. *See also* Dal Co, Francesco.

Talamo, Clara. 1979. *La Lidia Arcaica*. Bologna.

Tarde, Gabriel de. 1979. *Les Lois de l'Imitation: Etude Sociologique*. Paris.

Tarn, W. W. 1952. *Hellenistic Civilization*. London.

Tatarkiewicz, W. 1963. "Classification of Arts in Antiquity." *JHI* 24.

——. 1967. *Estetyka Nowozytna*. Wroclaw.

——. 1980. *A History of Six Ideas*. The Hague and Warsaw.

Tatlock, J. S. P. 1950. *The Legendary History of Britain*. Berkeley.

Tatton-Brown, V., ed. 1979. *Cyprus B.C.: 7000 Years of History*. London.

——. 1989. *Cyprus and the East Mediterranean in the Iron Age*. London.

Taylor, C. 1975. *Hegel*. Cambridge.

Taylour, Lord William. 1983. *The Mycenaeans*. London.

Teyssèdre, Bernard. 1957. *Roger de Piles et les Débats sur le Coloris au Siècle de Louis XIV*. Paris.

———. 1964. *L'Histoire de l'Art Vue du Grand Siècle*. Paris.

Théodorescu, Dinu. 1974. *Chapiteaux Ioniques de la Sicile Méridionale*. Naples.

———. 1985. "Le Comitium de Paestum, Projet et Réalisation." In *Le Dessin d'Architecture dans les Sociétés Antiques*. Strasbourg.

Thesiger, Wilfred. 1967. *The Marsh Arabs*. Harmondsworth.

Thiersch, H. 1935. *Artemis Ephesia: Eine Archeologische Untersuchung*. Berlin.

———. 1936. *Ependytes und Ephod: Gottesbild und Priesterkleid im Alten Vorderasien*. Stuttgart.

Thoenes, Christof, ed. 1989. *Sebastiano Serlio*. Milan.

Thomas, Rosalind. 1989. *Oral Tradition and Written Record in Classical Athens*. Cambridge.

Thomasset, Claude. 1982. *Une Vision du Monde à la Fin du XIIIe Siècle: Commentaire du Dialogue de Placides et Timéo*. Geneva.

Thompson, D'Arcy Wentworth. 1917. *On Growth and Form*. Cambridge.

Thompson, Homer A. 1940. "The Tholos of Athens and Its Predecessors." *Hesperia,* suppl. 4.

Thompson, Homer A., and R. E. Wycherley. 1972. *The Agora of Athens*. American School of Classical Studies at Athens, Athenian Agora 14. Princeton.

Thompson, Homer A., et al. 1976. *The Athenian Agora*. Princeton.

Thomson, George. 1946. *Aeschylus and Athens*. 1941. Reprint, London.

Thorndike, Lynn. 1923–1958. *A History of Magic and Experimental Science*. 8 vols. New York.

———. 1965. *Michael Scot*. London.

Thulin, Carl Olof. 1906–1909. *Die Etruskische Disciplin*. Göteborg.

Thumb, Albert. 1932–1959. *Handbuch der Griechischen Dialekte*. 2 vols. Ed. E. Kieckers and A. Scherer. Heidelberg.

Tigerstedt, E. N. 1965. *The Legend of Sparta in Classical Antiquity*. 3 vols. Stockholm.

Tiré, Claire, and Henri Van Effenterre. 1978. *Guide des Fouilles Françaises en Crète*. Paris.

Tobin, R. 1975. "The Canon of Polykleitos." *AJA* 79.

Toffanin, Giuseppe. 1943. *Storia dell'Umanesimo*. Bologna.

Toland, John. 1726. *Collection of Several Pieces*. London. (Reprinted Montrose, 1814.)

Tomberg, Friedrich. 1968. *Mimesis der Praxis und Abstrakte Kunst*. Neuwied and Berlin.

Torelli, Mario. 1971. "Il Santuario di Hera a Gravisca." *La Parola del Passato* 136–141. Rome.

———. 1975. *Elogia Tarquiniensia*. Studi e Materiali di Etruscologia e Antichità Italiche. Florence.

———. 1977. "Terracotte architettoniche arcaiche da Gravisca e una nota a Plinio XXXV, 151–152." *Nuovi Quaderni*.

———. 1987. *La Società Etrusca*. Rome.

Torge, P. 1902. *Aschera und Astarte*. Leipzig.

Torii, Takutoshi. 1983. *El Mundo Enigmático de Gaudí*. 2 vols. Madrid.

Tracy, Theodore. 1969. *Physiological Theory and the Doctrine of the Mean in Plato and Aristotle*. Chicago.

Trésors du Musée de Bagdad. 1977. Exhibition catalogue, Musée d'Art et d'Histoire, Geneva.

Treuil, René. 1983. *Le Néolithique et le Bronze Ancien Egéens*. Paris.

Trinkaus, C. 1970. *In Our Image and Likeness*. London.

Triomphe, Robert. 1989. *Le Lion, La Vierge et le Miel*. Paris.

Tunca, Önhan. 1984. *L'Architecture Religieuse Protodynastique en Mésopotamie*. 2 vols. Akkadica Suppl. Leuven.

Turchi, Nicola. 1939. *La Religione di Roma Antica*. Bologna.

Tuve, Rosemary. 1947. *Elizabethan and Metaphysical Imagery*. Chicago.

Tylor, E. B. 1903. *Primitive Culture*. London.

Tyrrell, W. B. 1984. *Amazons: A Study in Athenian Mythmaking*. Baltimore.

Tytler, Graeme. 1982. *Physiognomy in the European Novel: Faces and Fortunes*. Princeton.

Tzetzes, John. *See* Leone, P. A. M.; Harder, Carl Christian.

Tzonis, Alexander, and Liliane Lefaivre. 1988. *Classical Architecture: The Poetics of Order*. Cambridge, Mass.

Uberti, Fazio degli. 1952. *Il Dittamondo e le Rime*. 2 vols. Ed. G. Corsi. Bari.

Ucko, P. J., ed. 1977. *Form in Indigenous Art*. Canberra.

Ullmann, Walter. 1977. *Medieval Foundations of Renaissance Humanism*. London.

Usener, H. 1892. "Epikureische Inschriften auf Stein." *RhM* 47.

Uzielli, Gustavo. 1899. *Le Misure Lineari Medioevali e l'Effigie di Cristo*. Florence.

——. 1904. "Sulle Misure e sul Corpo di Cristo come Campione di Misura nel Medio Evo in Italia." In *Congresso Internazionale di Scienze Storiche.* Vol. 12. Rome.

Vagnetti, L., Ed. 1979. *Studi e Documenti di Architettura.* Nos. 9–10. Florence.

Valeriano of Belluno, Giovanni Pierio. 1625. *I Ieroglifici, overo Commentarii delle Occulte Significationi de gl'Egittii e Altre Nationi.* Venice.

Valesio, Paolo. 1960. "Un Termine della Poetica Antica." *Quaderni dell'Istituto di Glottologia* 5.

Vallois, René. 1944–1978. *L'Architecture Hellénique et Hellénistique à Delos.* 3 vols. Paris.

——. 1953. *Les Constructions Antiques de Délos: Documents.* Paris.

Van Buren, E. Douglas. 1923. *Archaic Fictile Revetments in Sicily and Magna Graecia.* London.

——. 1926. *Greek Fictile Revetments in the Archaic Period.* London.

——. 1931. *Foundation Figurines and Offerings.* Berlin.

——. 1933. *The Flowing Vase and the God with Streams.* Berlin.

——. 1945. *Symbols of the Gods in Mesopotamian Art.* Analecta Orientalia 23. Rome.

Van der Plas, Dirk. 1987. *Effigies Dei.* Leiden.

Vandier, J. 1954–. *Manuel d'Archéologie Egyptienne.* 4 vols. to date. Paris.

——. *See also* Jéquier, Gustave.

Vanoyeke, Violaine. 1992. *La Naissance des Jeux Olympiques et le Sport dans l'Antiquité.* Paris.

Vansteenberghe, E. 1920. *Le Cardinal Nicolas de Cues.* Paris.

Vasari, Giorgio. 1878–1906. *Vite.* 9 vols. Ed. G. Milanesi. Florence.

——. 1966–1976. *Vite.* 6 vols. Ed. Rosanna Bettarini and Paola Barocchi. Florence.

Vattimo, Gianni. 1961. *Il Concetto di Fare in Aristotele.* Turin.

——. 1989. *Al di la del Soggetto: Nietzsche, Heidegger e l'Ermeneutica.* Milan.

Vaux, Roland de. 1961. *Ancient Israel: Its Life and Institutions.* London.

Vellay, Charles. 1957. *Les Légendes du Cycle Troyen.* Monaco.

Verdenius, W. J. 1972. *Mimesis.* 1957. Reprint, Leiden.

Vermeule, Cornelius C., III. 1977. *Greek Sculpture and Roman Taste.* Ann Arbor.

Vermeule, Emily. 1972. *Greece in the Bronze Age.* 1964. Reprint, Chicago.

——. 1974. "Götterkult." *ArchHom* 3, fasc. 5.

——. 1979. *Aspects of Death in Early Greek Art and Poetry.* Berkeley.

Vernant, Jean-Pierre. 1985. *La Mort dans les Yeux.* Paris.

——. 1990. *Figures, Idoles, Masques.* Paris.

Vernant, Jean-Pierre, et al. 1981. *Le Sacrifice dans l'Antiquité.* Fondation Hardt 27. Geneva.

Verrall, Margaret de G., and J. E. Harrison. 1890. *Mythology and Monuments of Ancient Athens.* London.

Versenyi, Laszlo. 1974. *Man's Measure.* Albany.

Vesalius, Andreas. 1725. *Opera Omnia Anatomica et Chyrurgicae.* 2 vols. Ed. H. Borehaave and B. S. Albini. Leyden.

Vian, F. 1952. *La Guerre des Géants: Le Mythe avant l'Epoque Hellenistique.* Paris.

Vickers, Michael. 1985a. "Artful Crafts: The Influence of Metalwork on Athenian Painted Pottery." *JHS* 105.

——. 1985b. "Imaginary Etruscans: Changing Perceptions of Etruria since the Fifteenth Century." *Hephaistos* 7/8.

——. 1985c. "Persepolis, Vitruvius, and the Erechtheum Caryatids: The Iconography of Medism and Servitude." *RA* n.s., 1.

Vidal-Naquet, Pierre. 1986. *The Black Hunter: Forms of Thought and Forms of Society in the Greek World.* Baltimore.

Viel de Saint-Maux, Charles F. 1974. *Lettres sur l'Architecture.* 1787. Reprint, Geneva.

Vignola, Jacopo Barozzi da. 1596. *Regola delli Cinque Ordini d'Architettura.* 1562. Reprint, Venice.

Villani, Giovanni. 1559. *La Prima Parte delle Historie Universale de Suoi Tempi.* 2 vols. Venice.

Vincent of Beauvais. 1964. *Bibliotheca Mundi Vincentii Burgundi . . . Episcopi Bellovacensis: Speculum Quadruplex . . . op. et. st. Theologorum Benedictinorum.* 4 vols. 1624. Reprint, Graz.

Vincent, Howard P. 1965. *The Trying-out of Moby Dick.* Carbondale.

Viollet-le-Duc, E.-E. 1858. *Entretiens sur l'Architecture.* 2 vols. Paris.

——. 1877–1881. *Lectures on Architecture.* 2 vols. Trans. B. Bucknall. London.

Virel, André. 1980. *Ritual and Seduction: The Human Body as Art.* New York.

Virgilio, Biagio. 1981. *Il "Tempio Stato" di Pessi-*

nunte fra Pergamo e Roma nel II-I Secolo AC. Pisa.

Visser, R. P. W. 1985. *The Zoological Work of Petrus Camper.* Amsterdam.

Vlastos, Gregory. 1973. *Platonic Studies.* Princeton.

Voegelin, Eric. 1957–1974. *Order and History.* 4 vols. Baton Rouge.

Vollmann, B. 1965. *Studien zum Priszillianismus.* St. Ottilien.

Vries, Keith de, ed. 1980. *From Athens to Gordion: Memorial Symposium for Rodney S. Young.* Philadelphia.

Wace, A. J. B. 1949. *Mycenae: An Archaeological History and Guide.* Princeton.

Wace, A. J. B., and F. H. Stubbings. 1974. *A Companion to Homer.* 1962. Reprint, London.

Wace, A. J. B. and M. S. Thompson. 1912. *Prehistoric Thessaly.* Cambridge.

Walcot, P. 1966. *Hesiod and the Near East.* Cardiff.

Waldstein, C. 1902-1905. *The Argive Heraeum.* 2 vols. Boston.

——. *See also* Walston, Sir Charles.

Walker, D. Perkin. 1972. *The Ancient Theology.* London.

——. 1979. *Studies in Musical Science in the Late Renaissance.* London.

Wallace, R. W., and B. MacLachlan. 1991. *Harmonia Mundi.* Rome.

Walle, B. van de. 1954. "L'Erection du pilier Djed." *Nouvelle Clio* 6.

Wallis Budge, E. A. 1911. *Osiris and the Egyptian Resurrection.* 2 vols. London.

——. 1934. *From Fetish to God in Ancient Egypt.* Oxford.

——. 1969. *The Gods of the Egyptians.* 1904. 2 vols. Reprint, New York.

Walpole, Horace. 1828. *Anecdotes of Painting in England.* 3 vols. Ed. James Dallaway. London.

——. 1876. *Anecdotes of Painting in England.* 3 vols. Ed. Ralph Wornum. London.

Walston [Waldstein], Sir Charles. 1926. *Alcamenes and the Establishment of the Classical Type in Greek Art.* Cambridge.

Walter, Hans. 1976. *Das Heraion von Samos.* Zurich.

Walter-Karydi, Elena. 1980. "Die Entstehung der Griechischen Statuenbasis." *AK* 1.

Walton, Kendall L. 1990. *Mimesis as Make-Believe.* Cambridge, Mass.

Warburg, A. 1932. *Gesammelte Schriften.* 2 vols. Ed. Gertrud Bing. Berlin.

Ward, A. G., ed. 1970. *The Quest for Theseus.* London.

Warde Fowler, W. 1922. *The Religious Experience of the Roman People.* London.

Ware, William R. 1977. *The American Vignola.* Reprint with notes by John Barrington Bayley and Henry Hope Reed. New York.

Warren, Austin, and René Wellek. 1949. *Theory of Literature.* London.

Warren, Peter. 1975. *The Aegean Civilizations.* Oxford.

——. 1982–1983. "Knossos: Stratigraphical Museum Excavations, 1978–82. Part II." *Archaeological Reports* 29.

Warren, Peter, and Vronwy Hankey. 1989. *Aegean Bronze Age Chronology.* Bristol.

Washburn, Oliver. 1919. "The Origin of the Triglyph Frieze." *AJA* 23.

Watkin, D. W. 1974. *The Life and Work of C. R. Cockerell.* London.

Watts, Pauline Moffitt. 1982. *Nicolas Cusanus: A Fifteenth-Century Vision of Man.* Leiden.

Webb, Francis. 1815. *Panharmonicon.* London.

Webster, T. B. L. 1958. *From Mycenae to Homer.* London.

Wegner, M. 1949. *Das Musikleben der Griechen.* Berlin.

Wehrli, F. 1944–1959. *Die Schule des Aristoteles.* 10 vols. Basel.

Weidlé, Wladimir. 1963a. "Vom Sinn der Mimesis." *Eranos Jahrbuch* 31.

——. 1963b. *Der Mensch, Führer und Geführter im Werk.* Zurich.

Weinberger, Martin. 1967. *Michelangelo the Sculptor.* London.

Wellesz, E., ed. 1957. *Ancient and Oriental Music.* London.

Wendel, François, ed. 1963. *Les Sagesses du Proche-Orient Ancien.* Paris.

Weniger, Ludwig. 1904. "Das Hochfest des Zeus in Olympia." *Klio* 4.

Werbrouck, Marcelle. 1949. *Le Temple d'Hatshepsout à Deir-el-Bahari.* Brussels.

Wescott, Roger W., ed. 1974. *Language Origins.* Silver Spring, Md.

Wesenberg, Burkhardt. 1971. *Kapitelle und Basen: Beobachtungen zur Entstehung der Griechischen Säulenformen.* Düsseldorf.

——. 1972. "Kymation und Astragal." *Marburger Winckelmann Programm.*

——. 1982. "Thermos B1." *AA* 2.

——. 1983. *Beiträge zur Rekonstruktion Griechischer Architektur nach Literarischen Quellen.* Berlin.

West, E. W., ed. 1880–1897. *Pahlavi Texts.* 5 vols. Oxford.

West, M. L. 1971. *Early Greek Philosophy and the Orient.* London

——. 1992. Ancient Greek Music. Oxford.

Wetherbee, Winthrop. 1972. *Platonism and Poetry in the Twelfth Century.* Princeton.

——, ed. and trans. 1973. *The Cosmographia of Bernardus Silvestris.* New York and London.

Weyl, Hermann. 1949. *The Philosophy of Mathematics and Natural Science.* trans. Olaf Helmer. Princeton.

Wheatley, P. 1971. *The Pivot of the Four Quarters.* Edinburgh.

Wheelwright, P. 1959. *Heraclitus.* Princeton.

Whiston, William. 1725. *Astronomical Principles of Religion, Natural and Reveal'd.* 2nd ed. London.

——. 1753. *Memoir of My Life.* 2nd ed. London.

White, John. 1956. *On Perspective in Ancient Art.* Oxford.

Wide, S. C. A. 1893. *Lakonische Kulte.* Leipzig.

Widengren, G. 1965. *Die Religionen Irans.* Stuttgart.

Widmer, Bertha. 1955. *Heilsordnung und Zeitgeschehen in der Mystik Hildegards von Bingen.* Basel and Stuttgart.

Wiebenson, Dora, et al. 1982. *Architectural Theory and Practice from Alberti to Ledoux.* Chicago.

Wiegand, T. 1904. *Die Archaische Poros-Architektur des Akropolis zu Athen.* Cassel and Leipzig.

Wiemken, Helmut. 1972. *Der Griechische Mimus: Dokumente zur Geschichte des Antiken Volkstheater.* Bremen.

Wiener, Norbert. 1961. *Cybernetics.* Cambridge, Mass.

Wildung, D. 1977. *Imhotep und Amenhotep: Gottwerdung im Alten Aegypten.* Munich.

Wilkins, W. 1807. *The Antiquities of Magna Graecia.* London.

Will, Edouard. 1955. *Korinthiaka.* Paris.

——. 1956. *Doriens et Ioniens.* Paris.

Willey, Basil. 1934. *The Seventeenth-Century Background.* London.

William of Ockham. 1990. *Philosophical Writings: A Selection.* Trans. Philotheus Boehner. Rev. Stephen F. Brown. Indianapolis.

Williams, B. 1978. *Descartes: The Project of Pure Enquiry.* Harmondsworth.

Wilson-Jones, Mark. 1989. "Designing the Roman Corinthian Order." *JRA* 2.

——. 1991. "Designing the Roman Corinthian Capital." *Papers of the British School at Rome* 59.

——. 1994. "Designing Amphitheatres." *MDAIR* 100.

Winckelmann, J. J. 1964. *Anmerkungen ueber die Baukunst der Alten.* 1762. Reprint, Leipzig.

Wind, Jan, et al. 1992. *Language Origin: A Multidisciplinary Approach.* Dordrecht.

Winkler, John J., and Froma I. Zeitlin, eds. 1990. *Nothing to Do with Dionysos?* Princeton.

Winter, F. E. 1976. "Tradition and Innovation in Doric Design I: Western Greek Temples." *AJA* 80.

Wirth, Karl-August. 1967, September. "Bemerkungen zum Nachleben Vitruvs in 9. und 10. Jahrhundert und zu dem Schlettstädter Vitruv-Codex." *Kunstchronik* 9.

Wisse, Jakob. 1989. *Ethos and Pathos from Aristotle to Cicero.* Amsterdam.

Wissowa, Georg. 1904. *Gesammelte Abhandlungen zur Römischen Religions- und Stadtgeschichte.* Munich.

Wistrand, Erik. 1942. "Bemerkungen zu Vitruv und zur Antiken Architekturgeschichte." *Eranos* 40. Rpt. in his *Opera Selecta.* Stockholm, 1972.

Witt, R. E. 1971. *Isis in the Graeco-Roman World.* Oxford.

Wittgenstein, Ludwig. 1960–1982. *Schriften.* 8 vols. Ed. Rush Rhees. Frankfurt A/M.

——. 1967. "Bemerkungen über Frazers *Golden Bough.*" *Synthese* 17.

Wittkower, Rudolf, and H. Brauer. 1931. *Die Zeichnungen des Gianlorenzo Bernini.* 2 vols. Berlin.

——. 1975. *Studies in the Italian Baroque.* London.

Wölfflin, Heinrich. 1921. *Kunstgeschichtliche Grundbegriffe.* 5th ed. Munich.

Wollheim, Richard. 1987. *Painting as an Art.* London.

Wolska, Wanda. 1962. *La Topographie Chrétienne de Cosmas Indicopleustès.* Paris.

Wood, John. 1741. *The Origin of Building: or the Plagiarism of the Ancients Detected.* Bath.

——. 1747. *Choir Gaure, Vulgarly Called Stonehenge, on Salisbury Plain*. Oxford.

——. 1765. *A Description of Bath*. London.

Woolley, Sir Charles Leonard. 1935. *The Development of Sumerian Art*. London.

——. ed. 1952. *Carchemish: Report on the Excavations*. 3 vols. London.

——. 1953. *A Forgotten Kingdom*. Harmondsworth.

Woolley, Sir Charles Leonard, and Sir M. E. L. Mallowan. 1976. *Ur Excavations: The Old Babylonian Period*. Series ed. T. C. Mitchell. London.

Worsley, Giles. 1986, May. "The Baseless Doric Column in Mid-Eighteenth-Century English Architecture: A Study in Neo-Classicism." *Burlington Magazine*.

Wright, Frank Lloyd. 1943. *An Autobiography*. New York.

Wright, J. C. 1982. "The Old Temple at the Argive Heraeum and the Early Cult of Hera in the Argolid." *JHS* 102.

Wright, M. R. 1981. *Empedocles: The Extant Fragments*. New Haven.

Wundt, Wilhelm. 1908. *Voelkerpsychologie*. Leipzig.

Wurz, Erwin. 1906. *Plastische Dekoration des Stützwerkes in Baukunst und Kunstgewerbe des Altertums*. Strassburg.

Wyss, Beat. 1989. *Trauer der Vollendung*. Munich.

Xella, Paolo, et al. 1981. *La Religione Fenicia*. Colloquium of March 1979, Consiglio Nazionale delle Ricerche. Rome.

Yates, James. 1846. "On the Use of the Terms Acanthus, Acanthion etc. in the Ancient Classics." *CM* 3.

Yavis, C. G. 1949. *Greek Altars*. St. Louis.

Young, Rodney S. 1957. "Gordion 1956: Preliminary Report." *AJA* 61.

——. 1973. "The Phrygian Contribution." *ICCA* 10.

——. *See also* Vries, Keith de.

Zaehner, R. C. 1955. *Zurvan, a Zoroastrian Dilemma*. Oxford.

——. 1961. *The Dawn and Twilight of Zoroastrianism*. London.

Zafiropulo, Jean. 1961. *Apollon et Dionysos*. Paris.

Zahle, J. 1979. "Lykische Felsgräber mit Reliefs aus dem 4. Jahrhundert." *JDAI* 94.

Zakithinos [Zakuthēnos], D., et al. 1965. *Xaristērion éis Anastasion k. Orlandon*. Athens.

Zancani Montuoro, Paola. 1940. "La Struttura del Fregio Dorico." *Palladio* 18.2.

Zaner, Richard M. 1971. *The Problem of Embodiment*. The Hague.

Zeising, Adolph. 1854. *Neue Lehre von den Proportionen des Menschlichen Körpers*. Leipzig.

——. 1884. *Der Goldene Schnitt*. Leipzig.

Zeitlin, Froma. 1982. *Under the Sign of the Shield: Semiotics and Aeschylus' Seven against Thebes*. Rome.

Zeller, E. 1931. *Outlines of the History of Greek Philosophy*. Rev. W. Nestle. Trans. L. R. Palmer. London.

Zgusta, Ladislav. 1984. *Kleinasiatische Ortsnamen*. Heidelberg.

Zimmer, G. R. 1955. *The Art of Indian Asia: Its Mythology and Transformations*. 2 vols. Bollingen Series 39. New York.

Zimmermann, Albert, ed. 1983. *Mensura, Mass, Zahl, Zahlensymbolik im Mittelalter*. 2 vols. Berlin.

Ziomecki, Juliusz. 1975. *Les Représentations d'Artisans sur les Vases Attiques*. Warsaw.

Zoellner, Frank. 1985. "Agrippa, Leonardo and the Codex Huygens." *JWCI* 48.

——. 1987. *Vitruvs Proportionsfigur: Quellenkritische Studien zur Kunst-literatur im 15. und 16. Jahrhundert*. Worms.

Zografou, M. 1986. *Amazons in Homer and Hesiod*. Athens.

Zoubov, A. 1968. "Autour des Quaestiones super Geometriam Euclidis de Nicolas de Oreme." *MRS* 6.

Zuccaro, Federico. 1961. *Scritti d'Arte*. Ed. Detlef Heikamp. Florence.

Zuidervaart, Lambert. 1991. *Adorno's Aesthetic Theory: The Redemption of Illusion*. Cambridge, Mass.

Zuntz, Günther. 1971. *Persephone: Three Essays on Religion and Thought in Magna Graecia*. Oxford.

Zwerling, Charles, Frank H. Christensen, and Norman F. Goldstein. 1986. *Micropigmentation*. Thorofare, N.J.

Index

Page numbers in italic refer to illustration captions.

abacus, 181, 182, 311, 343
Abano, Pietro d', 39
Abelard, Peter, 415n16
Abu-Simbel (Egypt), 160, 314
Abydos, *160, 161,* 491n143, 493n158
acanthus
 and Corinthian capital, 317, *318,* 320, 343,
  345
 at Delphi, 327, *328,* 329–331, *331,* 334
 plant, 321, *324, 325,* 325, 327
 sculptured, *332, 333*
 Tarde on, 387
Achijawa, 485n92
Achilles, 342, 498n30
acroteria, *205,* 325–326
actio, 43
Adam, 65, 77, 82, *91*
Adamantius, 401n24
Adler, F., 453n41
Adonis, 487n113
Adorno, Theodor, 388–389, 391
 Aesthetic Theory, 389–390
aduton, 145
aedicules, 300
Aegina
 acroteria, 327
 Athena sculpture, 233
 sphinx, 267
 Temple of Aphaia, *174,* 217, *223,* 227,
  458n2, 467n2
Aeneas, 356
Aenianes, 499n30
Aeolic "order," 269, *270, 271,* 271, *272,* 290, *295,
 296, 298,* 302, 310, 315, 363–365, 474n12,
 475n14
Aeris, 488n113
Aeschylus, 337, 470n32
 Oresteia, 128
aesthetics, 30, 384–385
Agamedes, 193, 207
Agamemnon, 113, 128, 145–146, 498n30
Agapenor, King of Tegea, 292
Agatharcos of Samos, 470n32, 471n37
Ageladas, 104
Agesilaos II, King of Sparta, 175, *265, 266*
Aglosthenes, 481n52
agriculture, origins of, 337
Agrigento (Akragas, Sicily), temple of Olympian
 Zeus, 131, *132,* 133, 212, 458n70, 483n64

Agrippa, Cornelius, 82, 92
Ahab, King of Israel, 292
Ahhiyawa, 477n28
aisthēsis, 386
Akanthon (Thessaly), 331
Akkadian language, 274
Akragas. *See* Agrigento
Alaça Hüyük, 453n44, 486n105
Alalakh, 492n149
Alberti, Leon Battista
 and St. Antoninus, 420n50
 on Etruscan art, 367, 369
 and human body, 56, 59, 65–66, 89, 92
 on columns, 4, 54, 92, 460n25
Aldrich, Henry, 398n4
Alea, 340
Aleus, 503n67
Alexander the Great, 64, 110, 143, 265, 331,
 487n111, 504n84
Alexander of the Molossians, 499n30
Alison, Archibald, 15
Alkaios, 260
Alkamenes, 222, 224
Allers, R., 414n16
al-Muharrir, 415n18
Alpheios, river, 147
altars, Greek, 147–149
Alyattes of Lydia (or Sardis), 477n30, 483n72
Amathor Amathos, 488n113
Amathos (Cyprus), 314
Amazons
 and Artemis cult, 255
 statues of, *109,* 110–112, *112*
Ambrose, St., of Milan, 69
Amenhotep III, Pharaoh, 489n121
Amin-Berd-Erebani, near Yerevan, 485n93
Amrit (Marathus), 300
Amyklae, throne of Apollo, *238, 239,* 381, 386,
 465n72
anamnesis, 389
anathyrosis, *224,* 226
Anatolia
 arts and crafts, 258, 273–275
 timber from, 285
anatomy, artistic, 93. *See also* body, human
Anaxagoras, 470n32
Andania, mysteries at, 337
Andokides painter, 490n135
Andrae, Walter, 309–310
androgynes, 410n1
Androklos, 255

animals and birds, and human expression, *41, 41, 48, 49, 51*

Annianus, 505n4

Annio da Viterbo, 369

annuli (annulets), 178

anthropometry, 47

Antimachides, 504n90

Antinous, *111*

Antiochus, 349

Antistates, 504n90

Antium, battle of (338), 515n78

Antoninus Pius, Roman Emperor, 514n73

Antoninus, St., Bishop of Florence: *Summa Theologica*, 84–85

Anu (Hurrite deity), 487n112

Apelles, 441n40

Aphaia, Aegina temple, *174*

Aphrodisias of Carya, 257, 262, 478n39

Aphrodite. *See also* Astarte
  birth, 292
  and Corinth, 320
  cult, 257–258, *262*
  and Ionic order, *34*

Apollo, 200, 207–208, 227, 233–234, *238, 240, 241, 249, 255, 261, 264*, 267, 329, 331
  Belvedere, 92, *114*, 504n84
  Daphnephoros, 192, *194, 195*, 195, 212, 214
  Epikurios, 213, 217
  Karneios, 307
  Kereatas, 491n141
  Smintheus, 145, 149, *247*

Apollodorus, 487n113

Apollonius of Athens, *108*, 434n44

Appius Claudius Pulcher, 515n82

Apries, Pharaoh, 440n26

Apsas (Anatolian state), 275

Aquarossa (Italy), 514n66

araeostyle, 216

Aratus, 413n9

Arcadians, 502n60

Arcadius, Roman Emperor, 514n73

Arcesius, 467n93

Archanes, model, *191*, 468n14

Archilocus, 468n16

architrave, 182

Archytas of Tarentum, 427n11

Arctinus: *Iliupersis*, 498n29

Ares, 206

Argos, 113
  and Doric style, 249
  Heraion, 248–249
  temple model, *190*, 190, 212

Ariccia, Diana temple, 362

Aristophanes, 430n27

Aristotle, 36, 45–46, 69, 113
  on Athenian Olympeion, 347
  on harmony, 428n13
  on mimesis, 124–125, 127–128, 386
  on *poiēsis* and *praxis*, 226
  Poetics, 125, 127, 385, 435n1
  Rhetoric, 45

Aristoxenus of Tarentum, 102, 425n4, 429n16

Arkhanes (Crete), 156

Armenia, 279

Arrian, 485n83

Arsinoë II, Queen of Ephesus, 478n35

Arslan Tas (Köhnüs Valley), 283

Art Nouveau, 375, 377–378

Artemis
  cult, 255–260, 264
  statues and images of, 255, *256*, 257, 260, *262*
  temples, 110, 128, 200, 213, 247, 249, *250, 251, 252, 253, 254, 264*, 265–267, *266*, 269

Artemis Anassa, 257

Artemis Karneia (or Caryatis), 135

Artemis Leukophryene, 257

Artemisia of Halicarnassus, Caryan Queen, 129, 133

Arzawa (people), 275, 283, 285

Askelon, 487n112

Asklepios, 162, 343, *346*

asphodel, 321

Asplund, Gunnar, 18–19, *20, 21, 24*, 27

Assos, temple, 326, 469n22

Assurbanipal, King of Assyria, 269

Assur-nasir-pal, King of Assyria, 301

Assyria, 151–152, 156, 274–275, 278, 280, 290, 294, 301–302, 307, 326

Astarte/Ishtar, 258, 292, 294, 301, 479n39

Asterion (son of Minos), 146

astral man, 76

astrology/astronomy, 38–39, 65–66, 72, 76, 90. *See also* zodiac

Athena
  images and shrines, 113, 133, 213, 219, 224, 231, 233, 239, 257, 331, 340, 458n3
  saves Zagreos' heart, 337

Athenaeus, 133

Athens
  Academy, 385
  Acropolis, 135, 209, 217, *218*, 240, 267, 355, 474n9

Athens (cont.)
　　Dionysos theater, 443n57
　　Doric in, 212
　　Erechtheion, 133, *134*, 135, *242*, *246–247*,
　　　　467n1, 475n14
　　Hephaesteion, *223*
　　Ilissos (near), 246
　　Kerameikos, 468n12
　　Lysikrates monument, 327, 345, *347*, 495n3
　　Nike temple, 230, *240*, *241*, 246, 474n9
　　Odeion, 443n57
　　Olympieion (temple of Zeus), 347, *348*, 349,
　　　　360, 467n2
　　Parthenon (Athena Parthenos), 18, 27, 113,
　　　　188, 213, *222*, *223*, 231, 239, 246, 386,
　　　　473n53
　　Prytaneion, 342–343, 345
　　Theseion, 217
Atlanta, 503n70
atlantes (or telamones), 131, *132*, 133, 138. *See
　　also* caryatids
Atlas, 206, 369–370, 443n55
Atreidae, 153
Atreus, treasury, 151, 154, 156, 175
Attic order, 392n6
Attius Navius, 364, 515n75
Aubrey, John: *Brief Lives*, 43
Auerbach, Erich: *Mimesis*, 387
Augustine, St., 39, 46, 425n4, 428n12
Augustus (Gaius Octavianus Caesar), 228, 360–
　　361
avant-garde art (in Adorno and Lukács), 388, 391
Avebury (Wiltshire, England), 27
axes (tools), 175–177, *179*

Ba'al, 258
Baal-Shamin, 489n121
Baalbek, 479n40
Babylon, 152, 290, 326
Bacchos, 481n50. *See also* Dionysos
baetyls (holy stones), 156, 300
Bagehot, Walter, 387
Barbari, Jacopo de', 422n61
Barbaro, Daniele, 496n17
Barcelona, Parc Güell, 16–18, *16*, *17*, 381
Bar-Rekub of Zinçirli, 498n24
basileus, 146
Bassae
　　Apollo Epikurios, 213, 217, 246, 338, *339*,
　　　　340, 340, 342, 386, 467n2
　　Corinthian columns, 333–334, 338, 340,
　　　　343, 386, 495n3

Bataille, Georges, 122–123
Bath (England), 27, 29
Bathycles of Magnesia (throne of Apollo), *238*,
　　239
Batteux, Abbé Charles, 521n51
Baukunst, 375
Baumgarten, Alexander Gottlieb, 384
Beaux-Arts (Paris). *See* Ecole des Beaux-Arts
Becket, Thomas à, 38
Beckmann, Max, 518n6
Bede, Venerable, 69, *73*, 469n25
beehive tombs, 214
bees, 193, 479n43
Benda, Julien: *Trahison des Clercs*, 383–384
Beni-Hasan (Egypt), 158, 160
Benjamin, Walter, 373
Benn, Gottfried, 380–382
Berenguer, Francisco, 396n38
Berlin
　　Altes Museum, 11, 382
　　Neue Wache, 382
　　Philharmonic Hall, 389
　　Schauspielhaus, 11, 382
Bernini, Gian Lorenzo, 29, 31, 123, 370
Berry, Jean de France, Duke of: *Très Riches Heures*
　　(manuscript), 69, *70*
Beycesultan (Anatolia), 275, *280*, 283, 491n143
Billington, John, 395n30
bit-hilani, 280
Blake, William, 418n34
Blanc, Charles: *Grammaire des Arts du Dessin*, 8,
　　9, 11
Blom (or Bluom), Hans, 370
Blondel, François, 52
Blondel, Jacques-François, 34–36, *35*, *56*
blood, circulation of, 40–41, 97
Boaz, 294, 454n44
Boccaccio, Giovanni, 367
Boderie, Guy de la, 516n90
body, human
　　and built fabric, 93, 95, 97, 117
　　church plan as, 39, 61, 63–64, *63*
　　city as, 63–65, 67
　　and columns, 14, 29–35, 56, 113–115, 119,
　　　　122, 126, 128, 138, 373, 385
　　effect of passion on, 40–41
　　elements and humors, 73
　　female, *109*, 110, *112*, 112, 114–115, 117,
　　　　122
　　in God's image, 82–84
　　John Wood on, 29
　　as metaphor, 117–119, 122, 373

as microcosm, 69, 72–73, 75–77, *78, 79, 80, 81, 82,* 84, 112, 171
in motion, 90, 99
mutilation and marking of, 119
and numbers, 82, 101
proportions of, *56,* 59–61, *60, 62,* 63–64, 76, 82, 86–90, *87, 88, 89,* 92–93, 95, 97–99, 101, 110, 122, 217
and temple measures, 171
as temple of the Spirit, 27, 77, 82
Boetticher, Carl, 460n31
Boffrand, Charles, 53–54, 56
Boghazköy, 453n44
Bosse, Abraham, 47, 52, *52, 53, 54,* 399n12, 422n60
Bougainville, Louis de, 437n11
Boullant, Jean, 517n96
Boyle, Robert, 69
Bramante, Donato, 4, 370, 388
Brecht, Bertolt, 521n49
Bregna, Andrea, 420n50
Brethren of Purity. *See Ikhwan al-Safa*
Brittany, standing stones, *141*
bronze, 193, 230–231, 331
Brunelleschi, Filippo, 420n51
bucranium. *See* horns
Bülbül-Dagh (mountain), 255, 257
Bullant, Jean, 400n13
bulls, 271–272, 303–304, *305*
Bulwer, John, 404n44
Chirologia, 42
Burckhardt, Jacob, 380, 388
Cicerone, 10–11
burial, 320
Busiris (Abusir, Egypt), 160
Bussi, Giovanni Andrea de', 420n50
Butades (or Dibutades) of Sikyon, 208, 510n41
Büyükkale (Turkey), 275, 279
Byblos (*now* Ibrahim)
Ba'alat Gibal temple, 491n143
carvings, 297
coin from, *293*

Cabalists, 82, 85
Cabiroi sanctuary, near Thebes, 214
Cadiz (Gades), Melqart temple, 294
Caecina, Aulus, 506n6
Caere. *See* Cerveteri
Caesar, Julius, 360, 458n3, 509n34
Cailliaud mummy (Luxor), *74*
calendar, Mesopotamian, 75
Callescheros, 504n90
Callias the Eleusinian, 501n54

Callicrates, 246
Callimachus, 317, *319,* 320–321, 500n43
Callisto the Bear, 502n60
Calydon, 342
Cambyses, King of Persia, 273, 487n111
Camillus, 508n16, 510n39
Camper, Peter, 407n57
Canina, Luigi: *History of Ancient Architecture,* 18
canon, defined, 103
canonic man, 56, 76–77, 85, 90, 95, 97. *See also* body, human
Canosa, San Laucio church, 365
Carlyle, Thomas, 11, 518n7
Carneades of Cyrene, 402n29
carpenters, 126, *179*
Carthaginians, 131
Carya (Greece), 133, 135, 255, 283
caryatids, 31, 129–138, *130, 134, 136, 137,* 280, 333
Casal Marittimo, 513n64
Casaubon, Isaac, 43
Castalius, 499n33
Çatal Hüyük, 454n45, 491n143
Catherine de Médicis, Queen of France, 370
Cennini, Cennino, 56, 434n42
Censorinus: *De Die Natali,* 509n31
Ceres, 360–361, 363
Cerveteri (Caere), 354, 356
tombs, *360, 363, 364, 365,* 513n65
Cesariano, Cesare, 4, *88, 89, 90,* 368, *371,* 426n5
Cézanne, Paul, 388
Chaeronea, battle of (338), 343
Chalkis, 199
Chambray, Roland Fréart, Sieur de, 29, 31–32, 36, 47, *130,* 366, 461n33
Parallèle, 52, 319, 321
Champier, Symphorien, 516n90
Champollion, Jean François, 158
Chania Tekke (Crete), model, 190, *191,* 200, 455n48
Chantelou, Paul Fréart, Sieur de, 29–31, 123
Chapman, George, 176
character, defined, 35–36, 43, 45–46, 54, 56
Charlemagne, Emperor, 77, 79
Charles V, Emperor, 370
Cheops, Pyramid of, 455n48, 493n163
Chephren, Pyramid of, 314
Chersiphron of Knossos, 265
chiasmos (cross emphasis), 104
Chicago
Chicago Tribune Building, 19, *22, 23, 23,* 24
Columbian exhibition (1892), 15
Rookery Building, 396n46

Chinese "style," 8

Chipiez, Charles: *Critical History of the Greek Orders,* 6–8, 13

chirology. *See* hand

Chiusi, 354, 363, 513n63, 514n69, 515n80

Choisy, Auguste, 16, *185, 187,* 392n2

Christ. *See* Jesus Christ

Chronos, 249

Chryseis (daughter of Chryses), 145

Chryses (priest), 145, 149

Chrysippus, 109

church, body as plan model for, 39, 61, *63,* 63–64, *64, 65*

Cicero, 46, 228, 352, 361, 424n3, 425n5, 506n7
    On the Nature of the Gods, 83

Cimmerians, 278

Cinna, Gaius Helvius, 487n113

circular structures, 214, 342–343, *344, 345*

circumcision, 437n12

Cirò, Apollo Aleus temple, 494n164

city
    as a body, 63–64, 67
    as house, 65–66

Claros, 255, 264, 476n23

Claudius, Roman Emperor, 505n4, 506n7

Clytamnestra, 128
    treasury (Mycenae), 154, 175

Cnidos (now Tekir), treasury, 135, 333

Cockerell, C. R., 225

Cocles, Bartolomeus, 401n24

Colbert, Jean Baptiste, Marquis de Seignelay, 40

Coleridge, Samuel Taylor: *Philosophic Lectures,* 11

Colocci, Angelo, 392n6

Colombo, Realdo, 93

Columbia University, New York, 15

columen (*culmen*), 354

columns
    double spiral, 294, 297, 302
    Egyptian, 158, *159,* 160, 162–163, *166, 167,* 308–309, *309,* 311
    and the female, 110
    holy, 156–157
    and human body, 14, 29–35, *56,* 119, 122, 126, 128, 373, 385
    and the male, 129
    Mycenaean, 151–152, 154
    optical refinement and corrections of, 220–221, *223, 224,* 227, 229
    origins of, 9, 11–13, 113–114, 144, 148, 307–309
    and palm trunks, 303, *306*

sculptured figures as, 129–138, *130, 132, 134, 136, 137, 139, 140, 141,* 308, *309*
    as structural points, 12

Composite order, 3–4, *5, 30*

Constantine the Great, Roman Emperor, 499n34, 514n73

Constantinople
    "Burned Column," 514n73
    St. Sophia, 193

Copernicus, Nicolaus, 93

Cori, Hercules temple, 18, 473n65

Corinth
    and Aphrodite cult, 320
    Apollo temple, 208
    and Doric order, 208–209, 212, 469n22
    hipped temples, 189, 208
    social mores, 320
    and tholoi, 215
    treasury, 331

Corinthian genus
    capital, 317, *318*
    colors, 233
    design and proportions, 317, *341*
    and human body, 113, 217, 317, 386
    origins and development of, 317, *318, 319,* 320, 331, 333–334, 338, 340, 342–343, 345, 347, 349, 351
    as type, 3–4, *5,* 11, *31, 58*
    Vitruvius on, 239, 317, 320–321, 334

cornices
    as edging, 219–220
    and face profile, 35–36, *35, 57, 59*
    moldings of, *188,* 189, 190

corona. See cornice

Corsignano/Pienza, cathedral, 368

Cortona, 369

Corythus, 369

Cosa (Italy), temple, 359

cosmos
    fourfold division of, 76
    human body as, 69, 72–73, 75–77, *78, 79, 80, 81,* 82, 84

Cossutius, 349, 505n2

cow deities, 314

Creasy, Edward, 395n31

creativity, 386

Crete, 150, 190, 199, 275. *See also* Knossos; Minoans

Critias, 473n64

Croesus, King of Lydia, 265–266, 474n6

cube, in architecture, *58*

cult objects, 156, 160

Cumaean Sibyls, 510n38

cuneiform, 290

Cureau de la Chambre, Louis Marin, 40–41, 43

Cusanus, Nicholas. *See* Nicholas of Cusa

Cybele, 258, 455n47

cylinder seals, 151

cyma (capital molding), 189

cymatium, 247–248

Cyme, 260

Cyprus, 290, 292, 294, 297, *298, 299, 301, 303,*
303–304, 314

Cypsellus, 510n41

Cyrene, 483n68

Cyrus I, King of Persia, 265, 269

Cyrus II (the Great), King of Persia, tomb, 273,
274

Cytheria (Kythira), 320, 487n112

Da-u-Dukhtar, near Kurangun, Persia, 269

Daedalus, 150, 162, 472n41

Damon, 428n13

Damophilus, 361

Dante Alighieri
De Monarchia, 367
Inferno, 462n38
Purgatorio, 402n29

Daphne, 192, *194, 195*

Daphne-Harbie, House of Narcissus, *117*

Dardanus, 369, 509n34

Darius (the Great), King of Persia, 176, 272

David of Nerken, 414n16

decoration. *See also* polychromy
Aeolic, 294, *295, 296, 297, 298, 299,*
300–303
on columns, 182, 245–246

decorum, 53, 237, 239, 351

Deir-el-Bahari (Egypt), 158, *159,* 160, *313,* 315

de l'Orme, Philibert, 400n13, 516n86

Delos
monopteros, 345
Naxian *oikos,* 207
sphinx, 267
treasury, 331

Delphi, 135, 145, 193, 230
acanthus tripod, 327, *328,* 329–330, *331,*
334
Apollo temple, 192, *192, 193,* 234, 264, 327,
381, 471n39, 498n25
charioteer, 231
Hekation, 330, *330*
hestia-altar, 327
landscape, 331
Lesche, 498n30, 500n37

and origins of Doric, 249, 264, 267
sphinx, 267, *268*
Sybil, 249, 330–331
tholos, 343, *344, 345*
treasuries, 135, 138, *140,* 233, 234, 508n16

Delphos, 499n33

Demaratus, 510n41

Demeter, 334, *335, 337*–338, 361, 489n128

Demeter Thesmophora, 145, 217

Demetrios of Phalerum, 320

demiourgos, 124, 126, 150, 386. *See also poiēsis*

Democritus of Abdera, 414n16, 470n32, 472n41

Demophilus of Himera, 433n38, 513n58

Demosthenes, 460n30

Denderah (Egypt), temple of Hathor, *75, 313,* 314

Derham, William, 398n5

Désargues, Gérard (or Gaspard), 47

Descartes, René, 40–41, 43, *44,* 46–47

Deukalion, 327, 349

Dia Naxos (island), 483n68

Diderot, Denis, 520n38

Didyma, near Miletus, Apollo shrine, 215, 227,
241, *243, 246,* 255

Didymus, 429n16

Dietterlin, Wendel, 517n98

Dimeni (Thessaly), 452n34

Dinocrates, 63

Diocaesarea (Cilicia), 349

Diodorus Siculus, 131, 162, 331

Diogenes of Athens, 133

Diogenes Laertius, 472n41

Diomede, 443n61

Dionysius of Halicarnassus, 361, 440n32, 506n7

Dionysos (Bacchos), 255
cult, 261, 264, 329, 337–338, 345, 361,
381, 450n18
masked, 260–261
and Zagreos, 337

Diopos, 510n41

Dioscari, 365

Diospolis Parva (Egypt), 314

Diotima, 124, 222

Diulus, Caius, 515n78

Dix, Otto, 518n6

djed (or *tet*) "column," 156, 160, *161,* 162, 290,
308, *311*

Dodona, 249

Doesburg, Theo van, 519n26

Domitian, Roman Emperor, 360

Donatello, 420n51

Dorians
conquer Greece, 143
language, 115

Dorians (*cont.*)
mythology, 382
people, 381
Doric genus
and Apollo, 249, 264
base, 171–174, 176, *177*
Benn on, 380–381
Burckhardt praises, 11
capitals, 178–182, *184, 185, 185,* 186–189
column height, 216
contradictions and discrepancies in, 209, 211–212, 234
design, 171–182, 315
flutes, 177, *178*
and human (male) body, 112–113, 115, 122, 133, 138, 158, 165, 171, 386
idealized, 381
numbers, 215–218
origins and development, 143, 149, 158, 165, 208–209, 219, 249
shaft, 174–178
and statues, *32,* 131
in temple design, 172–173, *172, 173, 202*
territory, 249, 381
twentieth century examples, 16–24, 27, 378, 381
as type, 3–4, *5,* 11
Vitruvius on, 209, 211, 217, 237, 264, 386
wooden, 175–177
Dreros (Crete), temple, 149, *198,* 200, 202
Druids, 27, 399n6
Duchamp, Marcel, 397n51
Durand (Durandus), Guillaume, Bishop of Mende, 39
Durand, Jean-Nicolas-Louis, 11–14, 376
Dürer, Albrecht, 90, *91,* 92–93, 95, 471n35
Dvorak, Max, *24,* 388, 397n52

eagle, 208, *208*
Earle, John: *Microcosmographie,* 43
Ebstorf, *Mappa mundi, 79,* 82
echinus, 4, 10, 18, 178, 180, *181,* 182, 353
Ecole des Beaux-Arts (Paris), 11–14
Ecole Polytechnique (Paris), 11–12, 376
egg-and-dart ornament, 182, 245–246
Egypt
building methods, 311
columns, *140,* 158–159, *160,* 162–165, *166, 167,* 308, 314
cult objects, 156
and Hittites, 274
mythology, *74, 75*

and origins of Greek architecture and art, 14, 151, 158, 160, 162, 165
standards, 314
Ehrenburg, Ilya, 518n6
eikastikē, 125, 221
ekphora, 240
ekphrasis, 406n51
Elamites, 271, 278
Elea (city), 125–126, 260
elements, four (*stoicheia*), 69, 72–73, 109
Eleusis
capitals, 515n81
Demeter at, 334, 337
mysteries, 337
Telesterion (or Mystery Hall), 133, 166, 275, *336*
Temple G, 451n30
Elgin, Thomas Bruce, 7th Earl of, 453n41
emotion. *See* passion; pathos
Empedocles of Akragas, 69
Endoeos, 478n32
Enkomi (Cyprus), 491n141
Ennatum, King, 454n46
Ennius, 364, 508n21
entasis, 18, *221,* 226–227, 229
Epano Englianos (Navarino Bay), Palace of Nestor, 175
Ephesia, 475n22
Ephesos, 255
Ephesus
Acropolis of Ayusoluk, 255
Artemis temple and cult, 213, 247, 249, *254,* 255–257, 259–260, *264,* 265–267, *266,* 269, 381, 467n2, 474nn9,12, 507n12
Croesan building, 246, *264,* 265–266, 269
founding, 255
hekatompedon temple, 213
Prytaneion, 477n32
Epicurus, 103, 473n51
Epidauros, 327
Asklepeion, 343, 475n14, 495n3, 498n25
theater, 502n60
tholos, 343, *344, 346*
Epimetheus, 449n12
Epirus, 214
epistyle, 182–183, 220, 247
Epizephyrean Locri (Italy), 267
Erasmus, Georg Caspar, 517n98
Erechtheus, 135
Eretria
Daphnephoron, 168, 193, *194, 195,* 195–196, 202, 209, 212, 214

shrines, 213
tholos, 215, 343
Ergotimos, 462n46
Eros, 124
Eshnunna/Tell Asmar, 454n46
Espouy, Hector d': *Fragments d'Architecture Antique*, 18
ethikē, 46
ēthos, 45–46, 385
Etruscans. *See also* Tuscan arrangement
    architecture, 368
    cultural changes, 366–367
    language, 351, 367, 369
    origins, 506n7
    relations with Romans, 356–357, 369
    religion, 357
    temples, 172, 351, 353–356, 362, 364–365
    as term, 351–352, 357, 367
Eubea, 195, 198
Eucheir, 510n41
Euchidas of Platea, 498n29
Euclid, 227, 424n3, 427n12
Eudoxus of Cnidus, 413n9
Eugenius III, Pope, 37
Eugippius the Abbot, 416n24
Eugrammos, 510n41
Euphranor of Corinth (the Isthmian), 110
Euripides, 128–129, 177, 261, 462n47
    Iphigenia in Tauris, 186, 460n24, 462n47
eurythmy, 227–228, 272
Evans, Sir Arthur, 155
Expressionism, 375

face, human, *8, 9. See also* physiognomy
    and body proportions, *57,* 59–61, 86, 98
    on cornices and entablatures, 35–36, *35, 57, 59*
    division of, 86
    study of, 34–35, 36–37, 46–47, *48, 49, 50, 51*
Falerii, 364, 509n27
Fechner, Gustav, 394n13
female figure, 92, *109,* 110, 112, *112. See also*
    caryatids; korai
    and Ionic column, 114–115, 122, 133, 171
    Mycenaean shape, 156
    nudity, 447n4
Fethiye, tombs, *289*
Field, George, 394n13
Fiesole, 369
Figino, Ambrogio and Girolamo, 422n60
Filarete, Antonio Averlino, 4, 65
Firmicus Maternus, 501n51

first man, 72–73, 77
Flia (now Chalandri, Greece), 337
Florence
    Palazzo Rucellai, 368
    Piazza Santissima Trinita, 370
    Santa Maria Novella, 368
    Uffizi palace, 370
Foce del Sele, near Paestum, Heraion, 463n50
foot, human
    and body proportions, *62,* 86, 98
    as module, 248–249
Foucquet, Nicolas, Vicomte de Vaux, 40
Francesco di Giorgio. *See* Martini
François vase (Chiusi), 186, *187*
Fréart. *See* Chambray; Chantelou
Frederick II, Emperor, 39
free will, 76, 85
Freeman, Matthew L., 396n47
fresco, 232
frieze
    Doric, 182–183, 187, 189
    Ionic (zophoron), 247–148
Frontinus, S. Julius: *On Aqueducts,* 392n2
Fufidius of Arpinum, 505n2
Functionalism, 389
furniture and fabrics, 301–303

G (*Gestaltung;* periodical), 518n6
Galen, 109–110, 403n38
Gall, Franz Josef, 407n55
garlands, 271
Gassendi, Pierre, 40
Gaudí, Antoni, 16, *16, 17,* 18, 24, 27, 381
Gauricus, Pomponius
    and human proportions, 56, 410n87
    Dialogue on Sculpture (*De Sculptura*), 36, 90
Gea, 349
Gehlen, Arnold, 387
geison (coping), 189
Gellius, Aulus, 103, 512n50
Gelon of Syracuse, 131
genius, 386
Genseric (Vandal), 360
Geoffrey of Monmouth, 27
Gerbert, Martin, Abbot of St. Blasien, 429n17
Gerhardt, Paul, *22,* 396n47
Gestalt, 518n6
gesture, 40, 43, 45, 117
Ghiberti, Lorenzo, 426n5
Gilgamesh, 307, 491n140
Giocondo, Giovanni, 90, 507n9
Giotto, 39
Girard, René, 387–388

Giunta brothers (printers), 89–90

Gizeh, 314

Gnosticism, 82

goats, 271

God, man as likeness of, 82–84

gods, Greek. *See also individual deities*
    and cults, 260–261
    and kings, 146
    and temple types, 237, 239

Goethe, Johann Wolfgang von, 376, 385, 390, 394n16, 407n55

Gogh, Vincent van, 379, 388

Golem, 82

Gordion, megaron, *281,* 283, *284*

Gorgasos, 360

Gothic style, 13

Gozzoli, Benozzo, 417n30

Grapaldi, Francesco Maria, 4
    De Partibus Aedium, 393n7

Great Mother, 258–259, 264

Greece, ancient
    and canon, 103
    cults and gods, 260–261
    Dorians conquer, 143
    Egyptian influence on, 14, *151,* 158, 160, 162–163
    influence on Rome, 356–357
    languages, 114–115
    temples, 143–149, 171

Gribelin, Simon, 45

Guadet, Julien, 14

Gubbio tablets, 510n40

Gwilt, Joseph: *Encyclopaedia of Architecture,* 14–15

Gyges of Lydia, 468n16, 482n60, 486n94

Haçilar, 491n143

Hadad (Syrian deity), 258

Hades, 334, 337, 436n5

Hadrian, Roman Emperor, 207, 349

Haghia Triada, sarcophagus, *148,* 157

hair, 314–315

Halicarnassus, Mausoleum of, 23, 247, 258

Haller von Hallerstein, Karl, 502n65

Hambidge, Jay, 229

hand
    and body proportion, 86
    expressions and gesture (chirology), *42, 43*

Hannibal, 131

harmony
    musical, *53,* 100
    and numbers, 100–101

Harvey, William, 40–41, 97

Haselberger, Lothar, 227

Hathor, *299, 300, 312, 313,* 314–315

Hatshepsut, Egyptian Queen, 158, *159, 313, 315,* 494n164

Hattusas (Hittite city), 275, *276, 277*

Hay, David R., 438n20
    Science of Beauty, 394n13

Hazor, 294

head, human
    and body proportions, 56, 59–61, 86, 92
    seven openings in, 82

hearths, 150

Hecuba, Queen, 449n13

Hegel, Georg Wilhelm Friedrich
    and art history, 377
    and artistic creation, 384–385, 390
    on self-reflection, 435n2
    Aesthetics, 375–376, 390

Heidegger, Martin, 378–380, 382, 384–375
    "The Origin of the Work of Art," 379, 381

Hekate, 331

hekatompedon, 168, 213, 215, 266

Helike (Achaia), 475n22

Hellen, 113, 447n4

Hephaestus, 192, 459n18

Hera
    Argive temple, 113, 166
    cult of, 260
    Olympia temple, *172, 173, 202,* 203–206, *204, 205*
    Samos temple and altar, 147, 257, 260, 264
    and Zagreos, 337

Heraclides Ponticus, 431n29

Heraclitus, 260, 337, 381, 405n49, 476n27, 518n12

Herakles (Hercules). *See also* Melquart
    and Amazons, 255, 450n17
    ax wielding, 177, *179*
    battle with Apollo, 329
    columns, 370
    height, 104
    institutes Olympic games, 205, 343, 447n5
    labors, 342
    rescues Prometheus, 145
    sons (Heraklidae), 143, 146, 148, 205, 255
    statue of, *32*
    temples, 488n120, 513n56
    and Triptolemos, 501n54

heraldic figures, 151–153, *154,* 271

Hercules. *See* Herakles

Hermes, *102,* 491n140

Hermione, 498n30

Hermogenes, 213, 216, 327, 467n93, 482n58

Herodotus, 103, 294, 487n112, 488n120, 506n7

Herolt (or Heroldt), Georg, 392n2

heroon, 195, 198

Herostratus, 265

Herrade von Hohenberg (or Landsberg), 82

Hesiod, 115, 145, 149, 260, 487n112, 496n15

Hesychius, 174, 479n43, 497n18

Hierakonpolis (Egypt), 454n45

Hildegard of Bingen, St., 79, 80, 82

Himera, battle of (480), 131

Hippias, 347

Hippocrates of Cos, 109, 180

Hippodamia (daughter of Oenomaos), 206

Hippo, Amazon queen, 255, 257, 450n17, 477n29, 478n33

Hiram, King of Tyre, 489n121

Hiram (or Ahiram), 489n121

Hissarlik (Troy), 7

Hittites, 151–152, 271, 278–279, 283, 290, 326 royal seals, 278, 279

holy stones. See baetyls

Homer, 115, 145, 149, 213, 260, 320. See also Iliad; Odyssey Hymns, 345, 464n63

Honorius of Autun, 39, 61, 402n32

Hor, Pharaoh, 457n64

Horace, 356 Art of Poetry, 53–54

Horai (deities), 261

Horkheimer, Max, 388

horns (animal), 271, 303–304, 305, 307, 314

Hübsch, Heinrich, 377

Hugo, Victor, 376

Huizinga, Johan, 387

Humbaba (or Huwawa; deity), 307, 314

Humbert de Superville, David-Pierre, 8

humors (bodily), 73, 75–76, 92

Hunt, Philip, 453n41

Hunt, Richard Morris, 15

hupokritēs, 124

Husserl, Edmund Gustav Albrecht, 518n6

hut, primitive, 12–13

Huygens, Christian, 90

Hyperboreans, 192

hypotrachelion (hupotrachēlion), 178

Hypsipile, 342

Iamblichus, 501n53

ibex, 271

Ibn al-Sid of Badajoz, 409n76

Ictinus, 246, 338, 340

Idalion (Cyprus), 301, 301

Ikhwan al-Safa (Brethren of Purity), 409n76, 415n18

Iliad, 145–146, 482n57, 487nn112,113

Imhotep (Egyptian vizier), 162, 492n152

Imitation of Christ, as devotional work, 127

imitation. See mimēsis

inclination (of columns), 221, 223, 226

India, 72–73, 103

inlays, 231, 232

Innocent III, Pope: On the Misery of the Human Condition, 83

inscriptions, 234

Ion (son of Xuthos and Kreusa), 113

Ionian language, 115

Ionic genus
    base (plinth), 239–241, 241, 242, 243
    capitals, 242, 244, 244–246, 246, 247, 248, 250, 253, 266, 271, 297, 301, 315, 339, 340
    colors, 233, 244
    cornices, 246–248
    and gods, 264
    and human (female) body, 112–115, 133, 138, 171, 217, 317, 386
    idealized, 381–382
    and optical refinements, 220
    origins and development, 113–114, 211, 248–249, 267–269, 309, 315
    shaft and flutings, 241–242
    and sphinx, 267
    temple measure and dimensions, 216–217
    temples, 237, 239
    as type, 3–4, 5, 58
    Vitruvius on, 237, 239–242, 244–245, 247, 249, 269, 272, 303, 315

Iphigenia, 128, 186

Iphitus, King of Elis, 448n5

Ishtar. See Astarte

Isidore of Seville, 429n17

Isis, 160

"Isis knot," 290

Islam
    astrology, 76
    influence of, 37

Isocrates, 501n51

Isthmia, 189
    Poseidon temple, 208, 213, 498n25
    and tholoi, 215

"Italic" column, 368

Itba'al (or Etbaal), 292
Ixion on the wheel (relief), *98*

Jachin, 294, 454n44
Jacob (Biblical figure), 455n49
Janus, 369
Jerusalem, Temple, *6, 7,* 77, 79, 294–295
Jesus Christ
    body of, 77, 79, 82–83, 85, 397n2
    as judge, 37
Jezebel, 292
John of Salisbury, Bishop of Chartres (called Parvus): *Policraticus,* 37–39, 64
John Tzetzes. *See* Tzetzes
Jones, Owen, 377
Jupiter Heliopolitanus, 258
Jupiter Optimus Maximus, 349, 357, *358,* 359
Juventas, 511n48

Kahn, Louis, 383
Kalydon, 203
Kamilari, near Phaestos, 157
Kamir Blur, 498n24
kanēphorai (basket-bearers), 135
Kanofer (architect), 457n63
kanōn, 99–100, 103
Kant, Immanuel, 127, 378, 385–387, 390, 518n7
    Critique of Judgment, 384
Kapara ben Chadianu, 279
Karaindasch, Kassite king, 308
Karatepe, 279
Karkissa (people), 283
Karnak, 158, 160
Kekrops, first King of Athens, 135
Khadesh, battle of (1288 or 1275), 274
Khalabaseh (Egypt), 456n60
Kharkemish, 279, 290
Khattushilish III, Hittite Great King, 274
Khorsabad, 453n44, 490n134
kings, Greek, 146
Kinyras of Cyprus, 292
Kition (Cyprus), 292, 294, 304
Kleiton, 441n40
Kleomenes of Sparta, 445n69
Klitias, 462n46
Knossos, 146, 149, 155, 200, 214, 491n140
Kodros, King of Athens, 255
Kokoschka, Oskar, 397n52
korai (girl-statues), 115, 117, *121,* 133, 135, 355.
    See also caryatids
Korakou, 451n30
Kore. *See* Persephone

Koressos (son of Kaystris), 255
Korkyra (Corfu)
    Artemis temple, 180, 186, *233*
    Athena temple, 219
Kos, Aphrodite temple, 434n42
Kotchti (Cyprus), 490n138
kouroi (boy-statues), 115, 117, *120,* 162, 207, 447n4
Krauss, Karl, 389
krēpis, krēpidoma, 174
Kresilas, Amazon statue by, 110
Kronos, 292, 349
    Stone of, 331
Kubaba (Syro-Hittite deity), 258
Kuçuk Menderes (stream), 255
Kulka, Heinrich, 397n49
Kültepe (Anatolia), 452n34
Kumarbi (Hurrite deity), 487n112
Kunstgewerbe, 377, 388
Kurangun, 271
Kurash. *See* Cyrus
Kydon, Amazon statue by, 110
Kypsellos, Tyrant, 217

La Bruyère, Jean de: *Caractères,* 43, 45
Labarnash, Hittite Great King, 279
Labranda, Zeus Stratos statue, 258–259
Labrouste, Henri, 18, 376
Labys, 479n43
Lactantius, 83
Lagash, 455n46
Lakonia, Artemis temple, 477n29
Lane, Richard, 395n30
Langhans, Carl Gotthard, 519n25
language, nature of, 117–119, 122
Larissa, Aeolic "order," 269, *272, 273*
Lars Porsena of Clusium, 356, 364, 368, 516n90
lathes, 239–240
Lathuresa (Attica), 214
Latini, Brunetto, 367
Latona, 249, 258, 264
Laugier, Abbé Marc-Antoine: *Essai,* 12–13
Lavater, Johann Kaspar (or Gaspard): *La Physiognomie,* 407n55
Le Blon, Jean-Christophe, 400n22
Le Blond, Alexandre-Jean-Baptiste, 400n22
Le Blond, Jean, 36, 56
Le Brun, Charles, 36, 40–41, 43, 46–47, *48, 49, 50, 51,* 52–53, 56, 93
Le Corbusier, 27, 388
    Modulor series, 90, 390–391
Lebadia (Boeotia), 465n68, 471n40

lebēs, 330–331, *332*

Lecce, Palazzo Palmieri, 365

Lefkandi, "heroon," 195, *196,* 198–199, 209, 213

lekythoi, 321, *322, 323, 324*

Lenaea

    and Dionysos cult, 260

    vase, *258, 260*

Leochares (sculptor), 504n84

Leonardo da Vinci

    Codex Huygens, 82, 90

    and Nicholas of Cusa, 85

    reproduction of human proportions ("Vitruv-

        ian figure"), 86, 89–90, 93, 95, 97

    treatise on painting, 47

    and use of architrave, 393n6

Leoni, Leone, 517n100

Lerna (Argolid), 452n34

Lesbos, 269, *270,* 476n23

Leto, 200

Leto, Pomponio, 392n2

Lewerentz, Sigurd, 396n43

Liber, 360

Libera, 360

Lichtenberg, Georg Christoph, 407n55

Liknites, 329

Limbourg brothers, 69, 76

Limyra, 135

Lincoln Cathedral (England), 374

Linear B (script), 290

lions, 151–153

Lissitzky, El, 518n6

Livius Andronicus, 505n4

Livy, 367, 505n4, 512n51

Lomazzo, Gian Paolo, 92

Lombroso, Cesare, 407n55, 437n11

London

    Bedford Park, 16

    Cenotaph, 229

Loos, Adolf, 19, *22,* 23–24, *24,* 27, 378, 381, 389

Lorenzo de' Medici. *See* Medici, Lorenzo de'

lotus, *300, 326,* 387

Loudon, J. C.: *Encyclopaedia,* 15

Louis XIV, King of France, 29

Louis the Fat, Frankish Emperor, 79

Lucanians, 365

Lucian, 108

Lucius (Lucumo), King of Rome, 510n41

Lucretia, 367

Lucretius, 228

Luini, Aurelio, 422n60

Lukács, Georg, 388–389

Luqqa (people), 283

Luristan, 271

Lutyens, Sir Edwin, 15, 229

Luvian language, 285

Luxor

    Colossus of Rameses III, *166, 167*

    temple of Rameses II, *309*

Lycia

    people, 283, 285

    tomb, *7*

Lycosura, 337

Lycurgos, 176

Lydia, 271, 274, 283, 285, 506n7

Lydus, 506n7

Lysikrates, 327, 345

Lysimachos, Diadochos, 257

Lysippus of Sikyon, 110

    Eros (statue), *111*

McCormick, Robert Rutherford, 397n51

Macedonia, 214

Machaireos, 498n30

Machiavelli, Niccolò, 367

Maecenas, 509n34

Maenius, C., 515n78

Magnesia

    Artemis temple and cult, 213, *252, 253,* 257,

        327, 469n24, 482n58

    coins, *261*

    monopteros, 504n85

Malespini, Ricordano, 367

Mallia, "villa," 468n12

mandala, 73

Manetho (chronicler), 163

Mantinea, 207

Manto (daughter of Teresias), 476n28

Marathon, kouros, 231

Marcus Aurelius Antoninus, Roman Emperor,

    109, 363

Mardonius, 129

Marduk Nadin Ahze, Kassite king, 497n23

Mari, Zimrilim palace, 492n150

Marinatos, Spiridon, 200

Marmaria, 331

Mars, 351

Marseilles. *See* Massalia

Marsh Arabs, 310

Martini, Francesco di Giorgio, 4, *56, 57, 59*–61,

    60, 63, 63–65, *67,* 86, 368, 496n17

Marzabotto, Italy, 359, 514n66

masks, 260, 304, *305, 307,* 311

Massachusetts Institute of Technology, 15

Massalia (Marseilles), 267, 331, 333–334, *334,*

    343, 480n45

Matisse, Henri, 521n49

Mausolos of Halicarnassus, 258

Maximus, Praetor, 512n51

Medea, 331

Medici family, 369–370

Medici, Cosimo I de', 370, 516n90

Medici, Lorenzo de' ("the Magnificent"), 368

Medici Venus, *115*

medicine, and astrology, 76

Medinet-Habu (Egypt), 160

megabyzes, 479n43

Megakles, 498n26

Megalopolis, 337, 502n64

megaron, 144–145

Megiddo, 294, *295*

Melgunov scabbard, 498n24

Melquart/Herakles, 294, 301

Melville, Herman, 119

Menat-Khufu (Egypt), 158

Mendelsohn, Erich, 375

Menelaus, 327

Menes, Pharaoh, 314, 493n160

Mensicles, 503n68

Mersenne, Fr. Marin, 47, 429n18

Mesopotamia, 75, 151, 172, 193, 258, 271, 292, 294, 303, 307–308, 314

Messene, Asklepeion, 515n81

Metagenes, 265

metalsmiths, 150

metaphor, 13–14, 373–374
  body as, 117–119, 122, 373

methexis (inclusion), 126

metopes
  decoration, 233
  Doric, 182–186, 189, *201*, 203, 209, 217, 220
  Ionic, 248

Metrological Relief (Mensuration Slab), 99, *100*

Meyer, Hannes, 384

Michael the Scot, 39

Michelangelo Buonarroti, 92–93, 95, 388, 517n100
  Christ Holding the Cross (sculpture), *94*

Michelet, Jules, 11

microcosm, 56

Midas II, Phrygian king, 278, 283

Mies van der Rohe, Ludwig, 375, 382–384

Milan, 367
  cathedral, 393n6, 517n100

Miletus, 227, 241, *243, 246, 255,* 467n2

Milizia, Francesco, 56

Millawanda (Anatolian state), 275

mimēsis ("imitation"), 46, 123–128, 373, 386–391

Minias, treasury, 490n134

Minoans, 143, 149–153, 155, 160, 175, 182, 189, 214, 231, 303, 326

Minos, 146

Minotauros, 146, 304

Minturnae, battle of (340), 510n41

Minucius Augurinus, Lucius, 364, *364*

Mississippi Agricultural and Mechanical College, 396n47

Mittani (people), 274

Montagnola, 513n64

Montagu, Jennifer, 41

Montaigne, Michel Eyquem, seigneur de, 43

Monte Abetone, 509nn27,28

Monte Iato (Sicily), 445n72

Monte Sirai (Sardinia), 297

Montepulciano, 368

moon, and zodiac, 76

Mopsos, 476n28

Morris, Robert: *Lectures on Architecture, 54, 58*

Morris, William, 518n7

mosaics, *284, 285*

Mount Athos, 63
  Painters' Manual, 56

Mount Eryx (Sicily), 487n112

Mount Ida (Crete), 337

Mount Mykkale, Panionion, 249

Mucianus, 478n32

Mucius, Gaius, 505n3, 507n10

Müller, Carl Otfried, 381, 447n4

Mumford, Lewis, 374–375

Mundigak (Afghanistan), 454n45

Muret, Marc Antoine, 516n90

Mus, P. Decius, 511n42

Musasir (Urartu), 279, 294

Mushki (people), 283

music
  as imitative art, 127
  and number theory, 100–102
  teaching of, 226
  and visual harmony, 53

mutuli (modillions), 184, 189, 354–355

Muwatallish, Hittite Great King, 274

Mycenae, 7, 143, 146, 149–157, 160
  and Arzawa, 275
  beehive tombs, 214
  colonizes Cyprus, 290
  columns, 175–176
  friezes, 189
  grave circle, 320
  Lion Gate, 151–155, *152–153,* 175, 283

megaron, 150, 166
palace of Atreidae, 153
use of colors at, 231
Mykerinus, Pharaoh, 314
Myron (sculptor), 431n29
Myron, Sikyonian tyrant, 268
mystery cults, 337–338
mythology, 72. *See also individual deities*

Nabu-aplu-iddin, King of Babylon, *297, 301, 303*
nails and nailing, *304,* 307, 359
naos (Greek temple), 143–145, 171
Naqsh-i Rustem (Persia), 272
Narcissus, *117,* 118
Narmer (or Horus Nar), 314
Naucratis (Egypt), 267, 474n9
Naxos, 207, 267
Neandria (Asia Minor), 269, *270*
Nebel, Kay H., 518n6
Nebuchadnezzar II, King of Babylon, 302
Nefertary, Queen of Rameses II, 314
Neleos, 255
Nemea
lion, 342, 454n47
Zeus temple, 327, 333, 342
Nemi, 354
Neoplatonism, 37, 40–41, 127
Neoptolemos/Pyrrhus (son of Achilles), 327, 329
Neptune, 92
New York
Metropolitan building, 23
Union Trust building, 396n48
Woolworth building, 23
Niceron, J.-F., 47
Nicholas V, Pope, 83
Nicholas of Cusa (Nicholas Cusanus): *On Learned
Ignorance,* 83–85, 127
Nietzsche, Friedrich, 118, 122, 380–381, 384
The Birth of Tragedy, 447n4
Nikon, 498n27
Nîmes, Maison Carrée, 30, 507n11
Nimrud, *155,* 453n44
Nin-Astarte, 258
Ninos, 478n39
Niqmepa, King of Alalakh, 280
Nivelon, Claude, 404n40
Noah, 369
Nora (Sardinia), 515n82
Norchia, 513n64
Normand, Charles, 377
Parallèle, 7, 8, 13
Nortia, 512n51

Noyers, François Soublet des, 399n12
numbers
and humors, 92
and musical theory, 100–102
perfect, 60–61, 100–101
and proportions, 60–61
Nut (Egyptian deity), 74, *75*

Ocnus, 497n20
Octavian. *See* Augustus
Odysseus, 176–177
Odyssey, 176, 405n49, 487n112
Oenomaos, 206–207
oikos (Greek temple), 171
Olen (Hyperborean singer), 192
Olympia
Apollo, 233
Heraion, *172, 202,* 203–206, *204, 205,* 212–213, 502n62
House of Oenomaos, 206–207
Philippeion (tholos), 343, *345*
sacrifices, 147
treasury, 267–268, 331
Zeus statue, 231
Zeus temple, 113, 204, 206, *224,* 349, 458n2
Olympic games, 143, 193, 205, 343
Opheltes, 342
Opicinus de Canistris, 417n31
Oppenheim, Max von, 486n96
optical correction, 220–222, *223, 224, 224, 225,* 226–229, 234
Orchomenos, 214, 342, 490n134
order, defined, 27–29
Oresme, Nicolas, 402n29
Orestes, 128–129, 186, 327
original sin, 387
Orkos, 502n60
ornament, 377–378
Orpheus, 337
Orthos, 261
orthostats, 247, 290
Orvieto (Italy), 359, 508n22, 513n64
Tempio del Belvedere, 511n46
Osiris, 160
Oxford, Ashmolean Museum, 99, *100*
Oxylos (mythical king over Elis), 205

Pacioli, Fra Luca: *De Divina Proportione,* 56, 59, 90
Padua, Palazzo della Ragione, 38–39, *39*
Paestum (Poseidonia), temples, 10, 18, *173,* 180, *183,* 203, 212, 217, *218,* 365, 380–381, 495n4

Paleopaphos, 292

Palladio, Andrea, 31, *35, 35–36, 52,* 362, 366

Pallas Athena, 145, 441n37

palm trees, 301, *306,* 311, 314, 327, 334

palmette (ornament), 301, 325–327, *326, 332, 333, 333–334,* 343

Palmyra, 500n43

Pancrates of Argos, 499n31

Pandora, 145

Panofsky, Erwin, 397n2

Paphos (Cyprus), 292, *293,* 314

Paris

    Academy of Painting, 40

    Bibliothèque Sainte-Geneviève, 376

    Louvre palace, 29

    Opéra, 18

    Place Vendôme column, 23

    St. Eustache, 40

Parmigianino, 92

Parrhasios, 441n40

Pasargadae (Persia), 273, 453n44

Pasaryk, 285

Pascal, Blaise, 47

passion, effect on body, 40, 43, 46–47, *50, 51*

Patelski, Erich J., 396n47

pathos, 45, 47

Paul, St., 257

Pausanias, 129, 147, 162, 192–193, 204–207, 239, 268–269, 338, 342–343, 433n38, 487n112, 499n31

Paz, Octavio, 519n18

pediment. *See* tympanum

Peisistratos, 207, 347, 349

Pelias, King of Iolkos, 331

Pelops, 206, 343

Pennethorne, John, 229, 471n37

Penrose, Francis Cranmer, 471n37

Penthesilea, 477n29

Perachora, Hera temple, 190, *190,* 195, 200, 209

Pergamon, 147, 499n30

Pergamos, 499n30

Perge, Artemis Anassa, 257, *263,* 479n42

Pericles, King of Limyra, 135

Perrault, Charles, 52

Perrault, Claude, 15, 52–53, *55,* 95

Perret, Auguste, 16

Perrot, Georges: *History of Art in Antiquity,* 7–8, 387

Persephone (Proserpine), 334, *335,* 337–338, 361, 436n5, 489n128

Persepolis, 176, 181, 272, 453n44, 484n81

Persia and Persians

    column design, *130,* 129, 135, *139*

    goat and bull designs, 272–273

    and origins of Ionic, 269

Perseus, 369

perspective, 47, *52, 53, 54*

Perugia

    Arch of Augustus, 364

    Porta Augusta, *366*

    Porta Marzia, *365*

    Volumni tombs, 514n70

Peruzzi, Baldassare, 4, 370

Pessinus, 259

Petamenophis, *74*

Peter of Celle, Bishop of Chartres, 39, 61

Petrarch

    On Remedies against Either Fortune, 83

    sonnet (*Rime* 263), 462n38

Pevsner, Sir Nikolaus, 374–375, 385

Phaecia, 449n13

Phaestos, 200, 214

Phaidros, Archon, 443n57

phantastikē, 126, 221

Pheidias, 104, 113, 204, 222, 224, 338, 441n40, 511n43

    Amazon, 110

Phigaleia, 502n60

Philae (Egypt), Augustan temple, 495n4

Philandrier (or Philander), Guillaume, 33, 36

Philip II, King of Macedon, 110, 343

Philip V, King of Macedon, 203

Philo of Byzantium, 108, 226–228

    Belopoeika, 110

Philolaos of Croton, 428n16

Phoenicia and Phoenicians, 151, 292, 294, 297, *300,* 300, 356

Phoenix, 487nn112,113

Phradmon, Amazon sculpture by, 110

Phrygia, 182, 271, 274, 283, 285, 470n29

Phylakopi (Melos), 452n34

physiognomy, 34–35, 36–41, 43, 46–47, *48, 49, 50, 51, 56*

Piaget, Jean, 387

Picardo, Leon, 56

Picasso, Pablo, 518n6, 521n49

Pico della Mirandola, Giovanni, 76

Piero della Francesca: *De Perspectiva Pingendi,* 89

Pindar, 104, 115, 125, 193, 208, 217

Pindar, Tyrant of Ephesus, 265

pineal gland, 41, *44*

Pinhas ben Yair, 489n121

Piranesi, Giovanni Battista, 18

Pisa
    Camposanto, *81, 82*
    stadium, 104
Pistias, 441n40
Pius II, Pope, 83, 368
Pius IV, Pope, 517n100
Pius V, Pope, 516n90
Plataea, battle of (479), 129, 329
Platino (Piattino Piatti), 421n56
Plato
    on body and universe, 76
    and character, 46
    as a dog, *41*
    on laws, 473n51
    and *mimēsis*, 124–126
    on music, 428n13
    on musical skill, 226
    on optical correction, 221, 226–227
    and perfect numbers, 60–61, 101, 469n25
    on poets and painters, 126
    on *poiēsis*, 385
    on Polykleitos' canon, 433n38
    Laws, 125
    Phaedo, 338
    Phaedrus, 338
    Republic, 102, 125–126, 232
    Sophist, 125–127, 221
    Symposium, 386, 403n38
    Timaeus, 124, 386, 470n32
play, 117, 384, 387
Plessner, Helmut, 387
Pliny the Elder, 4, 104, 133, 176, 208, 216, 230
    231, 265, 352, 363–364, 366, 368, 425n3,
    446n73, 507nn12,13, 512n52
    Natural History, 433nn38,40,41
Plutarch, 38, 103–104, 108, 110, 175, 234,
    429n18, 493n158, 506n6, 509n31
    On Music (attrib.), 428n16
poets, as imitators, 126–127
Poggio a Caiano (Italy), Medici villa, 368, *369*
Poggio Civitate (Italy), 513n66
poiēsis, 124–125, 226, 384–385, 390
poiotēs, 227, 393n7
Pola, 362, 366
Poleni, Marchese Giovanni, 392n2
polis (Greek), 212
Poliziano, Angelo, as a rhinoceros, *41*
Pollaiuolo brothers, 93
Pollard, A. W., 392n2
Pollis (or Polis), 433n38
polychromy, on buildings and statues, *181, 182,*
    230–234, *230, 233,* 244–245, 271, 381

Polygnotos, 498n30
Polykleitos of Argos (or Sikyon), 100–101, 104–
    112, 113, 343, 502n60
    Amazon (sculpture), *109,* 110, *112*
    Doryphoros, 104, *105, 106, 107, 108,*
        110–112
Polykleitos the younger, 113
Polykrates, Tyrant, 347
Porinos, 504n90
Porphyry, 480n43
Porro, Riccardo, 439n23
Porta, Giambattista della, 36–37, 40, *41*
Portonaccio, 355
Poseidon
    Isthmia temple, 213
    Paestum temple, *173, 217, 217*
    Panionian shrine, 207, 249, 434n45
Poseidonia. *See* Paestum
posotēs, 227–228, 393n7
Post, George B., 396n48
Postel, Guillaume, 516n90
Poudra, Noël, 407n58
Poussin, Nicolas, 30, 43
Prado, Jeronimo, 27
Pratinas: *Karuatides*, 446n73
Prato, Gherardi da, 367
praxis, 46, 226
Praxiteles, 434n42
Priam, King of Troy, 327
Priene, temple at, *144,* 213, 247, 469n21
Prinias, temple at, 149, *199,* 199–200, 213
Priscillian, 402n29
Probus, Marcus Valerius, 443n61
Prometheus, 145
proportions. *See* body, human
proskephalaion. See pulvinar
Protagoras, 84
Prussia, as "Dorian" state, 11
Pteras, 192–193
Ptolemy (Claudius Ptolemaeus), 429nn16,18
Publius Septimius, 505n2
Pugin, Augustus, 377
Pulvillus, M. Horatius, 511n44
pulvinar (proskephalaion), 244–245
Puritanism, 27
pyknostyle, 216, 220
Pylades, 128, 186
Pylos, 150, *175, 177,* 214
pyramids, 158, 160, 162–163, *163, 164*
Pyrgi (Italy), 354, 359, 363, 508n24
Pythagoras (philosopher), 100–102, 104, 192,
    216, 260, 337, 425n4, 441n37, 496n15

Pythagoras of Rhegium, 472n41
Pythagoras, Tyrant of Ephesus, 482n60
Pytheios, 467n93
Pythia, 499n36
Pytho, 145, 267, 499n36

Quatremère de Quincy, Antoine-Chrysostome, 12–13, 56, 521n51
Quinet, Edgar, 11
Quintilian, 46

Ramat Rachel, 294, *296*
Rameses II, Pharaoh, 274, 314, 456n60
Ramesseum, 160
rams' heads, 269
Raphael Sanzio, 4, 47, 92
Ras Shamra, 468n14
Rasenna, 506n7
Ray, John, 398n5
Raymond of Marseilles, 37
reed bundles, 308–311, *311*
regula, 183
reification, 388, 518n6
Resep, 482n57
Restoro d'Arezzo, 77
Reynolds, Sir Joshua, 440n32
Rhabanus Maurus, Abbot of Fulda and Archbishop of Mainz, 77
Rhadamantes, 146
Rhea, 500n44, 501n50
rhetoric
        as "father of lies," 374
        and gesture, 43, 45, 47
Rhodians, 151
Rhoecos, 265, 267, 457n66, 476n28
rhuthmos, 224
Rig-Veda, 73
Rimini, Tempio Malatestiano, 367, 392n4
Rome
        Arch of the Argentarii, 392n5
        Arch of Titus, 4
        Basilica Aemilia, 461n33
        Baths of Diocletian, 392n5
        Capitolium and Temple of Jupiter Optimus Maximus, 357, *358*, 359, 492n146
        Colosseum, 4, 354, *355*, 363, 366
        column of Attus Navius, 364
        column of Marcus Aurelius, 363
        column of Minucius Augurinus, 364, *365*
        column of Trajan, 23, 363, 366
        Corinthian columns, popularity of, 349, 351
        use of Doric in, 173

        Etruscan and Greek influence, 356–357
        Forum of Augustus, 133, 473n58
        Palatine, 356
        Palazzo Caffarelli, 511n46
        Palazzo Massimi, 517n100
        Pantheon, 133, 193, 363
        St. John Lateran, 82, *84*
        St. Peter in the Vatican, forecourt, 370
        San Giovanni dei Fiorentini, 517n100
        San Pietro in Vincoli, 461n33
        Santa Costanza, 392n5
        Santa Maria degli Angeli, 392n5
        Santa Maria sopra Minerva, 93, *94*
        Tablinium, 461n33
        temple of Bacchus, 392n5
        temple of Ceres, 360
        temple of Diana and Apollo, 476n23
        temple of Fortuna Virilis, 507n11
        temple of Honor and Virtue, 507n10
        temple of Mars Ultor, 362
        temple of Mater Matuta, 353, 508n20, 516n84, 517n96
        Theater of Marcellus, 513n61, 516n86
Romulus, 506n6, 511n48
roofs, Doric, 182–184
Rossellino, Bernardo, 368
Rousseau, Jean-Jacques: *Essay on the Origin of Language*, 437n8
Rusas I, King of Urartu, 278
Ruskin, John, 11, 374, 395n32, 518n7

Sachlichkeit, 375–376, 382
sacrifice (to gods)
        and garlanding, 271
        in Greece, 145, 147–149, *148*, 168
        on Haghia Triada sarcophagus, 157
Sade, Donatien Alphonse François, Marquis de, 123
Sagazone, Pietro Paolo, 426n5
Sagredo, Diego de, 36, 56, 57, 425n5
        Medidas del Romano, 56
Sakjagözü, 280
Sakkara
        Heb-Sed pavilion, 311
        Zoser pyramid, 158, 160, 162–163, *163*, 308, 492n152
Sala Consilina (Lucania), 464n58
Salamis, battle of (480), 131
Salutati, Coluccio, 367
Samian brothers, 162
Samorna (or Smyrna), 255

Samos

and Aeolic capital, 269

coins, *261*

Heraion and shrine, 147, 168, 190, *190*, 209, 213, *255*, 257, 260, 267, *267*, 347, 467n2, 474n9

temples, 212

and tholoi, 215

Samothrace, 337

Arsinoeon, 345

San Giovenale, 356

San Giuliano, 513n64, 514n68

Sangallo, Giuliano da, 368

Sansovino, Andrea, 368

Sappho, 260

Sardis, Artemis temple, *250, 251*, 257, 474n9

Sargon II of Assyria, 278

Sarpedon, 146, 446n1

Sartre, Jean-Paul, 519n18

Satricum, Mater Matuta, 508n22

Saturnio, 365

Saumaise, Claude (Claudius Salmasius), 474n9

Scaevola, Mucius, 517n93

Scamozzi, Vincenzo, 33, 36, 104, 362

Schadow, Gottfried, 394n13

Scharoun, Hans, 389

Schelling, Friedrich Wilhelm Joseph von, 10, 520n47

Schiller, Johann Christoph Friedrich von, 387

Schindler, Rudolph M., 397n49

Schinkel, Karl Friedrich, 11, 381–382

Schlegel, Friedrich von, 394n16

Schliemann, Heinrich, 7, *168*, 453n41, 458n70

Schönberg, Arnold, 389

Schopenhauer, Arthur, 9–11, 123, 380

Schuyler, Montgomery, 396n48

sculpture, color and decoration of, 230–232

"Sea Peoples" (Eastern Mediterranean), 290, 292

seals, 151–153

seasons, and the elements, 73, 76, *78*

Sedlmayr, Hans, 388

Segesta, temple, *219*

Séguier, Pierre, Duc de Villemor, 40

Selçuk, 478n34

Seleucus Nicator, 349

Selinunte

Temple C, *230*

treasury, *188*

Selinus (Sicily), megaron temple, 451n30

Sellasia, 445n69

Semiramis, Queen, 478n39

Semper, Gottfried, 12–13, 377

Senmut, 314

Serlio, Sebastiano, 3–6, 33, *317*, 362, 366, 370

Servius Marius Honoratus, 369, 432n37

Servius Tullius, King of Rome, 480n45, 508n16

Sesostris I, Pharaoh, statue of, 454n45, 492n155

Severus, Septimius, 4

Sextus Empiricus, 429n16

Shaftesbury, Anthony Ashley Cooper, 3rd Earl of: *Characteristics,* 45, *45*

Shalmaneser III, 490n132

Shamash, statue of, 301, 303

Shub-Ad, Queen of Ur, 454n44

Shuppiluliumash II, 275

Shute, John: *First and Chief Grounds of Architecture,* 32–33, *32, 34*

Sibylline Books, 357

Sidon, 292

Sikyon, 268–269, 343, 497n20

Silamon, 433n38

Silber, Eucherius, 392n2

Silius Italicus: *Punica,* 488n119

Silvester, Bernard: *Cosmography* (or *De Mundi Universitate*), 37, 82

Simmel, Georg, 387

Sociology, 377–378, 383

Simonides, 482n57

Simplicius, 412n5, 413n9

Siphnos (?Siphanto), treasury, 135, 333

Sippur, 489n130

Skopas of Paros, 340

skulls, horned, 271, 303–304, *305, 307*

Sligo, John Denis Browne, 1st Marquess of, 453n42

Smyrna, 190, 255, 260, 269, 483n72

Smyrna (daughter of Kinyras), 487n113

Socrates, 124, 125–126, 222, 337, 470n32

Solomon, King of Israel, 27

Solon, 320

sophia, 431n29

Sovanna, Ildebranda cave, 365, 511n47

Sparta, 129, 133, 175, 433n42, 447n4

Artemis Orthia, 212–213

Spencer, Herbert, 387

sphinx, 152, 267, *268*, 297, *303*, 326

Spinoza, Benedict

Ethics, 404n42

Tractatus Theologico-Philosophicus, 397n3

Spintharos of Corinth, 193

Spurius Cassius (Vecellinus), 360–361

Staël, Anne Louise Germaine Necker, Baronne de Staël-Holstein, 394n16

Stanton Drew (near Bath, England), 27, *28*

statio (*thematismos*), 237

statues, as supporters, 31–33, *32, 33, 34. See also* caryatids; *and individual statues*

Stephen of Byzantium, 478n39

Stesichorus, 115, 498n29

Stockholm (Sweden)

Skandia cinema, 19

Woodland Chapel, 18–19, *20, 21*

stoicheia. See elements, four

Stoics, 69, 109, 228, 402n29

Stonehenge (England), 27, *28*

Strabo, 294, 453n41

Stratonice, 492n147

strife, and four elements, 69

stucco, 230–232, *230*

stylobate

Doric, 174, 176, 234

Ionic, 239

optical corrections to, *222, 223, 225, 226,* 229

Sugranes, Domingo, 396n42

sugraphē, 215

Sulla, 349, 360

Sullivan, Louis: "The Tall Office Building Architecturally Considered," 19, 24

Sulpizio, Giovanni, 392n2

Sumeria, 151, 156, *304,* 307–310, *310, 311*

Sumerian language, 274

post-knot signs, 309, *310*

Susa, 272, 484nn79,80

Sybaris, 365

symmetry, 109–110, 272

Syracuse, Apollo temple, 234

Syria and Syrians, 151, 172, 258, 271, 290, 292, 294, 304, 354

Ta'anit, 297, 300

Tabal, 283

Taccola, Pietro Mariano: *Liber de Ingeneis,* 86, 87

Tacitus, 488nn113,114

Tages, 357

Taine, Hippolyte, 380

tainia (fillet), 182, 189

Tamassos (now Politiko, Cyprus), 290, *291,* 294, 297

Tanaquil, 510n41

Taranto, 365

Tarde, Gabriel de, 387

Tarquinia, 354

Ara della Regina, 359

tombs, 362

Tarquinius Priscus, Lucius, 357, 364, 510n38

Tarquinius Superbus, 357

Tarsus, 279

Tasso, Bernardo, 393n6

tattooing, 119

Taurus, temple of Artemis, 128–129, 186

technē, 124–126, 128, 222, 390

technology, and aesthetics, 383–385

Tegea, Athena Alea temple, 213, 327, 338, 340, 342, 495n3, 498n25, 503n67

Teiresias, 436n5, 476n28

Teisheba, 294

Tel Tayanat, 280

Telamon, 443n55, 508n24

telamones. *See* atlantes

Tel-Balawat, 490n132

Tel-el Farah, 490n137

Telephos, 342

Tel-Halaf (Guzana), temple-palace, *140,* 279–280

Tell-el-Amarna, 158

Tell-el-Rimah (?Karana), 303, *306, 307*

temenos, 143–149, 171, 211

temperaments, 78

Teos (Sigaçik), temples, 467n93, 469n24

Termessos, 504n85

Terminus, 359

Tespes (Chishpish), King of Persia, 269

Tharros, temple, 300

Thebes (Boeotia), 150, 212

Thebes (Egypt), 158, 160, 314, 489n121

Thefarie Velianas, 508n23

Themis, 249, 474n2

Themistocles, 347, 434n43

Theodorus of Phocea, 504n80

Theodorus of Samos, 265, 457n66, 474n9

Theodosius, Roman Emperor, 514n73

Theon of Smyrna, 409n85

Theophrastus: *Characters,* 43, 45

Thera, 150

Thermon

temple excavations, *197, 200, 201,* 202–203, *206, 207,* 212, 463n50

and tholoi, 215

Theron of Akragas, Tyrant, 131

Theseus, 446n4, 477n29

Thesmophoria, 145

tholoi, 214, 342–343, *344, 345, 345*

Thomas of Celano, Fr., 37

Thorekos (Attica), 214

Thucydides, 103, 143

Thyades (maenads), 329

Thyia, 329

Tiberius, Roman Emperor, 361

Tilmen-Huruk, 279

Timotheos, 498n27

Tinia, 365

Tiryns
  circular citadels, 214
  megaron, 150, 166, *168,* 180, 207

Titans, 337

Tivoli, Hadrian's villa, 133, 445n72

Tlepolemos, 446n1, 450n20

Toland, John, 398n5, 399n6

torus (spira), 239–240

Toscanelli, Paolo, 420n50

trachēlion, 178, 181

Trajan, Roman Emperor, 23, 363, 366

Tralles, Asklepion temple, 467n93

treasuries, 267–269, 331, 333, 343. *See also individual sites*

trees, artificial, 156, 290

trees, sacred, 156, 290, 302, 315

triglyphs, 183–187, 189, 209, 217
  corner, 217–219

tripods, 329–331, 334, 345

Triptolemos, *335, 337*

Troad, Apollo Smintheus temple, *247*

trochilon (skotia), 239–240

Trophonios, 193, 207

Troy, 7, 113, 128, 131, 199

Tubertus Postumius, Aulus, 513n56

Tuby, Jean-Baptiste, 403n36

Türbe, *282*

Turianus of Fregene, 511n43

Tuscan arrangement
  base, 174, 352
  capital, 353, 363
  cornices, 35, *35, 354*
  design and proportions, 352, 353–357, *358,* 359–360
  entablatures, 35, *35*
  and Medici, 370
  origins, 351, 370
  temples, 351, 353–356, 362–365
  as type, 3, *5,* 370
  Vitruvius on, 351–352, 354, 356–357, 359, 363, 366, 368, 370

Tutankhamun, Pharaoh, 494n164

Tuthmosis III, Pharaoh, 158, 280

tympanum (or pediment), 217, 219, 233

tyrants, 215

Tyre, 292, 294

Tyrrhenus, 506n7

Tzetzes, John, 222, 226

Uadjit, Pharaoh, 491n143

Uni-Hera-Astarte, 354

Untashgal, 271

Uranos, 292

Urartu and Urartians, 271, 274, 275–276, 278–279, 283, 294, 326

Urban VIII, Pope, 464n67

Valeriano, Giovanni Pierio, 404n44

Valerius (Roman consul), 511n44

Varro, 104, 228, 364, 408n73, 424n3, 425n5, 441n45, 505n2, 508n21, 509n31, 515n77

Vasari, Giorgio, 368–369

Vastupurusamandala, 73

Vaux-le-Vicomte (France), 40

Vegoia (nymph), 357

Veii (Italy), 359, 510n39
  Ara della Regina, 362, 512n50

Veliano, near Perugia, Villa Danzetto, 514n72

Verdinglichung, 518n6

Verona, amphitheater, *355,* 362, 366

Verrius Flaccus, 505n4

Versailles, 29

Verus, Roman Emperor, 109

Vesalius, Andreas: *On the Fabric of the Human Body,* 93, 95

Veshch/Objet/Gegenstand (periodical), 518n6

Vespasian, Roman Emperor, 362, 507n11

Vetulonia, Pozzo dell'Abbate, 513n64

Vico, Giambattista, 11

Vienna, Goldmann and Salacz store ("Looshaus"), 19

Vignola, Jacopo Barozzi da, 7, 15, 31, 36, 52, 92, 366

Villalpando, Juan Bautista, 27

Villani, Giovanni, 367, 369

Villard d'Honnecourt, Master, notebook, 85–86, 89, 426n5

Vincent of Beauvais: *Speculum Quadruplex,* 77

Viola, Zanini (Gianni) Gioseffe, 400n13

Viollet-le-Duc, Eugène-Emmanuel, 7, 13, 15, 185, 377
  Entretiens, 13

Virgil, 233, 307, 367, 369

Vitruvius Pollio, Marcus
  on acanthus, 325
  and Athens Olympieion, 349
  and body as microcosm, 76–77
  on capitals, 178
  on Ceres shrine, 361
  on column numbers and building size, 211, 215–217

Vitruvius Pollio, Marcus (*cont.*)
   and Corinthian, 239, 317, 320–321, 334
   on corner metopes, 220
   and decorum, 53, 237
   on Doric, 209, 211, 217, 237, 264, 386
   and echinus, 181
   Ephesian Artemision, 257
   on eurythmy, 228–229
   on garland moldings, 272
   and human proportions, 56, 59–60, 64, 66,
      76, 79, 82–83, 86, *88, 89, 90,* 92–93, *95,*
      97–99, 110, 112, 165, 171, 217
   influence on Alberti, 66
   on Ionic, 237, 239–242, 244–245, 247, 249,
      269, 272, 303, 315
   on metopes, 365
   and number theory, 63, 101
   and optical refinements, 220; 226–228, 234
   on origins and types of columns, 113–114,
      144, 165, 174–175, 209, 213, 219, 237
   on ornaments and cornices, 183–187, 190
   and pagan form, 27
   and Polykleitos, 108, 110
   and principles of design, 97, 99–100
   on sculptured figures as columns, 129, 133,
      135
   on stone and timber construction, 9, 11–13,
      183–187, 189, 202, 315
   and surface colors, 230
   Taccola and, 86
   on temples, 171–172, 234, 237
   translations and editions of, 15, 52, 90
   on Tuscan, 351–352, 354, 356–357, *359,*
      363, 366, 368, 370
   on universal vision, 386
Volnius, 505n4
Volsci, 515n78
Volsinii
   Campanari tombs, 365
   Nortia temple, 512n51
Volterra, 364
volutes, Ionic, 242, *245,* 267, 269
Vrana, near Marathon, 214
Vredeman de Vries, Hans, 32
   Morning-Youth, 31
   Sunrise, 30
Vulca of Veii, 357
Vulcan, temples, 351

Ware, William R., 396n48
   The American Vignola, 15
Warka, temples, 308–309, *308,* 488n118

Waterhouse, Alfred, 395n30
Whiston, William, 397n4
White, Stanford, 15
wigs, 314–315, 494n164
Wittgenstein, Ludwig, 122
Wölfflin, Heinrich, 518n6
Wood, John, the Elder, 27, *28,* 29, 34, 422n60
work songs, 150

Xanthos
   Harpy tomb, 285, *287,* 490n135
   Lato shrine, 249
   Nereid monument, 133
   "sledded" tomb, *288*
   theater, *286*
Xenophanes of Colophon, 260, 501n51
Xenophon of Athens, 126, 479n43, 487n112
Xenophon of Corinth, 495n9
Xenophon of Ephesus, 479n43
Xeropolis, 199
xoana, 257, 259, 260

Yazilikaya, 479n41

Zagreos, 337
Zakim, 498n24
Zakros, 491n140
Zeising, Adolf, 394n13
Zernaki Tepe, 485n93
Zeus
   Agrigento temple, 131, *132,* 212
   Doric temples, 239
   and Doric territory, 249
   Nemea temple, 342
   Olympia temple, 113, 204, 206, *348, 349*
   and omphalos of world, 327
   and Persephone, 334, 337
   and Prometheus, 145
   sacrifices to, 147
   statues, images and cult, 258, 259, *262*
   Velchanos, 157
Zeuxis, 92, 441n40, 513n58
Zinçirli, 279–280
zodiac, 69, *70, 71, 72*–73, *72, 75,* 76
zophoron (frieze), 182, 247
Zoser, Pharaoh
   pyramid, 158, 162–163, *163,* 165, 308, 311